D1698166

Edited by
Moiz Mumtaz

Principles and Practice of Mixtures Toxicology

Related Titles

Bernhard, D. (ed.)

Cigarette Smoke Toxicity

Linking Individual Chemicals to Human Diseases

2011

ISBN: 978-3-527-32681-5

O'Brien, Peter J., Bruce, W. Robert (eds.)

Endogenous Toxins

Targets for Disease Treatment and Prevention

2010

ISBN: 978-3-527-32363-0

Salthammer, T., Uhde, E. (eds.)

Organic Indoor Air Pollutants

Occurrence, Measurement, Evaluation

2009

ISBN: 978-3-527-31267-2

Külpmann, W. R. (ed.)

Clinical Toxicological Analysis

Procedures, Results, Interpretation

2009

ISBN: 978-3-527-31890-2

Kumar, C. S. S. R. (ed.)

Nanomaterials - Toxicity, Health and Environmental Issues

2006

ISBN: 978-3-527-31385-3

Edited by
Moiz Mumtaz

Principles and Practice of Mixtures Toxicology

With a Foreword by John Doull

WILEY-VCH Verlag GmbH & Co. KGaA

The Editor

Dr. Moiz Mumtaz
ATSDR, Toxicology and
Environmental Medicine (F-62)
1600 Clifton Road
Atlanta, GA 30333
USA

Cover picture: Anne Christine Kessler

Library of Congress Card No.: applied for

British Library Cataloguing-in-Publication Data
A catalogue record for this book is available from the British Library.

Bibliographic information published by the Deutsche Nationalbibliothek
The Deutsche Nationalbibliothek lists this publication in the Deutsche Nationalbibliografie; detailed bibliographic data are available on the Internet at http://dnb.d-nb.de.

Cover Design Anne Christine Keßler, Karlsruhe
Typesetting Thomson Digital, Noida, India
Printing and Bookbinding betz-druck GmbH, Darmstadt

Printed in the Federal Republic of Germany
Printed on acid-free paper

ISBN: 978-3-527-31992-3

Dedicated to

My father, M.A. Raheem: an educationist and adviser, who instilled in me the significance of the word "read" and my mother, Khairunnisa Begum who personified patience.

My brothers and sisters who encouraged me and had the confidence that I could accomplish my goals.

My wife, Farzana, for a lifelong partnership and her everlasting support. My son, Nabeel, for his motivation and support.

Foreword

As toxicologists, we have two jobs. The first is to identify and characterize the adverse effects that chemicals and other agents can produce in biological systems. The second is to use this information to improve public health. We do this by making safety predictions, and the credibility and public support of our discipline depends on how well we do both jobs. Toxicology is what we do but risk assessment is why we do it. Historically, our ability to make accurate predictions for the adverse effects of mixtures has been limited by the difficulty of acquiring data for all the possible combinations of dose and time that exist even in simple mixtures. Such predictions are also compromised by our use of single-agent toxicity studies since most "real-world" exposures are to mixtures. This has resulted in a variety of approaches (models, protocols, techniques, etc.) to address these issues. These are described in detail in the two dozen chapters of this book along with case studies of mixtures that illustrate their use, advantages, limitations, and regulatory applications. After reviewing the exciting advances taking place in mixture toxicology, it seems to me that the greatest future impact will result from the use of high-throughput testing and the inclusion of emotional stress as a mixtures agent. The use of high-throughput testing as described in the NAS/NRC report "Toxicity Testing in the 21st Century" will enable us to resolve many of the data acquiring limitations of the past [1]. Before discussing emotional stress, I would like to describe how the approach that I use [2] to characterize toxicity can be adapted to the effects of mixtures.

The effects of chemical, biological, and physical agents (and combinations of these) have two things in common. First, all the effects that they produce are the result of an interaction between the agent(s) and a target and we define this interaction as the exposure. We can define each agent by the effects it is capable of producing and the target by its susceptibility to these effects. After doing this, we must then agree that if the exposure is sufficient to produce a specific effect in the target, we have defined a threshold for that effect even if we are not able to measure it. The bottom line here is that we should focus on the measurement problem rather than arguing about whether biological effects have thresholds since they all do.

The second thing that these agents have in common is that all their effects result from an action of the agent on a target (dynamics) or from the action of the target on the agent (kinetics). Thus, the key or rate-limiting events that we use to define the mode of action must occur in the dynamic or kinetic pathways. We focus on the

Principles and Practice of Mixtures Toxicology. Edited by Moiz Mumtaz

ISBN: 978-3-527-31992-3

mechanisms of injury and recovery for events in the dynamic pathway and on half-life (intake, distribution, metabolism, and elimination) for those in the kinetic pathway. Therefore, distinguishing the key events that occur in the dynamic pathway from those occurring in the kinetic pathway is a critical first step for defining the mode of action of the agent(s).

These two observations are applications of general biologic principles, but extending them to the "real-world" situation can be complicated. For example, the main factors of exposure are dose and time, but there are other factors such as the route and the presence of other ingredients (vehicles, mixtures). Dose can be a simple variable such as mass or surface area (as in nanotechnology), but time includes not only the duration and frequency but also the persistence of the agent (kinetics) and its effects (dynamics). Agents may be mixtures not only of chemicals but also of pathogens, radiation, and so on. Targets can range from genes to cells to organs and systems and from individuals to populations. Most agents exhibit multiple effects with increasing exposure (hormesis), and some exhibit adaptation or altered susceptibility with repeated exposure (allergens). For most chemical agents, recovery is slower (and therefore rate limiting) than injury in the dynamic pathway and elimination is slower than absorption in the kinetic pathway. In the dynamic pathway, injury can be reduced by adaptation and recovery is modulated by repair, reversibility, and adaptation. Distribution, biotransformation, and excretion are the major factors in the kinetic elimination pathway of chemicals. With pathogens, the rate of multiplication and the host defenses of the target can influence both the dynamic and the kinetic pathways. Exposure to physical agents (radiation, heat, vibration, noise, etc.) introduces new exposure units but the basic principles still apply.

With this approach, it is evident that there is no effect if the agent, target, or exposure is missing and it is equally evident that there are effect thresholds for each combination of agent(s) and target(s) where the exposure exceeds the homeostatic capability of the system. A log probit plot or a log-normal plot plus the Gaussian distribution can be used to identify the individual thresholds and estimate the relative risks in a population or an individual.

Emotional stress or psychosocial factors can certainly produce adverse effects, but including these as mixture agents in the above scheme will be a challenge since they not only have different exposure criteria but their mechanism of action and effects are also different. Furthermore, the proponents of the "exposome" concept are now suggesting that behavior, dietary, aging, lifestyle, and other environmental factors should also be incorporated as mixture components that influence human health and disease. Two things that would help us meet these challenges in mixture toxicology are, first, updating our definition of adversity and, second, developing methods for quantifying benefits to the same level of sophistication as currently exists for risks. This would make risk/benefit analysis a more attractive alternative for conventional (single-agent) and mixture toxicology.

John Doull
University of Kansas
Medical Center
Kansas City, Kansas

References

1 NRC (National Research Council) (2007) *Toxicity Testing in the 21st Century: A Vision and a Strategy*, National Academy Press, Washington, DC.

2 Rozman, K.K., Doull, J., and Hayes, W.J. (2010) Chapter 1: Dose and time determining and other factors influencing toxicity, in *Hayes Handbook of Pesticide Toxicology*, 3rd edn, vol. 1 (ed. R. Krieger), Elsevier, Amsterdam.

Contents

Principles and Practice of Mixtures Toxicology. Edited by Moiz Mumtaz

ISBN: 978-3-527-31992-3

Preface

On a daily basis we encounter chemicals in our lives. So throughout our lifetime, intentionally or unintentionally, we are exposed to chemicals in combination with one another [mixtures]. Exposures to a variety of mixtures occur through the food we eat, air we breathe, water we drink or through contact with soil. These mixtures, natural or synthetic, have the potential to cause adverse health effects under specific exposure scenarios. In the world of contaminants, it is especially important to identify and study significant mixtures in order to understand their mechanisms of action. From this we can develop methods to evaluate the risk they pose in the real world. Chemical mixtures toxicology is a rapidly developing sub-discipline of toxicology where advances are often based on using concepts and techniques developed through basic biomedical research. To provide a permanent platform for exchange of research and application of scientific advances in diverse fields, the Society of Toxicology has recently established a Mixtures Specialty section.

The risk assessment process involving mixtures uses the same National Academy of Sciences paradigm as for single chemicals, but incorporates mixtures issues in every aspect of exposure assessment, hazard identification, dose response assessment and risk characterization. Through this process new data needs have been recognized and research funded and performed to fill them. Mixtures research founded in molecular biology, statistical and mathematical modeling has led to routine use of models developed. For this field to remain current and continue making significant advances, the latest developments must translate basic research into usable methods and routine practice in the context of public health and protection of the environment. Recently several mixtures issues have become salient, and are deserving of a review that illustrates how new techniques have been applied to real life problems.

The goal of this book is to highlight basic concepts and new methods that may have major impact on general toxicology, as well as the field of safety and risk assessment of chemicals and their mixtures. National and international scholars and prominent toxicologists have provided timely perspectives on how the latest science can be applied to existing problems and provide overviews of areas where significant progress has been made. The target audiences for this book are practicing and future toxicologists in academia, government and industry. We anticipate the book may be

Principles and Practice of Mixtures Toxicology. Edited by Moiz Mumtaz

ISBN: 978-3-527-31992-3

especially useful to those individuals interested in the practical aspects of the risk assessment of chemical mixtures. It may also serve as a useful text for "special topics" courses for graduate curricula.

Each author was encouraged to write their chapter(s) in the style and format that suited their contribution, so the chapter formats may vary somewhat. The contents of these chapters do not represent the policy of any agency or organization, unless explicitly stated.

Atlanta, July 2010 *Moiz Mumtaz*

List of Contributors

S. Satheesh Anand
DuPont Haskell
Global Centers for Health and
Environmental Sciences
Newark, DE 19714
USA

Melvin E. Andersen
The Hamner Institutes for Health
Research
P.O. Box 12137
Research Triangle Park, NC 27709
USA

Subhash C. Basak
University of Minnesota Duluth
Center for Water and the Environment
Natural Resources Research Institute
5013 Miller Trunk Hwy.
Duluth, MN 55811
USA

Ronald E. Baynes
North Carolina State University
Center for Chemical Toxicology
Research and Pharmacokinetics
4700 Hillsborough Street
Raleigh, NC 27606
USA

Negash Belay
United States Food and Drug
Administration
Center for Food Safety and Applied
Nutrition
Office of Food Additive Safety
Division of Biotechnology and GRAS
Notice Review
5100 Paint Branch Parkway, HFS-255
College Park, MD 20740
USA

Linda S. Birnbaum
National Institute of Environmental
Health Sciences (NIEHS)
P.O. Box 12233
Mail Drop B2-01
Research Triangle Park, NC 27709
USA

Christopher J. Borgert
Applied Pharmacology and Toxicology,
Inc.
2250 NW 24th Avenue
Gainesville, FL 32605
USA

Laura K. Braydich-Stolle
Wright-Patterson AFB AFRL/HEPB
U.S. Air Force Research Laboratory
P.O. Box 31009
Dayton, OH 45437-0009
USA

Principles and Practice of Mixtures Toxicology. Edited by Moiz Mumtaz

ISBN: 978-3-527-31992-3

Jerry L. Campbell Jr.
The Hamner Institutes for Health Research
P.O. Box 12137
Research Triangle Park, NC 27709
USA

Ed Carney
The Dow Chemical Company
Toxicology & Environmental Research and Consulting
1803 Building
Midland, MI 48674
USA

Harvey J. Clewell III
The Hamner Institutes for Health Research
P.O. Box 12137
Research Triangle Park, NC 27709
USA

Alexander A. Constan
Infinity Pharmaceuticals, Inc.
780 Memorial Drive
Cambridge, MA 02139
USA

Rebecca P. Danam
United States Food and Drug Administration
Center for Food Safety and Applied Nutrition
Office of Food Additive Safety
Division of Biotechnology and GRAS Notice Review
5100 Paint Branch Parkway, HFS-255
College Park, MD 20740
USA

Christopher T. De Rosa
Agency for Toxic Substances and Disease Registry (ATSDR)
Division of Toxicology and Environmental Medicine
F-62, 1600 Clifton Road
Atlanta, GA 30033
USA

Kirby C. Donnelly†
Texas A&M University System Health Science Center
School of Rural Public Health
Department of Environmental and Occupational Health
1266 TAMU
College Station, TX 77843-1266
USA

Hisham El-Masri
U.S. Environmental Protection Agency
National Health and Environmental Effects Research Laboratory
Experimental Toxicology Division
109 T.W. Alexander Drive
Mail Drop B143-01
Research Triangle Park, NC 27711
USA

Mike Fay
Agency for Toxic Substances and Disease Registry (ATSDR)
Division of Toxicology and Environmental Medicine
F-62, 1600 Clifton Road
Atlanta, GA 30033
USA

†Deceased

Jeffrey W. Fisher
University of Georgia
College of Public Health
Department of Environmental Health
Science
102 Conner Hall
Athens, GA 30602
USA

Paulette M. Gaynor
United States Food and Drug
Administration
Center for Food Safety and Applied
Nutrition
Office of Food Additive Safety
Division of Biotechnology and GRAS
Notice Review
5100 Paint Branch Parkway, HFS-255
College Park, MD 20740
USA

Chris Gennings
Virginia Commonwealth University
Department of Biostatistics
1101 E. Marshall St.
Richmond, VA 23298-0032
USA

Panos Georgopoulos
UMDNJ/Rutgers University
Environmental and Occupational
Health Sciences Institute
107 Frelingshuysen Road
Piscataway, NJ 08854
USA

Brian D. Gute
University of Minnesota Duluth
Center for Water and the Environment
Natural Resources Research Institute
5013 Miller Trunk Hwy.
Duluth, MN 55811
USA

Laurie C. Haws
Tox Strategies
3420 Executive Center Drive
Suite 114
Austin, TX 78731
USA

Frank J. Hearl
U.S. Department of Health and Human
Services
Centers for Disease Control and
Prevention
National Institute for Occupational
Safety and Health
395 E Street, S.W.
Suite 9200
Patriots Plaza Building
Washington, DC 20201
USA

Saber M. Hussain
Wright-Patterson AFB AFRL/HEPB
USA
U.S. Air Force Research Laboratory
P.O. Box 31009
Dayton, OH 45437-0009
USA

Mark Johnson
Agency for Toxic Substances and
Disease Registry (ATSDR)
Division of Regional Operations
4770 Buford Hwy NE
Atlanta, GA 30341
USA

Kannan Krishnan
Université de Montréal
Département de santé
environnementale et santé au travail
C.P. 6128, Succ. Centre-ville
Montréal (Québec) H3C 3J7
Canada
UMDNJ/Rutgers University
Environmental and Occupational
Health Sciences Institute
170 Frelinghuysen Road
Piscataway, NJ 08854
USA

Jason C. Lambert
U.S. Environmental Protection Agency
National Center for Environmental
Assessment
Office of Research and Development
Chemical Mixtures Research Team
26 West Martin Luther King Drive
MC-A-110
Cincinnati, OH 45268
USA

Mary E. LaVecchia
United States Food and Drug
Administration
Center for Food Safety and Applied
Nutrition
Office of Food Additive Safety
Stakeholder Support Team
5100 Paint Branch Parkway, HFS-255
College Park, MD 20740
USA

John C. Lipscomb
U.S. Environmental Protection Agency
National Center for Environmental
Assessment
Office of Research and Development
Chemical Mixtures Research Team
26 West Martin Luther King Drive
MC-A-110
Cincinnati, OH 45268
USA

Michael A. Lyons
Colorado State University
Quantitative and Computational
Toxicology Group
1680 Campus Delivery
Fort Collins, CO 80523
USA
Colorado State University
Department of Environmental and
Radiological Health Sciences
1680 Campus Delivery
Fort Collins, CO 80523
USA

Margaret MacDonell
Argonne National Laboratory
Environmental Science Division
9700 South Cass Avenue, EVS/Bldg 240
Argonne, IL 60439
USA

Antonia Mattia
United States Food and Drug
Administration
Center for Food Safety and Applied
Nutrition
Office of Food Additive Safety
Division of Biotechnology and GRAS
Notice Review
5100 Paint Branch Parkway, HFS-255
College Park, MD 20740
USA

Arthur N. Mayeno
Colorado State University
Quantitative and Computational
Toxicology Group
1680 Campus Delivery
Fort Collins, CO 80523
USA
Colorado State University
Department of Environmental and
Radiological Health Sciences
1680 Campus Delivery
Fort Collins, CO 80523
USA

Eva D. McLanahan
University of Georgia
College of Public Health
Department of Environmental Health Science
102 Conner Hall
Athens, GA 30602
USA

Harihara M. Mehendale
University of Louisiana at Monroe
College of Pharmacy
Department of Toxicology
Bienville Building
700 University Avenue
Monroe, LA 71209
USA

Chander Mehta
Texas Southern University
College of Pharmacy
3100 Cleburn Street
Houston, TX 77004
USA

David Mellard
Agency for Toxic Substances and Disease Registry (ATSDR)
Division of Health Assessment and Consultation
1600 Clifton Road
Atlanta, GA 30033
USA

Nancy A. Monteiro-Riviere
North Carolina State University
College of Veterinary Medicine
Center for Chemical Toxicology Research and Pharmacokinetics
4700 Hillsborough Street
Raleigh, NC 27606
USA

Moiz Mumtaz
Agency for Toxic Substances and Disease Registry (ATSDR)
Computational Toxicology and Methods Development Laboratory
Division of Toxicology and Environmental Medicine
MS-F62, 1600 Clifton Road
Atlanta, GA 30333
USA

Richard C. Murdock
Applied Biotechnology Branch
Human Effectiveness Directorate
Air Force Laboratory
Wright-Patterson AFB Building 837
"R" Street, Area B
Dayton OH 45433-5705
USA

Ziad S. Naufal
Texas A&M University System Health Science Center
School of Rural Public Health
Department of Environmental and Occupational Health
1266 TAMU
College Station, TX 77843-1266
USA

Binu K. Philip
Indiana University
School of Medicine
1120 South Drive
Indianapolis, IN 46202
USA

Hana R. Pohl
Agency for Toxic Substances and Disease Registry (ATSDR)
Division of Toxicology and Environmental Medicine
F-62, 1600 Clifton Road
Atlanta, GA 30033
USA

Paul S. Price
The Dow Chemical Company
Toxicology & Environmental Research and Consulting
2030 Dow Center
Midland, MI 48674
USA

Brad Reisfeld
Colorado State University
Department of Chemical and Biological Engineering
1370 Campus Delivery
Fort Collins, CO 80523
USA

Glenn E. Rice
U.S. Environmental Protection Agency
Office of Research and Development
National Center for Environmental Assessment
Chemical Risk Assessment Branch
Chemical Mixtures Research Team
26 West Martin Luther King Drive
M.S. A 130
Cincinnati, OH 45268
USA

Jim E. Riviere
North Carolina State University
Center for Chemical Toxicology
Medicine Box 8410
4700 Hillsborough Street
Research and Pharmacokinetics
Raleigh, NC 27606
USA

Craig Rowlands
The Dow Chemical Company
Toxicology & Environmental Research and Consulting
2030 Dow Center
Midland, MI 48674
USA

Patricia Ruiz
Agency for Toxic Substances and Disease Registry (ATSDR)
Division of Toxicology and Environmental Medicine
1600 Clifton Road NE
Mailstop F62
Atlanta, GA 30333
USA

P. Barry Ryan
Emory University
Rollins School of Public Health
Department of Environmental and Occupational Health
1518 Clifton Road, NE,
Atlanta, GA 30322
USA

Alan Sasso
UMDNJ/Rutgers University
Environmental and Occupational Health Sciences Institute
170 Frelinghuysen Road
Piscataway, NJ 08854
USA

Jane Ellen Simmons
U.S. Environmental Protection Agency
Office of Research and Development
National Health and Environmental Effects Research Laboratory
109 T. W. Alexander Drive
Research Triangle Park, NC 27711
USA

Daniele F. Staskal
Tox Strategies
3420 Executive Center Drive
Suite 114
Austin, TX 78731
USA

William A. Suk
National Institute of Environmental Health Sciences
Division of Extramural Research and Training
P.O. Box 12233
Research Triangle Park, CO 80523-1681
USA

Linda K. Teuschler
U.S. Environmental Protection Agency
National Center for Environmental Assessment
Office of Research and Development
Chemical Mixtures Research Team
26 West Martin Luther King Drive
MC-A-110
Cincinnati, OH 45268
USA

Richard Y. Wang
Centers for Disease Control and Prevention
National Center for Environmental Health
Division of Laboratory Sciences
4770 Buford Highway, MS-F17
Atlanta, GA 30341
USA

Frank A. Witzmann
Indiana University School of Medicine
Biotechnology Research and Training Center
Department of Cellular and Integrative Physiology
1345 W. 16th Street
Indianapolis, IN 46202
USA

Kent Woodburn
Health and Environmental Sciences
Dow Corning
Midland, MI 48686
USA

and

The Dow Chemical Company
Toxicology & Environmental Research and Consulting
Midland, MI
USA

Raymond S. H. Yang
Colorado State University
Department of Environmental and Radiological Health Sciences
1680 Campus Delivery
Fort Collins, CO 80523
USA

1
Introduction to Mixtures Toxicology and Risk Assessment

M. Moiz Mumtaz, William A. Suk, and Raymond S.H. Yang

1.1
Chemical Mixtures Exposure

When humans are exposed to chemicals, they are not exposed to just one chemical at a time. A vast number of chemicals pervade our environment. Exposures, whether simultaneous or sequential, are to chemical mixtures. The standard definition of a chemical mixture is any set of multiple chemicals regardless of source that may or may not be identifiable that may contribute to joint toxicity in a target population [1, 2].

By some estimates, up to 6 billion tons of waste is produced annually in the United States. Several years ago, the US Office of Technology Assessment estimated 275 million of those tons were hazardous. Most waste finds its way to more than 30 000 toxic waste disposal sites across the United States, a majority of which the US EPA has categorized as uncontrolled hazardous waste sites [3]. Thus far, traditional risk assessment, even with its inherent shortcomings, has helped to control chemical exposures to that waste reasonably well, as evidenced by statistics on longevity, health status, and world population growth. Yet, new health and environment indicators have raised disquieting questions, and a consequent growing concern is that this success might be short-lived. One reason is an alarming, logarithmic increase in the synthesis, manufacture, and use of chemicals worldwide as "developed" and "developing" countries compete to provide their populations an improved quality of life. To help meet these concerns, the World Health Organization (WHO), as part of its harmonization of approaches project, recently published a report on methods and approaches for risk assessment of chemical mixtures [4, 5].

Former US Secretary of Defense Donald Rumsfeld once said with regard to intelligence reports

> There are known knowns. There are things we know we know. We also know there are known unknowns. That is to say, we know there are some things we do not know. But there are also unknown unknowns, the ones we don't know we don't know [6].

Principles and Practice of Mixtures Toxicology. Edited by Moiz Mumtaz

ISBN: 978-3-527-31992-3

Rumsfeld's wisdom also applies to the state of chemical toxicology, particularly to toxicology of chemical mixtures. Among the three categories, the *unknown unknowns* are, in the science of chemical mixture toxicology, the ones that cause the most worry.

Mixtures are one of the toxicology's huge unknowns. The concerns for chemical mixture toxicology's potential but unknown problems are illustrated by an examination of the presence of many chemicals, albeit at low levels, in our bodies. On December 10, 2009, the Centers for Disease Control and Prevention (CDC) released its Fourth National Report on Human Exposure to Environmental Chemicals [7]. This is the most comprehensive assessment to date of the exposure of the US population to chemicals in the environment. CDC has measured 212 chemicals in people's blood or urine, 75 of which have never before been measured in the US population. Similar to its three predecessors, but with expanded effort, this report contains exposure data for the US population for environmental chemicals monitored during 1999–2000, 2001–2002, and 2003–2004. The number 212 is not magic; it is merely the number of chemicals that could be identified and quantified per the established analytical laboratory protocol. The actual numbers of environmental pollutants in our body could be much higher than 212. The sample size ranged from hundreds to a few thousands, for example, a low of 1854 samples for 2,2′,3,3′,4,4′,5,5′,6,6′-decachlorobiphenyl (PCB 209) and a high of 8945 for cadmium or lead analyses. With such large sample sizes and the CDC staff's meticulous work, this report's results are widely viewed as a fair representation of those environmental chemicals and their respective concentrations that inhabit the general US population [7]. Recognizing, however, that associations are not causations, the CDC emphasizes in these reports that "... the measurement of an environmental chemical in a person's blood or urine does not by itself mean that the chemical causes disease."

Such a cautionary statement is understandable from a government agency responsible for public health. But toxicologically speaking, the 212 chemicals analyzed in the serum or urine samples were from a fairly large population sample. This then raises an important issue regarding the toxicological importance of a "mixture cocktail" in our bodies, albeit at very low concentrations. In many ways, this is the kind of *unknown unknowns* in chemical mixtures toxicology that should draw toxicologists to this real-life challenge: how do we assess the impact of these chemicals on current or future human health?

No one yet knows the answer for certain, but the question can be viewed from two different perspectives. A liberal perspective would hold that the presence of these chemicals in our bodies is merely a nuisance; they are the price paid for living in a modern, industrialized society that generates thousands of chemicals. These chemicals in our bodies are a necessary evil without any toxicological importance, particularly given such low levels. The average life span of our society is increasing, and the benefits derived from these xenobiotics outweigh their potential risks [8]. In fact, some scientists even believe that a small amount of any chemical might have certain beneficial effects [9–12].

But a second, more conservative and more cautionary perspective is that the presence of these chemicals in our bodies represents the toxicological *unknown unknowns*. These chemicals could potentially harm us. Thus, if we are to err, we should err on the safe side. Exposure to persistent chemicals such as metals, dioxins, and polychlorinated biphenyls (PCBs) could lead to their accumulation in our bodies and lead further to increasingly high tissue concentrations. Several lipophilic, persistent organic pollutants (POPs) can, for example, concentrate in breast milk and during pregnancy and through lactation expose the growing fetus and babies. Such chemical body burdens and their possible variations from person to person, together with the unlimited combinations of chemical mixtures that may be inherent in human populations, are beyond the capacity of traditional toxicity testing. The precautionary principle could be a solution, founded as it is on the use of comprehensive, coordinated research to protect human and environmental health. When the Collegium Ramazzini opened in 1983, Professor Irving Selikoff wrote, "Science is no stranger to uncertainty and incomplete data. The Collegium will utilize science to help unlock the rigidities of those fixed in legalistic and regulatory combats that prevent progress in environmental and occupational health" [13].

As environmental health hazards become increasingly complex and international in scope, this principle might play an increasingly important role. Its spirit is embodied in the 1992 Rio de Janeiro declaration: "Where there are threats of serious or irreversible damage, scientific uncertainty shall not be used to postpone cost-effective measures to prevent environmental degradation." That chemical mixtures or complex exposures are part of our lives is a subject of increased realization and awareness; to play a central role in environmental protection and public health, traditional risk assessment must accommodate such challenges [14]. The US Congress enacted, for example, clean air, clean water, and environmental laws because of concerns that contaminants in air and water and hazardous waste might cause adverse health effects before development of any antidotal, comprehensive body of biomedical science [15]. But such science is possible only by continued coordination and collaboration of efforts to develop alternative methods and transparent strategies; science that allows dynamic participation of data generators, data users, researchers, stakeholders, and decision makers.

Almost all applied and basic science underpinning current regulations tested one chemical at a time. Several US environmental laws including Superfund,[1)] the Safe Drinking Water Act Amendments (SDWAs), and the Clean Air Act (CAA) have acknowledged the significance of potential exposure to, and the health effects of, chemical mixtures. The Food Quality Protection Act (FQPA) even states that mixtures are the rule rather than the exception. Thus began the groundwork for a new approach to the study of chemical mixtures. Now, when calculating whether exposures exceed tolerance levels, compounds with similar mechanisms of action must receive joint consideration. FQPA has therefore initiated the cognitive transition – and the logical progression – from single to multiple chemical risk evaluation.

During the past two decades, quantitative risk assessment research has devised formulas, written documents, held workshops, and developed guidelines to address chemical mixtures issues [16, 17]. Recently, guidelines and guidance have incorpo-

1) The Comprehensive Environmental Response, Compensation and Liability Act of 1980, as amended by the Superfund Amendments Reauthorization Act (SARA) of 1986.

rated chemical interaction concepts and have suggested methods to evaluate the possible influence of such interactions on the overall joint toxicity of chemical mixtures [1, 2, 18].

The methods these guidelines propose have been, and could be applied to contrived mixtures, simple mixtures, complex mixtures, particulate matter, food additives, intended and unintended exposures of short-term and long-term episodic durations, and to environmental stressors. The evolution of the methods to evaluate the joint toxicity of mixtures and their success have provided the confidence to apply them to real-world exposures.

The past two decades have undoubtedly witnessed the gradual maturing of the toxicology of chemical mixtures. Symposia and major conferences dedicated to chemical mixtures have appeared with increasing frequency at annual meetings of major scientific societies, specifically the Society of Toxicology [19], the Society of Risk Analysis, the European Conference on Combination Toxicology, the ATSDR International Conference on Chemical Mixtures 2002 [20–24]. Concerns about chemical mixtures prompted the Society of Toxicology to establish a mixtures specialty in 2007, and mixtures were the theme of the 2009 annual meeting of Society of Toxicology, Canada. Recently, the ILSI Health and Environmental Sciences Institute (HESI) risk assessment methodologies technical committee convened a working group of academic, government, and industry representatives to explore and to improve methodologies available for assessing mixtures risk. The team elected to explore screening-level risk assessment methodologies that could address risks from low-dose exposure to mixtures and completed a critical analysis of chemical interactions and their magnitude [25]. As is true with any developing area, issues remain that are unique to chemical mixtures. Researchers must deal with an array of concerns about chemical mixtures, and many factors affect research and methods' development, including scientific advances, expert opinions, regulatory needs, administrative priorities, public interest, and legislative actions [26].

1.2
Superfund Research Program

The National Institute of Environmental Health Sciences (NIEHS) Superfund Research Program (SRP) was created as a network of multi- and interdisciplinary teams of researchers. Its purpose is to address broad, complex health and environmental issues that arise from the multimedia nature of hazardous waste sites – particularly, from both long-known and emerging environmental contaminants (http://www.niehs.nih.gov/sbrp) [27]. By creating multiproject and multidisciplinary programs, SRP encourages and fosters partnerships among diverse scientific disciplines. Recent technological advances have the capacity to stimulate interdisciplinary research in such disciplines as follows:

- Genomics, proteomics, and metabolomics technologies;
- Molecular, cellular, and whole-animal imaging methodologies;

- Miniaturized tools/technologies (i.e., at the micro- and nanolevel); and
- Improved cyber infrastructure and bioinformatic tools to gather, assimilate, and interrogate large diverse data sets.

Establishing multidisciplinary research programs provides a more comprehensive understanding of complex environmental issues. The knowledge gained through these research efforts has proven useful in supporting decisions made by state, local, and federal agencies, private organizations, and in industries involving the management of hazardous substances.

The mandates under which the SRP operates provide a framework that has, for example, allowed the NIEHS the latitude to address a wide array of scientific uncertainties facing the national Superfund Program (http://www.niehs.nih.gov/research/supported/srp/about/index.cfm). These mandates include the development of

1) Methods and technologies to detect hazardous substances in the environment;
2) Advanced techniques for the detection, assessment, and evaluation of the effect of hazardous substances on human health;
3) Methods to assess the risks to human health presented by hazardous substances; and
4) Basic biological, chemical, and physical methods to reduce the amount and toxicity of hazardous substances.

The methods grew out of Congress's recognition that the strategies for the cleanup of Superfund sites and the technologies available to implement these cleanups were inadequate to address the magnitude and complexity of the problems. Congress accordingly enacted the Superfund's 1986 SARA amendments.

To address the complex, interdependent, yet fundamental issues that arise in relation to hazardous waste integration, cooperation is needed from many disciplines. A holistic approach that borrows theories and methodologies from many diverse scientific disciplines is the future for integrated environmental health as it relates to the Superfund [28]. Ultimately, this approach will enable basic research findings to transition into epidemiological, clinical, ecological, and remediation studies, all of which are important for the public health decision-making process.

A central SRP premise recognizes the link between chemical exposure and health effects leading to disease outcome will assist in understanding, identifying, and establishing new or improved prevention and intervention modalities (Figure 1.1) [29]. Contributing factors that modulate the exposure–disease paradigm include

- Temporal factors (age and developmental stage);
- Spatial factors (geographic locations);
- Genetic factors (single-nucleotide polymorphisms (snps), methylation patterns); and
- Unique circumstances (e.g., comorbid conditions, nutritional status, etc.).

The considerable interplay between exposure and response results in "real-world" mixture exposures with widely varied biological effects. Ultimately, developing

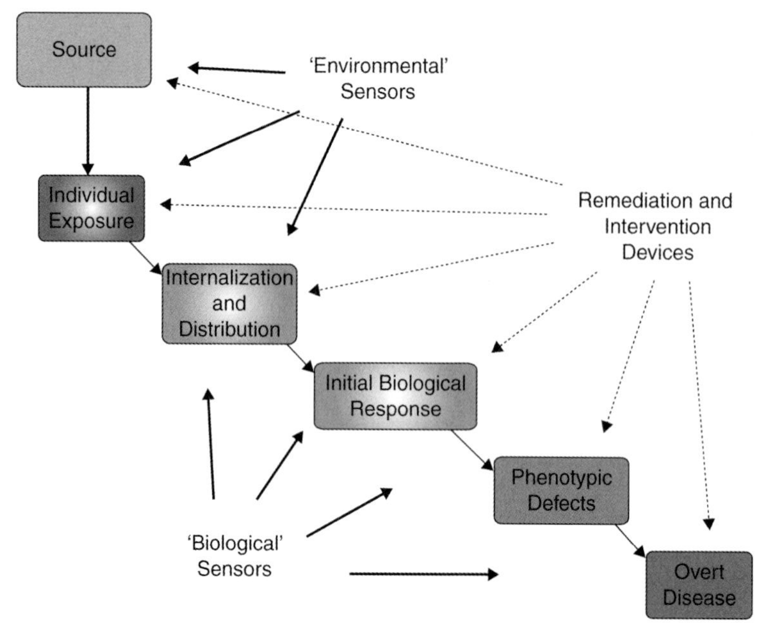

Figure 1.1 The environmental health sciences working paradigm of environmental exposures leading to disease through a cascade of events.

strategies must consider this foundational fact, not only to minimize the effect of exposure on disease risk but also to develop risk assessment models that incorporate these diverse parameters. Only in this way will science provide the biologically relevant information needed to make informed, human-health protective choices.

Thus, SRP-supported research is a continuum from basic to applied research. Its state-of-the-art techniques improve the sensitivity and specificity for detecting adverse effects on humans or on ecosystems exposed to hazardous substances. This research also promotes a better understanding of the underlying biology responsible for such adverse effects.

1.3 SRP and Mixtures Research

From a public health perspective, the inability to predict whether chemical agents act in an additive, synergistic, or antagonistic manner at concentrations encountered in the environment creates real problems for human health risk assessment. Many examples confirm that interactions of chemicals with each other or with other physical or biological agents affect health to a greater extent than would have been predicted given the toxicity of individual components. A critical issue related to hazardous waste sites for remediation or health effects research is that the concentrations at which chemicals occur in the environment are extremely low, and exposures are long-term and continual with simultaneous exposure to multiple chemicals. Thus, whether the subject is remediation strategies, exposure to humans or ecosystems, site characterization, bioavailability, or the development of risk assessment models, chemical mixtures are an issue of concern. Furthermore, real-life scenarios rarely reflect biomedical research, exposure assessments, or remediation strategies based on exposures to single substances in isolation. Indeed, such oversimplification fails to consider

- Prior exposure history and vulnerability (i.e., susceptibility);
- Interactions with other stressors of similar/dissimilar mechanisms of action;
- Potentiation or sensitization by chemicals not toxic in themselves; and
- Interaction of chemicals that could lead to synergistic or antagonistic effects [30].

In fact, the majority of diseases are the consequence of both environmental exposures and genetic factors [31]. Individual susceptibility to environmentally induced disease is a source of uncertainty. A better understanding of genetic influences on environmental response could lead to more accurate estimates of disease risks and could provide a basis for disease prevention. Researchers, environmental policy makers, and public health officials are challenged to design and implement strategies to reduce human disease and dysfunction resulting from exposure to chemical mixtures.

Interactions among mixture components and gene–environment interactions should be seen not as a limit to scientific progress but as a challenge to develop more complex and sensitive methods. To address the complexities and uncertainties surrounding human exposures to mixtures of chemical contaminants, researchers must fully utilize and integrate cellular and molecular biology methodologies, mechanistically based short-term toxicology studies, computational technology, and mathematical and statistical modeling [32, 33].

With continued development of and refinement in the available repertoire of advanced tools and approaches, the ability to better assess the effect of mixtures on human health is reachable. The types of research related to mixtures important to addressing issues within the SRP include the following:

- Development of computational toxicology approaches to understand dose/effect relationship in the context of chemical interactions;
- Application of high throughput functional assays to define critical mechanistic end points associated with potential adverse biological consequences of exposure to chemical mixtures;
- Integration of diverse data sets to develop biologically based predictive models for chemical mixtures;
- Application of metagenomics to understand the impact of chemical mixtures on the structure and function of microbial communities;
- Development of nanotechnologies to detect and measure individual components within complex mixtures in real time;
- Development of innovative approaches to remediate chemical mixtures in environmental media; and

- Adaptation and application of fate and transport models to predict and assess the influence of chemical mixtures on the efficiency and effectiveness of applied remediation approaches.

Multi-, inter-, and transdisciplinary research strategies are not easy to implement; many government, industry, and academic programs tend to foster and reward narrow approaches to problem solving. The NIEHS SRP, however, serves as a model of a successful program where biomedical researchers cooperate and collaborate with, for example, ecologists, engineers, and mathematicians. This results in creative synergisms and novel approaches to address complex problems, especially the problems of chemical mixtures at Superfund sites.

1.4 Drug–drug Interactions and Nanomaterials

For those interested in the toxicology of chemical mixtures, two areas of toxicological sciences – drug–drug interactions and nanomaterials – are particularly challenging. The former, though a long-time issue in the pharmaceutical industry, remains a toxicological *unknown unknown*. With regard to serious toxicological interactions, it has not received attention it deserves.

And nanomaterials are a completely new area. The challenges to toxicologists are particularly relevant from the perspective that many nanomaterials are "chemical mixtures" and their unique physicochemical properties would raise some highly unusual physiological, biochemical, and toxicological issues (see Chapter 21). These two areas warrant some special discussion.

Before prescribing multiple drugs, some physicians consider potential drug interactions. Physicians try to minimize these interactions by taking into consideration the time needed for each drug to reach maximum blood concentration, its half-life ($t_{1/2}$), its bioavailability, and its mode of action. Until recently, however, institutions would not allocate resources to study toxicological interactions from multiple drug intakes. In addition to combination therapy (i.e., polytherapy or polypharmacy), multiple drug intake could easily be realized when different doctors treat patients for multiple illnesses, particularly aging patients. As the exposure dose levels from drug intake are usually much higher than are doses of environmental chemicals, drug–drug toxicological interactions can become a serious problem. A number of case studies quoted below provide a glimpse of the seriousness of this issue.

Using a meta-analysis, in 1994 over 2.2 million cases of serious adverse drug reactions (ADRs) occurred among US hospital patients [34]. During hospital stays, the patients were given an average of eight drugs. Some 106 000 serious drug interaction cases were fatal, making ADRs the fourth to sixth leading cause of death for that year in the United States. In 1998, the US Food and Drug Administration (FDA) established the "Adverse Event Reporting System." Data analysis collected under this system revealed that from 1998 through 2005, serious adverse drug reactions increased 2.6 fold, from 34 966 to 89 842, and fatal adverse drug incidence increased 2.7 fold, from 5519 to 15 107 [35]. These results highlight the importance of ADRs as a public health problem.

An exposure situation and its related complications may also influence toxicological interactions. The anesthetic agent Fluroxene was safely used in clinical medicine for almost 20 years before the first fatal incident [36, 37]. In 1972, an epileptic surgical patient who was on a regimen of phenobarbital and diphenylhydantoin died within 36 h of operation due to massive hepatic necrosis. That fatal lesion was quickly confirmed with experimental animal toxicology studies; the cause of death was attributed to potentiation of fluroxene hepatotoxicity by phenobarbital and diphenylhydantoin through enzyme induction [36, 37].

In an experimental toxicology study, 7-day old infant rats were administered a combination of drugs commonly used in pediatric anesthesia (i.e., midazolam, nitrous oxide, and isoflurane) in doses sufficient to maintain a surgical plane of anesthesia for 6 h [38]. Researchers observed that such a common therapeutic practice combination in the infant rats caused widespread apoptotic neurodegeneration in the developing brain, deficits in hippocampal synaptic function, and persistent memory/learning impairments.

The intrinsic functions of the subject exposed to chemicals may modulate toxicological interactions. Renal dysfunction may change drug disposition such that the likelihood of drug–drug interactions would increase. A clinical example is the interaction between aminoglycoside antibiotics and penicillins in patients with impaired renal function [39]. In solution, these antibiotics bind to inactivate each other, but the reaction is slow. Because penicillin(s) is usually given in great molar excess to the aminoglycosides, the major consequence of such drug–drug interaction is inactivation of aminoglycoside to subtherapeutic concentration. This interaction, however, seems to occur only in patients with renal dysfunction. The reason has been attributed to the retention of both antibiotics in patients with impaired renal function, thereby allowing sufficient time for this interaction to take place.

Manufactured nanomaterials is the second area of science that toxicologists believe will pose a challenge to the study of chemical mixtures in the foreseeable future. The advancement of nanotechnology in the twenty-first century is probably so important that it represents yet another phase of the Industrial Revolution. Some estimates are that in few years, worldwide commerce involving nanomaterials will reach $1 trillion [40]. At present, more than 600 commercial products are known to contain nanomaterials [41]. Because these nanoparticles are invisible – usually under 100 nm in diameter – and because nothing much is known about their toxicities, concerns have been raised about their health effects on humans [42, 43]. Many of these nanomaterials have a core that consists of a number of metals [40, 44], hence arises the chemical mixtures issue. Moreover, nanomaterials have some unique physicochemical properties – some have rather persistent tissue pharmacokinetics [43, 45, 46]. In one of the first published physiologically based pharmacokinetic (PBPK) modeling papers on a nanoparticle, Quantum Dot 705 (QD705) in mice (Figure 1.2), the authors pointed out that such unique and worrisome pharmacokinetic nanoparticle properties might have a silver lining [43, 46]. That is, while the persistence of QD705 was of health concern specifically in the spleen, kidney, and

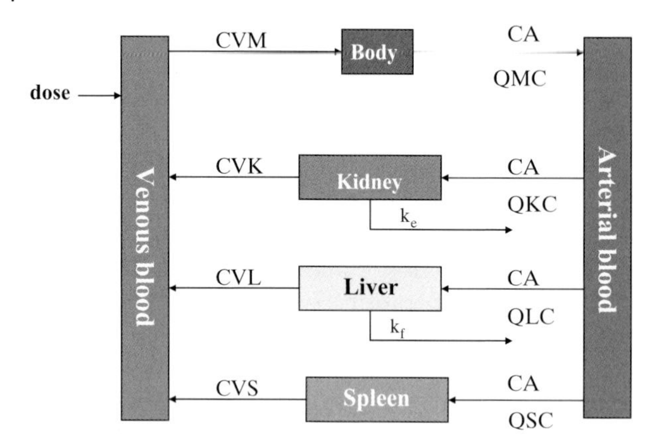

Figure 1.2 A conceptual PBPK model for QD705 in mice. CVM, CVK, CVL, and CVS represent QD 705 concentrations in venous blood, kidneys, liver, and spleen, respectively. CA is QD 705 concentration in arterial blood. QMC, QKC, QLC, and QSC represent blood flow to body, kidneys, liver, and spleen. (Reproduced with permission from Yang *et al.* (2008) *Environ. Sci. Technology*)

liver for up to an experimental duration of 6 months, the nanoparticles' affinity toward these tissues might be exploited to design drug delivery systems for potential targets in these same tissues. Thus, nanomaterials' unique properties will undoubtedly present an important future challenge for scientists in the environmental and occupational toxicology and risk assessment areas.

1.5 Waste Sites and Mixtures Risk Assessment

Communities near waste sites – particularly Superfund sites – can potentially be exposed to low levels of a wide range of chemicals originating from the site. Communities can also be exposed to various other environmental chemicals from nearby manufacturing, transportation, and other sources. At very low-level exposures, human populations do not show any observable health effects. Chemical(s) remain as body burdens showing no discernible effect on a person's overall health. Physiologically, the body adjusts to the presence of chemicals at this level through adaptive mechanisms. As the pollutant exposure levels increase, some effects may be observed.

But effects such as enzyme induction and certain biochemical and subcellular changes may be of uncertain importance. At this level of pollutant exposure, the body may have compensatory mechanisms [47]. Yet, as pollutant levels continue to increase, significant, readily observable adverse effects may ensue. At these higher pollutant levels, the body has exhausted its adaptive and compensatory mechanisms, and its functioning could be compromised. Such adverse effects could lead to organ function impairment through compromise of physiological processes, leading to pathophysiological changes such as fatty changes and necrosis resulting in significant organ function impairment. Exposure to higher levels of pollutants could lead to morbidity and ultimately to death. In this continuum of effects, exposures from multiple sources may cause some persons to cross the threshold for adverse health effects. Considering that the human population is heterogeneous and therefore lacks biochemical characteristic homogeneity, some persons within the population will be more susceptible than others to adverse effects. At the either end of the bell curve then, a small fraction of the population may be hypersensitive to pollutant burdens and exhibit adverse effects to levels of exposure that may otherwise be considered low. Moreover, as emphasized in a recent National Research Council report [48], both endogenous and exogenous background exposure and underlying disease processes contribute to population background risk by affecting the dose–response relationships of environmental chemicals.

The goal of waste-site risk assessment is to ensure "healthy people in healthy environment" through protecting the public from unintentional exposures to toxic substances. Determining the health risks of complex mixtures is daunting both to toxicologists using experimental approaches and to epidemiologists using observational approaches. Risk assessment is a four-step process that includes hazard identification, dose–response assessment, exposure assessment, and risk characterization [49]. Just as researchers often confront large data gaps, chemical mixture risk assessment of waste sites is often limited, incomplete, or inconclusive. Hazard assessment is the fundamental basis of the overall risk assessment process. If data were available on the whole mixture of concern, a toxicity index analogous to MRLs/RfDs would be calculated for the mixture [50]. Often, however, whole mixture data are not available; they often are available for some but not all mixture components. In such cases, the hazard index (HI) approach uses the doses of the individual mixture components after they have been scaled for toxic potency relative to each other. In practice, a screening level analysis is performed, summing across all target organs. If the HI value exceeds 1, this initial analysis is repeated by developing effect-specific HIs. Conceptually, this approach helps to construct the plausible toxicity index of a mixture that would have been calculated had the mixture itself been tested. Using this HI approach, if exposure or toxicity screening data are unavailable for all the components of the mixture, the risk could possibly be underestimated [51].

When using the HI approach, its limitations should be understood and special attention should be given to multiple target toxicities, the role of chemical interactions [52], and novel or new toxicity end points. Rarely does a chemical have single end point toxicity. Most chemicals cause multiple toxicities and cause them in multiple target organs in multiple cell types as a function of dose (Figure 1.3). Single chemicals can affect multiple organs/end points as a function of dose, and multiple chemicals can affect a single organ or system. For example, lead (Pb) can affect nervous, reproductive, and hematopoietic systems. On the other hand, arsenic, cadmium, chromium, and Pb can affect the nervous system, thus increasing the

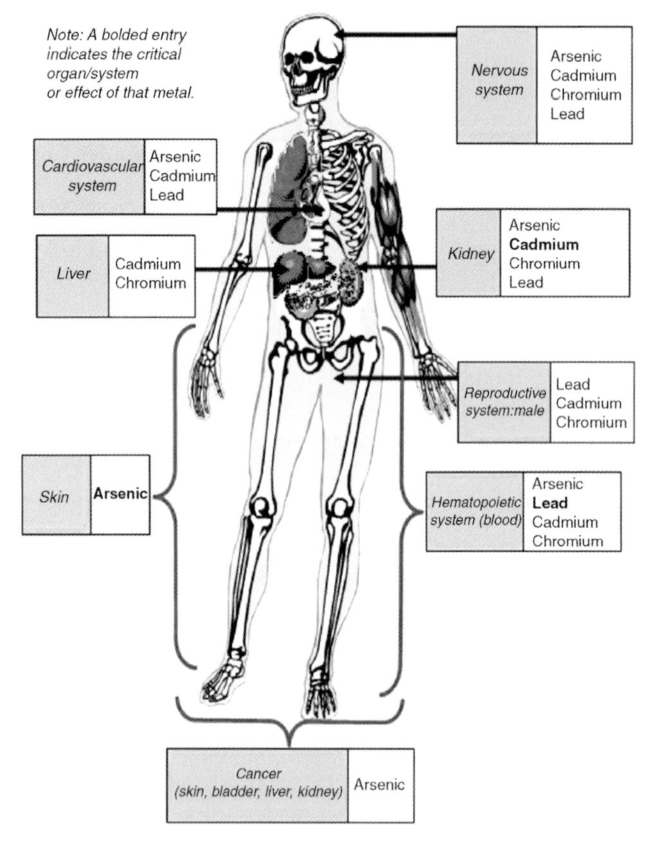

Figure 1.3 Single chemicals can affect multiple organs/end points and multiple chemicals can affect a single organ or system as a function of dose.

chances of chemical interactions and increasing overall joint toxicity. The point is that effects caused in the nervous system could be quite different from those caused in the reproductive system or in the liver or kidney, resulting in different disease outcomes. Often hazard assessment of chemical(s) is limited to critical effect or most sensitive effect. A full understanding of chemical mixtures' potential hazards is essential and a thorough evaluation of multiple end points is achievable. In this regard, a physiologically based, pharmacokinetic/pharmacodynamic model may serve as an integrator for all the relevant physiological and toxicological processes in the body – the essence of systems biology ([53–55], see Chapter 22). A full range of multiple toxicity values should be derived for all the secondary effects of a chemical component, analogous to its critical effect. Thus, if minimal risk levels (MRLs), reference doses (RfDs), threshold level values (TLVs), and other allowable levels are derived for hepatotoxicity as critical effect, then analogous values the target organ toxicity doses (TTDs) should be derived for all secondary effects such as nephrotoxicity, hematotoxicity, and immunotoxicity [56]. At times, in the absence of experimental toxicity data, computational tools such as structure–activity relationships (SAR) models can be used to derive such values.

The second aspect for consideration while using the HI approach is the role of potential of chemical interactions in the overall expression of chemical mixture joint toxicity. Ample studies demonstrate that chemicals can interact with one another and at times, by influencing the toxicity of other components of the mixture, can increase or decrease a mixture's overall toxicity. People are exposed to complex and highly variable mixtures of chemicals of naturally occurring and synthetic origin. The body in general disposes of all natural or synthetic chemicals by the same limited pathways. Thus, the probability arises of simple or complex interactions occurring at multiple levels in an organism. These interactions could be toxicokinetic (see Chapter 9) and toxicodynamic in nature (see Chapter 6); for realistic and accurate risk assessments, the interactions' consideration and, if needed, their integration into the overall toxicity assessment of a mixture, is important [57]. Often, this type of interaction assessment might lead to the conclusion that the interactions are insignificant, but it will serve the purpose by alleviating the concerns of communities cognizant of exposure to chemical mixtures.

Another more sophisticated approach – PBPK modeling – has also been used to study, validate, or verify interactions ([51, 53–55, 58, 59], see Chapter 20). Many early PBPK modeling efforts were based on the SimuSolv software. But at present, support for that is not forthcoming. More recently, the Advanced Continuous Simulation Language (ACSL; AEgis Technologies, Huntsville, AL) and Berkeley Madonna (the University of California, Berkley, CA) are being widely used. In addition to these dedicated computer software packages [60], the application of a spreadsheet program to support a PBPK model has also been demonstrated [61], and the Trent University (Peterborough, Ontario, Canada, updated 2003) has made available a spreadsheet program to run PBPK models.

PBPK models are mathematical representations of the animal or human body that group tissues or organs into compartments. Physiological and anatomical considerations of the sizes and blood flow of the organs they represent dictate the characteristics and links between these compartments. Thus, the model simulation is the resolving of a set of equations. These models were originally developed to understand the relationship between dose delivered to a target organ/tissue and its toxic response (s). Because of their increased biological relevance and reliability (fidelity), these models are now applied to study various aspects of toxicity of chemicals and interactions. During the past two decades, several mixtures, their mechanisms of interactions, and in some instances the threshold of such interactions, have been studied using PBPK models (see Chapter 7). Through these mixtures modeling exercises, great insights were acquired into the most commonly occurring mechanisms of interactions in biological systems such as competitive, noncompetitive, and noncompetitive enzyme inhibition or enzyme induction. With experience, increasing sophistication has been incorporated into new models for evaluating defined mixtures consisting of -2, -3, -4, -5 components and complex mixtures.

The third and last issue to consider for the HI approach is the possibility of new or novel toxicities the chemical mixtures might cause that individual components might not cause. This can happen when a shift occurs from chemical-specific, toxic responses to mixture-specific responses. If interactions occur (in the toxicokinetic or toxicodynamic phase), mixtures are likely to induce effects not seen in the individual chemicals. In both similar or dissimilar mechanisms of action, novel effects of mixtures are not likely to occur at low dose or no observed adverse effect levels of individual components. At high dose or adverse effect levels of the individual compounds, however, novel adverse effects of the mixture may occur and indeed have been observed [62, 63]. Such studies also show that some of the adverse effects seen with the individual chemicals are not found after exposure to the mixture at comparable dose levels.

In this challenging era of toxicology, application of transcriptomics,[2)] proteomics, and metabolomics are proving to be powerful tools. Transcriptomics using expression microarrays has provided increased insight into toxic actions and has led to the findings of gene expression signatures associated with types of action such as genotoxic or nongenotoxic carcinogenicity, peroxisome proliferators, oxidative stressors, and others [64–71]. These gene expression signatures are valuable in predicting the potential toxic actions of new compounds. Importantly, transcriptomics also provides information on the pathways and molecular processes affected by chemicals. While the number of publications on transcriptomics of individual chemicals is increasing, very few studies have applied transcriptomics to the effects of chemicals within mixtures. Transcriptomics can, however, address such important issues as

- Are profiles of the mixtures a simple sum of the profiles of the individual compounds or do one or two compounds dominate the effect of other toxins?
- Which processes and pathways are affected by the compounds?
- Do the mixtures affect genes or processes not affected by individual compounds?
- Which of the affected genes and processes can be linked to the pathology and clinical data?
- Can transcriptomics detect the initiation of potential harmful processes not detected by the classical toxicology methods?

Toward this end, recent transcriptomic studies with mixtures have shown several genes affected by the ternary mixture but not by single compounds or binary mixtures' synergistic action [72]. In the liver, the high-dose ternary mixture upregulated 57 genes, not significantly upregulated by any individual compound (Figure 1.4). Only 8 of these 57 genes were upregulated by at least one of the binary mixtures, leaving 49 genes uniquely upregulated by the ternary mixture. These 49 genes included those that influence cellular proliferation, apoptosis, and tissue-specific functions – a majority of these are stress genes not induced by individual chemical components [72]. The highly sensitive results from such new techniques need to be integrated into the hazard assessment step; the results allow detection and evaluation of end points undetected by classical toxicological testing. Thus, by integration of such molecular biomarkers into

2) The study of the transcriptome, that is, the complete set of RNA transcripts produced by the genome at any one time.

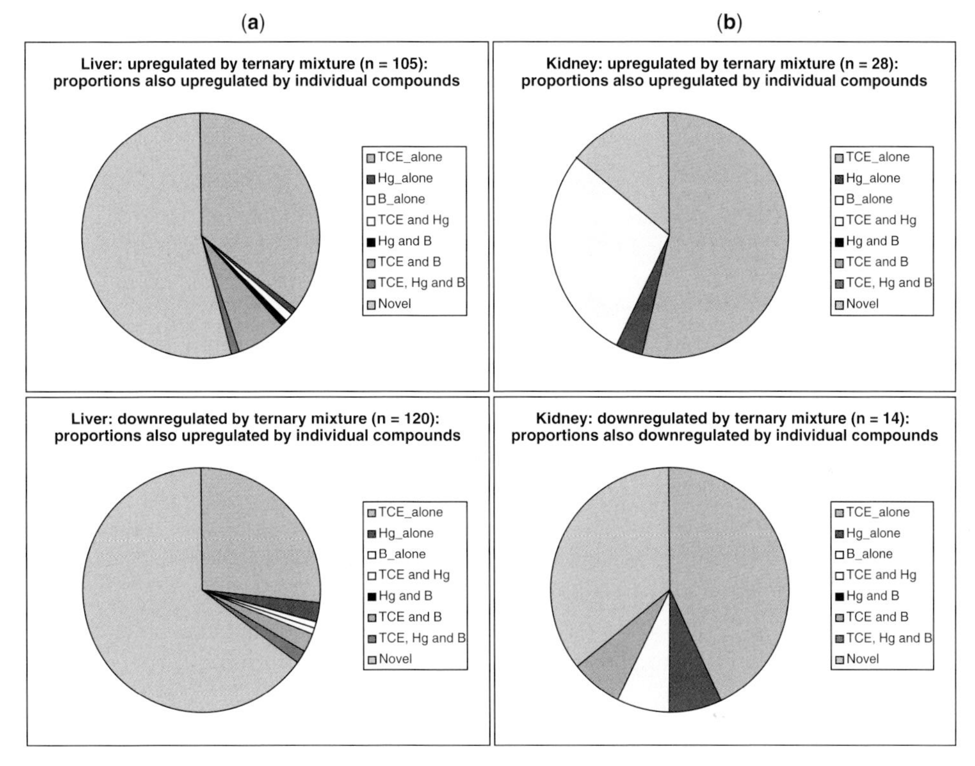

Figure 1.4 A majority of genes induced by a ternary mixture are stress genes not induced by individual chemical components.

the overall assessment process, unexpected outcomes following exposure to chemical mixtures can be avoided.

The first key step of the risk assessment process is accurate exposure assessment. For an accurate and realistic hazard assessment, the identification of all chemicals and stressors and their exposure assessment should be as complete as possible. This includes a thorough documentation of chemical mixtures and their compositions and concentrations, including bioavailability information (see Chapter 2). More recently, cumulative risk assessment – that is, the combined risk from aggregate exposures via multiple routes to multiple stressors, namely, chemical, biological, physical, and others – is gaining recognition as a pragmatic approach to characterize real-world risk [73] (see Chapter 10). Only with such advances can risk characterization of hazardous waste sites that integrates all the available information on toxicity of chemicals and their mixtures project, with some certainty, the frequency, as well as the severity of adverse health effects in potentially exposed populations. The fuller the risk characterization, the easier the comparison of the results of toxicological assessment with those of epidemiological studies to establish cause and effect relationships. Thus, the weight of evidence regarding human health effects of chemical mixtures should be derived from emerging evidence in the broad areas of toxicology and epidemiology. Once such a relationship(s) is established, steps can be taken for remediation of hazardous waste site and protection of public health – the fundamental goal of risk assessment.

1.6 Alternative Testing Methods

In August 2005, Hurricane Katrina brought unprecedented destruction to the Gulf Coast. Huge storm surges, widespread wind damage, and flooding of New Orleans displaced hundreds of thousands of people, damaged thousands of homes beyond repair, and disrupted thousands of lives and businesses. The City of New Orleans was particularly hard hit; its levees broke, flooding large parts of the city. Apart from the hundreds of lives lost, several chemicals were released from storage into the environment, chemicals that industries used as intermediates, reagents, and catalysts, mostly with unknown toxicity. Since Hurricanes Katrina and Rita, personnel from the multiagency unified command in Metairie, Louisiana, made important advances in the assessment, investigation, and oversight of the environmental cleanup efforts in southeast Louisiana. The unified command agencies, along with their local, state, and federal partners, have recovered millions of pounds of hazardous material/oil and disposed of large amounts of debris.

Such emergency situations and the shortcomings of the risk assessment community in providing solutions to real-life challenges brought a new awareness in the public both affected and unaffected by the happenings. This awareness has turned into demands for issue resolution pertaining to unknown chemical toxicities. In response, alternative testing protocols are being developed that could save time and could husband resources. The Interagency Coordination Committee for Validation of Alternative Methods (ICCVAMs) Authorization Act passed by the US Congress has authorized tests that will also achieve the goal of humane treatment of animals by reducing, refining, and replacing animal toxicity testing [74]. The ICCVAM has undertaken a mixture toxicity testing study of new products submitted for US EPA registration, 89% of which are chemical mixtures. They belong to a variety of chemical classes, including anti-inflammatory/analgesic agents, respiratory stimulants, barbiturates, pesticides (insecticides, herbicides, nematicides, algicides, and fungicides), and surfactants. This study was undertaken for the evaluation of the *in vitro* neutral red uptake (NRU) basal cytotoxicity test method. Researchers wanted to determine its usefulness for predicting the *in vivo* acute oral toxicity of chemical mixtures. One of the study's goals is to assess the relevance, including the accuracy, of NRU in an *in vitro* cytotoxicity assay for estimating rat oral LD_{50} values of mixtures representing the five Global Harmonization System (GHS) categories of acute oral toxicity [75].

Through these activities, toxicology adopts a more aggressive approach to overcome the data paucity – toxicology now accesses a broad panel of *in vitro* assays that the drug discovery industry has been using for years [76]. Several newly developed assays are now available that allow chemical bioactivity evaluation using an array of protein family pathways, critical cell signaling pathways, and cell health parameters. Building these assays into a series of screens to decipher the mechanisms of toxicity remains a formidable challenge in itself, but it can make the resulting information usable in risk assessment.

So much optimism surrounds this approach that several US government agencies and other organizations around the world are integrating these methods for high-throughput screening of chemicals. Noteworthy government-funded programs include ToxCast and Tox21. Registration, Evaluation, Authorization, and Restriction of Chemical substances (REACH) is a European initiative that will generate experimental data at a pace never duplicated in the history of toxicology. Tox21, a collaborative project of NIH, US EPA, and NIEHS combines advance automation and a growing assortment of *in vitro* assays and computational methods to reveal the interaction of chemicals with biological targets. The *in vitro* methods used an array of biochemical (e.g., metabolic kinase, multiple protein pathways, and protease) and cell-based assays (e.g., nuclear receptor, phenotypic, protease, signaling, and spicing). These assays can assess cell viability, nuclear receptors, pathways, and DNA damage.

ToxCast, a US EPA project, evaluates chemical properties and bioactivity profiles using a broad spectrum of gene assays, proteins, and metabolites that comprise the cellular "interactome" [77]. This data can help develop methods of prioritizing chemicals for further screening and testing to assist US EPA programs in the management and regulation of environmental contaminants and their mixtures. These pathways could serve as a good middle ground between biochemical or other target-focused assays and more phenomenological, phenotypic, or high-content assays. An important complement to ToxCast data will be that they are obtained from assays for detecting biotransformation and complex toxicities that use complex formats of human, nonhuman primate, or rodent cells. The ToxCast data can help identify overall patterns across many assays and data types that could be toxicity predictors. This type of testing takes advantage of HTS and toxicogenomic technologies for bioactivity profiling of environmental chemicals related in structure or in mechanism of action. Although the primary purpose is not to identify mechanisms of action of environmental toxicants per se, this might be a future benefit of the ToxCast program.

A scientific-method development norm is that whenever a new method is developed, it is compared with existing methods to show its advantages and define its limitations. The National Toxicology Program (NTP) has conducted methodical toxicological testing for the better part of this century. The results obtained using these new alternative methods should be compared and correlated with historical data, specifically those generated by the NTP. Similar correlative research would use data generated through those previous US EPA and FDA programs that guided toxicity testing for specific registration and regulation purposes.

Paralleling this newfound evolution of immense data generation openness and transparency is the emergence of data sharing. For example, ToxCast is making all of its data publicly available [78]. Establishment of databases such as ACToR, eChem-Portal, REACH, and Comparative Toxicogenomics Database (CTD) [79] will enhance data mining and interpretation. ACToR is a central database of toxicity information for thousands of chemicals that can be accessed to study chemical toxicity. Bioinformatics, the science of turning data into information, will play a critical role in experimental design and conduct of chemical toxicity studies.

Independent of these developments, in two separate reports, the National Research Council recently emphasized the use of toxicogenomics to link biological response indicators (biomarkers) to toxicity mechanisms, an approach susceptible of ready application to chemical mixtures [80]. These recommendations could enhance efforts to evaluate and remediate Superfund sites, and to reduce their effects on human health from exposure to chemical mixtures. The National Academy of Sciences (NAS) [81] has also recommended yet another alternative approach to traditional toxicity testing: the use of a complex array of animal/human studies and bioassays for the identification of toxicity pathways. This approach uses a systems understanding of the interconnected pathways composed of complex biochemical interactions operative in normal human and animal functioning. Following identification of these pathways, their qualitative and quantitative perturbations should be studied as a function of exposure to chemicals or their mixtures. This recommendation is predicated on the hypothesis that when sufficiently large biological perturbations of these pathways occur following exposures to chemicals or their mixtures' toxicity, adverse health effects or diseases result. For this to happen, the perturbations must be large enough that they exceed the adapting capacity of the host organism, namely, animals or humans. Thus, the degree of toxicity is host-specific and therefore dependant on a person's underlying health and disease status and on his or her individual ability to adapt [81].

Knowledge about the dose–response relationship (including its shape and slope) is a major factor in describing the toxicological characteristics of chemical(s) and their mixtures. The dose–response curve plays a key role in the assessment of health risks. Although toxicology deals with adverse effects and not with physiological changes, this latest NAS recommendation brings into focus the transitional research between physiological and harmful effects, assuming that nonadverse physiological changes (often controlled by homeostatic processes) seen at lower doses precede adverse effects observed at higher doses. Detailed knowledge of the relationship between biochemical changes at lower dose levels and adverse effects seen at higher doses will improve our understanding of mixture toxicology and will significantly contribute to a more knowledge-based risk assessment. Ultimately, this NAS recommendation will require standardization of magnitude of toxicity pathway perturbations and deviations from normal functioning of biological systems and will relate to toxicity and disease outcomes.

1.7
Translational Research

The scientific community increasingly recognizes that an understanding of risk requires consideration of the characteristics of the host population, the environmental chemical or chemical mixture, and the exposure milieu. If these factors are not adequately addressed, the shape and low-dose characteristics of dose–response relationships for environmental toxicants may be substantially misrepresented. Also, though recognized as scientifically important, site assessments give little if any consideration to nonchemical stressors, to population vulnerability, or to various background exposure and other risk factors. Affected stakeholder communities often question risk assessments as inadequate and point to their narrow focus and lack of comprehensive scope. Long-term basic research needs to generate the underlying scientific understanding to support assessments that would more realistically characterize low-dose risk. Any strategic planning exercise should consider promising lines of research to advance the ability to better predict risk from interactions of sensitive population groups and life stages and from chemical and nonchemical stressors.

In the near term, some advances are possible in the study of complex mixtures with high throughput, toxicogenomic studies. Complex mixtures of toxicants are a significant problem in Superfund sites as well as in other areas of toxicology and environmental health. These toxicant mixtures are known to interact in unexpected or poorly understood ways. Unfortunately, most toxicological studies use purified compounds, reconstructed mixtures, or both. Data from these studies are the basis for regulation of individual compounds and mixtures. As the number of components in the mixture increases, however, the study of reconstructed mixtures becomes more and more difficult and less and less valuable. This type of needed basic research might gain knowledge but will be of little value unless it can be applied in risk assessment to protect public health. And even if it is applied, it has to effectively bring about change in the decision-making process. Translational research is needed to transform basic and applied research into a risk assessment tool. Computational toxicology – a rapidly advancing discipline of toxicology that combines the modern-day computational power with the wisdom gained from conventional toxicity testing – is breathing optimism in this area of translational research and risk assessment tool development.

Once they are developed, computational tools could be made easily accessible, could decrease the cost of toxicity testing, and could meet the present demand for filling the fundamental knowledge gaps in chemical mixture studies. Modern computational chemistry and molecular and cellular biology tools allow researchers to characterize a broad spectrum of physical and biological properties for large numbers of chemicals [82].

Genomics, transcriptomics, proteomics, and metabolomics technologies are becoming integral components of the modern biology toolkit. Linking these molecular biology changes to adverse outcomes represents a significant research challenge that must be addressed before such data can provide information essential to support risk assessment. However, establishing a quantitative relationship between such changes and adverse responses will provide key information. Such information can be very relevant and, at times, critical to risk assessment by providing mechanistically oriented insight into the hazard identification, dose–response, and exposure portions of risk assessments [83]. Together with computational toxicology methods, researchers are using as biomarkers complimentary, alternative, *in vitro* methods in combination with -omics responses. As experience is gained through an increased use of such crosscutting science methods and technology, a more efficient approach to fill critical gaps in our knowledge base to support risk assessment will evolve. Public health and environmental medicine will then emerge together to solve chronic health problems such as obesity, diabetes, and other metabolic diseases linked to environmental factors. These changing perspectives have led to the evolution of the concepts of green chemistry, which has the potential to drastically reduce the synthesis, use, and production of hazardous chemicals, and largely limit the introduction of superfluous chemicals in our environment [84].

Exposure to environmental contaminants or toxicants is one of the many conditions or factors that compromise human quality of life. Toxic chemicals have been linked to deaths and to mortality increases from cancer, respiratory, and cardiovascular diseases [85]. The characteristics and patterns of exposures from waste sites, unplanned releases, and other sources of pollution need to be understood clearly to prevent potential adverse human health effects and diminished quality of life. Ideally, data from epidemiological findings supported by animal studies to verify mechanisms leading to the toxicity of chemical mixtures would be the most appropriate information needed for risk assessment [86–88]. Yet human and animal studies are costly and time-consuming and sometimes lead to inconclusive results. Available epidemiological studies that have examined the health effects of mixtures are usually based on retrospective epidemiological data, where exposure duration and concentrations can only be approximated. Apart from this, those epidemiological studies suffer from confounding factors such as genetic susceptibility, nutritional status, and lifestyle factors.

Looking at the science of toxicology holistically, a realization emerges as to how little is known about the millions of chemicals generally, or the over 80 thousand chemicals in commerce, let alone their mixtures [89]. The effects of chemical mixtures are extremely complex and vary as a function of the chemical composition of each mixture. This complexity is a major reason why mixtures have not been well studied. Thus, writings on chemical mixtures are more often a presentation of what we do not know than of what we do know. As with the expert, the more we understand, the more we realize how little we know. To quote,

> Learning is but an adjunct to ourself,
>
> And where we are, our learning likewise is.
>
> Shakespeare (*Love's Labor's Lost* 4,3)

Acknowledgment

We acknowledge the editorial review and suggestions of Wallace Sagendorph, Division of Creative Services, National Center for Health Marketing, Centers for Disease Control and Prevention (CDC).

References

1 ATSDR (2004) Guidance manual for the assessment of joint toxic action of chemical mixtures (final). Agency for Toxic Substances and Disease Registry, US Department of Health and Human Services, Atlanta, GA. Available at www.atsdr.cdc.gov/interactionprofiles/ipga.html.

2 US Environmental Protection Agency (2000) Supplementary guidance for conducting health risk assessment of chemical mixtures. Risk Assessment Forum. Office of Research and Development. Washington, D.C. EPA/630/R-00/002. Available at http://www.epa.gov/ncea/raf/pdfs/chem_mix/chem_mix_08_2001.pdf.

3 NRCEE (1991) *Environmental Epidemiology, Environmental Epidemiology, vol. 1, Public Health and Hazardous Wastes*, National Academy Press, Washington, DC.

4 Feron, V.J., Cassee, F.R., Groten, J.P., van Vliet, P.W., Job, A., and van Zorge, J.A. (2002) International issues on human health effects of exposure to chemical mixtures. *Environ. Health Perspect.*, **110** (Suppl. 6), 893–899.

5 World Health Organization (2009) Assessment of Combined Exposures to Multiple Chemicals: Report of a WHO/IPCS International Workshop. Available at http://www.who.int/ipcs/methods/hormonization/areas/aggregate/en/index.html.

6 Shermer, M. (2005) Rumsfeld's wisdom. *Sci. Amer.*, **293**, 38.

7 CDC (2009) Fourth National Report on Human Exposure to Environmental Chemicals. Available at http://www.cdc.gov/exposurereport/.

8 ATSDR (2004) Interaction profile for persistent chemicals found in fish (chlorinated dibenzo-*p*-dioxins, hexachlorobenzene, *p,p'*-DDE, methylmercury, and polychlorinated biphenyls). Agency for Toxic Substances and Disease Registry, US Department of Health and Human Services, Atlanta, GA. Available at www.atsdr.cdc.gov/interactionprofiles.

9 Lindsay, D.G. (2005) Nutrition, hermetic stress and health. *Nut. Res. Rev.*, **18**, 249–258.

10 Calabrese, E.J. and Baldwin, L.A. (2003) The hormetic dose–response model is more common than the threshold model in toxicology. *Toxicol. Sci.*, **71**, 246–250.

11 Cook, R. and Calabrese, E.J. (2007) The importance of hormesis to public health. *Cein. Saude Colet.*, **12**, 955.

12 Calabrese, E.J. (2008) Another California milestone: the first application of hormesis in litigation and regulation. *Int. J. Toxicol.*, **27**, 31–33.

13 Grandjean, P. (2005) Implications of the precautionary principle for public heath practice and research. *Human Ecol. Risk Assess.*, **11**, 3–15.

14 Landrigan, P. (2005) Opening remarks: Collegium Ramazzini. *Human Ecol. Risk Assess.*, **11**, 7–8.

15 Johnson, B.J. (2005) Editorial *Human Ecol. Risk Assessment*, **11**, 1–2.

16 Bucher, J. and Lucier, G.E. (1998) Current approaches toward chemical mixtures studies at the National Institute of Environmental Health Sciences and the U.S. National Toxicology Program. *Environ. Health Perspect.*, **106** (Suppl. 6), 1295–1298.

17 Yang, R.S.H., El-Masri, H.A., Thomas, R.S., Dobrev, I., Dennison, J.E. Jr., Bae, D.S., Campain, J.A., Liao, K.H., Reisfeld, B., Andersen, M.E., and Mumtaz, M.M. (2004) Chemical mixture toxicology: from descriptive to mechanistic, and going on to *in silico* toxicology. *Environ. Toxicol. Pharmacol.*, **18**, 65–81.

18 ACGIH (2006) TLV/BEI Resources. American Conference of Governmental and Industrial Hygienists. Available at www.acgih.org/TLV.

19 Mumtaz, M.M., Sipes, I.G., Clewell, H.J., and Yang, R.S.H. (1993) Risk assessment of chemical mixtures: biological and toxicologic issues. *Fundam. Appl. Toxicol.*, **21**, 258–269.

20 Combination Toxicology: Proceedings of a European Conference (1996) *Food Chem. Toxicol.*, **34**, 1025–1185.

21 Mason, A.M., Borgert, C.J., Bus, J.S., Mumtaz, M.M., Simmons, J.E., and Sipes, I.G. (2007) Improving the scientific foundation for mixtures joint toxicity and risk assessment: contributions from the SOT mixtures project – introduction. *Toxicol. Appl. Pharmacol.*, **223**, 99–103.

22 Groten, J.P., Heijne, W.H.M., Stierum, R.H., Freidig, A.P., and Feron, V.J. (2004) Toxicology of chemical mixtures: a challenging quest along empirical sciences. *Environ. Toxicol. Pharmacol.*, **18**, 185–192.

23 Mumtaz, M.M., De Rosa, C.T., Cibulas, W., and Falk, H. (2004) Seeking solutions to chemical mixtures challenges in public health. *Environ. Toxicol. Pharmacol.*, **18**, 55–63.

24 Andersen, M.E. and Dennison, J.E. (2004) Mechanistic approaches for mixtures risk assessments: present capabilities with simple mixtures and future directions. *Environ. Toxicol. Pharmacol.*, **16**, 1–11.

25 Boobis, A., Budinsky, R., Collie, S., Crofton, K., Embry, M., Felter, S., Hertzberg, R., Kopp, D., Mihlan, G., Mumtaz, M., Price, P., Solomon, K., Teuschler, L., Yang, R., and Zaleski, R. Critical analysis of literature on low dose synergy for use in screening chemical mixtures for risk assessment. *Critical Rev. Toxicol.*, in press.

26 Teuschler, L.K., Hertzberg, R.C., Rice, G.E., and Simmons, J.E. (2004) EPA project-level research strategies for chemical mixtures: targeted research for meaningful results. *Environ. Toxicol. Pharmacol.*, **18** (3), 193–199.

27 SRP (2010) National Institutes of Environmental Health Sciences, RTP, NC, USA. Available at http://www.niehs.nih.gov/research/supported/srp/about/index.cfm (accessed 2 March, 2010).

28 Suk, W.A., Anderson, B.E., Thompson, C.L., Bennett, D.A., and VanderMeer, D.C. (1999) Creating multidisciplinary research opportunities: a unifying framework model helps researchers to address the complexities of environmental problems. *Environ. Sci. Technol.*, **33** (11), 241A–244A.

29 Wilson, S.H. and Suk, W.A. (2005) Framework for environmental exposure research: the disease-first approach. *Mol. Interv.*, **5** (5), 262–267.

30 Suk, W.A., Olden, K., and Yang, R.S.H. (2002) Chemical mixtures research: significance and future perspectives. *Environ. Health Perspect.*, **110** (Suppl. 6), 891–892.

31 Suk, W.A. and Wilson, S.H. (2002) Overview and future of molecular biomarkers of exposure and early disease in environmental health, in *Biomarkers of Environmentally Associated Disease* (eds S.H. Wilson and W.A. Suk), CRC Press LLC/Lewis Publishers, Boca Raton, FL, pp. 3–15.

32 Yang, R.S., Thomas, R.S., and Gustafson, D.L. (1998) Approaches to developing alternative and predictive toxicology based on PBPK/PD and QSAR modeling. *Environ. Health Perspect.*, **106** (Suppl. 6), 1285–1293.

33 Suk, W.A. and Olden, K. (2004) Multidisciplinary research: strategies for assessing chemical mixtures to reduce risk of exposure and disease. *Euro. J. Oncol.*, **2**, 1–10.

34 Lazarou, J., Pomeranz, B.H., and Corey, P.N. (1998) Incidence of adverse drug reactions in hospitalized patients: a meta-analysis of prospective studies. *JAMA*, **279**, 1200–1205.

35 Moore, T.J., Cohen, M.R., and Furberg, C.D. (2007) Serious adverse drug events reported to the Food and Drug Administration, 1998–2005. *Arch. Intern. Med.*, **167** (16), 1752–1759.

36 Reynolds, E.S., Brown, B.R., Jr., and Vandam, L.D. (1972) Massive hepatic necrosis after fluroxene anesthesia: a case of drug interaction? *New Engl. J. Med.*, **286**, 530–531.

37 Harrison, G.G. and Smith, J.S. (1973) Massive lethal hepatic necrosis in rats anesthetized with fluroxene, after microsomal enzyme induction. *Anesthesiology*, **39** (6), 619–625.

38 Jevtovic-Todorovic, V., Hartman, R.E., Izumi, Y., Benshoff, N.D., Dikranian, K., Zorumski, C.F., Olney, J.W., and Wozniak,

D.F. (2003) Early exposure to common anesthetic agents causes widespread, neurodegeneration in the developing rat brain and persistent learning deficits. *J. Neurosci.*, **23**, 876–882.
39 Brater, D.C. (1990) *Toxic Interactions* (eds R.S. Goldstein, W.R., Hewitt, and J.B. Hook), Academic Press, San Diego, pp. 149–173.
40 Hardman, R. (2006) A toxicologic review of quantum dots: toxicity depends on physicochemical and environmental factors. *Environ. Health Perspect.*, **114** (2), 165–172.
41 Royal Commission on Environmental Pollution (2008) Novel Materials in the Environment: The Case of Nanotechnology, 27th Report, p. 147.
42 Maynard, A.D., Aitken, R.J., Butz, T., Colvin, V., Donaldson, K., Oberdorster, G., Philbert, M.A., Ryan, J., Seaton, A., Stone, V., Tinkle, S.S., Tran, L., Walker, N.J., and Warheit, D.B. (2006) Safe handling of nanotechnology. *Nature*, **444**, 267.
43 Yang, R.S.H., Chang, L.W., Yang, C.S., and Lin, P. (2010) Pharmacokinetics and physiologically-based pharmacokinetic modeling of nanoparticles. *J. Nanosci. Nanotech.*, **10**, 1–8.
44 Nel, A., Xia, T., Madler, L., and Li, N. (2006) Toxic potential of materials at the nanolevel. *Science*, **311** (5761), 622–627.
45 Yang, R.S., Chang, L.W., Wu, J.P., Tsai, M.H., Wang, H.J., Kuo, Y.C., Yeh, T.K., Yang, C.S., and Lin, P. (2007) Persistent tissue kinetics and redistribution of nanoparticles, Quantum Dot 705, in mice: ICP-MS quantitative assessment. *Environ. Health Perspect.*, **115** (9), 1339–1343.
46 Lin, P.P., Chen, J.W., Chang, L.W., Wu, J.P., Redding, L., Chang, H., Yeh, T.K., Yang, C.S., Tsai, M.H., Wang, H.J., Kuo, Y.C., and Yang, R.S.H. (2008) Computational and ultrastructural toxicology of a nanoparticle, Quantum Dot 705, in mice. *Environ. Sci. Technol.*, 42, 6264–6270.
47 De Rosa, C.T., Hansen, H., Wilbur, S., Pohl, H.R., El-Masri, H.A., and Mumtaz, M.M. (2001) Interactions, in *Patty's Toxicology*, 5th edn, vol. 1 (eds E. Bingham, B. Cohrssen, and C.H. Powell), John Wiley & Sons, Inc., pp. 233–284.
48 NRC (2008) *Science and Decisions: Advancing Risk Assessment, National Research Council*, The National Academy Press, Washington, DC.
49 NRC (1983 *Risk Assessment In The Federal Government: Managing The Process*, Committee on the Institutional Means for Assessment of Risks to Public Health, Commission on Life Sciences, National Research Council, National Academy Press, Washington, DC.
50 Mumtaz, M., Ruiz, P., and De Rosa, C. (2007) Toxicity assessment of unintentional exposures to multiple chemicals. *Toxicol. Appl. Pharmacol.*, **223**, 104–113.
51 Mumtaz, M.M., De Rosa, C.T., Groten, J., Feron, V.J., Hansen, H., and Durkin, P.R. (1998) Estimation of toxicity of chemical mixtures through modeling of chemical interactions. *Environ. Health. Perspect.*, **106** (Suppl. 6), 1353–1360.
52 Calabrese, E.J. (1991) *Multiple Chemical Interactions*, Lewis Publishers, Chelsea, MI.
53 Yang, R.S.H. (2010) Toxicologic interactions of chemical mixtures, in *Comprehensive Toxicology. Vol. 1. General Principles* (ed. J. Bond), Elsevier Ltd., Oxford, in press.
54 Yang, R.S.H. and Andersen, M.E. (2005) Physiologically based pharmacokinetic modeling of chemical mixtures, in *Physiologically Based Pharmacokinetics: Science and Applications* (eds M.B. Reddy, R.S.H. Yang, H.J. ClewellIII, and M.E. Andersen), John Wiley & Sons, Inc., New York, pp. 349–373.
55 Yang, R.S.H. and Lu, Y. (2007) The application of physiologically based pharmacokinetic (PBPK) modeling to risk assessment, in *Risk Assessment for Environmental Health* (eds M.G. Robson and W.A. Toscano), John Wiley & Sons, Inc., Hoboken, NJ, pp. 85–120.
56 Mumtaz, M.M., Poirier, K.A., and Coleman, J.T. (1997) Risk assessment for chemical mixtures: fine-tuning the hazard index approach. *J. Clean Technol. Environ. Toxicol. Occup. Med.*, **6**, 189–204.

57 Mumtaz, M.M. and Durkin, P.R. (1992) A weight-of-evidence scheme for assessing interactions in chemical mixtures. *Toxicol. Ind. Health*, **8**, 377–406.

58 Mumtaz, M.M., El-Masri, H., Chen, D., and Pounds, J. (2000) Joint toxicity of inorganic chemical mixtures: the role of dose ratios, in *Metal Ions in Biology and Medicine*, vol. 6 (eds J.A. Centeno, P.H. Collery, G. Vernet, R.B. Finkelman, H. Gibb and J.C. Etienne), John Libbey Eurotext, Paris, pp. 297–299.

59 El-Masri, H.A., Mumtaz, M.M., and Yushak, M.L. (2004) Application of physiologically based pharmacokinetic modeling to investigate the toxicological interaction between chlorpyrifos and parathion in the rat. *Environ. Toxicol. Pharmacol.*, **16**, 57–71.

60 Ray, M., Ritger, S.E., Mumtaz, M., Ruiz, P., Welsh, C., Fowler, D.A., Keys, D., and Fisher, J. (2009) Addressing public exposures to priority solvents using human PBPK models. *Toxicol. Sciences*, **108** (1), 471.

61 Haddad, S., Pelekis, M.L., and Krishnan, K. (1996) A methodology for solving physiologically-based pharmacokinetic models without the use of simulation software. *Toxicol. Letters*, **85**, 113–126.

62 Jonker, D., Jones, M.A., van Bladeren, P.J., Woutersen, R.A., Til, H.P., and Feron, V.J. (1993) Acute (24 hr) toxicity of a combination of four nephrotoxicants in rats compared with the toxicity of the individual compounds. *Food Chem. Toxicol.*, **31**, 45–52.

63 Jonker, D., Woutersen, R.A., and Feron, V.J. (1996) Toxicity of mixtures of nephrotoxicants with similar or dissimilar mode of action. *Food Chem. Toxicol.*, **34**, 1075–1082.

64 Heijne, W.H., Jonker, D., Stierum, R.H., van Ommen, O.B., and Groten, J.P. (2005) Toxicogenomic analysis of gene expression changes in rat liver after a 28-day oral benzene exposure. *Mutat. Res.*, **575** (1–2), 85–101.

65 Kienhuis, A.S., Wortelboer, H.M., Hoflack, J.C., Moonen, E.J., Kleinjans, J.C., van, O.B., van Delft, J.H., and Stierum, R.H. (2006) Comparison of coumarin-induced toxicity between sandwich-cultured primary rat hepatocytes and rats *in vivo*: a toxicogenomics approach. *Drug Metab. Dispos.*, **34** (12), 2083–2090.

66 McMillian, M., Nie, A.Y., Parker, J.B., Leone, A., Bryant, S., Kemmerer, M., Herlich, J., Liu, Y., Yieh, L., Bittner, A., Liu, X., Wan, J., and Johnson, M.D. (2004) A gene expression signature for oxidant stress/reactive metabolites in rat liver. *Biochem. Pharmacol.*, **68** (11), 2249–2261.

67 McMillian, M., Nie, A.Y., Parker, J.B., Leone, A., Kemmerer, M., Bryant, S., Herlich, J., Yieh, L., Bittner, A., Liu, X., Wan, J., and Johnson, M.D. (2004) Inverse gene expression patterns for macrophage activating hepatotoxicants and peroxisome proliferators in rat liver. *Biochem. Pharmacol.*, **67** (11), 2141–2165.

68 Natsoulis, G., El, G.L., Lanckriet, G.R., Tolley, A.M., Leroy, F., Dunlea, S., Eynon, B.P., Pearson, C.I., Tugendreich, S., and Jarnagin, K. (2005) Classification of a large microarray data set: algorithm comparison and analysis of drug signatures. *Genome Res.*, **15** (5), 724–736.

69 van Delft, J.H., van Agen, A.E., van Breda, S.G., Herwijnen, M.H., Staal, Y.C., and Kleinjans, J.C. (2005) Comparison of supervised clustering methods to discriminate genotoxic from non-genotoxic carcinogens by gene expression profiling. *Mutat. Res.*, **575** (1–2), 17–33.

70 Thomas, R.S., O'Connell, T.M., Pluta, L., Wolfinger, R.D., Yang, L., and Page, T.J. (2007) A comparison of transcriptomic and metabonomic technologies for identifying biomarkers predictive of two-year rodent cancer bioassays. *Toxicol. Sci.*, **96**, 40–46.

71 Thomas, R.S., Allen, B., Nong, A., Yang, L., Bermudez, E., Clewell, H.J. III, and Andersen, M.E. (2007) A method to integrate benchmark dose estimates with genomic data to assess the functional effects of chemical exposure. *Toxicol. Sci.*, **98**, 240–248.

72 Hendricksen, P., Freidig, A.P., Jonker, D., Thissen, U., Bogaards, J.J.P., Mumtaz, M.M., Groten, J.P., and Stierum, R.H.

(2007) Transcriptomics analysis of interactive effects of benzene, trichloroethylene, and methyl mercury within binary and ternary mixtures on the liver and kidney following subchronic exposure in the rat. *Toxicol. Appl. Pharmacol.*, **225**, 171–188.

73 Callahan, M.A. and Sexton, K. (2007) If cumulative risk assessment is the answer, what is the question? *Environ. Health Perspect.*, **115** (5), 799–806.

74 ICCVAM (2010) National Institutes of Environmental Health Sciences, RTP, NC, USA. Available at http://iccvam.niehs.nih.gov/ (accessed 2 March, 2010).

75 UN (2003) Skin corrosion/irritation. UN Globally Harmonized System of Classification and Labelling of Chemicals. ST/SG/AC.10/30, United Nations, New York, Geneva, pp. 123–135.

76 Houck, K.A. and Kavlock, R.J. (2008) Understanding mechanisms of toxicity: insights from drug discovery research. *Toxicol. Appl. Pharmacol.*, **277**, 163–178.

77 Dix, D.J., Houck, K.A., Martin, M.T., Richard, A.M., Setzer, R.W., and Kavlock, R.J. (2007) The ToxCast program for prioritizing toxicity testing of environmental chemicals. *Toxicol. Sci.*, **95**, 5–12.

78 Judson, R., Richard, A., Dix, D., Houch, K., Elloumi, F., Martin, M., Cathey, T., Transue, T., Spencer, R., and Wolf, M. (2008) ACToR: aggregated computational toxicology resource. *Toxicol. Appl. Pharmacol.*, **233**, 7–13.

79 Mattingly, C.J., Colby, G.T., Forrest, J.N., and Boyer, J.L. (2003) The comparative toxicogenomics database (CTD). *Environ. Health Perspect.*, **111**, 793–795.

80 NRC (2007) *Applications of Toxicogenomic Technologies to Predictive Toxicology and Risk Assessment*, The National Academy Press, Washington, DC.

81 NAS (2007) *Toxicity Testing in the 21st Century: A Vision and a Strategy*, The National Academy Press, Washington, DC.

82 Bredel, M. and Jacoby, E. (2004) Chemogenomics: an emerging strategy for rapid target and drug discovery. *Nat. Rev. Genet.*, **5** (4), 262–275.

83 US Environmental Protection Agency (2004) Potential Implications of Genomics for Regulatory and Risk Assessment Applications at EPA. Science Policy Council, U.S. Environmental Protection Agency, Washington, D.C. 20460.

84 NAS (2007) *Green Healthcare Institutions: Health, Environment, and Economics: Workshop Summary*, The National Academies Press, Washington, DC.

85 Mokadad, A.H., Marks, J.S., Stroup, D.F., and Gerberding, J. (2004) Actual causes of death in the United States, 2000. *JAMA*, **291** (10), 1238–1245.

86 Mumtaz, M.M., Ruiz, P., Whittaker, M., Dennison, J., Fowler, B.A., and De Rosa, C.T. (2006) Chemical mixtures risk assessment and technological advances, in *Biological Concepts and Techniques in Toxicology: An Integrated Approach* (ed. J.E. Riviere), Taylor & Francis, New York, pp. 177–204.

87 Gorell, J.M., Johnson, C.C., Rybicki, B.A., Peterson, E.L., Kortsha, G.X., Brown, G.G., and Richardson, R.J. (1997) Occupational exposure to metals as risk factors for Parkinson's disease. *Neurology*, **48**, 650–658.

88 Gorell, J.M., Johnson, C.C., Rybicki, B.A., Peterson, E.L., and Richardson, R.J. (1998) The risk of Parkinson's disease with exposure to pesticides, farming, well water, and rural living. *Neurology*, **50**, 1346–1350.

89 Pimentel, D., Tort, M., D'Anna, L., Krawic, A., Gerger, J., Rossman, J., Mugo, F., Doon, N., Shriberg, M., and Howard, E. (1995) Ecology of increasing disease: population growth and environmental degradation. *Bioscience*, **48**, 817–826.

2
Chemical Mixtures in the Environment: Exposure Assessment

Glenn Rice and Margaret MacDonell

2.1
Risk Assessment Paradigm: A Chemical Mixtures Context

During a time of increasing awareness of chemical contaminants in the environment, workplace, home, and diet, the NRC established an overarching paradigm for assessing human health risks [1]. This paradigm consisted of four components: hazard identification, exposure assessment, dose-response assessment, and risk characterization. As risk assessors and risk managers were implementing the paradigm, they eventually recognized the need for and usefulness of an initial planning phase; subsequently, a fifth, introductory component, problem formulation was added by the U.S. Environment Protection Agency (US EPA) [2].

The functions associated with each paradigm component follow; we focus on unique aspects of assessing health risks posed by chemical mixtures.

- **Problem Formulation**: Develop an initial understanding of key relationships between the types of health effects associated with the chemicals of interest and potential human exposures and stakeholder issues relevant to exposure.
- **Hazard Identification**: Identify the mixture(s) (i.e., combinations of chemicals) that could pose human health risks and associated health hazards based on information from the scientific literature. For contaminated sites, this includes defining the nature and extent of contamination.
- **Exposure Assessment**: Examine the sources of exposure, identifying who could be exposed to what chemicals, their locations when exposed, the timing of their exposures, how they are exposed, and to how much they are exposed. While not a component of every exposure assessment, some exposure assessments include an evaluation of the fate of chemical mixtures released into the environment (i.e., source to environmental concentrations) and others attempt

Principles and Practice of Mixtures Toxicology. Edited by Moiz Mumtaz

ISBN: 978-3-527-31992-3

to describe how a mixture that is already present in the environment has occurred. Following the initial step (e.g., analysis of environmental fate), exposure assessments next estimate "dose" by first examining how individuals and populations are exposed to chemical mixtures in their environments. To estimate dose, exposure assessments determine the pathways through which people are exposed to the chemical mixture and the routes by which the chemicals cross the body's outer boundary (i.e., exchange boundaries, which include the skin, gastrointestinal tract, and lungs). Exposure assessments estimate the potential dose of a mixture by quantifying the intensity, frequency, and duration of contact; sometimes, these assessments evaluate the rates at which chemicals cross the body's outer boundary and the amount of the chemicals that are absorbed (i.e., they might estimate an internal dose of the mixture or the dose to the toxicological target organ).

- **Dose-Response Assessment**: Evaluate the toxicity literature, including experimental animal data, other toxicological assays, and human epidemiological data, to quantitatively predict human responses (e.g., adverse human health effects) from exposures to those chemicals identified during the hazard identification component. There are two general approaches commonly used to estimate the body's response to a multiple chemical exposure: component-based approaches and whole-mixture approaches. Component-based approaches rely upon individual chemical toxicity data and an understanding of how the chemicals can collectively harm human health, whereas whole-mixture approaches rely on toxicity information about a mixture that has been evaluated in an epidemiological or toxicological study and how similar the encountered mixture (i.e., the mixture of concern in the environment) is to the tested mixture. Although infrequently used, a third approach, the use of dose-response information associated with specific fractions that comprise a mixture, is implemented for some mixtures (e.g., total petroleum hydrocarbons [3]). These three approaches are described in other chapters of this book.
- **Risk Characterization**: Integrate the results of the exposure and dose-response assessments to estimate the likelihood that adverse health effects might occur in the individuals or groups exposed to the mixture. To appropriately quantify risk, the results of the exposure and dose-response assessments must use the same dose metric.

Figure 2.1 presents these five components of the risk assessment paradigm [1, 2, 4–6].

2.2 Occurrence of Chemical Mixtures in the Environment

Humans routinely encounter environmental media such as air, soil, water, and biota that might be contaminated with multiple chemicals. Contact with contaminated environmental media can lead to human exposures to mixtures of chemicals

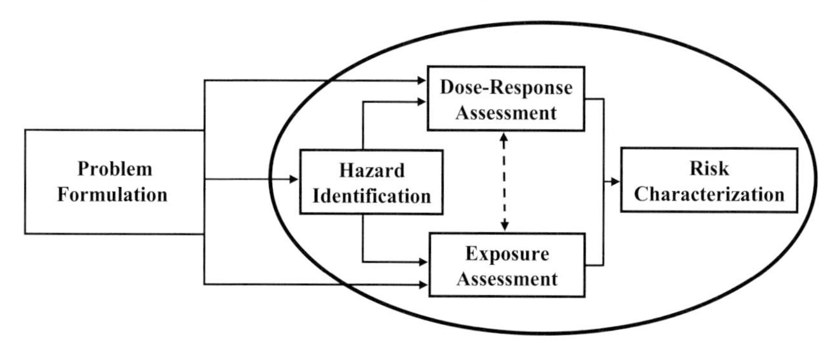

Figure 2.1 Paradigm for assessing human health risks from chemical exposures.

in these media. Mixtures of chemicals in these media typically arise from three broad circumstances: incidental release, intentional use, and coincidental occurrence [7].

Incidental releases describe mixtures released or formed during a process that is undertaken for another purpose. Examples of incidental releases include the occurrence of by-products in drinking water after chemical disinfection, and polycyclic aromatic hydrocarbons (PAHs) in emissions from hazardous waste combustors.

Intentional uses describe the purposeful use of a mixture in a product or application of a mixture. Examples of intentional releases include releases from the incorporation of polybrominated diphenyl ethers (PBDEs) as flame retardants in clothing and furnitures and the application of pesticides, which usually are chemical mixtures, to crops and lawns.

Coincidental occurrences describe chemicals that come about in an environmental medium due to their pattern of release and their physical and chemical properties. Examples of coincidental occurrences include combinations of diesel exhaust, polycyclic aromatic hydrocarbons, and volatile organic compounds (VOCs) in ambient urban air from multiple sources (e.g., vehicles and industrial facilities); and VOCs in indoor air released from sources such as household products (e.g., cleaning solutions).

2.3 Drivers for Assessing Exposures to Chemical Mixtures

Health concerns identified by the public or a public health agency often initiate an assessment of the risk posed by exposures to a mixture of chemicals. These concerns can stem from the presence in a local area of a specific source or combination of sources that release pollutants. Elevated measurements of mixtures in environmental media (including biota) can also initiate these assessments, as can a possible clustering of health effects in a community or workplace.

Public concern about exposures to chemical mixtures has resulted in specific legislation in the United States. The 1980 Comprehensive Environmental Response, Compensation, and Liability Act is among the earliest pieces of federal legislation requiring an examination of chemical mixture risks at Superfund sites. The 1996 Food Quality Protection Act required the US EPA and other federal agencies to evaluate human health risks associated with exposures to multiple chemicals.

As part of implementing these and other regulations, several different US federal agencies have published documents to guide assessments of various mixtures. These include the following:

- **Agency for Toxic Substances and Disease Registry (ATSDR)**: *Guidance Manual for the Assessment of Joint Toxic Actions of Chemical Mixtures* [7].
- **National Institute for Occupational Safety and Health (NIOSH)**: *Mixed Exposures Research Agenda, A Report by the NORA Mixed Exposures Team* [8].
- **Occupational Safety and Health Administration (OSHA)**: *Incorporation of General Industry Safety and Health Standards Applicable to Construction Work* [9], which includes a mandate for addressing airborne mixtures.
- **US Department of Energy (US DOE)**: *Radioactive Waste Management*, US DOE Order 435.1 [10], which requires low-level waste disposal facilities to account for potential exposures to multiple contaminants associated with their facilities.
- **US EPA**:

  –*Guidelines for the Health Risk Assessment of Chemical Mixtures* [11].

  –*Supplementary Guidance for Conducting Health Risk Assessment of Chemical Mixtures* [12].

  –*Risk Assessment Guidance for Superfund* [13].

  –*Guidelines for Exposure Assessment* [14].

  –*Methodology for Assessing Health Risks Associated with Multiple Pathways of Exposure to Combustor Emissions* [15].

  –*Guidance for Performing Aggregate Exposure and Risk Assessments* [16].

  –*General Principles for Performing Aggregate Exposure and Risk Assessments* [17].

  –*Framework for Cumulative Risk Assessment* [6].

  –*Concepts, Methods, and Data Sources for Health Risk Assessment of Multiple Chemicals, Exposures, and Effects* [18].

To address regulatory requirements and to evaluate exposures to specific mixtures in populations, several groups have conducted exposure assessments for mixtures. These mixture exposure assessments include dioxin-like compounds [19–21], polychlorinated biphenyls (PCBs) [22], PBDEs [23], drinking water DBPs [24], organophosphorous pesticides [17], and gasoline [25].

2.4
Using Conceptual Models to Guide the Development of Mixture Exposure Assessments

Conceptual models help organize key information and guide the exposure assessment process. Figure 2.2 presents a general form of a conceptual model for assessing exposures to mixtures released from anthropogenic pollution sources. This model can also be used to guide the assessment of human exposures to mixtures released from specific sources. During the assessment process, as data are identified and analyses are conducted, conceptual models typically are refined until they depict the relationships among the assessment components used in the final analysis [26].

Conceptual models are useful for clearly illustrating the exposure pathways relevant for the mixtures of concern. They ensure that key elements of the assessment are addressed and help the user understand the relationships among these key elements of the assessment. They can be nested (linked via a common element, e.g., an exposure point) or displayed spatially (e.g., via geographic information system plots) to integrate sources, affected media, and points of human exposure.

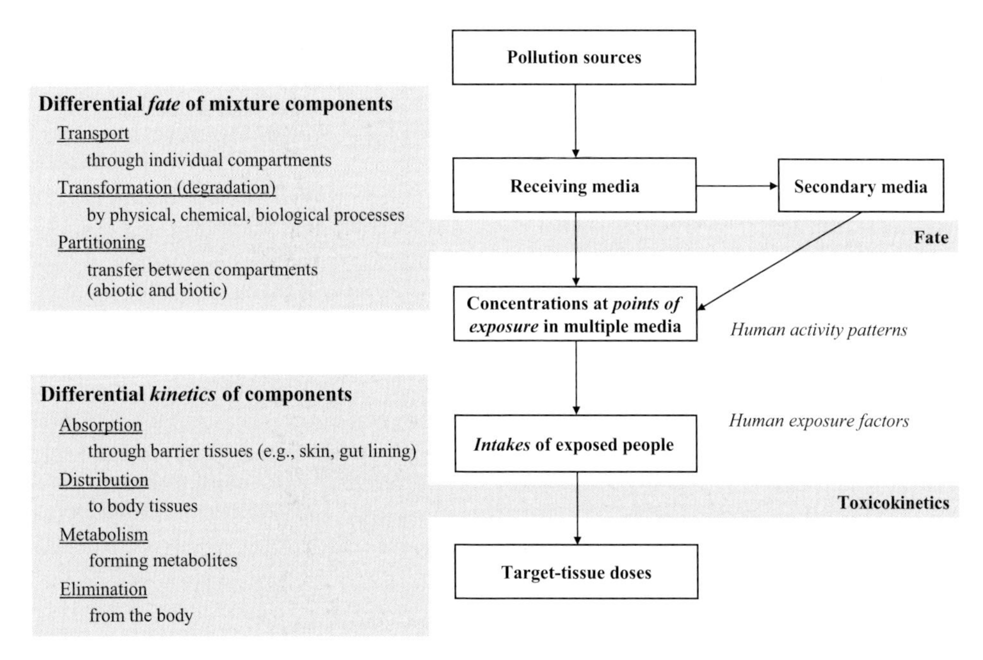

Figure 2.2 General conceptual model for assessing exposures to chemical mixtures.

2.5
Overview of Environmental Fate for Chemical Mixtures

Figure 2.2 illustrates that exposure assessments can include characterization of the sources and release mechanisms for the mixture of interest (e.g., emissions from a waste combustion facility), as well as the fate and transport of those chemicals. The transport of released chemicals results in concentrations in environmental media. People can be exposed when they encounter these contaminated media. We illustrate these concepts using a hypothetical facility. In this example, the dispersal of atmospheric releases from the waste combustion facility depends on local meteorological conditions (wind direction and speed and atmospheric stability), the release height of the source, and surrounding terrain. Thus, in this example, mixtures released from the combustor could be transported many kilometers through the atmosphere (the receiving medium[1]) before being deposited on soils or crops (secondary media) grown on farms downwind of that source. People having contact with the soil (e.g., during farming activities) or eating the crops might be exposed to the pollutants that were originally released by the combustion facility.

Concentrations of chemicals in a mixture typically change in the environment as the quantities released disperse in the receiving medium. Returning to our combustor example, concentrations of chemicals in the air at some distance from the facility differ from those at the point of release due to dilution in the air. (Similarly, concentrations of a mixture released in surface water will also be diluted.)

Natural physical, chemical, and biological processes can differentially transform the components of environmental mixtures. Processes such as hydrolysis, photolysis, atmospheric hydroxyl radical oxidation, and microbial dechlorination can affect components of a mixture in the environment.[2] Thus, the chemical *composition* of a mixture in the environment can substantially change over time because of differences in the fates of component chemicals, notably differential transport within the receiving medium, differential transformation to other chemicals, and differential partitioning to other media (e.g., volatilization from soil to air). Both the transport and partitioning of mixture components will depend, in part, on the physical–chemical properties of those components and the nature of the medium in which they are found.

Figure 2.2 indicates that humans encounter environmental mixtures through daily activity patterns, which can take place in their workplaces, homes, and communities. For mixtures, exposure pathways describe combinations of chemicals in media (or from sources), individuals who might be exposed (receptors), and the exposure routes through which these exposures might occur. Exposure pathways are comprised of four elements: (1) a source and mechanism by which the mixture is released; (2) a receiving medium and a mechanism(s) that transports the

1) The receiving medium is the environmental compartment (e.g., air, soil, or surface water) into which the mixture is initially released by a pollution source.

2) Information about these processes is readily available from peer-reviewed sources (e.g., the National Library of Medicine Hazardous Substances Data Bank) and recent ATSDR toxicological profiles [27].

chemicals either in the receiving medium or to secondary media; (3) a point of potential human contact with the contaminated medium; and (4) the likely route(s) of exposure. [3)]

Risk assessors typically quantify human exposures through one or more of the following three alternative approaches: point-of-contact measures, biomarker measures, and mathematical models of the exposure scenario.

1) While the exposure is occurring, risk assessors can measure the concentration of the chemical mixture at the point of contact (the outer boundary of the body). Risk assessors can also evaluate the intensity, duration, and frequency of the human contact with the contaminated medium. To quantify the exposure, risk assessors integrate the measured concentrations of chemicals and the contact information.
2) After the exposure has occurred, risk assessors can measure concentrations of pollutants or pollutant metabolites in tissues or body excretions (e.g., pollutant or metabolite levels in the blood, hair, or urine); they can also measure levels of chemical adducts in cells. These measures are sometimes referred to as "biomarkers" and the presence of these compounds in the body indicates that the individual has been exposed. To predict exposures based on measures of chemical concentrations in body fluids or tissues requires an understanding of toxicokinetics (absorption, distribution, metabolism, and elimination by the body) and the intensity, duration, and frequency of past exposures. Chapter 15 of this book addresses the topic of biomarkers for chemical mixtures [28].
3) Using mathematical models, risk assessors can simulate the exposure concentrations in the environmental media that individuals may contact and the intensity, duration, and frequency of contact. To predict exposures, risk assessors combine these data to simulate pollutant concentrations in environmental media at the point of assumed human contact [14] combined with relevant exposure factors to estimate the amount taken in.

Risk assessors develop exposure scenarios to describe how exposures occur and quantify human exposures [29]. Risk assessors construct these scenarios by integrating sets of facts and assumptions related to exposure events. (1) Early in the development of an exposure scenario, risk assessors determine concentrations of chemicals in contaminated media using either measurement data or models that predict these concentrations. (2) They identify individuals or population groups[4)] that likely come into contact with these contaminated media and the likely time, frequency, and, duration of those contacts by examining human activity patterns associated with locations of media affected by the chemicals of interest (e.g., workplaces, homes, and communities). The time, frequency and duration of the exposures with the affected media can be directly measured or estimated using given factor data

3) An exposure route is the way a pollutant enters the exposed person after contact. Ingestion, dermal absorption, and inhalation are the primary routes of exposure to environmental agents.

4) Some exposure scenarios also characterize the demographics of the people who might be exposed to specific mixtures.

(e.g., typical daily drinking water consumption rates) appropriate for the identified activity patterns and the exposed population in the exposure scenario. (3) Risk assessors can also estimate intake rates of the exposed individuals (e.g., inhalation rates, food consumption rates, and drinking water consumption rates). (4) Finally, the exposure scenario is completed by integrating these three sets of to estimates calculate exposures to for the individuals or population groups.[5)]

Risk assessors would then integrate the exposure assessment results – the estimated intake or dose of exposed or potentially exposed people – with the dose-response information for the mixture to characterize the risk. Chapter 10 addresses the topics of dose-response assessment and risk characterization for chemical mixtures [30].

2.6
Methods and Applications for Assessing Mixture Exposures

Rice *et al.* [31] suggest an approach for assessing chemical mixture exposures. The approach takes advantage of existing information for chemical mixtures while adhering to the basic concepts underlying conventional single-chemical approaches. This approach for assessing mixture exposures builds on previously published methods [14, 15, 29, 32, 33] and assessments of exposures to environmental chemical mixtures [17, 19–25]. While we describe the process linearly, exposure assessment requires the integration of a number of interdependent phases. Furthermore, the reader should be mindful that the exposure assessment itself is interdependent with other parts of the risk assessment paradigm.

Figure 2.3 identifies factors that risk assessors consider when evaluating exposures to mixtures, and it illustrates the iterative nature of the risk assessment process. In this process, targeted inputs from other risk assessment paradigm components (notably problem formulation and dose-response assessment) guide the quantitative evaluation of exposures to chemical mixtures. In this figure, we use shading to emphasize the components of the risk assessment process that are addressed in this section. This includes the information collected during the problem formulation component (see Section 2.6.1) that supports the detailed exposure assessment (see Section 2.6.2). We illustrate the approach for explicitly assessing chemical mixtures with several examples and information resources that can be used for various mixture assessments.

2.6.1
Problem Formulation: Exposure Context

The US EPA's *Ecological Risk Assessment Guidelines* [2] and the *Supplementary Guidance* for conducting health risk assessments for chemical mixtures [12] describe

5) Some define exposure scenarios to include descriptions of both the sources that release the pollutant mixtures and the environmental fate of the mixture in the receiving or secondary media.

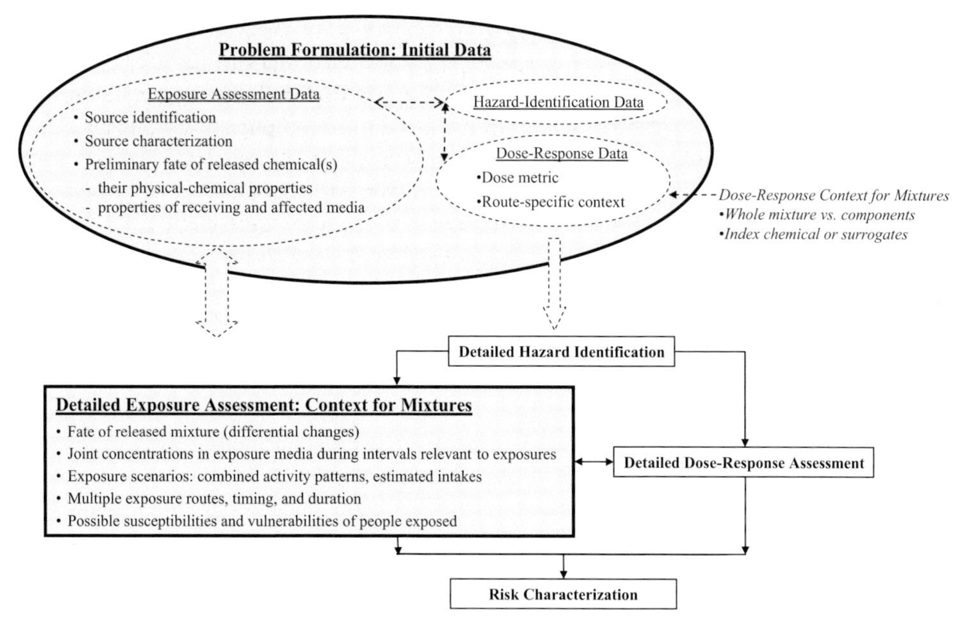

Figure 2.3 Approach for assessing chemical mixtures.

problem formulation as a process that generates and evaluates preliminary hypotheses about why health effects have occurred or could occur. Problem formulation provides a foundation for the risk assessment, wherein the nature of the problem is evaluated and the objectives for the risk assessment are defined and refined. This component culminates with the completion of a plan for analyzing the data and characterizing risk. To develop this plan, risk assessors undertake and then integrate three interdependent activities: (1) identification and integration of available information on the chemical mixture; (2) identification of the health end points of interest; and (3) development of a conceptual model (Section 2.4 of this chapter describes conceptual models) describing the relationship between the environmental mixture and the health end point(s). These activities are depicted inside the large oval in Figure 2.3.

The following discussion of the problem formulation component focuses on those aspects that guide the detailed exposure assessment. Section 2.6.1.1 describes the identification and characterization of mixture sources, Section 2.6.1.2 discusses the collection of data to analyze environmental fate, and Section 2.6.1.3 describes the importance of knowing the dose metric to be used in the dose-response analysis.

While not addressed explicitly because it is outside the scope of this chapter, risk assessors would assemble these available data during the development of the analytic plan. The analytic plan is the final product of the problem formulation component. It identifies the goals of the assessment and discusses how the analysis will be conducted [2].

2.6.1.1 Chemical Mixture Sources

The component of problem formulation that identifies and characterizes the source (s) of the chemicals of interest is an initial activity that will be incorporated into the exposure assessment. *Source identification* involves examination of the candidate sources including (1) whether they are anthropogenic or natural, (2) where they are located, and (3) whether they represent point, line, or area sources (notably for inhalation analysis).

Source characterization describes the nature of the release (e.g., amount, and continuous or episodic), its frequency and duration, and the receiving medium. Risk assessors quantify the composition of the mixture as it is released to the receiving medium if such data are available, and the quantities released.

Amounts or levels of these chemicals that are released can be either measured or simulated using process models. Because sampling and chemical analysis are resource intensive (i.e., they can be expensive and time consuming), when estimating chemical releases from a source, risk assessors generally use a dual approach involving (1) a limited number of measurements that are (2) supplemented by modeling techniques in order to fill measurement gaps. In other instances, risk assessors use surrogate measurements from similar processes, with adjustments, as needed, to align with the actual chemicals of interest in the assessment.

Characterizing variability in the quantity and composition of the mixture released is important. Seasonal and other process factors affecting the quantity and composition of the mixture as released, if known, are considered in the analysis.

2.6.1.2 Preliminary Evaluation of Environmental Fate

The risk assessor's next goal in problem formulation is to identify the likely pathways through which chemicals could reach individuals and factors that likely influence the composition of that mixture. Following the characterization of the source and the assessment of the mixture as it is released from a source to the receiving medium, risk assessors next develop an initial evaluation of the environmental fate of the chemical mixture following its release. These initial evaluations usually are based on physical–chemical properties of the chemicals comprising the mixture, of a chemical assumed to be representative of the mixture, or of the mixture as a whole. This evaluation should reflect the likely movement and transformation of the chemicals in the receiving medium (or media) and partitioning to other media. The media and locations where people might be exposed are also identified during this preliminary consideration of fate and transport.

The preliminary evaluation of fate begins with compiling a set of potentially useful environmental fate estimates or models and, if available, chemical concentration data in environmental media. Several mathematical models might be available to develop predictions of environmental fate. Model selection typically depends on the level of spatial and temporal resolution and the level of accuracy needed in the assessment.

Next, risk assessors determine the physical–chemical properties for the mixtures and chemicals of interest. These properties include physical form, valence state, vapor pressure, water solubility, solubility product, Henry's law constant, half-lives or degradation rates, and partitioning measures such as distribution coefficients and octanol/water partition coefficients (to predict organic partitioning). Appendix A of the protocol for hazardous waste combustion facilities [32] provides a list of properties for chemicals routinely emitted from such facilities – including several chemical mixtures. Such data help guide the evaluation of the partitioning of mixture components to specific media.

Existing information about the receiving medium and secondary media are commonly compiled as well, as are readily available data on climate/meteorology, topography, soil and geology, surface and groundwater hydrology and quality, ecological resources including vegetation, demographics and land use, and pollution sources if known. For example, risk assessors can download land use and land cover (LULC) maps from the US Geological Survey (USGS) [34]. Aerial photographs and remote sensing images of the local landscape can also be helpful.

Accurate soil type, geochemical, and atmospheric data are valuable inputs to the models used during the detailed exposure assessment phase to estimate contaminant fate. Site-specific data are preferred over default values and can substantially improve model accuracy. Medium-specific features include temperature, pH, and carbonate ion concentration (buffering capacity) for water, as well as texture, moisture, organic contents, coatings, colloids, and cation exchange capacities for soil and sediment. Similar for air, the nature and amount of adsorbing materials can also be important (e.g., levels of particulate matter). Risk assessors also identify pertinent models or relevant sets of measurement data that could be "tapped" for the detailed exposure assessment at this stage. In addition, risk assessors can review related data to determine whether information is available for potential surrogate chemicals (e.g., compounds with similar physical–chemical properties), or from similar sources or processes at other locations (see Box 2.1 for additional considerations).

2.6.1.3 **Influence of the Dose-Response Metric**

During problem formulation for mixture assessments, risk assessors review key toxicity studies relevant to the mixtures of interest to determine the types of available dose information. That is, as a practical matter, the risk assessors (i.e., those

Box 2.1: Evaluating Chemical Fate in a "Reverse Direction"

The evaluation of chemical fate, as described in the text, from sources to receiving media and receptors is traditionally conducted in a "forward" direction – that is, tracing the chemicals from their source(s) to potential receptors in the environment. As approaches for mixtures and cumulative risk analyses evolve, more assessments are expected to begin with the exposure points of interest – for example, where people are likely to contact the mixtures of interest. The evaluation would then proceed in the "reverse direction" to identify candidate sources (as described in US EPA [6]). Elevated levels of chemical mixtures in human blood or adipose tissue samples collected from individuals in a community might initiate a reverse analysis, as might an elevated incidence of a health end point in a community (e.g., elevated rate of leukemia).

characterizing dose-response data) should identify the dose terms reported in toxicity studies that they expect to utilize in the dose-response assessment (see link in Figure 2.3). In this way, every effort can be made during the exposure assessment to characterize the exposure conditions and provide intake or dose terms that are consistent with those used to describe the dose-response relationship. Achieving this concordance between dose-response and exposure assessments may be an iterative process occurring as these two elements of the risk assessment develop.

Two different types of exposure measures generally are used for environmental risk assessments: administered dose (the traditional metric) and internal dose. Administered dose refers to the mass of a substance contacted by an organism at the exchange boundary. Risk assessors generally express this dose as mass per unit body weight per unit time (e.g., mg/kg/day is the most common metric for oral route exposures). Internal doses include measures or predictions of mixture concentrations in body tissues (e.g., average daily blood concentrations of methylmercury and lead expressed as μg/ml or dl of blood) or the body burden of a mixture (e.g., dioxin-like compounds in the body expressed as ng/kg body weight). Text 2.2 highlights the importance of mixtures' considerations, when integrating the exposure assessment and dose-response assessment during risk characterization.

Box 2.2: Interdependence of the Dose-Response and Exposure Information in the Risk Assessment

During risk characterization, risk assessors will integrate the dose-response and exposure assessments to estimate the human health risks associated with the chemical mixture. Knowledge of the dose metric(s) to be used in the dose-response assessment and the reporting of the results of the exposure assessment using the same dose metric help ensure that the exposure assessment contributes to the best use of existing toxicity data in the risk assessment. Similarly, the risk assessor developing the exposure assessment needs to understand the nature of the dosing, including frequency and duration, utilized in the relevant dose-response studies.

- Did a single dose constitute the dosing in the relevant toxicological or epidemiological study?
- Did the doses leading to the health effects occur during a short span of time (e.g., a week)?
- Did the doses extend over a longer span of time (e.g., a number of months or years) or over a lifetime?

While the exposure assessment is not necessarily limited to the interpretation of the existing toxicity data (e.g., duration and dose metric used in the dose-response assessment), given that toxicity information commonly represents the limiting factor for health risk assessments, it is most practical to consider this information during development of the exposure assessment. If the exposure assessment and dose-response assessment results are not concordant, then the risk associated with exposure to a chemical mixture cannot adequately be evaluated.

A mixture's assessment can be more complicated than single-chemical assessments because risk assessors can utilize at least three different approaches to evaluate the "dose" of the mixture. Some toxicity studies measure the dose of the whole-chemical mixture [35], while others measure the dose of one or more of the mixture components and still others rely on measures of various fractions comprising the mixture [3]. Risk assessors need to determine whether whole-mixture, component, or mixture fractions data will be used in the dose-response assessment (and thus guide that element of the exposure assessment).

To summarize, the problem formulation elements that directly influence the mixture exposure assessment include (1) identification and characterization of the source(s) releasing the chemical mixture into the environment, if possible; (2) development of a preliminary analysis of environmental fate that uses the basic physical–chemical property data and estimates concentration where people might be exposed under scenarios; and (3) determination of what toxicity data exist in order to understand what types of exposure information will be most useful for guiding the ultimate risk characterization (i.e., the exposure and dose-response assessments must be concordant to characterize the risk).

2.6.2 The Detailed Exposure Assessment

The main exposure assessment phase involves evaluating human exposures to chemical mixtures in environmental media (i.e., "environmental concentration to exposure" phase). This phase builds on the preliminary data collected as part of problem formulation and should be described explicitly in the analytic plan developed during that step. Section 2.6.2.1 describes the complete characterization of the sources of the mixture components. Section 2.6.2.2 discusses the fate of mixture components in the environment, including the predicted or measured concentrations in the exposure media. Section 2.6.2.3 describes the development of an exposure scenario including the exposure setting and nature of human exposures. Section 2.6.2.4 addresses the development of quantitative exposure estimates to the mixture of chemicals. Each description of the steps in the method is followed by specific examples describing applications of those steps.

2.6.2.1 Sources

Methods For incidental and intentional releases of mixtures, risk assessors typically seek measurement data from the sources. If the quantity or composition of the mixture released varies, risk assessors generally seek additional data to estimate the variability of the emissions. For coincidental chemical mixtures, measured concentrations in various environmental media might be available, but the sources might be unknown. In those cases, environmental forensic tools can guide preliminary evaluations of source attribution, for example, by using an indicator for ambient conditions (see Box 2.3).

Box 2.3: Environmental Forensics Using Cesium-137 Soil Profiles

Cesium-137 was deposited on the earth's surfaces decades ago as atmospheric fallout from past weapon tests. Thus, the cesium-137 depth profile can be used to assess soil disturbance. For example, if groundwater was contaminated with a mix of chemicals and unreported waste burial was a suspected cause, then the soil profile for cesium-137 could be checked. A profile consistent with expectations for undisturbed soil would suggest that the source was not in overlying soil; on the other hand, A profile inconsistent with an undisturbed profile, suggests that the source of the contamination might be in the overlying soils. The isotopic ratio of uranium in groundwater can also be used as an indicator of contaminated versus uncontaminated conditions [36].

Applications Resources that provide information about chemicals released from anthropogenic sources include the Toxics Release Inventory (TRI) [37]. Begun in 1988, the TRI provides data for approximately 650 chemicals released from industries including manufacturing, metal and coal mining, electric utilities, and commercial hazardous waste treatment. Trial burn data provide information on air emissions for various hazardous waste combustors [32]. National emission inventory data are also available for a number of chemicals [38, 39].

Other public reporting requirements exist for releases of other mixtures in the United States. One example is the publication of quarterly average concentrations of four trihalomethanes (THMs) and five haloacetic acids (two classes of DBPs) by drinking water utilities. Risk assessors can use these data to estimate selected DBP concentrations in treated drinking waters in the United States [40].

Depending on what information is available for the combined chemicals of interest, risk assessors might use surrogate data for chemicals released from different sources or to different areas. Table 2.1 identifies several mobile sources that can release airborne chemicals and the information resources that can be used to identify and quantify the combined chemicals released from these sources. Source processes and seasonal factors can alter the composition of DBP mixtures initially released into drinking water distribution systems following water treatment. Such changes in DBP composition might alter the spectrum of health effects or the magnitude of the risks associated with exposures to these DBPs (see Ref. [41] and references cited therein).

2.6.2.2 Environmental Fate of Mixture Components

Methods In this phase of risk assessment, which evaluates the fate of a mixture in the environment, risk assessors use measured data or predictive models to characterize the fate of component chemicals or a whole mixture over the time period of interest. Typically, this involves evaluating the fate of a released mixture across multiple media, from the receiving medium to those into which the mixture chemicals or their transformation products may partition, and to which people

Table 2.1 Grouping airborne chemicals from multiple mobile sources.

Models used to estimate emissions	Chemicals estimated
On-road mobile MOBILE62 model [78]	Criteria pollutants (sulfur dioxide, nitrogen oxides, carbon monoxide, PM_{10}, $PM_{2.5}$, and lead); hydrocarbons; carbon dioxide; ammonia; and six organic compounds (benzene; methyl *tert*-butyl ether; 1,3-butadiene; formaldehyde; acetaldehyde; acrolein)
Nonroad mobile (vehicle/equipment engines): NONROAD model [79]	Criteria pollutants and hydrocarbons
Mobile, toxic fractions of hydrocarbons (e.g., engine exhaust)	Fraction-specific emissions for speciated hydrocarbons
(The 1999 inventory is available in US EPA [80]; for the most recent inventory, see US EPA [38])	(For the 1999 inventory, see supporting data in US EPA [81])

could be exposed. Naturally, measurement data are preferred because, if properly collected and evaluated, confidence in such data is higher than estimates from mathematical or empirical models. However, measurements at the scale typically needed for an exposure assessment of chemical mixtures in the environment are usually very expensive, so data availability is generally limited. Risk assessors more commonly use mathematical models to predict the fate of released chemical mixtures or the further fate of mixtures already present in the environment (e.g., coincidental mixtures). Limited measurement data can be used to augment or evaluate the modeling data. Identifying and quantifying accurately the changes in mixture compositions over time is important because such changes can significantly alter the predicted toxicity compared to that of the mixture originally released.

Differential transport is a key reason for the compositional changes in environmental mixtures over time. For example, certain mixture components will be advected away from a given source faster than others because differences in their physical–chemical properties lead to differences in their behaviors in the receiving medium. Differential degradation (i.e., some compounds are degraded into their breakdown products more rapidly than others) can also change the composition of the chemicals in a mixture following release. The timing of these changes can be important for predicting *when* and *where* people might be exposed to different combinations of chemicals following release of the mixture. Finally, mixture component chemicals can also partition differentially to other media depending on the physical–chemical properties of the chemicals and both the medium they initially enter and the media into which they may partition. It is useful to consider three types of possible transfers that can occur between environmental compartments: differential transfer between different abiotic media, differential transfer between abiotic and biotic media, and differential transfer between different biotic media (see Box 2.4).

Box 2.4: Differential Transfer of Technical-Grade Toxaphene Components

Technical grade toxaphene, a heavily used insecticide in the United States until the early 1980s, illustrates various cross-compartmental transfers of mixed chemicals:

Differential Transfer between Different Abiotic Media: Technical-grade toxaphene contains over 670 chemicals. Because some components of may volatilize to air and others are hydrophobic, the composition of the mixture will differ depending on where it is encountered, e.g. in soil at a contaminated site, the air around the site, or in nearby lake sediments [42].

Differential Transfer between Abiotic and Biotic Media: Although originally applied to soils, some components of technical-grade toxaphene have been measured in shellfish and fish, and they appear to be selectively retained by these organisms [42].

Differential Transfer between Different Biotic Media: After considering toxicokinetics for a dose-response analysis, Simon and Manning [43] suggested an approach for estimating cancer risk from eating fish contaminated with toxaphene congeners. This approach is based on concentrations of three congeners in fish that appear to be the primary forms selectively retained in humans after ingestion.

In the last part of this assessment phase, it can be useful to group the mixture components on the basis of (1) the media into which they partition and (2) the estimates of the timing of when the component chemicals would be expected to occur in the exposure media (e.g., at the points of human contact). For a given chemical mixture, most components might be measured in the receiving medium, while only those exhibiting certain physical–chemical properties or degrading to form certain chemicals would be expected to occur in other media. Some fraction of certain other mixtures (or individual chemical components) could exist in multiple media (e.g., soil, water, and air) at the same time. Identifying which components are likely to occur in what media over what time frames (e.g., considering field and laboratory data for persistence and fate product formation) is a useful way for risk assessors to organize the information developed in this phase.

Three elements are important to consider when assessing the timing of chemicals in various media and potential exposures. The first element addresses when, in the *future*, relative to the timing of the release, the chemical mixture or specific components will occur in a given medium through which exposures could occur. The second time-associated element addresses *overlaps* (i.e., whether the chemical mixture or some of the chemicals that comprise the mixture could concurrently occur in the same or different environmental media, and whether an individual could be exposed concurrently to multiple chemicals through contact with the media). The third element describes the *peak quantities* of a chemical mixture or its components in various media; this information is especially useful when assessing acute or subchronic effects and for bounding analysis where needed (e.g., for radioactive waste disposal sites).

Applications For this detailed evaluation, risk assessors commonly apply fate and transport models to fill gaps in measurement data. For example, models that estimate long-term steady-state concentrations, such as the multimedia models described in US EPA [15, 32], might be sufficient for evaluating the exposures that occur in some scenarios. Such scenarios include continuous emissions from an incinerator or long-term surveillance and maintenance after cleanup of a contaminated site. More advanced models are used to address episodic industrial releases or nonsteady-state exposures.

Advanced dispersion models can estimate nonsteady-state short-term concentrations to provide more realistic estimates of airborne exposures that serve as inputs to the inhalation models. For example, the Atmospheric Dispersion Modeling System, available from the Support Center for Regulatory Atmospheric Modeling [39, 44], can simulate short-term and average annual concentrations of chemicals released either continuously or discretely from a variety of source types. Risk assessors can program such models to include certain chemical reactions in the atmosphere and to account for physical decay. After risk assessors identify the environmental fate model(s) that are best suited to the given mixture exposure assessment, they can collect physicochemical and setting-specific data that will serve as model inputs to predict concentrations of the mixture or component chemicals in media and locations projected to be affected over time.

Table 2.2 illustrates how chemicals can be grouped based on typical primary release mechanisms that would be expected to control initial contamination and migration behavior in the environment. Some of these initial groupings might be developed during problem formulation. Chemicals released into the air can be dispersed by convective processes (e.g., through wind or surface water flow), while chemicals released into the soil (e.g., following waste placement) can leach downward to groundwater. Dominant sources and fate processes for a given setting will determine

Table 2.2 Exposure groupings by common release or migration behavior.

Release process	Organic chemicals	Inorganic chemicals and gases
Volatilization to air	Chlorinated solvents, petroleum-based solvents, fuels	Chlorine, ammonia, tritium, SO_2, NO_x, CO, CO_2
Permeation to groundwater	Chlorinated solvents (DNAPLs), aromatic hydrocarbons (BTEX) pesticides	Cations and anions
Particulate emissions from combustion (stacks)	Products of incomplete combustion (PICs) – PCBs, PAH, dioxins, furans	Heavy metals
Gaseous emissions from combustion (stacks)	Light hydrocarbons	SO_2, NO_x, CO, ammonia
Release of contaminated waters to surface waters	PCBs, PAH, dioxins, furans are sparingly soluble	Chlorine, ammonia, tritium, mercury

Heavy metals can include arsenic, cadmium, chromium, mercury, and nickel (e.g., see ATSDR [82]); BTEX: benzene, toluene, ethylbenzene, xylene; CO: carbon monoxide; CO_2: carbon dioxide; DNAPLs: dense nonaqueous phase liquids; NO_x: nitrogen oxides; PICs: products of incomplete combustion; SO_2: sulfur dioxide.

Table 2.3 Grouping chemicals by properties affecting environmental fate in soil and air.

Environmental compartment	Persistence environmental half-life	Environmental partitioning equilibrium-based[a]	Mobility dispersion-based
Organic matter in soil[b] and sediments, soil organisms	*High for* high K_{ow}/K_d[c], low biodegradability	Presence favored by high K_{ow}/K_d	*High binding for* high-K_{ow}/K_d organics and inorganics
	Low for high K_{ow}/K_d, high biodegradability	High persistence	*Low binding for* low-K_{ow}/K_d organics and inorganics
Soil inorganic phase	*High for* high-K_d inorganics, low-K_{sp} inorganics	*Presence favored by* high-K_d and low-K_{sp} inorganics	*High mobility for* cations, anions, water-soluble organics (low K_{ow}/K_d)
	Low for low-K_{ow}/K_d organics/inorganics		*Low mobility for* high-K_{ow}/K_d organics, high-K_{sp} solids
Air	*High for* low photodegradable, low reaction rate with hydroxyl radical and other free radicals, low washout rate (low K_H), gas phase	*Presence favored by* high volatility substances (gases and low boiling point liquids), high volatility from water (low K_H)	*High mobility for* gas phase, high persistence, small-particle bound
	Low for photodegradable, high reaction rates, high washout (high K_H), particulate phase		*Low mobility for* low persistence, large-particle bound

a) *Presence favored by* indicates that concentrations would be relatively higher compared to adjacent compartments, that is, activity coefficients for the substances are relatively low in the given compartment/medium.
b) Soil and sediment include the fraction of organic carbon (foc) and the clay content, which can reflect the quantity and types of sorption sites available.
c) K_d: soil/water partition coefficient, which is the amount of chemical associated with the solid phase (e.g., sorbed to soil particles) compared to the amount of chemical in pore water); K_H: Henry's constant (i.e., water/air distribution constant); K_{ow}: octanol/water partition coefficient (i.e., the extent to which a chemical partitions between octanol and water at equilibrium. Octanol is a surrogate for soil organic matter and lipids in the body); K_{sp}: solubility product constant for inorganic complexes.

what media are ultimately affected. Table 2.3 illustrates how mixture components can be grouped to assess fate in soil and air on the basis of properties such as volatility, water solubility, and partition coefficients. Identifying these groups of chemicals that co-occur in the same medium is important for mixture risk assessments because individuals could be coexposed to these combined chemicals through contact with those contaminated media. Thus, they potentially need to be considered as a chemical mixture in the assessment.

The components of these mixtures need to be identified for the risk assessor(s) developing the dose-response analysis. Those risk assessors can determine if health effects data exist for the mixture and convey important information regarding the duration of the doses analyzed, and how the mixture was analyzed, to the risk assessor developing the exposure assessment.

2.6.2.3 Characterizing the Exposure Scenario

Methods In characterizing the exposure scenario, the study area of interest is identified, including the boundary and the environmental characteristics that are relevant to the exposure assessment.[6] In addition, risk assessors identify people who are, or could be, exposed to multiple chemicals within that area, specifying where and how those exposures could occur. Like the previous discussion, this detailed assessment also builds on information collected during problem formulation. For example, identification of the chemical mixtures' source(s) and evaluation of environmental fate help define the study area boundary (see examples in Ref. [32]). An important consideration for chemical mixtures is whether sources of those chemicals of interest outside the study area might be contributing to ambient concentrations of mixture components in that targeted area.

During this phase, the risk assessors characterize the land uses in the study area, which may include residential, work, recreational, and garden/agricultural areas. Risk assessors also identify areas where certain population groups who might experience unique exposures might be located (including groups who could be differentially exposed). Risk assessors who are analyzing dose-response information could also be asked to identify those groups that might be vulnerable to exposures to the chemical mixture during this detailed assessment phase. For example, if schools and playgrounds are in the study area, the assessment may include an exposure evaluation for asthmatic children.

Thus, exposure assessments for chemical mixtures examine exposures for members of both "typical" populations and vulnerable subpopulations, with the latter being of key interest for cumulative health risk analyses. The US EPA *Framework for Cumulative Risk Assessment* [6] reflects "vulnerability" concepts that address receptor characteristics in four areas: differential exposure, susceptibility/sensitivity, differential preparedness, and differential ability to recover. Several other information resources also address this topic [45–51]. Clearly, close integration between the exposure and dose-response assessment components of the risk analysis process is crucial to effectively consider people who might be more susceptible or vulnerable to mixture exposures or effects.

Applications Basic geographic and soil survey data can be found through such information resources as the USGS and the US Department of Agriculture (USDA) Natural Resources Conservation Service. Climate and meteorology data are generally

6) While this text is developed for a site-specific exposure scenario, exposure scenarios can be developed through conjecture involving hypothetical sources and receptors.

Table 2.4 Information for susceptibility assessment.

Type of information	Example resources
Demographic data	US Census Bureau (www.census.gov)
Subpopulation groups	Sociodemographic data used for identifying potentially highly exposed populations [83]
Locations (e.g., schools, hospitals, and nursing homes)	Plat maps, county and city health department reports
Exposure data (e.g., blood lead levels)	State registries, county and city health department reports
Cancer registries	Centers for Disease Control (national data and links to state cancer registries, www.cdc.gov/cancer/npcr/statecon.htm)
Other health effect registries	State registries of birth defects, asthma

available from the National Weather Service and National Oceanic and Atmospheric Administration. Beyond the USGS resources noted earlier, risk assessors can obtain land use data from local agencies responsible for planning and zoning maps.

Table 2.4 identifies several sources of information that risk assessors can use to identify potentially vulnerable populations. Risk assessors can obtain values for the exposure factors associated with different potential receptors (e.g., quantity of various homegrown vegetables consumed daily from farming households in the West) from the *Exposure Factors Handbook* [52] and similar references.

The assessment might also address selected subgroups that could be differentially exposed due to proximity to a key source or unique activity patterns – such as subsistence fishing. In these cases, detailed recreational use and activity patterns can be developed from survey data, especially for fishing and hunting. These data may be available from state or county departments of environment, conservation, natural resources, or parks and recreation.

The assessor may also need to evaluate other sources of exposure to a chemical mixture in the population [18]. Such exposures could arise from releases from regional anthropogenic sources that contribute to levels of a chemical mixture at a given location under evaluation. Also, foods supplied from other regions might be contaminated with low levels of chemical mixtures. To assess whether other sources of chemical pollutants in a region could be contributing to chemical levels at a given location, risk assessors might examine whether relevant data could be available from US EPA including its regional offices (e.g., TRI data), state, county, or city environmental agencies, and state or local universities. For example, to assess oral exposures to dioxin-like compounds, US EPA [21] obtained data on concentrations in the US food supply from the USDA and other sources.

2.6.2.4 Exposure Quantification

Methods The last part of the exposure assessment for chemical mixtures involves predicting or measuring human exposures as doses. The exposure equations developed in US EPA [14] and also identified in Paustenbach [29] and other US EPA standard guidance documents, and the parameter values compiled in US EPA [52], and the subsequent reports provide a mathematical structure for predicting these doses from environmental concentrations.

Exposure rates for each exposure pathway are typically based on (1) the estimated concentration in a given medium, (2) the contact rate, (3) receptor body weight, (4) the frequency of exposure, and (5) the duration of exposure as well as (6) the exposure time if not accounted for in the contact rate (e.g., if the latter were m^3/hours, the exposure time would be hours/day). This calculation is repeated, e.g., for each component of the mixture (if a component-based approach is used) and e.g., for each exposure pathway included in an exposure scenario.

Equation (2.1) details the general form used in exposure equations. In the equation, intake (I) describes the amount of chemical at the exchange boundary. For oral route exposures, the units are mg/kg/day. Concentration (C) represents the concentrations of the individual mixture components in the environmental media people might come in contact with during the exposure period. Typically the units of mg/kg are used for contaminated foods and soils and mg/l for contaminated water. Contact rate (Cr) describes the amount of contaminated medium contacted per unit of time. Typically the of kg/day are used for foods and soils, and l/day for water. Exposure frequency (EF) describes the frequency at which the exposure event occurs; the unit is generally days/year. The exposure duration term (ED) describes the time period over which the exposures occur; the unit for this variable is typically years (yr). Body weight (Bw) describes the average body weight of exposed individuals. The unit is typically reported in kg. Finally, averaging time (AT) is the period over which the exposure is averaged (days).

For inhalation route exposures, this calculation typically is not used. Under the updated approach for assessing chronic exposures, which uses the C term directly with the relevant toxicity value. That is, risk assessors report the dose metrics for inhalation dose-response functions as concentrations in the air.

$$I = \frac{C \times \mathrm{Cr} \times \mathrm{EF} \times \mathrm{ED}}{\mathrm{Bw} \times \mathrm{AT}}. \quad (2.1)$$

Risk assessors routinely conduct uncertainty and variability analyses on the exposure estimates; these can be quite detailed [53–55]. In the relatively rare instances where tissue concentrations of chemical mixtures or mixture components have been measured, exposures can be reconstructed from those measurements using toxicokinetic information (which addresses the absorption, distribution, metabolism, and elimination of the chemicals) and information regarding the frequency, intensity, and duration of past exposures.

A key point for exposure assessments evaluating chemical mixtures is an emphasis on the identification of exposures to multiple chemicals that can occur over a period that is sufficiently short such that one chemical can influence the toxicity of others. Both toxicokinetics and toxicodynamics provide insight as to defining this length of time for various chemical mixtures. Toxicokinetic information can be useful in estimating the likelihood of co-occurring concentrations of multiple chemicals in body tissues

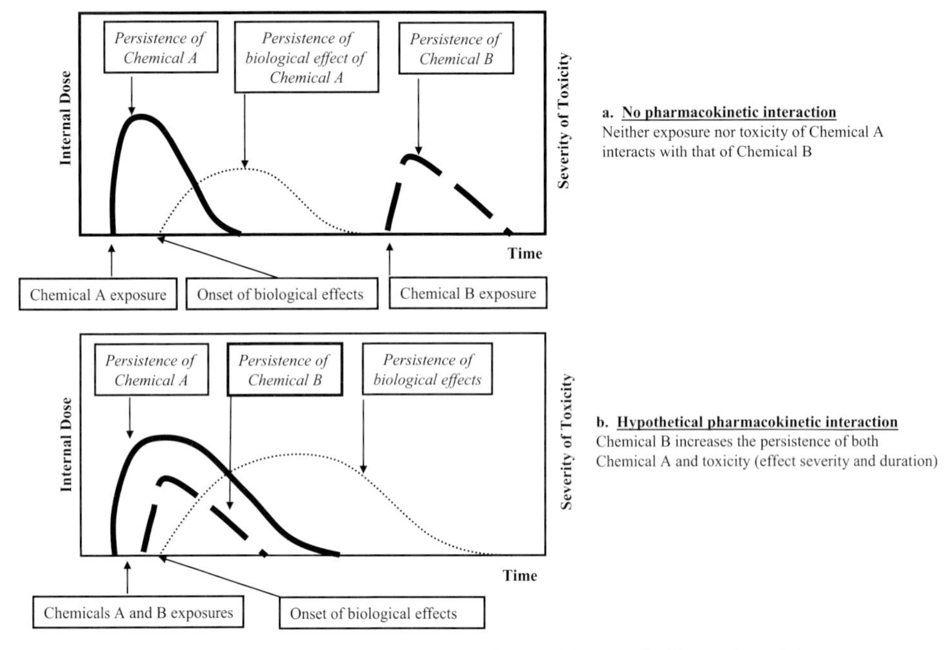

Figure 2.4 Assessing overlaps in timing of internal dose and effect. (adapted from [18]).

(i.e., overlapping internal doses of a mixture) and in quantifying the magnitude of the internal doses of each compound. Elevated concentrations of multiple similar compounds could lead to toxicity if detoxification pathways became overwhelmed. Similarly, toxicodynamics of multiple chemical exposures are considered because tissues targeted by the same chemicals may not have sufficient time to repair or regenerate prior to the subsequent insults leading to a toxic response. Figure 2.4 illustrates potential exposure and toxicity overlaps for a hypothetical two-chemical mixture.

Mixtures or various chemical components can coexist in the same medium or in different media to which people can be exposed during the same period. Internal doses of such chemicals that could overlap based on the timing of the exposures are grouped together. Toxicokinetic data for the compounds then are evaluated. For example, some chemicals have relatively long biological half-lives (e.g., highly chlorinated compounds), so internal doses could persist to extend the overall period of time during which exposures to other chemicals could exert a joint effect. To address this issue and to help guide the integrated evaluation of toxicity data, chemicals that comprise the mixture of interest can be grouped by exposure medium and timing.

Figure 2.4 illustrates how risk assessors can evaluate toxicokinetic information to assess the opportunity for internal dose overlaps. Highlighting discrepancies between the exposure assessment and the dose-response assessment can help guide the development of better information to support the next component of the risk assessment paradigm – risk characterization. These improvements may include refining information for the exposure durations of interest, the exposure pathways

Table 2.5 Exposure groupings by coexistence in media and time.

	Same medium	**Different media**
Same time	*Group* 1: Coexposures to a mixture of DBPs from consumption of unheated drinking water	*Group* 3: Coexposures to volatile and nonvolatile DBPs from consumption of unheated drinking water and via inhalation while showering
Different time	*Group* 2: Exposures to different pesticides with short environmental half-lives in contaminated drinking water	*Group* 4: Inhalation of PAHs from a mobile incinerator for limited-duration site remediation and, years later, ingestion of metals from well water contaminated by the site

and routes of interest, and the exposure metric of interest (such as an environmental concentration at the point of human contact or an internal tissue dose).

Tools and techniques that can support more detailed exposure assessments include software such as the "calendar approach" (which can develop information similar to that shown in Table 2.5) [16, 17]. The calendar approach uses physiologically based toxicokinetic modeling to estimate sequential, daily chemical exposures by linking episodic exposures (e.g., seasonal exposures to pesticides through surface water contact following residential lawn applications in spring and summer) with routine pesticide exposures (e.g., contaminants in the food supply). The calendar approach integrates these exposures by route using probabilistic input data to simulate daily exposures while allowing these to vary over the year.

Applications A number of studies use mathematical exposure models to estimate human mixture exposures. For example Lorber *et al.* [19] and US EPA [21] estimate US exposures to dioxin-like compounds. US EPA [17] uses mathematical models to estimate exposures to mixtures of organophosphorous pesticides.

Grouping chemicals on the basis of potential co-occurrence in affected compartments/media per location and time can be very useful for assessing chemical mixtures. Table 2.5 illustrates how grouping exposures can guide the analyses of multiple chemical exposures.

Group 1 includes chemical exposures that occur at the same time and through the same media. Risk assessors would assume such exposures to overlap. Someone drinking unheated drinking water would have overlapping exposures to DBPs in that water. Exposures to coincidental mixtures (e.g., via inhalation of urban air) occurring through a single pathway also illustrate group 1 (see Figure 2.4b).

Group 2 describes exposures occurring via the same medium at different times that would not be expected to overlap. Consider two pesticides that both partition to water at relatively low concentrations. These pesticides are applied during different seasons of the year and have relatively short environmental half-lives such that one has been eliminated from the water before the second is applied. While individuals could be exposed to both pesticides through their drinking water, it would be unlikely that the exposures would overlap. Also, if they are metabolized and eliminated rapidly, the likelihood of a toxicologically meaningful overlap is low. From the joint evaluation of environmental fate and toxicokinetics, exposures to these two chemicals would not be expected to overlap (see Figure 2.4a).

Group 3 describes exposures at the same time through different media. Risk assessors might assume logically that these exposures overlap. DBPs provide an example of such exposures through different media. An individual might be exposed by drinking the DBPs that occur in chlorinated drinking water and breathing volatile DBPs that partition to indoor air during showering. The extended example (based on the work by Teuschler *et al.* [24]) illustrates that although DBP exposures occur via different media (and routes), they can overlap. Determining whether these overlaps are significant requires toxicological evaluation.

Group 4 encompasses exposures that do not overlap because they occur via different media at different times. For example, such nonoverlapping exposures hypothetically could occur in a residential area near a contaminated site, where PAHs released by an onsite incinerator operating only during a limited cleanup period are inhaled during that short time, and, years later, metals from the site reach groundwater that is tapped by local wells. Consequently, the same residents that were exposed through inhalation to the PAHs eventually would be exposed to the contaminated groundwater by consuming drinking water. The PAHs released to the atmosphere could be transported away from the area fairly quickly following release from the incinerator (e.g., days to weeks), and those exposures would stop when the source was shut down. Groundwater transport of contaminants such as metals is typically substantially slower (e.g., decades) than atmospheric transport [56, 57]. The exposures to contaminated groundwater, such as via drinking or bathing, would occur many years after the inhalation exposures. According to the evaluation of environmental fate and kinetics, exposures to these two mixtures of chemicals would not be expected to overlap (see Figure 2.4a). These types of joint exposures might not be recognized in traditional assessments because the modeling needed to explicitly evaluate and integrate the information to assess such overlaps is not routinely conducted.

2.7
Illustrative Example: Assessing Exposures to DBP Mixtures in Drinking Water

Humans are daily exposed to complex mixtures of drinking water DBPs that form when water is chemically disinfected [58]. Although chemical disinfection of drinking water has greatly decreased the morbidity and mortality from waterborne microbial diseases, some epidemiological studies suggest that cancer and developmental effects are associated with exposures to DBP mixtures [59, 60].

These DBP mixtures likely consist of hundreds of chemicals [61]. Occurrence data also show that the mix of DBPs may considerably vary depending on the quality of the source water and water treatment process [62]. Factors influencing source water quality include season of the year [63], presence of contaminating sources, and geographic location [33]. Although critical for estimating exposures, the formation and transformations of DBPs in drinking water distribution systems is understudied.

Factors influencing DBPs in distribution systems include system and storage hydraulics, water temperature, pH, and time in the distribution system [62, 64–67].

We drew the materials for this illustrative example from two documents: Teuschler *et al.* [24] and US EPA [68]. For simplicity, many of the modeling details are not reported here. There are a number of approaches that these authors could have used to estimate exposures to DBPs – for example, direct measurement of compounds at the point of contact [52] or biomonitoring to analyze past exposures through internal indicators [69, 70]. Teuschler *et al.* [24] chose to use predictive mathematical models. The authors developed exposure estimates for several chemicals comprising these DBP mixtures through an exposure model, the total exposure model (TEM) [71], and internal dose estimates using a physiologically based pharmacokinetic (PBPK) model, the exposure-related dose-estimating model (ERDEM) [72].

The exposure modeling was based on

1) activity patterns affecting human contact time with drinking water;
2) building characteristics that influence indoor air concentrations of volatile DBPs; and
3) DBP physical–chemical properties that influence their fate in the indoor environment.

The use of internal dose estimates for the chemicals comprising the DBP mixture is an important choice because it allows integration of exposures to the DBPs across oral, dermal, and inhalation exposure routes. Teuschler *et al.* estimated internal doses for an adult female of reproductive age and a child (aged 6 years).

They elavuated four of the DBPs that are trihalomethanes [THMs]. These four DBPs-chloro- from, bromodichloro-methane, dibromo-chloromethane and bromo-form- are regulated by the US EPA under the 1998 *National Primary Drinking Water Regulations: Disinfectants and Disinfection Byproducts Final Rule* [73]. Because lifetime health effects data had been developed for these four chemicals, Teuschler *et al.* [24] combined the multiple daily dose estimates of individual chemicals to yield lifetime average daily doses of each chemical.

2.7.1 Problem Formulation

We do not present all aspects of problem formulation in this illustrative example. The phases of problem formulation relevant to exposure analysis include (1) the identification and characterization of chemical mixture source(s), (2) preliminary analysis of environmental fate, and (3) existing dose-response metrics for chemical mixtures, as well as receptor context. We highlight the relevant DBP exposure data compiled in each phase. We present a conceptual model in Figure 2.5. The model illustrates the exposure pathways relevant for the DBP mixtures. The figure identifies key elements of the assessment, specifically the determination of the concentrations of selected DBPs in drinking water estimation of concentrations of volatile DBPs in household air concentrations of the selected DBPs in the drinking water and indoor air estimated DBP intake via oral, inhalation, and dermal exposure routes and finally

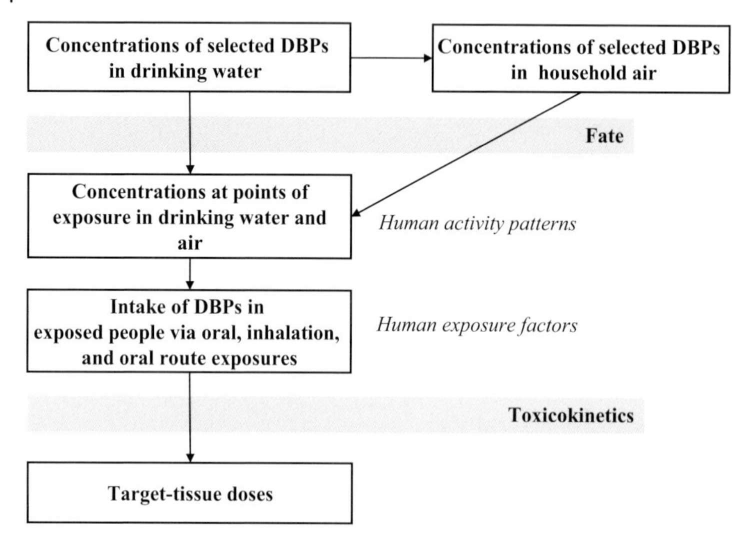

Figure 2.5 Conceptual model for assessing exposures to DBP mixtures.

doses in human target tissues. The figure also conveys the relationships among these key elements of the assessment to the reader.

2.7.1.1 Source Identification and Characterization

The primary source of human exposures to these four THMs is the use of treated drinking waters. Using the Ohio River as a source of water, Miltner *et al.* [62, 64] describe a series of studies in a pilot water treatment plant quantifying DBP concentrations in chlorinated waters. Importantly, for exposure assessment, Miltner *et al.* [62, 64] simulated the changes in DBP concentrations that occur in the distribution system by storing the treated waters in the dark. On the basis of Miltner's data, mean estimates of the DBP concentrations were generated; these are listed in Table 2.6.

2.7.1.2 Preliminary Analysis of Environmental Fate

The physical–chemical properties of the DBPs assembled in this exposure assessment included Henry's law constant, liquid-phase diffusivity, gas-phase diffusivity, octanol/water partition coefficient, molecular weight, boiling point, and volatility. The Henry's law constants indicate that all four DBPs can volatilize from the drinking

Table 2.6 DBP concentrations in chlorinated drinking water [62].

Chemical	Mean (μg/l)
Chloroform	55.50
Bromodichloromethane	24.40
Dibromochloromethane	10.20
Bromoform	0.35

water. The octanol/water partition coefficients suggest that these DBPs may cross human skin. Thus, human exposures may result from ingestion (drinking water consumption), inhalation (volatilization during drinking water uses), and dermal (during or after bathing or showering) exposures [74].

2.7.1.3 **Dose-Response Information for the Four DBPs**

This illustrative example focuses only on the cancer risk associated with these four DBPs. Teuschler et al. assumed that response addition, a component-based approach that assumes independence of the toxic mode of action (MOA), is appropriate for evaluating the carcinogenicity risk associated with this four-component mixture. US EPA [12] provides a detailed description of many component-based methods for chemical mixtures. Because a component-based method is used to evaluate human health risks, the risk assessors needed to develop individual chemical exposure information. (Simmons *et al.* [75] describe additional approaches for characterizing risks associated with DBP exposures.)

US EPA's Integrated Risk Information System (IRIS) lists cancer slope factors that estimate the potency of carcinogens (i.e., slope factors indicate the proportion of individuals statistically estimated to incur cancer per unit of dose or concentration for inhalation). In IRIS, the oral slope factor[7)] is expressed as a dose in mg/kg/day of chemical for three of the four DBPs: bromodichloromethane, dibromochloromethane, and bromoform (see Table 2.7) [76]. While the US EPA does not provide an oral slope factor for chloroform, an oral reference dose (RfD)[8)] for chloroform is listed. In this illustrative example, it is assumed that chloroform exposures below the RfD do not incrementally increase the human cancer risk.[9)]

2.7.2
Exposure Assessment

Exposure assessment consists of four key steps: (1) identifying the source of the exposure; (2) evaluating the fate of mixture components, including characterizing their concentrations in the exposure media; (3) characterizing the exposure setting (including potential receptors); and (4) estimating exposures and determining coexposures to mixture components in quantifying human exposures.

7) US EPA defines the oral slope factor as an upper bound, approximating a 95% confidence limit, on the increased cancer risk from a lifetime oral exposure to an agent. This estimate, usually expressed in units of proportion (of a population) affected per mg/kg/day, is generally reserved for use in the low-dose region of the dose-response relationship, that is, for exposures corresponding to risks less than 1 in 100.

8) US EPA defines the RFD as an estimate (with uncertainty spanning perhaps an order of magnitude) of a daily oral exposure to the human population (including sensitive subgroups) that is likely to be without an appreciable risk of deleterious effects during a lifetime. It can be derived from a no-observed adverse effect level, a lowest observed adverse effect level, or a benchmark dose, with uncertainty factors generally applied to reflect limitations of the data used. RfDs are generally used in noncancer health assessments.

9) Although not examined in this analysis, in cumulative risk assessments, the risk assessor may attempt to identify individuals who are potentially susceptible to the toxicants of concern [18, 77].

Table 2.7 Available cancer risk information for the four THMs.

Chemical	CAS number	Oral slope factor per (mg/kg/day)
Bromodichloromethane	75-27-4	6.2E-2 per (mg/kg/day)
Bromoform	75-25-2	7.9E-3 per (mg/kg/day)
Chloroform	67-66-3	No slope factor exists, the reference dose (noncancer end point) is 0.01 mg/kg/day
Dibromochloromethane	124-48-1	8.4E-2 per (mg/kg/day)

2.7.2.1 **Evaluating the Source of the Four Mixture Components**

In houses, these DBPs are associated with points where drinking water is released. These points include water faucets, showerheads, toilets, dishwashers, and clothes washers.

2.7.2.2 **Evaluating the Fate of the Four Mixture Components**

Using the TEM, Wilkes [71] estimated exposures to the four THMs. TEM input data include drinking water chemical concentrations, domestic water-use behaviors, building characteristics, and chemical volatilization. The authors used the mathematical models comprising the TEM to estimate daily concentrations of the four DBPs in the indoor air inside a hypothetical home.

2.7.2.3 **Characterizing the Exposure Setting**

The exposure setting was the inside of the hypothetical house. In this simulation, the authors assumed this house to be a collection of compartments where water-use zones were explicitly simulated and the remaining parts of the house were lumped into a common zone. Important building parameters for inhalation route exposures included whole-house volumes, volumes of the water-use zones, whole-house air exchange rates, and airflow between zones.

2.7.2.4 **Estimating Exposures and Determining Coexposures to Mixture Components**

The authors used the TEM to simulate human activity patterns, ingestion characteristics, and the physiological measurements that were used to estimate 24-h exposure time histories for the hypothetical individuals. Figure 2.6 lists the types of input data needed for the TEM and the types of output data generated by the TEM. The TEM depicts human water-use behaviors by querying databases that describe activities and behaviors affecting DBP releases from drinking water and subsequent human exposure. As detailed in US EPA [21], the authors focused on water-use activities: showering and bathing, and using the clothes washer, dishwasher, toilet, and faucet. They described the frequency of each water use (e.g., how often individuals shower) and the event duration (e.g., time spent in shower).

Exposure Quantification The authors used the TEM to simulate ingestion, inhalation, and dermal exposures. Here, a discussion of some of the data used to estimate

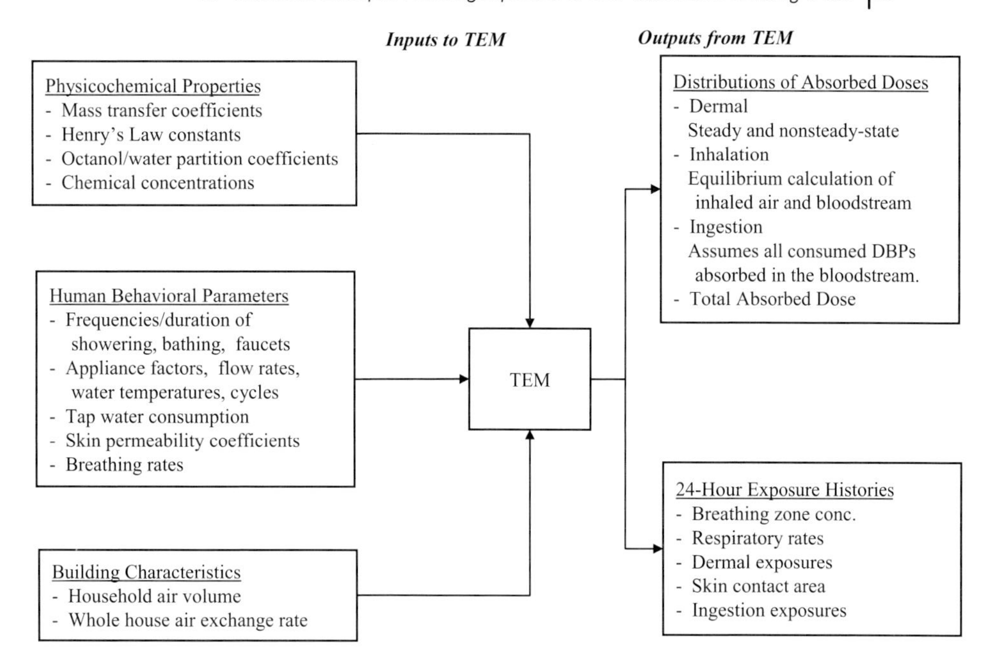

Figure 2.6 Overview of inputs and outputs to the Total Exposure Model.

ingestion route exposures follows a brief description of the inhalation and dermal route exposures. The authors estimated the inhalation exposures using the predicted indoor air concentrations of the four DBPs (based on water use and volatilization rates), human breathing rate information, and the time spent in various zones of the house. Dermal exposure estimates were based on chemical partitioning data for the DBPs and estimates of the time spent showering, bathing, and hand washing. These estimates were significantly less certain because the process of dermal absorption is not as well understood; consequently, there is less confidence in the models of this process compared to ingestion or inhalation routes (additional details are in Ref. [68]).

To assess oral DBP exposures, the authors identified the sources of water intake and estimated the quantities of water consumed. For direct consumption, they estimated the number of drinks and volumes consumed per day. They assumed either that the contaminant level remains constant from spigot to glass to body or that some fraction of the DBPs present in the drinking water volatilized during air contact. For indirect water consumption, such as via food or reconstituted drinks, the authors considered the quantity consumed, and they also determined whether the fraction of the contaminant remaining in the drink or food after volatilization and preparation was significant or inconsequential to the exposure calculation [52].

Because the timing of exposures for mixture exposure analyses could be critical, US EPA [68] describes an approach for estimating the distribution of water

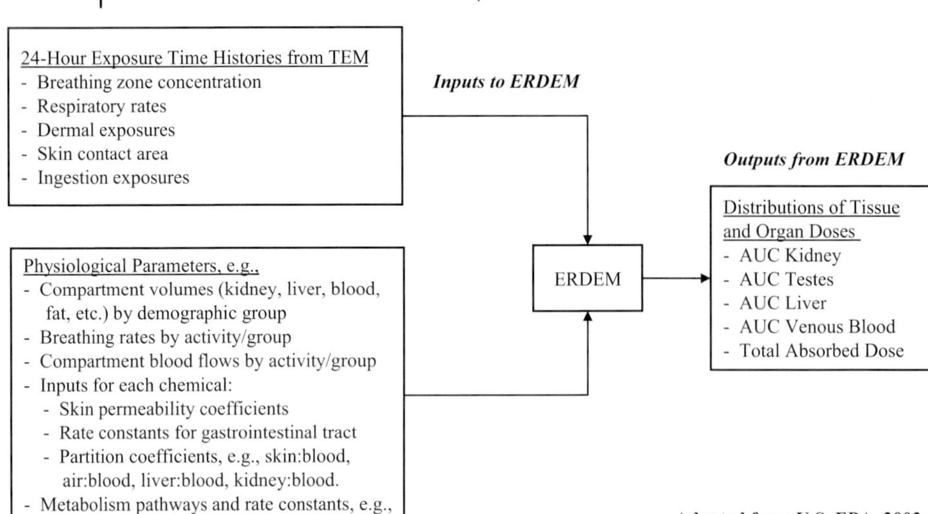

Figure 2.7 Integration of total exposure model and the exposure-related dose-estimating model. (AUC: area under the curve; V_{max}: maximum enzyme reaction rate; K_m: substrate concentration leading to one-half of the maximum rate)

consumption throughout the day. Because the authors identified no studies that quantified the manner in which water consumption is distributed throughout the day, the water consumption was distributed into a specified number of consumption events that occurred over the day (e.g., six consumption events over a 16-h period during the day).

Internal Dose Estimates To predict doses of the four DBPs experienced by relevant organs or target tissues, ERDEM (the PBPK model used) incorporated information on organ volumes, organ-specific blood flows, and metabolic capacity, and assumed no pharmacokinetic interactions occurred among the four chemicals (see Figure 2.7). The DBP exposures for oral, inhalation, and dermal routes that were predicted from the TEM simulation were used as inputs to ERDEM. Tissue doses for the four DBPs were estimated for liver, kidney, venous blood, ovaries, and testes. The dose metric used was area under the concentration–time curve averaged over 2 days.

2.7.3 Results

The TEM results include distributions of absorbed dose estimates for the dermal, ingestion, and inhalation exposure routes, and total absorbed dose. Teuschler *et al.* also examined absorbed doses for a 24-h period as a function of route, population group, and percentile of the population for each of the four DBPs (see Table 2.8). For each individual, a series of 24-h exposure simulations were combined to develop annual exposure estimates.

Table 2.8 Fiftieth percentile 24-h absorbed dose estimates (mg) output by TEM.

Chemical	Total[a)]	Dermal	Ingestion			Inhalation
			Direct	Indirect	Total[a)]	
Female[b)], age 15–45						
$CHCl_3$	3.00E-01	2.51E-02	2.09E-02	3.76E-03	2.52E-02	2.19E-01
BDCM	8.00E-02	2.70E-03	7.73E-03	1.71E-03	9.72E-03	6.12E-02
DBCM	5.12E-02	2.47E-03	5.33E-03	1.40E-03	7.03E-03	3.73E-02
$CHBr_3$	2.65E-02	1.60E-03	2.88E-03	3.00E-03	6.55E-03	1.63E-02
Child, age 6						
$CHCl_3$	1.56E-01	1.87E-03	1.09E-02	9.19E-04	1.26E-02	1.19E-01
BDCM	4.38E-02	2.66E-04	4.02E-03	1.07E-03	6.03E-03	3.36E-02
DBCM	2.91E-02	2.59E-04	2.77E-03	7.72E-04	4.18E-03	2.21E-02
$CHBr_3$	1.34E-02	1.73E-04	1.50E-03	7.42E-03	2.70E-03	8.77E-03

a) The total absorbed dose (by ingestion or by all three routes) is not equal to the sum of the doses in each row because each simulation provides a new data point to each of the dose estimates represented in the columns; the percentiles are then produced for each dose estimate (column) independent of one other. Furthermore, because the total absorbed dose is the sum of independent random variables, its variance is less than what is obtained when specific percentiles are summed.

b) $CHCl_3$–Chlorofom
BDCM–dibromodichloromethane
$CHBr_3$–bromoform.

2.8 Summary

Exposure assessments are an essential part of chemical mixture risk assessments. Approaches for these assessments can be improved by integrating (and specifically considering) mixture-specific information. With the aim of improving the knowledge and practice of assessing exposures to environmental chemical mixtures, we have described an approach comprised of interdependent phases that can be applied to strengthen mixture considerations during the problem formulation and exposure assessment components of the risk assessment paradigm. We identified points in the assessment process where an increased sharing of information between the risk assessors developing the dose-response and exposure assessments would improve the implementation of both risk assessment paradigm components. This approach also addresses whether people who might be exposed could be vulnerable or susceptible to exposures or effects for the mixture of interest.

The grouping of chemicals is central to our approach. We introduced grouping to account for common pathways of exposure (e.g., the same exposure medium contaminated with multiple chemicals) and to examine temporal relationships among exposures. Grouping helps organize the relevant information about potential sources that release mixtures of chemicals into the environment by focusing on the types and quantities of chemicals released. Grouping of chemicals that could co-occur in a given exposure medium also facilitates the generation of testable hypotheses and identifies mixtures of chemicals and their mixing ratios that could be the subject of toxicological tests. Furthermore, grouping of chemicals to which individuals could be exposed because they occur concurrently in different exposure media provides insight into additional relevant mixtures and their appropriate mixing ratios that could be studied toxicologically.

Finally, we illustrated our mixture exposure assessment approach using a DBP example that linked an exposure model and a toxicokinetic model. We expect that studies linking environmental fate, exposure, and toxicokinetic models will play an increasingly important role in assessing chemical mixture exposures because they can integrate exposures across multiple routes and because they can predict concentrations of multiple chemicals at toxicological target tissues of interest. They also can guide the targeted evaluation of potential exposure overlaps among multiple chemicals that can ultimately lead to a toxic response due to joint activity of the chemicals.

2.9
Future Directions

Further research is needed to improve the models used to evaluate both the environmental fate of chemical mixtures and the exposure to chemical mixtures. The evaluation model predictions (e.g., though comparison of model predictions with independently measured chemical mixture concentrations) is a particularly important and often overlooked aspect of model development and eventual improvement.

Research is also needed to fill the substantial gaps in the exposure assessment approaches – particularly those related to chemical mixture toxicokinetics (which bridges the exposure and dose-response assessments) with consideration of human susceptibilities. We expect the increased awareness of mixture exposure issues and ongoing advances in exposure assessment methods and exposure data to produce more realistic assessments of risks posed by multiple chemical exposures. Ultimately, these will lead to improved information that can be used to further inform public health decisions.

That said, many opportunities exist for improving exposure assessments of chemical mixtures. One important area needing attention is the evaluation of exposures to mixtures of *dissimilar* chemicals. One of the primary challenges with evaluating the risks posed by mixtures of dissimilar chemicals is the essentially limitless number of combinations and mixing ratios that could require analysis. This is a problem that requires the expertise of both exposure assessors and toxicologists to solve.

To make progress, area relevant studies are needed that document the levels of multiple pollutants across media overtime. Environmental studies need to report levels of dissimilar chemicals that co-occur in the same exposure medium (e.g., methylmercury, PCBs, and PAH in fish, or PAHs, chromium, and dioxins in air). Risk assessors conducting exposure assessments need to report levels of biomarkers for dissimilar chemicals when they are measured (e.g., levels of methylmercury, lead, and PCBs in the blood). Analysis of the NHANES data, which include a number of organic and inorganic pollutant levels measured in blood, could be used to identify some of the dissimilar chemicals that occur routinely in human tissues as a result of past exposures.

Identification of co-occurring compounds in environmental media and human tissues could help identify specific mixtures and mixing ratios comprised of dissimilar chemicals for toxicologists to analyze with an emphasis on common combinations. While conventional toxicological testing such as rodent bioassays could yield useful toxicity information these are typically at higher exposure levels than relevant to human environmental exposures, and, computational toxicology methods might yield the greatest gains for mixtures of dissimilar chemicals. Perhaps advances in computational toxicology will soon lead to a confident determination of whether these dissimilar chemicals might share a common toxicity pathway.[10)] The data that are needed to map a pathway would be obtained from assays that examine responses in a system as it is perturbed by a potentially toxic chemical or combination of chemicals. By identifying those chemicals that manifest their toxicity through the same toxicity pathway or through pathways comprising similar elements, perhaps a number of these chemical combinations of potential concern could be systematically identified and subjected to further toxicity testing (e.g., rodent bioassay) and additional evaluations that examine the potential for coexposure in human populations. For example, these mixtures could be those that affect common target tissues or systems (e.g., neurotoxic effects of methylmercury from ingesting contaminated fish, and lead from old paint).

Although researchers have studied the health effects of some low-level combinations of five or fewer chemicals, research on the health effects associated with more realistic combinations of environmental chemicals (e.g., more than 10 coincident chemicals) is virtually nonexistent. Categorizing typical and atypical exposure levels to multiple chemicals through targeted mixture exposure assessment studies would shed light on the composition of specific environmental chemical mixtures that warrant future toxicological or epidemiological analysis.

Acknowledgments

We appreciate the comments and suggestions of G. Suter and B. Hawkins. We thank L. Wood and S. Lewis for clerical support. We thank C. Broyles for editing this manuscript. With permission, we adapted the materials in Section 2.7 (illustrative example) from work published by L. Teuschler, C. Wilkes, J. Lipscomb, F. Power, and G. Rice.

References

1 National Research Council (1983) *Risk Assessment in the Federal Government: Managing the Process*, Committee on the Institutional Means for Assessments of, Risks to Public Health, Commission on Life Sciences, National Academy

10) Toxicity pathways describe the key molecular level-details of MOA, connecting chemical interactions with target tissues to adverse outcomes at a cellular, tissue, or organ level.

Press, Washington, DC. Available at http://www.nap.edu/openbook.php?isbn=0309033497.

2 U.S. Environmental Protection Agency (1998) Guidelines for ecological risk assessment. Risk Assessment Forum. *Fed. Reg.*, **63** (93), 26846–26924. Available at http://cfpub2.epa.gov/ncea/cfm/recordisplay.cfm?deid=12460.

3 Massachusetts Department of Environmental Protection (2002) Characterizing risks posed by petroleum contaminated sites: implementation of the MADEP VPH/EPH approach. Massachusetts Department of Environmental Protection, Boston, MA. Available at http://www.mass.gov/dep/cleanup/laws/02-411.pdf.

4 National Research Council (1994) *Science and Judgment in Risk Assessment*, Committee on Risk Assessment of Hazardous Air, Pollutants, Board on Environmental Studies and Toxicology, Commission on Life Sciences, National Academy Press, Washington, DC. Available at http://www.nap.edu/openbook.php?isbn=030904894X.

5 U.S. Environmental Protection Agency (1999) Issuance of final guidance: ecological risk assessment and risk management principles for Superfund sites. Memorandum from SD Luftig, Director, Office of Emergency and Remedial Response, to Superfund National Policy Managers, Regions1–10. OSWER Directive 9285.7-28P. Available at http://www.epa.gov/oswer/riskassessment/ecorisk/pdf/final99.pdf.

6 U.S. Environmental Protection Agency (2003) Framework for cumulative risk assessment. Risk Assessment Forum, Washington, DC. EPA/630/P-02/001F. Available at http://cfpub.epa.gov/ncea/raf/recordisplay.cfm?deid=54944.

7 Agency for Toxic Substances and Disease Registry (2004) Guidance manual for the assessment of joint toxic action of chemical mixtures. US Department of Health and Human Services, Public Health Service, Atlanta, GA. Available at http://www.atsdr.cdc.gov/interactionprofiles/IP-ga/ipga.pdf.

8 National Institute for Occupational Safety and Health (2004) Mixed exposures research agenda, a report by the NORA Mixed Exposures. National Institute for Occupational Safety and Health. NIOSH Publication No. 2005-106.

9 Occupational Safety and Health Administration (1993) Occupational Safety and Health Administration (OSHA) – Incorporation of general industry safety and health standards applicable to construction work, Correction. Final Rules. FRN 58:40468.

10 U.S. Department of Energy (2001) Radioactive waste management, DOE Order Number 435.1. Approved: 7-09-99. Review: 7-09-01. Change 1: 08-28-01.

11 U.S. Environmental Protection Agency (1986) Guidelines for the health risk assessment of chemical mixtures. Office of Research and Development, Washington, DC. EPA/630/R-98/002.

12 U.S. Environmental Protection Agency (2000) Supplementary guidance for conducting health risk assessment of chemical mixtures. Risk Assessment Forum, Office of Research and Development, Washington, DC. EPA/630/R-00/002. Available at http://www.epa.gov/ncea/raf/pdfs/chem_mix/chem_mix_08_2001.pdf.

13 U.S. Environmental Protection Agency (1989) Risk assessment guidance for superfund: Volume I. Human health evaluation manual (Part A). Office of Emergency and Remedial Response, Washington, DC. EPA/540/1-89/002.

14 U.S. Environmental Protection Agency (1992) Guidelines for exposure assessment. *Fed. Reg.*, **57** (104), 22888–22938. Available at http://rais.ornl.gov/homepage/GUIDELINES_EXPOSURE_ASSESSMENT.pdf.

15 U.S. Environmental Protection Agency (1998) Methodology for assessing health risks associated with multiple pathways of exposure to combustor emissions. National Center for Environmental Assessment, Cincinnati, OH. EPA/600/R-98/137.

16 U.S. Environmental Protection Agency (1999) Guidance for performing aggregate

exposure and risk assessments. Office of Pesticide Programs, Washington, DC. Available at http://www.epa.gov/EPA-PEST/2001/November/Day-28/p29386.htm.

17 U.S. Environmental Protection Agency (2001) General principles for performing aggregate exposure and risk assessments. Office of Pesticide Programs, Washington, DC. Available at http://www.epa.gov/pesticides/trac/science/aggregate.pdf.

18 U.S. Environmental Protection Agency (2007) Concepts, methods and data sources for health risk assessment of multiple chemicals, exposures and effects (external review draft). Office of Research and Development, National Center for Environmental Assessment, Cincinnati, OH in collaboration with U.S. Department of Energy, Argonne National Laboratory, Argonne, IL. EPA/600/R-06/013A.

19 Lorber, M., Cleverly, D., Schaum, J., Phillips, L., Schweer, G., and Leighton, T. (1994) Development and validation of an air-to-beef food chain model for dioxin-like compounds. *Sci. Total Environ.*, **156** (1), 39–65.

20 Lorber, M. and Phillips, L. (2002) Infant exposure to dioxin-like compounds in breast milk. *Environ. Health Perspect.*, **110** (6), A325–A332.

21 U.S. Environmental Protection Agency (2003) Exposure and human health reassessment of 2,3,7,8-tetrachlorodibenzo-*p*-dioxin (TCDD) and related compounds, NAS review draft. Exposure Assessment and Risk Characterization Group, National Center for Environmental Assessment (Washington Office), Office of Research and Development, Washington, DC. EPA/600/P-00/001Cb. Available at http://www.epa.gov/nceawww1/pdfs/dioxin/nas-review/.

22 Tsongas, T., Orlinskii, D., Priputina, I., Pleskachevskaya, G., Fetishchev, A., Hinman, G., and Butcher, W. (2000) Risk analysis of PCB exposure via the soil-food crop pathway, and alternatives for remediation at Serpukhov, Russian Federation. *Risk Anal.*, **20** (1), 73–79.

23 Lorber, M. (2007) Exposure of Americans to polybrominated diphenyl ethers. *J. Expo. Sci. Environ. Epidemiol.*, **18** (1), 2–19.

24 Teuschler, L.K., Rice, G.E., Wilkes, C.R., Lipscomb, J.C., and Power, F.W. (2004) A feasibility study of cumulative risk assessment methods for drinking water disinfection by-product mixtures. *J. Toxicol. Environ. Health A*, **67** (8–10), 755–777.

25 Foster, K.L., Mackay, D., Parkerton, T.F., Webster, E., and Milford, L. (2005) Five-stage environmental exposure assessment strategy for mixtures: gasoline as a case study. *Environ Sci. Technol.*, **39** (8), 2711–2718.

26 Suter, G.W. (1999) Developing conceptual models for complex ecological risk assessments. *Hum. Ecol. Risk Assess.*, **67** (2), 375–396.

27 Agency for Toxic Substances and Disease Registry (2007) Interaction Profiles for Toxic Substances, U.S. Department of Health and Human Services, Public, Health Service, Atlanta, GA. Available at http://www.atsdr.cdc.gov/interactionprofiles/.

28 Wang *et al.* (2009) Biomonetoring, in *Principles and Practice of Mixtures Toxicology* (Ed. M. Mumtaz), Wiley-VCH, Weinheim, 491–516.

29 Paustenbach, D.J. (2000) The practice of exposure assessment: a state-of-the-art review. *J. Toxicol. Environ. Health B Crit. Rev.*, **3** (3), 179–291.

30 Lipscomb *et al.* (2009) Chemical Mixtures and Cumulative Risk Assessment, in *Principle and Practice of Mixtures Toxicology* (Ed. M. Mumtaz), Wiley-VCH, Weinheim, 177–206.

31 Rice, G., MacDonell, M., Hertzberg, R.C., Teuschler, L., Picel, K., Butler, J., Chang, Y.S., and Hartmann, H. (2008) An approach for assessing human exposures to chemical mixtures in the environment. *Toxicol. Appl. Pharmacol.*, **233** (1), 126–136.

32 U.S. Environmental Protection Agency (2005) Human health risk assessment protocol for hazardous waste combustion facilities (final). Office of Solid Waste and Emergency Response, Washington, DC. EPA/530/R-05/006.

Available at http://www.epa.gov/epaoswer/hazwaste/combust/risk.htm.

33 James M. Montgomery Consulting, Metropolitan Water District of Southern California (1989) Disinfection by-products in United States drinking waters. Final report to the U.S. Environmental Protection Agency and the Association of Metropolitan Water Agencies. MWDSC Water Quality Division, La Verne, CA.

34 U.S. Geological Survey (2006) Land use and land cover (LULC). U.S. Department of Interior, Washington, DC. Available at http://eros.usgs.gov/products/landcover/lulc.html. Last updated 22 August 2008.

35 Cogliano, V.J. (1998) Assessing the cancer risk from environmental PCBs. *Environ. Health Perspect.*, **106** (6), 317–323.

36 Frederick, W.T., Keil, K.G., Rhodes, M.C., Peterson, J., and MacDonell, M. (2007) Utilizing isotopic uranium ratios in groundwater evaluations at FUSRAP sites. Proceedings of Waste Management, 2007 February 25–March 1, Tucson, AZ.

37 U.S. Environmental Protection Agency (2007) Toxic Release Inventory (TRI) Program, Washington, DC. Available at http://www.epa.gov/tri/chemical/index.htm#chemlist. Last updated 4 April 2008.

38 U.S. Environmental Protection Agency (2008) 2002 National Emissions Inventory data & documentation. Technology Transfer Network, Clearinghouse for Inventories & Emissions Factors, Washington, DC. Available at http://www.epa.gov/ttn/chief/net/2002inventory.html.

39 U.S. Environmental Protection Agency (2008) Air data: access to air pollution data. Office of Air and Radiation, Washington, DC. Available at http://www.epa.gov/oar/data/. Accessed 5 March 2008.

40 Greater Cincinnati Water Works (2006) 2006 Safe drinking water report. Available at http://www.cincinnati-oh.gov/water/downloads/water_pdf15886.pdf.

41 Rice, G.E., Teuschler, L.K., Bull, R.J., Simmons, J.E., and Feder, P.I. (2009) Evaluating the similarity of complex drinking-water disinfection by-product mixtures: overview of the issues. *J. Toxicol. Environ. Health A*, **72** (7), 429–436.

42 Agency for Toxic Substances and Disease Registry (1996) Toxicological Profile for Toxaphene, U.S. Department of Health and Human Services, Public Health Service, Atlanta, GA. Available at http://www.atsdr.cdc.gov/toxprofiles/tp94.html.

43 Simon, T. and Manning, R. (2006) Development of a reference dose for the persistent congeners of weathered toxaphene based on *in vivo* and *in vitro* effects related to tumor promotion. *Regul. Toxicol. Pharmacol.*, **44** (3), 268–281.

44 U.S. Environmental Protection Agency (2008) Alternative models. Technology Transfer Network, Support Center for Regulatory Atmospheric Modeling (SCRAM), Washington, DC. Available at http://www.epa.gov/scram001/dispersion_alt.htm#adms3. Accessed 8 February 2008.

45 Executive Order (1997) Executive Order 13045 – protection of children from environmental health risks and safety risks. Available at http://yosemite.epa.gov/ochp/ochpweb.nsf/content/whatwe_executiv.htm.

46 U.S. Environmental Protection Agency (2001) Summary report of the technical workshop on issues associated with considering developmental changes in behavior and anatomy when assessing exposure to children. Risk Assessment Forum, Washington, DC, EPA/630/R-00/005.

47 U.S. Environmental Protection Agency (2005) Supplemental guidance for assessing susceptibility from early-life exposure to carcinogens. Risk Assessment Forum, Washington, DC. EPA/630/R-03/003F. Available at http://www.epa.gov/ncea/iris/children032505.pdf.

48 U.S. Environmental Protection Agency (2005) Guidance on selecting age groups for monitoring and assessing childhood exposures to environmental contaminants. Risk Assessment Forum, Washington, DC. EPA/630/P-03/003. Available at http://cfpub.epa.gov/ncea/cfm/recordisplay.cfm?deid=146583.

49 Barton, H.A., Cogliano, V.J., Flowers, L., Valcovic, L., Setzer, R.W., and Woodruff, T.J. (2005) Assessing susceptibility from

early-life exposure to carcinogens. *Environ. Health Perspect.*, **113** (9), 1125–1133.

50 DeFur, P.L., Evans, G.W., Cohen Hubal, E.A., Kyle, A.D., Morello-Frosch, R.A., and Williams, D.R. (2007) Vulnerability as a function of individual and group resources in cumulative risk assessment. *Environ. Health Perspect.*, **115** (5), 817–824, Available at http://www.ehponline.org/members/2007/9332/9332.pdf.

51 Firestone, M., Moya, J., Cohen-Hubal, E., Zartarian, V., and Xue, J. (2007) Identifying childhood age groups for exposure assessments and monitoring. *Risk Anal.*, **27** (3), 701–714.

52 U.S. Environmental Protection Agency (1997) Exposure factors handbook (Volume I: general factors, Volume II: food ingestion factors, Volume III: activity factors). Office of Research and Development, National Center for Environmental Assessment, Washington, DC. Available at http://www.epa.gov/ncea/pdfs/efh/front.pdf.

53 Morgan, M.G. and Henrion, M. (1990) *A Guide to Dealing with Uncertainty in Quantitative Risk and Policy Analysis*, Cambridge University Press, Cambridge, UK.

54 Cooke, R.M. (1991) *Experts in Uncertainty: Opinion and Subjective Probability in Science*, Oxford University Press, New York.

55 Morgan, M.G., Dowlatabadi, H., Henrion, M., Keith, D., Lempert, R., McBride, S., Small, M., and Wilbanks, T. (2008) Best practice approaches for characterizing, communicating and incorporating scientific uncertainty in climate decision making. Synthesis and Assessment Product 5.2. U.S. Climate Change Science Program, Washington, DC. Available at http://www.climatescience.gov/Library/sap/sap5-2/sap5-2-draft3.pdf.

56 U.S. Environmental Protection Agency (1999) Understanding variation in partition coefficient, k_d, values. Volume II: review of geochemistry and available k_d values for cadmium, cesium, chromium, lead, plutonium, radon, strontium, thorium, tritium (^{3}h), and uranium. Office of Radiation and Indoor Air, Washington, DC. EPA 402/R-99/004B. Available at http://www.epa.gov/rpdweb00/docs/kdreport/vol2/402-r-99-004b.pdf.

57 U.S. Environmental Protection Agency (2000) Soil screening guidance for radionuclides: technical background document. Office of Radiation and Indoor Air, Washington, DC. OSWER No. 9355.4-16.

58 Richardson, S.D. (1998) Identification of drinking water disinfection by-products, in *John Wiley's Encyclopedia of Environmental Analysis and Remediation* (ed. R.A. Meyers), John Wiley & Sons, Inc., New York, pp. 1398–1421.

59 International Agency for Research on Cancer (2004) *Some Drinking Water Disinfectants and Contaminants Including Arsenic. Summary of Data Reported and Evaluation Last Updated.* IARC Monographs on the Evaluation of Carcinogenic Risks to Humans, vol. **84**. IARC/WHO, Lyon, France. Available at http://monographs.iarc.fr/ENG/Monographs/vol84/volume84.pdf.

60 Waller, K., Swan, S.H., DeLorenze, G., and Hopkins, B. (1998) Trihalomethanes in drinking water and spontaneous abortion. *Epidemiology*, **9** (2), 134–140.

61 Weinberg, H. (1999) Disinfection byproducts in drinking water: the analytical challenge. *Anal. Chem.*, **71** (23), 801A–808.

62 Miltner, R.J., Rice, E.W., and Stevens, A.A. (1990) Pilot-scale investigation of the formation and control of disinfection byproducts. 1990 Annual Conference Proceedings, AWWA Annual Conference, Cincinnati, OH, pp. 1787–1802.

63 Williams, D.T., LeBel, G.L., and Benoit, F.M. (1997) Disinfection by-products in Canadian drinking water. *Chemosphere*, **34** (2), 299–316.

64 Miltner, R.J., Rice, E.W., and Smith, B.L. (1992) Ozone's effect on assimilable organic carbon, disinfection byproducts and disinfection byproduct precursors. Proceedings of Water Quality Technology Conference, American Water Works Association, Orlando, Florida.

65 Clark, R.M. (1998) Chorine demand and TTHM formation: a second-order model. *J. Environ. Eng.*, **124** (1), 16–24.

66 Chen, W.J. and Weisel, C.P. (1998) Concentration changes of halogenated disinfection byproducts in a drinking water distribution system. *J. Am. Water Works Assoc.*, **90**, 151–163.

67 Pereira, V.J., Weinberg, H.S., and Singer, P.C. (2004) Temporal and spatial variability of DBPs in a chloraminated distribution system. *J. Am. Water Works Assoc.*, **96** (11), 91–102.

68 U.S. Environmental Protection Agency (2003) The feasibility of performing cumulative risk assessments for mixtures of disinfection by-products in drinking water. National Center for Environmental Assessment, Cincinnati, OH. EPA/600/R-03/051.

69 Wallace, L.A. (1997) Human exposure and body burden for chloroform and other trihalomethanes. *Crit. Rev. Environ. Sci. Technol.*, **27** (2), 113–194.

70 Weisel, C.P., Kim, H., Haltmeier, P., and Klotz, J.B. (1999) Exposure estimates to disinfection by-products of chlorinated drinking water. *Environ. Health Perspect.*, **107** (2), 103–110.

71 Wilkes, C. (1998) Case study, in *Exposure to Contaminants in Drinking Water: Estimating Uptake Through the Skin and by Inhalation* (ed. S. Olin), CRC Press, Washington, DC, pp. 183–224.

72 U.S. Environmental Protection Agency (2002) Exposure related dose estimating model (ERDEM) for assessing human exposure and dose. Office of Research and Development, Washington, DC. EPA/600/R00/011. Available at http://www.epa.gov/heasdweb/products/erdem/erdem.htm.

73 U.S. Environmental Protection Agency (1998) National primary drinking water regulations; disinfectants and disinfection by-products notice of data availability (proposed rule). *Fed. Reg.*, **63** (61), 15674–15692.

74 Weisel, C.P. and Jo, W.K. (1996) Ingestion, inhalation, and dermal exposures to chloroform and trichloroethene from tap water. *Environ. Health Perspect.*, **104** (1), 48–51.

75 Simmons, J.E., Teuschler, L.K., Gennings, C., Speth, T.F., Richardson, S.D., Miltner, R.J., Narotsky, M.G., Schenck, K.D., Hunter, E.S., III, Hertzberg, R.C., and Rice, G. (2004) Component-based and whole-mixture techniques for addressing the toxicity of drinking-water disinfection by-product mixtures. *J. Toxicol. Environ. Health A*, **67** (8–10), 741–754.

76 U.S. Environmental Protection Agency (2008) IRIS glossary/acronyms and abbreviations. Integrated Risk Information System, Washington, DC. Available at http://www.epa.gov/iris/help_gloss.htm. Last updated 10 January 2008.

77 U.S. Environmental Protection Agency (2001) Exploration of aging and toxic response issues. Risk Assessment Forum, Washington, DC. EPA/630/R01/003. Available at http://oaspub.epa.gov/eims/eimscomm.getfile?p_download_id=8298.

78 U.S. Environmental Protection Agency (2007) MOBILE6 vehicle emission modeling software. Office of Transportation and Air Quality, Washington, DC. Available at www.epa.gov/otaq/m6.htm.

79 U.S. Environmental Protection Agency (2007) NONROAD Model (nonroad engines, equipment, and vehicles). Office of Transportation and Air Quality, Washington, DC. Available at http://www.epa.gov/otaq/nonrdmdl.htm.

80 U.S. Environmental Protection Agency (2007) 1999 National Emission Inventory documentation and data – final version 3.0. Technology Transfer Network, Clearinghouse for Inventories & Emissions Factors, Washington, DC. Available at http://www.epa.gov/ttn/chief/net/1999inventory.html.

81 U.S. Environmental Protection Agency (2003) Documentation for aircraft, commercial marine vessel, locomotive, and other nonroad components of the National Emissions Inventory, Volume I – methodology. Emissions, Monitoring and Analysis Division, Emission Factor and Inventory Group (D205-01), Research Triangle Park, NC. Available at ftp://ftp.epa.gov/pub/EmisInventory/finalnei99ver3/criteria/documentation/nonroad/99nonroad_vol1_oct2003.pdf.

82 Agency for Toxic Substances and Disease Registry (2004) Interaction profile for

arsenic, cadmium, chromium and lead. U.S. Department of Health and Human Services, Public Health Service, Atlanta, GA. Available at http://www.atsdr.cdc.gov/interactionprofiles/ip04.htm.

83 U.S. Environmental Protection Agency (1999) Sociodemographic data used for identifying potentially highly exposed populations. Office of Research and Development, National Center for Environmental Assessment, Washington, DC. EPA/600/R-99/060. Available at http://www.epa.gov/NCEA/pdfs/frontmat.pdf.

3
Application of a Relative Potency Factor Approach in the Assessment of Health Risks Associated with Exposures to Mixtures of Dioxin-Like Compounds

Daniele F. Staskal, Linda S. Birnbaum, and Laurie C. Haws

For certain classes of related chemical compounds that are assumed to be toxicologically similar, regulatory and public health agencies have developed relative potency factor (RPF) approaches to address potential health risks posed by exposures to mixtures of compounds within the class [1]. Such approaches rely on the existence of toxicological dose-response data for at least one of the compounds in the mixture (referred to as the index compound or prototype) and scientific judgment as to the toxicity of the other individual compounds in the mixture relative to the index compound. To date, RPF approaches have been developed for two classes of compounds: (1) the dioxin-like chemicals [2–8] and (2) the polycyclic aromatic hydrocarbons (PAHs). Clearly, the most well-established and well-accepted approach is that for the dioxin-like compounds (DLCs). As such, the RPF approach developed for DLCs will be discussed in detail in this chapter to illustrate the application of the RPF approach in the assessment of health risks posed by mixtures of toxicologically similar compounds.

Assessment of the potential health risks associated with exposure to dioxin-like chemicals is complicated by the fact that these compounds are typically detected in the environment as part of complex mixtures of structurally related polyhalogenated aromatic hydrocarbons [4, 9]. Because the toxicity information is incomplete for this class of chemicals, a toxic equivalency factor (TEF) methodology has been developed to assess their potential health risks. The TEF methodology is a relative potency scheme, with the most toxic and well-studied member of the class, 2,3,7,8-tetrachlorodibenzo-*p*-dioxin (TCDD), serving as the index compound. A fundamental tenet of the TEF methodology is that DLCs act through a common mechanism of action initiated by binding to and activating the aryl hydrocarbon (Ah) receptor [10–15] to elicit a similar spectrum of biochemical and toxic responses [10, 16–19], and the combined effects of the different DLC compounds are assumed to be additive [7].

Principles and Practice of Mixtures Toxicology. Edited by Moiz Mumtaz

ISBN: 978-3-527-31992-3

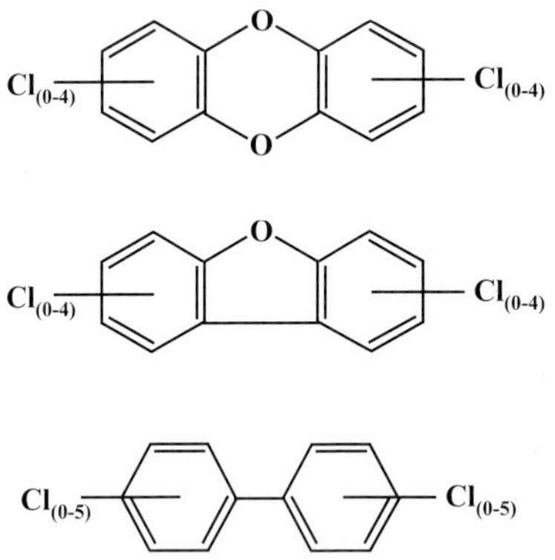

Figure 3.1 General chemical structures of dioxins, furans, and PCBs.

3.1 Dioxin-Like Chemicals

2,3,7,7-Tetrachlorodibenzo-*p*-dioxin is one of the most well-studied environmental compounds, as is evident by thousands of studies published in the peer-reviewed literature on the compound. TCDD, also commonly referred to as "dioxin," is part of a group of compounds that consists of the polychlorinated dibenzo-*p*-dioxins (PCDDs), polychlorinated dibenzofurans (PCDFs), and certain polychlorinated biphenyls (PCBs) that share similar chemical structures and biological characteristics. There are 75 different PCDD congeners, 135 different PCDF congeners, and 209 different PCB congeners. The individual congeners are identified on the basis of the number and position of chlorine atoms in the chemical structure (Figure 3.1). These compounds have been found to persist in the environment and to bioaccumulate in biological tissues [14]. Because of these characteristics, these compounds are ubiquitous in the environment.

PCDDs and PCDFs are not created intentionally; rather they are unwanted by-products of a number of anthropogenic combustion processes such as commercial or municipal waste incineration and burning wood, coal, or oil [20]. Other sources of these compounds include chlorine bleaching of pulp and paper, burning household trash, and cigarette smoke. Natural sources of emissions, such as forest fires, have also been cited [20]. Significant resources have been devoted to characterizing release of these compounds into the environment. In the United States, the US Environmental Protection Agency (USEPA) maintains an inventory estimates of annual release of dioxin-like compounds from known source activities, both historical and present, into air, water, and land. Though emissions have significantly decreased worldwide over the past several decades, low levels of these compounds still persist in the environment because of their slow rate of degradation and their continued release into the environment.

Unlike the PCDDs and PCDFs, PCBs were produced in relatively large quantities and used in a wide variety of industrial products including plasticizers, dielectric fluids for capacitors and transformers, heat transfer agents, and paint additives [20]. Though a minor source, PCBs are also generated through incomplete combustion of some waste products and as a by-product of some organic chemical manufacturing processes. Although they are no longer produced in the United States, these chemicals continue to be released into the environment given their extensive use in industrial products.

A subset of the 419 PCDDs, PCDFs, and PCBs is commonly referred to as "dioxin-like compounds" and comprises 17 laterally substituted (2,3,7,8-substituted) PCDD and PCDF congeners and 12 non-*ortho* and mono-*ortho* chlorine-substituted PCBs (other PCDD, PCDF, and PCB congeners not included in this subset are generally not considered to be dioxin-like as they are biologically inactive or act via different modes of action). Extensive research has been devoted to this class of compounds, and the criteria associated with classification as a DLC has evolved over time – a process that will be addressed in this chapter. Generally, in order for a compound to be considered dioxin-like, it must (a) have a similar chemical structure, (b) have similar physical–chemical properties, (c) act through a common mechanism of action, and (d) invoke a common battery of toxic responses relative to 2,3,7,8-tetrachlorodibenzo-*p*-dioxin. The spectrum of toxic effects associated with dioxin-like chemicals is generally characterized by severe weight loss, thymic atrophy, hepatotoxicity, edema, fetotoxicity, teratogenicity, reproductive toxicity, immunotoxicity, carcinogenicity, and enzyme induction in experimental animals [10, 14, 16–19, 21]. These common biological effects are mediated through a common mechanism of action that involves the binding of the Ah receptor [10–15]. In humans, numerous epidemiological studies have evaluated adverse health effects following exposure via industrial accidents or occupational exposure, although only chloracne has been conclusively demonstrated. The potential adverse effects of TCDD, other dioxins, and DLCs from long-term, low-level exposures are not directly observable and remain controversial [22].

3.2 Introduction of TEF Methodology

Scientists found that assessing the potential risk to both humans and wildlife resulting from exposure to these compounds was not straightforward for two main reasons. First, the dioxin-like congeners are typically not found in isolation but occur as part of a complex mixture of congeners in environmental media and biological tissues [4, 9]. Second, varying amounts of toxicity information for each of the dioxin-like congeners were available and thus incomplete for this group of compounds. As a result, the toxic equivalency factor methodology was proposed as an interim method of assessment for dioxin-like compounds. This methodology is based on a relative potency scheme in which all compounds are ranked against the most toxic and most thoroughly studied congener, 2,3,7,8-TCDD. For several decades, this methodology has been the topic of intense scientific research; an evolving process that has hinged primarily on six re-evaluations of the TEF approach in a number of scientific forums that are discussed in this section.

Throughout the evolution of this approach, scientists have consistently supported the application of a "toxic equivalency" approach using 2,3,7,8-TCDD as the prototype dioxin-like chemical. In such approach, each dioxin-like chemical is assigned an "order of magnitude" estimate of relative potency as compared to 2,3,7,8-TCDD. This value is defined as the TEF. Each congener is assigned a consensus-based value following a review of all available literature evaluating the relative potency (REP) of a given congener relative to TCDD. The TEF value for each congener is multiplied by the concentration of each specific congener, and the products are summed to give a single toxic equivalency (TEQ). This approach is demonstrated in Eq. (3.1) and an example TEQ calculation is provided in Table 3.1.

Table 3.1 Example calculation of TEQ using TEF methodology using theoretical data.

Dioxin-like congener	Amount measured (ppt lipid)	2005 WHO TEF	Congener TEQ
Polychlorinated dibenzo-*p*-dioxins			
2,3,7,8-TetraCDD	2.0	1	2.0
1,2,3,7,8-PentaCDD	4.8	1	4.8
1,2,3,4,7,8-HexaCDD	13.7	0.1	1.4
1,2,3,6,7,8-HexaCDD	41.8	0.1	4.2
1,2,3,7,8,9-HexaCDD	19.5	0.1	2.0
1,2,3,4,6,7,8-HeptaCDD	62.1	0.01	0.6
1,2,3,4,5,6,7,8-OctaCDD	1333.4	0.0003	0.4
Polychlorinated dibenzofurans			
2,3,7,8-TetraCDF	2.4	0.1	0.2
1,2,3,7,8-PentaCDF	2.1	0.03	0.1
2,3,4,7,8-PentaCDF	25.5	0.3	7.6
1,2,3,4,7,8-HexaCDF	10.9	0.1	1.1
1,2,3,6,7,8-HexaCDF	8.5	0.1	0.8
1,2,3,7,8,9-HexaCDF	1.5	0.1	0.2
2,3,4,6,7,8-HexaCDF	6.7	0.1	0.7
1,2,3,4,6,7,8-HeptaCDF	5.8	0.01	0.1
1,2,3,6,7,8,9-HeptaCDF	1.5	0.01	0.0
1,2,3,4,5,6,7,8-OctaCDF	12.3	0.0003	0.0
Coplanar polychlorinated biphenyls			
3,3′,4,4′-TetraCB (PCB 77)	34.2	0.0003	0.0
3,4,4′,5-TetraCB (PCB 81)	13.4	0.0003	0.0
3,3′,4,4′,5-PentaCB (PCB 126)	10.5	0.1	1.0
3,3′,4,4′,5,5′-HexaCB (PCB 169)	117.2	0.03	3.5
Mono-*ortho* substituted polychlorinated biphenyls			
2,3,3′,4,4′-PentaCB (PCB 105)	672.1	0.00 003	0.0
2,3,4,4′,5-PentaCB (PCB 114)	10 256.3	0.00 003	0.3
2,3′,4,4′,5-PentaCB (PCB 118)	6328.0	0.00 003	0.2
2′,3,4,4′,5-PentaCB (PCB 123)	7120.1	0.00 003	0.2
2,3,3′,4,4′,5-HexaCB (PCB 156)	15 782.3	0.00 003	0.5
2,3,3′,4,4′,5′-HexaCB (PCB 157)	4492.5	0.00 003	0.1
2,3′,4,4′,5,5′-HexaCB (PCB 167)	3252.1	0.00 003	0.1
2,3,3′,4,4′,5,5′-HeptaCB (PCB 189)	5723.8	0.00 003	0.2
Total dioxin toxic equivalent (ppt TEQ)			32.3

$$\text{Total toxic equivalency (TEQ)} = \sum_{n=1}^{k} C_n * \text{TEF}_n \quad (3.1)$$

3.3
Evolution of TEF Approach

3.3.1
Initial Proposal: The TEF Methodology, 1984

In 1984, the Ontario Ministry of the Environment (OME) issued a document entitled "Scientific Criteria Document for Standard Development. No. 4-84. Polychlorinated dibenzo-*p*-dioxins and polychlorinated dibenzofurans" recommending that DLCs should be evaluated using a toxic equivalency approach [3]. The OME recommended that this approach would provide an environmental standard designed specifically to protect human health from exposure to DLCs. Following review of the available research, this regulatory body concluded that there was sufficient evidence that DLCs were not only structurally related but also induced a similar spectrum of both biochemical and toxic responses. Furthermore, these effects were mediated through the Ah receptor, demonstrating a common mode of action relative to the most potent DLC, TCDD. As such, the OME recommended that toxicity factors be assigned to each of the PCDD and PCDF congeners on the basis of the potency of their homologue group (e.g., number of chlorine substitutions) relative to TCDD (Table 3.2), thus documenting the first official proposal for the use of TEF methodology.

Table 3.2 OME TEF values [3].

Isomer groups	Toxicity factor relative to 2,3,7,8-T_4CDD
DD	Nontoxic
M_1CDD	0.0001
D_2CDD	0.001
T_3CDD	0.01
T_4CDD	0.01
P_5CDD	0.1
H_6CDD	0.1
H_7CDD	0.01
O_8CDD	0.0001
DF	Nontoxic
M_1CDF	0.0001
D_2CDF	0.0001
T_3CDF	0.01
T_4CDF	0.5
P_5CDF	0.5
H_6CDF	0.1
H_7CDF	0.01
O_8CDF	0.0001

3.3.2
USEPA, 1987

Concurrently, the USEPA was evaluating a methodology to assess risks posed by PCDD and PCDF emissions from incinerators. It was their conclusion that the TEF methodology was the best available methodology, thus the agency further refined the OME's proposed TEF values and developed their own recommendations [5]. Specifically, because the agency was required to address limitations inherent in the selected methodology to assess risk from these compounds, the USEPA recommended TEF values be utilized for the individual congeners rather than homologue groups as recommended by the OME. The USEPA evaluated congener-specific data and assigned "order of magnitude" estimates of relative toxicity to TCDD. Actual TEF values were selected by expert scientific opinion (Table 3.3) and were based on review of both *in vitro* and *in vivo* toxicity studies [5, 23, 24]).

3.3.3
International TEFs, 1989

In conjunction with the USEPA's efforts, there was also an international effort underway to harmonize the TEF approach. This effort was spearheaded by the North Atlantic Treaty Organization Committee on the Challenges of Modern Society (NATO/CCMS) Dioxin Information Exchange Subcommittee in 1988. Following their 3-year study, the committee concluded that the TEF approach was the best available interim measure for PCDD and PCDF risk assessment. However, the committee refined the TEF values by adding two DLC congeners, octachlorodibenzofuran and octachlorodibenzodioxin, and removing TEF values for non-2,3,7,8-substituted congeners focusing on the 17 DLCs that were preferentially retained and bioaccumulated both in environment and in humans. In addition, the committee focused primarily on congener-specific *in vivo* studies when selecting their TEF values. The resulting recommendations were termed the "International TEFs/89" or the "I-TEFs/89" (Table 3.3). Shortly after the release of these values, several countries adopted the I-TEFs/89, though some countries accepted them with the caveat that the committee's proposed TEF methodology was interim and that revisions and updates should be included as new information becomes available.

3.3.4
Addition of PCBs, 1990–1993

The initial TEF approaches did not include PCBs; however, as the process evolved, a considerable amount of research was conducted to assess the structure–activity relationships between different PCB homologue groups [4, 14]. These investigations revealed that only the PCB congeners substituted in the meta and para positions were approximate isosteromers of TCDD. Subsequent toxicological studies investigating these PCB congeners demonstrated that the coplanar (PCBs 77, 81, 126, and 169) and mono-*ortho* coplanar PCBs (e.g., 1015, 118, 156, and 157) were Ah receptor agonists

Table 3.3 Evolution of TEF values.

Congener	EPA/87[a)]	NATO/89[b)]	WHO/94[c)]	WHO/98[d)]	WHO/05[e)]
PCDDs					
2,3,7,8-TCDD	1	1		1	1
1,2,3,7,8-PeCDD	0.5	0.5		1	1
1,2,3,4,7,8-HxCDD	0.04	0.1		0.1	0.1
1,2,3,7,8,9-HxCDD	0.04	0.1		0.1	0.1
1,2,3,6,7,8-HxCDD	0.04	0.1		0.1	0.1
1,2,3,4,6,7,8-HpCDD	0.001	0.1		0.01	0.01
1,2,3,4,6,7,8,9-OCDD	0	0.001		0.0001	0.0003
PCDFs					
2,3,7,8-TCDF	0.1	0.1		0.1	0.1
1,2,3,7,8-PeCDF	0.1	0.05		0.05	0.03
2,3,4,7,8-PeCDF	0.1	0.5		0.5	0.3
1,2,3,4,7,8-HxCDF	0.01	0.1		0.1	0.1
1,2,3,7,8,9-HxCDF	0.01	0.1		0.1	0.1
1,2,3,6,7,8-HxCDF	0.01	0.1		0.1	0.1
2,3,4,6,7,8-HxCDF	0.01	0.1		0.1	0.1
1,2,3,4,6,7,8-HpCDF	0.001	0.01		0.01	0.01
1,2,3,4,7,8,9-HpCDF	0.001	0.01		0.01	0.01
1,2,3,4,6,7,8,9-OCDF	0	0.001		0.0001	0.0003
PCBs					
77: 3,3′,4,4′-TCB			0.0005	0.0001	0.0001
81: 3,4,4′,5-TCB			—	0.0001	0.0003
105: 2,3,3′,4,4′-PeCB			0.0001	0.0001	0.0003
114: 2,3,4,4′,5-PeCB			0.0005	0.0005	0.0003
118: 2,3′,4,4′,5-PeCB			0.0001	0.0001	0.0003
123: 2′,3,4,4′,5-PeCB			0.0001	0.0001	0.0003
126: 3,3′,4,4′,5-PeCB			0.1	0.1	0.1
156: 2,3,3′,4,4′,5-HxCB			0.0005	0.0005	0.0003
157: 2,3,3′,4,4′,5′-HxCB			0.0005	0.0005	0.0003
1672,3′,4,4′,5,5′-HxCB			0.00 001	0.00 001	0.0003
169: 3,3′,4,4′,5,5′-HxCB			0.01	0.01	0.03
170: 2,2′,3,3′,4,4′,5-HpCB			0.0001		
180: 2,2′,3,4,4′,5,5′-HpCB			0.00 001		
189: 2,3,3′,4,4′,5,5′-HpCB			0.0001	0.0001	0.0003

a) Ref. [5].
b) Ref. [28].
c) Ref. [2].
d) Ref. [7].
e) Ref. [8].

and induced a variety of *in vitro* and *in vivo* effects that were similar to those of TCDD [25]. Research also demonstrated that substitution pattern directly influenced dioxin-like activity; PCBs with both para positions occupied, two or more meta positions occupied, and no ortho positions occupied demonstrated the greatest TCDD-like toxicological responses. However, even a single *ortho* substitution significantly decreases dioxin-like activity and increases nondioxin-like activity.

Di-*ortho*-substituted PCBs only induce very minor amounts (generally >5 orders of magnitude less than TCDD) of dioxin-like activity [14].

Given the consensus of these findings regarding dioxin-like activity of selected PCB congeners, the USEPA convened a workshop to evaluate PCBs with respect to inclusion in the TEF methodology for dioxin-like compounds [26]. This workshop was convened only shortly after TEFs were proposed for selected PCB congeners from research in the United States via publication in the peer-reviewed literature [14]. On the basis of the evaluation of this research, along with a variety of supporting literature, the EPA concluded a small subset of PCBs demonstrated dioxin-like activity and met the criteria for inclusion in the TEF methodology [23]. Criteria required for inclusion in the TEF methodology included (a) demonstrated need, (b) well-defined group of chemicals, (c) broad toxicological database, (d) consistency in relative toxicity for all end points, (e) demonstrated additivity between toxicity of individual congeners, (f) mechanistic rationale, and (g) consensus on values [26].

Following these events, international efforts to harmonize TEF schemes for dioxin-like compounds was initiated. Together, the WHO European Centre for Environmental Health (WHO-ECEH) and International Programme on Chemical Safety (IPCS) (referred to as the 1993 WHO-ECEH/IPCS expert panel) compiled a database of all relevant toxicological information on PCBs. In order to be included in the database, the study had to meet the following criteria: (1) TCDD or a reference coplanar PCB (PCB 77, 126, or 169) was used during the experiment (or results were available from similar previous studies), (2) at least one PCB congener was investigated, and (3) the end point was affected by both the reference compound and PCB in question [2]. Scientists from 8 different countries reviewed all available studies and ultimately assigned interim TEF values for 13 PCB congeners [27–29]. This group of experts also reported using the following criteria in their assessment of inclusion of each PCB congener in the TEF methodology: (1) the congener is structurally similar to the PCDDs and PCDFs, (2) the congener binds to the Ah receptor, (3) the congener elicits dioxin-like biochemical and toxic responses via Ah receptor mediation, and (4) the congener is persistent and accumulates in the food chain [2].

3.3.5 World Health Organization, 1998

Following their evaluation of dioxin-like PCBs, the 1993 WHO-ECEH/IPCS expert panel recommended that the database generated for the PCBs be expanded to include similar information for PCDDs, PDCFs, and other potential dioxin-like compounds that satisfied these above-mentioned four inclusion criteria. Over the next several years, significant efforts on behalf of regulators and independent scientists were made to compile information on all dioxin-like compounds and identify specific issues that needed to be addressed by an expert panel on TEF methodology. For example, various terminologies, definitions, and criteria were not transparent, including the use of “TEF” itself. A host of scientific experiments had evaluated dioxin-like chemicals in a variety of *in vitro* and *in vivo* systems; frequently, biological and/or biochemical end points were evaluated and were not necessarily considered "toxic." Therefore, relative potency values, or REPs, were assigned to describe the relative potency of a given compound as compared to TCDD from a single end point. These REP values were utilized by the expert panel in selecting a single point estimate TEF value, which was an order-of-magnitude estimate of toxicity of a compound relative to the toxicity of TCDD [7, 30, 31].

The 1993 REP database used for evaluation of PCBs was expanded to include all relevant studies of dioxin-like compounds. This work was spearheaded by scientists at the Institute of Environmental Medicine (IEM) at the Karolinska Institute (Stockholm, Sweden). This database became known as the "REP_{1997}" database and was used in the second WHO-ECEH/IPCS re-evaluation of TEFs when they convened in June of 1997 (note: the REP_{2007} database is discussed in Section 3.4.1). The expert panel's approach to the evaluation of TEFs was to (a) generally review the existing mammalian TEF values for PCDD/Fs [6, 27–29] as well as the existing TEFs for PCBs [2] and (b) determine if the TEFs should be changed on the basis of their evaluation of the studies in the REP_{1997} database. The WHO panel ultimately recommended consensus-based TEF values for 29 dioxin-like congeners, including 17 laterally substituted PCDDs/PCDFs, 4 non-*ortho* PCBs, and 8 mono-*ortho* PCBs. These recommendations were published in the peer-reviewed literature in 1998 [7], thus are known as the "WHO_{98} TEFs."

3.3.6
World Health Organization, 2006

The WHO-ECEH/IPCS expert panel again convened in June 2005 to re-evaluate the TEF methodology and values [8]. During this review, the panel re-evaluated the TEF values assigned in 1998 and also evaluated information for a number of other compounds for potential consideration in the TEF methodology. This reevaluation was based on a refined database of REPs developed by Haws *et al.* [30]. For the 29 DLCs that were previously assigned TEF values, new information published since the last expert panel meeting was evaluated. In addition, the expert panel utilized unweighted distributions of REP values to determine the depth of their review for each of the congeners (note: see discussion later in this chapter). As a result of their review, TEF values were changed for one PCDD, three PCDFs, and one non-*ortho*-substituted PCB (Table 3.3). In addition, all mono-*ortho*-substituted PCBs were assigned a common TEF value of 0.00 003.

Several chemicals, which demonstrated DLC activity, were evaluated by the expert panel for potential inclusion of the TEF concept. Though not ultimately included, data for PCB 37, polybrominated dibenzo-*p*-dioxins and dibenzofurans, mixed polyhalogenated dibenzo-*p*-dioxins and dibenzofurans, polyhalogenated napthalenes, and polybrominated biphenyls were reviewed by the panel. In general, these compounds did not meet all the criteria for inclusion in the TEF methodology, or there was not enough information to evaluate the compounds, or the compounds were potentially contaminated with low levels of more potent DLCs (i.e., observed DLC activity was likely due to these contaminants). The panel also considered nondioxin-like compounds that are AhR agonists and concluded that while these compounds may modulate the overall toxic potency of DLCs, these compounds would not impact the determination of individual REP or TEF values for DLCs because of their lack of persistence and lack of induction of the full spectrum of effects associated with DLCs. Therefore, risk from nondioxin-like AhR ligands needs to be investigated further.

Finally, the 2005 WHO-ECEH/IPCS expert panel discussed the application of the TEF methodology to abiotic matrices. However, the panel determined that the current methodology is specifically intended to estimate intake of DLCs, a limitation that is inherent in the REPs on which the TEF values are derived, given that the studies considered most relevant for TEF derivation are largely based on oral intake studies. Furthermore, the panel recommended that fate, transport, and bioavailability from each abiotic matrix be specifically considered when conducting a human risk assessment from abiotic matrices.

3.3.7
National Academy of Science, 2006

In the summer of 2004, the USEPA requested the National Research Council (NRC) to establish an expert committee to review USEPAs draft reassessment of the risks of dioxin and dioxin-like compounds (titled *Exposure and Human Health Reassessment of 2,3,7,8-Tetrachlorodibenzo-p-dioxin (TCDD) and Related Compounds,* referred to as the "Dioxin Reassessment" throughout this chapter). As a part of the charge to the NRC, the USEPA requested the expert committee to comment on the usefulness of TEFs and the uncertainties associated with their use for assessing health risks posed by mixtures of DLCs. As part of their review of the TEF methodology, the NAS committee considered a number of specific issues including the role of the Ah receptor in the toxic responses induced by DLCs, the role of AhR-independent mechanisms and interactions with other chemicals in the overall toxicity of DLCs, the need for analyses to address uncertainty in the TEF values, the consistency of the REP values across end points and exposure scenarios, the use of TEFs for determining human body burdens, the validity of the assumption that mixtures of DLCs exhibit additive toxicities, the use of REPs based on rodent studies to predict responses in humans, and the impact of natural and synthetic nondioxin-like AhR agonists on the toxicity of DLCs [22].

The expert committee released their findings in 2006 in a report titled *Health Risks from Dioxin and Related Compounds: Evaluation of the EPA Reassessment* [22]. Despite the assumptions, limitations, and uncertainties inherent in the TEF methodology, the NAS committee ultimately concluded that the TEF methodology is reasonable, scientifically justifiable, and widely accepted for the estimation of the relatively toxic potency of TCDD, other dioxins, and DLCs [22]. Another key conclusion concerning the TEFs was that there is a significant degree of uncertainty in the current consensus-based TEFs and, as a result, efforts should be made to develop an approach to better characterize uncertainty, including establishing a task force to build consensus probability density functions.

3.4 Relative Potency Estimates

Through the years, there has been some confusion about the terms TEF and relative potency estimate (REPs). The WHO has defined a TEF as an order of magnitude consensus estimate of the toxicity of a compound relative to TCDD based on all of the available evidence and scientific judgment [7]. In contrast, potency of a specific congener relative to TCDD for a single end point, in a single *in vivo* or *in vitro* study is referred to as a relative potency estimate. As mentioned previously, each dioxin-like congener is assigned a consensus-based TEF following review of all available literature evaluating the relative potency of a given congener as compared to TCDD. REP values are derived from many study types, different species, and different end points. Although thousands of studies have been published on DLCs, only a few of these studies were designed to specifically evaluate relative potencies. In fact, many of the studies that were utilized to derive TEF values were not studies designed to specifically evaluate relative potency, rather these studies were basic toxicological assessments investigating specific end points (e.g., reproductive toxicity, enzyme induction, etc.). However, as long as these studies met specific criteria for inclusion in the REP database [2], the 1993 WHO-ECEH/IPCS expert panel evaluated REP values based on these studies. In the original database, the authors of some studies provided REP values; however, in other cases, REP values had to be calculated by the database authors. This process was carried forward in the 2005 WHO-ECEH/IPCS evaluation of TEFs, as discussed in the following sections.

3.4.1 REP_{1997} Database

As already mentioned, prior to the 1997 WHO-ECEH/IPCS review of TEFs, a database of REP values was generated by scientists at the Karolinska Institute. When the REP_{1997} database was generated, a specific set of criteria was evaluated to determine if each study was suitable for inclusion. Specifically, the following criteria had to be met: (a) at lest one test congener (PCDD, PCDF, or PCB) and a reference compound (TCDD or PCB 126) were included in the study, or the reference compound data were from an identical experiment by the same authors, and (b) the end point used as the basis for the REP was Ah receptor-mediated and was affected by both the test congener and the reference compound. REP values, and associated study design information, from a wide range of studies were included because the TEF methodology was intended to address TEFs for mammals and wildlife; as such, information from studies that investigated fish and avian species were incorporated. When the REP_{1997} database was complete, there were a total of 976 mammalian REP values taken from 88 studies recorded in the database.

A host of information was contained in the database regarding the design of each study, such as duration of study, route of exposure, dosing regimen, chemical purity, cell line or species, and so on. The authors of the database not only recorded REP values calculated by the authors but also generated REP values for each study [7].

To do so, they applied one of the three general approaches: (a) comparison of dose-response curves or linear interpolation of log doses comparing the same effect level; (b) evaluation of effect ratios based on medial effective doses or concentrations, median lethal doses, tumor promotion indices, or dissociation constants for Ah receptor binding; or (c) estimation of REP values based on evaluation of figures presented in the publication. However, the authors of the database noted a number of issues with derivation of REP values using these approaches [30]. First, in a number of studies, some DLC congeners demonstrated a different maximal effect and therefore the REP values were based on comparison of different effect levels. Second, some REP values in the database had to be qualified with a "<" or ">" because some studies only included a single dose level of the test and/or reference compound. Third, original data were not provided in all studies, rather for some studies, only graphical data were provided. And, fourth, maximal responses and/or effects could not be confirmed in a number of studies as only a few of the studies evaluated high dose levels.

3.4.2
REP_{2005} Database

Prior to the 2005 WHO/IPCS meeting, the REP_{1997} database was re-examined in an effort to better characterize the range of observed REPs and updated to include new studies published since the last TEF evaluation [30]. The goal of this process was to create a structure that would facilitate quantitative analyses, which would ultimately allow a better characterization of uncertainty in TEF values. The refinement process was limited to review of studies associated with mammalian REP values for 28 of the 29 congeners for which WHO had established TEF values. The 29th congener was TCDD, which was treated as a reference compound in the database rather than a test congener. The authors first determined the critical study elements that they believed to be important metrics of study quality and relevance to humans: cell line, route of administration, chemical purity, exposure duration, delay between treatment and measurement of response, measurement end point, strain, tissue type, number of dose levels tested, attainment of maximal response, method of REP derivation, vehicle, animal age and sex, number of animals per treatment group, controls, and the reference compound. Four basic criteria were also developed to identify REP values that could not be utilized in quantitative analyses and therefore were excluded from the database: (1) REP repetitive studies (i.e., REPs from the same study presented in multiple publications), (2) repetitive end points (i.e., multiple REPs from a single study that used different methods to measure the same response), (3) incomplete dose-response data (e.g., studies that used only a single dose level of the test/and or reference), or (4) miscellaneous issues (e.g., lack of valid reference compound, lack of response, REP based on nonmammalian species, etc.).

Between the refinement of the REP_{1997} database and the addition of new studies, the overall number of studies in the revised database (REP_{2004} database) decreased from 88 to 83 and the total number of REPs decreased to 634. During the refinement process, it was noted that the volume of information available for each congener was vastly different; 1,2,3,7,8,9-hexachlorodibenzofuran had only a single study available for derivation of an REP where as other congeners, such as PCB 126, had over 20 studies available. The authors of the refinement process also noted the REP values for PCBs generally spanned three to six orders of magnitude whereas the REP values for the PCDD/Fs only spanned one to three orders of magnitude. Of particular interest, the authors note that for the congeners that account for the majority of TEQ in general US background population, the REP values tended to span two to five orders of magnitude (which was generally due to the PCBs). Because of this variability in REP values, the authors of the REP_{2004} database suggested the refined database could be used to improve risk assessment by facilitating quantitative approaches in the development of TEF values.

3.4.3 Variability in REP Values

There are many reasons for which REP values for the same congener can significantly vary. These differences are inherent in the underlying studies, specifically differences due to evaluation of an assortment of end points (e.g., enzyme induction, reproductive toxicity, tumor promotion, etc.), variations in study duration and dosing (i.e., acute versus chronic), and utilization of multiple species. In addition, for the mono-*ortho* PCBs, multiple-mechanism actions (e.g., mechanisms in addition to the AhR-mediated pathway) are involved. Finally, a wide variety of techniques were used to calculate REP values. Some techniques were very basic and crude in that REP values were estimated by evaluating graphs, whereas others were very sophisticated and involved dose-response modeling. The method used to derive an REP value is often limited by the type of data available. As such, some investigators have suggested that a more detailed database be developed that would include actual data sets, thereby allowing a comparative meta-analysis across studies [30].

As part of the 2005 WHO/IPCS evaluation of TEFs, the expert panel reviewed both the REP_{2004} database and the issues identified by the database authors. As a result, the expert panel derived the characteristics of an "ideally designed study" for evaluating relative potency from an *in vivo* study [8]. Such a study would include (a) a full dose-response curve for both TCDD and the congener of interest, (b) both the congener and TCDD would be administered by the same route to animals of the same species, strain, sex, and age – and these animals would be subjected to the same feeding, housing, and maintenance conditions, (c) a maximal response would be observed for both the congener and TCDD (and, ideally, the dose-response curves would be parallel), and (d) REP values would be calculated using an ED_{50} ratio of TCDD versus the congener. The panel also provided criteria for an "ideally designed" *in vitro* study used to evaluate relative potency [8]. For these studies, (a) at least four concentrations of both the congener and TCDD would be included, in addition to the vehicle; (b) out of the four concentration groups, three of the concentrations would elicit a response that falls between the EC_{20} and EC_{80}; (c) concentration–response curves would be parallel and at least one concentration would have elicited a maximal response; and (d) the REP was based on the EC_{50} ratio TCDD to the congener.

3.5 Derivation of TEF Values – Past, Present, and Future

3.5.1 TEF Derivation Process

Throughout the evolution of TEFs, expert panels have generally built upon previous efforts and refined TEF values based on updated information available in the peer-reviewed literature. As part of their review of new information, the 1993 and 1997 WHO-ECEH/IPCS expert panel thoroughly critiqued studies to evaluate quality and relevance. These critiques ultimately translated into the application of a qualitative "weighting" of REP values that were thought to have been derived from high-quality, relevant scientific studies. In assigning TEFs to each congener, the 1998 WHO-ECEH/IPCS expert panel employed scientific judgment and a qualitative weighting scheme whereby individual REPs from *in vivo* studies were given greater weight than *in vitro* studies and/or quantitative structure–activity relationship (QSAR) data; chronic studies were given greater weight than subchronic studies, which were given greater weight than subacute studies, which were given more weight than acute studies; and Ah receptor-mediated toxic responses were given more weight than biochemical responses. Resulting from the expert panel's review, a single, point-estimate TEF value for each congener was recommended on the basis of their scientific judgment.

In the 2005 WHO-ECEH/IPCS review [8], the expert panel further defined the qualitative "weighting" that was applied when setting the point estimate TEF values based on expert judgment. The panel noted that REP values from *in vivo* studies, which were actually designed to evaluate relative potency, were given highest priority by the WHO/IPCS panel because these studies inherently combine both toxicokinetic and toxicodynamic aspects of the experimental studies. However, the *in vitro* studies were still considered since they were useful in evaluating relative potency as they often aid in understanding the AhR-mediated mechanism of action for a given congener. In addition, *in vitro* studies significantly contributed to the characterization of differences in species sensitivity, particularly with respect to understanding the mechanism of action in humans.

For their evaluation, the 2005 WHO-ECEH/IPCS panel relied on the information in the REP_{2004} database. Specifically, the panel compared the WHO_{1998} TEF with the 75th percentile of the unweighted distribution of *in vivo* REP values for each congener. If the WHO_{1998} TEF was below the 75th percentile, all the data for that congener was extensively reviewed. However, if the WHO_{1998} TEF was above the 75th percentile, data was only briefly reviewed. For congeners in which the TEF was near or higher than the 90th percentile, particular attention was devoted to reviewing the underlying data for the REP values for that congener. In addition, studies conducted after the 1997 WHO expert panel meeting were reviewed to determine if the results would have influenced the WHO_{1998} TEF for that congener. Finally, the 2005 panel assigned TEF values based on half-orders of magnitude rather than a full order of magnitude in an effort to more fully characterize uncertainty in the TEF value.

Specifically, the panel noted that, by default, all TEF values are assumed to vary in uncertainty by at least one order of magnitude depending on the congener and its REP distribution [8]. Therefore, the assigned TEF values were representative of a central value with a degree of uncertainty that was at least a half-order of magnitude on both sides of the TEF value.

The 2005 WHO-ECEH/IPCS panel considered using REP distributions to determine TEF values for each congener. In doing so, a consistent percentile could be selected from the REP distributions, thereby instituting a transparent, conservative approach. However, the panel did not support this practice at the time because the REP values in the REP_{2004} database were not weighted [8, 30]. Without a method for weighting the REP values, the panel could not more heavily consider the studies to which they believed to be of highest quality and relevance. Therefore, the 2005 WHO-ECEH/IPCS panel utilized a combination of the unweighted REP distributions, expert judgment, and point estimates in their re-evaluation of the TEF values [8]. However, the panel concluded that if weighting criteria were developed, weighted REP distributions could be used for the derivation of TEF values in the future.

3.5.2 REP Weighting

In addition to the 2005 WHO-ECEH/IPCS panel's interest in developing weighting criteria for REPs, the NAS also noted such in their review of the Dioxin Reassessment [22]. Specifically, the NAS recommended that EPA should not only acknowledge the need for better uncertainty analyses for the TEF methodology but also establish a task force to develop "consensus probability density functions" [22]. As such, a number of investigators have proposed quantitative weighting schemes to weight REP values of the greatest quality and relevance [30, 32].

Most recently, a quantitative weighting scheme based on the Analytical Hierarchy Process was proposed [30, 33]. This framework was selected as it incorporates both multiple value comparison scales (different levels of better or worse) and a binary scale (better or not) and is well documented in the scientific literature. Using this framework, specific study elements were incorporated into a quantitative weighting scheme (Figure 3.2). Weighting factors were selected on the basis of study elements that most impacted REP quality and relevance and were based on the original qualitative criteria employed by the WHO expert panel in 1997. The proposed weighting scheme depicted in Figure 3.2 relies on a nonparallel framework where *in vivo* REP quality was determined based on consideration of pharmacokinetics, the quality of the underlying dose-response data ("REP derivation quality"), the perceived accuracy of the REP derivation method, and the nature of the end point on which the REP value was based. For the *in vitro* studies, REP quality was determined on the basis of the quality of the underlying dose-response data and the perceived accuracy of the REP derivation method. Using this framework, numerical values were assigned to each study element and the REP values were then compared against one another to determine the weight for each specific study element. The REP weights were then determined using an algebraic solution (i.e., the eigenvector) such that the ratios for

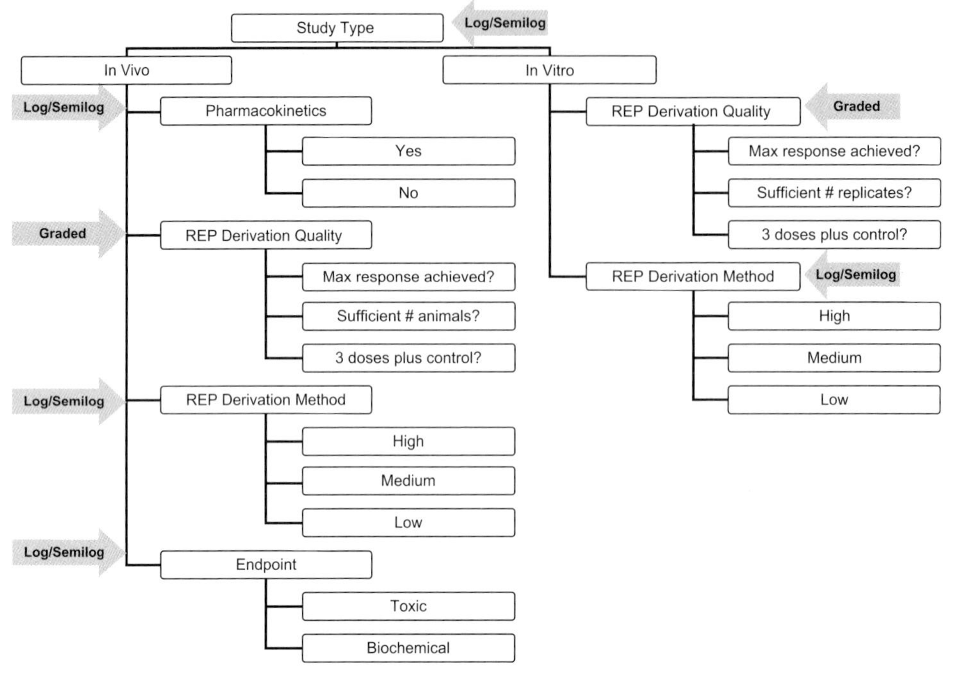

Figure 3.2 Proposed quantitative weighting scheme.

all paired comparisons were simultaneously taken into account. The overall weights for each individual REP using this weighting scheme can them be combined to prepare a cumulative distribution for each congener.

3.5.3
Use of REP Distributions to Characterize Uncertainty in Risk Assessment

As the practice of risk assessment has evolved over time, risk assessors have recognized the importance of quantitatively characterizing the uncertainty inherent in their estimates. Although a qualitative weighting scheme was applied by the WHO expert panel during the development of the TEFs, there is no way to quantitatively evaluate the uncertainty in the underlying REP values (which often span several orders of magnitude). One approach that has been proposed to address this limitation is the use of the full distribution of REP values for each congener in place of a single-point estimate TEF. Such an approach would allow better characterization of uncertainty and variability inherent in the risk estimates that are based on the TEFs [8, 22, 30, 32, 34]. Specifically, the REP distributions can be used in a probabilistic analysis (e.g., Monte Carlo techniques) to develop risk estimates. In 2003, the USEPA noted that "Future efforts by WHO or USEPA which develop guidelines and approaches to incorporating these weighting schemes into quantitative uncertainty analyzes are an important step in understanding the uncertainties of the TEF methodology" [20]. The need for a quantitative weighting scheme was echoed by the recent WHO-ECEH/IPCS expert panel charged with re-evaluating the TEF methodology [8]. Specifically, the WHO-ECEH/IPCS expert panel concluded that because all studies are not equal in terms of relevance and quality, it is important that a consensus-based weighting framework be developed and applied to the distributions of REP values for each congener prior to using those distributions to develop risk estimates [8]. A similar recommendation was made by the NAS expert committee charged with evaluating the USEPA Dioxin Reassessment [22].

Based on these recommendations by both WHO-ECEH/IPCS and NAS expert panels, efforts have been initiated to develop a consensus-based weighting framework [30]. Such an approach allows greater emphasis on those REP values that are believed to be more well suited for health risk assessment purposes. Applying such a framework to the REP values underlying each assigned TEF will allow development of weighted distributions of REP values that should ultimately facilitate better characterization of the variability and uncertainty inherent in the health risk estimates for this class of compounds. The use of distributions will also give risk managers the flexibility to tailor the desired level of protection to the specific situation under consideration and will facilitate establishment of a consistent level of protection for all congeners. Finally, the development and application of a quantitative weighting scheme will also yield a more transparent, reproducible, and consistent method for deriving TEFs from the underlying REP data.

The TEF methodology will continue to improve as the scientific community gains more understanding of dioxin-like compounds – particularly with respect to mode of action, additional techniques for evaluating relative potency, and additional research on relative potency. Throughout the history of TEFs, increased scientific rigor, clarity, and transparency have significantly improved the process.

3.6 Assumptions, Limitations, and Uncertainties of the TEF Approach

There are a number of assumptions and limitations inherent in the application of the TEF methodology for estimation of potential human health risks. Examples of some of the key assumptions and limitations include the following: (1) the spectrum of biological and toxic effects induced following exposure to TCDD is mediated through the Ah receptor; (2) all DLCs bind to the Ah receptor and induce a similar spectrum of dioxin-like effects; (3) the relative potency of a DLC is equivalent for all end points; (4) the relative potency of a DLC is equivalent across routes of exposure; (5) the relative potency of a DLC in humans is equivalent to that observed in rodents; (6) the toxicity of a mixture of DLCs is dose additive; (7) potential interactions among DLCs are ignored; (8) potential interactions with other chemicals are ignored; (9) naturally occurring (nonpersistent) chemicals that bind to the Ah receptor do not impact the potency estimates for the DLCs; and (10) non-Ah receptor-mediated effects are not accounted for. A number of these assumptions and limitations are discussed in greater detail in the sections that follow.

3.6.1
Role of the Ah Receptor

The TEF approach is based on the observation that the spectrum of dioxin-like effects induced following exposure to DLCs is mediated through a common mechanism of action involving binding to the Ah receptor [10, 11, 13, 14]. While binding to the Ah receptor is necessary for induction of biological and toxicological effects by this class of compounds, some investigators have shown that binding alone is not sufficient for generation of the wide variety of toxic effects [15, 35]. The ability of DLCs to produce their biochemical and toxicological effects results from downstream events regulated by the Ah receptor and AhR-dependent gene expression. In addition, though there are similarities in the expression and function of the AhR between species [36–38], which strengthens confidence in species extrapolation associated with the TEF methodology, the homeostatic physiological function of AhR is not completely understood. Significant efforts have been devoted to characterizing species differences in AhR affinity and subsequent differences in species sensitivities. In some species, such as mice, AhR binding affinity has been used to predict the sensitivity of both biological and toxicological effects of dioxin-like compounds. However, caution must be used as the relative binding affinity of TCDD and other dioxin-like compounds to the AhR is not consistent in all strains of mice – nor is it consistent between species [35, 39, 40]. Limited research has demonstrated that human AhR has a lower binding affinity than sensitive rodent strains; however, human populations express a wide range of variability in AhR binding affinity [41–43]. Despite the lack of complete understanding of the molecular mechanisms underlying the biological and toxic responses induced by DLC, and the potential species differences, the TEF approach is believed to be valid.

3.6.2
Consistency of REP Values Across End points and Routes of Exposure

The TEF methodology assumes that REPs for a given DLC are equivalent for all end points and for all exposure scenarios. While the evidence in literature generally supports this assumption, there are some studies that have shown wide variability in potency for some congeners across end points. For example, in a recent study by DeVito *et al.* [44], the investigators observed equivalent REPs for PCB126 based on hepatic EROD, dermal EROD, and ACOH, while the REP for pulmonary EROD was an order of magnitude lower. As a result of these and other findings, some have proposed developing end point-specific TEFs that could be used in cases where one is interested in assessing a particular end point. However, given that the concern in most cases is on the broad spectrum of effects induced by DLC, the TEFs developed considering all end points should be used to assess human health risk.

3.6.3
Additivity of DLCs

The assumption that the combined effects of mixtures of DLCs are additive has been debated for years. Examples of some of the arguments against the assumption of dose additivity that have been put forward include the presence of competing agonists or antagonists in the mixtures, nonadditive interactions with nondioxin-like chemicals, potential interactions among the DLCs in the mixtures (antagonism, synergism), the impact of nondioxin-like effects, and the observation of nonparallel dose–response curves for a given response across the different DLCs.

3.6.3.1 Studies with Mixtures of DLCs

The validity of the dose additivity assumption has been examined by numerous investigators over the years by testing mixtures containing DLCs [45–63]. In these studies, a host of cell culture systems and end points were examined and supported the assumption of additivity. The validity of the dose additivity assumption was revisited by the World Health Organization expert panel during their recent re-evolution of the TEF methodology [8]. The expert panel concluded that recent *in vivo* mixture studies supported the assumption of additivity [64–67]. Specifically, these studies were designed to evaluate if TEFs accurately predict response using potency-adjusted dose–additive mixtures in experimental studies. The results of the Walker *et al.* [67] study were heavily relied upon by the WHO/IPCS 2005 expert review panel. In this study, the investigators conducted a 2-year rodent cancer bioassay in which rats were administered three DLC congeners alone or in combination. Findings indicated that the observed dose-response relationships for several end points (i.e., hepatic, lung, and oral mucosal neoplasms) could be predicted using the potency-adjusted doses, and that the dose–response relationships of the three individual congeners were the same as the mixture. Most importantly, the authors demonstrated that the WHO_{1998} TEF values adequately predicted the tumor response induced by exposure to the mixture, lending strong support to the use of the TEF approach.

Several years before, Gao *et al.* [65] conducted a similar study investigating the validity of the TEF methodology with respect to endocrine disruption. These authors evaluated a number of reproductive end points in rats following exposure to individual congeners or a mixture of the congeners. The authors observed a common mode of action and that the dose-response relationships of the individual congeners were parallel both to each other and to the equipotent DLC mixture. In a similar study, Hamm *et al.* [66] evaluated developmental reproductive end points in rats following exposure to nine DLCs alone or in combination, the mixture of which approximated the relative abundance of DLCs in foodstuffs. Results indicated that the mixture of DLCs induced a similar spectrum of effects, though with a lesser potency, relative to TCDD. The authors concluded that the WHO_{1998} TEF values reasonably predicted the developmental toxicity end point evaluated in this study. This finding was also consistent with recent investigations [68].

Validation of the TEF methodology was also documented by Fattore *et al.* [64] in their evaluation of the effect of DLCs on hepatic enzymes. These authors investigated the effect of eight DLCs, either alone or in combination, on hepatic vitamin A levels in rats. In addition, the study was designed to evaluate the impact of dosing regimen (e.g., acute versus subchronic) on relative potency estimates using this end point. The findings of this study supported the additivity concept of the TEF methodology, though the effects of the mixture were approximately twofold lower than those predicted by the WHO_{1998} TEFs.

3.6.3.2 Nondioxin-Like Compounds
While the additivity of biochemical and toxic responses induced by DLCs has been shown in numerous mixture studies, there is also evidence in the literature indicating that nondioxin-like compounds can inhibit dioxin-like responses and thereby modulate their toxicity. For example, nondioxin-like PCBs and PCDFs have been observed to inhibit TCDD-induced CYP1A1 activity and immunotoxicity in mice [69–72]. In other studies, lower affinity synthetic PCDFs have been shown to inhibit CYP1A1, teratogenicity, immunotoxicity, and porphyria [73–76]. In contrast, certain nondioxin-like PCBs have been shown to synergize with dioxins [55–57, 77, 78].

3.6.3.3 Nonadditive Interactions
Although the PCDD and PCDF congeners included in the TEF methodology generally elicit response via the Ah receptor, it has been clearly demonstrated that for the mono-*ortho* PCBs, more than one mechanism of action (e.g., other mechanisms in addition to the AhR) are involved in the toxic responses observed. In fact, only a small subset of the PCBs has been classified as "dioxin-like" [79]; many of the nondioxin-like PCBs elicit a unique range of adverse effects, only some of which are also elicited by dioxin-like compounds. For example, several nondioxin-like compounds have demonstrated neurotoxicity, carcinogenicity, and endocrine disruption in rodent studies. Furthermore, several studies have indicated that these compounds may interact in a nonadditive fashion with dioxin-like chemicals. In some scenarios, the nondioxin-like PCBs demonstrate synergistic interactions whereas in others antagonistic interactions have been reported – and for some congeners, these interactions are dose specific (synergistic at low doses and antagonistic at high doses) [69], tissue specific, and end point specific [56, 57, 77, 80–82]. This situation is highlighted by PCB 153, which is a major environmental contaminant that is responsible for many nonadditive responses [83, 84].

3.6.3.4 Nonparallel Dose-Response Curves
Because derivation of REP values often depends on comparison of dose-response curves, situations in which the dose-response curves for some congeners and/or some effects are not parallel to the prototype TCDD curve contribute to the uncertainty associated with the TEF methodology [20, 85]. Demonstration of nonparallel dose-response curves have been observed for a number of congeners and end points, particularly when high doses are utilized. These observations primarily occur with PCB congeners, thus lending support to the theory that alternative mechanisms of action are involved for some congeners. For example, DeVito *et al.* [44] reported nonparallel dose-response curves for several individual PCBs at high doses, indicating a potential dose dependency for several end points. However, when these PCBs are administered in an environmentally relevant mixture, the assumption of additivity and parallelism of response is generally supported [45, 48, 53, 68, 78]. Starr *et al.* [85] have also indicated that the approach used to derive REP values also impacts the shape of the dose-response curve. These authors indicate that depending on the model selected to describe the dose-response relationship, the end point, and the species, a single order of magnitude of change in exposure at low doses can produce

responses for different end points that can vary by over 100 orders of magnitude. Such an observation would critically impact the selection of a TEF value and potentially bias such decisions to be based on specifically selected end points rather than the range of predicted REP values.

3.6.4 AhR-Independent Effects

As the TEF methodology is predicated on binding to the Ah receptor, AHR-independent effects are ignored. Examples of some of the types of AhR-independent effects that have been observed following exposure to DLCs include changes in gene expression [86], toxic responses in AhR knockout mice [87, 88], and effects on intracellular calcium levels [89]. However, current data are insufficient to determine the potential role of non-AhR-mediated effects in the overall toxicity of DLCS.

3.6.5 Other AhR Agonists

3.6.5.1 Natural Chemicals

Given the evolutionary conservation of the AhR, scientists have supported the hypothesis that there are endogenous ligands for this receptor although decisive evidence identifying such a ligand has not been demonstrated. However, several naturally offering compounds have been identified as "dioxin-like" in that they bind the Ah receptor and/or induce dioxin-like effects in both animals and humans. Examples include indole derivatives present in cruciferous vegetables, thermally treated meat protein (i.e., heterocyclic aromatic amines), oxidized amino acids, and bilirubin [90–96]. Research has demonstrated that these physiologically active AhR ligands can also act as antagonists by competing for AhR binding sites. However, in contrast to anthropogenic compounds included in the TEF-methodology, natural AhR ligands are rapidly metabolized and eliminated in biological systems (half-lives on the order of minutes to hours) and therefore are not persistent and do not bioaccumulate.

Several investigators have evaluated receptor binding and toxicity, and have estimated intake associated with these natural compounds [93, 97–108]. Together, these studies clearly indicate that there are many naturally occurring AhR agonists capable of inducing dioxin-like effects, such as induction of hepatic enzymes, both in experimental scenarios and in humans. Furthermore, several of these studies have suggested that human exposure to these compounds is much greater than to anthropogenic Ah receptor agonists. For example, Starr *et al.* [85] estimated that indole-3-carbinol (only one of the many natural compounds) alone contributes approximately 6 ppt TEQ, which was approximately the same TEQ estimated to be contributed by TCDD. Therefore, these authors concluded that although these natural compounds may not be persistent compared to the anthropogenic Ah receptor agonists, they significantly contribute to the daily intake. Similar findings were reported by Safe [14] following a comprehensive review of naturally occurring compounds that bind to the Ah receptor. In his report, Safe [14] compared the agonistic or antagonistic activity, relative potency, and dietary intake based on TEQ of several naturally occurring compounds and concluded that the mass intake of natural AhR agonists was much greater than that of the anthropogenic compounds. Although research has clearly demonstrated that the estimated intake of naturally occurring compounds that bind to the Ah receptor is significant, the relevance of such has not yet been well characterized.

Though there are very limited data characterizing the effects of mixtures of anthropogenic and natural AhR ligands, the TEF methodology relies on the assumption that these naturally occurring chemicals have short half-lives and also have varying degrees of affinity to the Ah receptor. The WHO expert panel indicated that while these compounds may impact the overall toxic effects associated with a defined TEQ, they will not impact the determination of an individual REP or TEF value for dioxin-like chemicals [8]. As such, naturally occurring Ah receptor agonists were not included in the TEF methodology.

3.6.5.2 **Anthropogenic Compounds**

A host of synthetic compounds have been shown to bind to AhR, though only a limited number is included in the TEF methodology. For example, within the dibenzofuran, dibenzo-*p*-dioxins, and biphenyl groups alone, there are hundreds of halogenated constituents; however, of the chlorinated compounds, only 7 of the PCDDs, 10 of the dibenzofurans, and 11 of the biphenyls have a planar confirmation that allows them to bind to the Ah receptor, thereby determining their "dioxin-like" status. In addition to these PCDDs, PCDFs, and PCBs, a number of industrial chemicals, such as polybrominated biphenyls, polychlorinated napthalenes, and chlorinated paraffins, pesticides, such as hexachlorobenzene, and other by-products associated with combustion or organic materials, such as polybrominated dioxins and dibenzofurans, as well as nonhalogenated polycyclic aromatic hydrocarbons, have demonstrated affinity for the Ah receptor. Several of the brominated congeners have similar affinities for the Ah receptor compared to the chlorinated congeners [14], as do some of the mixed bromo and chloro congeners. However, very little is known about the toxicity or relevance of human exposure given the large number of congeners [109]. As more information becomes available, these compounds will be considered for inclusion [8] given that several studies have documented Ah receptor binding and dioxin-like toxicities [110–114].

In the most recent evaluation of the TEFs, the WHO panel considered hexachlorobenzene for inclusion in the TEF methodology given that it fulfills the criteria; however, the panel noted that several studies investigating the dioxin-like activity of HCB were conducted with impure compound. Because high levels of dioxin-like contaminants were confirmed in some of the studies, the panel ultimately did not include HCB and recommended that HCB be re-evaluated for inclusion when studies using pure compound become available [8]. The expert panel also reached a similar conclusion regarding polychlorinated napthalenes and polybrominated naphthalenes; although there was a wealth of evidence demonstrating dioxin-like activity, the possible impurity of compounds used in the studies had to be addressed before they were included. However, recent studies using pure compounds that have been conducted since the most recent WHO expert panel meeting may influence this issue in the future. For example, recent research conducted by Puzyn *et al.* [113] reports REP values for several polychlorinated napthalenes that are in the same range as several of the dioxin-like compounds assigned TEF values by the WHO expert panel. PBDD and PBDF impurities present in commercial mixtures of polybrominated diphenyl ethers were also pinpointed by the expert panel; the PBDEs by themselves do not have AhR agonist properties and therefore were not included. As far as polybrominated biphenyls are concerned, the expert panel concluded that while these compounds generally met the criteria for inclusion in the TEF methodology, human exposure is extremely limited and therefore likely not relevant. In summary, while a number of anthropogenic compounds have been considered for inclusion in the TEF methodology, few have been assigned TEF values primarily because they do not meet all of the necessary criteria – specifically, they are readily metabolized (i.e., not persistent and/or bioaccumulative), they do not cause significant dioxin-like toxicological effects, or the available information regarding their dioxin-like activity is based on studies that utilized potentially contaminated compounds. However, given that volume of more recent research on these compounds supporting their inclusion in the TEF methodology, these anthropogenic compounds may be considered in future TEF evaluations.

3.6.6 Dose Metric

The current TEF methodology is based on administered dose studies and, as such, is most applicable to estimating health risks associated with intake of DLCs. As a result, some have raised concerns about applying TEFs to human tissue concentrations to develop body burden estimates. In fact, recent studies suggest that REPs for DLCs may differ when the REP is determined based on an administered dose basis versus tissue concentration [44, 66, 115]. While it would be ideal to develop TEFs based on tissue concentrations for purposes of assessing human body burdens, the most recent WHO expert panel concluded that the current data were insufficient for developing TEFs based on tissue concentrations [8].

3.7 Closing Remarks

As reviewed in this chapter, the relative potency factor approach for assessing mixtures of dioxin-like compounds has been the subject of intense scientific research for several decades. The TEF approach is still labeled as "interim" and has been subjected to periodic review. Clearly, the estimate of each TEF value is a critically important part of the risk assessment of dioxin-like compounds. As additional knowledge regarding the mode of action is understood and as more sophisticated quantitative tools become available, the uncertainty surrounding the TEF values will be further characterized and the overall TEF methodology will be strengthened. This has been demonstrated by each subsequent review of the TEFs, such that the scientific rigor, clarity, and transparency of the methodology have significantly improved in each review. Despite the inherent uncertainties, there is this general consensus that the TEF methodology provides a reasonable, scientifically justifiable, and widely accepted method to estimate the relative potency of dioxin-like compounds.

References

1 U.S. Environmental Protection Agency (U.S. EPA) (1996) PCBs: Cancer Dose-Response Assessment and Application to Environmental Mixtures. In National Center for Environmental Assessment. Office of Research and Development, Ed. U. S. Environmental Protection Agency.

2 Ahlborg, U., Becking, G.C., Birnbaum, L.S. *et al.* (1994) Toxic equivalency factors for dioxin-like PCBs: report on a WHO-ECEH and IPCS consultation, December 1993. *Chemosphere*, **28**, 1049–1067.

3 Ontario Ministry of the Environment (OME) (1984) Polychlorinated Dibenzop-Dioxins (PCDDs) and Polychlorinated Dibenzofurans (PCDFs), Scientific Criteria Document for Standard Development. No. 4-84.

4 Safe, S.H. (1994) Polychlorinated biphenyls (PCBs): environmental impact, biochemical and toxic responses, and implications for risk assessment. *Crit. Rev. Toxicol.*, **24**, 87–149.

5 U.S. Environmental Protection Agency (U.S. EPA) (1987) Interim Procedures for Estimating Risks Associated with Exposures to Mixtures of Chlorinated Dibenzo-*p*-Dioxins and -Dibenzofurans (CDDs and CDFs). In EPA/625/3-87/012, Ed.

6 U.S. Environmental Protection Agency (U.S. EPA) (1989) Interim Procedures for Estimating Risks Associated with Exposures to Mixtures of Chlorinated Dibenzo-*p*-Dioxins and -Dibenzofurans (CDDs and CDFs) And 1989 Update. In EPA/625/3-89/016, Ed.

7 van den Berg, M., Birnbaum, L., Bosveld, A.T., Brunstrom, B., Cook, P., Feeley, M., Giesy, J.P., Hanberg, A., Hasegawa, R., Kennedy, S.W., Kubiak, T., Larsen, J.C., van Leeuwen, F.X., Liem, A.K., Nolt, C., Peterson, R.E., Poellinger, L., Safe, S., Schrenk, D., Tillitt, D., Tysklind, M., Younes, M., Waern, F., and Zacharewski, T. (1998) Toxic equivalency factors (TEFs) for PCBs, PCDDs, PCDFs for humans and wildlife. *Environ. Health Perspect.*, **106**, 775–792.

8 Van den Berg, M., Birnbaum, L.S., Denison, M., DeVito, M., Farland, W., Feeley, M., Fiedler, H., Hakansson, H., Hanberg, A., Haws, L., Rose, M., Safe, S., Schrenk, D., Tohyama, C., Tritscher, A., Tuomisto, J., Tysklind, M., Walker, N., and Peterson, R.E. (2006) The 2005 World Health Organization reevaluation of human and mammalian toxic equivalency factors for dioxins and dioxin-like compounds. *Toxicol. Sci.*, **93**, 223–241.

9 Birnbaum, L.S. (1999) TEFs: a practical approach to a real-world problem. *Hum. Ecol. Risk Assess.*, **5**, 13–24.

10 Birnbaum, L.S. (1994) The mechanism of dioxin toxicity: relationship to risk assessment. *Environ. Health Perspect.*, **102** (Suppl. 9), 157–167.

11 Hankinson, O. (1995) The aryl hydrocarbon receptor complex. *Annu. Rev. Pharmacol. Toxicol.*, **35**, 307–340.

12 Martinez, J.M., DeVito, M.J., Birnbaum, L.S., and Walker, N.J. (2003) The toxicology of dioxins and dioxin-like compounds, in *Dioxins and Health*, 2nd edn (eds T. Gasiewiczand A. Schecter), John Wiley & Sons, Inc., Hoboken, NJ.

13 Okey, A.B., Riddick, D.S., and Harper, P.A. (1994) The Ah receptor: mediator of the toxicity of 2,3,7,8-tetrachlorodibenzo-*p*-dioxin (TCDD) and related compounds. *Toxicol. Lett.*, **70**, 1–22.

14 Safe, S. (1990) Polychlorinated biphenyls (PCBs), dibenzo-*p*-dioxins (PCDDs), dibenzofurans (PCDFs), and related compounds: environmental and mechanistic considerations which support the development of toxic equivalency factors (TEFs). *Crit. Rev. Toxicol.*, **21**, 51–88.

15 Sewall, C.H. and Lucier, G.W. (1995) Receptor-mediated events and the evaluation of the Environmental Protection Agency (EPA) of dioxin risks. *Mutat. Res.*, **333**, 111–122.

16 Birnbaum, L.S., and Tuomisto, J. (2000) Non-carcinogenic effects of TCDD in animals. *Food Addit. Contam.*, **17**, 275–288.

17 DeVito, M.J. and Birnbaum, L.S. (1994) Toxicology of the dioixins and related chemicals, in *Dioxins and Health* (ed. A. Schecter), Elsevier, New York, pp. 139–162.

18 McConnell, E.E., Moore, J.A., Haseman, J.K., and Harris, M.W. (1978) The comparative toxicity of chlorinated dibenzo-*p*-dioxins in mice and guinea pigs. *Toxicol. Appl. Pharmacol.*, **44**, 335–356.

19 Schwetz, B.A., Norris, J.M., Sparschu, G.L., Rowe, U.K., Gehring, P.J., Emerson, J.L., and Gerbig, C.G. (1973) Toxicology of chlorinated dibenzo-*p*-dioxins. *Environ. Health Perspect.*, **5**, 87–99.

20 US Environmental Protection Agency (USEPA) (2003) Exposure and Human Health Reassessment of 2,3,7,8-Tetrachlorodibnzo-*p*-Dioxin (TCDD) and Related Compounds.

21 Kociba, R.J., Keyes, D.G., Beyer, J.E., Carreon, R.M., Wade, C.E., Dittenber, D.A., Kalnins, R.P., Frauson, L.E., Park, C.N., Barnard, S.D., Hummel, R.A., and Humiston, C.G. (1978) Results of a two-year chronic toxicity and oncogenicity study of 2,3,7,8-tetrachlorodibenzo-*p*-dioxin in rats. *Toxicol. Appl. Pharmacol.*, **46**, 279–303.

22 National Academy of Sciences (NAS) (2006) *Health Risks from Dioxin and Related Compounds: Evaluation of the EPA Reassessment*, National Academies Press, Washington, D.C.

23 Barnes, D.G. (1991) Toxicity equivalents and EPA's risk assessment of 2,3,7,8-TCDD. *Sci. Total Environ.*, **104**, 73–86.

24 Barnes, D.G., Bellin, J., and Cleverly, D. (1985) Interim procedures for estimating risks associated with exposures to mixtures of chlorinated dibenzodioxins and -dibenzofurans (CDDs and CDFs). *Chemosphere*, **15**, 1895–1903.

25 Leece, B., Denomme, M.A., Towner, R., Li, S.M., and Safe, S. (1985) Polychlorinated biphenyls: correlation between *in vivo* and *in vitro* quantitative structure–activity relationships (QSARs). *J. Toxicol. Environ. Health*, **16**, 379–388.

26 Barnes, D., Alford-Stevens, A., Birnbaum, L., Kutz, F.W., Wood, W., and Patton, D. (1991) Toxicity equivalency factors for PCBs? *Qual. Assur.*, **1**, 70–81.

27 North Atlantic Treaty Organization/Committee on the Challenges of Modern Society (NATO/CCMS) (1988) International Toxicity Equivalency Factor (I-TEF) Method of Risk Assessment for Complex Mixtures of Dioxins and Related Compounds. Report No. 176.

28 North Atlantic Treaty Organization/Committee on the Challenges of Modern Society (NATO/CCMS) (1988) Scientific Basis for the Development of International Toxicity Equivalency Factor (I-TEF) Method of Risk Assessment for Complex Mixtures of Dioxins and Related Compounds. Report No. 178.

29 North Atlantic Treaty Organization/Committee on the Challenges of Modern Society (NATO/CCMS) (1988) Inventory of Regulations/Statutes Concerning Dioxins and Related Compounds. Report No. 169.

30 Haws, L.C., Su, S.H., Harris, M., Devito, M.J., Walker, N.J., Farland, W.H., Finley, B., and Birnbaum, L.S. (2006) Development of a refined database of mammalian relative potency estimates

for dioxin-like compounds. *Toxicol. Sci.*, **89**, 4–30.

31 van Leeuwen, F. (1997) Derivation of toxic equivalency factors (TEFs) for dioxin-like compounds in humans and wildlife. *Organohalogen Compd.*, **34**, 237.

32 Finley, B.L., Connor, K.T., and Scott, P.K. (2003) The use of toxic equivalency factor distributions in probabilistic risk assessments for dioxins, furans, and PCBs. *J. Toxicol. Environ. Health A*, **66**, 533–550.

33 Scott, P.K., Haws, L., Staskal, D.F., Birnbaum, L., Walker, N.J., DeVito, M., Harris, M.A., Farland, W.H., Finley, B.L., and Unice, K.M. (2006) An alternative method for establishing TEFs for dioxin-like compounds. Part 1. Evaluation of decision analysis methods for use in weighting relative potency data. *Organohalogen Compd.*, **68**, 2519–2522.

34 Finley, B., Kirman, C., and Scott, P. (1999) Derivation of probabilistic distributions for the W.H.O. mammalian toxic equivalency factors. *Organohalogen Compd.*, **42**, 225–228.

35 DeVito, M.J. and Birnbaum, L.S. (1995) Dioxins: model chemicals for assessing receptor-mediated toxicity. *Toxicology*, **102**, 115–123.

36 Abbott, B.D., Birnbaum, L.S., and Perdew, G.H. (1995) Developmental expression of two members of a new class of transcription factors: I. Expression of aryl hydrocarbon receptor in the C57BL/6N mouse embryo. *Dev. Dyn.*, **204**, 133–143.

37 Dohr, O., Li, W., Donat, S., Vogel, C., and Abel, J. (1996) Aryl hydrocarbon receptor mRNA levels in different tissues of 2,3,7,8-tetrachlorodibenzo-*p*-dioxin-responsive and nonresponsive mice. *Adv. Exp. Med. Biol.*, **387**, 447–459.

38 Dolwick, K.M., Schmidt, J.V., Carver, L.A., Swanson, H.I., and Bradfield, C.A. (1993) Cloning and expression of a human Ah receptor cDNA. *Mol. Pharmacol.*, **44**, 911–917.

39 Birnbaum, L.S., McDonald, M.M., Blair, P.C., Clark, A.M., and Harris, M.W. (1990) Differential toxicity of 2,3,7,8-tetrachlorodibenzo-*p*-dioxin (TCDD) in C57BL/6J mice congenic at the Ah Locus. *Fundam. Appl. Toxicol.*, **15**, 186–200.

40 Poland, A. and Glover, E. (1990) Characterization and strain distribution pattern of the murine Ah receptor specified by the Ahd and Ahb-3 alleles. *Mol. Pharmacol.*, **38**, 306–312.

41 Micka, J., Milatovich, A., Menon, A., Grabowski, G.A., Puga, A., and Nebert, D.W. (1997) Human Ah receptor (AHR) gene: localization to 7p15 and suggestive correlation of polymorphism with CYP1A1 inducibility. *Pharmacogenetics*, **7**, 95–101.

42 Okey, A.B., Giannone, J.V., Smart, W., Wong, J.M., Manchester, D.K., Parker, N.B., Feeley, M.M., Grant, D.L., and Gilman, A. (1997) Binding of 2,3,7,8-tetrachlorodibenzo-*p*-dioxin to AH receptor in placentas from normal versus abnormal pregnancy outcomes. *Chemosphere*, **34**, 1535–1547.

43 Rowlands, J.C. and Gustafsson, J.A. (1995) Human dioxin receptor chimera transactivation in a yeast model system and studies on receptor agonists and antagonists. *Pharmacol. Toxicol.*, **76**, 328–333.

44 DeVito, M.J., Menache, M.G., Diliberto, J.J., Ross, D.G., and Birnbaum, L.S. (2000) Dose-response relationships for induction of CYP1A1 and CYP1A2 enzyme activity in liver, lung, and skin in female mice following subchronic exposure to polychlorinated biphenyls. *Toxicol. Appl. Pharmacol.*, **167**, 157–172.

45 Birnbaum, L.S. and DeVito, M.J. (1995) Use of toxic equivalency factors for risk assessment for dioxins and related compounds. *Toxicology*, **105**, 391–401.

46 Birnbaum, L.S., Harris, M.W., Crawford, D.D., and Morrissey, R.E. (1987) Teratogenic effects of polychlorinated dibenzofurans in combination in C57BL/6N mice. *Toxicol. Appl. Pharmacol.*, **91**, 246–255.

47 Birnbaum, L.S., Weber, H., Harris, M.W., Lamb, J.C.T., and McKinney, J.D. (1985) Toxic interaction of specific polychlorinated biphenyls and 2,3,7,8-tetrachlorodibenzo-*p*-dioxin: increased incidence of cleft palate in mice. *Toxicol. Appl. Pharmacol.*, **77**, 292–302.

48 DeVito, M.J., Diliberto, J.J., Ross, D.G., Menache, M.G., and Birnbaum, L.S. (1997) Dose-response relationships for polyhalogenated dioxins and dibenzofurans following subchronic treatment in mice. I. CYP1A1 and CYP1A2 enzyme activity in liver, lung, and skin. *Toxicol. Appl. Pharmacol.*, **147**, 267–280.

49 Lipp, H.P., Schrenk, D., Wiesmuller, T., Hagenmaier, H., and Bock, K.W. (1992) Assessment of biological activities of mixtures of polychlorinated dibenzo-*p*-dioxins (PCDDs) and their constituents in human HepG2 cells. *Arch. Toxicol.*, **66**, 220–223.

50 Nagao, T., Golor, G., Hagenmaier, H., and Neubert, D. (1993) Teratogenic potency of 2,3,4,7,8-pentachlorodibenzofuran and of three mixtures of polychlorinated dibenzo-*p*-dioxins and dibenzofurans in mice. Problems with risk assessment using TCDD toxic-equivalency factors. *Arch. Toxicol.*, **67**, 591–597.

51 Schrenk, D., Buchmann, A., Dietz, K., Lipp, H.P., Brunner, H., Sirma, H., Munzel, P., Hagenmaier, H., Gebhardt, R., and Bock, K.W. (1994) Promotion of preneoplastic foci in rat liver with 2,3,7,8-tetrachlorodibenzo-*p*-dioxin, 1,2,3,4,6,7,8-heptachlorodibenzo-*p*-dioxin and a defined mixture of 49 polychlorinated dibenzo-*p*-dioxins. *Carcinogenesis*, **15**, 509–515.

52 Schrenk, D., Lipp, H.P., Wiesmuller, T., Hagenmaier, H., and Bock, K.W. (1991) Assessment of biological activities of mixtures of polychlorinated dibenzo-*p*-dioxins: comparison between defined mixtures and their constituents. *Arch. Toxicol.*, **65**, 114–118.

53 Smialowicz, R.J., Williams, W.C., and Riddle, M.M. (1996) Comparison of the T cell-independent antibody response of mice and rats exposed to 2,3,7,8-tetrachlorodibenzo-*p*-dioxin. *Fundam. Appl. Toxicol.*, **32**, 293–297.

54 Stahl, B.U., Kettrup, A., and Rozman, K. (1992) Comparative toxicity of four chlorinated dibenzo-*p*-dioxins (CDDs) and their mixture. Part I: Acute toxicity and toxic equivalency factors (TEFs). *Arch. Toxicol.*, **66**, 471–477.

55 van Birgelen, A., Visser, T., Kaptein, E. *et al.* (1997) Synergistic effects on thyroid hormone metabolism in female Sprague Dawley rats after subchronic exposure to mixtures of PCDDs, PCDFs and PCBs. *Organohalogen Compd.*, **34**, 370–375.

56 Van Birgelen, A.P., Van der Kolk, J., Fase, K.M., Bol, I., Poiger, H., Brouwer, A., and Van den Berg, M. (1994) Toxic potency of 3,3′,4,4′,5-pentachlorobiphenyl relative to and in combination with 2,3,7,8-tetrachlorodibenzo-*p*-dioxin in a subchronic feeding study in the rat. *Toxicol. Appl. Pharmacol.*, **127**, 209–221.

57 Van Birgelen, A.P., Van der Kolk, J., Fase, K.M., Bol, I., Poiger, H., Van den Berg, M., and Brouwer, A. (1994) Toxic potency of 2,3,3′,4,4′,5-hexachlorobiphenyl relative to and in combination with 2,3,7,8-tetrachlorodibenzo-*p*-dioxin in a subchronic feeding study in the rat. *Toxicol. Appl. Pharmacol.*, **126**, 202–213.

58 Viluksela, M., Stahl, B.U., Birnbaum, L.S., and Rozman, K.K. (1997) Subchronic/chronic toxicity of 1,2,3,4,6,7,8-heptachlorodibenzo-*p*-dioxin (HpCDD) in rats. Part II. Biochemical effects. *Toxicol. Appl. Pharmacol.*, **146**, 217–226.

59 Viluksela, M., Stahl, B.U., Birnbaum, L.S., and Rozman, K.K. (1998) Subchronic/chronic toxicity of a mixture of four chlorinated dibenzo-*p*-dioxins in rats. II. Biochemical effects. *Toxicol. Appl. Pharmacol.*, **151**, 70–78.

60 Viluksela, M., Stahl, B.U., Birnbaum, L.S., Schramm, K.W., Kettrup, A., and Rozman, K.K. (1998) Subchronic/chronic toxicity of a mixture of four chlorinated dibenzo-*p*-dioxins in rats. I. Design, general observations, hematology, and liver concentrations. *Toxicol. Appl. Pharmacol.*, **151**, 57–69.

61 Weber, H., Harris, M.W., Haseman, J.K., and Birnbaum, L.S. (1985) Teratogenic potency of TCDD, TCDF and TCDD-TCDF combinations in C57BL/6N mice. *Toxicol. Lett.*, **26**, 159–167.

62 Weber, L.W., Lebofsky, M., Stahl, B.U., Kettrup, A., and Rozman, K. (1992) Comparative toxicity of four chlorinated dibenzo-*p*-dioxins (CDDs) and their mixture. Part II: Structure–activity relationships with inhibition of hepatic phosphoenolpyruvate carboxykinase, pyruvate carboxylase, and gamma-glutamyl transpeptidase activities. *Arch. Toxicol.*, **66**, 478–483.

63 Weber, L.W., Lebofsky, M., Stahl, B.U., Kettrup, A., and Rozman, K. (1992) Comparative toxicity of four chlorinated dibenzo-*p*-dioxins (CDDs) and their mixture. Part III: Structure–activity relationship with increased plasma tryptophan levels, but no relationship to hepatic ethoxyresorufin *o*-deethylase activity. *Arch. Toxicol.*, **66**, 484–488.

64 Fattore, E., Trossvik, C., and Hakansson, H. (2000) Relative potency values derived from hepatic vitamin A reduction in male and female Sprague-Dawley rats following subchronic dietary exposure to individual polychlorinated dibenzo-*p*-dioxin and dibenzofuran congeners and a mixture thereof. *Toxicol. Appl. Pharmacol.*, **165**, 184–194.

65 Gao, X., Son, D.S., Terranova, P.F., and Rozman, K.K. (1999) Toxic equivalency factors of polychlorinated dibenzo-*p*-dioxins in an ovulation model: validation of the toxic equivalency concept for one aspect of endocrine disruption. *Toxicol. Appl. Pharmacol.*, **157**, 107–116.

66 Hamm, J.T., Chen, C.Y., and Birnbaum, L.S. (2003) A mixture of dioxins, furans, and non-ortho PCBs based upon consensus toxic equivalency factors produces dioxin-like reproductive effects. *Toxicol. Sci.*, **74**, 182–191.

67 Walker, N.J., Crockett, P.W., Nyska, A., Brix, A.E., Jokinen, M.P., Sells, D.M., Hailey, J.R., Easterling, M., Haseman, J.K., Yin, M., Wyde, M.E., Bucher, J.R., and Portier, C.J. (2005) Dose-additive carcinogenicity of a defined mixture of "dioxin-like compounds". *Environ. Health Perspect.*, **113**, 43–48.

68 Smialowicz, R.J., Devito, M.J., Williams, W.C., and Birnbaum, L.S. (2008) Relative potency based on hepatic enzyme induction predicts immunosuppressive effects of a mixture of PCDDS/PCDFS and PCBS. *Toxicol. Appl. Pharmacol*, **227**, 477–484.

69 Bannister, R. and Safe, S. (1987) Synergistic interactions of 2,3,7,8-TCDD and 2,2′,4,4′,5,5′-hexachlorobiphenyl in C57BL/6J and DBA/2J mice: role of the Ah receptor. *Toxicology*, **44**, 159–169.

70 Biegel, L., Harris, M., Davis, D., Rosengren, R., Safe, L., and Safe, S. (1989) 2,2′,4,4′,5,5′-Hexachlorobiphenyl as a 2,3,7,8-tetrachlorodibenzo-*p*-dioxin antagonist in C57BL/6J mice. *Toxicol. Appl. Pharmacol.*, **97**, 561–571.

71 Chen, G. and Bunce, N.J. (2004) Interaction between halogenated aromatic compounds in the Ah receptor signal transduction pathway. *Environ. Toxicol.*, **19**, 480–489.

72 Davis, D. and Safe, S. (1988) Immunosuppressive activities of polychlorinated dibenzofuran congeners: quantitative structure–activity relationships and interactive effects. *Toxicol. Appl. Pharmacol.*, **94**, 141–149.

73 Astroff, B., Zacharewski, T., Safe, S., Arlotto, M.P., Parkinson, A., Thomas, P., and Levin, W. (1988) 6-Methyl-1,3,8-trichlorodibenzofuran as a 2,3,7,8-tetrachlorodibenzo-*p*-dioxin antagonist: inhibition of the induction of rat cytochrome P-450 isozymes and related monooxygenase activities. *Mol. Pharmacol.*, **33**, 231–236.

74 Bannister, R., Biegel, L., Davis, D., Astroff, B., and Safe, S. (1989) 6-Methyl-1,3,8-trichlorodibenzofuran (MCDF) as a 2,3,7,8-tetrachlorodibenzo-*p*-dioxin antagonist in C57BL/6 mice. *Toxicology*, **54**, 139–150.

75 Harris, M., Zacharewski, T., Astroff, B., and Safe, S. (1989) Partial antagonism of 2,3,7,8-tetrachlorodibenzo-*p*-dioxin-mediated induction of aryl hydrocarbon hydroxylase by 6-methyl-1,3,8-trichlorodibenzofuran: mechanistic studies. *Mol. Pharmacol.*, **35**, 729–735.

76 Yao, C. and Safe, S. (1989) 2,3,7,8-Tetrachlorodibenzo-*p*-dioxin-induced porphyria in genetically inbred mice: partial antagonism and mechanistic studies. *Toxicol. Appl. Pharmacol.*, **100**, 208–216.

77 De Jongh, J., DeVito, M., Nieboer, R., Birnbaum, L., and Van den Berg, M.

(1995) Induction of cytochrome P450 isoenzymes after toxicokinetic interactions between 2,3,7,8-tetrachlorodibenzo-*p*-dioxin and 2,2′,4,4′,5,5′-hexachlorobiphenyl in the liver of the mouse. *Fundam. Appl. Toxicol.*, **25**, 264–270.

78 van Birgelen, A.P., Fase, K.M., van der Kolk, J., Poiger, H., Brouwer, A., Seinen, W., and van den Berg, M. (1996) Synergistic effect of 2,2′,4,4′,5,5′-hexachlorobiphenyl and 2,3,7,8-tetrachlorodibenzo-*p*-dioxin on hepatic porphyrin levels in the rat. *Environ. Health Perspect.*, **104**, 550–557.

79 U.S. Environmental Protection Agency (U.S. EPA) (1991) Workshop Report on Toxicity Equivalency Factors for Polychlorinated Biphenyls Congeners. In EPA/625/3-91/020, Ed.

80 Hornung, M.W., Zabel, E.W., and Peterson, R.E. (1996) Toxic equivalency factors of polybrominated dibenzo-*p*-dioxin, dibenzofuran, biphenyl, and polyhalogenated diphenyl ether congeners based on rainbow trout early life stage mortality. *Toxicol. Appl. Pharmacol.*, **140**, 227–234.

81 Morrissey, R.E., Harris, M.W., Diliberto, J.J., and Birnbaum, L.S. (1992) Limited PCB antagonism of TCDD-induced malformations in mice. *Toxicol. Lett.*, **60**, 19–25.

82 Zabel, E.W., Walker, M.K., Hornung, M.W., Clayton, M.K., and Peterson, R.E. (1995) Interactions of polychlorinated dibenzo-*p*-dioxin, dibenzofuran, and biphenyl congeners for producing rainbow trout early life stage mortality. *Toxicol. Appl. Pharmacol.*, **134**, 204–213.

83 Safe, S. (1997) Limitations of the toxic equivalency factor approach for risk assessment of TCDD and related compounds. *Teratog. Carcinog. Mutagen.*, **17**, 285–304.

84 van den Berg, M. (2000) Human risk assessment and TEFs. *Food Addit. Contam.*, **17**, 347–358.

85 Starr, T.B. (1997) Concerns with the use of a toxic equivalency factor (TEF) Approach for risk assessment of "Dioxin-Like" compounds. *Organohalogen Compd.*, **34**, 91–94.

86 Oikawa, K., Ohbayashi, T., Mimura, J., Iwata, R., Kameta, A., Evine, K., Iwaya, K., Fujii-Kuriyama, Y., Kuroda, M., and Mukai, K. (2001) Dioxin suppresses the checkpoint protein, MAD2, by an aryl hydrocarbon receptor-independent pathway. *Cancer Res.*, **61**, 5707–5709.

87 Fernandez-Salguero, P.M., Hilbert, D.M., Rudikoff, S., Ward, J.M., and Gonzalez, F.J. (1996) Aryl-hydrocarbon receptor-deficient mice are resistant to 2,3,7,8-tetrachlorodibenzo-*p*-dioxin-induced toxicity. *Toxicol. Appl. Pharmacol.*, **140**, 173–179.

88 Lin, T.M., Ko, K., Moore, R.W., Buchanan, D.L., Cooke, P.S., and Peterson, R.E. (2001) Role of the aryl hydrocarbon receptor in the development of control and 2,3,7,8-tetrachlorodibenzo-*p*-dioxin-exposed male mice. *J. Toxicol. Environ. Health A*, **64**, 327–342.

89 Puga, A., Hoffer, A., Zhou, S., Bohm, J.M., Leikauf, G.D., and Shertzer, H.G. (1997) Sustained increase in intracellular free calcium and activation of cyclooxygenase-2 expression in mouse hepatoma cells treated with dioxin. *Biochem. Pharmacol.*, **54**, 1287–1296.

90 Bradfield, C.A. and Bjeldanes, L.F. (1984) Effect of dietary indole-3-carbinol on intestinal and hepatic monooxygenase, glutathione S-transferase and epoxide hydrolase activities in the rat. *Food Chem. Toxicol.*, **22**, 977–982.

91 Bradfield, C.A. and Bjeldanes, L.F. (1987) Structure–activity relationships of dietary indoles: a proposed mechanism of action as modifiers of xenobiotic metabolism. *J. Toxicol. Environ. Health*, **21**, 311–323.

92 Gillner, M., Bergman, J., Cambillau, C., Fernstrom, B., and Gustafsson, J.A. (1985) Interactions of indoles with specific binding sites for 2,3,7,8-tetrachlorodibenzo-*p*-dioxin in rat liver. *Mol. Pharmacol.*, **28**, 357–363.

93 Michnovicz, J.J. and Bradlow, H.L. (1991) Altered estrogen metabolism and excretion in humans following consumption of indole-3-carbinol. *Nutr. Cancer*, **16**, 59–66.

94 Sinal, C.J. and Bend, J.R. (1997) Aryl hydrocarbon receptor-dependent induction of cyp1a1 by bilirubin in mouse hepatoma hepa 1c1c7 cells. *Mol. Pharmacol.*, **52**, 590–599.

95 Sinha, R., Rothman, N., Brown, E.D., Mark, S.D., Hoover, R.N., Caporaso, N.E., Levander, O.A., Knize, M.G., Lang, N.P., and Kadlubar, F.F. (1994) Pan-fried meat containing high levels of heterocyclic aromatic amines but low levels of polycyclic aromatic hydrocarbons induces cytochrome P4501A2 activity in humans. *Cancer Res.*, **54**, 6154–6159.

96 Wattenberg, L.W., and Loub, W.D. (1978) Inhibition of polycyclic aromatic hydrocarbon-induced neoplasia by naturally occurring indoles. *Cancer Res.*, **38**, 1410–1413.

97 Bjeldanes, L.F., Kim, J.Y., Grose, K.R., Bartholomew, J.C., and Bradfield, C.A. (1991) Aromatic hydrocarbon responsiveness-receptor agonists generated from indole-3-carbinol *in vitro* and *in vivo*: comparisons with 2,3,7,8-tetrachlorodibenzo-*p*-dioxin. *Proc. Natl. Acad. Sci. USA*, **88**, 9543–9547.

98 Connor, K.T., Harris, M.A., Edwards, M.R., Budinsky, R.A., Clark, G.C., Chu, A.C., Finley, B.L., and Rowlands, J.C. (2008) AH receptor agonist activity in human blood measured with a cell-based bioassay: evidence for naturally occurring AH receptor ligands *in vivo*. *J. Expo. Sci. Environ. Epidemiol*, **18** (4), 369–380.

99 Crowell, J.A., Page, J.G., Levine, B.S., Tomlinson, M.J., and Hebert, C.D. (2006) Indole-3-carbinol, but not its major digestive product 3,3′-diindolylmethane, induces reversible hepatocyte hypertrophy and cytochromes P450. *Toxicol. Appl. Pharmacol.*, **211**, 115–123.

100 Guengerich, P.F., Martin, M.V., McCormick, W.A., Nguyen, L.P., Glover, E., and Bradfield, C.A. (2004) Aryl hydrocarbon receptor response to indigoids *in vitro* and *in vivo*. *Arch. Biochem. Biophys.*, **423**, 309–316.

101 Henry, E.C., Bemis, J.C., Henry, O., Kende, A.S., and Gasiewicz, T.A. (2006) A potential endogenous ligand for the aryl hydrocarbon receptor has potent agonist activity *in vitro* and *in vivo*. *Arch. Biochem. Biophys.*, **450**, 67–77.

102 Liu, H., Wormke, M., Safe, S.H., and Bjeldanes, L.F. (1994) Indolo[3,2-*b*]carbazole: a dietary-derived factor that exhibits both antiestrogenic and estrogenic activity. *J. Natl. Cancer Inst.*, **86**, 1758–1765.

103 Menzie, C.A., Potocki, B.B., and Santodonato, S. (1992) Exposure to carcinogenic PAHs in the environment. *Environ. Sci. Technol.*, **26**, 1278–1284.

104 Moon, Y.J., Wang, X., and Morris, M.E. (2006) Dietary flavonoids: effects on xenobiotic and carcinogen metabolism. *Toxicol. In Vitro*, **20**, 187–210.

105 Popp, J.A., Crouch, E., and McConnell, E.E. (2006) A Weight-of-evidence analysis of the cancer dose-response characteristics of 2,3,7,8-tetrachlorodibenzodioxin (TCDD). *Toxicol. Sci.*, **89**, 361–369.

106 Reed, G.A., Peterson, K.S., Smith, H.J., Gray, J.C., Sullivan, D.K., Mayo, M.S., Crowell, J.A., and Hurwitz, A. (2005) A phase I study of indole-3-carbinol in women: tolerability and effects. *Cancer Epidemiol. Biomarkers Prev.*, **14**, 1953–1960.

107 Tiwari, R.K., Guo, L., Bradlow, H.L., Telang, N.T., and Osborne, M.P. (1994) Selective responsiveness of human breast cancer cells to indole-3-carbinol, a chemopreventive agent. *J. Natl. Cancer Inst.*, **86**, 126–131.

108 Poellinger, L. (2000) Mechanistic aspects – the dioxin (aryl hydrocarbon) receptor. *Food Addit. Contam.*, **17**, 261–266.

109 Birnbaum, L.S., Staskal, D.F., and Diliberto, J.J. (2003) Health effects of polybrominated dibenzo-*p*-dioxins (PBDDs) and dibenzofurans (PBDFs). *Environ. Int.*, **29**, 855–860.

110 Behnisch, P.A., Hosoe, K., and Sakai, S. (2001) Bioanalytical screening methods for dioxins and dioxin-like compounds a review of bioassay/biomarker technology. *Environ. Int.*, **27**, 413–439.

111 Mason, G., Zacharewski, T., Denomme, M.A., Safe, L., and Safe, S. (1987) Polybrominated dibenzo-*p*-dioxins and related compounds: quantitative *in vivo* and *in vitro* structure–activity relationships. *Toxicology*, **44**, 245–255.

112 Olsman, H., Engwall, M., Kammann, U., Klempt, M., Otte, J., Bavel, B., and Hollert, H. (2007) Relative differences in

aryl hydrocarbon receptor-mediated response for 18 polybrominated and mixed halogenated dibenzo-*p*-dioxins and -furans in cell lines from four different species. *Environ. Toxicol. Chem.*, **26**, 2448–2454.

113 Puzyn, T., Falandysz, J., Jones, P.D., and Giesy, J.P. (2007) Quantitative structure–activity relationships for the prediction of relative *in vitro* potencies (REPs) for chloronaphthalenes. *J. Environ. Sci. Health A* **42**, 573–590.

114 Weber, L.W. and Greim, H. (1997) The toxicity of brominated and mixed-halogenated dibenzo-*p*-dioxins and dibenzofurans: an overview. *J. Toxicol. Environ. Health*, **50**, 195–215.

115 Chen, C.Y., Hamm, J.T., Hass, J.R., Albro, P.W., and Birnbaum, L.S. (2002) A mixture of polychlorinated dibenzo-*p*-dioxins (PCDDs), dibenzofurans (PCDFs), and non-ortho polychlorinated biphenyls (PCBs) changed the lipid content of pregnant Long Evans rats. *Chemosphere*, **46**, 1501–1504.

4
Statistical Methods in Risk Assessment of Chemical Mixtures

Chris Gennings

Rigorous statistical methods are especially important in risk assessment of chemical mixtures that may include the detection and characterization of toxicological interaction (i.e., departure from additivity) among the components in the mixture. As the complexity of the design and data structure increases, so does the need for sound statistical methods. This chapter provides an overview of various statistical methods that may be useful in describing the effect and possible interactions among the components in a chemical mixture. We begin with a review of general statistical principles and concepts in Section 4.1. Section 4.2 includes a discussion with two examples of classical methods for describing multidimensional relationships through a response surface; here, the response is a function of the dose combination of the chemicals in the mixture. Estimation of the surface depends on the support from data through experimental design points. Two examples are provided that include a full factorial design in a 5^3 study (i.e., a mixture of 3 chemicals with 5 dose groups each for a total of 125 dose combination groups) and a fractionated 2^9 design (i.e., a study of 9 chemicals with 2 levels of each – but only 16 dose groups). A description of an interaction threshold model concludes Section 4.2 that allows dose-dependent regions of additivity and of interaction. As the number of chemicals in the mixture gets large, the experimental support for classical methods becomes impractical. An alternative approach is to focus the experimental effort and resulting inference on environmentally relevant mixing ratios of chemicals in a mixture through the use of ray designs. Two approaches are described with examples in Section 4.3 – the "single chemical required" (SCR) method and the "single chemical not required" (SCNR) method. Section 4.4 includes a description of an approach that focuses on testing *for* additivity in a low-dose region. That is, the approach uses equivalence testing methods to require the mixture data to provide significant evidence of additivity while controlling the type I error rate. Finally, Section 4.5 describes recent work on the concept of sufficient similarity in dose responsiveness. The chapter concludes with a short summary in Section 4.6.

Principles and Practice of Mixtures Toxicology. Edited by Moiz Mumtaz

ISBN: 978-3-527-31992-3

4.1 Principles of Statistics

Analysis of data begins with basic underlying assumptions about characteristics of the data. There is generally a response variable and a set of covariates that may be related to the response variable. The corresponding analysis strategy depends on the type of response variable; it may be on a continuous scale (serum enzyme levels, blood pressure, etc.), ordinal scale (mild, moderate, and severe muscle weakness), binary (alive/dead), count variable (motor activity), or time to response variable (time to death). The covariates may include dose/exposure/concentration levels, gender, age, and location (row/column or center/boundary) on a 96-well plate and other variables describing characteristics of the experimental unit. Dose-response data are generally analyzed using analysis of variance (ANOVA) or regression models for the mean with various variance assumptions based on the type of response variable. Normally distributed error terms are often assumed with continuous response variables; the assumption of a multinomial or binomial distribution is common for ordinal or binary data, respectively; a Poisson distribution is common for count data. An important unifying generalization of these approaches is found in generalized linear model methodology [1, 2] and quasi-likelihood estimation [3] where only an assumption on the mean and variance is required and not the full underlying distribution. Time-to-response data often include information about censoring that should be accounted for in the analysis. Popular methods for these types of data include Kaplan–Meier estimation, logrank tests, and proportional hazards models.

Another important characteristic of the data is whether the response variables are independent or dependent. When repeated measures are taken on the same experimental subject, the data are generally considered correlated. But observations on different animals are generally considered independent. An exception is when the data may be clustered, for example, in litters. The point is that the assumption of independence also determines the type of analysis to be conducted relating to the variance–covariance assumptions about the data. A common assumption for independent data is that the variance in response is constant across the covariates. This is analogous to assuming the variance does not change with the mean. More generally, the assumption may be made that the variance is a function of the mean (e.g., proportional to the mean) and that the covariance across subject responses is zero. When the data are clustered (e.g., based on littermates or experimental labs), the covariance structure may be quite complex by allowing intracluster covariances to follow some pattern of relationship while intercluster observations may be assumed independent with zero covariance. The appropriate analysis strategy of a given data set depends on characteristics of the data and should appropriately account for the relationship between the mean and the covariates, and the underlying variance–covariance structure in the data. The examples in this chapter assume the observations are independent.

Once an analysis strategy is determined, focus is on the use of efficient statistical methods for testing hypotheses of interest. When analyzing mixtures data, that hypothesis is often based on additivity:

H_0 : additivity
H_1 : interaction

In testing a hypothesis, we assume the null hypothesis (H_0) and look for evidence in the data to reject the null hypothesis in favor of the alternative (H_1). Thus, there are two types of errors that can be made: (i) a type I error is rejecting the null hypothesis when it is true; (ii) a type II error is not rejecting the null hypothesis when the alternative hypothesis is true. Hypothesis testing generally fixes the probability of a type I error (called the significance level or α) and tries to minimize the probability of a type II error (called β) by increasing the power of the study (defined as $1 - \beta$). Ways of increasing the power of a study include using statistics with minimum variance and by increasing sample size. It is important to point out that failure to reject the null hypothesis does not imply that the null hypothesis is true. This is analogous to the US court system where a defendant is "assumed innocent until proven guilty." When a defendant is found "not guilty," it does not prove innocence; instead it indicates lack of evidence to prove guilt whether the defendant is innocent or guilty. In testing the hypothesis of additivity as stated above, we assume additivity and look for evidence in the mixture data to prove interaction. When that evidence is not found, it should not necessarily be assumed that the mixture is additive.

When planning a statistical analysis strategy, it is important to avoid multiple testing. For example, if many mixture points are compared with additivity, then the probability of a type I error is inflated. There are many post hoc tests used for correcting multiple testing. We refer to a Bonferroni correction and Hochberg's correction in the examples in this chapter.

Compared to the typical hypothesis setup for testing for interaction, Stork *et al.* [4] used a bioequivalence testing framework and switched the hypotheses. That is, in testing *for* additivity in a low-dose region, their null hypothesis was that of interaction at mixture points of interest, and they looked for evidence from the data that the mixture is additive. The null hypothesis is rejected when a confidence interval(s) on the ratio of means under additivity to the sample means is contained within a prespecified boundary. More details and an example are provided in Section 4.4.

Many analysis strategies are based on representing data with statistical models. That is, in a dose-response study, in addition to summary statistics on responses at each dose group, a model can be used to represent the full dose-response curve. Such models are parsimonious with fewer parameters making them more powerful. However, it is important to demonstrate the adequacy of the model's representation of the data before model-based inference should be conducted. Goodness-of-fit criteria may include a formal test of goodness of fit or simply a demonstration of adequacy of fit by plots. It is important to consider the impact of sample size on goodness-of-fit tests. When the sample size is large, even slight and biologically inconsequential departure from the model may result in a statistically significant lack of fit. In our experience, measures of fit (e.g., Akaike's information criterion (AIC) and Bayesian information criterion (BIC) [2]) are adequate to distinguish which model from a set of models best represents the data instead of formal hypothesis testing. In fact, the US EPA guidelines for determining benchmark doses [5] suggest the use of AIC. In lower dimensional analyses, plots of the dose-response data overlaid with the predicted model provide overall evidence of the quality of the fit and also an indication of interpolated prediction from the model.

4.2
Statistical Approaches for Evaluating Mixtures

It is important to distinguish the discussion of mixtures herein as being associated with dose responsiveness where both the amount of the mixture and the mixing ratios may impact response. This is in contrast to the literature [6] on mixtures where the mixing ratio (or blend) of the mixture is of primary importance and dose (level of exposure) does not impact response. For example, an automobile engine may perform differently with different blends of gasoline, and the amount of gas in the tank is inconsequential (assuming there is at least some gas in the tank!). Here, we are interested in dose responsiveness where the dose (exposure) amount and mixture ratio are associated with response. Of particular interest in the analysis of chemical mixtures is whether the chemicals in the mixture interact and thereby possibly increase the toxic effect from what would be expected on the basis of single-chemical data. Thus, it is important to clearly describe what would be the "zero interaction" or additivity relationship among the chemicals in the mixture.

4.2.1
Definition of Additivity

A basic concept of an interaction is that the slope (i.e., steepness) of a dose-response curve of a chemical changes in the presence of one or more other components in a mixture. Conversely, if the slope of the dose-response curve of a chemical is not altered in the presence of another chemical, then the chemicals are said to exhibit no interaction or they are said to combine *additively* (i.e., zero interaction) [7, 8]. This concept of interaction as a change in slope is well grounded in the pharmacology literature. The earliest designations of receptors (receptive substance) and drug–receptor interactions occurred in the late nineteenth century when Langley suggested that these interactions were based on the law of mass action [9]. This hypothesis was theoretically and experimentally extended by A. J. Clark in the 1920s, Ariëns and Beld [10], and many others [11]. The equation that was used to describe this interaction between a drug and its receptor was that of a simple rectangular hyperbola (e.g., Michaelis–Menten). It is from these initial studies that terms for drugs with different activities (e.g., potency, partial agonist, and antagonist) were coined and various types of drug interactions (e.g., "additivity" as no interaction) were described [9].

A definition of additivity, which is often used to test for interactions among components in a mixture, is given by Berenbaum [11] and is based on the classical isobologram for the combination of two chemicals [12, 13]. In fact, Berenbaum [11, 14] refers to this definition as a "general solution" that is "mechanism-free" with the advantage of being based on empirical information. In a combination of c chemicals, let E_i represent the concentration/dose of the ith component alone that yields a fixed response, and let x_i represent the concentration/dose of the ith component in combination with the c agents that yields the same response. According to this definition of additivity, if the substances combine with *zero interaction*, then

$$\sum_{i=1}^{c} \frac{x_i}{E_i} = 1. \tag{4.1}$$

If the left-hand side of (4.1) is less than 1, then a greater than additive response (i.e., *synergism*) can be claimed at the combination of interest. If the left-hand side of (4.1) is greater than 1, then a less than additive response (i.e., *antagonism*) can be claimed at the combination. As (4.1) is the equation of a plane in c dimensions, this definition of additivity implies that under additivity contours of constant response are planar. It is important to note that Berenbaum's general definition as given in Eq. (4.1) places no constraint on the single-chemical slopes, and the mixture may include active and inactive compounds. Furthermore, the chemicals in the mixture do not need to have similar shaped dose-response curves, a requirement for applications of dose addition that use an index chemical to estimate risk. An example of the use of an index chemical to assess mixture risk is the toxic equivalency factor (TEF) approach to dose addition for dioxins that assumes common slopes across the chemicals under study [15, 16]; Berenbaum's definition of additivity in (4.1) does not require such an assumption.

Gennings *et al.* [8] show an algebraic equivalence of Berenbaum's definition of additivity given in (4.1) to statistical additivity models. They also demonstrate that these statistical additivity models satisfy the fundamental concept of zero interaction for additivity. Their argument begins with a general discussion of the justification for using empirical statistical models, which approximate an underlying relationship using a Taylor series argument; these models are not based on mechanistic assumptions/knowledge. This approach relates the mean of the response variable to a set of covariates and is applicable to many types of responses, dose-response shapes, and distributional assumptions. That is, the methods may be applied to continuous response values, proportional responses, count data, and so on. These are additivity models when they are parameterized to include an intercept and linear terms only. Otherwise, higher order cross-product terms are associated with interactions according to Berenbaum's definition. An advantage of this framework is that the important hypotheses of additivity can be tested with statistical rigor.

For this model development to be useful, it is necessary to demonstrate the adequacy of the approximation. Since the observed data contain information about the underlying dose-response relationship, comparisons of these data to the predictions of the model are important in assessing the model. Such comparisons can be accomplished with varying levels of statistical rigor. Often simple plots of observed and predicted results are sufficient. In other cases, it may be necessary to test the null hypothesis of model adequacy. While testing model adequacy is an activity that can occur only after the data collection and analysis phases of a study, it should be noted that experimental designs have been developed to maximize the power of the test of model adequacy [17]. For models that provide an adequate representation of the data, the appropriateness of Box's [18] observation, "all models are wrong, but some are useful," is readily understood.

For a mixture of c agents, an additivity model can be written as

$$g(\mu) = \beta_0 + \beta_1 x_1 + \beta_2 x_2 + \cdots + \beta_c x_c, \tag{4.2}$$

where $x_1, x_2, \ldots, x_c$ are the doses of the c individual components, $g(\mu)$ is a user-specified link function ([1]; p. 375), and $\beta_0, \beta_1, \ldots, \beta_c$ are unknown parameters. At a fixed response, μ_0, $g(\mu) - \beta_0 = \beta_1 x_1 + \beta_2 x_2 + \cdots + \beta_c x_c$, and

$$1 = \frac{x_1}{(g(\mu_0) - \beta_0)/\beta_1} + \cdots + \frac{x_c}{(g(\mu_0) - \beta_0)/\beta_c} = \frac{x_1}{\mathrm{ED}_{\mu_0}^{(1)}} + \cdots + \frac{x_c}{\mathrm{ED}_{\mu_0}^{(c)}}$$

$$= \text{interaction index.}$$

This follows, since $(g(\mu_0) - \beta_0)/\beta_i = \mathrm{ED}_{\mu_0}^{(i)}$, the dose associated with the response μ_0 for the ith component of the mixture. Berenbaum [14] related the interaction index to the isobologram and showed that when the interaction index is not equal to 1, an interaction, that is, a departure from additivity, is present. The interaction index is important since it does not have the graphical limitations associated with producing plots of multidimensional isobolograms when the number of agents in the mixture is greater than 3. It is important to note that the additivity model in (4.2) may be estimated with support only from single-chemical dose-response data. Thus, the additivity model for a mixture of c chemicals may be estimated with approximately $4c$ to $6c$ judiciously chosen (i.e., within the active range of the dose-response curve) single-chemical dose groups.

Carter *et al.* [19] and Gennings *et al.* [8] showed that when model parameters associated with interaction in a generalized linear model are different from zero, the interaction index is different from one. For example, consider a two-agent mixture with interaction, that is,

$$g(\mu) = \beta_0 + \beta_1 x_1 + \beta_2 x_2 + \beta_{12} x_1 x_2.$$

From this, it follows that

$$1 - \frac{\beta_{12} x_1 x_2}{g(\mu) - \beta_0} = \frac{x_1}{\mathrm{ED}_{\mu}^{(1)}} + \frac{x_2}{\mathrm{ED}_{\mu}^{(2)}} = \text{interaction index.}$$

When $\beta_{12} = 0$, the interaction index is equal to 1. For $g(\mu) > \beta_0$, if the estimate of $\beta_{12} > 0$, the interaction index is less than 1, which is indicative of a synergism. Similarly, when $\beta_{12} < 0$, an antagonistic interaction can be claimed. The significance and interpretation of cross-product terms in a response surface model are illustrated in the next section.

4.2.2
Testing for Departure from Additivity: Analysis of a Response Surface

To improve the interpretation of slope and interaction parameters in a response surface model, it may be advantageous to imbed the slope and cross-product terms in a nonlinear function that accommodates the general sigmoidal or exponential shape of the relationship. In general, we may define a generalized nonlinear model for the mean of Y as

$$\begin{aligned} g(\mu;\omega) &= \mathbf{X}\boldsymbol{\beta} \\ &= \beta_0+\beta_1x_1+\beta_2x_2+\cdots+\beta_cx_c+\sum_{i=1}^{c-1}\sum_{j=i+1}^{c}\beta_{ij}x_ix_j \\ &\quad +\sum_{i=1}^{c-2}\sum_{i=j+1}^{c-1}\sum_{k=i+2}^{c}\beta_{ijk}x_ix_jx_k+\cdots+\beta_{12\ldots c}x_1x_2\ldots x_c. \end{aligned} \tag{4.3}$$

where $\boldsymbol{\omega}$ is a vector of nonlinear parameters. For example, $\mu=\alpha+\gamma\exp(\mathbf{X}\boldsymbol{\beta})$ allows the dose-response relationship to change exponentially, and with response at 0, $\mu_0=\alpha+\gamma$. It is often the case in dose-response data that the variance of the response changes with the mean, that is,

$$\mathrm{Var}(Y)=\tau V(\mu),$$

where τ is a scale parameter and $V(\cdot)$ is a user-specified function of the mean. In such a case, the method of quasi-likelihood [1, 3] may be used to find estimates for model parameters. Using a quasi-likelihood estimation criterion, it is of interest to maximize $\sum_i Q_i(\mu_i;y_i)$, where

$$Q_i(\mu_i;y_i)=\int_y^{\mu}\frac{y-t}{\tau V(t)}dt.$$

When the variance is proportional to the mean, that is,

$$\mathrm{Var}(Y)=\tau V(\mu)=\tau\mu, \tag{4.4}$$

the quasi-likelihood objective function is

$$\int_y^{\mu}\frac{y-t}{\tau V(t)}dt=\int_y^{\mu}\frac{y-t}{\tau t}dt=\frac{1}{\tau}(y_i\log(\mu_i)-\mu_i).$$

Estimation of τ is based on a moment estimator

$$\hat{\tau}=\frac{1}{(n-p)}\sum_i\frac{(y_i-\hat{\mu}_i)^2}{\hat{\mu}_i}.$$

To implement the maximum quasi-likelihood criterion for estimation in PROC NLIN in SAS, the loss and weight functions can be used. The loss function, negative log quasi-likelihood, for the ith observation can be defined as $-(y_i\log(\mu_i)-\mu_i)/w_i$ where the weight $w_i=1/\mu_i$ (which is the inverse of $V(\mu)$ in general). Using these weight and loss statements in PROC NLIN results in maximizing the quasi-likelihood using the nonlinear mean model in (4.3) and the variance assumption in (4.4). The estimate for the (weighted) mean square error is the estimate for τ.

Define the $p \times 1$ vector of model parameters as

$$\boldsymbol{\theta} = \begin{bmatrix} \alpha \\ \gamma \\ \beta_1 \\ \beta_2 \\ \\ \beta_{12\ldots c} \end{bmatrix},$$

which is conveniently partitioned into two submatrices

$$\boldsymbol{\theta}_{p\times 1} = \begin{bmatrix} \boldsymbol{\theta}_1 \\ \boldsymbol{\theta}_2 \end{bmatrix} \begin{matrix} q\times 1 \\ p-q\times 1 \end{matrix},$$

where we may assume that the q parameters in $\boldsymbol{\theta}_1$ are of primary interest and those in $\boldsymbol{\theta}_2$ are considered "nuisance parameters." Suppose it is of interest to test the hypothesis

$$\mathrm{H}_0 : \boldsymbol{\theta}_1 = \mathbf{0} \quad \text{versus} \quad \mathrm{H}_1 : \boldsymbol{\theta}_1 \neq \mathbf{0}$$

Various tests may be used to test this hypothesis. The quasi-likelihood ratio test, for example, is evaluated by comparing the quasi-likelihood (i.e., negative loss function) for a (full) model that includes θ_1 with that of the reduced model where θ_1 is removed. In particular, for large samples,

$$\mathrm{LR} = \frac{2[Q(\hat{\boldsymbol{\theta}}; y) - Q(\boldsymbol{\theta}_1 = 0; \tilde{\boldsymbol{\theta}}_2; y)]}{g\hat{\tau}}$$

may, conservatively, be compared with an $F_{q,n-p}$ distribution.

4.2.3 Example 1: Analysis of a Mixture of Three Chemicals from a 5^3 Study

Moser *et al.* [20] experimentally evaluated neurobehavioral effects of a mixture of heptachlor (HEPT), di(2-ethylhexyl)phthalate (DEPH), and trichloroethylene (TCE) in young adult female Fischer-344 rats. A full 5^3 factorial design was used, requiring 125 treatment groups, with $n = 10$/group (total sample size was 1250). The logistics of conducting the study required 5 replicates of 125 treatment groups, with 2 rats in each group, and each replicate was further broken into 5 blocks of 50 rats each (the maximum that could be tested at a time). Numerous response variables were measured including continuous (e.g., grip strength, motor activity counts), ordinal (e.g., gait abnormality, handling reactivity score), and binary (e.g., piloerection, pupil response) end points. To illustrate the approach, we describe the analysis of the gait score using an ordinal logistic model. A gait score was assigned from 1 to 4, where 1 = a normal, unaffected gait; 2 = a slightly affected gait (foot splay, slight hindlimb weakness and spread); 3 = a moderately affected gait (foot splay, moderate hindlimb weakness, moderate limb spread during ambulation); 4 = a severely affected gait (foot splay, severe hindlimb weakness, dragging hindlimb, and inability to rear).

It was assumed that the response count in each category at each dose combination has a multinomial distribution and that the multinomial counts at different dose combinations are independent. Preliminary analysis of using a cumulative logits model included evaluation of the proportional odds assumption that was not satisfied. Instead, continuation-ratio logits were used [21].

For gait score, three continuation-ratio logits were defined [2] as

$$
\begin{aligned}
L_1 &= \text{logit}(\text{Prob}(\text{normal gait})) = \text{logit}(P(Y=1)),\\
L_2 &= \text{logit}(\text{Prob}(\text{slightly abnormal gait given abnormal gait}))\\
&= \text{logit}(P(Y=2|Y\geq 2)),\\
L_3 &= \text{logit}(\text{Prob}(\text{moderatley abnormal gait given at least}\\
&\quad \text{a moderately abnormal gait}))\\
&= \text{logit}(P(Y=3|Y\geq 3)).
\end{aligned}
$$

It can be shown that the conditional probabilities associated with these continuation-ratio logits are independent indicating that they can be separately analyzed using available software for binary responses. Let π_j be the probability for the jth logit, $j=1, 2, 3$, which is a function of dose. Each logistic model was parameterized as follows:

$$\pi(x) = \frac{1}{1+\exp(-g(x,\beta,\alpha))},$$

where

$$
\begin{aligned}
g(x,\beta,\alpha) = {}& \beta_0+\beta_1x_1+\beta_2x_2+\beta_3x_3+\beta_{12}x_1x_2+\beta_{13}x_1x_3+\beta_{23}x_2x_3+\beta_{123}x_1x_2x_3\\
&+\alpha_1z_1+\alpha_2z_2+\alpha_3z_3+\alpha_4z_4
\end{aligned}
$$

and $x_1=$ dose of TCE (mg/kg), $x_2=$ dose of DEHP (mg/kg), $x_3=$ dose of HEPT (mg/kg), $z_i=$ indicator variable designating whether the observation is from the ith block, $i=1, \ldots, 4$, with the constraint that $\alpha_5 = -(\alpha_1+\alpha_2+\alpha_3+\alpha_4)$. The "linear terms" are associated with the effect of each chemical on the response and the cross-product terms are associated with either two- or three-way interactions among the chemicals. Parameter terms for individual chemicals and their interactions were tested for significance ($p<0.05$). When an interaction was identified as significantly different from additivity, the interaction was characterized as synergistic (greater than additive) or antagonistic (less than additivity). The signs of the parameter estimates for the interaction and main effects were compared. An interaction parameter with the same algebraic sign as the main effect parameters indicates synergism, and the opposite sign indicates antagonism. For example, in the analysis of gait score [20], none of the parameters associated with DEHP was significant (i.e., $\beta_2=\beta_{12}=\beta_{23}=\beta_{123}=0$). The slope parameters for TCE and HEPT were both positive and significant (i.e., β_1 and β_3). Since β_{13} was negative and borderline significant ($p=0.07$), there was an indication of an antagonistic relationship between TCE and HEPT. Thus, the contours of constant response (isobols) from this model curve were slightly curved upward.

This example illustrates a typical analysis strategy for estimating and interpreting a response surface. It is important to note that the response surface is estimated with support points from the experimental design. In this example with 3 chemicals in combination, there were 125 treatment combinations using a full 5^3 factorial design. To reduce the number of design points, a fractionated factorial design may be used that confounds higher order interaction terms with lower order terms, thereby reducing the number of required design points. In the next example, we illustrate the approach from a study with a two-level factorial design with 9 factors (i.e., 9 chemicals) in 16 experimental groups, which is a 1/32 fraction of a complete study [22].

4.2.4
Example 2: Analysis of a Mixture of Nine Chemicals from a Fractionated 2^9 Study

Groten *et al.* [22] designed a study "intended to determine whether simultaneous administration of nine compounds at a concentration equal to the NOAEL for each of them will result in a NOAEL for the combination." Nine chemicals were selected to represent a combination of compounds highly relevant to the general population in terms of their use pattern and level and frequency of exposure (drugs, food additives, food contaminants, bioagenic amines, and industrial solvents). The compounds were not chosen on the basis of their similar mode of action. The study included 20 groups of male albino Wistar outbred rats: 4 main groups ($n = 8$/group) and 16 satellite groups ($n = 5$/group). Rats in the main groups were simultaneously exposed to the nine chemicals at concentrations equal to the minimum observed adverse effect level (MOAEL), NOAEL, or 1/3 NOAEL. The fourth main group was a control group. In the 16 satellite groups, the rats were simultaneously exposed to various combinations of chemicals, all at the MOAEL. The combinations were selected such that the results would allow analysis of the two-way interactions among the nine chemicals. Higher order interactions were assumed to be negligible. Except for two of the chemicals, all test substances were administered *ad libitum* in the diet, at constant dietary concentrations, for 4 weeks. Animals were exposed to the other two chemicals through multitiered inhalation chambers.

Groten *et al.* [22] present the analysis for body weight, food intake, and selected clinical and hematological parameters using a general linear model parameterized as

$$\mu = \alpha_0 + \sum_{i=1}^{9} \alpha_i z_i + \sum_{i=1}^{8} \sum_{j=i}^{9} \alpha_{ij} z_i z_j,$$

where α_0 is the overall mean effect, α_i is half the additional effect due to the ith compound, α_{ij} is the interactive effect between the ith and jth compounds, and z_i have values of either 1 or -1 (presence or absence) of the ith compound. For example, the estimated equation ([22], Table 8) for prothrombin time (PTT) was additive with main effects due to butyl hydroxyanisol (BHA), DEHP, methylene chloride (MC), and aspirin (Asp). Of the 29 end points evaluated, 17 (59%) included significant pairwise-interaction parameters. Groten *et al.* report that only a few adverse effects were observed in the NOAEL group (hyperplasia and metaplasia of the nasal epithelium, hepatocellular hypertrophy, decreased plasma triglyceride concentrations, altered ALP enzyme activities, and increased relative kidney weight), demonstrating that combined exposure to compounds at their individual NOAELs can result in a significant effect of the combination. Except for a slight increase in the relative kidney weight, no toxic effects were observed with combined exposure at 1/3 NOAEL. Groten *et al.* conclude that in general combined exposure to the single chemicals does not "constitute an evidently increased hazard, provided the exposure level of each chemical is similar to or lower than its own NOAEL."

4.2.5
Interaction Threshold Model

An assumption often made by mixture toxicologists [23] is that interaction among chemicals is generally a high dose phenomenon and that the assumption of additivity in a low-dose region generally holds. To explore this assumption, Hamm *et al.* [24] developed an empirical response surface model for an interaction threshold. That is, the model includes a boundary between a low-dose region of additivity and a high-dose region of interaction. They demonstrated their approach with four parameterizations for the boundary shape: (i) segmented line boundary; (ii) four-parameter nonlinear boundary; (iii) inverse cubic boundary; and (iv) elliptical boundary. Selection of a boundary for a given data set was based on the consideration of a goodness-of-fit statistic. They demonstrated their model on the combination of ethanol and chloral hydrate.

Gessner and Cabana [25] studied the mixture of ethanol and chloral hydrate, using the loss of righting reflex in mice as the response. These authors compared confidence interval estimates of experimentally determined ED_{50} values at 19 different dose combinations with lines formed by connecting the upper and lower confidence limits of the ED_{50} values of individual drugs. They interpreted these lines as some form of a confidence bound about the line of additivity formed by connecting the ED_{50} values determined for each drug. Evidence for a departure from additivity was provided whenever the confidence bounds for the experimentally determined ED_{50} for a combination did not overlap with the "confidence" region for the line of additivity. Although the statistical approach used to analyze their data is crude, their work is important because it provides an empirical estimate of the ED_{50} contour of the underlying dose response surface associated with these two agents. It is noteworthy that the experiment used a total of 1681 animals in 21 treatment groups, the 19 combinations of the two agents mentioned previously, and two single-agent groups [26].

It is interesting to compare the contours of the fitted dose response surface for the interaction threshold model with the isobologram presented by Gessner and Cabana (Figure 4.1) using an independently observed data set. The contours associated with various ED_{100} values are found in Figure 4.1b. The similarity between Gessner and Cabana's ED_{50} contour and that estimated by the interaction threshold model [25] would appear to lend support to the concept of interaction

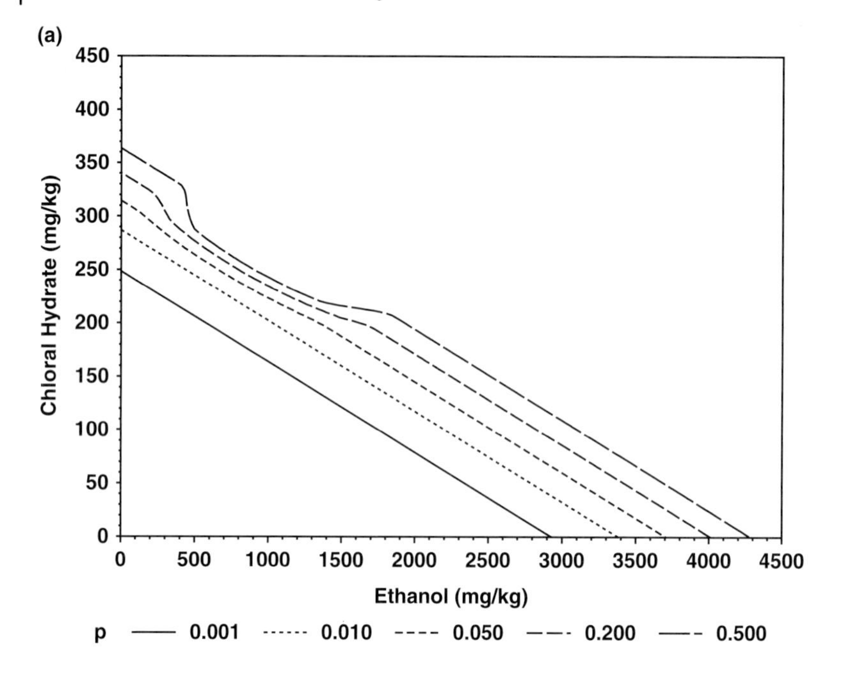

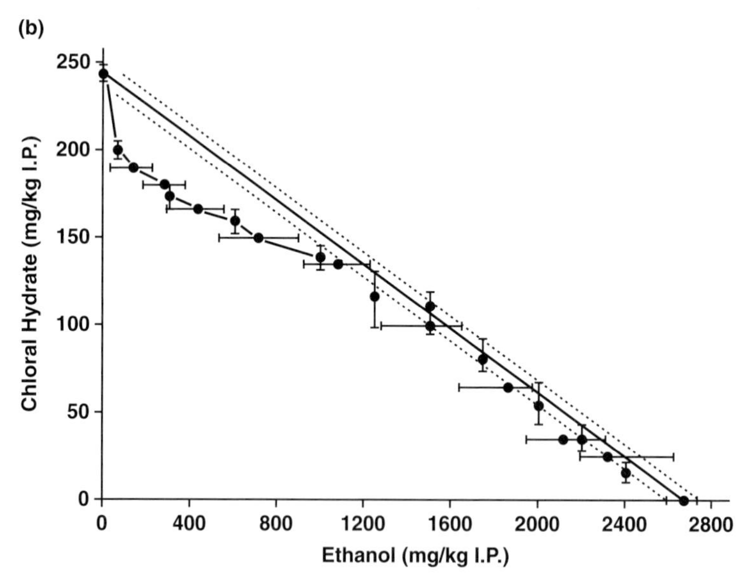

Figure 4.1 Predicted contours of constant response from (a) the interaction threshold model described by Hamm *et al.* [24] and (b) Gessner and Cabana's [25] isobologram in the study of ethanol and chloral hydrate.

thresholds. Although response surface models may be theoretically used in a general case of c compounds, there are practical limitations to the supporting experimental effort for c greater than 2 or 3. An alternative approach to a response surface analysis for detecting and characterizing chemical interactions, which is economically more feasible, is to focus the mixture data on relevant mixing ratios of the chemicals.

4.3
Alternative Approach: Use of Ray Designs with Focus on Relevant Mixing Ratios

A general strategy for testing for interactions among chemicals in a mixture is to use an additivity model to define the "no interaction" case and to use mixture data to describe the so-called "unrestricted" or the general case, as described by the mixture data. Only single-chemical dose-response data are necessary to estimate an additivity model. Selection of mixture points in regions of environmental or biological relevance (defined by fixed-ratio mixtures of chemicals in the mixture) results in economical and practical designs for use in testing for interactions when the number of components in the mixture is large. Examples of the use of additivity models in testing the additivity hypothesis follow.

Gennings *et al.* [26, 50] compared mean responses from an additivity model with that observed at a mixture point of interest. In particular, these authors describe 100 $(1-\alpha)$% prediction intervals at each mixture point of interest using the additivity model. If the observed sample mean from the mixture point is included outside of the prediction interval, then they conclude evidence of departure from additivity. As the number of mixture points increases, multiple comparison corrections (e.g., Bonferroni corrections) become important. Dawson *et al.* [27] compared the dose locations at specified responses under an additivity model to that observed using the interaction index. These authors estimate the interaction index at each mixture point of interest and develop a statistical test of whether the index equals 1. They used Hochberg corrections for multiple testing.

More recently, several authors used a ray design to compare predicted responses from an additivity model with a mixture model along one or more fixed-ratio mixture rays [28–31]. Figure 4.2 depicts a ray design for a combination of two chemicals with two mixture rays. Casey *et al.* [30] developed methodology for testing the hypothesis of additivity in a mixture of c chemicals and for testing whether subsets of the chemicals interact with the remaining chemicals. Let a_i be the proportion of the ith chemical in the fixed mixture ratio, $i = 1, \ldots, c$, where $\sum_{i=1}^{c} a_i = 1$. Gennings *et al.* [28] and others have pointed out that the slope in terms of total dose along the fixed-ratio ray under additivity is given by $\boldsymbol{\theta}_{\text{add}} = \sum_{i=1}^{c} a_i\beta_i$. These authors develop a test of additivity by testing whether the slope for the dose-response curve of the mixture in terms of total dose is equivalent to θ_{add}. Although this inference is limited to the mixing ratio used in the experiment, it results in experimentally feasible studies of mixtures of many chemicals.

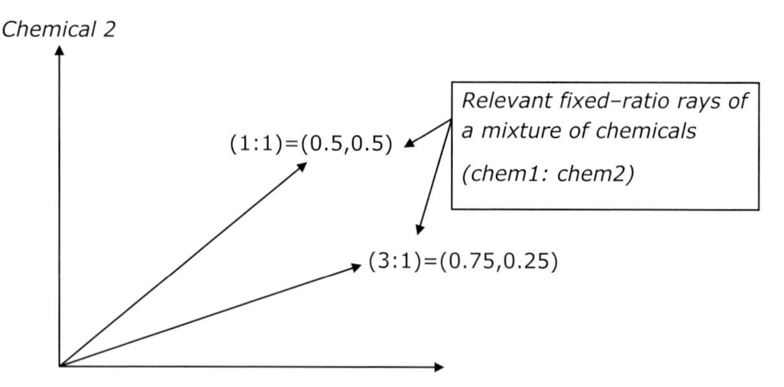

Figure 4.2 Schematic of a ray design depicted by dose-response data for each single chemical alone and mixture data in terms of total dose along two fixed mixing ratio rays: (1 : 1) and (3 : 1).

4.3.1
Example of SCR Method: Analysis of Functional Effects of a Mixture of Five Pesticides Using a Ray Design

Gennings *et al.* [32] demonstrated the SCR and SCNR analysis using data from a mixture of five OP pesticides (acephate, diazinon, chlorpyrifos, malathion (MAL), and dimethoate). The study included dose-response data for each single chemical and at a fixed relative proportion of the pesticides. Their relative proportions in the mixture were based on the relative dietary human exposure estimates of each chemical as projected by the US EPA Dietary Exposure Evaluation Model (DEEM). Dose-response data in terms of total dose for the mixture were generated along two fixed-ratio rays. The "full ray" included all five pesticides at the specified mixing ratio; the "reduced ray" omitted MAL (82.5% of the mixture in the "full ray") and the remaining four chemicals were at the same relative proportions as used in the "full ray." The biological end point was based on the evidence of presence or absence of a gait abnormality in healthy adult male Long–Evans rats following acute oral exposure. An additivity model was parameterized as

$$\log\left(\frac{\pi_{\text{add}}}{1-\pi_{\text{add}}}\right) = \beta_0 + \sum_{i=1}^{5} \beta_i x_i,$$

where π is the probability of gait abnormality, β_0 is an unknown intercept term, β_i are the unknown slope parameters for the pesticides, and x_i is the dose of each pesticide. It is important to note that this model may be estimated with only single-chemical -response data. The additivity model along a specified fixed-ratio ray defined with mixing ratios $[a_1, a_2, \ldots, a_c]$ where $\sum_{i=1}^{c} a_i = 1$ is given by

$$\begin{aligned}\log\left(\frac{\pi_{\text{add}}}{1-\pi_{\text{add}}}\right) &= \beta_0 + \sum_{i=1}^{5} \beta_i a_i t \\ &= \beta_0 + \boldsymbol{\theta}_{\text{add}} t.\end{aligned}$$

For comparison with the additivity model along the fixed-ratio rays, the mixture data were fit to a "mixture model" along each ray in terms of total dose:

$$\log\left(\frac{\pi_{\text{mix}}}{1-\pi_{\text{mix}}}\right) = \beta_0 + \boldsymbol{\theta}_{\text{mix}}t.$$

To achieve adequate fit to the data, higher order terms in total dose are added to the mixture model when necessary. The overall model included a common intercept term, linear terms for each of the single chemicals (excluding MAL), a linear and quadratic term in total dose for the full ray, and a linear term in total dose for the reduced ray (described above), which adequately represented the data ($p = 0.900$, i.e., no indication of lack of fit). Plots of observed and predicted responses along both fixed-ratio mixture rays are provided (Figure 4.3) based on the mixture model and under the assumption of additivity. A test of additivity is a test of coincidence of the dose-response curve under additivity to the mixture model, that is,

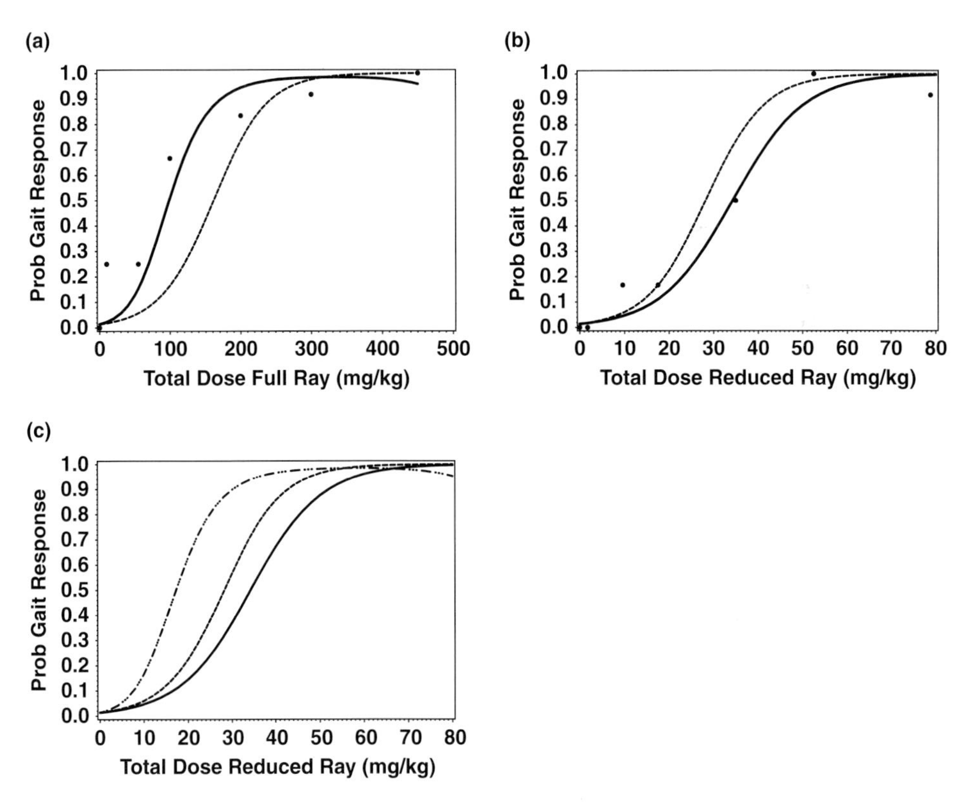

Figure 4.3 Observed (*) and predicted responses based on the *SCR mixture model* (solid line) and assuming additivity (dotted line) as a function of total dose (mg/kg) for the (a) full ray with all five chemicals and (b) for the reduced ray with MAL omitted. Part (c) includes the predicted response under additivity (dotted line) and based on the reduced ray mixture model (solid line) without MAL as given in (b). In addition, part (c) includes an adjusted curve (dashed line) from the full ray mixture data (including MAL).

$$H_0: \quad \text{additivity} \quad \Leftrightarrow \quad H_0: \quad \boldsymbol{\theta}_{\text{add}} = \boldsymbol{\theta}_{\text{mix}}$$

Following the approach in Gennings *et al.* [28], as the curve predicted using the mixture model for the full mixture ray falls significantly above the predicted dose-response curve under the assumption of additivity (Figure 4.3a) ($p = 0.053$), it can be inferred that the overall effect of the specified fixed ratio of the five pesticides is associated with a greater than additive, or synergistic, relationship. Furthermore, there is not a significant difference (Figure 4.3b) in the dose-response curves from the mixture model and that predicted under the assumption of additivity for the mixture (omitting MAL) along the reduced ray ($p = 0.314$). Following the logic of Casey *et al.* [30], this indicates that the interaction (i.e., synergy) in the full ray may be associated with the presence of MAL. Casey *et al.* [30] developed a method for combining the results from multiple rays with the same relative ratios (like the full and reduced rays here) in the same figure. If the adjusted dose-response curve from the full ray is identical to the dose-response curve from the reduced ray, then it indicates the chemical or subset of chemicals that were removed in the reduced ray do not interact with the remaining chemicals. For these data, testing for the difference in the corrected and corresponding parameters from the full and reduced rays [30] indicates that the presence of MAL significantly ($p = 0.014$) changes the shape of the dose-response curve in the full ray. This is elucidated in Figure 4.3c where the dashed curve corresponds to the adjusted dose-response curve from the five-pesticide mixture plotted with the dose-response curve without MAL (solid curve). These dose-response curves are significantly different due to the presence of MAL. The total dose that is associated with a 50% response (ED_{50}) for the four pesticides in the absence of MAL is about 34 mg/kg. In the presence of MAL, the total dose of the other four pesticides associated with a 50% response is roughly 17 mg/kg. This twofold difference indicates the magnitude of the effect of the synergism at a 50% response level. It is apparent from Figure 4.3c that the magnitude of the synergism depends on the level of the effect (i.e., 10%, 50%, etc.).

4.3.2
Example of "Single Chemicals Not Required" Method Using a Mixture of Five Pesticides

A comparison analysis strategy for the SCR method [29, 30, 32] is based on an assumption of a general parameterization of the underlying six-dimensional (five chemicals and one response dimension) response surface. Meadows *et al.* [29] showed that polynomial terms of degree 2 or greater for the model along a fixed-ratio ray are associated with interactions among the chemicals in a mixture. Casey *et al.* [30] generalized this result to the case of multiple mixture rays and termed this approach the "single chemicals not required" method. As the name suggests, the method is applicable to the case where single agent data are not available for estimating the additivity relationship. This information is "replaced" with the assumption of the parametric form of the underlying response surface. However, if single-agent data are available, they can be used to support the estimation of the corresponding parameters in the model.

In comparison to the mixture model used in the SCR analysis, here higher degree terms are allowed and interpreted as being associated with interactions. The interaction model for a mixture of c chemicals is given by

$$g(\mu_{\text{mix}}) = \beta_0 + \sum_{i=1}^{c} \beta_i t^i.$$

For example, for data from the full ray with five pesticides, a polynomial model with up to a fifth-degree term was considered; for the data from the reduced ray with four pesticides in the mixture, a polynomial model with up to a fourth-degree term was considered. The significance of the ith-degree term is interpreted as evidence of an ith-way interaction ($i = 2, \ldots, c$), but it is not suggestive of which i components are interacting. For example, in the mixture analysis of five pesticides (for ease of notation denoted as A, B, C, D, and E), there are five possible four-way interactions: ABCD, ABCE, ABDE, ACDE, and BCDE. The significance of the fourth-degree term in the SCNR model indicates that one or more of these four-way interactions are present. Further experimentation is needed in order to infer which components are interacting [29, 30]. Estimation of model parameters is similar to that described in the previous section.

The initial model fit to the gait score data had a common intercept term across the experiments (five single-chemical and two mixture experiments), linear terms for each single-chemical experiment for estimating the additivity response, a fifth-degree polynomial for the full ray data, and a fourth-degree polynomial for the reduced ray data. All the data were used to estimate the additive part of the model, but only the mixture data were used to estimate the corresponding higher degree terms. From this model, the simultaneous test for the significance of the higher degree terms is equivalent to a test of additivity. For these binary gait score data, there was an indication of departure from additivity ($p = 0.058$, with 7 df). To make the model more parsimonious, Gennings *et al.* [32] used a backward elimination approach to delete terms that were not statistically significant. The model was reduced for the full ray first while keeping the reduced ray fully parameterized. Once the model for the full ray was reduced to having only significant terms, then the model for the reduced ray was simplified. None of the higher order terms for the reduced ray was significant; however, the pure quadratic term was kept for flexibility in the model.

The overall test of additivity was rejected indicating the significance of at least one higher degree term ($p = 0.021$, 4 df). Starting with the highest degree term, since the fourth-degree term was significant for the full ray, it was concluded that at least one four-way interaction exists. Without further information, it is not evident which four chemicals are involved. By comparison, none of the higher degree terms was significant along the reduced ray ($p = 0.098$). Therefore, there was no evidence of departure from additivity among the four pesticides (ACE, DIA, CPF, and DIM). Recall the relative ratios of these pesticides are equivalent to those used in the full ray. Since there was an indication of at least one four-way interaction on the full ray where MAL was present, and no indication of departure from additivity on the reduced ray where MAL was absent, there is evidence to suggest that MAL interacts with at least some or all the other four pesticides. Following the work of Casey *et al.* [30], the hypothesis of no MAL interaction with the other pesticides was rejected ($p = 0.019$, 3 df).

4.3.3
Experimental Designs

One of the primary advantages of using fixed-ratio ray designs is the savings in terms of experimental resources required to test additivity hypotheses. In general, the estimation of the additivity model requires only suitable single-chemical dose-response data. Our experience suggests that four to six dose groups spanning the active region of each of the single-chemical dose-response curves is sufficient to predict the additivity surface. Similarly, if a single fixed-ratio mixture is of interest, then a target of about six total dose groups along the ray spanning the active part of the dose-response curve is suggested. Thus, with c single chemicals and one mixture ray, such a design includes about $6(c + 1)$ dose groups. By contrast, a factorial design with c chemicals and d dose groups per chemical has d^c dose groups. Another advantage of using statistical models is their connection to statistical experimental designs. A vast experimental design literature [17, 33–35] has developed that can be exploited to provide estimates with desirable properties, such as estimates with minimized variance. Meadows *et al.* [29] and Casey *et al.* [31] developed optimal experimental design strategies for testing interaction using fixed-ratio ray designs. That is, these designs specify dose locations and sample size allocations that are associated with desirable statistical properties of the model parameters. The approach taken by Meadows *et al.*, Casey *et al.*, and Coffey *et al.* [36] was to determine the experimental designs associated with minimizing the variance of the test statistic associated with the test of additivity. By reducing its variance, the resulting test statistic has increased power for rejecting the hypothesis of additivity.

4.4
Testing for Additivity in the Low-Dose Region

Based on heuristic arguments [23, 37], it is often assumed that low-dose regions of chemical mixtures, particularly individual chemical subthreshold doses, behave additively. However, the US EPA Risk Assessment Guidance for Superfund [16] acknowledges that "simultaneous subthreshold exposure to several chemicals could result in an adverse health effect," whereby estimates based on single-chemical data may underestimate the overall risk. According to the US EPA [38, 39], "additivity assumptions are expected to yield generally neutral risk estimates (i.e., neither conservative nor lenient)." However, the US EPA argues that if an antagonistic interaction occurs, an assumption of additivity could produce an overestimate of risk. Similarly, synergistic interactions might result in an underestimate.

Stork *et al.* [40] applied statistical equivalence testing methodology [41, 42] to mixtures toxicology to rigorously test for additivity in the sense of Berenbaum [11] at selected mixture groups of interest, particularly in the low-dose region. The equivalence testing framework is useful in this context because it allows one to claim additivity with a false positive rate chosen a priori by the investigator to be some acceptably small value. In addition, claims of additivity as defined by prespecified additivity margins are based on biologically meaningful deviations such that small deviations from additivity that are not considered to be biologically important are not statistically significant. The investigator or risk assessor often has insight into the magnitude of differences that are important, on the basis of their expert judgment. The choice of boundaries may be discipline related. The method is proposed as an alternative to assuming additivity based on heuristic arguments.

In short, the ratio of the mean response of the mth mixture group predicted under additivity to the mean response at the mth mixture group is formed [40] as

$$r_m = \frac{\mu_{m,\text{add}}}{\mu_m}, \quad m = 1, 2, \ldots, M. \tag{4.5}$$

A ratio is chosen for inference to detect a shift of the mth mixture group from that predicted under additivity. When the ratio in (4.5) is equal to 1, the overall effect of the chemicals at the mth mixture group is additive (i.e., zero interaction). When decreasing dose-response relationships are observed, if the ratio (4.5) is greater than 1, the chemicals at the mth mixture group synergistically interact (i.e., greater than additive toxicity), and when the ratio is less than 1, the chemicals at the mth mixture group interact antagonistically (i.e., less than additive toxicity). Without loss of generality, when increasing dose-response curves are considered, the reciprocal of the mean ratio in (4.5) is evaluated and the interpretations are the same.

To illustrate the approach, we again use the mixture of five OP pesticides with neurotoxicity measured in terms of motor activity in healthy adult male Long–Evans rats following acute oral exposure [4]. The study involved single-pesticide exposure of acephate, diazinon, chlorpyrifos, dimethoate, and malathion. Four low-dose mixture groups from this study were evaluated to test for additivity in the low-dose region (Figure 4.4a).

For this example, a log-linear threshold additivity model was fit to the single-chemical data [4]. The parameter associated with the slope for malathion was not significant and therefore was removed from the additivity model. The malathion dose-response data were thus used to estimate the intercept and threshold parameters. The parameters associated with the slopes of the four biologically active chemicals were all negative and significant ($p < 0.001$) indicating that as the dose of each chemical increases, the mean motor activity decreases. The dose thresholds of the four active chemicals were also significantly different from zero [4]. It is interesting to note that the doses of the individual chemicals in the mixture groups (Figure 4.4a) were all below their respective dose threshold estimates ([40], Table 2).

In the absence of specific additivity margins defining plausible biological regions of additivity, for the sake of this example, Stork *et al.* [39] adopted the equivalence margins recommended for clinical trials in the FDA 1992 guidance that define the margins to be (0.80, 1.25), which are symmetric about unity in the ratio scale since $0.80 = 1/1.25$. The mean motor activity responses from each of the four observed mixture groups and those predicted under the assumption of additivity are included

(a)

Mixture group	Dose of pesticide in the mixture (mg/kg)					Mean motor activity (STD)	Predicted mean under additivity
	ACE	DIA	CPF	DIM	MAL		
1	0.40	0.02	0.31	1.02	8.25	200.92 (27.1)	201.76[a]
2	2.20	0.11	1.71	5.61	45.38	167.92 (37.6)	189.97
3	0.40	0.02	0.31	1.02	0.00	189.58 (30.8)	201.76[a]
4	2.20	0.11	1.70	5.60	0.00	186.92 (26.9)	190.07

[a]Mixture group evaluated at a subthreshold total dose as predicted under additivity.

(b)

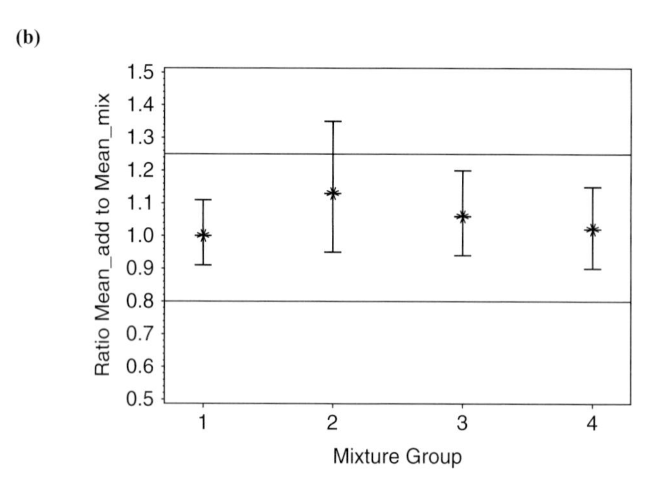

Figure 4.4 (a) Summary statistics for motor activity associated with four low-dose mixture groups ($n = 12$ at each). (b) Point estimates (asterisk) for mean ratios of four mixture groups and familywise 95% Bonferroni confidence intervals, plotted with additivity margins (0.80, 1.25).

in Figure 4.4a. Mixture groups (1) and (3) are subthreshold as predicted by the additivity model, with the same response as background.

The point estimates for the ratio of the mean response predicted under additivity to the observed sample mean response for each mixture along with corresponding Bonferroni adjusted familywise 95% confidence intervals about the true mean ratio are plotted in Figure 4.4b. Additivity can be claimed for mixture groups (1), (3), and (4) since their familywise confidence intervals are completely contained within the preselected additivity interval (0.80, 1.25). There is no sufficient evidence to claim additivity for mixture group (2). It is interesting to note that group (2) has only approximately a 6% mean response change from control as predicted under additivity, yet the observed mixture indicates that the shift from additivity is much larger. The results obtained here by applying the equivalence testing approach depend on the selection of the prespecified additivity margins. In the absence of expert biological judgment, the additivity interval (0.80, 1.25) is chosen here for illustrative purposes only.

4.5 Sufficient Similarity in Dose Responsiveness

An important complicating factor when studying complex chemical mixtures is the dynamic process that changes the mixture (and mixing ratio) over time (e.g., due to seasonal changes) and location (e.g., application of pesticides, location of source water for disinfection). The study of chemical mixtures is challenging because of the large number of combinations that are possible, even with simple mixtures [43]. The US EPA [38, 39] has developed a framework for conducting risk assessments of chemical mixtures, when toxicity data are not available on the mixture of concern, on the basis of observed data from a "sufficiently similar" mixture as a surrogate for the mixture of concern. The EPA guidelines define a similar mixture on the basis of similar chemical composition and similar component proportions, based on expert biological judgment. However, as noted by Seed *et al.* [44] and Stork *et al.* [4], a challenge with the use of this method is associated with determining how similar an observed mixture is to an unobserved mixture of interest.

Since the publication of the original 1986 EPA guidance document, the concept of "sufficiently similar" mixtures has received recent attention in the toxicology literature by many authors [4, 43–48]. Many authors suggest that the use of professional judgment is imperative in conducting health risk assessments by defining sufficiently similar mixtures [39, 43, 45]. Using an equivalence testing framework, chemical mixtures may be determined to be sufficiently similar in dose responsiveness if the variability in the parameter estimates (i.e., as determined by a confidence ellipsoid that accounts for biological variability and variability in the model parameter estimates) is contained within predetermined "similarity bounds" [40, 49]. These authors point out that if dose-response data were available on chemical mixtures at random samples of a dynamic process, a nonlinear mixed-effects model could be used to determine sufficient similarity in dose responsiveness. In most cases, such data are not available.

Stork *et al.* [40] develop an approach to determine sufficient similarity for complex chemical mixtures based on empirical methods and expert biological judgment. Extensions and characterization of this approach is the topic of ongoing funded research (#R01ES015276). For illustration, Crofton *et al.* [50] describe a mixture of 18 different polyhalogenated aromatic hydrocarbons (PHAHs: 2 dioxins, 4 dibenzofurans, and 12 polychlorinated biphenyls) in a mixing ratio based on a rough average of concentrations found in breast milk, fish, and other food sources of human exposure. The end point studied for these prototypic thyroid-disrupting chemicals was serum total thyroxine (T_4) (% of vehicle control). A fit of a threshold model to the dose-response mixture data is provided in Figure 4.5a. Suppose that expert judgment suggests that a 50% shift in the ED_{50} and a fold shift in the ED_{10} is associated with

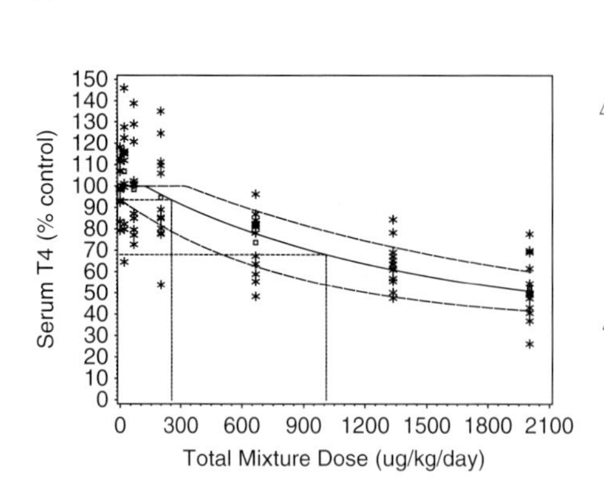

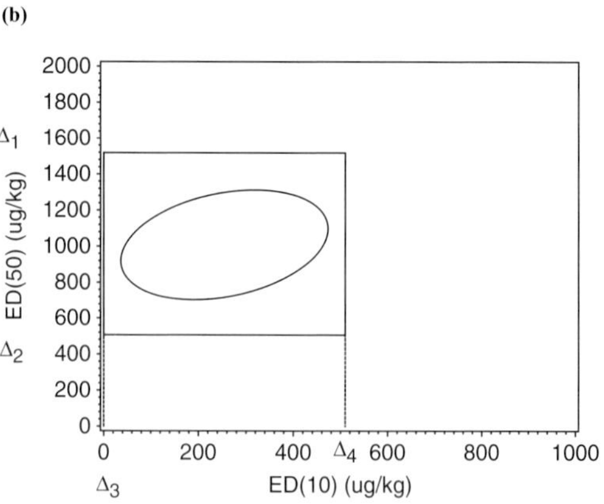

Figure 4.5 (a) Observed mixture responses (asterisk) for a mixture of 18 PHAHs [50], mixture sample means (square), model predicted responses in terms of total dose for the threshold model (solid line). Conditional on the maximum and minimum response parameters, the dose-response curve is completely characterized by the ED_{10} and ED_{50}. Supposed upper and lower mixture boundary curves (dashed lines) defining the region of similarity for sufficiently similar mixtures that are based on the ED_{10} and ED_{50} (elucidated by the dotted lines). (b) Conservative 95% confidence region for the PHAH reference mixture conditional on the maximum and minimum response parameters. Horizontal axis is the ED_{10} (in terms of total dose) of the mixture. Vertical axis is ED_{50} (in terms of total dose) of the mixture. Similarity margins are Δ_1, Δ_2, Δ_3, and Δ_4 (solid lines) that define the region of similarity associated with sufficiently similar mixtures in terms of dose responsiveness.

inappreciable differences. The boundary curves chosen here for illustration of the method are provided in Figure 4.5a as dashed curves. The boundary curves define a region of similarity where the mean dose-response relationship of the mixture described by the ED_{10} and ED_{50} of candidate mixtures is sufficiently similar to that of the reference mixture. A graphical representation of the confidence region on the ED_{10} and ED_{50} and the similarity margins is provided in Figure 4.5b. When D parametric functions are considered, the similarity margins will form a D-dimensional hyper-rectangle. The rectangle does not necessarily need to be symmetric about the confidence region. However, the rectangle defining the region of similarity is assumed to be defined such that it completely contains the confidence region from the dose-response data from the reference mixture of interest. Thus, variability associated with the reference mixture is accounted for when determining sufficiently similar candidate mixtures.

In reality, complex mixtures are often the result of a dynamic process that change temporally or due to changes in the process/location. We can think of these dose-response curves as resulting from random exposure levels (or random coefficients) that should be accounted for in the volume of the confidence region. It is of interest to determine how much additional variability due to the random nature of the process may be associated with sufficiently similar mixtures in dose responsiveness. The maximum tolerable shifts of the ED_{50} and the ED_{10} that define the boundary curves

(Figure 4.5a) and equivalently the similarity margins (Figure 4.5b) are considered to be associated with added variability due to the random dilution/exposure effect, *h*, for candidate mixtures from a mixed-effects model. It is of interest to determine how much additional variability due to the random dilution effect, *h*, is tolerable such that the volume of the confidence region increases and comes "close" to the similarity margins but remains completely contained within the rectangle defining the region of similarity. If the additional variability associated with the random dilution/exposure effect is such that the confidence region for a candidate mixture is contained within the similarity margins, then the candidate mixture will be defined as sufficiently similar in the mean dose-response relationship to that of the reference mixture. More details are described in Stork *et al.* [40].

4.6 Summary

One of the overall objectives of this chapter is to demonstrate the variety of statistical designs and analysis methods that may be useful in empirical analysis of chemical mixtures. The chapter begins with a general description of classical methods for analyzing combinations of chemicals. These methods are based on a basic form of an additivity model that is linked to the classical isobologram through the interaction index. But, alternative approaches are warranted as the number of chemicals in the mixture increases. Examples are provided for the SCR and SCNR methods. These approaches are framed to allow the data to describe departure from additivity, so that the null hypothesis of additivity is rejected with evidence of departure from additivity (interaction). When the null and alternative hypotheses are switched, the data can be used to provide evidence for additivity while controlling the type I error. This approach is described while focusing on mixtures in a low-dose region. Finally, recent work on an empirical assessment of sufficient similarity in dose responsiveness of chemical mixtures is provided.

References

1 McCullagh, P. and Nelder, J.A. (1989) *Generalized Linear Models*, 2nd edn, Chapman & Hall, London.

2 Agresti, A. (2002) *Categorical Data Analysis*, 2nd edn, John Wiley & Sons, Inc., New York.

3 McCullagh, P. (1983) Quasi-likelihood functions. *Ann. Stat.*, **11**, 59–67.

4 Stork, L.G., Gennings, C., Carchman, R., Carter, W.H., Jr., Pounds, J., and Mumtaz, M. (2006) Testing for additivity at select mixture groups of interest based on statistical equivalence testing methods. *Risk Anal.*, **26**, 1601–1612.

5 U.S. EPA (United States Environmental Protection Agency) (2000) Benchmark Dose Technical Guidance Document. EPA/630/R-00/001. Available at www.epa.gov/ncea/pdfs/bmds/BMD-External_10_13_2000.pdf.

6 Cornell, J.A. (1990) *Experiments with Mixtures: Designs, Models and Analysis of Mixture Data*, 2nd edn, John Wiley & Sons, Inc., New York.

7 Teuschler, L., Klaunig, J., Carney, E., Chambers, J., Connolly, R., Gennings, C., Giesy, J., Hertzberg, R., Klaassen, C., Kodell, R., Paustenbach, D.,and Yang, R.

(2002) Support of science-based decisions concerning the evaluation of the toxicology of mixtures: a new beginning. *Regul. Toxicol. Pharm.*, **36**, 34–39.

8 Gennings, C., Carter, W.H., Jr., Carchman, R.A., Teuschler, L.K., Simmons, J.E., and Carney, E.W. (2005) A unifying concept for assessing toxicological interactions: changes in slope. *Toxicol. Sci.*, **88**, 287–297.

9 Goodman, L.S. and Gilman, A.G. (2001) *Goodman and Gilman's The Pharmacological Basis of Therapeutics*, 10th edn (eds A.G. Gilman, L.S. Goodman, T.W. Rall, and F. Murad), MacMillan Publishing Co., New York.

10 Ariëns, E.J. and Beld, A.J. (1977) The receptor concept in evolution. *Biochem. Pharm.*, **26**, 913–918.

11 Berenbaum, M.C. (1985) The expected effect of a combination of agents: the general solution. *J. Theor. Biol.*, **114**, 413–431.

12 Loewe, S. and Muischnek, H. (1926) Uber kombinationswirkunger. I. Mitteilung: Hiltsmittel der gragstellung. *Naunyn-Schmiedebergs Arch. Pharmacol.*, **114**, 313–326.

13 Loewe, S. (1953) The problem of synergism and antagonism of combined drugs. *Arzneimittle Forshung*, **3**, 285–290.

14 Berenbaum, M.C. (1989) What is synergy? *Pharmacol. Rev.*, **41**, 93–141.

15 Safe, S.H. (1998) Hazard and risk assessment of chemical mixtures using the toxic equivalency factor approach. *Environ. Health Perspect.*, **106** (Suppl. 4), 1051–1058.

16 U.S. EPA (United States Environmental Protection Agency) (1989) Interim Procedures for Estimating Risks Associated with Exposures to Mixtures of Chlorinated Dibenzo-*p*-Dioxins and -Dibenzofurans (CDDs and CDFs) and 1989 Update. Risk Assessment Forum. EPA/625/3-89/016.

17 Atkinson, A.C. and Donev, A.N. (1996) *Optimum Experimental Designs*, Oxford Science Publications, Clarendon Press, Oxford.

18 Box, G.E.P. (1979) Robustness in the strategy of scientific model building. In R.L. Launer and G.N. Wilhinson (eds.) Robustness in statistics. New York: Academic Press.

19 Carter, W.H., Jr., Gennings, C., Staniswalis, J.G., Campbell, E.D., and White, K.L., Jr. (1988) A statistical approach to the construction and analysis of isobolograms. *J. Am. Coll. Toxicol.*, **7**, 963–973.

20 Moser, V.C., MacPhail, R.C., and Gennings, C. (2003) Neurobehavioral evaluations of mixtures of trichloroethylene, heptachlor, and di(2-ethylhexyl)phthalate in a full-factorial design. *Toxicology*, **188**, 125–137.

21 Agresti, A. (1990) Categorical data analysis. New York: Wiley.

22 Groten, J.P., Schoen, E.D., van Bladerenm, P.J., Kuper, C.F., van Zorge, J.A., and Feron, V.J. (1997) Acute toxicity of a mixture of nine chemicals in rats: detecting interactive effects with a fractionated two-level factorial design. *Fundam. Appl. Toxicol.*, **36**, 15–29.

23 Carpy, S.A., Kobel, W., and Doe, J. (2000) Health risk of low-dose pesticides mixtures: a review of the 1985–1998 literature on combination toxicology and health risk assessment. *J. Toxicol. Environ. Health B*, **3**, 1–25.

24 Hamm, A.K., Carter, W.H., Jr., and Gennings, C. (2005) Analysis of an interaction threshold in a mixture of drugs and/or chemicals. *Stat. Med.*, **24**, 2493–2507.

25 Gessner, P.K. and Cabana, B.E. (1970) A study of the interaction of the hypnotic effects and of the toxic effects of chloral hydrate and ethanol. *J. Pharmacol. Exp. Ther.*, **174**, 247–259.

26 (a) Gennings, C., Schwartz, P., Carter, W.H., Jr., and Simmons, J.E. (1997) Detection of departures from additivity in mixtures of many chemical with a threshold model. *J. Agric. Biol. Environ. Stat.*, **2**, 198–211.
(b) Gennings, C., Schwartz, P., Carter, W.H., Jr., and Simmons, J.E. (2000) Erratum: Detection of departures from additivity in mixtures of many chemical with a threshold model. *J. Agric. Biol. Environ. Stat.*, **5**, 257–259.

27 Dawson, K.S., Carter, W.H., Jr., and Gennings, C. (2000) A Statistical test for

detecting and characterizing departures from additivity in drug/chemical combinations. *J. Agric. Biol. Environ. Stat.*, **5**, 342–359.

28 Gennings, C., Carter, W.H., Jr., Campain, J.A., Bae, D., and Yang, R.S.H. (2002) Statistical analysis of interactive cytotoxicity in human epidermal keratinocytes following exposure to a mixture of four metals. *J. Agric. Biol. Environ. Stat.*, **7**, 58–73.

29 Meadows, S.L., Gennings, C., Carter, W.H., Jr.,and Bae, D.-S. (2002) Experimental designs for mixtures of chemicals along fixed ratio rays. *Environ. Health Perspect.*, **110** (Suppl. 6), 979–983.

30 Casey, M., Gennings, C., Carter, W.H., Jr., Moser, V.C., and Simmons, J.E. (2004) Detecting interaction(s) and assessing the impact of component subsets in a chemical mixture using fixed-ratio mixture ray designs. *J. Agric. Biol. Environ. Stat.*, **9**, 339–361.

31 Casey, M., Gennings, C., Carter, W.H., Jr., Moser, V.C., and Simmons, J.E. (2005) D_S-optimal designs for studying combinations of chemicals using multiple fixed-ratio ray experiments. *Environmetrics*, **16**, 129–147.

32 Gennings, C., Carter, W.H., Jr., Casey, M., Moser, V., Carchman, R., and Simmons, J.E. (2004) Analysis of functional effects of a mixture of five pesticides using a ray design. *Environ. Toxicol. Pharmacol.*, **18**, 115–125.

33 Abdelbasit, K.M. and Plackett, R.L. (1983) Experimental design for binary data. *J. Am. Stat. Assoc.*, **78**, 90–98.

34 Minkin, S. (1987) Optimal designs for binary data. *J. Am. Stat. Assoc.*, **82**, 1098–1103.

35 Kalish, L.A. (1990) Efficient design for estimation of median lethal dose and quantal dose-response curves. *Biometrics*, **46**, 737–748.

36 Coffey, T., Gennings, C., Simmons, J.E., and Herr, D.W. (2005) D-optimal experimental designs to test for departure from additivity in a fixed-ratio mixture ray. *Toxicol. Sci.*, **88**, 467–476.

37 Monosson, E. (2005) Chemical mixtures: considering the evolution of toxicology and chemical assessment. *Environ. Health Perspect.*, **113**, 383–390.

38 U.S. EPA (United States Environmental Protection Agency) (1986) Guidelines for the Health Risk Assessment of Chemical Mixtures. EPA/630/R-98/002. Available at http://www.epa.gov/NCEA/raf/pdfs/chem_mix/chemmix_1986.pdf.

39 U.S. EPA (United States Environmental Protection Agency) (2000) Supplementary Guidance for Conducting Health Risk Assessment of Chemical Mixtures. EPA/630/R-00/002. Available at http://www.epa.gov/NCEA/raf/chem_mix.htm.

40 Stork, L.G., Gennings, C., Carter, W.H., Jr., Teuschler, L., and Carney, E.W. (2008) Empirical evaluation of sufficient similarity in dose-response for environmental risk assessment of chemical mixtures. *J. Agric. Biol. Environ. Stat.*, **13** (3), 313–333.

41 Blackwelder, W.C. (1982) "Proving the null hypothesis" in clinical trials. *Control. Clin. Trials*, **3**, 345–353.

42 Berger, R.L. and Hsu, J.C. (1996) Bioequivalence trials, intersection–union tests and equivalence confidence sets. *Stat. Sci.*, **11**, 283–319.

43 Nair, R.S., Dudek, B.R., Grothe, D.R., Johannsen, F.R., Lamb, I.C., Martens, M.A., Sherman, J.H., and Stevens, M.W. (1996) Mixture risk assessment: a case study of Monsanto experiences. *Food. Chem. Toxicol.*, **34**, 1139–1145.

44 Seed, J., Brown, R.P., Olin, S.S., and Foran, J.A. (1995) Chemical mixtures: current risk assessment methodologies and future directions. *Regul. Toxicol. Pharmacol.*, **22**, 76–94.

45 Mumtaz, M.M. (1995) Risk assessment of chemical mixtures from a public health perspective. *Toxicol. Lett.*, **82/83**, 527–532.

46 Suk, W.A., Olden, K., and Yang, R.S.H. (2002) Chemical mixtures research: significance and future perspectives. *Environ. Health Perspect.*, **110** (Suppl. 6), 891–892.

47 Cizmas, L., McDonald, T.J., Phillips, T.D., Gillespie, A.M., Lingenfelter, R.A., Kubena, L.F., Phillips, T.D., and Donnely, K.C. (2004) Toxicity characterization of complex mixtures using biological and chemical analysis in preparation for assessment of mixture similarity. *Environ. Sci. Technol.*, **38**, 5127–5133.

48 Teuschler, L.K., Rice, G.E., Feder, P., Bull, R., and Simmons, J.E. (2004) Toxicological and statistical criteria for defining sufficient similarity of complex chemical mixtures. Society for Risk Analysis Annual Meeting, W3.4 (abstract).

49 Stork, L.G. (2005) A statistical equivalence testing approach to testing for additivity and determining sufficiently similar mixtures of many chemicals based on dose response. PhD dissertation, Department of Biostatistics, Virginia Commonwealth University, Richmond, VA.

50 Crofton, K.M., Craft, E.S., Hedge, J.M., Gennings, C., Simmons, J.E., Carchman, R.A., Carter, W.H., Jr., and DeVito, M.J. (2005) Thyroid-hormone-disrupting chemicals: evidence for dose-dependent additivity or synergism. *Environ. Health Perspect.*, **113** (11), 1549–1554.

5
Modeling Kinetic Interactions of Chemical Mixtures

Jerry L. Campbell Jr., Kannan Krishnan, Harvey J. Clewell III, and Melvin E. Andersen

5.1
Pharmacokinetic Modeling

Pharmacokinetics is the study of the time course for the absorption, distribution, metabolism, and excretion of a chemical substance in a biological system [1, 2]. In pharmacokinetic modeling, established descriptions of chemical transport and metabolism are employed to simulate observed kinetics *in silico* [3]. Implicit in any application of pharmacokinetics to toxicology or risk assessment is the assumption that the toxic effects on a particular tissue can be related in some way to the concentration time course of an active form of the substance in that tissue. Moreover, in the absence of pharmacodynamic differences between animal species, it is expected that similar responses will be produced at equivalent tissue exposures regardless of species, exposure route, or experimental regimen [4, 5]. Of course the actual nature of the relationship between tissue exposure and response, particularly across species, may be quite complex.

Classic compartmental modeling is largely an empirical exercise where data on the time course of the chemical of interest in blood (and perhaps other tissues) are collected. On the basis of the behavior of the data, a mathematical model is selected that possesses a sufficient number of compartments (and therefore parameters) to describe the data. The compartments do not generally correspond to identifiable physiological entities but rather are abstract concepts with meaning only in terms of a particular calculation. The advantage of this modeling approach is that there is no limitation to fitting the model to the experimental data. If a particular model is unable to describe the behavior of a particular data set, additional compartments can be added until a successful fit is obtained. Since the model parameters do not possess any intrinsic meaning, they can be freely varied to obtain the best possible fit, and different parameter values can be used for each data set in a related series of experiments. Once developed, these models are useful for interpolation and limited extrapolation of the concentration profiles that can be expected as experimental conditions are varied. They are also useful for statistical evaluation of a chemical's apparent kinetic complexity [6]. However, since the compartmental model does not

Principles and Practice of Mixtures Toxicology. Edited by Moiz Mumtaz

ISBN: 978-3-527-31992-3

possess a physiological structure, it is often not possible to incorporate a description of these nonlinear biochemical processes in a biologically appropriate context. For example, without a physiological structure it is not possible to correctly describe the interaction between blood transport of the chemical to the metabolizing organ and the intrinsic clearance of the chemical by the organ.

Physiologically based pharmacokinetic (PBPK) models differ from the conventional compartmental pharmacokinetic models in that they are based to a large extent on the actual physiology of the organism [7]. Figure 5.1 illustrates the structure of a simple PBPK model for the volatile lipophilic compound, styrene. The model equations represented by the diagram are described in the original publication [8]. Instead of compartments defined solely by mathematical analysis of the experimental

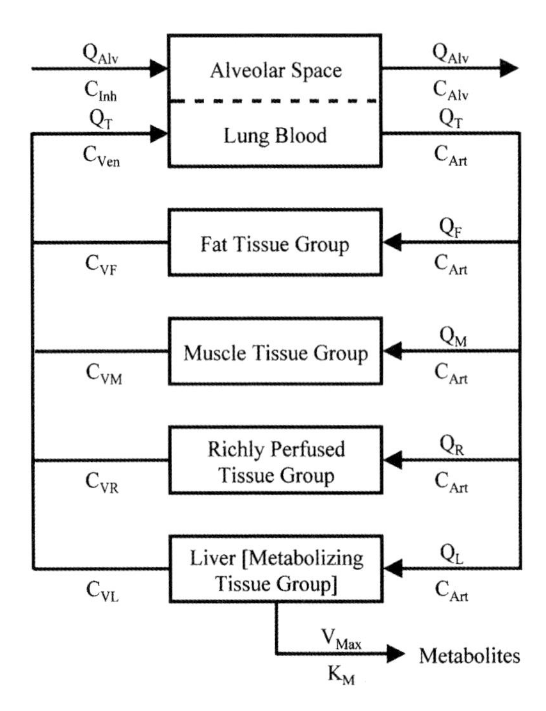

Figure 5.1 Diagram of a physiologically based pharmacokinetic model for styrene. In this description, groups of tissues are defined with respect to their volumes, blood flows (Q), and partition coefficients for the chemical. The uptake of vapor is determined by the alveolar ventilation (Q_{alv}), cardiac output (Q_t), blood:air partition coefficient (P_b), and the concentration gradient between arterial and venous pulmonary blood (C_{art} and C_{ven}). The dashed line reflects the fact that the lung compartment is described by a steady-state equation assuming that diffusion between the alveolar air and the lung blood is fast compared to ventilation and perfusion. Metabolism is described in the liver with a saturable pathway defined by a maximum velocity (V_{max}) and affinity (K_m). The mathematical description assumes equilibration between arterial blood and alveolar air as well as between each of the tissues and the venous blood exiting from that tissue. (Adapted from Ref. [8].)

kinetic data, compartments in a PBPK model are based on realistic organ and tissue groups, with weights and blood flows obtained from experimental data. Moreover, instead of compartmental rate constants determined solely by fitting data, actual physicochemical and biochemical properties of the compound, which can be experimentally measured or estimated by quantitative structure–property relationships, are used to define parameters in the model. To the extent that the structure of the model reflects the important determinants of the kinetics of the chemical, the result of this approach is a model that can predict the qualitative and quantitative behavior of an experimental time course without having been based directly on it. In recent years, there has been an enormous expansion of uses of PBPK modeling in areas related to environmental chemicals and drugs [9].

5.2 PBPK Modeling of Individual Chemicals

5.2.1 Theoretical Considerations

PBPK modeling involves the development of mathematical descriptions of the processes determining the uptake and disposition of chemicals as a function of their physicochemical (e.g., tissue:blood partition coefficients), biochemical (e.g., rate constants for metabolism and binding), and physiological (e.g., tissue volume, blood flow rates, and breathing raises) characteristics. Typically, a mammalian PBPK model consists of a series of tissue compartments, each receiving the chemical via the arterial blood and loses the free chemical via the venous blood leaving the tissue (Figure 5.1). These biologically relevant compartments, arranged in an anatomically accurate manner, are defined with appropriate physiological characteristics. The compartments may represent a single tissue or a group of tissues with similar blood flow and solubility characteristics (e.g., adrenals, kidney, thyroid, brain, heart, and other organs are pooled into one compartment, referred to as "richly perfused tissues"). The processes of absorption, distribution, metabolism, and excretion of chemicals are described with a series of mass balance differential equations. A brief account of some of the descriptions frequently employed in PBPK modeling follows.

5.2.1.1 Absorption

The concentration of volatile organic chemicals in arterial blood following pulmonary equilibration has been calculated by accounting for the breathing rate, cardiac output, blood:air partition coefficient, and concentration of the chemical in the inhaled air and in the mixed venous blood [8]:

$$C_a = \frac{Q_{alv} C_{inh} + Q_c C_v}{\left(Q_C + \dfrac{Q_{alv}}{P_b} \right)}. \tag{5.1}$$

This algebraic expression is derived from the mass conservation equation (5.2) for the lung, which specifies that the loss of chemical in the air is balanced by an identical gain of the chemical in the pulmonary blood:

$$Q_{\mathrm{alv}}(C_{\mathrm{inh}}-C_{\mathrm{alv}}) = Q_{\mathrm{c}}(C_{\mathrm{a}}-C_{\mathrm{v}}). \tag{5.2}$$

Since the lung equilibrates vapor between air and blood, $C_{\mathrm{alv}} = C_{\mathrm{a}}/P_{\mathrm{b}}$. This relationship assumes rapid equilibrium of the chemical across the alveolar walls, there is no significant pulmonary metabolism, and there is negligible storage capacity in lungs.

The dermal absorption of vapors has been described by introducing a diffusion-limited compartment to represent skin as a portal of entry [10]. Accordingly, the rate of change in the amount of the chemical in the skin compartment during exposures is given by

$$V_{\mathrm{s}}\frac{dC_{\mathrm{s}}}{dt} = K_{\mathrm{p}}A\left[C_{\mathrm{air}}-\left(\frac{C_{\mathrm{s}}}{P_{\mathrm{s/a}}}\right)\right]+Q_{\mathrm{s}}\left[C_{\mathrm{a}}-\left(\frac{C_{\mathrm{s}}}{P_{\mathrm{s}}}\right)\right]. \tag{5.3}$$

The oral uptake of chemicals has been described by introducing a first-order absorption rate constant [8, 11]:

$$\frac{dA_{\mathrm{a}}}{dt} = K_{\mathrm{a}}D_0\,e^{-K_{\mathrm{a}}t}. \tag{5.4}$$

In this case, the compound is assumed to appear in the liver after absorption. If gastrointestinal absorption occurs at a constant rate, it is described as a zero-order process.

The entry of a chemical via the intravenous route is described by including a term to represent the rate of infusion into the mixed venous blood [8]:

$$C_{\mathrm{v}} = \frac{\left[K^0+\left(\sum_t^n Q_t C_{\mathrm{vt}}\right)\right]}{Q_{\mathrm{c}}}. \tag{5.5}$$

The infusion rate can be adjusted for very long or very short periods by adjusting the infusion time parameter, T, where

$$K^0 = \begin{bmatrix} k^0; 0 < t \leq T \\ 0; t > T \end{bmatrix}. \tag{5.6}$$

5.2.1.2 **Distribution**

Reversible binding of a chemical to plasma albumin can be described by accounting for the binding capacity, the binding affinity constant, and the concentration of the free chemical in blood. Thus, the relationship between bound and free concentrations of a chemical in blood is represented by a Langmuir isotherm depicting the interaction with a single class of noninterfering binding sites [12]:

$$C_{\mathrm{b}} = \frac{nBK_{\mathrm{b}}C_{\mathrm{f}}}{(1+K_{\mathrm{b}}C_{\mathrm{f}})}. \tag{5.7}$$

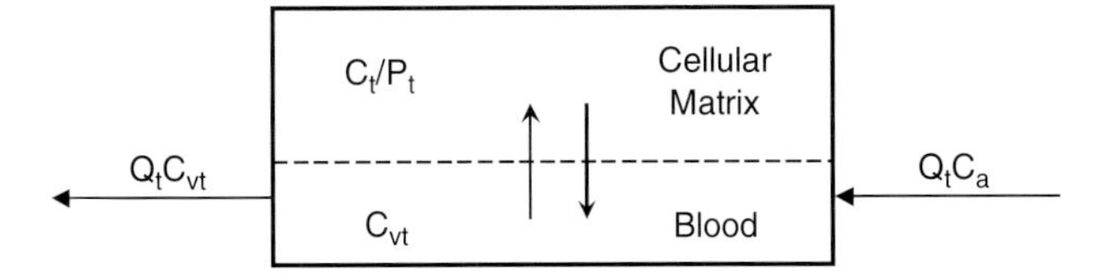

Figure 5.2 Schematic of a tissue compartment. Q_t is tissue blood flow rate, C_a is the arterial blood concentration of the chemical, C_{vt} is the concentration of the chemical in the venous blood leaving the tissue, C_t is the tissue concentration, and P_t is the tissue:blood partition coefficient of the chemical.

Covalent binding to various amino acid residues of hemoglobin and tissue DNA has been described as a second-order process with respect to the concentration of hemoglobin/DNA and the reactive form of the chemical. Thus, the rate of change in the amount of a chemical bound to hemoglobin at a specific site (e.g., N-terminal valine, cysteine, histidine) is represented as [13]

$$\frac{dC_{bh}}{dt} = K_h C_h C_a V_a + K_h C_h C_v V_v. \tag{5.8}$$

In chronic exposure scenarios, rate constant(s) for adduct repair/removal and red blood cell turnover rate should be taken into consideration to accurately simulate the changes in adduct levels [14].

The tissue distribution of chemicals is described as either a perfusion-limited or a diffusion-limited process. If diffusion is rate limiting, the tissue subcompartments (i.e., tissue blood and cellular matrix; Figure 5.2) are described separately with a set of mass balance differential equations of the following form:

$$\text{for cellular matrix:} \quad \frac{dA_{cm}}{dt} = PA_t \left[C_{vt} - \left(\frac{C_t}{P_t} \right) \right] \tag{5.9}$$

$$\text{and for tissue blood:} \quad \frac{dA_{tb}}{dt} = Q_t (C_a - C_{vt}) - PA_t \left[\left(\frac{C_t}{P_t} \right) - C_{vt} \right]. \tag{5.10}$$

If the diffusion from blood to tissue is slow with respect to total tissue blood flow, both equations are necessary. On the other hand, if blood flow is slow with respect to diffusion, the tissues are described as homogeneous well-mixed compartments. In this case, the rate of change in the amount of chemical in the tissue is described with a single equation for the whole tissue mass (cellular matrix plus tissue blood):

$$\frac{dA_t}{dt} = \frac{dA_{cm}}{dt} + \frac{dA_{tb}}{dt} = Q_t (C_a - C_{vt}). \tag{5.11}$$

Time integration of this mass balance differential equation provides the total amount of the chemical in the tissue.

5.2.1.3 Metabolism

Metabolism in individual tissues can be described by adding one or more terms to account for the rate of loss of the parent chemical due to particular biotransformation pathways. These equations may be first order (5.12), second order (5.13), or saturable (5.14):

$$\frac{dA_{met}}{dt} = K_f C_{vt} V_t, \tag{5.12}$$

$$= K_s C_{vt} C_{cf} V_t, \tag{5.13}$$

$$= \frac{V_{max} C_{vt}}{(K_m + C_{vt})}. \tag{5.14}$$

Thus, in a perfusion-limited description, the rate of change in the amount of the chemical in metabolizing organs (e.g., liver) is described as

$$\frac{dA_t}{dt} = Q_t(C_a - C_{vt}) - \frac{dA_{met}}{dt}. \tag{5.15}$$

5.2.1.4 Excretion

Renal clearance of unbound chemicals is modeled by accounting for the excretory process(es) of interest. For example, glomerular filtration of unbound chemical that is neither secreted nor reabsorbed can be described as [15]

$$\frac{dA_{rc}}{dt} = C_p \mathrm{GFR}. \tag{5.16}$$

Reabsorption can be modeled by assuming that a fixed portion of a chemical, which is either filtered and/or secreted, is reabsorbed [16] or by accommodating the known dependence of reabsorption on urine flow [17–19].

5.2.2 Experimental Approach

The parameters included in Eqs. (4.1–4.16) for PBPK modeling of individual chemicals can be grouped into three types: (i) physiological, (ii) physicochemical, and (iii) biochemical. The physiological parameters, such as alveolar ventilation rate, cardiac output, and tissue blood flow rates, can be obtained from biomedical literature [20–22]. A compilation of this information from a variety of sources is also available [23].

The physicochemical parameters, such as partition coefficients representing the ratio of the concentration of a chemical between two media (blood and air, blood and tissue) at equilibrium, can be estimated *in vivo* [24, 25] or *in vitro* by vial equilibration [26, 27], ultrafiltration [11], or equilibrium dialysis [28].

The biochemical parameters, such as the rates of absorption, biotransformation, and macromolecular binding, can be determined *in vivo* by conducting time course analyses or can be estimated from *in vitro* studies. One strategy for accurate estimation of specific biochemical parameters *in vivo* is conducting experiments under conditions in which pharmacokinetic behavior of a chemical is related to one or two dominant factors and, thereby, deriving estimates of these parameters. For example, the dermal absorption rate of volatile chemicals has been determined by conducting body-only exposures of animals by providing them with a latex face mask that prevents inhalation of the vapor [29]. The skin permeability constant required to simulate dermal absorption during exposures is obtained by fitting the set of blood concentration–time course curves to a PBPK model that has all parameters except K_p defined [10].

The rate constant for gastrointestinal absorption of volatile organic chemicals has been estimated by analyzing the time course of exhaled concentration of volatile metabolite or the parent chemical with a PBPK model [30]. The rate constant of gastrointestinal absorption for hydrophilic chemicals ($P_b > 90$) may have to be estimated by determining a time course of their blood concentration followed by data analysis using a PBPK model that has all parameters except K_a defined.

Pulmonary absorption can be estimated in nose-only exposure experiments, in which chemical concentration at the inlet and outlet ports of the exposure tower containing a known number of animals is measured [31]. Pulmonary absorption can also be determined from gas uptake exposures if the data analysis is performed with a PBPK model [32].

For volatile organic chemicals, gas uptake studies have been employed to determine metabolic rate constants [32]. Gas uptake experiments involve exposure to a volatile organic chemical of a group of rodents placed in a closed chamber with recirculated atmosphere. The temporal decline in the chamber concentration of the chemical in a certain number of animals is a function of the rate of primary metabolic pathway(s), represented by K_m, V_{max}, and/or K_f, once the chemical has equilibrated within the body. This exposure system is less suitable for some hydrophilic chemicals with a prolonged equilibration phase that renders the rapid decline in chamber concentration of chemicals insensitive to their metabolic rates. In such cases, the metabolic rate constants have been assessed using an exhaled breath chamber. In this procedure, an animal previously administered a chemical is placed in a chamber and the temporal change in the concentration of the chamber effluent is monitored [33]. The data analysis is performed using a PBPK model in which the temporal change in the effluent concentration of the chemical is quantitatively related to physiological parameters, chemical solubility, and rates of metabolism [30].

The rate constants for intermediary metabolic pathways and metabolite kinetics cannot be determined using data from gas uptake studies with parent chemicals alone. In such cases, additional experiments are required to determine the rate constants for clearance of each of the metabolites. These PBPK models for metabolites then can be interconnected to provide an integrated model for the simulation of the transformation of parent chemical to a secondary metabolite [34]. Frequently, because of the polarity of the metabolites, gas uptake or exhaled breath systems cannot be used to determine the rates of their metabolism. Accordingly, the use of *in vitro* systems must be considered.

Hepatocytes, microsomes, and liver slices are all potentially useful for estimating metabolic rate constants *in vitro*. The relevance of rate constants determined *in vitro* to the intact animal is not clear in all cases [35]. However, several studies using microsomal and cytosolic preparations, or hepatocytes, to determine metabolic rate constants for direct incorporation into PBPK models have described and predicted the kinetics of volatile and nonvolatile chemicals successfully [28, 36–44]. The reaction rates of chemicals with tissue proteins, DNA, and hemoglobin can also be estimated *in vitro* or can be estimated from *in vivo* dose-response studies and then used in the respective physiological models of chemical disposition [35].

PBPK models, once formulated by combining information on animal physiology, partition coefficients, and biochemical rate constants, can be used to simulate the disposition kinetics of a chemical in the test animal species. Simulation requires simultaneously solving of a set of mass balance differential equations by numerical integration. Using these pharmacokinetic simulation models, instantaneous and integrated amounts of a chemical in specific tissue compartments can be calculated with a computer and readily accessible simulation software.

Simulations of any variable in the PBPK model (e.g., percentage dose exhaled, amount of metabolites produced, level of hepatic and extrahepatic glutathione depletion, and tissue and blood concentration of parent chemical and its metabolite) can be generated for a variety of exposure scenarios. The a priori model predictions of various end points for exposure conditions different from the ones used to derive the model parameters are compared with experimental data to validate the PBPK model. The failure of model validation with experimental data implies an incomplete understanding of the mechanistic determinants of the process by the investigator. In such cases, additional model-directed experiments have provided insight into the possible mechanisms involved [45].

5.2.3
Examples

To date, a number of PBPK models describing the disposition kinetics of a variety of chemicals in one or more animal species have been developed. These chemicals include common air pollutants (ozone, nicotine), metals (lead, mercury, nickel, and zinc), solvents (acetone, benzene, 1,3-butadiene, 2-butoxyethanol, carbon tetrachloride, chloroform, dihalomethanes, 1,4-dioxane, ethylene dichloride, methanol, methyl chloroform, methylene chloride, toluene, trichloroethylene (TCE), tetrachloroethylene, vinylidine chloride, and *m*-xylene), industrial intermediates (ethyl acrylate, ethylene oxide), pesticides (dieldrin, hexachlorobenzene, and parathion), and other pollutants of concern (dioxins, furans, and polychlorinated biphenyls) [46]. The validated PBPK models can be used to estimate target tissue exposure to the toxic moiety of a chemical. The mechanistic basis of these models allows more accurate extrapolation of tissue dosimetry from high to low dose, between different routes of exposure, and among various species. This ability is particularly important since human health risk is often estimated on the basis of the response seen in animal toxicity studies, in which high doses of chemicals are administered by routes that may differ from those anticipated for human exposures. Using PBPK models, the response seen in animal studies can be related to the tissue dose of the toxic moiety instead of the environmental levels of the parent chemical, thus reducing the uncertainties associated with the extrapolation of tissue dosimetry in conducting quantitative risk assessment [47].

5.3 PBPK Modeling of Binary Chemical Mixtures

5.3.1 Theoretical Considerations

When biota are exposed to two chemicals, simultaneously or sequentially, an interaction of toxicological significance may occur. Toxic interactions refer to significant deviations of the combined effect of two chemicals from the simple summation of their individual effects. Interactions result from the modulation of the toxicokinetics and/or toxicodynamics of one chemical by another. Toxicokinetic interactions involve the modulation of the absorption, distribution, metabolism, and/or excretion of one chemical by another, which alters the physiological and/or biochemical parameters. PBPK models of binary chemical mixtures must account for changes in the critical biological determinants during combined exposures since these changes directly influence the extent and rate of absorption, distribution, metabolism, and excretion of one or both of the chemicals.

5.3.1.1 Absorption

The enhancement or reduction of the absorption of one chemical as a result of the presence of another chemical is often caused either by interference with an active uptake process or, perhaps, by modulation of the critical biological determinants of uptake. Modulation of the rate of absorption of one chemical by another can be quantitatively described with a mathematical relationship that is consistent with the pattern and magnitude of change in the critical determinant(s) as a function of the exposure concentration of the modifier chemical.

5.3.1.2 Distribution

During the distribution phase, the chemicals and/or their metabolites may interact with various macromolecules such as hemoglobin, albumin, and metallothionein. Using the example of binding of two chemicals with hemoglobin, the rate of change in the amount of a chemical bound to hemoglobin can be described as a second-order process that accounts for the total amount of the macromolecule (C_hV_b) available for binding by both chemicals. This term C_hV_b becomes a common factor in the description of macromolecular binding for both chemicals ((5.17) and (5.18)); so, depending on the rate constant for reaction, greater quantities of one chemical than the other may remain unbound at high exposure concentrations:

$$dA_{bh1} = K_{h1} C_h C_{c1} V_b, \tag{5.17}$$

$$dA_{bh2} = K_{h2} C_h C_{c2} V_b. \tag{5.18}$$

Competition for plasma protein binding has been modeled by accounting for the number of binding sites, total protein level, dissociation constants, and concentration of the unbound form of both the substrate and the inhibitor [38].

5.3.1.3 Metabolism

Induction Prior exposures to certain chemicals induces specific isoenzymes, causing an increase in the capacity of the enzyme to metabolize other chemicals. The effect of enzyme induction may be modeled by determining both the quantitative relationship between V_{max} or K_f for a particular chemical and the pretreatment dose of the inducer. Enzyme induction does not alter the substrate affinity for metabolism (K_m) if the same isoform is induced.

Inhibition Combined exposure to two chemicals may result in competition between them for enzyme-mediated biotransformation, causing an inhibition of the metabolism of one chemical by the other. In simple cases, enzyme inhibition may be one of three kinds, namely, competitive, noncompetitive, and uncompetitive. Competitive inhibition is the consequence of two substrates (S_1, S_2) requiring the same site for metabolism. In other words, one of the two substrates acting as the inhibitor (S_2) combines with the form of the enzyme (E) as the substrate of interest (S_1); hence, less E is available to S_1. The same consideration applies to the metabolism of S_2, that is, the combination of S_1 with the same form of E as S_2 results in less E being available to S_2. The equilibria describing competitive inhibition, with S_1 as the substrate and S_2 as the inhibitor, are shown here [48, 49].

$$\begin{array}{l} \mathbf{E} + \mathbf{S_1} \underset{}{\overset{K_{m1}}{\rightleftharpoons}} \mathbf{ES_1} \xrightarrow{K_{p1}} \mathbf{E} + \mathbf{P_1} \\ + \\ \mathbf{S_2} \\ K_{i2,1} \updownarrow \\ \mathbf{ES_2} \end{array} \tag{5.19}$$

where

$$K_{m1} = \frac{[E][S_1]}{[ES_1]},$$

$$K_{i2,1} = \frac{[E][S_2]}{[ES_2]},$$

and K_{pi} is the rate constant for the breakdown of ES_1 to E and P_1.

The expression relating v_1, V_{max1}, S_1, K_{m1}, S_2, and $K_{i2,1}$ can be derived using rapid equilibrium or steady-state assumptions. Thus, the relative velocity in the presence of a competitive inhibitor can be derived as [48]

$$v_1 = K_{p1}[ES_1], \tag{5.20}$$

$$\frac{v_1}{E_t} = \frac{K_{p1}[ES_1]}{[E] + [ES_1] + [ES_2]}, \tag{5.21}$$

where the total enzyme level (E) is represented by the sum of free enzyme (E) and that which reacted with chemical 1 (or the substrate ES_1) and with chemical 2 (or the inhibitor ES_2).

Rearranging,

$$\frac{v_1}{K_{p1} E_t} = \frac{[ES_1]}{[E] + [ES_1] + [ES_2]}, \tag{5.22}$$

since

$$K_{p1} E_t = V_{max1} \tag{5.23}$$

then

$$\frac{v_1}{V_{max1}} = \frac{[ES_1]}{[E] + [ES_1] + [ES_2]}. \tag{5.24}$$

Furthermore, since

$$[ES_1] = \frac{[E][S_1]}{K_{m1}}$$

and

$$[ES_2] = \frac{[E][S_2]}{K_{i2,1}}$$

then

$$\frac{v_1}{V_{max1}} = \frac{\frac{[E][S_1]}{K_{m1}}}{[E] + \frac{[E][S_1]}{K_{m1}} + \frac{[E][S_2]}{K_{i2,1}}}. \tag{5.25}$$

Dividing the numerator and denominator on the right-hand side by [E],

$$\frac{v_1}{V_{max1}} = \frac{\frac{[S_1]}{K_{m1}}}{1 + \frac{[S_1]}{K_{m1}} + \frac{[S_2]}{K_{i2,1}}}. \tag{5.26}$$

Multiplying both denominator and numerator in Eq. (5.26) by K_{ml}

$$\frac{v_1}{V_{max1}} = \frac{[S_1]}{K_{m1} + [S_1] + \frac{[S_2] K_{m1}}{K_{i2,1}}}. \tag{5.27}$$

Rewriting the right-hand side of Eq. (5.27) gives

$$\frac{v_1}{V_{\max 1}} = \frac{[\mathrm{S}_1]}{K_{\mathrm{m}1}\left(1 + \frac{[\mathrm{S}_2]}{K_{\mathrm{i}2,1}}\right) + [\mathrm{S}_1]} \tag{5.28}$$

or

$$v_1 = \frac{[\mathrm{S}_1]V_{\max 1}}{[\mathrm{S}_1] + K_{\mathrm{m}1}\left(1 + \frac{[\mathrm{S}_2]}{K_{\mathrm{i}2,1}}\right)}. \tag{5.29}$$

Thus, the competitive inhibitor (S_2) increases the apparent K_m for the substrate (S_1) since a portion of the enzyme exists in the ES_2 form (which has no affinity for S_1). As the concentration of S_2 increases, the apparent K_{m1} increases. V_{max1} remains unchanged since, at very high $[S_1]$, all the enzyme can be driven to the ES_1 form. In the presence of a competitive inhibitor, a much greater concentration of the substrate is needed to attain any given fraction of V_{max1}. An increase in $[S_1]$ at constant $[S_2]$ decreases the degree of inhibition. An increase in $[S_2]$ at constant $[S_1]$ increases the degree of inhibition. The lower the value of $K_{i2,1}$, the greater the degree of inhibition at any given $[S_1]$ and $[S_2]$. $K_{i2,1}$ is the constant corresponding to K_{m2}, which reflects the concentration of S_2 (inhibitor) required to slow the reaction of the enzyme with S_1 to half the rate observed in its absence.

Noncompetitive inhibition is thought to result from the inhibitor (S_2) binding to both the free enzyme (E) and the enzyme–substrate complex (ES_1). In the classical description of noncompetitive inhibition, we assume that the binding of S_1 to the enzyme does not affect the binding of S_2 [50]. The equilibria describing noncompetitive inhibition, with S_1 as the substrate and S_2 as the inhibitor, are shown here [48, 49]:

$$\begin{array}{ccccc}
\mathbf{E + S_1} & \underset{}{\overset{K_{m1}}{\rightleftharpoons}} & \mathbf{ES_1} & \xrightarrow{K_{p1}} & \mathbf{E + P_1} \\
+ & & + & & \\
\mathbf{S_2} & & \mathbf{S_2} & & \\
K_{i2,1}\updownarrows & & \updownarrows K_{i2,1} & & \\
\mathbf{ES_2 + S_1} & \rightleftharpoons & \mathbf{ES_1S_2} & &
\end{array}$$

An expression of the relative velocity in the presence of a noncompetitive inhibitor can be derived as follows [48]:

$$v_1 = K_{\mathrm{p}1}[\mathrm{ES}_1], \tag{5.20}$$

$$\frac{v_1}{E_{\mathrm{t}}} = \frac{K_{\mathrm{p}1}[\mathrm{ES}_1]}{[\mathrm{E}] + [\mathrm{ES}_1] + [\mathrm{ES}_2] + [\mathrm{ES}_1\mathrm{S}_2]}. \tag{5.30}$$

Rearranging,

$$\frac{v_1}{V_{\max 1}} = \frac{\frac{[\mathrm{S}_1]}{K_{\mathrm{m}1}}}{1 + \frac{[\mathrm{S}_1]}{K_{\mathrm{m}1}} + \frac{[\mathrm{S}_2]}{K_{\mathrm{i}2,1}} + \frac{[\mathrm{S}_1]}{K_{\mathrm{m}1}}\frac{[\mathrm{S}_2]}{K_{\mathrm{i}2,1}}}. \tag{5.31}$$

Simplifying,

$$\frac{v_1}{V_{max1}} = \frac{[S_1]}{K_{m1}\left(1+\frac{[S_2]}{K_{i2,1}}\right)+[S_1]\left(1+\frac{[S_2]}{K_{i2,1}}\right)}. \tag{5.32}$$

Dividing by $(1 + [S_2]/K_{i2,1})$ on either side,

$$\frac{v_1}{V_{max1}\Big/\left(1+\frac{[S_2]}{K_{i2,1}}\right)} = \frac{[S_1]}{K_{m1}+[S_1]}. \tag{5.33}$$

Rearranging,

$$v_1 = \frac{[S_1]\left(\frac{V_{max1}}{1+\frac{[S_2]}{K_{i2,1}}}\right)}{K_{m1}+[S_1]}. \tag{5.34}$$

Thus, the decrease in V_{max1} is a consequence of a decrease in the total amount of available sites for reaction with the substrate to yield ES_1 in the presence of a noncompetitive inhibitor. However, K_{m1} remains unchanged because the remaining enzyme functions normally. The decrease in V_{max1} denotes that the steady-state $[ES_1]$ is decreased but that the breakdown of ES to E and P_1 is not altered. At any $[S_1]$ and $[S_2]$, the enzyme–substrate complex is present as a mixture of productive ES_1 and nonproductive ES_1S_2 forms. Therefore, the degree of inhibition in the presence of a noncompetitive inhibitor solely depends on $[S_1]$ and $K_{i2,1}$. The factor $(1 + [S_2]/K_{i2,1})$ in Eq. (5.34) may be considered an $[S_2]$-dependent statistical factor describing the distribution of enzyme–substrate complexes between ES_1 and ES_1S_2.

Uncompetitive inhibition results from the inhibitor (S_2) binding to the enzyme–substrate complex (ES_1) to produce an inactive ES_1S_2 complex. The inhibitor does not bind to the free enzyme [48, 50]. Classical uncompetitive inhibition is described by these equilibria [48, 49]:

$$\mathbf{E} + \mathbf{S_1} \overset{K_{m1}}{\rightleftharpoons} \mathbf{ES_1} \overset{K_{p1}}{\longrightarrow} \mathbf{E} + \mathbf{P_1}$$

$$\mathbf{ES_1} + \mathbf{S_2} \overset{K_{i2,1}}{\rightleftharpoons} \mathbf{ES_1S_2}$$

The equilibria show that, at any $[S_2]$, some nonproductive ES_1S_2 complex will be present, indicating that the V_{max} of S_1 in the presence of an uncompetitive inhibitor (S_2) will be lower than that obtained in its absence. Unlike noncompetitive inhibition, the apparent K_m of S_1 is also decreased because the reaction $ES_1 + S_2 \rightarrow ES_1S_2$ removes some ES_1, thus driving the reaction $E + S_1 \leftrightarrow ES_1$ to the right. An expression of the relative velocity in the presence of an uncompetitive inhibitor, with S_1 as the substrate and S_2 as the inhibitor, is derived as [48]

$$v_1 = K_{p1}[ES_1], \tag{5.20}$$

$$\frac{v_1}{E_t} = \frac{K_{p1}[ES_1]}{[E] + [ES_1] + [ES_1 S_2]}. \tag{5.35}$$

Rearranging,

$$\frac{v_1}{V_{max1}} = \frac{\dfrac{[S_1]}{K_{m1}}}{1 + \dfrac{[S_1]}{K_{m1}} + \dfrac{[S_1]}{K_{m1}} \dfrac{[S_2]}{K_{i2,1}}}. \tag{5.36}$$

Simplifying,

$$\frac{v_1}{V_{max1}} = \frac{[S_1]}{K_{m1} + [S_1] + \dfrac{[S_1][S_2]}{K_{i2,1}}}. \tag{5.37}$$

Rearranging,

$$\frac{v_1}{V_{max1}} = \frac{[S_1]}{K_{m1} + [S_1]\left(1 + \dfrac{[S_2]}{K_{i2,1}}\right)}. \tag{5.38}$$

Dividing by $(1 + [S_2]/K_{i2,1})$ on either side,

$$\frac{v_1}{V_{max1} \Big/ \left(1 + \dfrac{[S_2]}{K_{i2,1}}\right)} = \frac{[S_1]}{K_{m1} \Big/ \left(1 + \dfrac{[S_2]}{K_{i2,1}}\right) + [S_1]}. \tag{5.39}$$

Rearranging,

$$v_1 = \frac{[S_1]\left(\dfrac{V_{max1}}{1 + \dfrac{[S_2]}{K_{i2,1}}}\right)}{[S_1] + \left(\dfrac{K_{m1}}{1 + \dfrac{[S_2]}{K_{i2,1}}}\right)}. \tag{5.40}$$

Under certain conditions, a mixed type of inhibitor can produce the same effects as an uncompetitive inhibitor. Although these relatively simple inhibitions are easily identified because of their kinetic characteristics, describing these relationships with various mixed-type inhibitions can be more complicated, and the discrimination between mechanistic descriptions can become more difficult. For example, when a chemical is biotransformed to varying extents by several isoenzymes and a second chemical inhibits the various isoenzymes to varying degrees, the kinetic description becomes more complicated. These various inhibitory interactions are limited not only to the P450-mediated reactions but also to those involving the phase II conjugation processes.

5.3.2 Experimental Approach

5.3.2.1 Absorption

Quantitative changes in the critical biological determinants of absorption during combined exposures can be determined as outlined for single substances. These experiments repeated in the presence of two substances will indicate any possible change in physiological parameters that might affect the uptake of one or both chemicals. For example, changes in the breathing rate caused by one chemical, thus altering the uptake of both chemicals, can be determined by monitoring the animals with a flow plethysmograph during exposures [31]. These measurements may also be carried out during gas uptake studies [32]. Changes in the dermal absorption of one compound by another can be determined similarly by examining the rate of absorption of one chemical in the absence and then in the presence of another chemical. A change in the skin permeability constant (K) can be determined by analyzing the pharmacokinetic data with a PBPK model that has all parameters defined except K. The same principle applies to the determination of K to describe changes in the gastrointestinal absorption of one chemical in the presence of another chemical.

5.3.2.2 Distribution

Plasma protein binding of chemicals can be determined by estimating Scatchard constants or stepwise constants. Scatchard plots of chemical binding, most often used by pharmacologists in drug-receptor binding studies, are easier to calculate using graphical procedures than the stepwise constants that require complex computer programs. One major advantage of stepwise constants is that those of different chemicals can be directly compared. Another advantage is that allosteric effects, if present, are accounted for in the calculation of these constants. For competitive displacement reactions, the stepwise constants appear to be distinctly advantageous, especially for competitors with different binding capacities because the stepwise constants for different ligands can be directly compared. For a detailed discussion of the experimental approaches and computer programs available for estimation of parameters required to evaluate protein-binding interactions, the reader may refer to Luecke and Wosilait [51] and Wosilait and Luecke [52].

5.3.2.3 Metabolism

In vivo analysis of metabolic interactions between two volatile organic chemicals can be performed by conducting a series of gas uptake studies.

First, a set of gas uptake studies is conducted for each substance separately and the metabolic rate constants are determined (V_{max}, K_m, K_f). Then, similar gas uptake studies are conducted by introducing both chemicals at different initial concentrations. By keeping the initial chamber concentration of one chemical constant and changing the concentration of the other chemical, and vice versa, the temporal changes in the chamber concentrations of both chemicals are determined and the data are analyzed with a PBPK model that has all parameters defined except for the interaction term. The type of inhibitory interaction can also be determined by conducting quantitative kinetic analysis *in vitro* [53]. The mass balance differential equation of the metabolizing tissue (e.g., liver) in PBPK models for individual chemicals then is modified to accommodate the hypothesis of the different types of interactions.

5.3.2.4 Excretion

The inhibition or enhancement of the excretion of a parent compound or its metabolite can be modeled by obtaining information on the kinetic parameter of interest, as in the case of single chemicals (e.g., glomerular filtration, urine flow rate, and rate of reabsorption). The difference is that such parameters are obtained in both the absence and the presence of the second chemical, so interaction at the excretory level can be described within a predictive modeling framework (Table 5.1).

5.3.3 Examples

Only a few examples of PBPK modeling of binary chemical mixtures are available. With environmental chemical mixtures, the most common single mechanism of interaction investigated in detail is inhibition or induction of hepatic cytochrome P450 [54]. A brief analysis of the PBPK modeling studies of binary chemical mixtures

Table 5.1 List of metabolic inhibition hypotheses and corresponding equations used in the physiologically based pharmacokinetic models[a].

Hypotheses	Equations[b]
No metabolic interaction	$\dfrac{V_{max1} \cdot C_{vl1}}{K_{m1} + C_{vl1}}$
Competitive interaction	$\dfrac{V_{max1} \cdot C_{vl1}}{K_{m1}\left(1 + \dfrac{C_{vl2}}{K_{i2,1}}\right) + C_{vl1}}$
Noncompetitive interaction	$\dfrac{V_{max1} \cdot C_{vl1}}{(K_{m1} + C_{vl1})\left(1 + \dfrac{C_{vl2}}{K_{i2,1}}\right)}$
Uncompetitive interaction	$\dfrac{V_{max1} \cdot C_{vl1}}{K_{m1} + C_{vl1}\left(1 + \dfrac{C_{vl2}}{K_{i2,1}}\right)}$

a) Reproduced from Krishnan *et al.* [92], with permission of the National Institute of Environmental Health Sciences.

b) These equations are written by designating chemical 2 as inhibitor and chemical 1 as substrate. Chemical 1 is assumed to be metabolized only in the liver by a saturable pathway.

in which induction or inhibition of hepatic P450-mediated metabolism was examined as the mechanistic basis of interaction follows.

5.3.3.1 Dibromomethane and Isofluorane

This study by Clewell and Andersen [55] evaluated the potential metabolic interaction between dibromomethane (DBM) and isofluorane (ISO). A PBPK model was first developed for DBM by determining the rate constants for hepatic metabolism of this compound by a saturable and a first-order process. The P450-mediated saturable metabolism of DBM produces carbon monoxide, leading to the induction of carboxyhemoglobinemia. On coexposure to ISO, the carboxyhemoglobin levels produced by DBM were significantly reduced. This interaction was successfully predicted by a competitive metabolic interaction model (Figure 5.3).

5.3.3.2 Trichloroethylene and 1,1-Dichloroethylene

1,1-Dichloroethylene (DCE) is a potent hepatotoxicant that exerts acute toxicity when reactive metabolites (probably chloroacetylchloride, CAC) are formed faster than they can be detoxified by glutathione or other cellular antioxidants. In naïve rats, formation of 30 mg of reactive metabolites per kg body weight in 2 h is lethal to 50% of exposed rats. This metric for toxicity was called LQ_{50}, that is, the lethal quantity metabolized causing death of 50% of a group of exposed rats. The presence of some other chlorinated hydrocarbons, such as vinyl chloride or trichloroethylene, reduces the toxicity of 1,1-DCE. For instance, the degree of toxicity at a given 1,1-DCE exposure is lower and/or higher levels of DCE exposure are required to produce equivalent hepatic injury. In rats with depleted hepatic GSH, the LQ_{50} is reduced to about 5 mg metabolized/kg body weight.

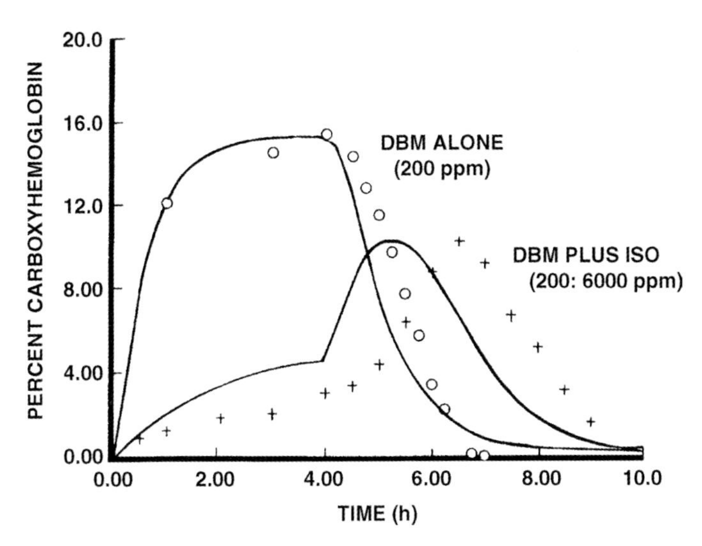

Figure 5.3 Model predictions (solid lines) and experimental observations (circles) of carboxyhemoglobin levels during and following combined exposure to DBM ISO. In this PBPK model, the hepatic metabolism of both chemicals was described as a competitive inhibition process.

Today, we know that 1,1-DCE is metabolized by cytochrome P4502E1 (CYP2E1) to an epoxide intermediate (Figure 5.2). This metabolism primarily occurs in the liver, where about 90% of the body's CYP2E1 enzyme resides. The epoxide is conjugated with glutathione (GSH) or spontaneously rearranges to an acyl chloride, CAC. If CAC is not detoxified by GSH or other antioxidants, it will cause liver injury, leading to the release of liver enzymes such as aspartate transaminase. Damage to hepatocytes leads to release of aspartate transaminase (SGOT) from liver into blood. Increases in plasma SGOT is used as a measure of damage.

Andersen *et al.* [47, 56] investigated the basis for decrease in 1,1-DCE toxicity when animals were coexposed to trichloroethylene. Like 1,1-DCE, TCE is primarily metabolized in the liver by CYP2E1 but does not tend to deplete GSH or cause acute liver toxicity. Since the two compounds are both metabolized by the same enzyme, they were expected to compete for metabolism. To characterize the pharmacokinetics of TCE and 1,1-DCE in rats, Andersen *et al.* developed a PBPK model for the two chemicals. The PBPK model describes the animal as composed of four compartments (liver, fat, rapidly perfused tissues, and slowly perfused tissues), with blood flows, and gas exchange with the exposure atmosphere. Metabolism in the liver is described using standard Michaelis–Menten kinetics, with competition occurring between the two compounds. Gas uptake pharmacokinetic data were used to determine the rates of metabolism of each of the individual chemicals. Experimental data for uptake of chemicals when coexposure to TCE and DCE occurred were also used to determine the nature and extent of inhibition. After verifying by modeling of the uptake curves that the metabolic inhibition was competitive, the PBPK model was used to determine the amounts of 1,1-DCE metabolized in constant concentration exposures to 1,1-DCE or to combinations of 1,1-DCE and TCE. This calculation permits extrapolation across concentration and across exposure regimen (i.e., from closed chamber exposures to constant concentration inhalation exposure).

In Figure 5.4, the degree of liver injury, estimated by increases in serum aspartate transaminase, is plotted against either the concentration of 1,1-DCE in the 6 h exposures or against the calculated amount of 1,1-DCE metabolized determined with the PBPK model. In the left panel, toxicity of 1,1-DCE is seen to be decreased by coexposure to 500 ppm of TCE, causing the coexposure dose-response curve to shift to the right compared to 1,1-DCE without TCE coexposure. However, these shifts in the dose-response curves by themselves do not shed light on the nature of the interaction. The close correspondence between toxicity measurements when the dose is considered on the basis of the dose metric "amount DCE metabolized" supports the idea that the production of metabolites is responsible for toxicity and that the only change with TCE coexposure is alteration of the amount of 1,1-DCE metabolized per unit of exposure.

5.3.3.3 Benzene and Toluene

This study by Purcell *et al.* [57] was an extension of previous investigations in which the rate of metabolism of benzene and toluene had been separately determined (benzene: $V_{max} = 3.3$ mg/h, $K_m = 0.3$ mg/l; toluene: $V_{max} = 7.5$ mg/h, $K_m = 0.3$ mg/l). In a series of gas uptake studies, both chemicals in combination were introduced

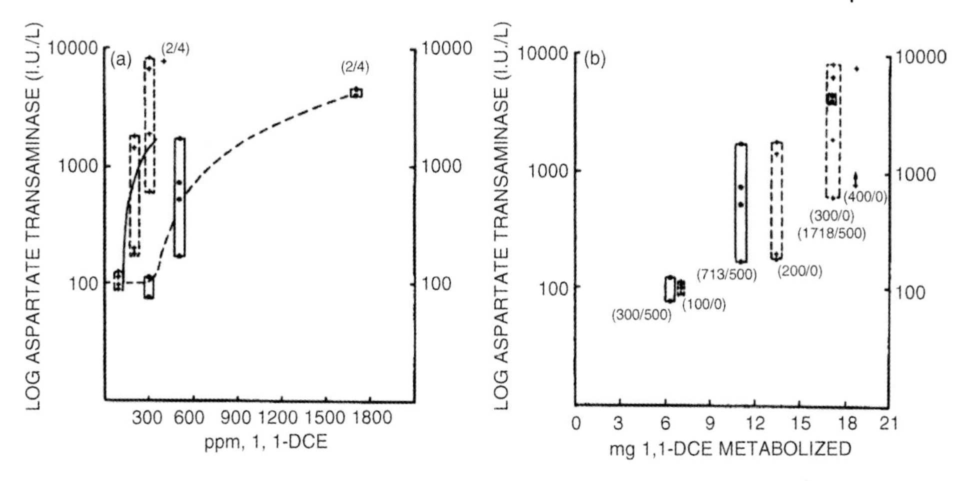

Figure 5.4 Increased serum aspartate transaminase after 6 h exposures of male F344 rats to DCE with or without simultaneous coexposure to 500 ppm of TCE. Serum aspartate transaminase was measured 24 after beginning of the exposure. Dotted boxes show the range of response data for DCE exposures; solid boxes show the range of response data for coexposure to 500 ppm TCE with DCE at level indicated (panel a). When coexposure to TCE occurs, the dose-response curve shifts to the right, indicating decreased toxicity. Higher levels of DCE were required to produce the same responses (b). The numbers in parentheses indicate the exposure level for DCE/TCE. The PBPK model for DCE and TCE was used to calculate the amount of DCE metabolized during the experiment. Panel b shows a close correspondence between toxicity for the experiments shown in Panel a and the amount of DCE metabolized. (Reprinted with permission).

at different initial concentrations (200, 1000, or 2000 ppm) and the temporal change in their chamber concentrations was analyzed with a PBPK model that had the metabolic interaction terms defined in the liver compartment. The noncompetitive interaction model provided adequate fit to all the experimental data. Toluene was a better inhibitor of benzene metabolism $K_{mi} = 0.28$ mg/l) than benzene was of toluene metabolism ($K_{mi} = 2.47$ mg/l) (Figure 5.5).

It is important to note that the observation in this study (that the kinetics of the interaction between benzene and toluene was most consistent with noncompetitive inhibition) does not, in itself, support a conclusion that the interaction is truly by the noncompetitive mechanism described earlier in this chapter. Additional studies would be required to support such a conclusion. In fact, other studies of the interaction between benzene and toluene [58] concluded that the kinetics could be well described using a description of competitive inhibition. The discrepancy between these two studies may reflect contributions to the metabolism of these compounds from multiple P450 isozymes.

5.3.3.4 **Mirex, Phenobarbital, or Chlordecone and Bromotrichloromethane**

This study by Thakore *et al.* [59] examined the effect of dietary pretreatment with several chemicals on the metabolism of bromotrichloromethane ($BrCCl_3$). Rats were fed with either normal Purina pelleted diet or a pelleted diet that contained mirex

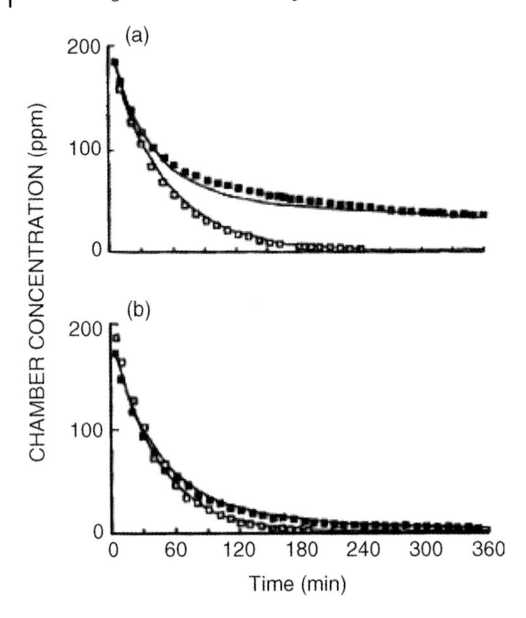

Figure 5.5 (a) Temporal change in benzene concentration in gas uptake chamber for an exposure of 200 ppm benzene alone and for a coexposure of 200 ppm benzene with 1000 ppm toluene. The solid lines are generated with the PBPK model, assuming a noncompetitive interaction between the two solvents. (b) Temporal change in chamber toluene concentration for an exposure of 200 ppm toluene alone and for a coexposure of 200 ppm toluene with 1000 ppm benzene. The solid lines were generated with the PBPK model assuming a noncompetitive interaction between the two solvents. (Reprinted from Purcell *et al.* [57], with permission of Elsevier Press.)

(10 ppm), phenobarbital (225 ppm), or chlordecone (10 ppm) for 15 days. Following this pretreatment regimen, each group of animals was placed in a gas uptake chamber at initial concentrations of 30, 200, 700, or 3000 ppm of $BrCCl_3$. The data on the decline in the chamber concentration of $BrCC1_3$ were analyzed with a PBPK model that had all parameters defined except metabolic rate constants. The V_{max}, K_m (for the saturable pathway yielding the reactive CCl_3), and K_f (for the glutathione conjugation pathway) initially were obtained for the control groups and then compared with those for other pretreated groups. In modeling the gas uptake data from the enzyme inducer treatments, the K_m was kept constant and the fit to the experimental data was obtained only by changing V_{max} and K_f. Pretreatment with phenobarbital increased the V_{max} and decreased the K_f significantly, whereas both V_{max} and K_f were increased in mirex-pretreated rats. The mild enhancement of $BrCCl_3$ toxicity by phenobarbital and mirex correlated with an increase in metabolism via the saturable pathway; however, the marked potentiation seen after chlordecone pretreatment could not be attributed to the induction of $BrCCl_3$ metabolism. Additional experimental evidence indicated that this potentiation phenomenon is a result of chlordecone interfering with the initial tissue repair process that follows $BrCC1_3$-induced liver injury [60].

5.3.3.5 Ethanol and Trichloroethylene

This study by Sato *et al.* [61] utilized the predictive capacity of PBPK models to determine the magnitude of presumed toxicokinetic interactions between TCE and ethanol in humans. The induction of TCE metabolism by ethanol was modeled by assuming that workers consume ethanol the evening before the workshift. The inhibitory effect of ethanol on TCE metabolism was simulated by assuming the consumption of ethanol 15 min before the start of the workshift. Metabolic rates and inhibition constants determined previously in both *in vitro* and human studies were used to construct the model [62–65]. The model simulations showed that the effect of ethanol on TCE metabolism would be marked if it were consumed just before work (inhibitory effect) rather than the evening before the workshift (induction effect). When TCE concentration is below 100 ppm (i.e., at which level hepatic blood flow limits metabolism), the effect of enzyme induction by ethanol on TCE metabolism can be expected to be negligible. On the basis of this analysis, these authors concluded that enzyme induction caused by ethanol would not affect the pharmacokinetic behavior of TCE and other organic chemicals as much as animal and *in vitro* studies might suggest.

5.3.3.6 Toluene and *m*-Xylene

In this study, Tardif *et al.* [66] determined the rate constants for the metabolism of toluene (TOL) ($V_{max} = 4.8$ mg/h/kg, $K_m = 0.55$ mg/l) and *m*-xylene (XYL) ($V_{max} =$ 8.4 mg/h/kg, $K_m = 0.2$ mg/l) by gas uptake technique. Then, in additional gas uptake studies, rats were exposed to initial chamber concentrations of 500 ppm TOL + 1000 ppm XYL, 1000 ppm TOL + 500 ppm XYL, and 1000 ppm TOL + 1000 ppm XYL. The data on temporal change in the chamber concentrations of both chemicals were analyzed with a PBPK model. The competitive metabolic interaction model adequately described the altered uptake of these two substances in the gas uptake studies. The validated rat model then was used to predict the magnitude of inhibitory interactions between TOL and XYL at untested exposure concentrations. The simulations of the amount of one chemical metabolized in the presence of the second chemical suggested that during a 5 h exposure to mixtures of equiconcentrations, interaction becomes important when the exposure concentrations exceed approximately 50 ppm of each solvent. Interestingly, in human volunteer studies, an interaction between TOL and XYL was significant after a 4 h exposure to 95 and 80 ppm, respectively, whereas the interaction was not demonstrable at lower exposure concentrations (50 ppm TOL + 40 ppm XYL, 7 h) [67]. These observations are consistent with both the occurrence of competitive metabolic interaction between TOL and XYL in humans and the notion that the interaction becomes important only at exposure concentrations that are somewhat higher but within the threshold limit value (TLV) of the individual chemicals (100 ppm).

5.3.3.7 Carbaryl and Chlorpyrifos

Carbamate and organophosphorus (OP) compounds are two major types of insecticides used for agriculture and domestic or commercial pest control. These two types of pesticides both act by inhibiting acetylcholinesterase (AChE), a class of enzymes that catalyzes the hydrolysis of the neurotransmitter acetylcholine (ACh). The inhibition

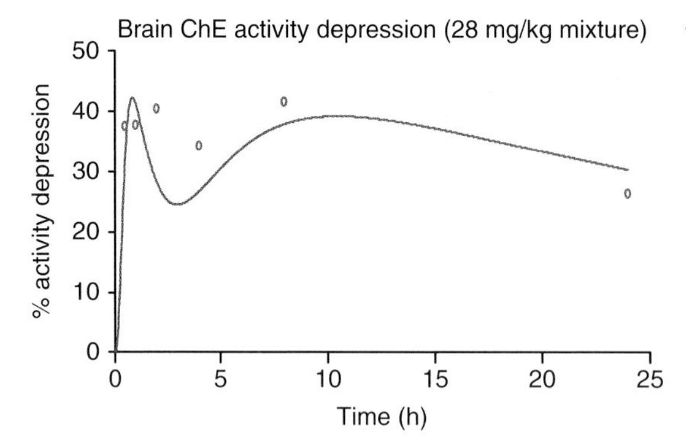

Figure 5.6 Predicted (curve) and measured time course for brain AChE activity following dosing with 28 mg/kg of both carbaryl and chlorpyrifos. (Data from Ref. [91].)

causes an accumulation of ACh at nerve synapses and a prolonged stimulation of ACh receptors on postsynaptic cells, leading to neurotoxic effects. On the other hand, both carbamate and OP insecticides are metabolized by cytochrome P450-mediated monooxygenases [68, 69]. Therefore, it is possible that they interact both in metabolism and at target sites. Carbaryl and chlorpyrifos are widely used carbamate and OP pesticides, respectively, for which PBPK models have been developed [70, 71], providing an opportunity to investigate the interactions of compounds that compete both for metabolism and for activity. An important distinction between the compounds is that while the interaction of carbaryl with AChE is rapidly reversible, the interaction of CPF-oxon with AChE is essentially irreversible.

To study carbaryl–chlorpyrifos mixtures, the PBPK models for the individual compounds were linked through descriptions of competitive inhibition at sites of metabolism and at AChE. In both individual chemical models, metabolism was described as a Michaelis–Menten process, and the interaction of each chemical with AChE was described as a second-order processes. In the mixture model, the compounds carbaryl and CPF are assumed to compete with each other at CYP450; while carbaryl and CPF-oxon compete with each other for AChE. The resulting mixture model was able to successfully predict the time course for AChE in the brain following mixed exposures to the two compounds (Figure 5.6). The complex time course reflects the combination of the rapid but reversible binding of carbaryl to AChE together with the slower, irreversible binding of CPF-oxon.

5.4
PBPK Modeling of Complex Chemical Mixtures

5.4.1
Theoretical Considerations

With a multichemical mixture, some components may act independently, neither interfering with nor being modified by other chemicals, whereas some components might interfere with and modify the disposition of others. Approaches for the PBPK modeling of both independently acting and interacting binary chemical mixtures have been presented in the earlier sections. In modeling complex mixtures, complications arise from the possibility that a third or a fourth chemical might alter further the kinetics of interacting binary chemicals but may not alter the disposition kinetics of some other components in the mixture. With environmental chemicals, metabolic interference appears to be the most common single mechanism by which toxicokinetic interactions occur. Inhibitory metabolic interactions among the components of a chemical mixture can be modeled by approaches similar to those detailed in the section for binary chemical mixtures.

5.4.2 Experimental Approach

The two alternative approaches used for modeling chemical mixtures are bottom-up and top-down approaches. The bottom-up approach first collects kinetic information about individual components in the mixture, and accordingly, constructs individual models. These individual models are then complied into one mixture model by accounting for the binary and/or higher order inactions occurring within the mixture. Ultimately, the complied mixture model would be capable to predict the responses of interest for the entire mixture.

For this approach, information such as tissue:blood partition coefficients, metabolic pathways, and rate constants must be available for each individual chemical to establish a reasonable PBPK model. These individual chemical models then must be linked at some point such as the blood, lung, or liver compartment. For example, in case of competitive interaction among several chemicals, the K for hepatic metabolism of each chemical (substrate) will be increased (by a factor equal to $1 + I/K_i + X/K_x + Y/K_y \ldots$) by the other chemicals (i.e., inhibitors I, X, Y . . .). The inhibition constant representing the interaction between two components, then the influence of a third component, and so on should be obtained and incorporated into the model to establish adequate predictability.

The bottom-up approach has been successfully demonstrated in a series of modeling efforts on mixtures of *m*-xylene (M), toluene (T), ethyl benzene (E), dichloromethane (D), and benzene (B) [58, 72–75]. In these studies, individual PBPK model for each mixture component was first developed and then interconnected at the binary level in the liver compartment. Inhibitive metabolism was incorporated within the models since all five chemicals are metabolized by cytochrome P4502E1 [66, 76–80]. Using the bottom-up approach, a ternary mixture model of T-X-E was built, followed by addition of a fourth chemical (B) and later a fifth chemical (D) [58, 72, 73].

Another bottom-up example is an interactive PBPK model that predicts individual kinetics of trichloroethane, perchloroethylene (PERC), and methylchloroform (MC) [81, 82] in a ternary mixture. Individual PBPK models were linked via the saturable metabolism term in the liver compartment based on the assumption that competitive inhibition occurs when TCE, PERC, and MC coexist. The mixture model was used to predict the PK profiles of two common biomarkers of exposure, peak TCE blood concentrations and total amount of TCE metabolites were generated, over a range of PERC and MC concentrations. Using these predicted dose metrics and assuming their increase of 10% or more corresponds to significant health effects, interaction thresholds for binary and ternary mixtures were calculated to provide a quantitative measure of interactions at occupationally relevant levels of exposure.

The top-down approach, or lumping approach, starts with considering the mixture as a single entity and proceeds to splitting out chemicals about which specific information is required, such as the pharmacokinetics of a chemical that causes an effect of interest. This approach is appropriate when it is unnecessary to distinguish one component in the mixture from another and the relevant properties of each component are similar enough to be described as a whole. This is also an effective approach when the number of binary interactions in the mixture of interest becomes too large to assess as in the case of gasoline exposure. The gasoline mixture contains several hundred chemicals derived from crude petroleum and fuel additives. Using the top-down approach, a PBPK model was developed for two blends (winter and summer) of gasoline with PK descriptions of specific compounds (*n*-hexane, benzene, toluene, ethylbenzene, and *o*-xylene) and a lumped component that represented the rest of the gasoline mixture [83, 84]. The lumped component was described using a single set of parameters that depended on the blend of gasoline. In a subsequent study, the model was extended to simulate the evaporative fractions of gasoline, which would be inhaled from exposure scenarios such as emissions from gas tank refilling [84].

When modeling mixture exposure, one must consider the characteristic of the mixture. The top-down approach is most effective when the mixture is composed of similar components, such as polycyclic aromatic hydrocarbons and gasoline. Although gasoline is not a homogeneous mixture, all gasoline components are primarily metabolized in the liver. When the mixture is composed of different components, for example, disinfection by-products that may contain trihalomethanes and bromate and air pollutants that may contain metals and organics, a combination of top-down and bottom-up approaches could be useful.

5.4.3 Examples

In reality, each individual chemical, on biotransformation, yields a complex mixture of metabolites. Multiple metabolic interactions involving the parent chemical and its metabolites can occur that together determine the kinetics of each of them. Parent chemical–metabolite interactions have been described for *n*-hexane, benzene, and hexachlorobenzene [85–88]. The quantitative aspect of the mechanism of interaction has been studied only for *n*-hexane [87].

5.4.3.1 *n*-Hexane

n-Hexane is metabolized by a saturable pathway yielding methyl *n*-butyl ketone (MnBK). MnBK is metabolized further by w-1 oxidation to 2,5-hexanedione (2,5-HD), the neurotoxic metabolite, and by a-oxidation and decarboxylation to pentanoic acid. After a 6 h exposure of F344 rats to 500, 1000, 3000, or 10 000 ppm *n*-hexane, the blood concentrations of *n*-hexane and MnBK increased linearly but 2,5-HD showed anomalous behavior [89]. The concentration of 2,5-HD at the end of hexane exposure substantially increased from 500 to 1000 ppm, remained fairly constant between 1000 and 3000 ppm, and was lower for the 10 000 ppm exposure than for the lower exposure concentrations. At higher hexane exposure levels, the blood concentration of 2,5-HD immediately after exposure was lower than the eventual peak, indicating that inhibition of 2,5-HD production during hexane exposure ceased with the termination of the exposure. In addition, the time at which the peak 2,5-HD concentration occurred was immediately after exposure at the lowest exposure level but not until 16 h later for the highest exposure concentration. In modeling this complex kinetic behavior of 2,5-HD resulting from hexane exposure, Andersen and Clewell [87] considered multiple competitive interactions among *n*-hexane, MnBK, and 2,5-HD. This PBPK description, consistent with the observed kinetics of *n*-hexane, MnBK, and 2,5-HD, predicts a peak toxicity (corresponding to the increased AUC of 2,5-HD) for *n*-hexane at 1000 ppm (Figure 5.7). The model also suggests that, at higher concentrations, *n*-hexane actually might be less toxic because of less conversion to 2,5-HD.

This example of hexane shows that multichemical exposures can ensue after exposure to a single chemical, that is, exposure to hexane is equivalent to exposure to a

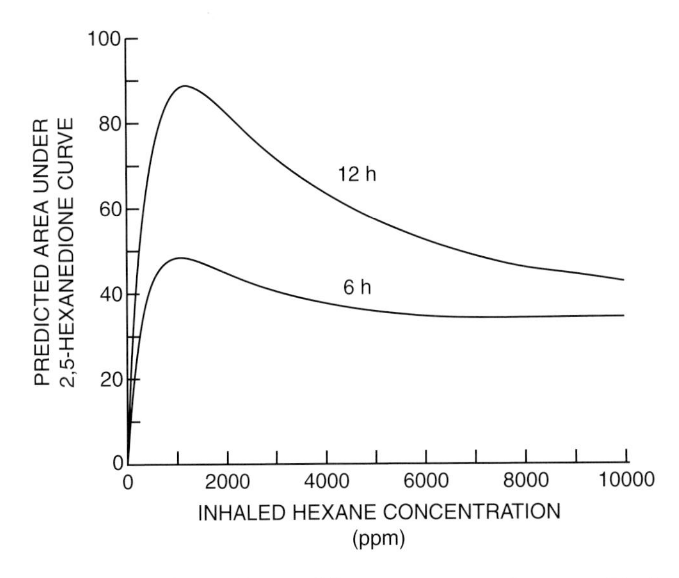

Figure 5.7 Model simulations of the area under the blood 2,5-hexanedione concentration versus time curve for 6 and 12 h inhalation exposures of up to 10 000 ppm *n*-hexane. The lines were generated with a PBPK model assuming competitive interactions among *n*-hexane, 2,5-hexanedione, and methyl *n*-butyl ketone. (Reprinted from Krishnan *et al.* [92] with permission of the National Institute of Environmental Health Sciences.)

mixture of related chemicals. Modeling of metabolite interactions for more chemicals may offer an opportunity to develop experimental approaches and strategies for PBPK modeling of complex mixtures. The approaches useful for studying toxicokinetic interactions among metabolites should be applicable to multichemical interactions as well. Evidently, in multichemical mixture modeling, toxicokinetic interactions among their metabolites might further complicate the predicted outcome of such studies.

5.4.3.2 **Toluene, *m*-Xylene, and Ethylbenzene**

Building upon the binary mixtures work discussed previously, Tardif *et al.* [72] assessed the metabolic interaction of a ternary mixture of alkyl benzenes with the bottom-up approach. The objective of the 1996 study was to develop a PBPK model for a ternary mixture of alkyl benzenes (T, X, and E) in rats and humans. This approach involved initial development of the mixture PBPK model in the rat followed by extrapolation to humans. The development of the ternary mixture PBPK model in the rat was accomplished by initially validating or refining the existing PBPK models for T, X, and E and linking the individual chemical models via the hepatic metabolism term. As with the previous binary mixtures work, the Michaelis–Menten equation for each solvent was modified to test four possible mechanisms of metabolic interaction (i.e., no interaction, competitive inhibition, noncompetitive inhibition, and uncompetitive inhibition). The metabolic inhibition constant (K_i) for each binary pair of alkyl benzenes was estimated by fitting the binary chemical PBPK model simulations to previously published data on blood concentrations of T, X, and E in rats exposed for 4 h to a binary combination of 100 or 200 ppm of each of these solvents. This assessment determined competitive metabolic inhibition to be the most plausible mechanism of interaction at relevant exposure concentrations for all binary mixtures of alkyl benzenes in the rat (K_i, T-X = 0.17; K_i, T-E = 0.79; K_i, X-T = 0.77; K_i, X-E = 1.50; K_i, E-T = 0.33; K_i, E-X = 0.23 mg/l). The K_i values obtained with the binary chemical mixtures were incorporated into the PBPK model for the ternary mixture. This approach resulted in adequate simulation of the venous blood time course concentrations of T, X, and E in rats exposed to a mixture containing 100 ppm each of these solvents. The ternary mixture model in the rat was then scaled to predict the kinetics of T, X, and E in blood and alveolar air of human volunteers exposed for 7 h to a combination of 17, 33, and 33 ppm of these solvents, respectively. Model simulations and experimental data obtained in humans indicated that exposure to atmospheric concentrations of T, X, and E remaining within the permissible concentrations for a mixture would not result in biologically significant modifications of their pharmacokinetics.

5.4.3.3 **Benzene, Toluene, *m*-Xylene, Ethylbenzene, and Dichloromethane**

Haddad *et al.* [58, 73] developed physiologically based toxicokinetic (PBTK) models to describe the interactions of both quaternary (BTEX) mixtures and BTEX with dichloromethane (D) substituted for one of the BTEX components or in addition to BTEX. The objective of the 1999 study was to develop and validate an interaction-based PBTK model for simulating a quaternary mixture (i.e., BTEX) in the rat. The primary focus was on obtaining and refining validated individual chemical PBTK models from the literature, interconnecting the individual chemical PBTK models at the level of liver based on the mechanism of binary chemical interactions (e.g., competitive, noncompetitive, or uncompetitive metabolic inhibition), and comparing the a priori predictions of the interaction-based model to corresponding experimental data on venous blood concentrations of B, T, E, and X during mixture exposures. Analysis of the blood kinetics data from inhalation exposures (4 h, 50–200 ppm each) of rats to all binary combinations of B, T, E, and X was suggestive of competitive metabolic inhibition as was similar to the previous work. The metabolic inhibition constant (K_i) for each binary combination was quantified and incorporated into the mixture PBTK model. The binary interaction-based PBTK model was able to predict the inhalation toxicokinetics of all four components in rats following exposure to mixtures of BTEX (50 ppm each of B, T, E, and X, 4 h; 100 ppm each of B, T, E and X, 4 h; 100 ppm B + 50 ppm each of T, E, and X, 4 h). Thus, it was concluded that the binary interaction or bottom up-approach was adequate for predicting more complex mixtures when all binary combinations are accounted for.

The follow-up to this work was to investigate the incorporation of another P4502E1 substrate (i.e., D) into the PBTK model framework in an attempt to validate the bottom-up approach. First, D was used to replace one of the components in the BTEX mixture and then D was added to the BTEX mixture. The binary interactions between D and the BTEX components were also explored. The analysis of rat blood kinetic data (4 h inhalation exposures, 50–200 ppm each) to all binary combinations (D-B, D-T, D-E, and D-X) investigated in the present study were again suggestive of competitive metabolic inhibition as the plausible interaction mechanism. By incorporating the newly estimated values of metabolic inhibition constant (K_i) for each of these binary combinations within the five-chemical PBTK model (i.e., for the DBTEX mixture), the model adequately predicted the venous blood kinetics of chemicals in rats following a 4 h inhalation exposure to various mixtures (mixture 1: 100 ppm of D and 50 ppm each of T, E, and X; mixture 2: 100 ppm each of D, T, E, and X; mixture 3: 100 ppm of D and 50 ppm each of B, T, E, and X; mixture 4: 100 ppm each of D, B, T, E, and X).

5.4.3.4 **Gasoline and JP-8 Jet Fuel**

In some cases, the mixture is too complex to afford the bottom-up approach such as with fuel mixtures. These mixtures contain thousands of individual chemical components and vary between batches due to variability in refining processes and source crude oil. In these cases, it is more prudent to apply a bottom-up or lumping approach, as the number of binary exposures that would be necessary to describe the entire fuel is unattainable.

Dennison *et al.* [83, 84] used the initial approach of Tardif to investigate the metabolic interaction of gasoline with several individual components including BTEX and *n*-hexane (H) using gas uptake to determine kinetic parameters of V_{max} and K_m. The primary assumption was that the whole gasoline mixture shared a common metabolic pathway (i.e., P4502E1). Thus, the gas uptake experiments with the fuel were fit with individual interaction model descriptions for each component and then the remaining measured concentration of the fuel was accounted for with a single interaction model description. The model was able to adequately predict the metabolic interaction of both winter and summer blends of gasoline that slightly differs in composition due to variations in refinement and additives. Dennison *et al.* [84] used the same basic approach to investigate the metabolic interaction of evaporative fractions of gasoline on individual and lumped component kinetics. The evaporative fractions (i.e., 1/3 and 2/3 cuts) were expected to more accurately describe the composition of real-world exposure to gasoline (e.g., during auto refueling). Again, the model was able to describe both the individual BTEXH components and the different fractions of gasoline.

Campbell and Fisher [90] used a similar approach to investigate the metabolic interaction between the components of JP-8 jet fuel and the EX components of BTEX. Jet fuel is kerosene based and contains a larger range of hydrocarbon components from C8 to C18. Thus, it was not amenable to gas uptake experiments. These researchers used constant concentration and direct analysis of blood and liver concentrations to determine the metabolic interaction of a lumped aromatic fraction. The fraction was determined based on the best available knowledge concerning components of JP-8 vapor that would be expected to share similar metabolic pathways with EX. This fraction comprised approximately 18–25% of the total hydrocarbon exposure concentration. The lumping approach was able to describe both the kinetics of the individual components from the previous work of Haddad and the kinetics of the lumped fraction.

5.5 Summary and Future Directions

Multichemical exposure is the rule rather than an exception in both the general and the occupational environment. Simultaneous or sequential exposure to multiple chemicals may cause alterations in the toxicokinetics and/or toxicodynamics of the individual chemicals, resulting in a change in the toxicity predicted based on the summation of the effects of the components. Toxicokinetic interactions occur as a result of one chemical altering the absorption, metabolism, distribution, and/or excretion of other chemical(s). Such interactions often result in modification of the internal or target dose of one component of a mixture by another component. In such cases, conducting quantitative risk assessment for chemicals present as a mixture is difficult. The uncertainties that arise from changes in the toxicokinetics of the components can be addressed by developing physiologically based descriptions of the disposition of chemical mixtures that can be used for dose, route, and interspecies extrapolations of the target tissue concentration of the toxic moieties. Furthermore, biologically based models for mechanisms of action and tissue response should be developed so the toxicodynamics of the interacting chemicals are also integrated into a predictive modeling framework. Such a quantitative mechanistic approach to the study of chemical interactions is imperative to achieve the ultimate goal of assessing the health risks associated with human exposure to complex chemical mixtures.

References

1 Teorell, T. (1937) Kinetics of distribution of substances administered to the body. I. The extravascular modes of administration. *Arch. Int. Pharmacodyn.*, **57**, 205–225.

2 Teorell, T. (1937) Kinetics of distribution of substances administered to the body. II. The intravascular modes of administration. *Arch. Int. Pharmacodyn.*, **57**, 226–240.

3 Andersen, M.E., Clewell, H.J., and Frederick, C.B. (1995) Applying simulation modeling to problems in toxicology and risk assessment – a short perspective. *Toxicol. Appl. Pharmacol.*, **133**, 181–187.

4 Andersen, M.E. (1981) Saturable metabolism and its relation to toxicity. *Crit. Rev. Toxicol.*, **9**, 105–150.

5 Andersen, M.E., Clewell, H.J., and Krishnan, K. (1995) Tissue dosimetry, pharmacokinetic modeling, and interspecies scaling factors. *Risk Anal.*, **15**, 533–537.

6 O'Flaherty, E.J. (1987) Modeling: An introduction. In: (2nd Edn. ed.), *Pharmacokinetics in Risk Assessment. Drinking Water and Health*, Volume. 8. National Academy Press, Washington DC, pp. 27–35.

7 Clewell, R.A. and Clewell, H.J. (2008) Development and specification of physiologically based pharmacokinetic models for use in risk assessment. *Reg. Toxicol. Pharmacol.*, **50**, 129–143.

8 Ramsey, J.C. and Andersen, M.E. (1984) A physiologically based description of the inhalation pharmacokinetics of styrene in rats and humans. *Toxicol. Appl. Pharmacol.*, **73**, 159–175.

9 Reddy, M.B., Yang, R.S.H., Clewell, H.J. III, and Andersen, M.E. (2005) *Physiologically Based Pharmacokinetic Modeling: Science and Applications*, John Wiley & Sons, Inc., Hoboken, New Jersey.

10 McDougai, J.N., Jepson, G.W., Clewell, H.J. III, MacNaughton, M.G., and Andersen, M.E. (1986) A physiological pharmacokinetic model for dermal absorption of vapors in the rat. *Toxicol. Appl. Phamacol.*, **85**, 286–294.

11 Fisher, J.W., Whittaker, T.A., Taylor, D.H., Clewell, H.J. III, and Andersen, M.E. (1990) Physiologically based pharmacokinetic modeling of the pregnant rat: multiroute exposure model for trichloroethylene and trichloroacetic acid. *Toxicol. Appl. Pharmacol.*, **99**, 395–414.

12 Travis, C.C. and Bowers, J.C. (1989) Protein binding of benzene under ambient exposure conditions. *Toxicol. Ind. Health*, **5**, 1017–1024.

13 Krishnan, K., Gargas, M.L., Fennell, T.R., and Andersen, M.E. (1992) A physiologically based description of ethylene oxide dosimetry in the rat. *Toxicol. Ind. Health*, **8**, 121–140.

14 Fennell, T.R., Sumner, S., and Walker, V.E. (1992) A model for the simulation of the formation and degradation of hemoglobin adducts. *Cancer Epidemiol. Biomark. Prev.*, **1**, 213–219.

15 Rowland, M. (1985) Physiological pharmacokinetic models and interanimal species scaling. *Pharmacol. Ther.*, **29**, 49–68.

16 Levy, G. (1980) Effect of plasma protein binding on renal clearance of drugs. 1. *Pharmaceut. Sci.*, **69**, 482–483.

17 Wesson, L.G. (1954) A theoretical analysis of urea excretion by the mammalian kidney. *Am. J. Physiol.*, **179**, 364–371.

18 Hall, S. and Rowland, M. (1983) Relationship between renal clearance, protein binding and urine flow for digitoxin, a compound of low clearance in isolated perfused rat kidney. *J. Pharmacol. Exp. Ther.*, **227**, 174–180.

19 Tangliu, D.D., Tozer, T.N., and Riegelman, S. (1983) Dependence of renal clearance on urine flow: a mathematical model and its applications. *J. Pharmaceut. Sci.*, **72**, 154–158.

20 Caster, W.O., Poncelet, J., Simon, A.B., and Armstrong, W.D. (1956) Tissue weights of the rat. 1. Normal values determined by dissection and chemical methods. *Proc. Soc. Exp. Biol. Med.*, **91**, 122–126.

21 International Commission on Radiation Protection (1975) *Report of the Task Group*

on Reference Man, ICRP Publication No. 23, Pergamon Press, New York.
22 Delp, M.D., Manning, R.O., Bruckner, J.V., and Armstrong, R.B. (1991) Distribution of cardiac output during diurnal changes of activity in rats. *Am. J. Physiol.*, **261**, H1487–H1493.
23 Arms, A.D. and Travis, C.C. (1988) Reference Physiological Parameters in Pharmacokinetic Modeling. NTIS PB 88-196019, Office of Health and Environmental Assessment, U.S. Environmental Protection Agency, Washington, D.C.
24 Chen, H.S.G. and Gross, J.F. (1979) Estimation of tissue to plasma partition coefficients used in physiological pharmacokinetic models. *Pharmacokinet. Biopharmaceut.*, **7**, 117–125.
25 Lin, J.H., Sugiyama, Y., Awazu, S., and Hanano, M. (1982) *In vitro* and *in vivo* evaluation of the tissue to blood partition coefficients for physiological pharmacokinetic models. I. *Pharmacokinet. Biophamaceut.*, **10**, 637–647.
26 Sato, A. and Nakajima, T. (1979) Partition coefficients of some aromatic hydrocarbons and ketones in water, blood and air. *Br. J. Ind. Med.*, **36**, 231–234.
27 Gargas, M.L., Burgess, R.J., Voisard, D.J., Cason, G.H., and Andersen, M.E. (1989) Partition coefficients of low molecular weight volatile compounds in various liquids and tissues. *Toxicol. Appl. Pharmacol.*, **98**, 87–99.
28 Sato, H., Sugiyama, Y., Sawada, Y., Iga, T., and Hanano, M. (1987) Physiologically-based pharmacokinetics of radioiodinated human endorphin in rats: An application of the i capillary membrane-limited model. *Drug. Metab. Disp.*, **15**, 540–550.
29 McDougal, J.N., Jepson, G.W., Clewell, H.J. III, and Andersen, M.E. (1985) Dermal absorption of dihalomethane vapors. *Toxicol. Appl. Pharmacol.*, **79**, 150–158.
30 Gargas, M.L. (1990) An exhaled breath chamber system for assessing rats of metabolism and rats of gastrointestinal absorption with volatile chemicals. *J. Am. Coll. Toxicol.*, **9**, 447–453.
31 Mauderly, J.L. (1990) Measurement of respiration and respiratory responses during inhalation exposures. *J. Am. Coll. Toxicol.*, **9**, 397–406.
32 Gargas, M.L., Andersen, M.E., and Clewell, H.J. III (1986) A physiologically based simulation approach for determining metabolic constants from gas uptake data. *Toxicol. Appl. Pharmacol.*, **86**, 341–352.
33 Gargas, M.L. and Andersen, M.E. (1989) Determining the kinetic constants of chlorinated ethane metabolism in the rat from rates of exhalation. *Toxicol. Appl. Pharmacol.*, **99**, 344–353.
34 Deitz, F.K., Rodriguez-Giaxoia, M., Traiger, G.J., Stella, V.J., and Himmelstein, K.J. (1981) Pharmacokinetics of 2-butanol and its metabolites in the rat. *Pharmacokinet. Biopharmaceut.*, **9**, 553–576.
35 Krishnan, K. and Andersen, M.E. (1993) *In vitro* toxicology and risk assessment, in *Alternative Methods in Toxicology*, vol. 10 (ed. A.M. Goldberg), Liebert, New York, pp. 113–124.
36 Lin, J.H., Hayashi, M., Awazu, S., and Hanano, M. (1978) Correlation between *in vitro* and *in vivo* drug metabolism rate: oxidation of ethoxybenzamide in rat. *J. Pharmacokinet. Biophamaceut.*, **6**, 327–337.
37 Lin, J.H., Sugiyama, Y., Awazu, S., and Hanano, M. (1982) Physiological pharmacokinetics of ethoxybenzamide based on biochemical data obtained *in vitro* as well as on physiological data. *J. Phamacokinet. Biophamaceut.*, **10**, 649–661.
38 Sugita, O., Sawada, Y., Sugiyama, Y., Iga, T., and Hanano, M. (1982) Physiologically based pharmacokinetics of drug–drug interaction: a study of tolubutamide-sulfonamide interaction in rats. 1. *Pharmacokinet. Biopharmaceut.*, **10**, 297–312.
39 Sato, A. and Nakajima, T. (1979) A vial equilibration method to evaluate the drug metabolizing enzyme activity for volatile hydrocarbons. *Toxicol. Appl. Pharmacol.*, **47**, 41–46.
40 Sato, H., Sugiyama, Y., Miyauchi, S., Sawada, Y., Iga, T., and Hanano, M. (1986)

A simulation i study on the effect of a uniform diffusional barrier across hepatocytes on drug metabolism ! by evenly or unevenly distributed unienzyme in the liver. *J. Pharmaceut. Sci.*, **75**, 3–8.

41 Reitz, R.H., Mendrala, A.L., Park, C.N., Andersen, M.E., and Guengerich, F.P. (1988) Incorporation of *in vitro* enzymatic data into the physiologically-based pharmacokinetic (PBPK) model for methylene chloride: implications for risk assessment. *Toxicol. Lett.*, **43**, 97–116.

42 Gearhart, J.M., Jepson, G.W., Clewell, H.J. III, Andersen, M.E., and Conolly, R.B. (1990) Physiologically based pharmacokinetic and pharmacodynamic model for the inhibition of acetylcholinesterase by diisopropyl fluorophosphate. *Toxicol. Appl. Pharmacol.*, **106**, 295–310.

43 Sultatos, L.G. (1990) A physiologically based pharmacokinetic model of parathion based on chemical specific parameters determined *in vitro*. *J. Am. Coll. Toxicol.*, **9**, 611–617.

44 Carfagna, M.A. and Kedderis, G.L. (1992) Isolated hepatocytes as *in vitro* models for the biotransformation and toxicity of chemicals *in vivo*. *CIIT Activities*, **12** (9–10), 1–6.

45 Clewell, H.J. III, and Andersen, M.E. (1987) Dose, species, and route extrapolations using physiologically based pharmacokinetic models. *Drink. Water Health*, **8**, 159–184.

46 Krishnan, K. and Andersen, M.E. (1994) Physiologically-based pharmacokinetic modeling in toxicology, in *Principles and Methods of Toxicology*, 3rd edn (ed. A.W. Hayes), Raven Press, New York, pp. 149–188.

47 Andersen, M.E., Clewell, H.J., III, Gargas, M.L., Smith, F.A., and Reitz, R.H. (1987) Physiologically based pharmacokinetics and the risk assessment process for methylene chloride. *Toxicol. Appl. Pharmacol.*, **87**, 185–205.

48 Segel, I.H. (1974) *Enzyme Kinetics: Behavior and Analysis of Rapid Equilibrium and Steady-State Enzyme Systems*, John Wiley & Sons, Inc., New York.

49 Kuhy, S.A. (1991) *A Study of Enzymes*, vol. I, CRC Press, Boca Raton, Florida, pp. 20–26.

50 Cunningham, E.B. (1978) *Biochemical Mechanisms of Metabolism*, McGraw Hill, New York.

51 Luecke, R.H. and Wosilait, W.D. (1986) Estimation of drug binding parameters. *Pharmacokinet. Biopharmaceut.*, **14**, 65–78.

52 Wosilait, W.D. and Luecke, R.H. (1990) Displacement interactions resulting from competition for binding sites on proteins, in *Toxic Interactions* (eds R. Goldstein, W.R. Hewitt, and J. Hook), Academic Press, New York, pp. 115–148.

53 Sato, A. and Nakajima, T. (1979) Dose-dependent metabolic interaction between benzene and toluene *in vivo* and *in vitro*. *Toxicol. Appl. Pharmacol.*, **48**, 249–256.

54 Krishnan, K. and Brodeur, J. (1991) Toxicological consequences of combined exposure to environmental pollutants. *Arch. Complex Environ. Stud.*, **3** (3), 1–106.

55 Clewell, H.J., III, and Andersen, M.E. (1985) Risk assessment extrapolations and physiological modeling. *Toxicol. Ind. Health*, **1**, 111–131.

56 Andersen, M.E., Gargas, M.L., Clewell, H.J. III, and Severyn, K.M. (1987) Quantitative evaluation of the metabolic interaction between trichloroethylene and 1,l-dichloroethylene *in vivo* using gas uptake methods. *Toxicol. Appl. Pharmacol.*, **89**, 149–157.

57 Purcell, K.J., Cason, G.H., Gargas, M.L., Andersen, M.E., and Travis, C.C. (1990) *In vivo* metabolic interactions of benzene and toluene. *Toxicol. Lett.*, **52**, 141–152.

58 Haddad, S., Tardif, R., Charest-Tardif, G., and Krishnan, K. (1999) Physiological modeling of the toxicokinetic interactions in a quaternary mixture of aromatic hydrocarbons. *Toxicol. Appl. Pharmacol.*, **161**, 249–257.

59 Thakore, K.N., Gargas, M.L., Andersen, M.E., and Mehendale, H.M. (1991) PBPK derived metabolic constants, hepatotoxicity, and lethality of $BrCCl_3$ in rats pretreated with chlordecone, phenobarbital or mirex. *Toxicol. Appl. Pharmacol.*, **109**, 514–528.

60 Mehendale, H.M. (1991) Role of hepatocellular regeneration and

hepatolobular healing in the final outcome of liver injury: a two stage model of toxicity. *Biochem. Pharmacol.*, **42**, 1155–1162.

61 Sato, A., Endoh, K., Kaneko, T., and Johanson, G. (1991) Effects of consumption of ethanol on the biological monitoring of exposure to organic solvent vapor: a simulation study with trichloroethylene. *Br. J. Ind. Med.*, **48**, 548–556.

62 Westerfeld, W.W., Schulman, M.P., and Syracuse, N.Y. (1959) Metabolism and clearance rate of alcohol. *J. Am. Med. Assoc.*, **170**, 197–203.

63 Fernandez, J.G., Droz, P.O., Humbert, B.E., and Capers, J.R. (1977) Trichloroethylene exposure: simulation of uptake, excretion and metabolism using a mathematical model. *Br. J. Ind. Med.*, **34**, 43–55.

64 Sato, A., Nakajima, T., and Koyama, Y. (1981) Dose-related effects of a single dose of ethanol on the metabolism of some aromatic and chlorinated hydrocarbons. *Toxicol. Appl. Pharmacol.*, **60**, 8–15.

65 Koizumi, A. (1989) Potential of physiological pharmacokinetics to amalgamate kinetic data of trichloroethylene and tetrachloroethylene obtained in rats and man. *Br. J. Ind. Med.*, **46**, 239–249.

66 Tardif, R., Lapare, S., Krishnan, K., and Brodeur, J. (1993) Physiologically based modeling of the toxicokinetic interaction between *m*-xylene and toluene. *Toxicol. Appl. Phamacol.*, **120**, 266–273.

67 Tardif, R., Lapar, S., Plaa, G.L., and Brodeur, J. (1991) Effect of simultaneous exposure to toluene and xylene on their respective biological exposure indices in humans. *Int. Arch. Occup. Environ. Health*, **63**, 279–284.

68 Matsumura, F. (1975) Metabolism of carbamate insecticides, in *Toxicology of Insecticides* (ed. F. Matsumura), Plenum Press, New York, pp. 228–239.

69 Chambers, H.W. (1992) Organophosphorus compounds: an overview, in *Organophosphates: Chemistry, Fate, and Effects* (eds J.E. Chambersand P.E. Levi), Academic Press, San Diego, pp. 3–17.

70 Timchalk, C., Nolan, R.J., Mendrala, A.L., Dittenber, D.A., Brzak, K.A., and Mattsson, J.L. (2002) A physiologically based pharmacokinetic and pharmacodynamic (PBPK/PD) model for the organophosphate insecticide chlorpyrifos in rats and humans. *Toxicol. Sci.*, **66**, 34–53.

71 Nong, A., Tan, Y.M., Krolski, M.E., Wang, J. Lunchick, C., Conolly, R.B., and Clewell, H.J. 3rd. (2008) Bayesian calibration of a physiologically based pharmacokinetic/ pharmacodynamic model of carbaryl cholinesterase inhibition. *J. Toxicol. Environ. Health A*. **71**, 413–426.

72 Tardif, R., Charest-Tardif, G., Brodeur, J., and Krishnan, K. (1997) Physiologically based pharmacokinetic modeling of a ternary mixture of alkyl benzenes in rats and humans. *Toxicol. Appl. Pharmacol.*, **144**, 120–134.

73 Haddad, S., Charest-Tardif, G., Tardif, R., and Krishnan, K. (2000) Validation of a physiological modeling framework for simulating the toxicokinetics of chemicals in mixtures. *Toxicol. Appl. Pharmacol.*, **167**, 199–209.

74 Haddad, S., Beliveau, M., Tardif, R., and Krishnan, K. (2001) A PBPK modeling-based approach to account for interactions in the health risk assessment of chemical mixtures. *Toxicol. Sci.*, **63**, 125–131.

75 Krishnan, K., Haddad, S., Beliveau, M., and Tardif, R. (2002) Physiological modeling and extrapolation of pharmacokinetic interactions from binary to more complex chemical mixtures. *Environ. Health Perspect.*, **110** (Suppl. 6), 989–994.

76 Toftgard, R. and Nilsen, O.G. (1982) Effects of xylene and xylene isomers on cytochrome P-450 and *in vitro* enzymatic activities in rat liver, kidney and lung. *Toxicology*, **23**, 197–212.

77 Guengerich, F.P., Kim, D.H., and Iwasaki, M. (1991) Role of human cytochrome P-450 IIE1 in the oxidation of many low molecular weight cancer suspects. *Chem. Res. Toxicol.*, **4**, 168–179.

78 Liira, J., Elovaara, E., Raunio, H., Riihimaki, V., and Engstrom, K. (1991) Metabolic interaction and disposition of methyl ethyl ketone and *m*-xylene in rats at single and repeated inhalation exposures. *Xenobiotica*, **21**, 53–63.

79 Nakajima, T., Wang, R.S., Elovaara, E., Park, S.S., Gelboin, H.V., Hietanen, E., and Vainio, H. (1991) Monoclonal antibody-directed characterization of cytochrome P450 isozymes responsible for toluene metabolism in rat liver. *Biochem. Pharmacol.*, **41**, 395–404.

80 Tassaneeyakul, W., Birkett, D.J., Edwards, J.W., Veronese, M.E., Tukey, R.H., and Miners, J.O. (1996) Human cytochrome P450 isoform specificity in the regioselective metabolism of toluene and *o*-, *m*- and *p*-xylene. *J. Pharmacol. Exp. Ther.*, **276**, 101–108.

81 Dobrev, I.D., Andersen, M.E., and Yang, R.S. (2001) Assessing interaction thresholds for trichloroethylene in combination with tetrachloroethylene and 1,1,1-trichloroethane using gas uptake studies and PBPK modeling. *Arch. Toxicol.*, **75**, 134–144.

82 Dobrev, I.D., Andersen, M.E., and Yang, R.S. (2002) In *silico* toxicology: simulating interaction thresholds for human exposure to mixtures of trichloroethylene, tetrachloroethylene, and 1,1,1-trichloroethane. *Environ. Health Perspect.*, **110**, 1031–1039.

83 Dennison, J.E., Andersen, M.E., and Yang, R.S. (2003) Characterization of the pharmacokinetics of gasoline using PBPK modeling with a complex mixtures chemical lumping approach. *Inhal. Toxicol.*, **15**, 961–986.

84 Dennison, J.E., Andersen, M.E., Clewell, H.J., and Yang, R.S. (2004) Development of a physiologically based pharmacokinetic model for volatile fractions of gasoline using chemical lumping analysis. *Environ. Sci. Technol.*, **38**, 5674–5681.

85 Saito, F.U., Kocsis, J.J., and Snyder, R. (1973) Effect of benzene on hepatic drug metabolism and ultrastructure. *Toxicol. Appl. Pharmacol.*, **26**, 209–217.

86 Debets, F.M., Strik, J.J.T.W.A., and Olie, K. (1980) Effects of pentachlorophenol on rat liver changes induced by hexachlorobenzene, with special reference to porphyria and alterations in mixed function oxygenases. *Toxicology*, **15**, 181–195.

87 Andersen, M.E. and Clewell, H.J. (1983) Pharmacokinetic interaction of mixtures. Proceedings of the Fourteenth Annual Conference on Environmental Toxicology, AFAMRL-TR-83-099, Dayton, Ohio, pp. 226–238.

88 Gilmour, S.K., Kalf, G.F., and Snyder, R. (1986) Comparison of the metabolism and benzene and its metabolite phenol in rat liver microsomes. *Adv. Exp. Med. Biol.*, **197**, 223–235.

89 Baker, T. and Rickert, D. (1981) Dose-dependent uptake, distribution and elimination of inhaled n-hexane in the Fischer-344 rat. *Toxicol. Appl. Pharmacol.*, **61**, 414–422.

90 Campbell, J.L. Jr.and Fisher, J.W. (2007) A PBPK modeling assessment of the competitive metabolic interactions of JP-8 vapor with two constituents, *m*-xylene and ethylbenzene. *Inhal. Toxicol.*, **19**, 265–273.

91 Gordon, C.J., Herr, D.W., Gennings, C., Graff, J.E., McMurray, M., Stork, L., Coffey, T., Hamm, A., and Mack, C.M. (2006) Thermoregulatory response to an organophosphate and carbamate insecticide mixture: testing the assumption of dose-additivity. *Toxicology*, **217** (1), 1–13.

92 Krishnan, K., Clewell, H.J. III, and Andersen, M.E. (1994) Physiologically-based pharmacokinetic analysis of simple mixtures. *Environ. Health Perspect.*, **102** (Suppl. 9), 151–155.

6
Toxicodynamic Interactions

Binu K. Philip, S. Satheesh Anand, and Harihara M. Mehendale

6.1
Introduction

Humans and animals are exposed concurrently or sequentially to multiple chemicals via multiple exposure routes from multiple environmental media, including the workplace (solvents, particles, dusts, etc.), environment (air pollutants, cigarette smoke, etc.), home (drinking water contaminants, cleaning agents, and solvents), leakage and distribution from hazardous waste sites, and drinking water containing disinfection by-products in addition to other contaminants. Indeed, all the living organisms contain complex mixture of biosynthesized or man-made chemicals. The nature of chemical mixtures varies widely in the specific chemicals present and their concentrations. In most cases, many of the components and concentrations of complex mixtures such as those found at hazardous waste sites (hundreds of chemicals and identity of most of them is unknown) and their toxicities are unknown. Determining the risk that such mixtures may pose to human populations is one of the most daunting tasks for toxicological research and risk assessment. These realities have heightened the need for exposure assessment, hazard identification, and risk characterization of chemical mixtures.

It has long been known that the presence of multiple chemicals within a biological system increases the potential for interactions that could enhance or diminish the toxicity of each other. Bliss [1] defined seven decades ago three main categories of joint chemical actions: (*a*) independent joint action, which refers to chemicals that act independently and have different modes and mechanisms of action, such that the presence of one chemical will not impact the toxicity of another and the combined toxicity can be predicted from the knowledge of the individual chemicals; (*b*) similar joint action, which refers to chemicals that cause similar effects often through similar mechanisms, and in this case how the presence of one chemical may affect the toxicity of another chemical (e.g., if two chemicals, A and B, act by combining with the same receptor in the body, the impact of B will depend on how much chemical A is present – its effect might be lessened or heightened in the presence of A). Therefore, as with independent joint action, toxicity can be predicted with the knowledge of the individual chemicals; and (*c*) synergistic or antagonistic action, where the effect of the

Principles and Practice of Mixtures Toxicology. Edited by Moiz Mumtaz

ISBN: 978-3-527-31992-3

mixture cannot be predicted from the knowledge of the individual ingredients because it may depend upon their combined toxicity when used in different proportions. From public health perspective, a major toxicological issue is the possibility of unusual toxicity due to interaction of two or more toxic chemicals at individually harmless levels.

Despite the awareness that exposure to mixtures is a reality and toxic interactions could cause exaggerated toxicity, safety evaluation studies are primarily based on the toxicity of individual chemicals. A cursory review of toxicological journals by Dr. Raymond Yang in 1992 revealed that nearly 95% of the toxicology studies are devoted to single chemicals [2]. Toxicity studies with single chemicals are conducted for the sake of simplicity, to understand the mechanisms of toxicity with minimum confounding factors, and to satisfy regulatory requirements on individual chemicals. Of the 5% of toxicological studies that do address mixtures, the great majority focus on binary mixtures (or two chemicals at a time), employ relatively high chemical concentrations, and observe relatively crude end points.

There has been a greater emphasis on the development of both scientific and regulatory methodology to improve our ability to evaluate the human and environmental health impacts of chemical mixtures for the last two decades [3–5]. The importance of addressing chemical interactions was highlighted by regulatory agencies, US EPA and ATSDR, in their recommendations for evaluating risk associated with chemical mixtures [6, 7]. Public awareness of chemical interactions of biological significance also heightened the need for studying chemical mixture toxicity. Consequently, the area of chemical mixture toxicology has been advancing rapidly [8–10].

6.2
Historical Perspective of Chemical Mixtures

People working in industrial settings and living in the vicinity of landfill sites are the best examples of exposures to complex chemical mixtures. The presence of large quantities of potentially hazardous mixtures of toxic chemicals in Love Canal (upstate New York) and Times Beach (Missouri) attracted a great deal of attention to the complexity of exposure to chemical mixtures. Other well-known incidents include human consumption of rice bran oil contaminated with polychlorinated biphenyls (PCBs) in Japan and Taiwan and flame retardant (polybrominated biphenyls (PBBs)) contamination of cattle feed in Michigan. These incidents increased the public awareness of mixture toxicity and led to various regulations such as Comprehensive Environmental Response, Compensation, and Liability Act (CERCLA). These incidents are briefly discussed in the following sections.

6.2.1
Love Canal

Love Canal began with a dream of model industrial city and the availability of abundant, inexpensive hydroelectric power around Niagara River in 1892 and turned into a nightmare and became a poster child for chemical mixture exposure. William T. Love envisioned a 7-mile-long transport and hydropower canal that would be helped by 300-ft drop of Niagara Falls. Love Canal project began with the support of local government, private investors, and manufacturers. In 1910, the project was abandoned due to fading financial support and discovery of a new method of transport of electric power, leaving behind a partially completed and empty 16-acre canal as its only legacy. The canal served as chemical waste disposal site. In 1953, the site was covered with 1–4 ft of earth and sold to Niagara School Board and an elementary school was built. A blue-collar neighborhood developed in the area. In 1976, the New York State Department of Environmental Conservation (NY DEC) suspected mirex contamination in the site. However, no mirex was found, rather more than 200 organic compounds with a labyrinth of toxic properties were found. Most of the compounds were chlorinated hydrocarbons, but many of which could not be identified and toxicity of only a small percentage of the compounds had been described in the literature. No information on the toxicity of the combination of the chemicals in the mixture was available. The New York State Health Commissioner stated Love Canal site to be "an extremely serious threat and danger to the health, safety, and welfare of those using it, living near it, or exposed to it" and directed that "all appropriate and necessary corrective actions be taken to abate the public health nuisance." Intensive chemical analysis of the site, adjacent homes, and residents was initiated. Two epidemiological studies [11, 12] reported contradicting results: one study reported significant chromosomal damage in human populations, while the other found no such evidence. Subchronic exposure of mice to soil collected from the site caused hepatic lesions [13]. Vianna and Polan [14] reported a transient, but statistically significant, increase in the incidence of low birth weights in people living near Love Canal from 1940 through 1978. Subsequent animal studies with organic extract of soil supported this finding [15]. Finally, Superfund legislation was signed, authorizing the expenditure of federal money to aid in the cleanup of waste dumps and the recovery of these costs from the companies responsible for the contamination. Thus, Love Canal became an example of notorious chemical mixture problem and provided the impetus for the development of new regulatory and scientific approaches to study the issues of mixture toxicology. In addition, this incident led to the realization that very little toxicity information was available for complex mixtures, which was important in assessing the health problems created or perceived by the presence of industrial dumps such as Love Canal.

6.2.2 Times Beach

The Times Beach was a resort town occupying a 0.8 square mile area on the Meramec River with a population of 2000 people in Missouri. From 1972 to 1976, due to lack of city funds, the city contracted with a waste oil hauler to spray oil on unpaved roads for dust control [16]. The contractor mixed waste oil with wastes from a pharmaceutical and chemical company that contained PCBs to pave the road. He sprayed the oil containing dioxins at dozens of sites in Missouri, including his own farm, and created the biggest toxic waste disaster since Love Canal. The dioxins ended up permeating the soil in dangerous amounts, and this was further exacerbated by the flood of the Meramec River in 1982. Many illnesses, miscarriages, and animal deaths occurred in the city attributed to the dioxins. Polychlorinated biphenyls are a family of more than 200 chemical compounds (congeners), which degrade very slowly in the environment, can build up in the food chain, and are known to cause acne, birth defects, immune effects, and cancer [17]. An anonymous caller tipped EPA about the presence of PCBs in the waste oil application. A site investigation was conducted in the town of Times Beach in 1982. In February 1983, under the advice of the Center for Disease Control (CDC), EPA transferred funds to the Federal Emergency Management Agency (FEMA) to permanently relocate Times Beach residents and businesses [18]. The government bought out the land and relocated its citizens and businesses. Times Beach has been devoid of human life for about two decades as a result of the contamination. As part of the nearly 20-year cleanup, Syntex was granted a contract by the EPA in 1990 to install an incinerator and thermal treatment facility, which began operation in 1995. The new facility was utilized to destroy all dioxin-contaminated soil. Two years later, after 265 000 tons of soil were burned, the incinerator was shut down and dismantled, as the authorities believed the cleanup to be finished. The former town has since been reborn as Route 66 State Park, which opened in 1999. This incident remains as the worst man-made disaster in American history. The former town is a chilling reminder of the harm we are capable of inflicting on our environment. Times Beach, along with the Love Canal section of Niagara Falls, New York, share a special place in the history of the United States environmental policy as the two sites that in large part led to the Comprehensive Environmental Response, Compensation, and Liability Act (CERCLA). CERCLA is much more commonly referred to as "Superfund" because of the fund established within the act to help the cleanup of locations like Times Beach.

6.2.3
Rice Bran Oil Contaminated with PCBs in Japan and Taiwan

The problems of PCBs and polychlorinated dibenzofurans (PCDFs) have aroused a great deal of concern following the two episodes of poisoning due to contamination of rice bran oil with these chemicals. These events took place in Japan in 1968 (Yusho incident) and in Taiwan in 1979 (Yu-Cheng incident). Victims of the poisoning experienced a range of symptoms, including chloracne, fatigue, nausea, numbness in the arms and legs, headache, liver disorders, brown pigmentation of the skin and nails, excessive eye discharge, swelling of the eyelids, and distinctive hair follicles [19, 20]. Some babies born to mothers who had consumed the contaminated oil showed skin hyperpigmentation and behavioral and cognitive problems [21, 22]. Studies of children born to poisoned mothers showed developmental delays at all stages, persistent delays in growth, behavioral problems at age 3–9, and reduced penile length in boys aged 11–14 [22]. The rice bran oil that many people consumed in these two countries was unknowingly contaminated during the heat transfer process from the pipes that carried PCBs as the heat transfer fluid. Small holes developed in these pipes allowing PCBs from the heat transfer fluid to leak into the oil flowing in a countercurrent direction in the pipe surrounding the pipe carrying rice oil extract. The fluid that contaminated the oil was partially heat degraded, so in addition to PCBs, high levels of chlorinated dibenzofurans and chlorinated quaterphenyls also escaped into the rice bran oil [17, 19]. Since the thermal degradation products of PCBs are highly toxic, there has been some controversy over whether the adverse health outcomes experienced by people who consumed food cooked in the oil are due to PCBs or the products of heated PCB degradation: dibenzofurans and quaterphenyls [17].

6.2.4 Cattle Feed Contamination with Polybrominated Biphenyls

Polybrominated biphenyls have been widely used as flame retardants in industrial and consumer products. In 1973, PBBs accidentally entered the food chain in Michigan, when Firemaster FF-1, a commercial flame retardant, was inadvertently substituted for magnesium oxide (Foodmaster) in filling the Foodmaster bags used as a supplement in the formulation of cattle feed [23]. Contaminated Feedmaster bags were shipped for use as mineral supplement for cattle feed. Dairy cattle were the first to show signs of intoxication (decreased milk production, anorexia, alopecia, and abnormal growth), which were attributed to the ingestion of PBB-contaminated feed. Subsequently, other farm animals were also found to have been contaminated and adversely affected. For about 9 months following the accident, contaminated dairy and poultry products were consumed by Michigan farmers and residents. In order to minimize further human exposure, at least 29 800 cattle, 5900 hogs, 1500 sheep, and 1.5 million chickens were destroyed along with large amounts of animal feed, cheese, butter, dry milk products, and eggs [24]. Until the Michigan accident, very little was known about the toxicity of PBBs. This incident triggered a series of studies to investigate the health effects of PBBs. Toxicological assessments of Firemaster FF-1 have been complicated because this commercial preparation contains a mixture of PBBs as well as traces of impurities (brominated naphthalenes), which have their own toxic properties.

6.2.5 Great Lakes

The Great Lakes is the largest body of surface freshwater on earth and the basin is a major industrial and agricultural region of North America. The health and integrity of the Great Lakes ecosystem has been the subject of much public and scientific interest and debate for decades [25]. By the mid-1980s, over 800 distinct chemical substances from a variety of industrial, agricultural, and municipal sources had been identified in the Great Lakes basin, of which only half of them are well known such as PCBs, DDT, toxaphene, mirex, dieldrin, hexachlorobenzene, the heavy metals (lead, cadmium, nickel, copper, etc.) and benzo[*a*]pyrene. In 1978, the United States and Canada signed Great Lakes Water Quality Agreement and subsequent efforts have led to the virtual elimination of all persistent toxic substances from the Great Lakes.

Because of the efforts, levels of critical pollutants have dropped significantly [26]. While laboratory animal studies and human epidemiologic studies of accidental and occupational exposures have demonstrated that high concentrations of certain heavy metals (lead, cadmium, nickel, copper, etc.) and persistent chlorinated contaminants present in Great Lakes can cause a range of adverse health effects in humans [26], interactive effects of these chemicals are unknown. A wide range of effects were observed in the region, including reproductive failure and population declines in mink, lake trout, bald eagles, snapping turtles, double-crested cormorants, herring gulls, and common terns. However, whether these effects are caused by single chemicals or combination of chemicals is still unclear.

6.3
Current Status

The need for improved toxicological data and methodology for assessment of health effects of chemical mixtures to which the public is exposed has resulted in several initiatives. US EPA and ATSDR developed guidance documents for chemical mixtures. Some of the commonly used approaches in the field of mixture risk assessment are as follows: (1) Treat mixtures as simple mixtures (two or three known chemicals), deriving the combined toxicity primarily from single-chemical studies. (2) Prioritize chemical mixtures and collect toxicity data. (3) Competitive metabolic inhibition as way of interaction. (4) Mixtures cause additive toxicity. The additivity assumption is valid only when there is no interaction among the components of the mixture through toxicokinetics (TK) or toxicodynamics (TD) mechanisms. Several environmental chemicals interact with each other by various mechanisms that are dependent on the dose, dosing regimen (i.e., single or repeated exposure), exposure pattern (i.e., simultaneous, staggered, pretreatment, coadministration, or sequential), and/or exposure route of one or all chemicals [27]. Several studies with chemical mixtures at concentrations near or below NOAEL have reported potentially harmful biologic responses [28–30]. We have well-documented examples of synergistic actions of environmental agents in humans; decay products of radon and smoking have synergistic effects on the incidence of lung cancer [31] and smoking and exposure to asbestos exert synergistic effects on incidence of lung cancer [32]. Obviously, smoking, considered alone, represents an exposure to a very complex chemical mixture, but this does not alter the clear evidence that the addition of a second environmental factor, either radon or asbestos, results in more than an additive risk for injury. Because humans are exposed to a wide variety of environmental compounds, it is necessary to consider the potential for interactions that could result in a cumulative toxicity different from the default assumption of additivity.

In order to effectively address the mixture toxicity problem, all the available integrated scientific evidence should be considered. Each component of a mixture may possess a unique toxic potential and may influence the toxicity of another component chemical by affecting absorption, distribution, metabolism, and excretion (toxicokinetics) [33]; or by altering inherent susceptibility of the cell or tissue to damage; or by affecting the ability of the cell or tissue to repair damage (toxicodynamics) [8, 34–36]. Both TK and TD interactions can significantly impact the toxicity of chemical mixtures. While it is widely accepted that understanding TK and TD interactions of chemical mixtures is critical, the present mixture toxicity studies measure only toxicokinetic interactions and toxicodynamic responses such as tissue repair (TR) receive scant recognition. But, it has been shown that the opposing biological dynamic events characterized by cell division and TR play a decisive role in the outcome of hepatotoxicity [8, 34, 37–41]. Toxicokinetic interactions will permit us to predict infliction of injury satisfactorily but do not allow us to predict the ultimate outcome of toxic injury. The ultimate outcome of toxicity hinges on whether and how much cellular or tissue level protective mechanisms are permitted to occur.

In this chapter, we review experimental evidence showing that TD response, TR, is a dose-dependent event regardless of the number of chemicals involved in the exposure and any alteration in this response markedly influences the final toxic outcome. In addition, the implications of stimulated or inhibited TR for risk assessment and public health, mechanisms, and factors affecting this biological response are discussed. While TD interaction is discussed extensively, TK interaction is addressed only to the extent of supporting the importance of TD.

6.4 Tissue Repair

The ability to overcome injury from physical or chemical agents encountered in the environment is a remarkable attribute of all living organisms. In an attempt to enhance survival from any noxious injury, organisms have developed several lines of defense mechanisms. By and large, such mechanisms may be categorized into two classes: (1) biochemical mechanisms that enable the organism to prevent or minimize initiation of injury after a noxious insult and (2) biological response intended to overcome injury, by promoting tissue healing after the fact. The extraordinary ability of liver to regenerate upon surgical resection (surgical injury) or other types of tissue injury has been known since prehistoric times [42]. Liver regeneration has been studied in detail in a variety of models with two-thirds partial hepatectomy (PH) in rodents serving as the principal model system [43, 44]. Investigations during the last quarter century have revealed that a similar dynamic regeneration response or TR occurs following cell death and tissue injury after exposure to toxic chemicals in liver and other tissues [38, 40, 45–52]. Consideration of TR as the biologic event opposing injury may result in two sets of dose-response curves in the classic dose-response paradigm. At lower doses, upon initiation of toxic injury, a cascade of distress signals triggered in a dose-response fashion stimulate TR by promoting surrounding healthy cells to divide in order to replace the dead cells [45, 53–57]. Predictably, these animals will suffer from injury, but are rescued from progression of injury and death. As the dose increases, a threshold is reached where such promitogenic signaling is inhibited resulting in delayed or inhibited TR culminating in unrestrained progression of injury and animal death (Figure 6.1) [38, 53, 54, 58].

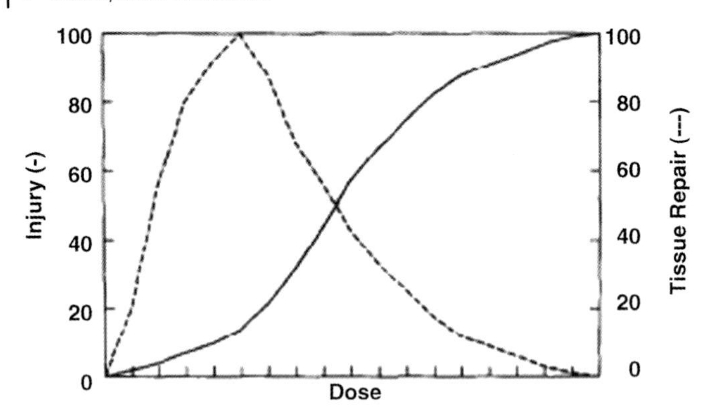

Figure 6.1 A typical dose-response relationship between the two opposing forces of inflicted injury and stimulated tissue repair upon exposure to a toxic chemical. As the dose of the toxicant increases, tissue repair is increased, facilitating recovery from tissue injury. When the dose exceeds the threshold, tissue repair is attenuated and delayed, allowing injury to progress in an unrestrained manner and leading to organ or tissue failure and animal death. Quantifying both injury as well as stimulated tissue repair simultaneously as a dose-response relationship is important in assessing the outcome of the interaction between these two opposing forces.

6.4.1 Studies That Brought TR to the Forefront

Three decades ago, when researchers were reluctant to investigate the effects of chemical mixtures let alone studying the importance of TD response, Dr. Mehendale's research group was the first to show the importance of TR in the outcome of exaggerated toxicity of chemical mixture at individually nontoxic levels.

First, Dr. Mehendale and associates [59, 61] showed that a simple binary mixture (sequential exposure) of toxicants caused highly exaggerated toxicity of one of the toxicants. Because the toxicity of only one component was increased, they introduced the term "amplification" of toxicity. Using combinations of chlordecone (CD), CCl_4, chloroform ($CHC1_3$), bromotrichloromethane ($BrCCl_3$), and so on, they showed that prior exposure to 10 ppm CD for 15 days in the diet sensitized experimental animals (all species and strains investigated thus far except gerbils) to highly exaggerated toxicity of the halomethanes. Their work showed that hepatotoxicity of a normally nontoxic dose of CCl_4 was amplified 67-fold higher in male SD (Sprague–Dawley) rats [59, 64, 92, 93]. Also, bioactivation of CCl_4 was not increased [8, 59]. Existing literature at the time suggested that increased toxicity of CCl_4 might be due to induction of hepatic P450 enzymes bioactivating CCl_4 to $^{\bullet}CCl_3$. However, Mehendale and associates [8, 38, 40, 46, 48, 59] went on to establish that this was not the mechanism of 67-fold amplification of CCl_4 by CD. The toxicity was amplified because the ability of the liver to replace the dead cells by the necrogenic action of the low dose of CCl_4 was inhibited by the interactive toxicity of $CD + CCl_4$.

6.4.2
CD + CCl_4 Toxicity

The importance of TR in interactive toxicity of chemical mixtures can be explained well with the classical potentiation of CCl_4 toxicity by pretreatment with Kepone® (chlordecone). Dietary exposure to a nontoxic dose of CD (10 ppm, 15 days) results in a 67-fold amplification of hepatotoxicity and lethality of an ordinarily nontoxic dose of CCl_4 (100 μl/kg, i.p.). Neither the close structural analogues of CD, mirex, and photomirex nor the phenobarbital (PB) exhibit this property [59–62]. Although the PB and halomethane combination results in highly increased liver injury, even more than that observed with chlordecone and halomethanes, this injury is of no consequence to animal survival [63]. CD also potentiates the hepatotoxicity and lethality of $CHC1_3$ and $BrCCl_3$. Although the toxicity of these closely related halomethanes is potentiated by such low levels of CD, the toxicity of structurally and mechanistically dissimilar compounds like trichloroethylene (TCE) and bromobenzene is not potentiated [8, 35, 39, 64]. The mechanism underlying the CD-amplified toxicity of halomethanes was the subject of extensive investigations during the 1980s. Studies showed that this remarkable capacity to potentiate halomethane hepatotoxicity is not related to CD-induced cytochrome P450 or associated enzymes, enhanced bioactivation of $CC1_4$, increased lipid peroxidation, or decreased glutathione. These and other candidate mechanisms were considered carefully in experiments designed to verify their adequacy and was found inadequate; additional experiments revealed that inhibition of TR plays an important role in the progression of toxicity [8].

Figure 6.2 illustrates the mechanism of recovery from injury after administration of a low dose of CCl_4 alone and the mechanism of progression of liver injury following exposure to CD + CCl_4 mixture. Within 6 h after the administration of a low dose of CCl_4 (100 μl/kg, i.p.), limited hepatocellular necrosis accompanied by ballooned cells and steatosis inflicted by the same widely accepted mechanisms of CCl_4 bioactivation followed by lipid peroxidation occurs. Simultaneously, the liver tissue responds by stimulating the early phase of hepatocellular regeneration at 6 h even though centrilobular necrosis only begins to manifest at that time [8, 64]. It is clear that the early burst of cell division is due to the mobilization of a small number of hepatocytes that are present normally in the liver in G_2 phase [65, 66]. Although the molecular events responsible for the stimulation of hepatocellular division have not been fully explored, glycogen, the principal form of hepatic energy resource, is mobilized prior to cell division [46, 47]. Glycogen levels are restored after cell division has been adequately completed [46, 47]. The limited hepatocellular necrosis enters the progressive phase between 6 and 12 h, while the hepatocellular regeneration and tissue healing processes continue [46–48]. By 24 h, no significant liver injury is evident. Any remaining level of tissue injury is overcome by the second phase of hepatocellular division that occurs at 36–48 h [67–69]. These observations indicate that stimulation of hepatocellular regeneration is a protective response of the liver, occurs very early after the administration of a low dose of $CC1_4$, and leads to replacement of dead cells, thereby restoring the hepatolobular architecture and liver function [59, 70].

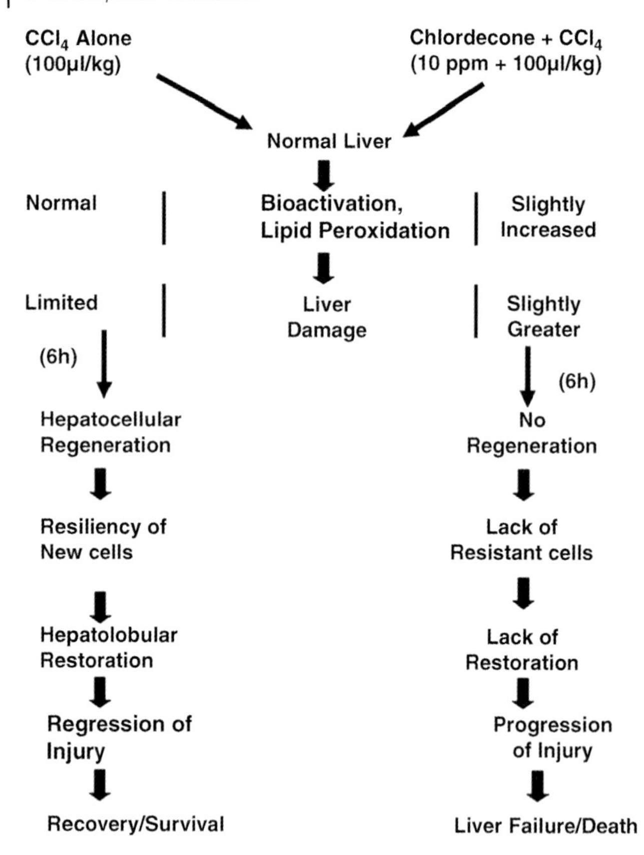

Figure 6.2 Mechanism for the highly amplified interactive toxicity of chlordecone + CCl_4. The scheme depicts the concept of suppressed hepatocellular regeneration, simply permitting what is normally limited liver injury caused by a subtoxic dose of CCl_4 to progress in the absence of hepatolobular repair and healing mechanisms stimulated by the limited injury. The limited hepatotoxicity from a low dose of CCl_4 is normally controlled and held in check owing to the hepatocellular regeneration and hepatolobular healing. The chlordecone + CCl_4 combination treatment results in unabated progression of injury owing to lack of tissue repair obtunded due to lack of cellular energy. These events lead to complete hepatic failure, culminating in animal death.

Furthermore, this remarkable biological event results in another important protective action. It is known that newly divided cells are relatively resistant to toxic chemicals [48, 71]. Therefore, in addition to the restoration of the hepatolobular architecture by cell division, by virtue of the relatively greater resistance of the new cells, the liver tissue is able to overcome the imminence of continued expansion of injury during the progressive phase (6–12 h), preventing the spread of injury on the one hand and speeding up the process of recovery through tissue healing on the other. By 6 h, over 75% of the administered CCl_4 is eliminated in the expired air leaving less than 25% in the animal. At later time points (12 h and onward), most of the CCl_4 will have been eliminated by the animal, thereby preventing additional infliction of injury. Continued cellular regeneration during this time period and at later time points allows the complete restoration of the hepatolobular architecture during and after the progressive phase of injury [67–69]. Relative resiliency of the newly divided cells at this critical time frame, as the animal continues to exhale the remaining CCl_4, is an added critical defense mechanism easily available through cell division.

If the animals had been exposed to food contaminated with 10 ppm CD for 15 days prior to the administration of a single low dose of CCl_4 (100 μl/kg, i.p.), the early phase of cell division would have abolished (Figure 6.2). It is known that during cell division, a number of growth factors and protooncogenes are overexpressed and these products of gene expression profoundly facilitate division of other cells in the tissue [72–76]. Therefore, if the early phase of cell division does not occur, the mechanism and signals necessary to facilitate or "prime" the neighboring cells would be unavailable. This results in progression of injury, given the case that there is no other biological mechanism to restrain injury [77]. By 24–36 h, while the second phase of cell division does occur, in the face of the unrestrained progression of injury this cell division and TR becomes far too little and too late to restrain the accelerated progression of injury [67–69]. It is now known that the reason for failed early-phase cell division in animals receiving the combination of CD and CCl_4 is due to lack of sufficient cellular energy [77–82]. At 6 h after CCl_4 administration, the available ATP is precipitously decreased in the livers of rats receiving CD and CCl_4 combination, thereby incapacitating the cells from dividing [83–85]. The concept of insufficient hepatocellular energy being linked to failure of hepatocellular regeneration and TR has gained support from experiments in which the administration of external source of energy (Catechin (cyanidanol)) resulted in augmented ATP levels and significant protection against the lethal effect of CD + CCl_4 [80, 81]. Protection by catechin is accompanied by a restored stimulation of hepatolobular repair and tissue healing [80]. The most interesting aspect of catechin protection against the interactive toxicity of CD + CCl_4 is that protection does not appear to be the result of decreased infliction of hepatic injury, as evidenced by a lack of difference in injury up to 24 h after CCl_4 administration [80, 81].

The aforementioned studies showed that the animals' inability to mount sufficient TR led to the progression of injury and death when exposed to CD + CCl_4. To support this conclusion, CD + CCl_4 mixture was tested in experimental models (partial hepatectomy and developing rats) where regeneration is active. When CCl_4 was administered to CD-fed rats 2 days after partial hepatectomy at a time of maximally stimulated hepatocellular division, remarkable protection was observed [86]. At 7 days after partial hepatectomy, when the stimulated cell division phases out, the interactive toxicity becomes fully manifested again [86]. Moreover, protection was not observed when a large dose of CCl_4 was administered since the early-phase cell division normally stimulated by a low dose of CCl_4 was inhibited entirely by the large dose of CCl_4. These findings indicate that the real difference between a low and a high dose of CCl_4 is the presence or absence of hormetic response in the form of stimulated early-phase cell division and tissue repair. Newborn and young developing rats (2, 5, 20, and 35 days) were completely resilient to CD potentiation of $CC1_4$ toxicity. The resiliency of younger rats to CD potentiation of CCl_4 toxicity is more likely related to ongoing hepatocellular regeneration during early development rather than to differences in the bioactivation of $CC1_4$ [87]. These studies clearly demonstrate the role of insufficient and suppressed hepatic TR as the mechanism of potentiation of CCl_4 toxicity by pretreatment with CD.

6.4.3
CD and Other Halomethanes

Importance of TR has been demonstrated with other models of CD-potentiated toxicities. Plaa and colleagues [88, 89] have demonstrated the capacity of CD to potentiate $CHC1_3$ hepatotoxicity in mice. These observations have been extended to demonstrate that, in addition to the hepatotoxic effects of CCl_4, the lethal effect of $CHC1_3$ is also potentiated by exposure to 10 ppm dietary CD [90] and that this is also associated with suppressed repair of the liver tissue [91]. Preexposure to CD also leads to potentiation of hepatotoxicity and lethality of $BrCCl_3$ [60, 92].

6.5
Interactions Leading to Increased Liver Injury, But Not Death

Prior exposure to PB and subsequent exposure to very low levels of CCl_4 (Figure 6.3), $CHCl_3$, or $BrCCl_3$ result in highly increased liver injury [60, 61, 90, 92–97]. However, this highly exaggerated liver injury does not lead to increased lethality in exposed animals [61, 70]. Studies showed that methanol, ethanol, and isopropanol also potentiate CCl_4 liver injury without having any impact on the survival [98–100]. The potentiation of CCl_4 hepatotoxicity by prior exposure to isopropanol observed by Harris and Anders [98] did not cause animal death. The reason behind this perplexing observation remained uninvestigated until the investigations of Rao *et al.* [101]. They reported that the reason for survival of rats despite the highly potentiated hepatotoxicity was the stimulated liver tissue repair, just as observed with PB pretreatment [101]. Depletion of ATP does not occur in these livers [83, 84]. Therefore, the only consequence of this highly toxic liver injury is to postpone the early-phase cell division until 24 h. However, the second phase of cell division is greatly stimulated [83]. In combination, this wave of highly stimulated cell division and TR leads to systematic restoration of hepatolobular structure and function, followed by full animal recovery.

The above-mentioned examples clearly show that the biological events that occur beyond infliction of injury determine whether the progression or regression of that injury will occur, which in turn determines the ultimate outcome. Thus, if TR is permitted to occur, injury is restrained, tissue structure and function are restored, and animal recovery occurs. On the other hand, if TR is blocked, unrestrained progression of injury leads to further deterioration of the tissue and ability to survive.

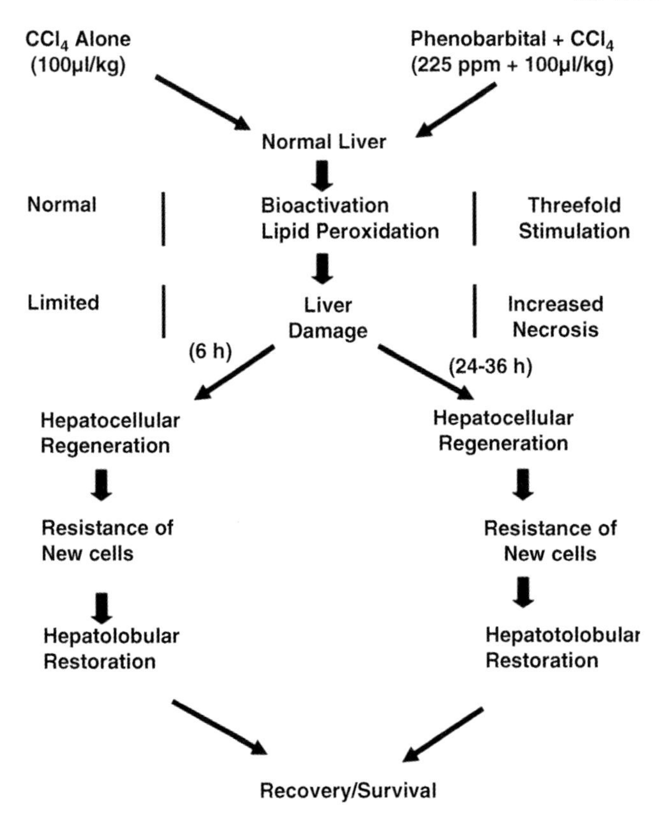

Figure 6.3 Mechanism for phenobarbital-induced potentiation of CCl_4 hepatotoxicity without any increase in lethality. Normal liver response to a low-dose CCl_4 injury is not abrogated by phenobarbital + CCl_4 interaction. Instead, the early phase of cell division is postponed from the normal 6–24 h. Enhanced putative mechanisms such as increased bioactivation of CCl_4 and resultant increased lipid peroxidation are responsible for the increased infliction of stage 1 injury. Because hepatocellular regeneration and tissue repair processes continue albeit a bit later than normal, these hormetic mechanisms permit tissue restoration resulting in recovery from the enhanced liver injury. This mechanism explains the remarkable recovery from phenobarbital-induced enhancement of CCl_4 liver injury. Despite a remarkably enhanced liver injury by phenobarbital, this is of no real consequence to the animal's survival because depletion of cellular energy does not occur with this interaction, which permits hormetic mechanisms to restore hepatolobular architecture resulting in complete recovery.

6.6 Two-Stage Model of Toxicity

The studies described above show the determining effect of compensatory TR in the final outcome of toxicity, that is, progression or regression of injury. Hormetic mechanisms are activated upon exposure to toxicants either individually or as mixtures. Although the mechanisms responsible for triggering mobilization of

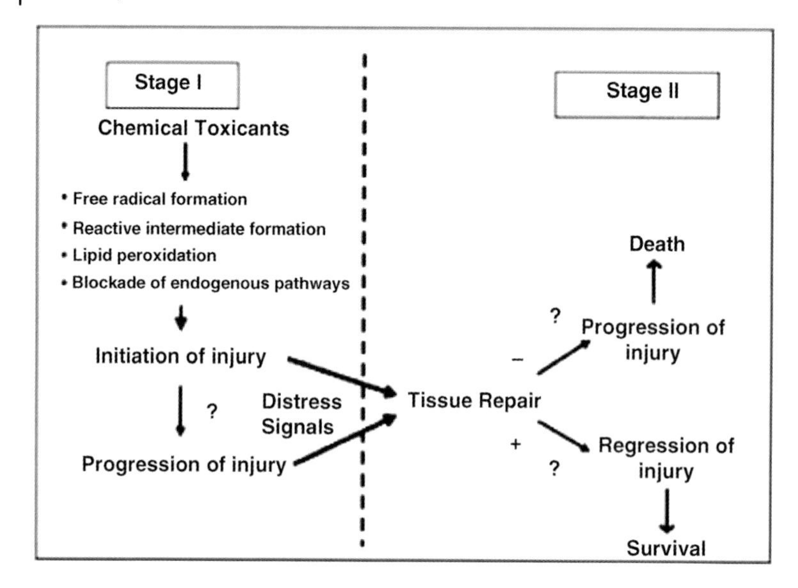

Figure 6.4 Mechanistic two-stage model of toxicity. During stage 1, toxic chemicals initiate tissue injury via the well-established bioactivation-based events. The injury further progresses due to some unknown mechanisms. During stage 2, infliction of injury stimulates compensatory tissue repair response after treatment with low to moderate doses of the toxicant (+ tissue repair) leading to prompt regression of injury permitting animal survival. In contrast, high dose of the toxicants inhibit tissue repair (– tissue repair) further leading to unrestrained progression of injury and animal death.

biochemical events leading to cellular proliferation after the administration of a toxicant are not fully understood, it is clear that early events involved in TR are the critical determinants for the final outcome of toxicity, that is, progression or regression of injury [102]. These studies emphasize the existence of two distinct stages of toxicity (Figure 6.4). Stage 1 is the inflictive stage in which toxic chemicals initiate injury through well-established mechanisms tempered via the net effect of bioactivation and detoxification processes, while stage 2 is the progression/regression phase of injury corresponding to the absence/presence of compensatory TR, respectively. Cell replacement and TR stimulated after the low to moderate doses of the toxicants restrain injury resulting in recovery [37, 58], while the high doses of toxicants inhibit compensatory TR leading to unrestrained progression of liver injury and animal death [54]. The two-stage model of toxicity emphasizes the critical role of opposing interplay of progression and regression of acute toxic injury in determining the final outcome.

6.7
Tissue Repair Follows a Dose Response After Exposure to Chemical Mixtures

The characteristic of exposure to noxious agents and the spectrum of toxic effects come together in a correlative relationship, customarily referred to as a dose-response relationship, the most fundamental and pervasive concept of toxicology. This concept is used in predictive toxicology and is a very basic principle justifiably used in risk assessment. In developing dose-response relationships for toxic chemicals, toxicologists have used only toxic injury measured against a series of increasing doses. This information is incomplete and leads to erroneous predictions because it does not take into consideration the opposing and dynamic reparative and tissue restoration response. Including dose response of TR would be more precise and accurate in predicting the final outcome of toxic injury.

TR is a dose-dependent phenomenon up to a threshold. Dose dependency of TR has been well established using a number of individual toxicants such as CCl_4 [58], thioacetamide (TA) [38, 103], chloroform ($CHCl_3$) [104], 1,2,dichlorobenzene [105, 106], trichloroethylene, and allyl alcohol (AA) [39, 107] and their binary, ternary, and quaternary mixtures [40, 41]. These compounds are common environmental contaminants and cause regiospecific hepatotoxicity. Brief description on the results of individual exposure to TA, CCl_4, and various mixtures are presented below. These studies clearly show that regardless of the number of chemicals, TR is a dose-dependent event and plays major role in the toxic outcome.

6.7.1 Individual Exposure

TA and CCl_4 are two classical hepatotoxicants and a wealth of background information is available on the mechanisms by which cellular and tissue injury is initiated. Corn oil was used as vehicle for CCl_4 and saline for thioacetamide. Both compounds were administered by intraperitoneal route in the interest of greater accuracy and reproducibility. The main objective of these studies was to test the hypothesis that tissue repair response is a dose-dependent event regardless of the mechanism of action of toxicants and any alteration in this response markedly influences the final toxic outcome [108].

6.7.1.1 Thioacetamide

Rats were injected with a 12-fold dose range of thioacetamide (50, 150, 300, and 600 mg/kg) and both liver injury and TR were measured over a time course [38]. Liver injury was assessed by serum alanine aminotransferase (ALT) elevations and by histopathology. Tissue regeneration response was measured by [^{3}H]-thymidine incorporation into hepatocellular DNA and by proliferating nuclear cell antigen (PCNA). Serum ALT did not show any dose response over a 12-fold dose range up to 24 h (Figure 6.5). With 50, 150, and 300 mg/kg, ALT levels increased initially and then declined to normal, indicating liver injury followed by recovery from injury. With the highest dose (600 mg/kg), though the initial injury (24 h) was similar to other three doses, after 48 h injury progressed and all rats died between 4 and 7 days. On the other hand, TR showed dose response (Figure 6.5). Tissue regeneration peaked at 36 h after administration of a low dose of thioacetamide (50 mg/kg). With increment in doscs up to sixfold (150 and 300 mg/kg), a dose-dependent increase in stimulation of TR was observed [38]. Peak TR response was delayed by several hours with each

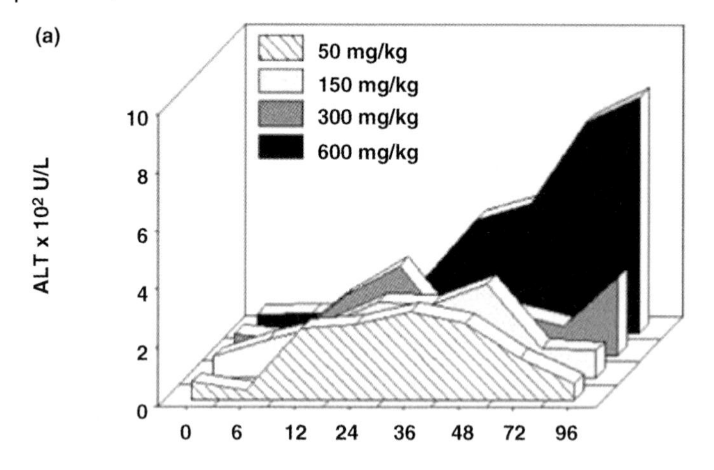

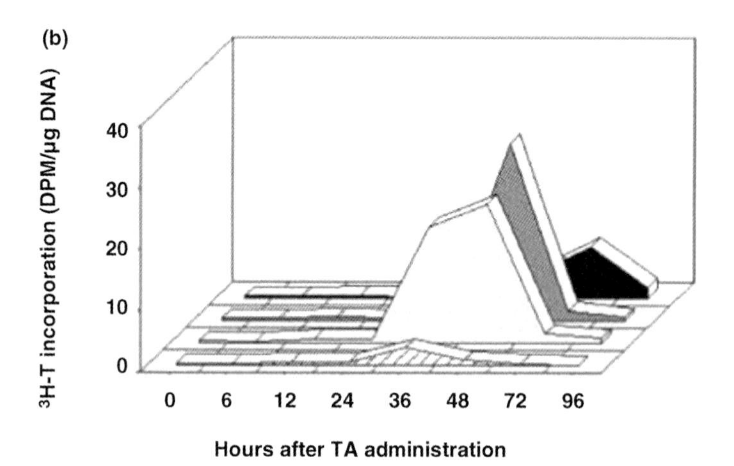

Figure 6.5 Dose response for liver injury (a) and tissue repair (b) after thioacetamide administration to rats. Male SD rats were divided into four groups. At time zero, groups received i.p. injection of 50, 150, 300, and 600 mg/kg thioacetamide. Controls received normal saline. Plasma ALT measured (a) as a marker of liver injury and [^{3}H]-thymidine incorporation into hepatocellular nuclear DNA measured as a marker of hepatocellular regeneration (b) over a time course of 0–96 h.

increment in the TA dose, suggesting that TR is delayed corresponding to the increments by which doses are escalated. With further increase in the dose to 600 mg/kg, the TR response was substantially delayed and suppressed leading to accelerated progression of injury and 100% mortality at this dose [38].

6.7.1.2 CCl_4

A dose-response relationship for hepatic injury and TR has also been observed with CCl_4 in rats [58]. Following administration of 0.1 ml CCl_4/kg, peak liver regeneration was observed at 36 h. With increase in dose to 1 ml CCl_4/kg, a corresponding increase in the magnitude of TR was observed, but peak was delayed to 48 h [58]. With a further increase in the CCl_4 dose to 2 ml/kg, although an increase in S-phase synthesis was observed, this response was delayed further to 72 h. Up to 2 ml CCl_4/kg (threshold dose), TR increased in a dose-dependent fashion, leading to recovery from injury. An increase in doses to 3 and 4 ml CCl_4/kg resulted in significantly delayed and diminished TR, resulting in unrestrained progression of injury and 60 and 80% mortality, respectively. Although 3 and 4 ml CCl_4/kg caused 60–80% lethality, TR response in the surviving rats was increased by about fivefold, aptly demonstrating the critical role of TR in overcoming injury and enabling these animals to survive [58]. These findings suggest that tissue injury and repair are two simultaneous and opposing dynamic responses consistent with dose response and timely TR response is critical in restraining the progression of injury, thereby determining the ultimate outcome of toxicity.

6.7.2 Mixture Exposure

Doses for mixture studies described below were chosen from individual toxicity studies. Since our objective was to examine whether tissue repair elicits dose dependency, lethal doses of the respective chemicals were omitted and the three lower doses were employed. In order to assess whether these individually nonlethal doses cause any mortality when administered in combination, lethality studies were conducted with all three dose combinations. Like the individual compounds, mixtures were also administered intraperitoneally. Chemicals were administered simultaneously one after the other on opposite sides of the abdomen. TA and AA were administered in saline and TCE and $CHCl_3$ were administered in corn oil (maximum 0.5 ml/kg). The doses, vehicle, and route of exposure employed in this study were environmentally irrelevant. But, this approach was appropriate because the central objective of this work was to test the concept that tissue repair plays a pivotal role in the outcome of liver damage regardless of the number of chemicals administered.

6.7.2.1 $CHCl_3 + AA$

The liver injury increased in a dose-dependent manner following the administration of three combinations of binary mixture (BM) of $CHCl_3$ and AA [41]. At the lowest combination (74 mg $CHCl_3$ + 5 mg AA/kg), injury peaked at 48 h (Figure 6.6). With the increase in dose (185 mg $CHCl_3$ + 20 mg AA/kg), the time of maximum injury did not change, but the magnitude of injury was higher compared to the preceding combination. Further increase in dose (370 mg $CHCl_3$ + 35 mg AA/kg) caused much higher injury and, interestingly, the peak injury was observed 24 h earlier than the two lower combinations. With all three combinations, liver injury regressed thereafter, diminishing undetectable injury by 72 h [41]. Compared to individual treatments of $CHCl_3$ and AA [40, 104], the injury was supraadditive (higher than twofold) after exposure to highest combination, whereas at lower combinations, injury was only additive. The enhanced liver damage was due to the increased absorption and

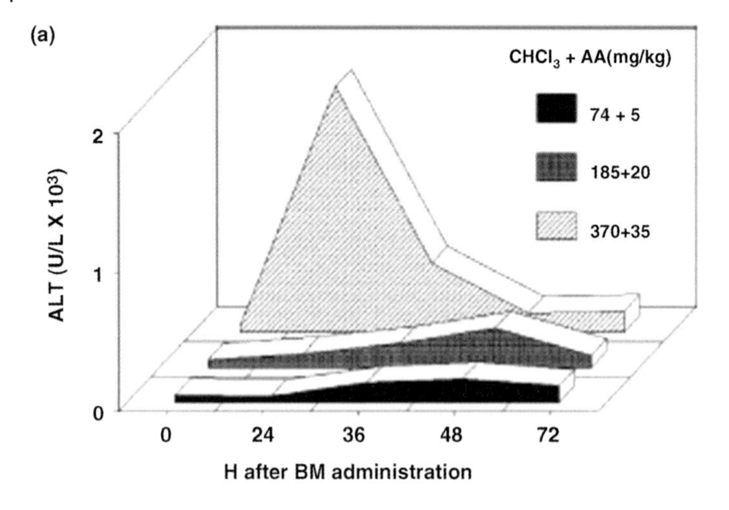

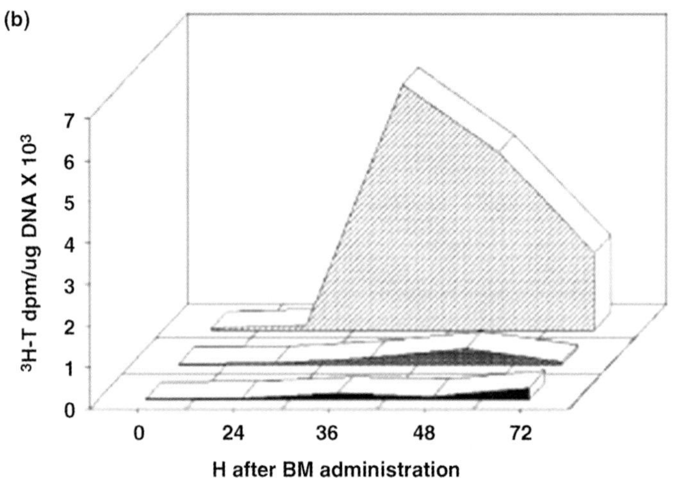

Figure 6.6 Plasma ALT levels (a) as a measure of liver injury after exposure to binary mixture of $CHCl_3$ and AA. Male SD rats were divided into three groups. At time zero, the respective groups received single i.p. injections of indicated doses of $CHCl_3$ (corn oil) and AA (distilled water). Plasma ALT activity was measured over a time course of 0–72 h. Hepatic tissue repair (b) measured by [^{3}H]-T incorporation into hepatocellular nuclear DNA. [^{3}H]-T (35 mCi) was administered 2 h prior to termination at each time point.

metabolism of $CHCl_3$ in BM at the highest dose level. Liver regeneration was also stimulated in a dose-related fashion following various doses of BM (Figure 6.6). The two lower dose combinations (74 mg $CHCl_3$ + 5 mg AA and 185 $CHCl_3$ + 20 mg AA/kg) yielded a small, but statistically significant, increase in liver regeneration at 72 and 48 h, respectively [41]. This minimal liver regeneration response correlated with the lower liver injury occurring at these doses. With highest dose combination (370 mg $CHCl_3$ + 35 mg AA/kg), the TR was robust and peaked at 36 h, which gradually declined thereafter over the time course. The early and augmented TR at highest combination explains why peak injury occurred 24 h earlier than the low-dose combinations and did not progress despite supraadditive injury [41]. Indeed, the peak injury observed at 24 h is not the maximum injury that would have occurred had TR not intervened with the progressive development of injury.

6.7.2.2 $CHCl_3$ + TA

Liver injury and TR were measured over a time course of 0–72 h following three combinations of $CHCl_3$ + TA BM (74 + 50, 185 + 150, and 370 + 300 mg $CHCl_3$ + TA/kg) [109]. The doses were selected from individual toxicity studies [38, 104]. With all three combinations, the liver injury was evident at 24 h. While the two lower combinations (74 + 50 and 185 + 150 mg/kg) showed peak at 36 h, the injury progressed to a maximum at 48 h with the high-dose combination (370 + 300 mg/kg). With all three dose combinations, the ALT levels returned to normal by 72 h, indicating recovery from injury. The injury from the BM was no more than additive compared to individual compounds. Liver regeneration was evident at 24 h with all three combinations. With the lowest dose combination (74 + 50 mg/kg), peak TR occurred at 36 h. At moderate dose combination (185 + 150 mg/kg), increased TR reached maximum 12 h later. Although the TR was significantly higher than the control with the high-dose combination (370 + 300 mg/kg), this response was significantly lower compared to preceding dose combination (185 + 150 mg/kg) at 24 and 36 h. However, the TR was robust at 48 h, indicating that the animals overcame the early suppression of TR. At lower dose combinations, the TR declined to normal quiescent level by 72 h, whereas it was still active at the highest dose combination [109]. Both injury and repair showed dose response and TR was adequate to prevent the injury from progressing at all three combinations.

6.7.2.3 $CHCl_3$ + TCE

Three dose combinations of $CHCl_3$+TCE BM (74 + 250, 185 + 500, and 370 + 1250 mg $CHCl_3$ + TCE/kg) were administered to rats [110]. The blood and liver $CHCl_3$ levels after the administration of binary mixture were similar compared to the administration of $CHCl_3$ alone. The blood and liver TCE levels after the BM were significantly lower compared to TCE alone due to higher elimination in the presence of $CHCl_3$, resulting in decreased production of metabolites [110]. The antagonistic toxicokinetics resulted in lower liver injury than the summation of injury caused by the individual components at all three dose levels. On the other hand, TR elicited by the binary mixture was dose dependent. The interactive toxicity of this mixture resulted in subadditive initial liver injury because of a combined effect of higher elimination of TCE and mitigated progression of liver injury by timely dose-dependent stimulation of compensatory TR.

6.7.2.4 $CHCl_3$+ TCE + AA

In this study, rats were treated with three combinations of $CHCl_3$ + TCE + AA (74 + 250 + 5, 185 + 500 + 20, and 370 + 1250 + 35 mg/kg) and the hepatotoxicity and liver regeneration were studied over a time course [111]. Exposure to a ternary

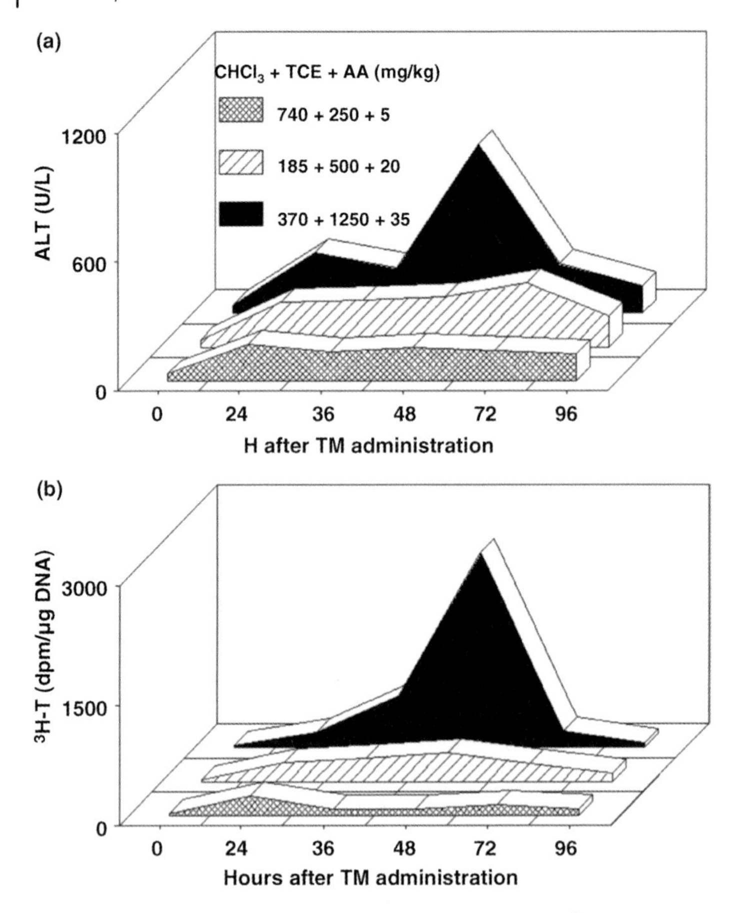

Figure 6.7 Plasma ALT activity (a) measured as a biomarker of liver injury after exposure to the ternary mixture (TM) of $CHCl_3$, TCE, and AA. Male SD rats were divided into three groups. At time zero, the respective groups received single i.p. injections of indicated doses of $CHCl_3 + TCE + AA$. Hepatic cell division as measured by ^{3}H-thymidine incorporation into hepatocellular nuclear DNA after TM (b). [^{3}H]-T (35 μCi) was administered 2 h prior to termination at each time point. The ALT activity and tissue repair was measured over a time course of 0–96 h.

mixture (TM) resulted in lower initial liver injury owing to higher elimination of TCE, and compensatory liver TR was stimulated in a dose-dependent manner mitigating progression of injury (Figure 6.7). Stimulation of a dose-dependent liver TR suggests that this compensatory biological response was unaffected by any interaction among the toxicant components of TM. As a result, new cells were available to replace the dying cells, imparting resistance to the mechanism of progression of injury [112], leading to complete recovery [113, 114]. Although the initial liver injury with this TM was lower, adequate and timely TR was important in preventing the injury from progressing, because it is known that regardless of the extent of initial injury, inhibition of TR results in progression of injury, liver failure, and morbidity even after normally inconsequential injury [34, 102, 103]. Even though the injury after TM was lower compared to individual treatments, peak ALT observed with the high-dose TM is comparable to that observed with lethal dose (740 mg/kg) of $CHCl_3$ alone [104]. Paradoxically, 90% of the rats receiving the highest dose of $CHCl_3$ alone died because of inhibited TR, whereas all the rats receiving the TM containing same dose of $CHCl_3$ survived due to adequate TR, underscoring the significance of TR in the final outcome of liver injury. Although the liver injury was lower and progression was contained by timely TR, 50% mortality (between days 1 and 2) occurred with the high-dose combination. This could be due to extrahepatic (CNS) toxicity resulting from exposure to chemicals ($CHCl_3$ and TCE) that are known to be CNS depressants at high doses. At the respective doses of individual treatments, however, these compounds did not cause any mortality. Although the doses used in this study were several fold higher than environmental levels, unusual toxicity observed in this study underscores the uncertainty associated with mixture exposures.

6.7.2.5 **$CHCl_3$ + AA + TA + TCE**

Tissue repair and injury induced by a ternary mixture of mechanistically dissimilar chemicals ($CHCl_3$, AA, TA, and TCE) showed a dose response as with the case with individual chemicals [40]. All four combinations of mixtures (Mixture I: 62.5 + 12.5 + 1.25 + 62.5 mg; Mixture II: 125 + 25 + 2.5 + 125 and 250 + 50 + 5 + 250 mg; Mixture III: 250 + 50 + 5 + 250 mg; and Mixture IV: 500 + 150 + 20 + 500 mg TCE + TA + AA + $CHCl_3$/kg, respectively) showed dose response for liver injury (Figure 6.8). With lower and medium dose mixtures, the dose-dependent injury peaked at 24 h and declined thereafter, whereas after high-dose mixture, the substantially higher injury peaked at 36 h and regressed subsequently in surviving animals. Like injury, TR also was dose dependent after lower and medium dose mixtures (Figure 6.8). With the low-dose mixture, TR peaked at 48 h, whereas with medium-dose mixtures, peak regeneration occurred between 36 and 48 h. With further increase in dose, the liver regeneration was delayed and was substantially lower compared to low and medium doses [40]. The lack of cell division and TR explains the progressive liver injury and death after the highest dose combination. The results of these studies indicate that irrespective of the number of chemicals to which exposure occurs, the opposing biological phenomenon of compensatory TR plays a critical role in the outcome of liver injury [108]. This shows that TR is the biological response to injury that counters and overcomes the toxic response.

6.8
Tissue Repair Determines the Outcome of Toxicity

The fascinating outcome of the dose-response studies is the finding of the dynamic relationship between the TR response and the progression or regression of the injury. The injury regresses when animals can mount adequate TR, whereas it progresses when animals fail to elicit a prompt TR response that culminates in liver failure and

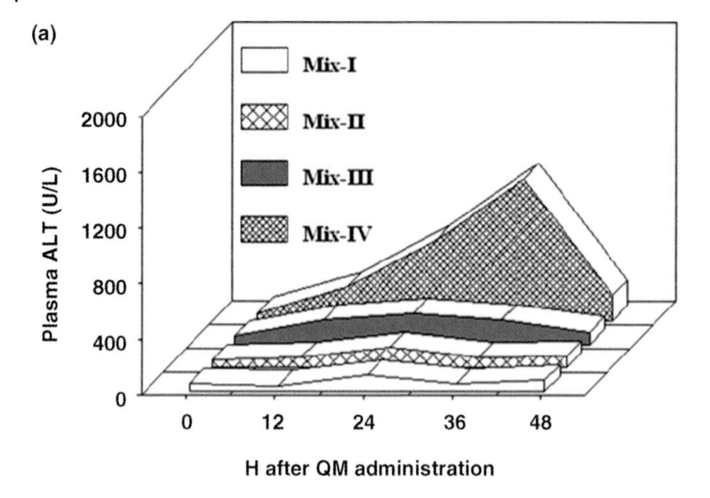

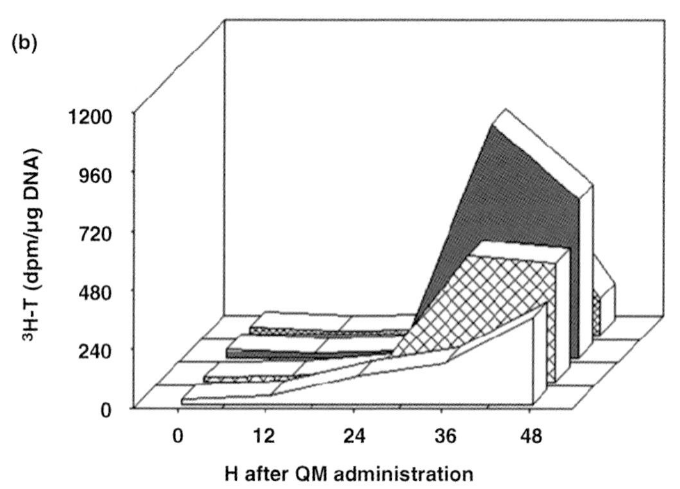

Figure 6.8 Dose response for liver injury (a) and tissue repair (b) after administration of mixture of TCE, TA, AA, and $CHCl_3$ to rats. Male Sprague–Dawley rats were divided into four groups. At time zero, the respective group received intraperitoneal injection of mixture. (a) Plasma alanine aminotransferase (ALT) measured as a marker of liver injury. (b) [^{3}H]-thymidine incorporation into hepatocellular nuclear DNA measured as a marker of S-phase stimulation. Results are expressed as means for four rats in each group. Mixture I: 62.5 + 12.5 + 1.25 + 62.5; Mixture II: 125 + 25 + 2.5 + 125; 250 + 50 + 5 + 250; Mixture III: 250 + 50 + 5 + 250; and Mixture IV: 500 + 150 + 20 + 500 mg TCE + TA + AA + $CHCl_3$/kg, respectively).

death. Although time course studies on TR following various doses of toxicants indicated that TR plays a determining role in the final outcome, that is, survival versus death [40, 64, 66], conclusive evidence comes from interventional studies employing two opposing strategies: (1) antimitosis studies where TR was deliberately inhibited, and (2) preplacement of TR in auto- and heteroprotection studies. One very successful strategy to demonstrate the critical importance of compensatory TR in the recovery from liver injury is to intervene with cell division and TR that oppose progression of injury. Colchicine (CLC) is an antimitotic agent that inhibits cell division by two separate mechanisms: by inhibiting DNA synthesis [115] and by disrupting microtubular formation [116]. Antimitotic intervention with CLC (1 mg/kg) at crucial time points well after toxicant-initiated injury (150 and 300 mg TA/kg) but before or during TR resulted in complete inhibition of cell proliferation and TR. This resulted in conversion of these normally nonlethal doses of TA (150 and 300 mg/kg) into 100% lethal [103]. Analysis of TR indicated that CLC inhibited cell proliferation and TR. Consequently, the injury progressed leading to liver failure and animal death. Treatment with CD (10 ppm for 15 days in the diet) + CCl_4 (100 μl/kg)-exposed male SD rats with CLC (1 mg/kg) resulted in an increase in mortality from 25 to 85%, with a significant decrease in TR in the CLC-treated group [117]. Similar increase in lethality was observed in Fischer 344 (F344) rats treated with *o*-DCB [106]. Taken together, these data highlight the importance of TR in the final outcome of toxicity.

Another strategy to study the role of TR in the final outcome of toxicity is by preplacement of TR using auto- and heteroprotection models [37, 45, 49, 97]. By administration of low dose of compound "A," TR is stimulated that further protects against a subsequently administered lethal dose of the same compound "A" (autoprotection) or entirely different compound "B" (heteroprotection). The first small dose of the toxicant initiates promitogenic cellular signals and essentially preplaces TR, which serves to inhibit progression of injury initiated by the subsequently administered normally lethal dose and protects the animals. Autoprotection has been studied using CCl_4 (Figure 6.9) [97], TA [38], and APAP [45, 49] while heteroprotection has been investigated using TA and APAP combination [118]. Preplacement of TR can also be achieved by surgical two-thirds resection of liver by partial hepatectomy before toxicant treatment [119]. Liver regeneration after 70% partial hepatectomy protects the animals from a lethal challenge of CCl_4 or CD + CCl_4 due to attenuation of progression phase of injury [67, 68, 86, 87]. In these models of auto-and heteroprotection, it should be noted that liver injury initiated by the high dose of toxicants is not diminished by the prior administration of the priming agents [37, 118]. Even though the same massive and normally lethal liver injury is reached, and is lethal in unprimed animals, the primed animals overcome this injury as a result of sustainable and early onset of TR stimulated by the priming dose.

In essence, in an acute toxicity paradigm, absence or presence of TR response leads to either progression or regression of injury, respectively. Injury regresses upon the onset of timely and robust TR because the dividing/newly divided cells are resilient to progression of injury [68, 69, 71, 86, 87, 113, 114, 117, 120, 121]. Recently we have shown that subchronic chloroform exposure leads to adaptive tolerance from liver and kidney injury in mice [122] due to increased tissue regeneration. Further these animals were protected from a subsequently administered lethal dose of $CHCl_3$ [123] due to upregulated compensatory TR.

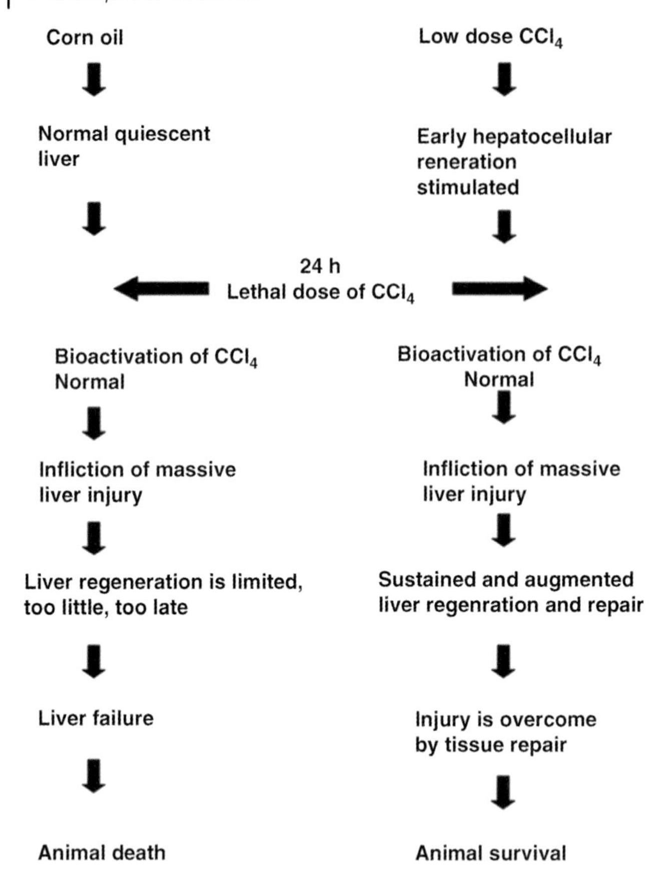

Figure 6.9 Proposed mechanism of CCl_4 autoprotection. Same massive liver injury occurs when a large dose of CCl_4 is administered to rats regardless of whether a protective dose of CCl_4 was given 24 h previously. This suggests that even though cytochrome P-450 is decreased in rats treated with the protective dose, bioactivation of CCl_4 and infliction of liver injury are not compromised. The dramatic difference in the final outcome and consequence of liver injury is a reflection of whether cell division and tissue repair are sustained. Augmented and sustained tissue repair in livers of rats receiving the protective dose enable these rats to overcome liver injury through tissue healing, thereby allowing survival.

6.8.1 Factors Affecting Tissue Repair

A number of physiological factors, including species, strain, age, nutrition, caloric restriction, and disease, affect the TR response. Difference in responses among the general public is a serious problem in assessing risk of exposure to drugs and toxicants. While many factors are considered as underlying causes of such unpredictable diversity, the role of TR in such variation has not been considered. Factors that influence the TR response may help in explaining the wide range of interindividual differences in responses to drugs and toxicants.

6.8.1.1 Species, Strain, and Age Difference

Marked differences in the toxicity of chemicals are known to occur. While many factors are considered as underlying causes of such differences, the role of tissue repair as a substantial factor has not been considered. Mehendale and associates [8, 105, 106, 117, 124–130] examined this issue closely. For example, Mongolian gerbils are highly susceptible (35-fold lower LD50) to CCl_4-induced hepatotoxicity than SD rats due to extremely sluggish TR [131]. Gerbils are also remarkably resistant to CD-amplified toxicity of CCl_4 [131]. CD-amplified CCl_4 toxicity is known to be due to inhibition of CCl_4-induced increase in compensatory tissue repair [8], and since TR is minimal in gerbils, the interactive toxicity does not occur. Similar species difference has been shown between rats and mice under disease conditions [124–126]. Streptozotocin (STZ)-induced type 1 diabetic (DB) rats were found to be highly sensitive to TA-induced liver injury where even a normally nonlethal dose of TA is lethal in DB rats because of compromised TR response [126, 127]. However, streptozotocin-induced type 1 DB mice were completely refractory to liver injury induced by a lethal dose of TA due to their ability to mount effective TR response [125]. A classic example of strain difference in TR is observed between F344 and SD rats exposed to 1,2 dichlorobenzene (*o*-DCB) [132]. F344 rats exhibited significantly higher liver injury following exposure to *o*-DCB as compared to SD rats treated with the same doses. However, the mortality induced by *o*-DCB is not higher in F344 rats, since these rats are capable of mounting a much stronger TR compared to SD rats [105]. The significantly higher TR in F344 rats enables them to prevent the progression of liver injury and thereby escape *o*-DCB-induced liver failure even though they suffer 10-fold higher initiation of liver injury than the SD rats [105, 106].

Age is another important factor that determines the extent of TR following toxicant exposure. In general, newborn animals are capable of mounting faster and efficient TR during early developing age compared to adults. This has been demonstrated with CCl_4- and CD + CCl_4-amplified liver injury in 20-day-old neonatal and 2-month-old young adult SD rats [117, 128, 129]. Twenty-day old neonates were found to be resistant to the CCl_4- and CD + CCl_4-induced liver injury as compared to the 2-month-old young adult rats [117, 128, 129]. Neonatal rats have growing liver that undergoes rapid cell proliferation that affords protection from toxicant exposure. Moreover, protooncogenes such as transforming growth factor-α (TGF-α), c-fos, H-ras, and K-ras were expressed at much higher levels and at much earlier time points following toxicant exposure [128]. This ability is lost in the adults at 2 months of age when most of the hepatocytes are in quiescence and fail to divide and mount an effective TR. Another interesting phenomenon is the resiliency of old F344 rats (24 months) to CD + CCl_4 toxicity [130, 133]. The 14- and 24-month-old rats exhibited higher liver TR and survival as compared to the young adult (3-month-old) rats to CD + CCl_4 toxicity. Mehendale *et al.* [133] reported that protection against CD + CCl_4-amplified toxicity was also evident in SD rats, suggesting that this remarkable resiliency due to very high compensatory TR in liver is not strain dependent.

Sanz *et al.* [134] reported that severity of TA-induced liver injury was significantly lower in the old rats (30 months) than young rats (6 months) and the authors attributed it to difference in antioxidant enzyme system. These data suggest that the TR response is not only intact in the old animals but it is also surprisingly enhanced. Taken together, these data indicate that ability to mount TR following toxicant exposure varies among different species, strains, and age groups and may have a significant impact on the drug development and risk assessment process.

6.8.1.2 **Nutritional Status**

Numerous studies have demonstrated that modulation of the nutritional factors can directly affect the final outcome of toxicity by modulating TR response. Glucose loading (15% glucose supplementation in drinking water for 8 days) inhibits compensatory TR in rats following exposure to centrilobular hepatotoxicants such as TA, $CHCl_3$, and CCl_4 with a concomitant increase in toxicity. Glucose loading did not affect the bioactivation of these compounds nor did the insulin levels change in the rats. Supplementation of diet with palmitic acid, a primary or preferred source of energy to the periportal hepatocytes, along with its mitochondrial carrier, L-carnitine protected the rats from a lethal dose of TA [135]. These studies suggest that excessive glucose in the body can adversely affect TR, an observation further supported by findings that DB rats exhibit inhibited TR following exposure to TA and CCl_4. The mechanisms underlying these two models, however, may be different since glucose loading did not increase blood glucose as in the case of DB rats [136].

6.8.1.3 **Diet Restriction**

Caloric or diet restriction (DR) has extensive health benefits. Numerous studies have shown the beneficial effects of DR by decreasing age-related diseases, including cancer, and also protecting from chemical-induced toxicity [54, 137–142]. Ramaiah *et al.* [143] have shown that moderate DR (65% of *ad libitum* (AL) for 21 days) protects male SD rats from lethal challenge of the model hepatotoxicant TA [143, 144]. Timely and robust liver cell proliferation and TR protect DR rats in spite of twofold higher liver injury than AL rats. This enhanced TR is due to upregulation of promitogenic signaling via interleukin-6 (IL-6)-mediated and MAPK pathways [54]. DR rats treated with a low dose of TA (50 mg TA/kg) exhibited sixfold higher initial bioactivation-based liver injury than AL rats treated with the same dose. In spite of higher initial injury, DR rats survived due to the enhanced TR response. The increased liver injury of TA in DR rats was attributed to the induction of CYP2E1, the primary enzyme involved in bioactivation of TA [126, 145].

The mechanism of higher survival in DR rats exposed to TA is stimulated TR, and this TR is not dependent on the extent of initial liver injury as demonstrated by an equihepatotoxic study with TA [53]. DR rats treated with a low dose of TA (50 mg/kg) and AL rats treated with a 12-fold higher dose of TA (600 mg/kg) produced equal initial liver injury. Even though the initial liver injury was the same, AL rats experienced a progressive increase in liver injury leading to 90% lethality, while DR rats experienced complete survival due to prompt TR, which in the case of the AL rats was delayed. Moreover, in DR rats, the number of apoptotic cells increases rapidly and remains high until 60 h after TA administration. This early and sustained increase in apoptosis rapidly removes damaged cells that would otherwise die later via necrosis, permitting more rapid replacement, and also prevents leakage of degradative death proteins like calpain that would have otherwise happened in case of necrosis. AL feeding regimen does not promote the same rate of apoptosis of weaker hepatocytes and timely compensatory TR. The early upregulation of cell division by prompt and elevated expression of IL-6, TGF-α, HGF, and iNOS in the DR rats appears to be the key for early onset of TR [53]. These growth factors also increase in AL rats, but the increase is delayed and diminishes very rapidly leading to diminished cell division in the AL rats upon toxic challenge. In DR rats, the TR was not inhibited even after massive initial liver injury indicating that DR rats have a mechanism that can operate under extreme stress of injury to promptly stimulate the TR. The signaling mechanisms in DR rats are viable, even during the massive liver injury, regardless of whether it is from a low dose due to induction of CYP2E1 or due to a high dose [53].

6.8.1.4 **Disease Condition: Diabetes**

More than 18 million Americans are suffering from diabetes (DB), and it is a predominant cause of morbidity and mortality. DB is known for inducing secondary complications, including renal and cardiovascular problems and impaired wound healing. DB have nearly twofold higher risk of liver failure due to drug-induced toxicities and chronic liver disease [146, 147]. Concerns about hepatotoxic sensitivity in DB suddenly increased when the use of the highly effective anti-DB drug troglitazone led to severe and in some cases fatal hepatotoxicity in a number of DB patients resulting in the withdrawal of the drug from the US and UK markets [136]. Other incidents of hepatotoxicity have been reported in DB patients on drug therapy with methotrexate, acarbose, and metformin [148–151]. The question of whether DB increases the sensitivity to drugs and other potential hepatotoxicants still remains uninvestigated. Historically, investigations of hepatotoxicity of drugs or toxicants have focused on biotransformation. Factors other than the mechanisms that initiate injury as being involved in determining the final outcomes of the initiated injury have received little attention. Hence, the literature regarding liver TR in the presence of DB condition is sparse as compared to altered biotransformation in DB. Various animal models are available to study the etiology of DB. DB animal models fall into two categories, type 1 DB (insulin-dependent DB mellitus) and type 2 DB (noninsulin-dependent DB mellitus), consistent with the etiologies of the two major types of DB in humans. Alloxan or streptozotocin-induced DB has been the most extensively employed and studied animal model of type 1 DB [152–154]. However, the experimental type 1 DB induced by alloxan has fallen behind the STZ-induced DB model because of its severe toxicity [155]. In addition, various genetic models of type 1 DB, including nonobese DB (NOD) mice and BB Wistar rat model are also available. Leptin and leptin-receptor deficient yellow obese mice, Zucker DB Rats, and fat-fed/STZ-induced type 2 DB rats and mice models are the various type 2 DB animal models. High-fat diet has been recognized to lead to insulin resistance and is considered as a major predisposing factor for type 2 DB [156, 157]. Rats fed with a 40% fat (by weight) diet become hyperinsulinemic, similar to the pre-DB state characterized by resistance to insulin-mediated glucose disposal and compensatory hyperinsulinemia in humans. A subsequent administration of a single dose of STZ (50 mg/kg, i.v.) to these animals lead to decreased insulin levels and hyperglycemia resembling the DB condition in humans. However, these fat-fed/STZ-induced type 2 DB models yielded only slight hyperglycemia, and it was not characterized for persistent and irreversible type 2 DB and associated chronic complications. Sawant *et al.* [158] refined and characterized this type 2 DB model so as to develop robust DB in 24 days by feeding rats with 20% fat (by weight) diet for 2 weeks followed by a single dose of STZ (45 mg/kg, i.p.). These rats developed marked hyperglycemia, normoinsulinemia, high free fatty acids, and hypertriglyceridemia by day 24. This DB state could be controlled by metformin treatment in the animals, further confirming the type 2 DB status in this model. Subsequent 6-month study showed that the hyperglycemic rats developed cataract and DB nephropathy, indicating that the type 2 DB condition was persistent and irreversible.

Remarkable species difference exists in terms of effects of DB on the ultimate outcome of hepatotoxicity. Regardless of type 1 or type 2 DB, rats are sensitive to hepatotoxicants, whereas mice are resistant to the hepatotoxicity induced by toxicants [136, 159]. Streptozotocin-induced type 1 (insulin-dependent) DB rats exhibit inhibited TR response followed by significantly higher liver injury after treatment with a normally nonlethal dose of TA (300 mg/kg), leading to 100% mortality [126, 127, 160]. The increased liver injury is partly due to induction of TA-bioactivating enzyme, CYP2E1, in the liver. However, inhibition of CYP2E1 in DB rats by a relatively specific inhibitor, diallyl sulfide (DAS), failed to protect from TA-induced liver failure and mortality. Although DAS treatment decreased initial bioactivation-based injury (stage 1 of toxicity) to the same level as seen in the non-DB rats, only DB rats failed to stimulate an effective TR response, resulting in progression of initial injury, massive liver failure, and death [127]. Interestingly, type 1 DB mice were resilient to TA- bromobenzene-, CCl_4-, and acetaminophen (APAP)-induced hepatotoxicity, due to increased TR, indicating a species difference in TR response in type 1 DB animals. The increased TR in DB mice was partly explained by timely signaling via PPAR-α, stimulating cell cycle genes such as cyclin D1 [56, 124, 125].

Studies with type 2 DB rats revealed inhibition of TR after hepatotoxicity by diverse group of toxicants such as AA, TA, and CCl_4 [158, 161]. A subsequent mechanistic study showed that bioactivation of CCl_4 was unchanged and liver TR was inhibited in the type 2 DB rats. Compromised TR caused unabated progression of CCl_4-induced liver injury, leading to the conclusion that potentiation of CCl_4 hepatotoxicity is due to compromised liver TR. In the type 1 diabetic rats, TR is inhibited in the absence of insulin; and in the type 2 diabetic rats, TR is inhibited even in the presence of insulin [126, 158]. Leclercq *et al.* [162] showed that CCl_4-induced liver injury was equally severe in leptin-deficient *ob/ob* type 2 DB and control non-DB mice at 48 h following CCl_4 injection. These authors also reported profound impairment of liver regeneration in the *ob/ob* DB mice due to exaggerated activation of nuclear factor-kappaB (NF-κB) and signal transducer and activators of transcription 3 (STAT3) during the priming phase, abrogation of tumor necrosis factor (TNF)-α and

IL-6release at the time of G_1/S transition, and failure of hepatocyte induction of cyclin D1 and cell cycle entry. However, whether liver injury initiated by CCl_4 bioactivation progressed or regressed after 48 h was not reported nor was the ultimate outcome of that injury reported. Sawant *et al.* [159] reported decreased hepatotoxicity of APAP, CCl_4, and bromobenzene in the fat-fed/STZ-induced type 2 DB mice. This study indicated that hepatotoxicity of all three toxicants was decreased in the DB mice allowing substantial survival after treatment with LD_{80} doses of these model hepatotoxicants. A subsequent time-course study with a normally LD_{80} dose of APAP revealed substantially lower liver injury at 6, 12, and 24 h in the type 2 DB mice as assessed by plasma ALT and AST, while 80% of the non-DB mice died of liver failure. Subsequent studies showed that after initial bioactivation-mediated liver injury, liver TR was enhanced in the type 2 DB mice, leading to resiliency to hepatotoxicity. Increased expression of calpastatin (the endogenous inhibitor of calpain) in type 2 DB mice appeared to be at least in part responsible for the restricted progression of injury following administration of APAP [114, 159].

Taken together, these data provide substantial evidence that DB animals are highly susceptible to toxicant-induced liver injury due to inhibition of compensatory TR response and not due to lower insulin or higher bioactivation hepatotoxicants.

6.9 Molecular and Cellular Mechanisms of Tissue Repair

Numerous studies have shown that following liver damage, quiescent hepatocytes are primed to reenter from G_0 into the G_1 phase of the cell cycle. Cytokines like TNF-α and IL-6 released from nonparenchymal cells bind with their receptors on hepatocytes and activate downstream DNA binding of preexisting transcription factors such as nuclear factor (NF-κB), STAT3 (signal transducers and activators of transcription 3), activator protein-1, and CAAT enhancer binding proteins (C/EBPs) [42, 163–170] leading to activation of immediate early genes, including c-fos, c-Jun, and c-myc [23, 72]. After G_0 hepatocytes progress to G_1 phase, several growth factors, such as epidermal growth factor, transforming growth factor-α, and hepatocyte growth factor, act predominantly in the progressive phase of liver regeneration to stimulate hepatocyte DNA synthesis [42, 43, 53, 167, 171–173]. These growth factors signal through their receptors that have intrinsic tyrosine kinase activities and activate downstream targets, including mitogen-activated protein kinases (MAPKs), leading to DNA synthesis. Parallel pathways of Jun nuclear kinase (JNK) and protein kinase C (PKC) are also necessary for hepatocyte proliferation [174–176]. Transient expression of protooncogenes during liver regeneration follows the order of c-fos, c-myc, p53, and ras. Increase in p53 mRNA corresponds to the G_1/S transition, and mRNAs from c-ras genes are increased later, coinciding with DNA replication and mitosis [172, 173]. In contrast, TGF-β inhibits cell cycle progression in hepatocytes via its action on membrane-bound TGF-β receptors I and II. Phosphorylation of these receptors leads to activation of cellular proteins, which enter the nucleus and inhibit progression of the cell cycle from $G_1$to S phase and terminate DNA synthesis [43, 177].

Hyperglycemia in DB induces oxidative stress and thereby activates p38 MAPK and other stress pathways [178–180]. Several studies have shown that p38 negatively and ERK1/2 MAPK in turn positively regulates cyclin D1 [181–183], a critical G_1/S checkpoint gatekeeper that binds with cdk4 and cdk6 to form complexes and allows G_1/S transition by phosphorylation of retinoblastoma protein (ppRB). Expressions of ERK1/2 MAPK, cyclin D1, cdk4, cdk6, and p-pRB were decreased with a resultant reduction in hepatic DNA synthesis after acute CCl_4 challenge of type 2 DB rats [184] suggesting that p38 MAPK may downregulate cell cycle regulatory molecules, resulting in inhibited S-phase DNA synthesis. These findings are consistent with compromised liver TR leading to progression of injury initiated by CCl_4, liver failure, and death of the type 2 DB rats. Administration of TA (300 mg/kg) to non-DB rats led to sustained NF-ϰB-regulated cyclin D1 signaling and prompts compensatory TR and survival, whereas inhibited NF-ϰB-regulated cyclin D1 signaling resulting in inhibited TR, liver failure, and death was observed in type 1 DB rats challenged with the same dose of TA. Administration of a 10-fold lower dose of TA (30 mg/kg) to DB rats led to remarkably higher NF-ϰB-DNA binding, but transient downregulation of cyclin D1 expression, explaining delayed TR in the DB rats. These data suggest that NF-ϰB-regulated cyclin D1 signaling might explain inhibited versus delayed TR observed in the DB rats receiving higher (300 mg/kg) and lower dose (30 mg/kg) of TA, respectively. Sawant *et al.* [184] have shown that NF-ϰB signaling is concomitantly downregulated along with compromised liver TR in the type 2 DB rats.

As described earlier, TR responses of DB rats and mice to toxicant challenge are species specific. Shankar *et al.* [56] demonstrated a novel role for the nuclear receptor PPAR-α in mediating DB-induced resistance against hepatotoxicity in mice. Pretreatment with peroxisome proliferators, such as the prototypical clofibrate and others, protects against lethal hepatotoxicity of APAP and other hepatotoxicants to DB [185–189]. DB state in mice has been shown to induce peroxisome proliferation, PPAR-α activation, and increased oxidative stress [190–192]. Resistance to lethal challenge of APAP was substantially decreased in type 1 DB PPAR-α knockout mice than in wild-type DB mice [56] further consistent with the role of PPAR-α activation as the mechanism of resistance observed in DB mice. After liver injury, NF-ϰB is activated in the DB mice in a PPAR-α-dependent manner. Cyclin D1 gene and protein expressions, which were found to be increased in DB mice, can be influenced directly via NF-ϰB or other regulators. p38-MAPK, a negative regulator of cyclin D1, is significantly decreased in DB. Higher cyclin D1 expression is consistent with earlier progression of hepatocytes from G_0 to G_1 cell cycle phases after initial liver injury. Hepatic DNA synthesis was significantly decreased in the PPAR-$\alpha^{-/-}$ type 1 DB mice. Microarray analyses also identified the key cell cycle regulator, cyclin D1, to be induced in DB mice in a PPAR-α-dependent manner. Consistently, literature indicated that PPAR-α activation could increase cell proliferation and decrease apoptosis, both of which can increase the ability of the liver to repair tissue after chemical injury. Anderson *et al.* [193] demonstrated that peak of S-phase DNA synthesis is delayed almost by 24 h in PPAR-$\alpha^{-/-}$ mice after partial hepatectomy [193]. In another recent study, PPAR-$\alpha^{-/-}$ mice challenged with CCl_4 developed acute liver failure with 50% greater mortality compared to wild-type controls [194].

Recent studies have revealed that PPAR-α dependency of hepatoprotection of DB mice against toxicants may be dependent on the type of DB. In type 2 DB mice, hepatoprotection occurs in the absence of PPAR-α stimulations [159]. Even though DB rats also demonstrate markers of PPAR-α activation, they are unable to translate this into a protective response in terms of increased cell proliferation. A marked difference between the responses of DB rats and mice is also evident at the level of cell division regulator cyclin D1, which is preferentially increased in DB mice and not in DB rats.

6.9.1
How Does Injury Progress?

Liver injury initiated by hepatotoxicants such as CCl_4, APAP, and TA progresses even after the offending toxicant has been eliminated from the body [112, 113]. Following exposure to a toxic dose of hepatotoxicants, injury is initiated through bioactivation-based mechanisms or via generation of free radicals [195]. Upon initiation of injury, low to moderate doses of toxicants stimulate compensatory cell division and TR, promptly replacing the lost cells and tissue structure to restore tissue function, whereas high doses of toxicants inhibit cell division, and the lack of new cells to replace lost tissue facilitates unabated progression of injury [39, 54, 102]. This raises two questions: (1) How does injury progress? (2) Why newly divided cells are resistant to progression of injury? The following sections address these questions. Recent studies have shown that hydrolytic enzymes, such as calpain, leaking out of toxicant-induced necrotic cells is activated by high extracellular concentrations of calcium and acts as a death protein destroying neighboring cells to spread tissue injury regardless of how the injury is initiated (Figure 6.10) [113].

With increasing numbers of dying hepatocytes, injury progresses in self-perpetuated fashion. Calpain escaped from necrotic cells is involved in the progression of injury initiated by CCl_4. Intervention with calpain inhibitors, CBZ (Cbz-Val-Phe-methyl ester) and E64 (cell impermeable calpain inhibitor), administered 1 h after the administration of the toxicant (LD_{80} dose) markedly attenuates the progression of liver injury with a concomitant decrease in the release of calpain and increased animal survival by 75–80%. These observations indicate that calpain leaks out of dying cells and this extracellular calpain mediates the progression of injury by destroying neighboring healthy cells. Further, a significant breakdown of calpain-specific substrate, α-fodrin, was observed in the CCl_4-treated rats, whereas no such breakdown was observed in the rats that received CBZ intervention after CCl_4 administration. Incubation of normal rat hepatocytes with calpain led to cell death in a Ca^{2+}-dependent fashion, confirming the role of extracellular calpain in mediating cell death [113]. Intervention with calpain inhibitor CBZ also protected mice from progression of liver injury and lethality induced by a lethal (LD_{80}) dose of APAP indicating that calpain mediates progression of liver injury initiated by APAP in mice [113]. Protection afforded by calpain inhibitor CBZ is not due to interference with the bioactivation mechanisms of the two toxicants since neither CYP2E1 activity nor covalent binding was affected. In addition to its role in progression of

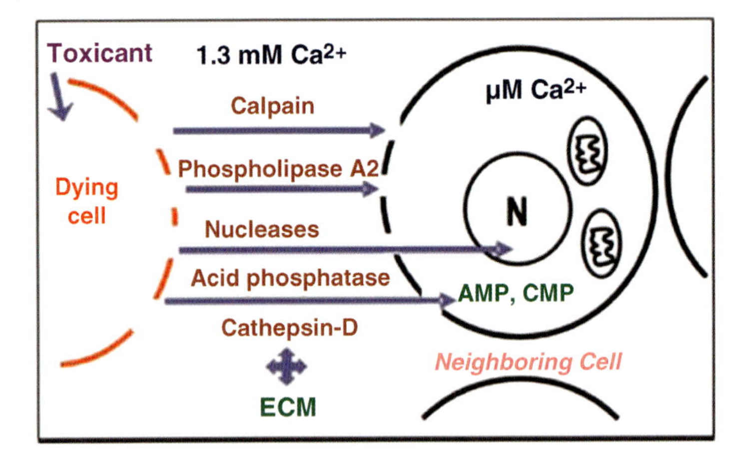

Figure 6.10 Role of death proteins in progression of injury. Schematic representation of the lytic action of death proteins released from dying hepatocytes in the extracellular space on the neighboring hepatocytes. Upon infliction of cellular injury by bioactivation-based events, the injured cells eventually are lysed releasing their contents into the extracellular space. Thus, the potent degradative enzymes (death proteins) enter the extracellular environment rich in Ca^{2+}, are activated, and hydrolyze their substrates contained in the plasma membrane of the neighboring healthy cells or cells partly affected by toxicants, causing progression of injury until the augmented tissue repair is mounted or is severely inhibited.

injury, calpain could also play a role in the initiation of injury. Increase in intracellular calcium concentration during necrotic liver injury would activate intracellular calpain, which then works as a secondary mechanism after initiation of injury [112]. Secretory phopholipase 2 (sPLA2) is another member of death proteins that leaks out of the necrosed hepatocytes or is secreted in the intercellular space by vesicular emptying following drug/toxicant-initiated injury [112, 113, 196]. sPLA2 hydrolyzes the ester bond at the sn-2 position of glycerophospholipids in the presence of Ca^{2+} with a broad fatty acid and phospholipid specificity [197–199]. Bhave *et al.* [200] reported that sPLA2 causes progression of liver injury via hydrolysis of membrane phospholipids in the perinecrotic areas in the absence of inadequate induction of cyclooxygenase (COX-2). Adequate induction and copresence of COX-2 with sPLA2 upregulate prostaglandin production that affords the following hepatoprotective effects: (i) by acting as endogenous ligands for signal transduction pathways, thus enabling structural and functional restoration of the liver [201–203]; (ii) by upregulating anti-inflammatory cytokines [204, 205]; and (iii) by regulating sPLA2 transcription through the anti-inflammatory effects of PGE2 and PGJ2 [206–208]. Like calpain, extracellular sPLA2 has been shown to cause death of hepatocytes in a time- and dose-dependent manner. Napirei *et al.* [209] have reported that deoxyribonuclease-1 (Dnase-1) leaking out of necrosed hepatocytes causes progression of injury initiated by APAP by destroying DNA of cells surrounding the necrotic cells [113]. Numerous reports have linked activation of calpain in the pathogenesis of various ailments such as neurodegenerative disorders, including Alzheimer's disease, epilepsy, cerebral ischemia/excitotoxicity, demyelination, cataracts, ischemic liver and ischemic renal injury, toxic renal injury, and muscular dystrophy [113].

6.9.2 Why Newly Divided Cells Are Resistant to Progression of Injury?

Substantial research has shown that timely liver regeneration can prevent progression of injury leading to a favorable prognosis [8, 49, 210]. However, the mechanism by which compensatory regeneration prevents progression of injury was not known. Recent studies have shown that proliferating/newly proliferated hepatocytes overexpress calpastatin (CAST) preventing calpain-mediated progression of injury [114]. CAST is a specific endogenous inhibitor of calpain and is known to inhibit both activated as well as a proenzyme form of calpain [211–215].

Expression of CAST has been correlated with resistance to hepatotoxicity in three liver cell division models [114]: rats challenged with a nonlethal dose of CCl_4 that stimulates compensatory liver TR, 70% partial hepatectomy, and 20-day-old newborn rat pups challenged with APAP. In all three models, CAST was upregulated in the dividing/newly divided hepatocytes and declined to normal levels with the cessation of cell proliferation. Interestingly, CAST expression pattern coincided not only with the gradual increase/decrease in cell division but also with the time course of resistance/loss of resistance against hepatotoxicity in these models. These data strongly indicate that upregulation of CAST and recovery from progression of injury are closely and inversely related. Also, mice made to overexpress CAST using recombinant adenoviral (Ad/CAST) transfection exhibited markedly attenuated progression of liver injury and 57% survival following challenge with a lethal dose of APAP. Although APAP-bioactivating enzymes and covalent binding of the APAP-derived reactive metabolites were unaffected, degradation of calpain-specific target substrates such as fodrin was significantly reduced in these mice [114].

Moreover, CAST was found to be localized to the plasma membrane and cytoplasm of hepatocytes. This strategic location would be very effective in inhibiting the degradative actions of extracellular calpain. CAST localized in nucleus may not be critical in protection against degradative action of extracellular calpain because nuclear CAST staining disappeared with time after CCl_4 administration, and moreover, CAST was primarily localized in the cytoplasm and membranes of adenoviral Ad/CAST transfected mice livers [114]. The upregulated CAST may help in preventing activation of intracellular calpain by subsequent entry of extracellular Ca^{2+} inside the cells on initiation of extracellular calpain-mediated injury. Collectively, these observations suggest that CAST overexpression may be a generalized protective mechanism against destruction by proteases such as calpain. A similar finding has been reported in spinal cord injury [216]. Twenty-one-day-old rats showed resiliency against spinal cord injury due to upregulation of spinal cord calpastatin. Furthermore, following spinal cord injury, calpain expression was qualitatively less in the 21-day-old rats than in adult rats. Decreased levels of TUNEL positive neurons were also noted in juvenile rat spinal cord, indicating that the developing cord may have an increased resistance to injury [217].

6.10
Implications for Risk Assessment

Interaction of chemical mixtures can result in modulation of either stage 1 or stage 2 of toxicity. Studies dealing with the modulation of infliction of injury during stage 1 are abundant [218–220]. For example, the presence of drug-metabolizing enzyme inducers or inhibitors can result in increased or decreased stage 1 of toxicity. The resulting consequences of toxicity are well known and well described in the toxicological literature. Although inhibition or augmentation of stage 2 response is known to alter the toxic outcome [63, 97, 103, 118], interactive interference of chemical mixtures at this stage has been only sparsely investigated [39, 48, 66]. The studies described in this chapter clearly show the following: (1) TR is dose dependent. (2) It is a simultaneous, but opposing response to injury. (3) Timely and adequate stimulation of TR prevents continued injury. (4) Delayed and inhibited TR leads progression of injury and lethality. (5) Interaction of chemical mixture is inconsequential to the toxicity outcome at stage 1 regardless of the magnitude of the injury unless that injury progresses. (6) Experimental modulation of TR mechanisms has been shown to have significant impact on the outcome of toxicity.

Despite the evidence that TR plays a decisive role in the progression or regression of injury, this response has not been considered in the risk assessment process. Measuring dose-response paradigm of TR in addition to injury in response to exposure to individual chemicals or mixtures will lead to more precise and accurate prediction of the final outcome of toxic injury. Inclusion of TR stimulation as a biological event opposing injury, in the classic dose-response paradigm, can result in two sets of dose-response curves. At lower doses, as injury begins, there is a simultaneous but opposing TR response allowing the animals to overcome that injury preventing death. As the dose progresses, a threshold level is reached where any additional increase in the dose results in a delay in the stimulation of TR and a decrease in the amplitude of the TR response resulting in unrestrained progression of injury and animal death. Another fascinating observation of these dose-response studies is the dynamic relationship between the TR response and the progression of the injury. Accelerated progression of liver injury becomes evident only after failure to elicit a prompt TR response and leads to liver failure and death. Approaches proposed in the literature to date for describing the dose-response relationship for cytotoxic chemicals such as chloroform have implicitly assumed that injury as measured by cell death or enzyme leakage is coupled in a one-to-one relationship with repair, as measured by cell division [221–223]. However, such an approach accounts neither for the restrained injury nor for the acceleration of recovery through division of cells. Both are time-dependent dynamic events working in direct and mutual opposition. The true functional relationship can be investigated by the application of statistical methods to test various hypotheses regarding the temporal causative interactions between injury and repair. This dynamic interplay can be stated in a biologically based, empirical mathematical model:

$$\text{Outcome} = f(\text{repair}, \text{injury}), \text{ which can be mathematically represented as}$$

$$\text{Outcome} = {}^{t}f_{0}\{\text{repair}(t) - \text{injury}(t)\}dt, \text{ which after weightage for simultaneous injury and repair becomes}$$

$$\text{Outcome} = {}^{t}f_{0}\{\text{repair}(t)XW_{\text{r}}(t) - \text{injury}(t)XW_{\text{i}}(t)\}dt$$

where $W_r(t)$ is the weight given to repair (cell division) and $W_i(t)$ is the weight given to injury (cell death at time t). We suggest that for a given dose, the ultimate outcome of injury is a function of the net difference between repair and injury and that this difference can be integrated. This model may provide greater precision in risk assessment because this model is biologically based and it takes into account the dynamic nature of TR and tissue healing processes. If experimentally validated, this concept might be more useful in improving the current paradigms of predictive toxicology and the science behind it.

6.11 Conclusions

There has been a greater emphasis on development of both scientific and regulatory methodology to improve our ability to evaluate the human and environmental health impacts of chemical mixtures. From a public health perspective, exposure to single chemicals is seldom relevant. Multiple exposures to environmental chemicals, drugs, food additive, and recreational or abused substances such as ethanol, tobacco, and so on simultaneously, intermittently, or sequentially are almost always the case. A large body of evidence suggests that chemicals can interact with each other and enhance or inhibit the toxicity of one another. As a consequence, the studies investigating chemical mixture toxicity have increased in recent years. Most of the current mixture studies investigate the interactions of chemicals with the events that initiate injury (TK phase) and disregard the interactions that might occur with the events that alter the ability of cell or tissue to repair the damage (TD phase). The studies reviewed in this chapter show a pivotal role for TR as a definitive determinant of the ultimate outcome of toxicity regardless of number of chemicals exposed. These studies establish the existence of a two-stage threshold, one for infliction of injury and the other for repair mechanisms whether the exposure is to single chemicals or mixture of chemicals. Generally speaking, the threshold for stage 1 of toxicity must lie in the cytoprotective mechanisms (cellular hormesis). The threshold for stage 2 of toxicity appears to be in the ability of the tissue to respond promptly by augmenting tissue healing mechanisms. Of greater interest from a public health perspective is the finding that the tissue repair mechanisms, which are dissociable from the mechanisms that initiate injury and constitute the threshold for physical or chemical toxicity, can be mitigated by other chemical and physical agents, resulting in highly accentuated toxicity. Therefore, the recognition of the existence of compensatory TR response may offer more accurate and reliable risk assessment strategies for the assessment of risk from exposure to mixtures.

In addition to risk assessment of chemicals, establishing that the initial toxic or injurious events, regardless of how they are caused, can be separated from the subsequent events that determine the ultimate outcome of injury offers promising opportunities for developing new avenues of therapeutic intervention, with the aim of restoring the hormetic tissue repair mechanisms. Such a development will open up avenues for two types of measures to protect public health. The current approach is to decrease the injury by interfering with stage 1 of toxicity by treatment with an antidote, which either prevents further injury or decreases already inflicted injury. A second option is to enhance tissue repair and healing mechanisms, not only to obstruct progression of injury but also to augment recovery from that injury simultaneously. Consideration of this endogenous compensatory response may also offer explanation for interindividual variation in adverse drug/toxicant effects and provide additional mechanistic information extremely valuable for drug development process.

Taken together, these data indicate that assessment of TR initiated by toxicants upon exposure can have enormous impact on public health. The evidence presented in this chapter justifies the necessity for incorporating TR as a component in the decision tree of chemical mixture risk. Given the magnitude of the challenge of mixture toxicity, it is clear that substantial enhancement of experimental and risk assessment methods are needed. Improvements can be achieved by using organism-based uptake, distribution, and elimination modeling coupled with data from well-defined, model-based *in vivo* and *in vitro* experiments analyzed with improved statistical and mathematical protocols.

References

1 Bliss, C.I. (1939) The toxicity of poisons applied jointly. *Ann. Appl. Biol.*, **26**, 585–615.

2 Yang, R. (1994) Introduction to the toxicology of chemical mixtures, in *Toxicology of Chemical Mixtures*, Academic Press, Inc., pp. 1–10.

3 De Rosa, C.T., Cibulas, W., El-Masri, H.E., Pohl, H., Hansen, H., and Mumtaz, M.M. (2002) Chemical mixtures in public health practice: ATSDR's strategic approach. International Conference on Chemical Mixtures, The Agency for Toxic Substances and Disease Registry, Atlanta, p. 99.

4 Robinson, P.J. and MacDonell, M.M. (2002) Priorities for mixtures health effects research. International Conference on Chemical Mixtures 2002, The Agency for Toxic Substances and Disease Registry, Atlanta, p. 85.

5 MacDonell, M., Robinson, P., and Hertzberg, R. (2002) Approaches for assessing risks of mixed exposures. International Conference on Chemical Mixtures 2002, The Agency for Toxic Substances and Disease Registry, Atlanta, p. 86.

6 USEPA (2000) Supplementary guidance for conducting health risk assessment of chemical mixtures. United States Environmental Protection Agency, Washington, DC.

7 ATSDR (2002) Guidance manual for the assessment of joint toxic action of chemical mixtures. Agency for Toxic Substances and Disease Registry, US Department of Health and Human Services, Atlanta.

8 Mehendale, H.M. (1994) Amplified interactive toxicity of chemicals at nontoxic levels: mechanistic

considerations and implications to public health. *Environ. Health. Perspect.*, **102** (Suppl. 9), 139–149.

9 Feron, V.J. and Groten, J.P. (2002) Toxicological evaluation of chemical mixtures. *Food Chem. Toxicol.*, **40**, 825–839.

10 Yang, R.S.H. (2002) Chemical mixture toxicology: from descriptive to mechanistic, and going on to in *silico* toxicology. International Conference on Chemical Mixtures 2002, The Agency for Toxic Substances and Disease Registry, Atlanta, p. 15.

11 Wolff, S. (1984) Love Canal revisited. *J. Am. Med. Assoc.*, **251**, 1464.

12 Heath, C.W., Jr., Nadel, M.R., Zack, M.M., Jr., Chen, A.T., Bender, M.A., and Preston, R.J. (1984) Cytogenetic findings in persons living near the Love Canal. *J. Am. Med. Assoc.*, **251**, 1437–1440.

13 Silkworth, J.B., McMartin, D.N., Rej, R., Narang, R.S., Stein, V.B., Briggs, R.G., and Kaminsky, L.S. (1984) Subchronic exposure of mice to Love Canal soil contaminants. *Fundam. Appl. Toxicol.*, **4**, 231–239.

14 Vianna, N.J. and Polan, A.K. (1984) Incidence of low birth weight among Love Canal residents. *Science*, **226**, 1217–1219.

15 Silkworth, J.B., Tumasonis, C., Briggs, R.G., Narang, A.S., Narang, R.S., Rej, R., Stein, V., McMartin, D.N., and Kaminsky, L.S. (1986) The effects of Love Canal soil extracts on maternal health and fetal development in rats. *Fundam. Appl. Toxicol.*, **7**, 471–485.

16 USEPA (1983) NPL site narrative for Times Beach site. Federal Register Notice, September 8. Available at www.epa.gov/superfund/sites/npl/nar833.htm.

17 ATSDR (2000) Polychlorinated biphenyl (PCB) toxicity. Available at http://www.atsdr.cdc.gov/HEC/CSEM/pcb/physiologic_effects.html.

18 USEPA (1988) Times Beach record of decision signed. EPA press release, September 30. Available at www.epa.gov/cgi-bin/epaprintonly.cgi.

19 ACSH (2003) What's the story? – PCBs. Available at http://www.acsh.org/publications/story/pcb.

20 Lloyd, R.J.W. *et al.* (1975) Current Intelligence Bulletin 7. National Institute for Occupational Safety and Health, November 3. Available at http://www.cdc.gov/niosh/78127_7.html.

21 Hsu, S.T., Ma, C.I., Hsu, S.K., Wu, S.S., Hsu, N.H., Yeh, C.C., and Wu, S.B. (1985) Discovery and epidemiology of PCB poisoning in Taiwan: a four-year followup. *Environ. Health. Perspect.*, **59**, 5–10.

22 NYSOAG (2001) Fact sheet on polychlorinated biphenyls (PCBs). Available at http://www.oag.state.ny.us/pcb/pcb_shorter.pdf.

23 Kay, M.A. and Fausto, N. (1997) Liver regeneration: prospects for therapy based on new technologies. *Mol. Med. Today*, **3**, 108–115.

24 Carter, L.J. (1976) Michigan's PBB incident: chemical mix-up leads to disaster. *Science*, **192**, 240–243.

25 Tremblay, N.W. and Gilman, A.P. (1995) Human health, the Great Lakes, and environmental pollution: a 1994 perspective. *Environ. Health. Perspect.*, **103** (Suppl. 9), 3–5.

26 HWC (1992) A vital link: health and the environment in Canada. Cat. No. H21-112/1992E Ottawa: Minister of Supply and Services Canada.

27 Krishnan, K. and Brodeur, J. (1991) Toxicological consequences of combined exposure to environmental pollutants. *Arch. Complex Environ. Studies*, **3** (3), 1–106.

28 Cavieres, M.F., Jaeger, J., and Porter, W. (2002) Developmental toxicity of a commercial herbicide mixture in mice: I. Effects on embryo implantation and litter size. *Environ. Health. Perspect.*, **110**, 1081–1085.

29 Welshons, W.V., Thayer, K.A., Judy, B.M., Taylor, J.A., Curran, E.M., and vom Saal, F.S. (2003) Large effects from small exposures. I. Mechanisms for endocrine-disrupting chemicals with estrogenic activity. *Environ. Health. Perspect.*, **111**, 994–1006.

30 Rajapakse, N., Silva, E., and Kortenkamp, A. (2002) Combining xenoestrogens at levels below individual no-observed-effect concentrations dramatically

enhances steroid hormone action. *Environ. Health. Perspect.*, **110**, 917–921.

31 Morrison, H.I., Villeneuve, P.J., Lubin, J.H., and Schaubel, D.E. (1998) Radon-progeny exposure and lung cancer risk in a cohort of Newfoundland fluorspar miners. *Radiat Res.*, **150**, 58–65.

32 Erren, T.C., Jacobsen, M., and Piekarski, C. (1999) Synergy between asbestos and smoking on lung cancer risks. *Epidemiology*, **10**, 405–411.

33 Andersen, M.E. and Dennison, J.E. (2004) Mechanistic approaches for mixture risk assessments: present capabilities with simple mixtures and future directions. *Environ. Toxicol. Pharmacol.*, **16**, 1–11.

34 Mehendale, H.M. (1995) Injury repair as opposing forces in risk assessment. *Toxicol. Lett.*, **82–83**, 891–899.

35 Mehendale, H.M., Roth, R.A., Gandolfi, A.J., Klaunig, J.E., Lemasters, J.J., and Curtis, L.R. (1994) Novel mechanisms in chemically induced hepatotoxicity. *FASEB J.*, **8**, 1285–1295.

36 Mehendale, H.M., Thakore, K.N., and Rao, C.V. (1994) Autoprotection: stimulated tissue repair permits recovery from injury. *J. Biochem. Toxicol.*, **9**, 131–139.

37 Mangipudy, R.S., Chanda, S., and Mehendale, H.M. (1995) Hepatocellular regeneration: key to thioacetamide autoprotection. *Pharmacol. Toxicol.*, **77**, 182–188.

38 Mangipudy, R.S., Chanda, S., and Mehendale, H.M. (1995) Tissue repair response as a function of dose in thioacetamide hepatotoxicity. *Environ. Health. Perspect.*, **103**, 260–267.

39 Soni, M.G. and Mehendale, H.M. (1998) Role of tissue repair in toxicologic interactions among hepatotoxic organics. *Environ. Health. Perspect.*, **106** (Suppl. 6), 1307–1317.

40 Soni, M.G., Ramaiah, S.K., Mumtaz, M.M., Clewell, H., and Mehendale, H.M. (1999) Toxicant-inflicted injury and stimulated tissue repair are opposing toxicodynamic forces in predictive toxicology. *Regul. Toxicol. Pharmacol.*, **29**, 165–174.

41 Anand, S.S., Murthy, S.N., Vaidya, V.S., Mumtaz, M.M., and Mehendale, H.M. (2003) Tissue repair plays pivotal role in final outcome of liver injury following chloroform and allyl alcohol binary mixture. *Food Chem. Toxicol.*, **41**, 1123–1132.

42 Michalopoulos, G.K. and DeFrances, M.C. (1997) Liver regeneration. *Science*, **276**, 60–66.

43 Fausto, N., Laird, A.D., and Webber, E.M. (1995) Liver regeneration. 2. Role of growth factors and cytokines in hepatic regeneration. *FASEB J.*, **9**, 1527–1536.

44 Taub, R. (1996) Liver regeneration 4: transcriptional control of liver regeneration. *FASEB J.*, **10**, 413–427.

45 Dalhoff, K., Laursen, H., Bangert, K., Poulsen, H.E., Anderson, M.E., Grunnet, N., and Tygstrup, N. (2001) Autoprotection in acetaminophen intoxication in rats: the role of liver regeneration. *Pharmacol. Toxicol.*, **88**, 135–141.

46 Lockard, V.G., Mehendale, H.M., and O'Neal, R.M. (1983) Chlordecone-induced potentiation of carbon tetrachloride hepatotoxicity: a morphometric and biochemical study. *Exp. Mol. Pathol.*, **39**, 246–255.

47 Lockard, V.G., Mehendale, H.M., and O'Neal, R.M. (1983) Chlordecone-induced potentiation of carbon tetrachloride hepatotoxicity: a light and electron microscopic study. *Exp. Mol. Pathol.*, **39**, 230–245.

48 Mehendale, H.M. and Cai, Z. (1991) Role of ongoing versus stimulated hepatocellular regeneration in resiliency to amplification of CCl_4 toxicity by chlordecone. *FASEB J.*, **5**, A1248.

49 Shayiq, R.M., Roberts, D.W., Rothstein, K., Snawder, J.E., Benson, W., Ma, X., and Black, M. (1999) Repeat exposure to incremental doses of acetaminophen provides protection against acetaminophen-induced lethality in mice: an explanation for high acetaminophen dosage in humans without hepatic injury. *Hepatology*, **29**, 451–463.

50 Sivarao, D.V. and Mehendale, H.M. (1995) 2-Butoxyethanol autoprotection is due to resilience of newly formed

erythrocytes to hemolysis. *Arch. Toxicol.*, **69**, 526–532.

51 Vaidya, V.S., Shankar, K., Lock, E.A., Bucci, T.J., and Mehendale, H.M. (2003) Role of tissue repair in survival from s-(1,2-dichlorovinyl)-L-cysteine-induced acute renal tubular necrosis in the mouse. *Toxicol. Sci.*, **74**, 215–227.

52 Barton, C.C., Bucci, T.J., Lomax, L.G., Warbritton, A.G., and Mehendale, H.M. (2000) Stimulated pulmonary cell hyperplasia underlies resistance to alpha-naphthylthiourea. *Toxicology*, **143**, 167–181.

53 Apte, U.M., Limaye, P.B., Desaiah, D., Bucci, T.J., Warbritton, A., and Mehendale, H.M. (2003) Mechanisms of increased liver tissue repair and survival in diet-restricted rats treated with equitoxic doses of thioacetamide. *Toxicol. Sci.*, **72**, 272–282.

54 Apte, U.M., Limaye, P.B., Ramaiah, S.K., Vaidya, V.S., Bucci, T.J., Warbritton, A., and Mehendale, H.M. (2002) Upregulated promitogenic signaling via cytokines and growth factors: potential mechanism of robust liver tissue repair in calorie-restricted rats upon toxic challenge. *Toxicol. Sci.*, **69**, 448–459.

55 Gardner, C.R., Laskin, J.D., Dambach, D.M., Chiu, H., Durham, S.K., Zhou, P., Bruno, M., Gerecke, D.R., Gordon, M.K., and Laskin, D.L. (2003) Exaggerated hepatotoxicity of acetaminophen in mice lacking tumor necrosis factor receptor-1. Potential role of inflammatory mediators. *Toxicol. Appl. Pharmacol.*, **192**, 119–130.

56 Shankar, K., Vaidya, V.S., Corton, J.C., Bucci, T.J., Liu, J., Waalkes, M.P., and Mehendale, H.M. (2003) Activation of PPAR-alpha in streptozotocin-induced diabetes is essential for resistance against acetaminophen toxicity. *FASEB J.*, **17**, 1748–1750.

57 Tomiya, T., Ogata, I., and Fujiwara, K. (1998) Transforming growth factor alpha levels in liver and blood correlate better than hepatocyte growth factor with hepatocyte proliferation during liver regeneration. *Am. J. Pathol.*, **153**, 955–961.

58 Rao, P.S., Mangipudy, R.S., and Mehendale, H.M. (1997) Tissue injury and repair as parallel and opposing responses to CCl_4 hepatotoxicity: a novel dose-response. *Toxicology*, **118**, 181–193.

59 Mehendale, H.M. (1984) Potentiation of halomethane hepatotoxicity: chlordecone and carbon tetrachloride. *Fundam. Appl. Toxicol.*, **4**, 295–308.

60 Klingensmith, J.S. and Mehendale, H.M. (1981) Potentiation of brominated halomethane hepatotoxicity by chlordecone in the male rat. *Toxicol. Appl. Pharmacol.*, **61**, 378–384.

61 Klingensmith, J.S. and Mehendale, H.M. (1982) Potentiation of CCl_4 lethality by chlordecone. *Toxicol. Lett.*, **11**, 149–154.

62 Mehendale, H.M. (1984) Pulmonary disposition and effects of drugs on pulmonary removal of endogenous substances. *Fed. Proc.*, **43**, 2586–2591.

63 Thakore, K.N. and Mehendale, H.M. (1994) Effect of phenobarbital and mirex pretreatments on CCl_4 autoprotection. *Toxicol. Pathol.*, **22**, 291–299.

64 Mehendale, H.M. (1991) Role of hepatocellular regeneration and hepatolobular healing in the final outcome of liver injury. A two-stage model of toxicity. *Biochem. Pharmacol.*, **42**, 1155–1162.

65 Calabrese, E.J., Baldwin, L.A., and Mehendale, H.M. (1993) G_2 subpopulation in rat liver induced into mitosis by low-level exposure to carbon tetrachloride: an adaptive response. *Toxicol. Appl. Pharmacol.*, **121**, 1–7.

66 Calabrese, E.J. and Mehendale, H.M. (1996) A review of the role of tissue repair as an adaptive strategy: why low doses are often non-toxic and why high doses can be fatal. *Food Chem. Toxicol.*, **34**, 301–311.

67 Kodavanti, P.R., Joshi, U.M., Mehendale, H.M., and Lockard, V.G. (1989) Chlordecone (Kepone)-potentiated carbon tetrachloride hepatotoxicity in partially hepatectomized rats: a histomorphometric study. *J. Appl. Toxicol.*, **9**, 367–375.

68 Kodavanti, P.R., Joshi, U.M., Young, R.A., Bell, A.N., and Mehendale, H.M. (1989) Role of hepatocellular regeneration in chlordecone potentiated hepatotoxicity of carbon tetrachloride. *Arch. Toxicol.*, **63**, 367–375.

69 Kodavanti, P.R., Joshi, U.M., Young, R.A., Meydrech, E.F., and Mehendale, H.M. (1989) Protection of hepatotoxic and lethal effects of CCl_4 by partial hepatectomy. *Toxicol. Pathol.*, **17**, 494–505.

70 Mehendale, H.M. (1989) Mechanism of the lethal interaction of chlordecone and CCl_4 at non-toxic doses. *Toxicol. Lett.*, **49**, 215–241.

71 Ruch, R.J., Klaunig, J.E., and Pereira, M.A. (1985) Selective resistance to cytotoxic agents in hepatocytes isolated from partially hepatectomized and neoplastic mouse liver. *Cancer Lett.*, **26**, 295–301.

72 Thompson, N.L., Mead, J.E., Braun, L., Goyette, M., Shank, P.R., and Fausto, N. (1986) Sequential protooncogene expression during rat liver regeneration. *Cancer Res.*, **46**, 3111–3117.

73 Fausto, N. and Mead, J.E. (1989) Regulation of liver growth: protooncogenes and transforming growth factors. *Lab Invest.*, **60**, 4–13.

74 Sasaki, Y., Hayashi, N., Morita, Y., Ito, T., Kasahara, A., Fusamoto, H., Sato, N., Tohyama, M., and Kamada, T. (1989) Cellular analysis of c-Ha-ras gene expression in rat liver after CCl_4 administration. *Hepatology*, **10**, 494–500.

75 Michalopoulos, G.K. (1990) Liver regeneration: molecular mechanisms of growth control. *FASEB J.*, **4**, 176–187.

76 Herbst, H., Milani, S., Schuppan, D., and Stein, H. (1991) Temporal and spatial patterns of proto-oncogene expression at early stages of toxic liver injury in the rat. *Lab Invest.*, **65**, 324–333.

77 Rao, S.B. and Mehendale, H.M. (1993) Halomethane-chlordecone (CD) interactive hepatotoxicity: current concepts on the mechanism. *Indian J. Biochem. Biophys.*, **30**, 191–198.

78 Rao, S.B. and Mehendale, H.M. (1989) Protective role of fructose 1,6-bisphosphate during CCl_4 hepatotoxicity in rats. *Biochem. J.*, **262**, 721–725.

79 Rao, S.B. and Mehendale, H.M. (1989) Protection from chlordecone (Kepone)-potentiated CCl_4 hepatotoxicity in rats by fructose 1,6-diphosphate. *Int. J. Biochem.*, **21**, 949–954.

80 Soni, M.G. and Mehendale, H.M. (1991) Protection from chlordecone-amplified carbon tetrachloride toxicity by cyanidanol: regeneration studies. *Toxicol. Appl. Pharmacol.*, **108**, 58–66.

81 Soni, M.G. and Mehendale, H.M. (1991) Protection from chlordecone-amplified carbon tetrachloride toxicity by cyanidanol: biochemical and histological studies. *Toxicol. Appl. Pharmacol.*, **108**, 46–57.

82 Soni, M.G. and Mehendale, H.M. (1993) Hepatic failure leads to lethality of chlordecone-amplified hepatotoxicity of carbon tetrachloride. *Fundam. Appl. Toxicol.*, **21**, 442–450.

83 Kodavanti, P.R., Kodavanti, U.P., Faroon, O.M., and Mehendale, H.M. (1992) Pivotal role of hepatocellular regeneration in the ultimate hepatotoxicity of CCl_4 in chlordecone-, mirex-, or phenobarbital-pretreated rats. *Toxicol. Pathol.*, **20**, 556–569.

84 Kodavanti, P.R., Kodavanti, U.P., and Mehendale, H.M. (1990) Altered hepatic energy status in chlordecone (Kepone)-potentiated CCl_4 hepatotoxicity. *Biochem. Pharmacol.*, **40**, 859–866.

85 Kodavanti, P.R., Kodavanti, U.P., and Mehendale, H.M. (1991) Carbon tetrachloride-induced alterations of hepatic calmodulin and free calcium levels in rats pretreated with chlordecone. *Hepatology*, **13**, 230–238.

86 Bell, A.N., Young, R.A., Lockard, V.G., and Mehendale, H.M. (1988) Protection of chlordecone-potentiated carbon tetrachloride hepatotoxicity and lethality by partial hepatectomy. *Arch. Toxicol.*, **61**, 392–405.

87 Cai, Z. and Mehendale, H.M. (1993) Resiliency to amplification of carbon tetrachloride hepatotoxicity by chlordecone during postnatal development in rats. *Pediatr. Res.*, **33**, 225–232.

88 Hewitt, L.A., Palmason, C., Masson, S., and Plaa, G.L. (1990) Evidence for the involvement of organelles in the mechanism of ketone-potentiated chloroform-induced hepatotoxicity. *Liver*, **10**, 35–48.

89 Hewitt, W.R., Miyajima, H., Cote, M.G., and Plaa, G.L. (1979) Acute alteration of chloroform-induced hepato- and nephrotoxicity by Mirex and Kepone. *Toxicol. Appl. Pharmacol.*, **48**, 509–527.

90 Purushotham, K.R., Lockard, V.G., and Mehendale, H.M. (1988) Amplification of chloroform hepatotoxicity and lethality by dietary chlordecone (Kepone) in mice. *Toxicol. Pathol.*, **16**, 27–34.

91 Mehendale, H.M., Purushotham, K.R., and Lockard, V.G. (1989) The time course of liver injury and [3*H*]thymidine incorporation in chlordecone-potentiated $CHCl_3$ hepatotoxicity. *Exp. Mol. Pathol.*, **51**, 31–47.

92 Agarwal, A.K. and Mehendale, H.M. (1982) Potentiation of bromotrichloromethane hepatotoxicity and lethality by chlordecone preexposure in the rat. *Fundam. Appl. Toxicol.*, **2**, 161–167.

93 Klingensmith, J.S., Lockard, V., and Mehendale, H.M. (1983) Acute hepatotoxicity and lethality of CCl_4 in chlordecone-pretreated rats. *Exp. Mol. Pathol.*, **39**, 1–10.

94 Klingensmith, J.S. and Mehendale, H.M. (1983) Hepatic microsomal metabolism of CCL_4 after pretreatment with chlordecone, mirex, or phenobarbital in male rats. *Drug Metab. Dispos.*, **11**, 329–334.

95 Faroon, O.M. and Mehendale, H.M. (1990) Bromotrichloromethane hepatotoxicity. The role of stimulated hepatocellular regeneration in recovery: biochemical and histopathological studies in control and chlordecone pretreated male rats. *Toxicol. Pathol.*, **18**, 667–677.

96 Faroon, O.M., Henry, R.W., Soni, M.G., and Mehendale, H.M. (1991) Potentiation of $BrCCl_3$ hepatotoxicity by chlordecone: biochemical and ultrastructural study. *Toxicol. Appl. Pharmacol.*, **110**, 185–197.

97 Thakore, K.N. and Mehendale, H.M. (1991) Role of hepatocellular regeneration in CCl_4 autoprotection. *Toxicol. Pathol.*, **19**, 47–58.

98 Harris, R.N. and Anders, M.W. (1981) 2-Propanol treatment induces selectively the metabolism of carbon tetrachloride to phosgene. Implications for carbon tetrachloride hepatotoxicity. *Drug Metab. Dispos.*, **9**, 551–556.

99 Ueng, T.H., Moore, L., Elves, R.G., and Alvares, A.P. (1983) Isopropanol enhancement of cytochrome P-450-dependent monooxygenase activities and its effects on carbon tetrachloride intoxication. *Toxicol. Appl. Pharmacol.*, **71**, 204–214.

100 Ray, S.D. and Mehendale, H.M. (1990) Potentiation of CCl_4 and $CHCl_3$ hepatotoxicity and lethality by various alcohols. *Fundam. Appl. Toxicol.*, **15**, 429–440.

101 Rao, P.S., Dalu, A., Kulkarni, S.G., and Mehendale, H.M. (1996) Stimulated tissue repair prevents lethality in isopropanol-induced potentiation of carbon tetrachloride hepatotoxicity. *Toxicol. Appl. Pharmacol.*, **140**, 235–244.

102 Mehendale, H.M. (2005) Tissue repair: an important determinant of final outcome of toxicant-induced injury. *Toxicol. Pathol.*, **33**, 41–51.

103 Mangipudy, R.S., Rao, P.S., and Mehendale, H.M. (1996) Effect of an antimitotic agent colchicine on thioacetamide hepatotoxicity. *Environ. Health. Perspect.*, **104**, 744–749.

104 Anand, S.S., Soni, M.G., Vaidya, V.S., Murthy, S.N., Mumtaz, M.M., and Mehendale, H.M. (2003) Extent and timeliness of tissue repair determines the dose-related hepatotoxicity of chloroform. *Int. J. Toxicol.*, **22**, 25–33.

105 Kulkarni, S.G., Duong, H., Gomila, R., and Mehendale, H.M. (1996) Strain differences in tissue repair response to 1,2-dichlorobenzene. *Arch. Toxicol.*, **70**, 714–723.

106 Kulkarni, S.G., Warbritton, A., Bucci, T.J., and Mehendale, H.M. (1997) Antimitotic intervention with colchicine alters the outcome of o-DCB-induced hepatotoxicity in Fischer 344 rats. *Toxicology*, **120**, 79–88.

107 Soni, M.G., Mangipudy, R.S., Mumtaz, M.M., and Mehendale, H.M. (1998) Tissue repair response as a function of dose during trichloroethylene hepatotoxicity. *Toxicol. Sci.*, **42**, 158–165.

108 Anand, S.S. and Mehendale, H.M. (2004) Liver regeneration: a critical toxicodynamic response in predictive toxicology. *Environ. Toxicol. Pharmacol.*, **18**, 149–160.

109 Anand, S.S., Murthy, S.N., Mumtaz, M.M., and Mehendale, H.M. (2004) Dose-dependent liver tissue repair in chloroform plus thioacetamide acute hepatotoxicity. *Environ. Toxicol. Pharmacol.*, **18**, 143–148.

110 Anand, S.S., Mumtaz, M.M., and Mehendale, H.M. (2005) Dose-dependent liver tissue repair after chloroform plus trichloroethylene binary mixture. *Basic Clin. Pharmacol. Toxicol.*, **96**, 436–444.

111 Anand, S.S., Mumtaz, M.M., and Mehendale, H.M. (2005) Dose-dependent liver regeneration in chloroform, trichloroethylene and allyl alcohol ternary mixture hepatotoxicity in rats. *Arch. Toxicol.*, **79**, 671–682.

112 Mehendale, H.M. and Limaye, P.B. (2005) Calpain: a death protein that mediates progression of liver injury. *Trends Pharmacol. Sci.*, **26**, 232–236.

113 Limaye, P.B., Apte, U.M., Shankar, K., Bucci, T.J., Warbritton, A., and Mehendale, H.M. (2003) Calpain released from dying hepatocytes mediates progression of acute liver injury induced by model hepatotoxicants. *Toxicol. Appl. Pharmacol.*, **191**, 211–226.

114 Limaye, P.B., Bhave, V.S., Palkar, P.S., Apte, U.M., Sawant, S.P., Yu, S., Latendresse, J.R., Reddy, J.K., and Mehendale, H.M. (2006) Upregulation of calpastatin in regenerating and developing rat liver: role in resistance against hepatotoxicity. *Hepatology*, **44**, 379–388.

115 Tsukamoto, I. and Kojo, S. (1989) Effect of colchicine and vincristine on DNA synthesis in regenerating rat liver. *Biochim. Biophys. Acta*, **1009**, 191–193.

116 Fitzgerald, P.H. and Brehaut, L.A. (1970) Depression of DNA synthesis and mitotic index by colchicine in cultured human lymphocytes. *Exp. Cell Res.*, **59**, 27–31.

117 Dalu, A. and Mehendale, H.M. (1996) Efficient tissue repair underlies the resiliency of postnatally developing rats to chlordecone + CCl_4 hepatotoxicity. *Toxicology*, **111**, 29–42.

118 Chanda, S., Mangipudy, R.S., Warbritton, A., Bucci, T.J., and Mehendale, H.M. (1995) Stimulated hepatic tissue repair underlies heteroprotection by thioacetamide against acetaminophen-induced lethality. *Hepatology*, **21**, 477–486.

119 Uryvaeva, I.V. and Faktor, V.M. (1976) Resistance of regenerating liver to hepatotoxins. *Biull. Eksp. Biol. Med.*, **81**, 283–285.

120 Abdul-Hussain, S.K. and Mehendale, H.M. (1992) Ongoing hepatocellular regeneration and resiliency toward galactosamine hepatotoxicity. *Arch. Toxicol.*, **66**, 729–742.

121 Roberts, E., Ahluwalia, M.B., Lee, G., Chan, C., Sarma, D.S., and Farber, E. (1983) Resistance to hepatotoxins acquired by hepatocytes during liver regeneration. *Cancer Res.*, **43**, 28–34.

122 Anand, S.S., Philip, B.K., Palkar, P.S., Mumtaz, M.M., Latendresse, J.R., and Mehendale, H.M. (2006) Adaptive tolerance in mice upon subchronic exposure to chloroform: increased exhalation and target tissue regeneration. *Toxicol. Appl. Pharmacol.*, **213**, 267–281.

123 Philip, B.K., Anand, S.S., Palkar, P.S., Mumtaz, M.M., Latendresse, J.R., and Mehendale, H.M. (2006) Subchronic chloroform priming protects mice from a subsequently administered lethal dose of chloroform. *Toxicol. Appl. Pharmacol.*, **216**, 108–121.

124 Shankar, K., Vaidya, V.S., Apte, U.M., Manautou, J.E., Ronis, M.J., Bucci, T.J., and Mehendale, H.M. (2003) Type 1 diabetic mice are protected from acetaminophen hepatotoxicity. *Toxicol. Sci.*, **73**, 220–234.

125 Shankar, K., Vaidya, V.S., Wang, T., Bucci, T.J., and Mehendale, H.M. (2003) Streptozotocin-induced diabetic mice are resistant to lethal effects of thioacetamide hepatotoxicity. *Toxicol. Appl. Pharmacol.*, **188**, 122–134.

126 Wang, T., Fontenot, R.D., Soni, M.G., Bucci, T.J., and Mehendale, H.M. (2000) Enhanced hepatotoxicity and toxic outcome of thioacetamide in

streptozotocin-induced diabetic rats. *Toxicol. Appl. Pharmacol.*, **166**, 92–100.
127 Wang, T., Shankar, K., Bucci, T.J., Warbritton, A., and Mehendale, H.M. (2001) Diallyl sulfide inhibition of CYP2E1 does not rescue diabetic rats from thioacetamide-induced mortality. *Toxicol. Appl. Pharmacol.*, **173**, 27–37.
128 Dalu, A., Cronin, G.M., Lyn-Cook, B.D., and Mehendale, H.M. (1995) Age-related differences in TGF-alpha and proto-oncogenes expression in rat liver after a low dose of carbon tetrachloride. *J. Biochem. Toxicol.*, **10**, 259–264.
129 Dalu, A., Warbritton, A., Bucci, T.J., and Mehendale, H.M. (1995) Age-related susceptibility to chlordecone-potentiated carbon tetrachloride hepatotoxicity and lethality is due to hepatic quiescence. *Pediatr. Res.*, **38**, 140–148.
130 Murali, B., Korrapati, M.C., Warbritton, A., Latendresse, J.R., and Mehendale, H.M. (2004) Tolerance of aged Fischer 344 rats against chlordecone-amplified carbon tetrachloride toxicity. *Mech. Ageing Dev.*, **125**, 421–435.
131 Cai, Z.W. and Mehendale, H.M. (1990) Lethal effects of CCl_4 and its metabolism by Mongolian gerbils pretreated with chlordecone, phenobarbital, or mirex. *Toxicol. Appl. Pharmacol.*, **104**, 511–520.
132 Stine, E.R., Gunawardhana, L., and Sipes, I.G. (1991) The acute hepatotoxicity of the isomers of dichlorobenzene in Fischer-344 and Sprague-Dawley rats: isomer-specific and strain-specific differential toxicity. *Toxicol. Appl. Pharmacol.*, **109**, 472–481.
133 Mehendale, H.M., Ramaiah, S.K., Dalu, A., and Soni, M.G. (1999) Older rats are resilient to the hepatotoxicity of CCl_4 and chlordeone + CCl_4. *Toxicologist*, **48**, 281.
134 Sanz, N., Diez-Fernandez, C., and Cascales, M. (1998) Aging delays the post-necrotic restoration of liver function. *Biofactors*, **8**, 103–109.
135 Chanda, S. and Mehendale, M. (1994) Role of nutritional fatty acid and l-carnitine in the final outcome of thioacetamide hepatotoxicity. *FASEB J.*, **8**, 1061–1068.
136 Wang, T., Shankar, K., Ronis, M.J., and Mehendale, H.M. (2007) Mechanisms and outcomes of drug- and toxicant-induced liver toxicity in diabetes. *Crit. Rev. Toxicol.*, **37**, 413–459.
137 Berg, T.F., Breen, P.J., Feuers, R.J., Oriaku, E.T., Chen, F.X., and Hart, R.W. (1994) Acute toxicity of ganciclovir: effect of dietary restriction and chronobiology. *Food Chem. Toxicol.*, **32**, 45–50.
138 Duffy, E.H., Feuers, R.J., Pipkin, J.L., Berg, T.F., Leakey, J.E.A., Turturo, A., and Hart, R.W. (1995) The effect of dietary restriction and aging on the physiological response to drugs, in *Dietary Restriction: Implications for the Design and Interpretation of Toxicity and Carcinogenicity Studies* (eds R.W. Hart, D.A., Neumann, and R.T. Robertson), ILSI Press, Washington, DC, pp. 127–140.
139 Hass, B.S., Lewis, S.M., Duffy, P.H., Ershler, W., Feuers, R.J., Good, R.A., Ingram, D.K., Lane, M.A., Leakey, J.E., Lipschitz, D., Poehlman, E.T., Roth, G.S., Sprott, R.L., Sullivan, D.H., Turturro, A., Verdery, R.B., Walford, R.L., Weindruch, R., Yu, B.P., and Hart, R.W. (1996) Dietary restriction in humans: report on the Little Rock Conference on the value, feasibility, and parameters of a proposed study. *Mech. Ageing Dev.*, **91**, 79–94.
140 Masaro, E.F. (1998) Mini review: food restriction in rodents: an evaluation of its role in aging. *J. Gerontol.*, **43**, B59–64.
141 Ramaiah, S.K., Apte, U.M., and Mehendale, H.M. (2000) Diet restriction as a protective mechanism in non-cancer toxicity outcomes: a review. *Int. J. Toxicol.*, **19**, 1–13.
142 Weindruch, R. and Walford, R.L. (1982) Dietary restriction in mice beginning at 1 year of age: effect on life-span and spontaneous cancer incidence. *Science*, **215**, 1415–1418.
143 Ramaiah, S.K., Soni, M.G., Bucci, T.J., and Mehendale, H.M. (1998) Diet restriction enhances compensatory liver tissue repair and survival following administration of lethal dose of thioacetamide. *Toxicol. Appl. Pharmacol.*, **150**, 12–21.
144 Ramaiah, S.K., Bucci, T.J., Warbritton, A., Soni, M.G., and Mehendale, H.M. (1998) Temporal changes in tissue repair permit

survival of diet-restricted rats from an acute lethal dose of thioacetamide. *Toxicol. Sci.*, **45**, 233–241.

145 Ramaiah, S.K., Apte, U., and Mehendale, H.M. (2001) Cytochrome P4502E1 induction increases thioacetamide liver injury in diet-restricted rats. *Drug Metab. Dispos.*, **29**, 1088–1095.

146 El-Serag, H.B. and Everhart, J.E. (2002) Diabetes increases the risk of acute hepatic failure. *Gastroenterology*, **122**, 1822–1828.

147 El-Serag, H.B., Tran, T., and Everhart, J.E. (2004) Diabetes increases the risk of chronic liver disease and hepatocellular carcinoma. *Gastroenterology*, **126**, 460–468.

148 Andrade, R.J., Lucena, M., Vega, J.L., Torres, M., Salmeron, F.J., Bellot, V., Garcia-Escano, M.D., and Moreno, P. (1998) Acarbose-associated hepatotoxicity. *Diabetes Care*, **21**, 2029–2030.

149 Carrascosa, M., Pascual, F., and Aresti, S. (1997) Acarbose-induced acute severe hepatotoxicity. *Lancet*, **349**, 698–699.

150 Gentile, S., Turco, S., Guarino, G., Sasso, F.C., and Torella, R. (1999) Aminotransferase activity and acarbose treatment in patients with type 2 diabetes. *Diabetes Care*, **22**, 1217–1218.

151 Hsiao, S.H., Liao, L.H., Cheng, P.N., and Wu, T.J. (2006) Hepatotoxicity associated with acarbose therapy. *Ann. Pharmacother.*, **40**, 151–154.

152 Dulin, W.E. and Wyse, B.M. (1969) Reversal of streptozotocin diabetes with nicotinamide. *Proc. Soc. Exp. Biol. Med.*, **130**, 992–994.

153 Fischer, L.J. and Rickert, D.E. (1975) Pancreatic islet-cell toxicity. *CRC Crit. Rev. Toxicol.*, **3**, 231–263.

154 Like, A.A. and Rossini, A.A. (1976) Streptozotocin-induced pancreatic insulitis: new model of diabetes mellitus. *Science*, **193**, 415–417.

155 Mansford, K.R. and Opie, L. (1968) Comparison of metabolic abnormalities in diabetes mellitus induced by streptozotocin or by alloxan. *Lancet*, **1**, 670–671.

156 Kraegen, E.W., James, D.E., Storlien, L.H., Burleigh, K.M., and Chisholm, D.J. (1986) *In vivo* insulin resistance in individual peripheral tissues of the high fat fed rat: assessment by euglycaemic clamp plus deoxyglucose administration. *Diabetologia*, **29**, 192–198.

157 Lillioja, S., Mott, D.M., Spraul, M., Ferraro, R., Foley, J.E., Ravussin, E., Knowler, W.C., Bennett, P.H., and Bogardus, C. (1993) Insulin resistance and insulin secretory dysfunction as precursors of non-insulin-dependent diabetes mellitus. Prospective studies of Pima Indians. *N. Engl. J. Med.*, **329**, 1988–1992.

158 Sawant, S.P., Dnyanmote, A.V., Shankar, K., Limaye, P.B., Latendresse, J.R., and Mehendale, H.M. (2004) Potentiation of carbon tetrachloride hepatotoxicity and lethality in type 2 diabetic rats. *J. Pharmacol. Exp. Ther.*, **308**, 694–704.

159 Sawant, S.P., Dnyanmote, A.V., Mitra, M.S., Chilakapati, J., Warbritton, A., Latendresse, J.R., and Mehendale, H.M. (2006) Protective effect of type 2 diabetes on acetaminophen-induced hepatotoxicity in male Swiss-Webster mice. *J. Pharmacol. Exp. Ther.*, **316**, 507–519.

160 Wang, T., Shankar, K., Ronis, M.J., and Mehendale, H.M. (2000) Potentiation of thioacetamide liver injury in diabetic rats is due to induced CYP2E1. *J. Pharmacol. Exp. Ther.*, **294**, 473–479.

161 Sawant, S.P., Dnyanmote, A.V., Warbritton, A., Latendresse, J.R., and Mehendale, H.M. (2006) Type 2 diabetic rats are sensitive to thioacetamide hepatotoxicity. *Toxicol. Appl. Pharmacol.*, **211**, 221–232.

162 Leclercq, I.A., Field, J., and Farrell, G.C. (2003) Leptin-specific mechanisms for impaired liver regeneration in ob/ob mice after toxic injury. *Gastroenterology*, **124**, 1451–1464.

163 Bruccoleri, A., Gallucci, R., Germolec, D.R., Blackshear, P., Simeonova, P., Thurman, R.G., and Luster, M.I. (1997) Induction of early-immediate genes by tumor necrosis factor alpha contribute to liver repair following chemical-induced hepatotoxicity. *Hepatology*, **25**, 133–141.

164 Devi, S.S. and Mehendale, H.M. (2005) The role of NF-kappaB signaling in

impaired liver tissue repair in thioacetamide-treated type 1 diabetic rats. *Eur. J. Pharmacol.*, **523**, 127–136.

165 Devi, S.S. and Mehendale, H.M. (2005) Disrupted G_1 to S phase clearance via cyclin signaling impairs liver tissue repair in thioacetamide-treated type 1 diabetic rats. *Toxicol. Appl. Pharmacol.*, **207**, 89–102.

166 Devi, S.S. and Mehendale, H.M. (2006) Microarray analysis of thioacetamide-treated type 1 diabetic rats. *Toxicol. Appl. Pharmacol.*, **212**, 69–78.

167 Michalopoulos, G.K. and DeFrances, M. (2005) Liver regeneration. *Adv. Biochem. Eng. Biotechnol.*, **93**, 101–134.

168 Yamada, Y. and Fausto, N. (1998) Deficient liver regeneration after carbon tetrachloride injury in mice lacking type 1 but not type 2 tumor necrosis factor receptor. *Am. J. Pathol.*, **152**, 1577–1589.

169 Yamada, Y., Kirillova, I., Peschon, J.J., and Fausto, N. (1997) Initiation of liver growth by tumor necrosis factor: deficient liver regeneration in mice lacking type I tumor necrosis factor receptor. *Proc. Natl. Acad. Sci. USA*, **94**, 1441–1446.

170 Zimmers, T.A., McKillop, I.H., Pierce, R.H., Yoo, J.Y., and Koniaris, L.G. (2003) Massive liver growth in mice induced by systemic interleukin 6 administration. *Hepatology*, **38**, 326–334.

171 Fausto, N. (2000) Liver regeneration. *J. Hepatol.*, **32**, 19–31.

172 Fausto, N., Mead, J.E., Braun, L., Thompson, N.L., Panzica, M., Goyette, M., Bell, G.I., and Shank, P.R. (1986) Proto-oncogene expression and growth factors during liver regeneration. *Symp. Fundam. Cancer Res.*, **39**, 69–86.

173 Fausto, N. and Webber, E.M. (1993) Control of liver growth. *Crit. Rev. Eukaryot. Gene Expr.*, **3**, 117–135.

174 Brenner, D.A. (1998) Signal transduction during liver regeneration. *J. Gastroenterol. Hepatol.*, **13** (Suppl.), S93–S95.

175 Davis, R.J. (1994) MAPKs: new JNK expands the group. *Trends Biochem. Sci.*, **19**, 470–473.

176 Schwabe, R.F., Bradham, C.A., Uehara, T., Hatano, E., Bennett, B.L., Schoonhoven, R., and Brenner, D.A. (2003) c-Jun-N-terminal kinase drives cyclin D1 expression and proliferation during liver regeneration. *Hepatology*, **37**, 824–832.

177 Diehl, A.M. and Rai, R.M. (1996) Liver regeneration 3: regulation of signal transduction during liver regeneration. *FASEB J.*, **10**, 215–227.

178 Duzgun, S.A., Rasque, H., Kito, H., Azuma, N., Li, W., Basson, M.D., Gahtan, V., Dudrick, S.J., and Sumpio, B.E. (2000) Mitogen-activated protein phosphorylation in endothelial cells exposed to hyperosmolar conditions. *J. Cell Biochem.*, **76**, 567–571.

179 Pascal, M.M., Forrester, J.V., and Knott, R.M. (1999) Glucose-mediated regulation of transforming growth factor-beta (TGF-beta) and TGF-beta receptors in human retinal endothelial cells. *Curr. Eye Res.*, **19**, 162–170.

180 Tsiani, E., Lekas, P., Fantus, I.G., Dlugosz, J., and Whiteside, C. (2002) High glucose-enhanced activation of mesangial cell p38 MAPK by ET-1, ANG II, and platelet-derived growth factor. *Am. J. Physiol. Endocrinol. Metab.*, **282**, E161–E169.

181 Albrecht, J.H. and Hansen, L.K. (1999) Cyclin D1 promotes mitogen-independent cell cycle progression in hepatocytes. *Cell Growth Differ*, **10**, 397–404.

182 Awad, M.M., Enslen, H., Boylan, J.M., Davis, R.J., and Gruppuso, P.A. (2000) Growth regulation via p38 mitogen-activated protein kinase in developing liver. *J. Biol. Chem.*, **275**, 38716–38721.

183 Lavoie, J.N., L'Allemain, G., Brunet, A., Muller, R., and Pouyssegur, J. (1996) Cyclin D1 expression is regulated positively by the p42/p44MAPK and negatively by the p38/HOGMAPK pathway. *J. Biol. Chem.*, **271**, 20608–20616.

184 Sawant, S.P., Dnyanmote, A.V., and Mehendale, H.M. (2007) Mechanisms of inhibited liver tissue repair in toxicant challenged type 2 diabetic rats. *Toxicology*, **232**, 200–215.

185 Manautou, J.E., Emeigh Hart, S.G., Khairallah, E.A., and Cohen, S.D. (1996) Protection against acetaminophen hepatotoxicity by a single dose of clofibrate: effects on selective protein arylation and glutathione depletion. *Fundam. Appl. Toxicol.*, **29**, 229–237.

186 Manautou, J.E., Hoivik, D.J., Tveit, A., Hart, S.G., Khairallah, E.A., and Cohen, S.D. (1994) Clofibrate pretreatment diminishes acetaminophen's selective covalent binding and hepatotoxicity. *Toxicol. Appl. Pharmacol.*, **129**, 252–263.

187 Manautou, J.E., Silva, V.M., Hennig, G.E., and Whiteley, H.E. (1998) Repeated dosing with the peroxisome proliferator clofibrate decreases the toxicity of model hepatotoxic agents in male mice. *Toxicology*, **127**, 1–10.

188 Nicholls-Grzemski, F.A., Calder, I.C., and Priestly, B.G. (1992) Peroxisome proliferators protect against paracetamol hepatotoxicity in mice. *Biochem. Pharmacol.*, **43**, 1395–1396.

189 Nicholls-Grzemski, F.A., Calder, I.C., Priestly, B.G., and Burcham, P.C. (2000) Clofibrate-induced *in vitro* hepatoprotection against acetaminophen is not due to altered glutathione homeostasis. *Toxicol. Sci.*, **56**, 220–228.

190 Asayama, K., Sandhir, R., Sheikh, F.G., Hayashibe, H., Nakane, T., and Singh, I. (1999) Increased peroxisomal fatty acid beta-oxidation and enhanced expression of peroxisome proliferator-activated receptor-alpha in diabetic rat liver. *Mol. Cell Biochem.*, **194**, 227–234.

191 Engels, W., van Bilsen, M., Wolffenbuttel, B.H., van der Vusse, G.J., and Glatz, J.F. (1999) Cytochrome P450, peroxisome proliferation, and cytoplasmic fatty acid-binding protein content in liver, heart and kidney of the diabetic rat. *Mol. Cell Biochem.*, **192**, 53–61.

192 Kroetz, D.L., Yook, P., Costet, P., Bianchi, P., and Pineau, T. (1998) Peroxisome proliferator-activated receptor alpha controls the hepatic CYP4A induction adaptive response to starvation and diabetes. *J. Biol. Chem.*, **273**, 31581–31589.

193 Anderson, S.P., Yoon, L., Richard, E.B., Dunn, C.S., Cattley, R.C., and Corton, J.C. (2002) Delayed liver regeneration in peroxisome proliferator-activated receptor-alpha-null mice. *Hepatology*, **36**, 544–554.

194 Dharancy, S., Malapel, M., Perlemuter, G., Roskams, T., Podevin, P., Conti, F., Canva, V., Gambiez, L., Calmus, Y., Mathurin, P., Pol, S., Brechot, C., Auwerx, J., and Desreumaux, P. (2002) Impaired liver expression of the PPAR-α modulates liver steatosis and inflammation during hepatitis C virus infection. *Hepatology*, **36**, 264A.

195 Freeman, B.A. and Crapo, J.D. (1982) Biology of disease: free radicals and tissue injury. *Lab Invest.*, **47**, 412–426.

196 Ito, M., Ishikawa, Y., Kiguchi, H., Komiyama, K., Murakami, M., Kudo, I., Akasaka, Y., and Ishii, T. (2004) Distribution of type V secretory phospholipase A2 expression in human hepatocytes damaged by liver disease. *J. Gastroenterol. Hepatol.*, **19**, 1140–1149.

197 Murakami, M. and Kudo, I. (2004) Secretory phospholipase A2. *Biol. Pharm. Bull.*, **27**, 1158–1164.

198 Winstead, M.V., Balsinde, J., and Dennis, E.A. (2000) Calcium-independent phospholipase A(2): structure and function. *Biochim. Biophys. Acta*, **1488**, 28–39.

199 Wolf, M.J., Wang, J., Turk, J., and Gross, R.W. (1997) Depletion of intracellular calcium stores activates smooth muscle cell calcium-independent phospholipase A2. A novel mechanism underlying arachidonic acid mobilization. *J. Biol. Chem.*, **272**, 1522–1526.

200 Bhave, V.S., Donthamsetty, S., Latendresse, J.R., Muskhelishvili, L., and Mehendale, H.M. (2008) Secretory phospholipase A2 mediates progression of acute liver injury in the absence of sufficient cyclooxygenase-2. *Toxicol. Appl. Pharmacol.*, **228**, 225–238.

201 Beraza, N., Marques, J.M., Martinez-Anso, E., Iniguez, M., Prieto, J., and Bustos, M. (2005) Interplay among cardiotrophin-1, prostaglandins, and vascular endothelial growth factor in rat liver regeneration. *Hepatology*, **41**, 460–469.

202 Casado, M., Callejas, N.A., Rodrigo, J., Zhao, X., Dey, S.K., Bosca, L., and Martin-Sanz, P. (2001) Contribution of cyclooxygenase 2 to liver regeneration

after partial hepatectomy. *FASEB J.*, **15**, 2016–2018.

203 Fernandez-Martinez, A., Callejas, N.A., Casado, M., Bosca, L., and Martin-Sanz, P. (2004) Thioacetamide-induced liver regeneration involves the expression of cyclooxygenase 2 and nitric oxide synthase 2 in hepatocytes. *J. Hepatol.*, **40**, 963–970.

204 Demeure, C.E., Yang, L.P., Desjardins, C., Raynauld, P., and Delespesse, G. (1997) Prostaglandin E2 primes naive T cells for the production of anti-inflammatory cytokines. *Eur. J. Immunol.*, **27**, 3526–3531.

205 Oppenheimer-Marks, N., Kavanaugh, A.F., and Lipsky, P.E. (1994) Inhibition of the transendothelial migration of human T lymphocytes by prostaglandin E2. *J. Immunol.*, **152**, 5703–5713.

206 Beck, S., Lambeau, G., Scholz-Pedretti, K., Gelb, M.H., Janssen, M.J., Edwards, S.H., Wilton, D.C., Pfeilschifter, J., and Kaszkin, M. (2003) Potentiation of tumor necrosis factor alpha-induced secreted phospholipase A2 (sPLA2)-IIA expression in mesangial cells by an autocrine loop involving sPLA2 and peroxisome proliferator-activated receptor alpha activation. *J. Biol. Chem.*, **278**, 29799–29812.

207 Lappas, M., Permezel, M., and Rice, G.E. (2006) 15-Deoxy-delta(12,14)-prostaglandin J(2) and troglitazone regulation of the release of phospholipid metabolites, inflammatory cytokines and proteases from human gestational tissues. *Placenta*, **27**, 1060–1072.

208 Piraino, G., Cook, J.A., O'Connor, M., Hake, P.W., Burroughs, T.J., Teti, D., and Zingarelli, B. (2006) Synergistic effect of peroxisome proliferator activated receptor-gamma and liver X receptor-alpha in the regulation of inflammation in macrophages. *Shock*, **26**, 146–153.

209 Napirei, M., Basnakian, A.G., Apostolov, E.O., and Mannherz, H.G. (2006) Deoxyribonuclease 1 aggravates acetaminophen-induced liver necrosis in male CD-1 mice. *Hepatology*, **43**, 297–305.

210 Horn, K.D., Wax, P., Schneider, S.M., Martin, T.G., Nine, J.S., Moraca, M.A., Virji, M.A., Aronica, P.A., and Rao, K.N. (1999) Biomarkers of liver regeneration allow early prediction of hepatic recovery after acute necrosis. *Am. J. Clin Pathol.*, **112**, 351–357.

211 Inomata, M., Kasai, Y., Nakamura, M., and Kawashima, S. (1988) Activation mechanism of calcium-activated neutral protease. Evidence for the existence of intramolecular and intermolecular autolyses. *J. Biol. Chem.*, **263**, 19783–19787.

212 Maekawa, A., Lee, J.K., Nagaya, T., Kamiya, K., Yasui, K., Horiba, M., Miwa, K., Uzzaman, M., Maki, M., Ueda, Y., and Kodama, I. (2003) Overexpression of calpastatin by gene transfer prevents troponin I degradation and ameliorates contractile dysfunction in rat hearts subjected to ischemia/reperfusion. *J. Mol. Cell Cardiol.*, **35**, 1277–1284.

213 Nishiura, I., Tanaka, K., Yamato, S., and Murachi, T. (1978) The occurrence of an inhibitor of $Ca^{2}+$-dependent neutral protease in rat liver. *J. Biochem.*, **84**, 1657–1659.

214 Suzuki, K., Imajoh, S., Emori, Y., Kawasaki, H., Minami, Y., and Ohno, S. (1987) Calcium-activated neutral protease and its endogenous inhibitor. Activation at the cell membrane and biological function. *FEBS Lett.*, **220**, 271–277.

215 Waxman, L. and Krebs, E.G. (1978) Identification of two protease inhibitors from bovine cardiac muscle. *J. Biol. Chem.*, **253**, 5888–5891.

216 Wingrave, J.M., Sribnick, E.A., Wilford, G.G., Matzelle, D.D., Mou, J.A., Ray, S.K., Hogan, E.L., and Banik, N.L. (2004) Higher calpastatin levels correlate with resistance to calpain-mediated proteolysis and neuronal apoptosis in juvenile rats after spinal cord injury. *J. Neurotrauma*, **21**, 1240–1254.

217 Wingrave, J.M., Sribnick, E.A., Wilford, G.G., Matzelle, D.D., Mou, J.A., Ray, S.K., Hogan, E.L., and Banik, N.L. (2004) Relatively low levels of calpain expression in juvenile rat correlate with less neuronal apoptosis after spinal cord injury. *Exp. Neurol.*, **187**, 529–532.

218 Bornheim, L.M. (1998) Effect of cytochrome P450 inducers on cocaine-mediated hepatotoxicity. *Toxicol. Appl. Pharmacol.*, **150**, 158–165.

219 Conney, A.H. (1967) Pharmacological implications of microsomal enzyme induction. *Pharmacol. Rev.*, **19**, 317–366.

220 Whysner, J., Ross, P.M., Conaway, C.C., Verna, L.K., and Williams, G.M. (1998) Evaluation of possible genotoxic mechanisms for acrylonitrile tumorigenicity. *Regul. Toxicol. Pharmacol.*, **27**, 217–239.

221 Beck, B.D., Conolly, R.B., Dourson, M.L., Guth, D., Hattis, D., Kimmel, C., and Lewis, S.C. (1993) Improvements in quantitative noncancer risk assessment. Sponsored by the Risk Assessment Specialty Section of the Society of Toxicology. *Fundam. Appl. Toxicol.*, **20**, 1–14.

222 Conolly, R.B., Reitz, R.H., Clewell, H.J., and Andersen, M.E. (1988) Biologically structured models and computer simulation. *Commun. Toxicol.*, **2**, 305–319.

223 Reitz, R.H., Mendrala, A.L., Corley, R.A., Quast, J.F., Gargas, M.L., Andersen, M.E., Staats, D.A., and Conolly, R.B. (1990) Estimating the risk of liver cancer associated with human exposures to chloroform using physiologically based pharmacokinetic modeling. *Toxicol. Appl. Pharmacol.*, **105**, 443–459.

7
Toxicological Interaction Thresholds of Chemical Mixtures

Hisham El-Masri

7.1
Introduction

Toxicity mechanism of many single chemicals can possibly change with respect to dose. As dose levels increase, some key events in the principal mechanism of toxicity may shift to other mechanisms. Several chemicals exhibit this dose-dependent behavior, including acetaminophen, butadiene, ethylene glycol, formaldehyde, manganese, methylene chloride (MC), peroxisome proliferator-activated receptor (PPAR), progesterone/hydroxyflutamide, propylene oxide, vinyl acetate, vinyl chloride, vinylidene chloride, and zinc [1]. The possibility of mechanism shifts for single chemicals can also be observed for interaction mechanism of chemical mixtures. For instance, interactions (synergism or antagonism) taking place at high individual doses of a mixture may not be significant at low levels. One of the early experiments indicating a change in the interaction mechanism with dose was conducted using chloral hydrate and ethanol [2]. The binary mixture was investigated by fixing response to 50% of the mice losing the righting reflex (ED_{50}). The authors compared the experimentally determined ED_{50}'s at 19 different dose combinations to lines of additivity formed by connecting the upper and lower confidence limits of the ED_{50}'s of the individual chemicals. Evidence for a departure from additivity was provided whenever the confidence bounds for the experimentally determined ED_{50} for a combination did not overlap with the additivity confidence region. Inspection of the experimentally determined lines of equal ED_{50} effects (isobolograms) shows additivity to be prevalent for a combination of chloral hydrate less than 125 mg/kg and ethanol greater than 1200 mg/kg for the ED_{50} response. When chloral hydrate exceeds 125 mg/kg and ethanol is less than 1200 mg/kg in combination, there is a synergistic interaction (Figure 7.1).

In 1996, the idea of "interaction thresholds" as the minimal level of change in tissue dosimetry associated with a significant health effect was first introduced [3]. The interaction threshold is broadly defined as "the dose region where the magnitude of toxicological interactions between components of a chemical mixture significantly

Principles and Practice of Mixtures Toxicology. Edited by Moiz Mumtaz

ISBN: 978-3-527-31992-3

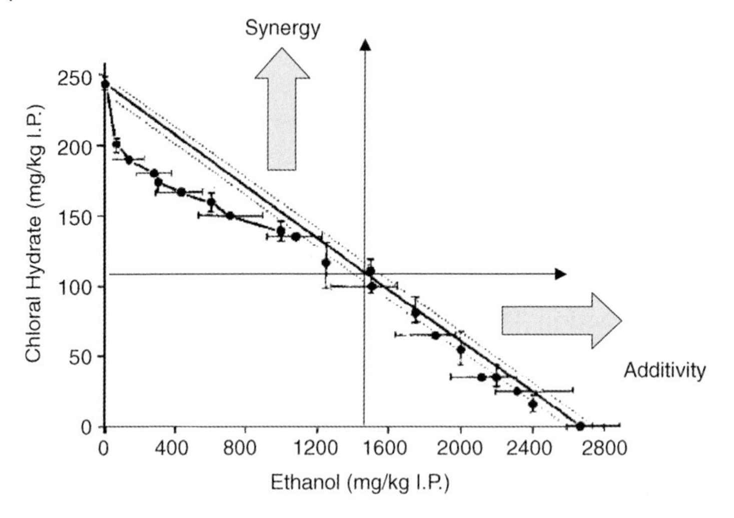

Figure 7.1 A graph depicting ED50 isobologram of the righting reflex in mice in response to a mixture of ethanol and chloral hydrate. The graph was adopted with modifications from Cabana and Gessner [16]. The mixture appears to be additive for a combination of chloral hydrate less than 125 mg/kg and ethanol greater than 1200 mg/kg. When chloral hydrate exceeds 125 mg/kg and ethanol is less than 1200 mg/kg in combination, there is a synergistic interaction. This suggests the possibility of an interaction threshold when the drugs are combined.

deviated from additivity." This definition is based on three concepts that are essential for quantitatively determining an interaction threshold level.

1) An interaction threshold is a region in the dose-response relationship space encompassing several levels of individual doses. In the simplest binary case, a graphical representation of the doses response of a mixture is a surface in the x-y-z domain, where x, y, and z are dose A, dose B, and response axes, respectively. The plane containing levels of dose A and dose B below which interactions are not significant comprises the interaction threshold surface. The importance of this concept stems from the need to identify interaction threshold in relation to health hazards of a chemical mixtures. Therefore, identifying a set of exposure levels for components in a mixture where interactions are not observed may not necessarily be an interaction threshold. However, once experimentally observed, any determined "no interaction" level can be compared with exposure levels. If exposure levels are less than the observed "no interaction" levels, then additivity can be considered for assessing health risks of the mixture.
2) Defining an interaction threshold necessitates establishing criteria for additivity. A general definition for additivity is given by Berenbaum [4] as follows:

$$\sum_{i}^{c} \frac{x_i}{E_i} = 1. \tag{7.1}$$

In a combination of c chemicals, E_i represent the concentration/dose of the ith component alone that yields a fixed response, and x_i represent the concentration/dose of the ith component in combination with the c agents that yields the same response. According to this definition, if the left-hand side of Eq. (7.1) is less than 1, then a greater than additive response (i.e., *synergism*) can be claimed at the combination of interest. If the left-hand side of Eq. (7.1) is greater than 1, then a less than additive response (i.e., *antagonism*) can be claimed at the combination [4, 5].

3) Interactions at doses above threshold levels are statistically significant. In general, interactions may occur at the cellular level but may still be numerically insignificant. For example, consider enzymatic competitive inhibition as a possible interaction mechanism between chemical A and B. The mathematical formula describing the metabolic rate of a chemical (A) is given in Eq. (7.2) below when a Michaelis–Menten mechanism is employed.

$$\mathrm{RMA} = \frac{V_A \cdot C_A}{K_{\mathrm{mA}} + C_A}, \tag{7.2}$$

where RMA is the rate of metabolism of chemical A, V_A is the maximum Michaelis–Menten metabolism velocity for chemical A, C_A is the level of chemical A, and K_{mA} is the concentration of chemical A when metabolic rate is at half V_A.

In the presence of chemical B,

$$\mathrm{RMA/B} = \frac{V_A \cdot C_A}{K_{\mathrm{mA}} \cdot \left(1 + \frac{C_B}{K_{\mathrm{mB}}}\right) + C_A}, \tag{7.3}$$

where RMA/B is the rate of metabolism of chemical A in the presence of chemical B, C_B is the level of chemical B, and K_{mB} is the concentration of chemical B when metabolic rate is at half its maximum Michaelis–Menten velocity (V_B).

It can be seen from Eqs. (7.2) and (7.3) that a no-interaction effect will be observed numerically only if C_B is zero. The difference in metabolic rate of chemical A alone (RMA) from the one in combination with chemical B (RMA/B) is diminished when C_B is too small or when K_{mB} is very large. A very large value of K_{mB} compared to C_B will derive their ratio to zero and hence have a smaller effect on any change to RMA. It is at this point where the change in metabolic rate is too small to be statistically determined that a quantitative threshold is determined. Equations (7.2) and (7.3) both represent a mathematical description of a biological process for metabolism. The basic biological mechanism affecting the rate of metabolism is based on the availability of enzyme sites. A situation may occur where free enzymatic sites are available to metabolize both chemicals. Another situations may arise when the enzymatic binding affinity and concentration of one chemical are much lower than the other (driving $C_B/K_{mB} \rightarrow 0$) causing the interactive impact of the first chemical to be negligible. In both situations, biological interaction may occur but can be at a very small scale to be numerically distinguished from thresholds. The absence of significant metabolic interaction was demonstrated for the binary mixture of trichloroethylene (TCE) and 1,1-dichloroethylene (1,1-DCE) [3]. Earlier interaction PBPK model showed competitive inhibition to be the mode of interaction between TCE and 1,1-DCE [6]. To predict the range at which the interaction threshold was determined, the PBPK model was modified to include mathematical descriptions of the percentage of enzyme sites occupied by either chemical in the presence or the absence of the other [3]. By comparing the percentage of occupied sites by one chemical, in the presence of the other, with those sites occupied in the absence of the latter, the PBPK model predicted a range of concentrations (100 ppm or less) of either chemical where the competitive inhibition interaction would not be observed. Consequently, gas uptake experiments were designed where the initial concentration was selected at 2000 ppm for one chemical while the other chemical was set at 100 ppm in one experiment and 50 ppm in the other. Under these conditions, the best stimulation to the concentration depletion curves in the gas uptake system of the chemical in the higher concentration was obtained when the PBPK model was run under the assumption of no-interaction. This substantiated the model predictions of the presence of observable interaction only at concentrations higher than 100 ppm for both chemicals.

7.2 Statistical Analysis for Interaction Thresholds

Statistical analysis of interactions and hypothesis testing for the presence of an interaction threshold starts with the Berenbaum additivity criteria given for multiple chemicals as shown in the equation below:

$$1 = \frac{x_1}{\mathrm{ED}_1} + \frac{x_2}{\mathrm{ED}_2} + \cdots + \frac{x_i}{\mathrm{ED}_i} + \cdots + \frac{x_c}{\mathrm{ED}_c}, \tag{7.4}$$

where x_i is the level of chemical i in a mixture of c chemicals, ED_i is the effective dose of chemical i.

Equation (7.4) can be derived from a general function describing the dose-response relationship for a mixture when additivity is considered [7, 8]. The form of the general function for multiple chemicals is shown as

$$g(\mu) = \beta_0 + \beta_1 x_1 + \beta_2 x_2 + \cdots + \beta_c x_c, \tag{7.5}$$

where $g(\mu)$ is a function describing the dose-response relationship, μ is a response, β_0 is the background response in the absence of chemicals, and β_1, β_2, and β_c are the individual slopes of dose-response curve for chemicals 1, 2, and, c, respectively.

A general formula describing interactions can be generated by adding terms to Eq. (7.5) depicting combination effects. In the simple binary mixture, the interaction general equations is presented as

$$g(\mu) = \beta_0 + \beta_1 x_1 + \beta_2 x_2 + \beta_{12} x_1 x_2. \tag{7.6}$$

For $\beta_{12} = 0$, additivity is the case. If $\beta_{12} > 0$, then synergism can be claimed, and if $\beta_{12} < 0$ antagonism can be claimed [9]. The task for a statistical analysis to show deviation from additivity or to test the presence of an interaction threshold by determining that β_{12} is significantly different from zero. Equation (7.6) can be extended to mixtures larger than binary by adding multiplicative interaction terms and testing the significance of their associated constants [9].

The dose region where interaction-signaling deviation from additivity is not significant is described as the interaction threshold dose region. Graphically, the boundary that separates the dose space into regions of interaction and additivity is called the interaction threshold boundary. Hamm *et al.* [8] applied statistical analysis to experimental results of the chloral hydrate and ethanol interaction effects on righting reflex in mice to identify the presence and shape of the threshold boundary [8, 10].

7.3
Predictive Modeling of the Interaction Threshold

Predictive modeling (such as PBPK/PD models) along with mode-designed experiments provides an efficient methodology for the identification of interaction thresholds (Figure 7.2). On the basis of mechanistic consideration of enzyme inhibitions, the presence of an interaction threshold between binary chemicals was predicted using PBPK models for two different sets of chemicals: volatiles (TCE and 1,1-DCE) and pesticides (chlorpyrifos and parathion). In the first example, an earlier interaction PBPK model assuming competitive inhibition between the chemicals was modified to include mathematical descriptions of the percentage of enzyme sites occupied by either chemical in the presence or the absence of the other [3, 6]. In the

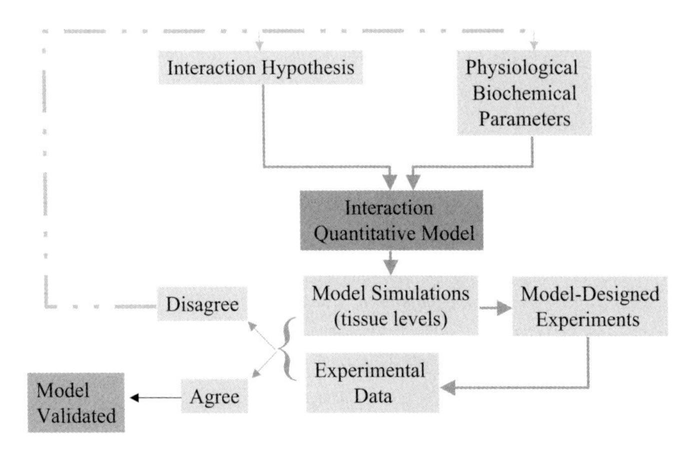

Figure 7.2 A graph showing a methodology of testing interaction hypotheses (such as presence or absence of a threshold) using model-designed experiments.

second example, a binary interaction PBPK model was developed to estimate an interaction threshold for the joint toxicity between chlorpyrifos and parathion in the rat [11]. Initially, individual PBPK models were developed for parent chemicals and their oxon-metabolites to estimate blood concentrations of their respective metabolites. The metabolite concentrations in blood were then linked to acetylcholinesterase kinetic submodel. The resulting overall PBPK model described interactions between these pesticides at two levels: (i) the P450 enzymatic bioactivation site and (ii) acetylcholinesterase binding sites. Using the overall model, a response surface was constructed at various dose levels of each chemical to investigate the mechanism of interaction and to calculate interaction threshold doses.

Interaction thresholds of three common volatile organic solvents, trichlorethey-lene, perchlorate (PERC), and methylene chloride, under different dosing conditions were investigated using PBPK modeling [12, 13]. Briefly, an interactive PBPK model was built where PERC and MC were competitive inhibitors for TCE. The model was developed and validated by gas uptake pharmacokinetic studies in F344 rats at relatively high doses of single chemicals, binary mixtures, and the ternary mixture. Computer simulations were used to extrapolate from high to low concentrations to investigate the toxicological interactions at occupational exposure levels, specifically at threshold limit value/time weighted average (TLV/TWA). Using a 10% elevation in parent compound blood level as a criterion for significant interaction, interaction thresholds were predicted with two of the three chemicals held at constant concentrations.

7.4 "No Interaction" Exposure Levels

Statistical methods can be useful in identifying a threshold when existing data cover a wide enough range of experiments along the dose-response continuum, specifically in the low-dose region. However, estimating low-dose interaction thresholds experimentally is costly and challenging because of the need to use a large number of laboratory animals. For instance, the chloral hydrate/ethanol experiments used a total of 1681 animals in 21 treatment groups [8, 10]. Predictive modeling and model-designed experimentations can be effective in determining threshold levels for a mixture. However, development, calibration, and evaluation of predictive models are usually challenging and can be experimentally tasking. In most cases, biological descriptions of interaction mechanisms are not readily available to be mathematically incorporated into predictive models. In view of the experimental and computational difficulties, estimations of interaction threshold levels per se may not be feasible for health risk assessments of mixtures. Instead, studies can be conducted to test the presence of "no interaction" at environmentally relevant low doses. One such study was conducted to find out whether simultaneous administration of nine chemicals (dichloromethane, formaldehyde, aspirin, di(2-ethylhexyl)phthalate, cadmium chloride, stannous chloride, butyl hydroxyanisol, loperamide, and spermine) at a concentration equal to the "no-observed adverse effect level" (NOAEL) for each of them would result in a NOAEL for the combination [14]. This 4-week oral/inhalation study on male Wistar rats was performed in which the toxicity (clinical chemistry, hematology, biochemistry, and pathology) of combinations of the nine compounds was examined. It was concluded that simultaneous exposure to these nine chemicals does not constitute an evidently increased hazard compared to exposure to each of the chemicals separately, provided the exposure level of each chemical in the mixture is at most similar to or lower than its own NOAEL [15].

References

1 Slikker, W. Jr., Andersen, M.E., Bogdanffy, M.S., Bus, J.S., Cohen, S.D., Conolly, R.B., David, R.M., Doerrer, N.G., Dorman, D.C., Gaylor, D.W., Hattis, D., Rogers, J.M., Setzer, R.W., Swenberg, J.A., and Wallace, K. (2004) Dose-dependent transitions in mechanisms of toxicity: case studies. *Toxicol. Appl. Pharmacol.*, **201**, 226–294.

2 Gessner, P.K. and Cabana, B.E. (1967) Chloral alcoholate: reevaluation of its role in the interaction between the hypnotic effects of chloral hydrate and ethanol. *J. Pharmacol. Exp. Ther.*, **156**, 602–605.

3 El-Masri, H.A., Tessari, J.D., and Yang, R.S. (1996) Exploration of an interaction threshold for the joint toxicity of trichloroethylene and 1,1-dichloroethylene: utilization of a PBPK model. *Arch. Toxicol.*, **70**, 527–539.

4 Berenbaum, M.C. (1985) The expected effect of a combination of agents: the general solution. *J. Theor. Biol.*, **114**, 413–431.

5 Berenbaum, M.C. (1991) Concepts for describing the interaction of two agents. *Radiat. Res.*, **126**, 264–268.

6 Andersen, M.E., Gargas, M.L., Clewell, H.J. 3rd, and Severyn, K.M. (1987) Quantitative evaluation of the metabolic interactions between trichloroethylene and 1,1-dichloroethylene *in vivo* using gas uptake methods. *Toxicol. Appl. Pharmacol.*, **89**, 149–157.

7 Gennings, C. (1996) Economical designs for detecting and characterizing departure from additivity in mixtures of many chemicals. *Food Chem. Toxicol.*, **34**, 1053–1058.

8 Hamm, A.K., Hans Carter, W. Jr., and Gennings, C. (2005) Analysis of an interaction threshold in a mixture of drugs and/or chemicals. *Stat. Med.*, **24**, 2493–2507.

9 Carter, W.H. Jr. (1995) Relating isobolograms to response surfaces. *Toxicology*, **105**, 181–188.

10 Gessner, P.K. and Cabana, B.E. (1970) A study of the interaction of the hypnotic effects and of the toxic effects of chloral hydrate and ethanol. *J. Pharmacol. Exp. Ther.*, **174**, 247–259.

11 El-Masri, H.A., Mumtaz, M.M., and Yushak, M.L. (2004) Application of physiologically-based pharmacokinetic modeling to investigate the toxicological interaction between chlorpyrifos and parathion in the rat. *Environ. Toxicol. Pharmacol.*, **16**, 57–71.

12 Dobrev, I.D., Andersen, M.E., and Yang, R.S. (2001) Assessing interaction thresholds for trichloroethylene in combination with tetrachloroethylene and 1,1,1-trichloroethane using gas uptake studies and PBPK modeling. *Arch. Toxicol.*, **75**, 134–144.

13 Dobrev, I.D., Andersen, M.E., and Yang, R.S. (2002) *In silico* toxicology: simulating interaction thresholds for human exposure to mixtures of trichloroethylene, tetrachloroethylene, and 1,1,1-trichloroethane. *Environ. Health Perspect.*, **110**, 1031–1039.

14 Groten, J.P., Schoen, E.D., van Bladeren, P.J., Kuper, C.F., van Zorge, J.A., and Feron, V.J. (1997) Subacute toxicity of a mixture of nine chemicals in rats: detecting interactive effects with a fractionated two-level factorial design. *Fundam. Appl. Toxicol.*, **36**, 15–29.

15 Groten, M., Girthofer, S., and Probster, L. (1997) Marginal fit consistency of copy-milled all-ceramic crowns during fabrication by light and scanning electron microscopic analysis *in vitro*. *J. Oral Rehabil.*, **24**, 871–881.

16 Cabana, B.E. and Gessner, P.K. (1967) Determination of chloral hydrate, trichloroacetic acid, trichloroethanol, and urochloralic acid in the presence of each other and in tissue homogenates. *Anal. Chem.*, **39**, 1449–1452.

8
Characterization of Toxicoproteomics Maps for Chemical Mixtures Using Information Theoretic Approach[1)]

Subhash C. Basak, Brian D. Gute, Nancy A. Monteiro-Riviere, and Frank A. Witzmann

8.1
Introduction

In the postgenomic era, the emerging technologies of genomics, proteomics, and metabolomics are taking an increasingly important role in predicting the biological activities and toxicities of chemicals. While microarray studies provide an assessment of cellular transcriptional processes, proteomics provides a better understanding of cell function because the intracellular proteins are the workhorses of biological systems. At present, researchers need to evaluate real and virtual chemicals both as novel beneficial agents (pharmaceuticals and agrochemicals are two notable examples) and as potential threats to environmental and human health. Chemical evaluation strategies can be divided into three main approaches: (i) *in vivo*, (ii) *in vitro*, and (iii) *in silico*. The difficulty with this scheme arises from the extremely large number of candidate chemicals. The number of industrial chemicals in the United States of America today, maintained in the Toxic Substances Control Act (TSCA) Chemical Substance Inventory of the United States Environmental Protection Agency (USEPA), is approximately 75 000 and is increasing each year [1], and the TSCA inventory does not include food additives and pharmaceutical agents regulated by the US Food and Drug Administration (FDA). All these chemicals, and their possible mixtures, cannot be tested *in vivo* because of the lack of testing facilities, need of enormous resources, and the astronomical number of animals that would need to be sacrificed for such a project. Meanwhile, modern combinatorial chemistry is producing very large libraries of real and hypothetical structures that need to be evaluated both for their toxic effects and therapeutic action. We need to understand the toxic effects and modes of action (MOA) of these chemicals. Calculated chemodescriptors, which can be directly computed from molecular structure without the input of any other experimental data, have found considerable success in predicting

1) The views and conclusions contained herein are those of the authors and should not be interpreted as necessarily representing the official policies or endorsements, either expressed or implied, of the Air Force Research Laboratory or the US Government.

Principles and Practice of Mixtures Toxicology. Edited by Moiz Mumtaz

ISBN: 978-3-527-31992-3

the toxicity and toxic modes of action of chemicals [2, 3]. While chemical structure can be easily characterized through chemodescriptors, toxicoproteomics gives us some unique and complementary information on the finer biological processes underlying chemical toxicity. These biological data provide a global snapshot of the most abundant cellular proteins that can be used to identify a complex biosignature related to the toxic insult or an individual protein that may serve as specific biomarker of toxicity [2, 4–19].

Mixture toxicity is an important problem in toxicology. Real-world chemical exposures are hardly ever to pure substances. Instead, people are exposed to a soup of chemicals consisting of a mixture of parent toxicants and a variety of derivatives resulting from environmental degradation and metabolic processes. Researchers have attempted to predict mixture toxicity for chemicals with the same mode of action using the concept of concentration addition [20–24]. However, there is no general methodology for predicting mixture toxicity when the mixture's constituents act by multiple known and unknown mechanisms. Toxicoproteomics data, which represent holistic, protein-level responses of the cellular machinery to toxicants, can be employed in understanding toxicity and toxic mode of action in such cases. In this chapter, we describe our work on mathematical proteomics that can characterize the proteomic perturbation resulting from both individual chemicals, such as the peroxisome proliferators, and complex mixtures, like jet propellant formulation #8 (JP-8).

8.2
Current Proteomics Technologies

Current technologies for analysis and characterization of cellular proteins (proteomics patterns or profiles) include two-dimensional gel electrophoresis (2DE), matrix-assisted laser desorption ionization (MALDI), surface-enhanced laser desorption ionization (SELDI), and isotope-coded affinity tagging (ICAT). Many authors have used 2DE gel technology in understanding the molecular basis of chemical toxicity [25, 26]. In this method, the tissue or cell exposed to the toxicant is homogenized and the proteins are separated by charge and mass through two-dimensional electrophoresis. The magnitude of each spot after separation gives the abundance of a particular type of protein or a closely related set of proteins that comprise an individual spot. A typical 2DE gel can result in the isolation and identification of 1500–2000 proteins. To date, our work in developing mathematical biodescriptors for proteomics data has been restricted to data from 2DE gels. As such, this chapter will focus entirely on biodescriptors derived from such proteomic maps.

8.3
Mathematical Proteomics Approaches

As mentioned above, an important goal of toxicoproteomics is to study the effect of toxins on protein expression in various tissues. Characterizing patterns composed of

1500 or more objects is a daunting task and cannot be accomplished simply through visual inspection; rigorous mathematical and statistical methods are required for a thorough and objective analysis of such patterns. Our group has been involved in characterizing toxicoproteomic patterns using four different techniques: (i) invariants of graphs associated with proteomics maps [4], (ii) spectrum-like representation based on projections of the three-dimensional space (mass, charge, and abundance) onto three two-dimensional planes [12], (iii) robust statistical selection of toxicologically relevant spots [14], and (iv) information theoretic characterization of protein spot patterns [27, 28]. While this chapter will touch on all four of these approaches, the primary focus is on the application of the information theoretic method.

In each gel, the proteins are separated by charge and mass (see Figure 8.1a). In addition, each protein spot is characterized by its relative abundance, quantifying the amount of that particular protein or closely related class of proteins gathered in one spot. Mathematically, the data generated by 2DE may be looked upon as points in a three-dimensional space, with the axes described by charge, mass, and spot abundance (see Figure 8.1b).

8.3.1 The Spectrum-Like Approach

The three-dimensional 2DE gel data can be "deconstructed" and projected onto three planes (*xy*, *yz*, and *xz*). An example of these three projections using the human keratinocyte data from this study are depicted in Figure 8.2. The spectrum-like data derived from this approach can be converted into vectors, and the similarity of a set of proteomics maps can be computed based on these map descriptors. We have attempted to use this class of model on the data generated by Anderson *et al.* [25] and Witzmann [4] on a set of peroxisome proliferators. Animals treated with these chemicals show an increase in the number of liver cell peroxisomes (a cellular organelle). This peroxisome proliferation is considered to be the basis for the occurrence of cancer in these animals. Results using spectrum-like descriptors show that this method can differentiate maps (2DE patterns of liver proteins) derived from animals treated with different peroxisome proliferators.

8.3.2 The Graph Invariant Approach

In the graph invariant biodescriptor approach, different types of embedded graphs are associated with the proteomics maps. For instance, zigzag graphs or neighborhood graphs can be associated with a proteomics map, with proteins representing the vertices of such graphs.

In the zigzag approach, one considers the spot with the highest (or lowest) abundance to be the first point (vertex) and draws a line (edge) between it and the next most abundant (or next least abundant) spot and continues the process until all points on the map are connected by a single zigzag "curve." This curve is then

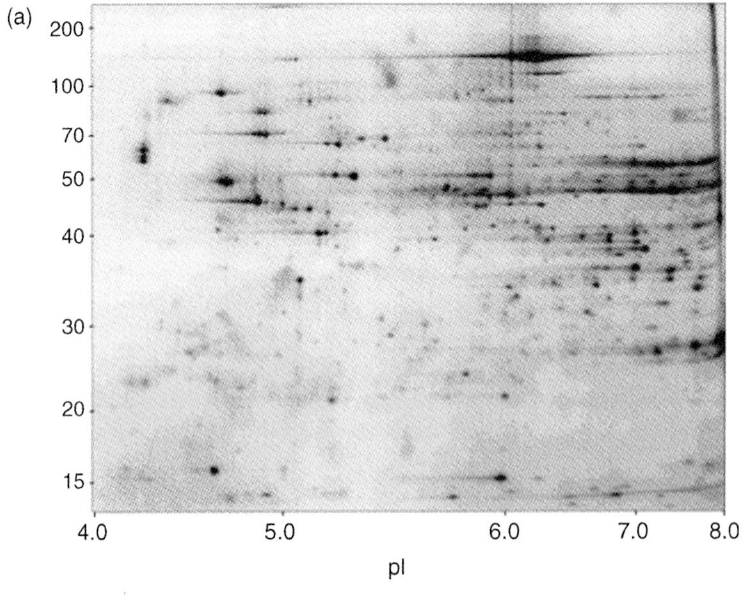

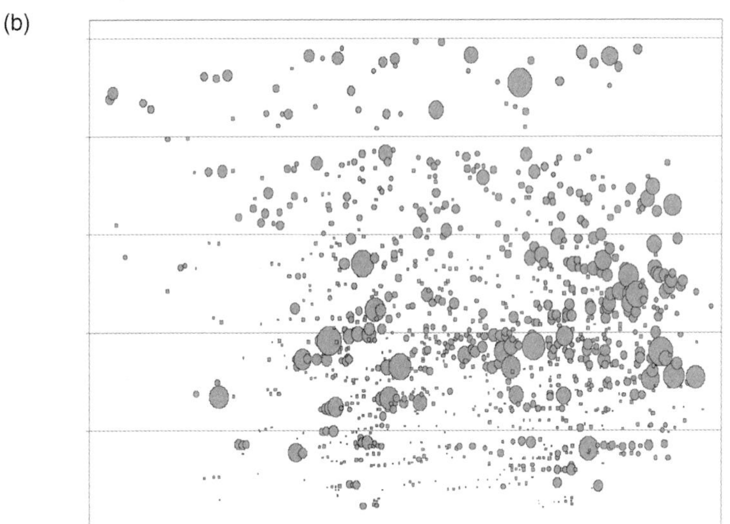

Figure 8.1 (a) Two-dimensional polyacrylamide gel (2DE gel) pattern of primary rat hepatocyte lysate obtained after monolayer culture and (b) computer-derived map based on coordinate data derived from the original gel. Protein quantification and *x*, *y* coordinate position determined using image analysis on each detected spot in the pattern.

converted into a D_E/D_G matrix. This is a special form of distance matrix where the (i, j) entries for such a matrix are the quotient of the Euclidean distance between points i and j and the (pathwise or through-bond) distance (see Figure 8.3) between them. The leading eigenvalue pattern for such D_E/D_G matrices and their higher order

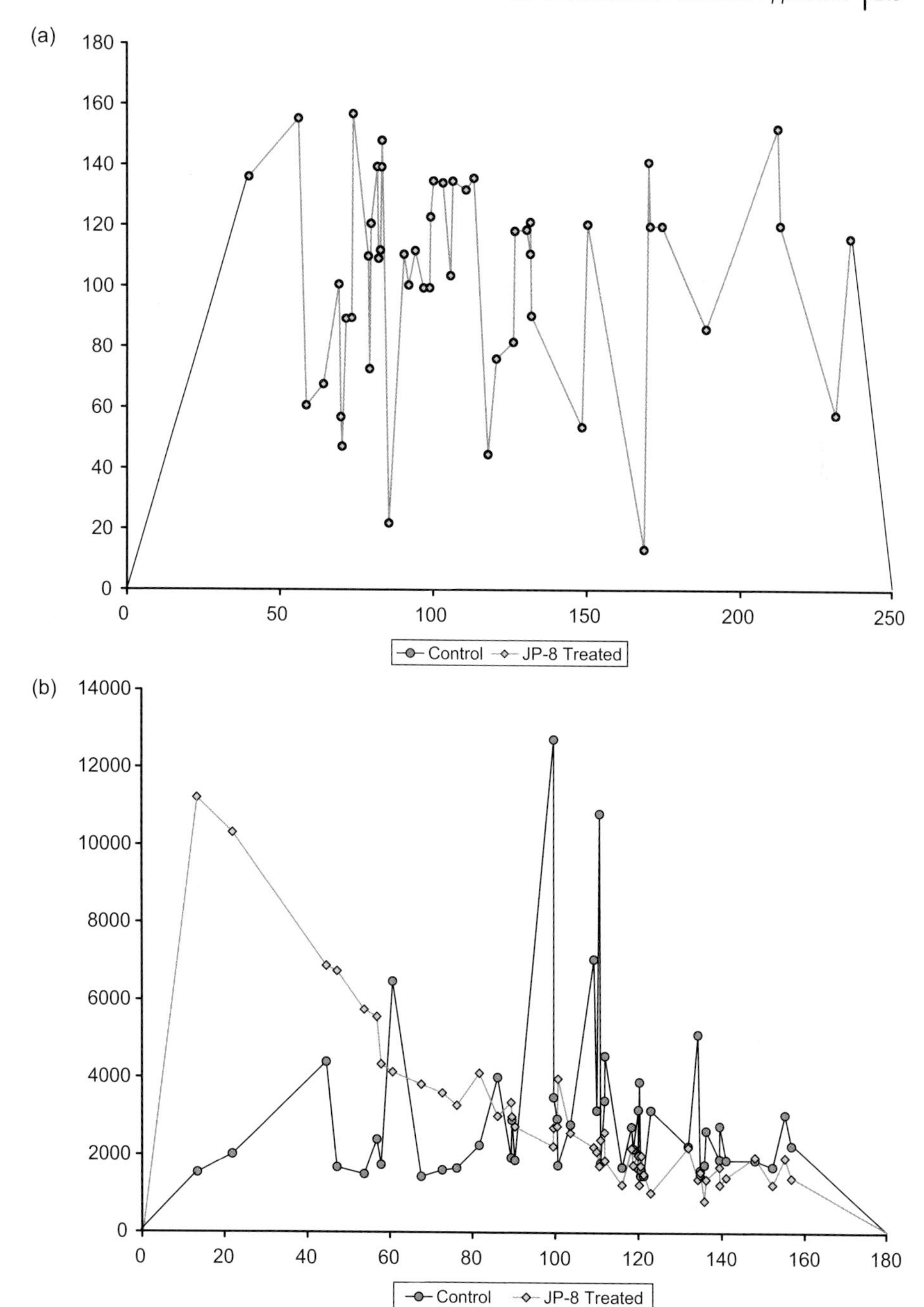

Figure 8.2 Spectrum-like representations of proteomics map data for a map consisting of the 50 most abundant proteins in the control and JP-8-treated HEK. (a) Shows the *xy* plots for control and treatment that overlap completely since the spots were sorted by control abundance, (b) shows the *yz* plots, and (c) shows the *xz* plots.

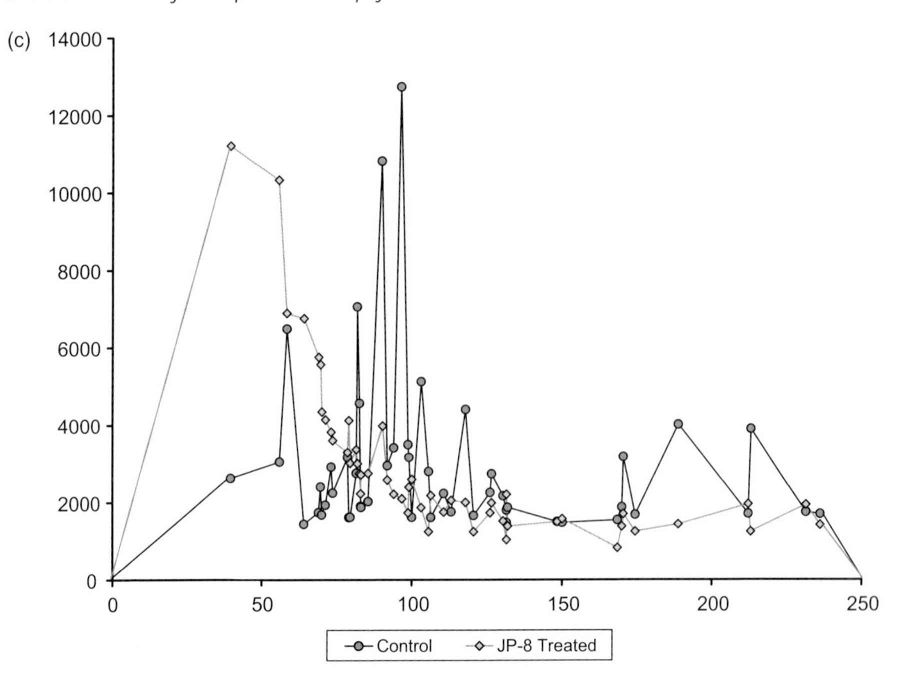

Figure 8.2 (*Continued*)

Kronecker products derived from proteomics maps of peroxisome proliferator-treated liver cells are shown in Figure 8.4. A perusal of the figure shows that the leading eigenvalues for the zigzag graph are capable of discriminating among the 2DE patterns of closely related peroxisome proliferators. This gives us confidence that such biodescriptors might be useful in QSAR modeling for such compounds.

Another graph invariant approach defines a neighborhood of graphs corresponding to the neighborhood of protein spots on the map [10–12], similar to the topological neighborhood of atoms in molecules defined by Basak *et al.* [29, 30]. Invariants derived in this way have been used to measure the similarity/dissimilarity of maps of peroxisome proliferators.

8.3.3
The Protein Biodescriptor Approach

The spots of proteomics maps can be looked upon as independent variables, and statistical methods can be applied to find which of the spots are related to the particular situation. For example, specific proteins may be up- or downregulated as a result of cellular stress, disease, or the effects of drugs or xenobiotics on biological systems. Unfortunately, we need a reasonable amount of data to conduct such a study.

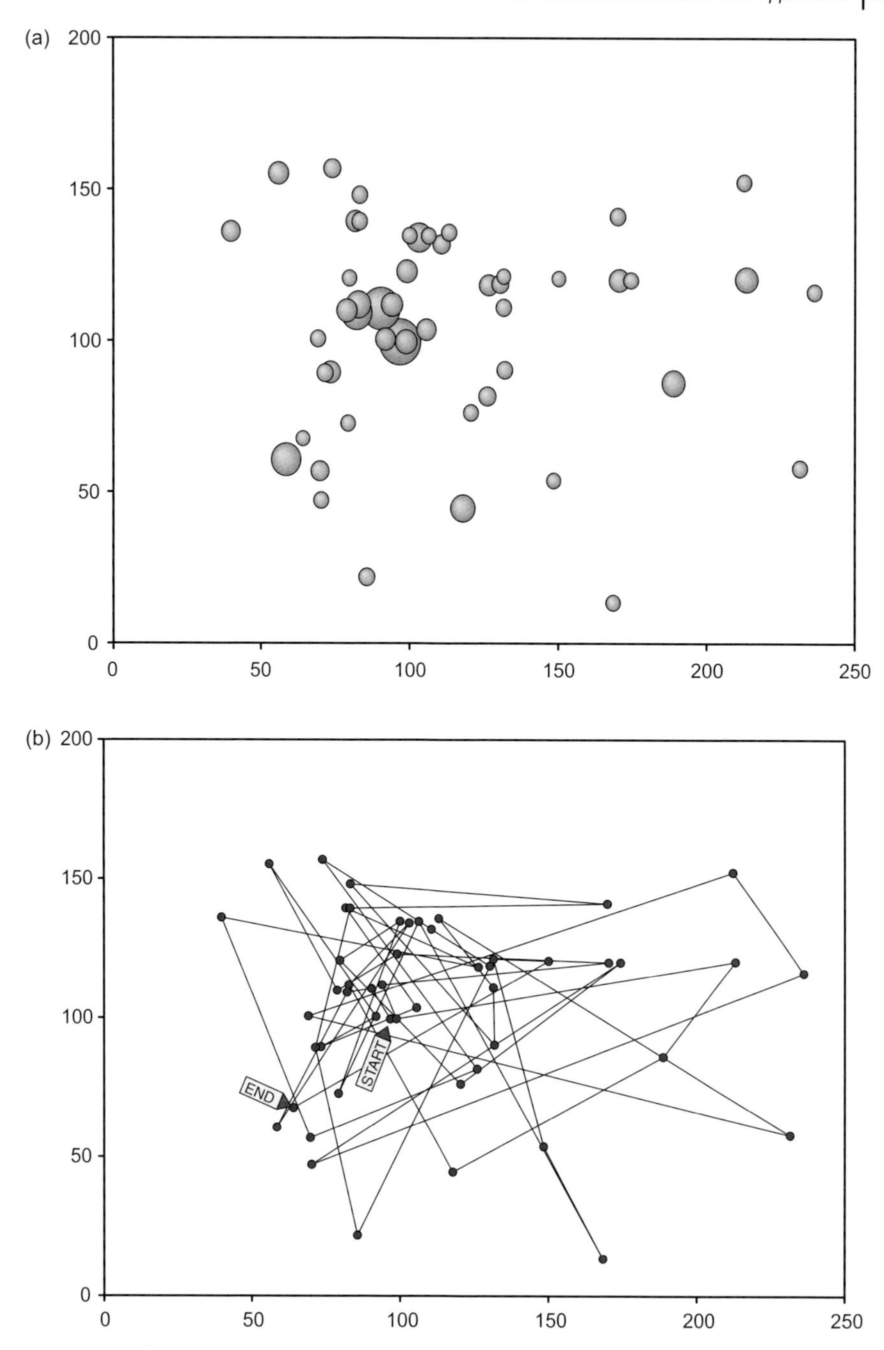

Figure 8.3 (a) The bubble diagram of the 50 most abundant map spots for the control treatment keratinocytes and (b) the embedded zigzag curve for the same proteomics map.

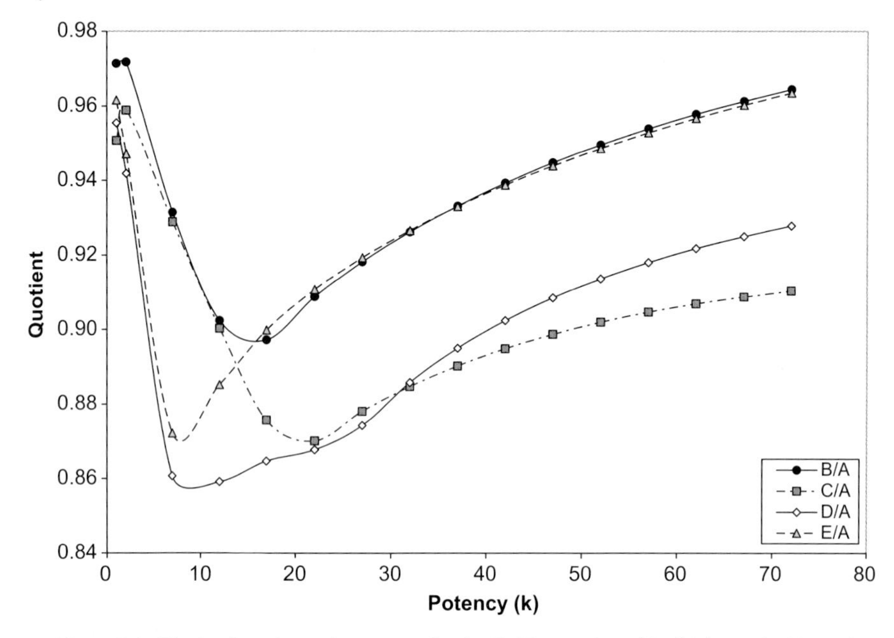

Figure 8.4 The leading eigenvalue pattern for the D_E/D_G matrix and its higher order Kronecker products for proteomics maps derived from peroxisome proliferator-treated liver cells.

The data on the four peroxisome proliferators are too small for this kind of study; however, a preliminary statistical analysis of 1401 spots derived from exposing primary hepatocytes to 14 halocarbon toxicants was able to order the spots on the basis of their ability to discriminate among the toxicants [14]. The most important spots so derived can be used as biodescriptors in predictive toxicology and can lead to a much more mechanistic understanding of the toxic mode of action. Alternatively, when key proteins are identified in the course of this type of analysis, the information can be used to validate current theories regarding the mechanistic basis of toxicity.

8.3.4
The Information Theoretic Method

Finally, a proteomics map may be looked upon as a pattern of protein mass distributed over a plane. The distribution may vary depending on the functional state of the cell both under various developmental and pathological conditions and under the influence of therapeutics and xenobiotics. One means of examining spot distribution is to characterize the map's center of mass. Preliminary work has been initiated in this area, but further research is needed to test the utility of this approach in characterizing proteomics maps.

The discussion in this chapter will focus on our work on applying information theoretic methods to characterize toxicoproteomics maps following the exposure of

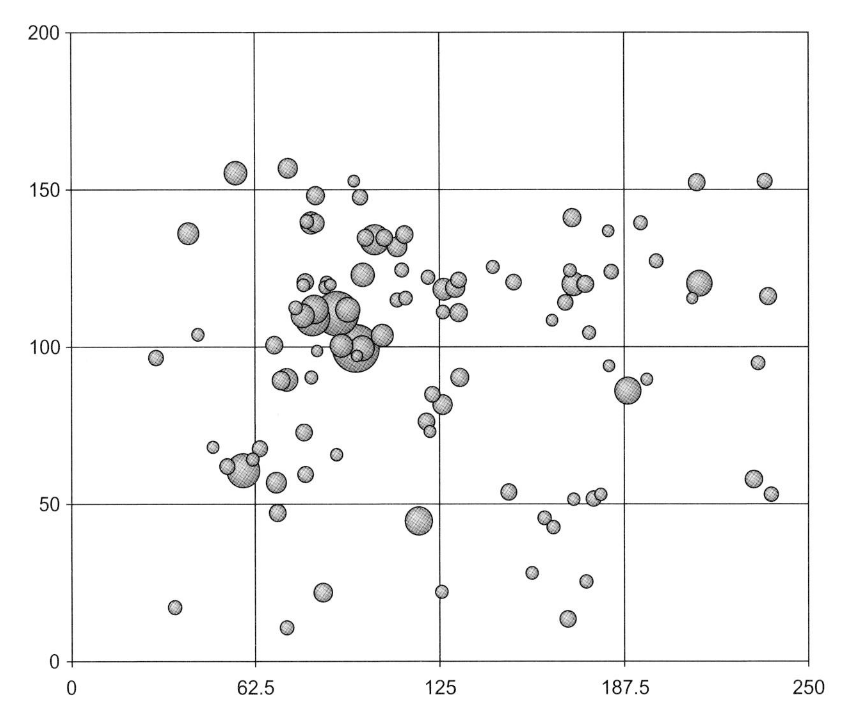

Figure 8.5 A 4×4 grid superimposed on the control keratinocyte proteomics map of the 100 most abundant proteins.

mice to four peroxisome proliferators, namely, perfluoro octanoic acid (PFOA), perfluoro decanoic acid (PFDA), clofibrate, and diethylhexyl phthalate (DEHP), as well as the exposure of cultured keratinocytes to JP-8. This particular information theoretic approach divides the proteomics map into an $n \times n$ grid (where $n \geq 2$) and applies Shannon's formula to calculate map complexity (see Figure 8.5). While this method has been discussed in detail in several earlier publications [27, 28], a cursory explanation of the calculations is given in Section 8.5.

8.4 Experimental Methods

8.4.1 Cell Culture and JP-8 Exposure

Cryopreserved human neonatal epidermal keratinocytes (approximately 260 K cells/vial) were purchased from Cambrex BioScience (Walkerville, MD) and plated onto three 75 cm^2 culture flasks, each containing 15 ml of serum-free keratinocyte growth media (KGM-2, from keratinocyte basal media, supplemented with 0.1 ng/ml human epidermal growth factor, 5 mg/ml insulin, 0.4% bovine pituitary extract, 0.1% hydrocortisone, 0.1% transferrin, 0.1% epinephrine, and 50 mg/ml gentamicin/50 ng/ml amphotericin-B). The culture flasks were maintained in a humidified incubator at 37 °C with a $95\%O_2/5\%CO_2$ atmosphere. After reaching approximately 50% confluency, the keratinocytes were passed into eight 75 cm^2 culture flasks and grown in 15 ml of KGM-2. After that they were harvested and plated in 6-well culture plates in 2 ml of media at a concentration of approximately 96 000 cells per well. Cells were dosed with 0.1% JP-8 (100 μl fuel in 900 μl EtOH stock; 250 μl of stock added to 25 ml prewarmed KGM-2, mixed well, and dosed 2 ml per well). The cells were exposed to JP-8 for 24 h and cell media collected and frozen immediately at −80 °C for IL-8 determination, in triplicate, using a human IL-8 cytoset (Biosource International, Camarillo, CA). Previous studies in our lab have demonstrated no ethanol effect on either protein or RNA transcription in this model [31]; therefore, no EtOH controls were run. Other plates containing the cells ((i) control plate; (ii) JP-8 plates, media removed) were quickly frozen at −80 °C and shipped overnight for proteomic analysis.

8.4.2 Sample Preparation

HEK cells were directly solubilized in well (*in situ*) after removal of the medium [32]. Four hundred microliters of lysis buffer containing 9 M urea, 4% Igepal CA-630 ([octylphenoxy]polyethoxyethanol), 1% DTT, and 2% carrier ampholytes (pH 8–10.5) were directly added to each well. The culture plates were then placed in a 37 °C incubator for 1 h with intermittent manual agitation. After 1 h, the entire volume was removed from each well and placed in 2 ml Eppendorf tubes. Each sample was then sonicated with a Fisher Sonic Dismembranator using 3 × 2 s bursts. Sonication was carried out every 15 min for 1 h after which the fully solubilized samples were transferred to a cryotube for storage at −80 °C until thawed for analysis.

8.4.3 Two-Dimensional Electrophoresis

Proteins were resolved by two-dimensional electrophoresis using the ISO-DALT System in which up to 20 first- and second-dimension gels can be simultaneously run. Solubilized protein samples were placed on each of the 18 first-dimension isoelectric focusing (IEF) tube gels (23.5 cm × 1.5 mm) containing 3.3% acrylamide, 9 M urea, 2% Igepal CA-630, and 2% ampholyte (BDH pH 4–8), and isoelectrically focused for 28 000 V h at room temperature. Each IEF gel was then placed on a second-dimension slab gel (20 cm × 25 cm × 0.15 cm) containing a linear 11–19% acrylamide gradient. Second-dimension slab gels were run for approximately 18 h at 160 V and 8 °C and later stained with colloidal Coomassie brilliant blue [33].

8.4.4
Image Analysis

After staining, the 2DE protein patterns were scanned under visible light at 200 μm/pixel resolution using the Fluor-S MAX MultiImager System (Bio-Rad, Hercules, CA, USA). Image data were analyzed using PDQuestTM software (Bio-Rad). Background was subtracted and peaks for the protein spots located and counted. Because total spot counts and the total optical density are directly related to the total protein concentration, individual protein quantities were thus expressed as parts per million (ppm) of the total integrated optical density. A reference pattern was constructed and gel matching was performed with image analysis software matching each gel in the match set to the reference gel. Numerous proteins that were uniformly expressed in all patterns were used as landmarks to facilitate rapid gel matching. Raw data for each protein spot was exported to Excel for statistical analysis.

8.4.5
Statistical Analysis

In a preprocessing step, all zero and negative protein expression values derived from PDQuest were converted to missing values. Protein spots that were not present at least 75% in either group were filtered out, reducing the number of analyzable spots from 929 to 749. The remaining percentage of missing values was calculated (11%) and their values imputed using the KNN algorithm (which is robust to about 20%) [34]. The integrated densities in all gel patterns were then normalized using nonlinear quantile normalization [35] and outlier gels were sought using a Euclidian distance bootstrap method. Using $p < 0.001$, no outlier gels were observed; therefore, all 18 gel patterns were included in subsequent analyses.

For purposes of this study, only the protein spots present in every gel pattern and completely matched to the reference gel were considered. This screening process reduced the total number of spots further from the 749 analyzable spots to a set of 298 protein spots present in all 18 gel patterns.

8.5
Theoretical Calculation of Information Theoretic Biodescriptors

A proteomics map can be looked upon as a two-dimensional pattern representing the cellular mass of identifiable proteins based on charge (x) and mass (y). Figure 8.1a presents a typical 2DE pattern (proteomic map). When a cell is exposed to a toxin, a whole host of transcriptional and translational processes are perturbed, resulting in up- or downregulation of particular proteins and the activation or deactivation of other proteins. These changes ultimately impact the abundance of proteins within the proteome, leading to changes in the spots on the proteomics map. Information theory is a suitable mathematical tool for characterizing such complex patterns. As was mentioned earlier, we have previously applied information theory to characterize the neighborhood complexity of atomic bonding patterns within molecules [29, 30, 36–39]. However, several studies have been published on the application of information theory in characterizing the proteomic patterns of cells [27, 28], and in this chapter we will apply map information content to HEK cells exposed to JP-8.

In the information theoretic formalism, a set A of N objects is partitioned into subsets A_i with cardinalities N_i; $\Sigma N_i = N$. A probability scheme is then associated with the distribution:

$$A_1, A_2, \ldots, A_h,$$

$$p_1, p_2, \ldots, p_h,$$

where $p_i = N_i/N$.

The complexity of the system consisting of N objects is computed by Shannon's formula:

$$\text{Complexity} = -\sum_{i=1}^{h} p_i \log_2 p_i. \qquad (8.1)$$

Of the 18 2DE gels, 6 were control gels while the other 12 were exposure gels. The abundance values for the gels were averaged to determine the relative abundance value to be used in our analyses. Maximum gel dimensions for determining grid size were set at $x = 250$ and $y = 200$, with the actual charge (x) values varying from 26.73 to 237.19 and the actual mass (y) values varying from 10.67 to 181.03.

Proteomics maps were created as bubble plots, using the same x- and y-coordinates as the original gels, charge and mass, respectively, while abundance is represented by the relative size of the bubble (Figure 8.1b). The maps were then divided into 2×2, 3×3, 4×4, and 5×5 grids, and spot abundances were assigned to the grid square in which the spot's center was located. Once the abundance values were determined for each grid cell, information content was calculated using the method described previously [27, 28]. Figure 8.6 shows the control proteomics map divided into 2×2, 3×3, 4×4, and 5×5 grids.

8.6 Results

8.6.1 Peroxisome Proliferators

The magnitudes of the map information content indices calculated using a 4×4 grid (MIC(4)) for the most abundant 200 and 500 spots, and for the entire set of 1054 protein spots reported by Witzmann for peroxisome proliferators [40] are presented in Table 8.1 and Figure 8.7. It may be noted that the magnitude of MIC(4) decreases for the proteomics maps of cells exposed to toxicants.

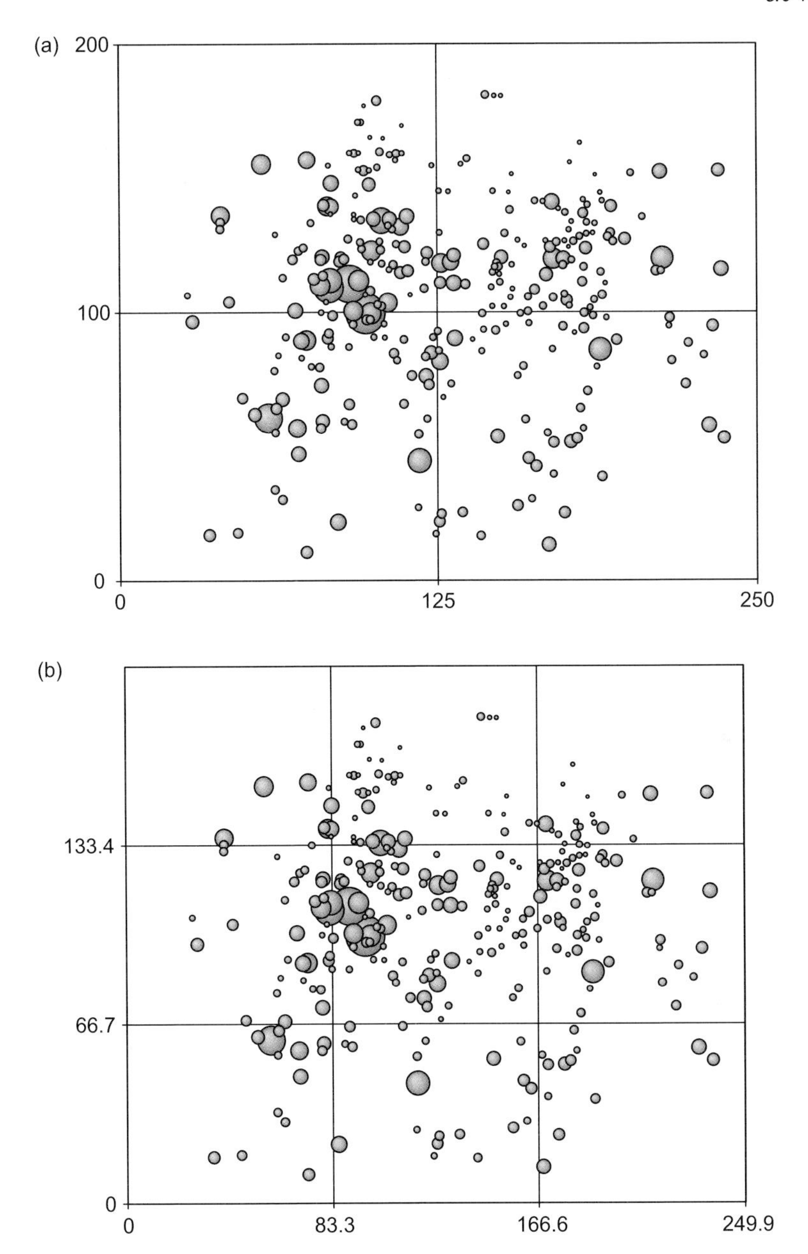

Figure 8.6 Control HEK proteomics map of all 298 spots shown as (a) 2 × 2 grid, (b) 3 × 3 grid, (c) 4 × 4 grid, and (d) 5 × 5 grid superimposed on the map.

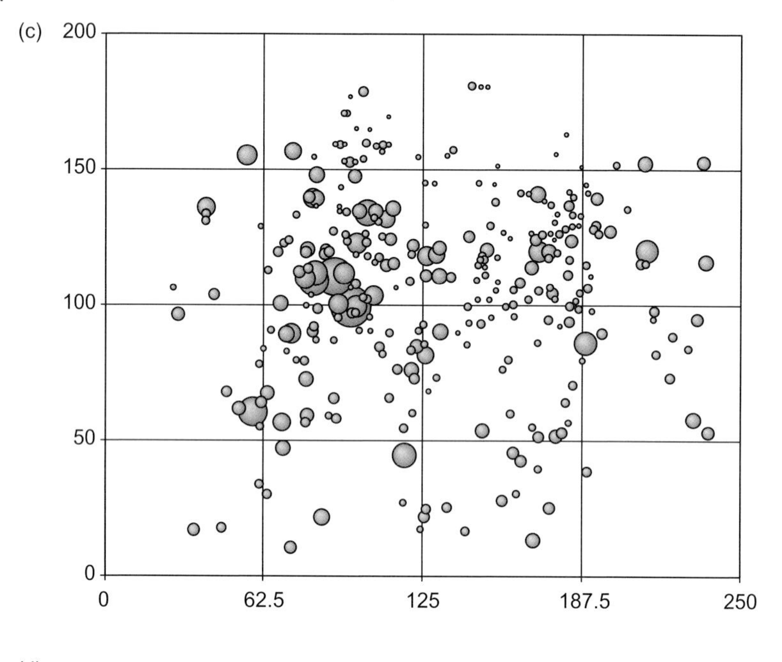

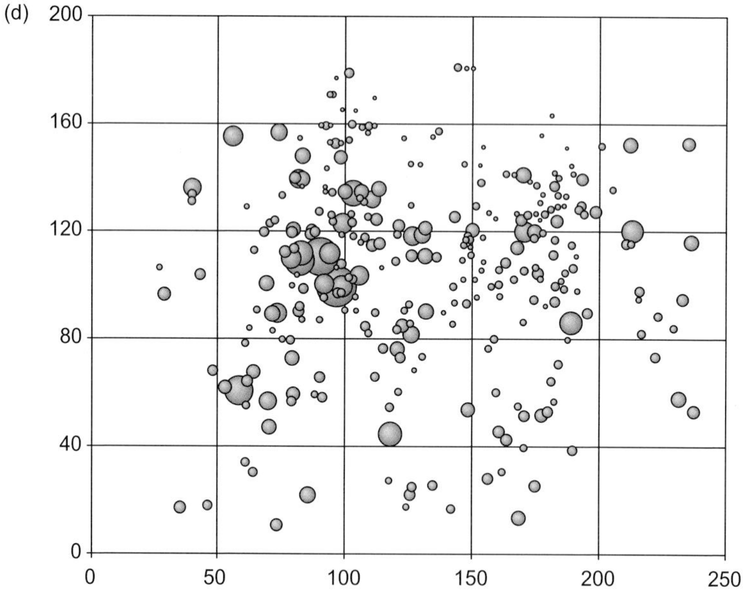

Figure 8.6 *(Continued)*

A toxicologically interesting fact is that the two structurally related peroxisome proliferators PFOA and PFDA, along with clofibrate, have relatively similar MIC values compared to the control and DEHP. This indicates that the MIC biodescriptor reported here may be capable of characterizing toxicity and toxic mode of action.

Table 8.1 Calculated values of MIC(4) for the five treatments using the most abundant 200, most abundant 500, and entire set of 1054 proteins.

	200	500	1054
Control	3.8390	3.9486	3.9892
PFOA	3.7865	3.9112	3.9519
PFDA	3.7702	3.8861	3.9289
Clofibrate	3.7769	3.9051	3.9584
DEHP	3.7033	3.8387	3.8923

One desirable property of chemodescriptors is the ability to discriminate among closely related chemical structures. Analogously, we would expect that biodescriptors derived from biological systems would be able to discriminate among closely related biochemical processes or responses; for example, perturbations in proteomics maps as a result of exposure to similar toxicants. MIC(4) not only discriminates among maps derived from different structural classes of peroxisome proliferators but also discriminates between closely related compounds, for example, PFOA and PFDA. It is expected that the MIC index will find applications in pattern recognition for

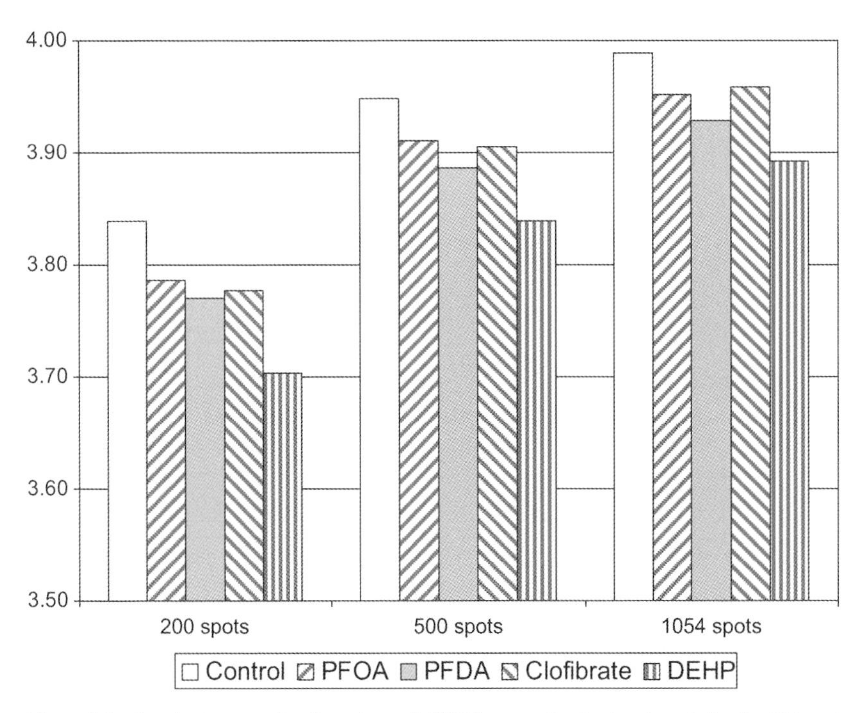

Figure 8.7 Bar chart comparing the values of MIC(4) across the five treatments for the 200-spot, 500-spot, and 1054-spot proteomics maps.

Table 8.2 Calculated values of MIC for the two treatments using the most abundant 100, most abundant 200, and entire set of 298 proteins.

	50	100	150	200	298
MIC(2)					
Control	1.7640	1.8658	1.8929	1.9138	1.9258
JP-8	1.7849	1.8735	1.8930	1.9115	1.9225
MIC(3)					
Control	2.6449	2.7695	2.7863	2.7611	2.7796
JP-8	2.6480	2.7648	2.7790	2.7507	2.7705
MIC(4)					
Control	2.7597	2.9684	3.0654	3.0916	3.1350
JP-8	2.7531	2.9522	3.0370	3.0574	3.1035
MIC(5)					
Control	3.0135	3.4153	3.5475	3.5752	3.6142
JP-8	3.0584	3.4316	3.5701	3.5701	3.6076

proteomics maps pertinent to biomedicinal chemistry, pharmacology, pathology, and toxicology.

8.6.2 JP-8 Exposure

The magnitudes of the map information content indices MIC(2)–MIC(5) for the most abundant 50, 100, 150, 200, and 298 spots for control and JP-8 exposure of keratinocytes are presented in Table 8.2, while Figure 8.8 presents a plot of the MIC(4) values for the same sets of spots. As with the peroxisome proliferators, the magnitude of MIC(4) decreases for the JP-8-exposed proteomics maps compared to the control values.

It should be noted that even at this low level of JP-8 exposure, any change in protein expression and proteomics map patterns can be detected using the information content biodescriptors presented in this study.

8.7 Discussion and Conclusion

The primary objectives of this chapter were twofold: (i) to give a general overview of recent advances in mathematical proteomics and (ii) to discuss the utility of an information theoretic approach in toxicoproteomics. The modern "omics" sciences, namely, genomics, proteomics, and metabolomics, are generating an enormous amount of raw data at an ever-increasing speed. However, in many cases, the utility and meaning of such data are not clear. One approach to tackling this data-rich, but meaning-poor, quagmire could be the use of robust and proven mathematical

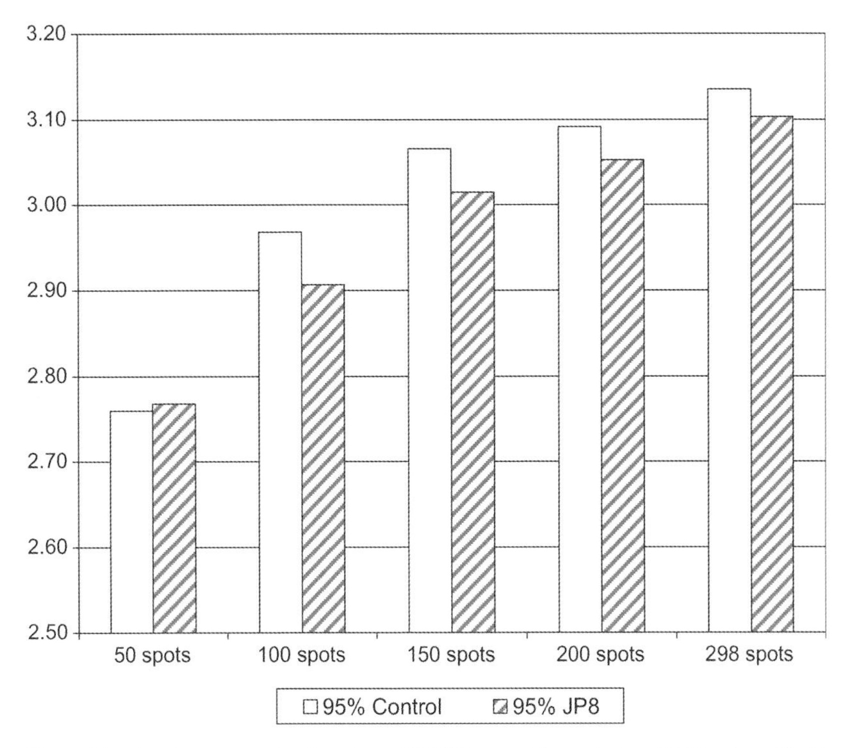

Figure 8.8 Bar chart comparing the values of MIC(4) across the control and JP-8 treatment for the 50-spot, 100-spot, 150-spot, 200-spot, and 298-spot proteomics maps.

methods for pattern recognition. The four different mathematical proteomics techniques developed by Basak and collaborators fall in this category.

The information theoretic method gives a compact measure of complexity based on the distribution of the various protein spots on the 2DE gel. In spite of the large number of proteins perturbed after treatment with PFOA, PFDA, clofibrate, and DEHP, the calculated information theoretic indices have close magnitudes for the two most similar chemicals, PFOA and PFDA. Also, in the case of the complex jet fuel mixture, JP-8, the calculated indices discriminate between the control and the fuel-exposed maps. Further research on the utility of such information theoretic indices in proteomics pattern recognition is necessary to test the overall utility of these mathematical biodescriptors in toxicology.

Acknowledgments

This manuscript is contribution number 455 from the Center for Water and the Environment of the Natural Resources Research Institute and is based on research sponsored by the Air Force Research Laboratory, under agreements F49 620-02-1-0138 (SCB), F49 620-03-1-0089 (FAW), and F49 620-01-1-0080 (JER). The US

Government is authorized to reproduce and distribute reprints for official purposes notwithstanding any copyright notation thereon.

References

1 What is the TSCA Chemical Substance Inventory? United States Environmental Protection Agency, http://www.epa.gov/oppt/newchems/pubs/invntory.htm, accessed 8/30/2007.

2 Basak, S.C., Mills, D., and Gute, B.D. (2007) Predicting bioactivity and toxicity of chemicals from mathematical descriptors: a chemical-cum-biochemical approach, in *Advances in Quantum Chemistry* (eds D.J. Klein and E. Brandas), Elsevier.

3 Basak, S.C., Grunwald, G.D., Host, G.E., Niemi, G.J., and Bradbury, S.P. (1998) A comparative study of molecular similarity, statistical and neural network methods for predicting toxic modes of action of chemicals. *Environ. Toxicol. Chem.*, **17**, 1056–1064.

4 Randic, M., Witzmann, F., Vracko, M., and Basak, S.C. (2001) On characterization of proteomics maps and chemically induced changes in proteomes using matrix invariants: application to peroxisome proliferators. *Med. Chem. Res.*, **10**, 456–479.

5 Randic, M. and Basak, S.C. (2002) A comparative study of proteomics maps using graph theoretical biodescriptors. *J. Chem. Inf. Comput. Sci.*, **42**, 983–992.

6 Randic, M., Zupan, J., Novic, M., Gute, B.D., and Basak, S.C. (2002) Novel matrix invariants for characterization of changes of proteomics maps. *SAR QSAR Environ. Res.*, **13**, 689–703.

7 Bajzer, Z., Randic, M., Plavsic, D., and Basak, S.C. (2003) Novel map descriptors for characterization of toxic effects in proteomics maps. *J. Mol. Graph. Model.*, **22**, 1–9.

8 Gute, B.D., Basak, S.C., Balasubramanian, K., Geiss, K., and Hawkins, D.M. (2004) Chemodescriptors versus biodescriptors for toxicity prediction on halocarbons. *Environ. Toxicol. Pharmacol.*, **16**, 121–129.

9 Randic, M. and Basak, S.C. (2004) On similarity of proteome maps. *Med. Chem. Res.*, **13**, 800–811.

10 Randic, M., Lers, N., Plavsic, D., and Basak, S.C. (2004) Characterization of 2-D proteome maps based on the nearest neighborhoods of spots. *Croat. Chem. Acta*, **77**, 345–351.

11 Randic, M., Lers, N., Plavsic, D., and Basak, S.C. (2004) On invariants of a 2-D proteome map derived from neighborhood graphs. *J. Proteome Res.*, **3**, 778–785.

12 Vracko, M. and Basak, S.C. (2004) Similarity study of proteomic maps. *Chemometr. Intell. Lab. Syst.*, **70**, 33–38.

13 Randic, M., Lers, N., Vukicevic, D., Plavsic, D., Gute, B.D., and Basak, S.C. (2005) Canonical labeling of proteome maps. *J. Proteome Res.*, **4**, 1347–1352.

14 Hawkins, D.M., Basak, S.C., Kraker, J.J., Geiss, K.T., and Witzmann, F.A. (2006) Combining chemodescriptors and biodescriptors in quantitative structure–activity relationship modeling. *J. Chem. Inf. Model.*, **46**, 9–16.

15 Balasubramanian, K., Khokhani, K., and Basak, S.C. (2006) Complex graph matrix representations and characterizations of proteomic maps and chemically induced changes to proteomes. *J. Proteome Res.*, **5**, 1133–1142.

16 Basak, S.C., Mills, D., and Gute, B.D. (2006) Quantitative structure–toxicity relationships using chemodescriptors and biodescriptors, in *Biological Concepts and Techniques in Toxicology: An Integrated Approach* (ed. J.E. Riviere), Taylor & Francis, New York.

17 Randic, M., Witzmann, F.A., Kodali, V., and Basak, S.C. (2006) On the dependence of a characterization of proteomics maps on the number of protein spots considered. *J. Chem. Inf. Model.*, **46**, 116–122.

18 Vracko, M., Basak, S.C., Geiss, K., and Witzmann, F. (2006) Proteomics maps-toxicity relationship of halocarbons studied with similarity index and genetic algorithm. *J. Chem. Inf. Model.*, **46**, 130–136.

19 Bajzer, Z., Basak, S.C., Vracko, M., and Randic, M. (2007) Use of proteomics based biodescriptors in the characterization of chemical toxicity, in *Genomic and Proteomic Applications to Toxicity Testing* (ed. M.J. Cunningham), Humana Press, Totowa, NJ.

20 Anderson, P.D. and Weber, L.J. (1975) The toxicity to aquatic populations of mixtures containing certain heavy metals. Symposium Proceedings, International Conference on Heavy Metals in the Environment, vol. 2, pp. 933–954.

21 Mumtaz, M.M. and Durkin, P.R. (1992) A weight-of-evidence approach for assessing interactions in chemical mixtures. *Toxicol. Ind. Health*, **8**, 377–406.

22 Gennings, C. (1996) Economical designs for detecting and characterizing departure from additivity in mixtures of many chemicals. *Food Chem. Toxicol.*, **34**, 1053–1059.

23 Mumtaz, M.M., de Rosa, C.T., Groten, J., Feron, V.J.H., Hansen, H., and Durkin, P.R. (1998) Evaluation of chemical mixtures of public health concern: estimation versus experimental determination of toxicity. *Environ. Health Perspect.*, **106**, 1353–1361.

24 Cassee, F.R., Groten, J.P., Van Bladeren, P.J., and Feron, V.J. (1998) Toxicological evaluation and risk assessment of chemical mixtures. *Crit. Rev. Toxicol.*, **28**, 73–101.

25 Anderson, N.L., Esquer-Blasco, R., Richardson, F., Foxworthy, P., and Eacho, P. (1995) The effects of peroxisome proliferators on protein abundances in mouse liver. *Toxicol. Appl. Pharmacol.*, **137**, 75–89.

26 Witzmann, F.A. (2002) Proteomic applications in toxicology, in *Comprehensive Toxicology, Vol. XIV: Cellular and Molecular Toxicology* (eds J.P. Vanden Heuvel, W.F. Greenlee, G.H. Perdew, and W.B. Mattes), Elsevier.

27 Basak, S.C., Gute, B.D., and Witzmann, F.A. (2005) Information-theoretic biodescriptors for proteomics maps: development and applications in predictive toxicology. *WSEAS Trans. Inf. Sci. Appl.*, **7**, 996–1001.

28 Basak, S.C., Gute, B.D., Geiss, K.T., and Witzmann, F.A. (2007) Information-theoretic biodescriptors for proteomics maps: application to rodent hepatotoxicity, in *Computation in Modern Science and Engineering*, vol. 2, Part A (eds T.E. Simos and G. Maroulis), American Institute of Physics.

29 Roy, A.B., Raychaudhury, C., Ray, S.K., Basak, S.C., and Ghosh, J.J. (1983) Information-theoretic topological indices of a molecular graph and their applications in QSAR, in *Proceedings of the Fourth European Symposium on Chemical Structure–Biological Activity: Quantitative Approaches*, Elsevier.

30 Roy, A.B., Basak, S.C., Harriss, D.K., and Magnuson, V.R. (1984) Neighborhood complexities and symmetry of chemical graphs and their biological applications, in *Mathematical Modelling in Science and Technology* (eds X.J.R. Avula, R.E. Kalman, A.I. Liapis, and E.Y. Rodin), Pergamon Press, New York.

31 Allen, D.G., Riviere, J.E., and Monteiro-Riviere, N.A. (2001) Cytokine induction as a measure of cutaneous toxicity in primary and immortalized porcine keratinocytes exposed to jet fuels, and their relationship to normal human epidermal keratinocytes. *Toxicol. Lett.*, **119**, 209–217.

32 Witzmann, F.A., Clack, J.W., Geiss, K., Hussain, S., Juhl, M.J., Rice, C.M., and Wang, C. (2002) Proteomic evaluation of cell preparation methods in primary hepatocyte cell culture. *Electrophoresis*, **23**, 2223–2232.

33 Witzmann, F.A., Strother-Robinson, W.N., McBride, W.J., Hunter, L., Crabb, D.W., Lumeng, L., and Li, T.-K. (2003) Innate differences in protein expression in the nucleus accumbens and hippocampus of inbred alcohol-preferring (iP) and -nonpreferring (iNP) rats. *Proteomics*, **3**, 1335–1344.

34 Troyanskaya, O., Cantor, M., Sherlock, G., Brown, P., Hastie, T., Tibshirani, R.,

Botstein, D., and Altman, R.B. (2001) Missing value estimation methods for DNA microarrays. *Bioinformatics*, **17**, 520–525.

35 Bolstad, B.M., Irizarry, R.A., Astrand, M., and Speed, T.P. (2003) A comparison of normalization methods for high density oligonucleotide array data based on variance and bias. *Bioinformatics*, **19**, 185–193.

36 Basak, S.C., Roy, A.B., and Ghosh, J.J. (1979) Study of the structure–function relationship of pharmacological and toxicological agents using information theory. Proceedings of the Second International Conference on Mathematical Modelling, University of Missouri-Rolla, Rolla, Missouri (eds X.J.R. Avula, R. Bellman, Y.L. Luke, and A.K. Rigler).

37 Magnuson, V.R., Harriss, D.K., and Basak, S.C. (1983) Topological indices based on neighborhood symmetry: chemical and biological applications, in *Chemical Applications of Topology and Graph Theory* (ed. R.B. King), Elsevier.

38 Basak, S.C. (1987) Use of molecular complexity indices in predictive pharmacology and toxicology: a QSAR approach. *Med. Sci. Res.*, **15**, 605–609.

39 Basak, S.C. (1999) Information theoretic indices of neighborhood complexity and their applications, in *Topological Indices and Related Descriptors in QSAR and QSPR* (eds J. Devillers and A.T. Balaban), Gordon and Breach Science Publishers, The Netherlands.

40 Witzmann, F.A., Jarnot, B.M., Parker, D.N., and Clack, J.W. (1994) Modification of hepatic immunoglobulin heavy chain binding protein (BiP/Grp78) following exposure to structurally diverse peroxisome proliferators. *Fundam. Appl. Toxicol.*, **23**, 1–8.

9
Pharmacokinetic Mechanisms of Interactions in Chemical Mixtures

Kannan Krishnan, Alan Sasso, and Panos Georgopoulos

9.1
Introduction

Mixed exposure to chemicals in the general environment and occupational settings may lead to interactions at the exposure, pharmacokinetic, or pharmacodynamic levels (Figure 9.1). While the exposure phase interactions result in a change in the potential dose per unit environmental concentration of mixture components, the pharmacokinetic interactions alter the internal dose of a chemical per unit exposure concentration. Pharmacodynamic interactions are said to occur when tissue response associated with a unit tissue dose of the active chemical is altered during mixed exposures in comparison with single exposures [1]. The impact of interactions on the toxicological outcome during mixed exposures depends, among other factors, on the relative concentrations and mode of action of mixture components as well as mechanisms of and thresholds for interactions.

In most experimental studies of interactions, chemicals have been evaluated as binary mixtures [2, 3]. Such studies have been particularly useful in understanding both the mechanisms of interactions and the thresholds of interactions as a function of dose and exposure sequence to chemicals [3–8]. However, in real-life and occupational settings, chemicals co-occur as mixtures of more than two, in which some components may act independently whereas others might interact with each other. Recent research has indicated that binary interaction data can be used to predict the internal dose of chemicals in increasingly complex mixtures [7, 9, 10], underpinning the pivotal importance of mechanistic understanding of interactions in chemical mixtures. This chapter describes the basis and mechanisms of interactions occurring during mixed chemical exposures at the level of pharmacokinetics (i.e., absorption, distribution, metabolism, and excretion).

Principles and Practice of Mixtures Toxicology. Edited by Moiz Mumtaz

ISBN: 978-3-527-31992-3

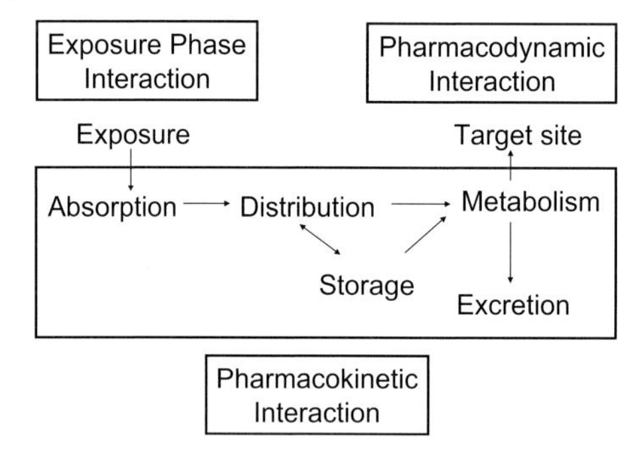

Figure 9.1 Schematic depicting the possible levels at which chemicals may interact with each other in the environment and in the organism.

9.2
Absorption-Level Interactions

9.2.1
Pulmonary Absorption

Combined exposures may alter the rate and the extent of absorption (i.e., the process by which a chemical moves from the environment across the biological barrier to reach systemic circulation) of one or more chemicals. Mixed exposure to chemicals, in comparison with single-chemical exposures, can change the rate of pulmonary uptake if they modify one or more key determinants, namely, blood solubility, breathing rate, or cardiac output. Blood solubility, as determined by the blood:air partition coefficient of volatile chemicals, is a function of their solubility in blood lipids, protein, and water [11]. During combined exposure to chemicals, the concentration of lipid-like components in blood might increase (as a function of the number and concentration of mixture constituents) [12], thus leading to an increase in the uptake of lipophilic chemicals into systemic circulation. For highly blood-soluble chemicals (e.g., methanol), the limiting determinant of pulmonary uptake is the alveolar ventilation rate. Thus, when mixed exposures lead to a significant change in alveolar ventilation rate, it can be expected to alter the inhaled dose of other substances in the atmosphere; hydrogen cyanide, for example, at low concentrations increases the pulmonary ventilation rate and the uptake of other air-borne substances [13]. For inhalants exhibiting low blood solubility, however, cardiac output is the factor limiting their uptake; thus, mixed exposure situation altering the cardiac output in exposed individuals is likely to influence the pulmonary uptake of poorly soluble chemicals (e.g., carbon tetrachloride, perchloroethylene, and so on).

9.2.2
Dermal Absorption

Interactions at the dermal absorption level might result from alteration of the permeability of the skin or the rate of diffusion through the lipid or protein pathways.

For example, exposure to an organic chemical, such as dimethyl sulfoxide, leads to the disruption of dermal layers, allowing greater quantity of other chemicals to enter the systemic circulation [14–20]. Increased dermal absorption of several chemicals has been reported to occur following exposure to highly irritant skin-allergens and surfactants [21–23]. Fuel (gasoline) mixtures containing lower percentages of methanol showed an enhancement of dermal penetration, indicating that the increasingly nonpolar nature of the mixture facilitated the movement of the polar product (methanol) out of the vehicle and into the stratum corneum of the skin [24]. More recent studies have reported the enhanced dermal absorption of pesticides and metals in the presence of organic solvents (e.g., triazines in the presence of trichloroethylene [25, 26]; *N*,*N*-diethyl-*m*-toluamide in the presence of ethanol [27]; and arsenic and nickel in organic solvent mixtures [28]).

9.2.3 Gastrointestinal Absorption

Interactions during oral absorption result from changes in pH, motility, and bioavailability as well as competition for transporters. There are examples both from the pharmaceutical and from the toxicological literature regarding enhancement or reduction of oral absorption of substances during mixed exposure situations [3, 29]. Of particular interest is the extensive literature on metal–metal and metal–nutrient interactions at the absorption level [30–35]. In general, low essential element status has been found to increase toxic metal absorption (See Table 9.1 for examples). This is due to the fact that absorption of many divalent cations (Cd, Co, Cu, Fe, Mn, Pb, and Zn) through the gastrointestinal (GI) tract depends on common metal-binding transporters such as DMT1 (divalent metal transporter, also known as DCT1 or Nramp-2 [36–42]). Deficiency in essential elements leads to an increase in the expression of DMT1 and such metal-binding transporters, resulting in increased absorption of not only essential nutrients but also toxic metals [43–45]. Some toxic metals, in turn, may reduce the absorption of essential nutrients leading to symptoms associated with their deficiency (e.g., selenium and mercury, lead and

Table 9.1 Examples of essential nutrient–toxic metal interactions (based on Refs. [31–33, 35, 49, 106]).

Deficiency	Increased absorption
Iron	Lead
	Cadmium
	Arsenic
	Manganese
Zinc	Lead
	Cadmium
	Arsenic
Calcium	Lead
	Cadmium

iron [44, 46, 47]). Absorption-level interactions involving several cations (Fe^{3+}, Zn^{2+}, Hg^{2+}, and Cu^{2+}) and ethylene bis-thiocarbamate anion (of the fungicide Maneb) have also been reported [48].

Interactions occurring during the transport of toxic metals are considered to be a consequence of molecular mimicry or ionic mimicry [49]. Molecular mimicry refers to the phenomenon by which the bonding of metal ions to nucleophilic groups on certain biomolecules results in the formation of organometallic complexes that can behave or serve as a structural and/or functional homologue of other endogenous biomolecules. For example, amino acid transporters and organic anion transporters (OAT1, OAT3) have been implicated in the transport of organic and inorganic forms of mercury in kidney cells [49]. Ionic mimicry, on the other hand, refers to the ability of an unbound cationic species of a metal to mimic an essential element. For example, cadmium can use the ion channels and certain membrane transporters such as the divalent metal transporter 1 (DMT1/DCT1/Nramp1) [49].

9.3 Distribution-Level Interactions

Distribution-level interactions occur when the rate of delivery or the unbound concentration of a chemical is altered by another chemical during mixed exposures. These interactions result primarily from one chemical or its metabolite altering or competing with another chemical for transport in and out of tissues (diffusion or perfusion-limited process), blood protein binding (e.g., hemoglobin, albumin), or tissue macromolecular binding.

The number of accessible plasma-protein binding sites and the relative affinity for the binding sites are the primary determinants of the extent to which a chemical is retained in bound state. In case of mixed exposure to two toxicants, one might displace another from a binding site, resulting in an increase in the free form of the latter chemical, which might either cause immediate effects or become available to be metabolized. The ability of one chemical to preferentially bind to a site over another chemical is determined by its relative concentration at the site, binding affinity and availability of alternative binding site(s). This kind of interaction, particularly at the serum protein level, has been more commonly observed with pharmaceutical substances than environmental agents [3, 29].

During combined exposure scenarios, one chemical may alter the solubility and distribution characteristics of another chemical. For example, cyanide forms complexes with essential metals changing their tissue concentrations and distribution pattern (reviewed in Ref. [3]). Similarly, various dithiocarbamates form lipophilic complexes with inorganic lead, leading to a greater accumulation in the lipid-rich brain compartments [50, 51]. Chemical reactions may also occur during the distribution phase resulting in the formation of compounds that are markedly less or more toxic than the reacting chemicals. The formation of nitrosamines from nitrite and secondary amines is a classical example of activation-type reaction. An example of reduction of internal dose and toxicity resulting from such reaction would be the chelation of certain metal ion(s) with ethylene diamine tetraacetic acid (EDTA), 1,2-cyclohexylene amine tetraacetic acid (CDTA), and diethylene triamine pentaacetic acid (PTDA) [52].

Interactions among metals and between metals and nutrients, involving competition or induction of metallothionein (a low molecular weight protein containing –SH groups), have been reported (reviewed in Ref. [31, 53]). The toxicological consequence of such an interaction would depend on whether the host is at the deficient or excess state for the specific metals or metalloids. In general, induction of metallothionein in hepatic and extrahepatic tissues has been shown to afford protection against a variety of toxicants including toxic metals [31, 54].

9.4 Metabolism-Level Interactions

The single most common mechanism of interaction among environmental chemicals appears to be the alteration of metabolism (e.g., phase I metabolism (CYP-mediated oxidation, hydroxylation, reduction, etc.) or phase II reactions such as GSH conjugation, glucuronidation, and sulfation) [3, 4]. Metabolic interactions, occurring in liver or another organ, may lead to a change in qualitative profile of metabolites or an alteration of the rate or magnitude of specific metabolic pathways. Alteration of affinity or capacity (due to enzyme induction, inhibition, or destruction) and change in the rate of delivery to metabolizing organ figure among the most common mechanisms of metabolic interaction.

The importance of reversible or irreversible inhibition of metabolism of one chemical by another chemical during mixed exposures depends upon the relative affinity of the chemicals, their concentrations, and their relationship with the internal dose of relevance to risk assessment. Table 9.2 lists metals that have been shown to bind to the prosthetic group of metabolizing enzymes or cause irreversible inhibition

Table 9.2 Metal–organic chemical interactions: implication of hepatic cytochrome P450 (CYP).

Metals	Effects on hepatic CYP	Potential substrates affected by the interaction	References
Cadmium	Induction of CYP2A6, CYP2C9, CYP2E1	Carbamates, drugs, halogenated hydrocarbons, VOCs, organophosphates, triazines	[107]
Lead	Inhibition of CYP2A6, CYP1A2	Drugs, arylamines, organophosphates, triazines, VOCs	[108–111]
Arsenic	Increase of CYP1A1	PAHs, triazines, VOCs	[71, 112]
Metal mixtures	Alteration of CYP1A1/2	PAHs, triazines, VOCs, drugs	[113–116]

or destruction of enzymes. Reversible inhibition, which can occur at lower exposure levels (compared to irreversible inhibition) mostly involving organic chemicals, is one of the following types: competitive, noncompetitive, and uncompetitive.

Competitive inhibition is the consequence of substrates competing for the same binding site for metabolism. In the context of environmental and occupational exposure to chemical mixtures, each chemical can simultaneously play the role of both a substrate and an inhibitor to metabolism of another chemical. As a function of the concentration of I and its affinity for metabolism, referred to as inhibition constant, it influences the metabolism of another chemical (i.e., substrate) in the mixture. The equilibria describing competitive inhibition, with C as the substrate chemical and I as the inhibitor chemical, are as follows [55, 56]:

$$\begin{array}{ccccc} E + C & \underset{}{\overset{K_m}{\rightleftharpoons}} & EC & \xrightarrow{K_p} & E + P \\ + & & & & \\ I & & & & \\ K_i \updownarrow & & & & \\ EI & & & & \end{array}$$

where

$$K_m = \frac{[E][C]}{[EC]}, \tag{9.1}$$

$$K_i = \frac{[E][I]}{[EI]}, \tag{9.2}$$

and K_p is the rate constant for the breakdown of EC to E and P.

In this case, the rate of metabolism of the substrate C, in the presence of the inhibitor I, is computed as follows:

$$v = \frac{[C]\, V_{max}}{[C] + K_m\left(1 + \frac{[I]}{K_i}\right)}, \tag{9.3}$$

where V_{max} is the maximal velocity, K_m denotes affinity constant, and K_i is the inhibition constant.

The above relationship, as depicted in Figure 9.2, indicates that the competitive inhibitor I increases the apparent K_m of the substrate chemical (C) since a portion of the enzyme exists in the EI form (which has no affinity for the substrate). Therefore, in the presence of a competitive inhibitor, a greater dose of the substrate would be required to achieve the same level of internal dose at the target site.

Classical examples of competitive metabolic inhibition include interactions among substrates of hepatic cytochrome P450 and alcohol dehydrogenase [57–60]. Several volatile organic chemicals that are known or potential substrates of CYP2E1 have been shown to interact via competitive metabolic inhibition in

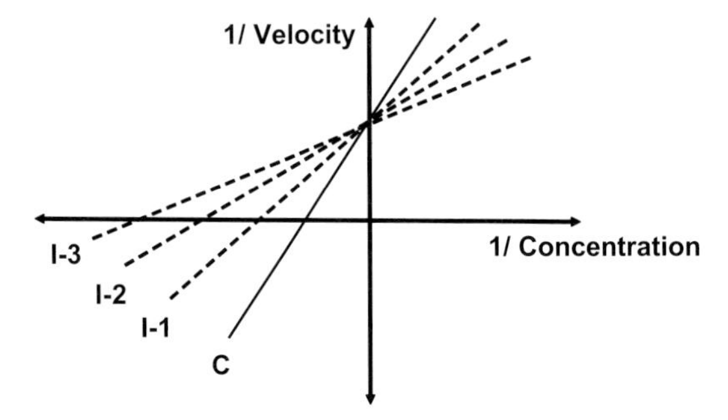

Figure 9.2 Impact of changing concentrations of an inhibitor (I) on the metabolism of a substrate (c) at a single concentration. Note that the maximal velocity remains unchanged whereas the affinity is altered.

mixtures of varying complexity and composition [7, 9, 61]. In the pesticide literature, there are reports of competitive metabolic interactions between chlorpyriphos and diazinon [62], as well as chlorpyriphos-oxon and malathion [63]. The toxicological consequence of metabolic inhibition, of course, depends on the nature of the toxic moiety (i.e., parent chemical, metabolite) and the importance of hepatic metabolism relative to whole body clearance of the interacting chemicals.

Noncompetitive inhibition is thought to result from the inhibitor (I) binding to both the free enzyme (E) and the enzyme–substrate complex (EC) [55, 56]:

$$\begin{array}{ccccc} \mathrm{E+C} & \underset{}{\overset{K_m}{\rightleftharpoons}} & \mathrm{EC} & \xrightarrow{K_p} & E+P \\ + & & + & & \\ \mathrm{I} & & \mathrm{I} & & \\ K_i \upharpoonleft\downharpoonright & & \upharpoonleft\downharpoonright K_i & & \\ \mathrm{EI\ +C} & \rightleftharpoons & \mathrm{ECI} & & \end{array}$$

The rate of metabolism of a substrate in the presence of a noncompetitive inhibitor can be calculated as follows [55]:

$$v = \frac{[\mathrm{C}]\dfrac{V_{\mathrm{max}}}{\left(1+\frac{[\mathrm{I}]}{K_{\mathrm{i}}}\right)}}{K_{\mathrm{m}}+[\mathrm{C}]}. \tag{9.4}$$

The above relationship, depicted in Figure 9.3, indicates that the V_{max} of the substrate is decreased in the presence of a noncompetitive inhibitor as a consequence of a decrease in the amount of available sites for binding. However, K_m remains unchanged because the remaining enzyme is intact and exhibits the same affinity for the substrate as it would in the absence of an inhibitor.

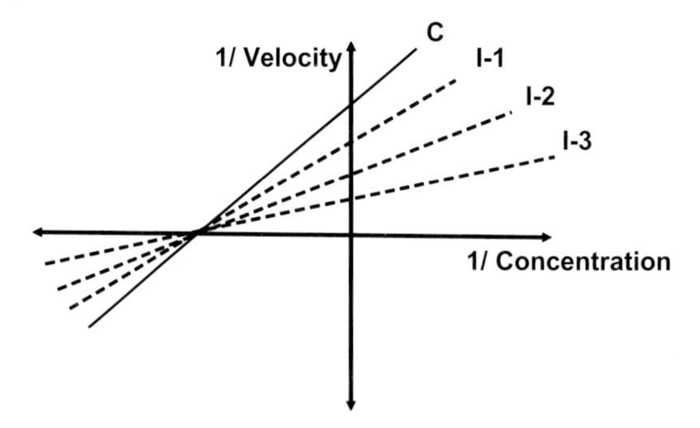

Figure 9.3 Impact of changing concentrations of an inhibitor (I) on the metabolism of a substrate (C) at a single concentration. Note that the affinity remains unchanged whereas the maximal velocity is altered.

Uncompetitive inhibition, as depicted below, leads to a reduction in V_{max} and K_m of the substrate because the inhibitor binds to the enzyme–substrate complex [55]:

$$\mathrm{E + C} \underset{}{\overset{K_m}{\rightleftharpoons}} \mathrm{EC} \xrightarrow{K_p} E + P$$

$$\begin{array}{c} \mathrm{EC} \\ + \\ \mathrm{I} \\ K_i \updownarrow \\ \mathrm{ECI} \end{array}$$

Uncompetitive metabolic inhibition among environmental chemicals has been observed/reported less frequently compared to competitive or noncompetitive inhibition [6, 9, 64]. The various inhibitory interactions described above are limited not only to the cytochrome P450-mediated reactions but also to those involving phase II conjugation processes [65]. The existence of polymorphic forms of several metabolizing enzymes and parallel metabolic pathways impact the predictability and outcome of such interactions, however [66].

Metabolic inhibition as the interaction mechanism is restricted not only to chemicals belonging to the same class or family but also to chemicals belonging to multiple classes (e.g., metals and organics). For example, lead, not only diminishes heme synthesis but also changes the rate of heme degradation, leading to a change in P450 levels [67]. Kadiiska and Stoytchev [68] reported the inhibition of aniline hydroxylase by copper, cobalt, cadmium, lead, nickel, and arsenic. Inhibition of specific isoforms of cytochrome P450 by metals has also been reported: CYP1A1 by mercury [69], CYP2E and CYP3A by cadmium [70], and CYP1A1 by arsenic [71]. It should not, however, be assumed that the inhibitors or inducers would act similarly under all conditions. For example, chronic ethanol treatment enhances hepatic CYP2E1 activity, but following an acute administration, it acts as an inhibitor [72–77]. SK&F 525A (2-(dimethyl-amino)ethyl-2,2-diphenyl pentanonate) and imidazole, known as effective inhibitors of hepatic CYP after a single oral dose, have been reported to induce drug metabolism after repetitive administrations [78]. Similarly,

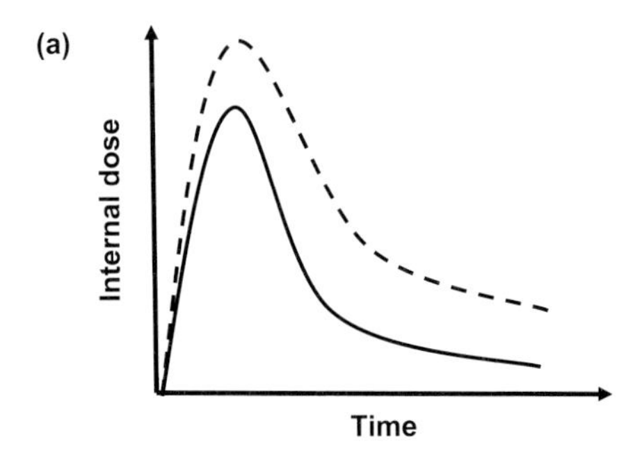

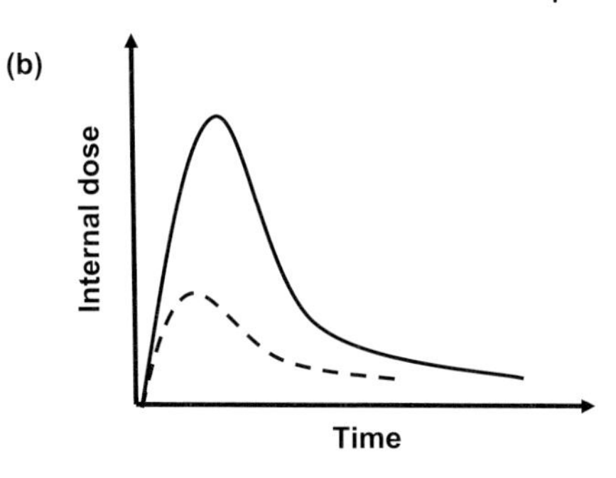

Figure 9.4 (a) Comparison of the time course of the internal dose of a hypothetical chemical during mixed exposures (dotted line) and single exposure (solid line). In this case, the alteration of kinetics during mixed exposure is caused by metabolic inhibition. (b) Comparison of the time course of the internal dose of a hypothetical chemical during mixed exposures (dotted line) and single exposure (solid line). In this case, the alteration of kinetics during mixed exposure is caused by metabolic induction.

certain organophosphate pesticides have been shown to inhibit drug metabolism after repeated administration but induce hepatic CYP activity after acute exposures [79–81].

Inhibition or induction of the metabolism of one chemical by another often leads to increase or decrease, respectively, in the circulating concentration of the unchanged (parent) form (Figure 9.4a and b). The impact of metabolic inhibition is likely to be significant only in the case of highly extracted chemicals (E greater than 0.7). For such chemicals, enzyme inhibition will shift the blood flow-limited metabolism to the capacity-limited metabolism, the impact being comparable to overexpression versus underexpression of metabolizing enzymes. In contrast, the occurrence of metabolic inhibition is unlikely to have any impact on the whole body pharmacokinetics or absorbed dose of poorly extracted chemicals. Since metabolic extraction is already low (i.e., $E = 0 - 0.3$), any further reduction in metabolic clearance in this case is unlikely to impact the circulating concentration of the parent chemical. In the case of poorly metabolized chemicals (e.g., certain PCBs, dioxin), however, enzyme induction is likely to have a marked effect [82, 83]. Here, increased capacity of the metabolizing enzymes, resulting either from increased synthesis rate or decreased degradation rate, will augment the rate of metabolism resulting in reduced parent chemical concentration coupled with detectable increase in the metabolite levels. Correspondingly then, enzyme induction is likely to reduce the toxicity of direct-acting chemicals but increase the toxicity associated with metabolites. The interpretations get more complicated when one of the multiple pathways is inhibited during combined exposures. In such cases, inhibition or induction of one metabolic pathway might result in an increase in the flux through other uninhibited pathways. In the case of dichloromethane and toluene, for example, competition between these chemicals for CYP2E1-metabolism leads to a decrease in the amount of DCM oxidized; however, it is compensated by an increase in the amount of DCM conjugated with glutathione,

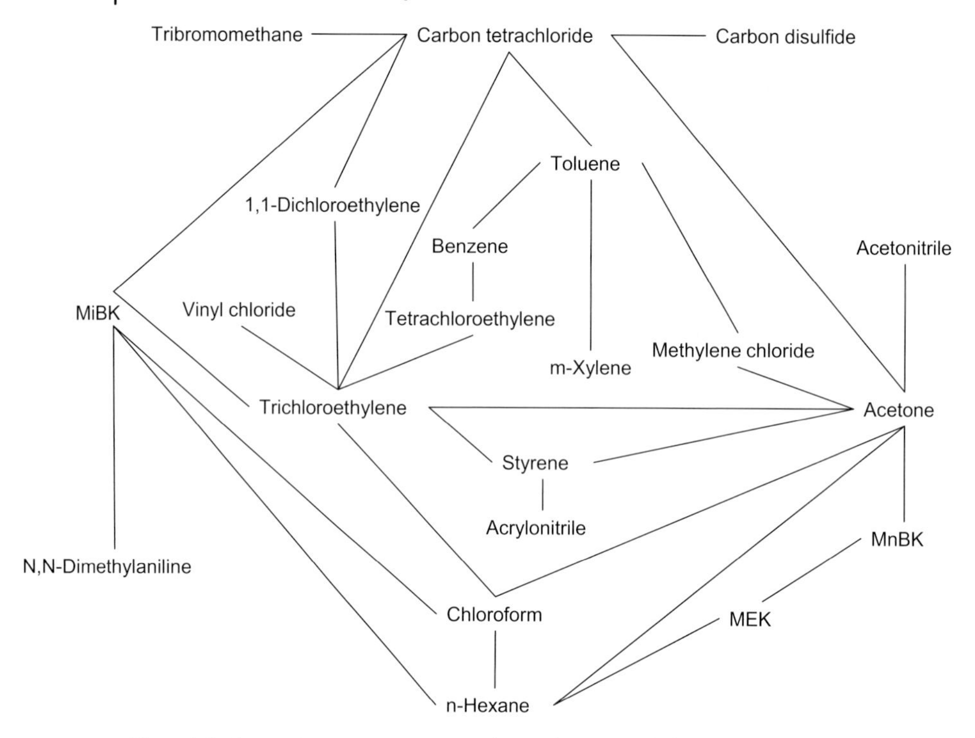

Figure 9.5 Interactions among organic solvents due to metabolic interactions. Any two solvents connected by a solid line signify the existence of literature data on the occurrence of interactions either in an experimental model or in humans (Based on [4]).

the pathway responsible for the DCM carcinogenicity [1, 7, 84]. Figure 9.5 summarizes the available information on the metabolic interactions among organic solvents, most of which are metabolized by CYP2E1.

9.5 Elimination-Level Interactions

Changes in physiological clearance or unbound concentration of chemicals occurring during mixed exposures can lead to an enhancement or impairment of the elimination of toxicants from systemic circulation. Accordingly, changes in rates of glomerular filtration, biliary flow, and alveolar ventilation can have a direct impact on the amount excreted in feces, urine, and expired air, respectively [85]. Furthermore, the reabsorption, recirculation, and urinary elimination of a chemical and/or its metabolite(s) can be altered by another chemical that causes a change in the urinary pH [86]. Modulation of biliary excretion has been suggested to be the basis of arsenic–selenium and copper–TCDD interactions. Arsenic and selenium, during combined exposures, increase the biliary excretion of each other following the formation of (seleno-bis (*S*-glutathionyl)arsenium ion) that is subsequently excreted in bile (reviewed in Ref. [87]). In case of copper–TCDD interaction, the hepatic injury caused by TCDD leads to an impairment of biliary excretion and concomitant change in the copper content of tissues [88]. There are a number of other reports in the literature describing the alteration of the rate and/or extent of biliary, pulmonary, and urinary elimination of one substance by another during combined exposures [85, 86, 89–99].

9.6 Pharmacokinetic Interactions and Impact on Internal Dose

The interactions and mechanisms operating at the absorption, distribution, metabolism, and excretion levels should not be evaluated in isolation, rather these processes should be considered together in the context of their impact on internal dose and health risk associated with chemicals during mixed exposures. In this regard, physiologically based pharmacokinetic (PBPK) models are useful tools that facilitate the integration of mechanistic data on interactions with animal/human physiology to predict the ensuing kinetics of chemicals during mixed exposures [5, 6, 100]. The PBPK models represent a unique framework for forecasting the impact of interactions occurring at various levels on the internal dose of one or more of the mixture components [6–8, 61, 101–103, 117].

The key feature of the PBPK modeling framework is the ability to predict the impact of interactions on the internal dose of a specific chemical as a function of interaction mechanisms. For example, Figure 9.6 depicts the impact of a noncompetitive inhibitor on the arterial blood concentration of toluene, during mixed exposures. The model simulations presented in this figure represent a priori predictions of the extent of modulation of internal dose (i.e., arterial blood concentration of toluene) as a function of the extent of interaction (i.e., extent of reduction in V_{max}). Using this simulation approach makes it possible to determine the interaction threshold, that is, the exposure concentration of an interacting chemical at which the modulation of internal dose of the primary toxicant does not exceed 10% compared to individual chemical exposures [102, 104]. Several studies have investigated the interaction threshold as a function of mechanism and exposure concentrations of interacting chemicals [7, 9, 61, 100, 102–105]. Further descriptions of PBPK modeling of chemical mixtures can be found in Chapter 5 of this book.

9.7 Conclusions

Interactions occurring among chemicals at the absorption, distribution, metabolism, and excretion levels can lead to qualitative and quantitative changes in the pharmacokinetics during mixed exposures compared to single-chemical exposures. Pharmacokinetic interactions change the internal dose per unit of exposure con-

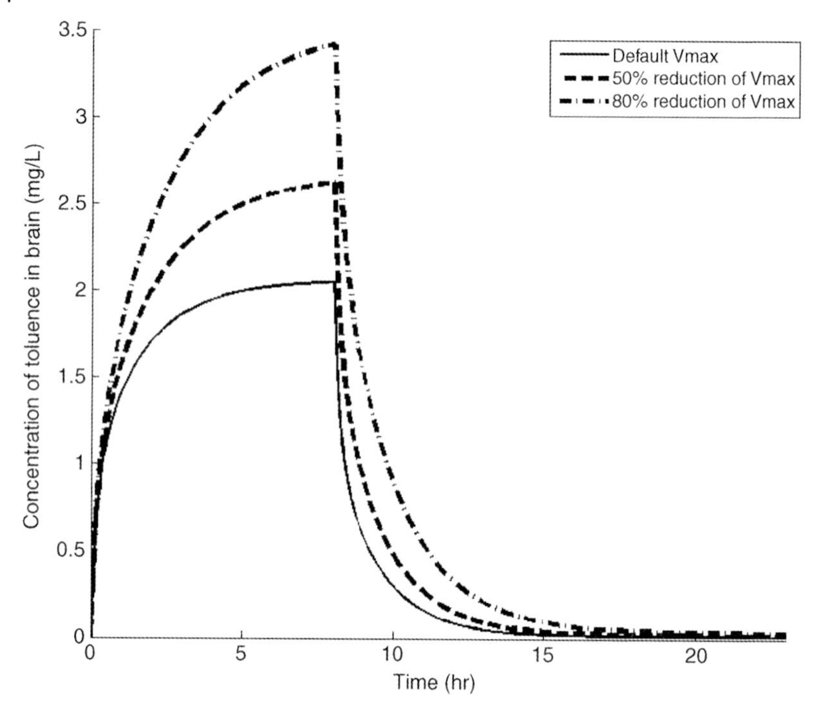

Figure 9.6 Inhalation PBPK model simulations of the tissue kinetics of toluene in humans experiencing a reduction of maximal velocity (V_{max}).

centration, and it depends primarily on the interaction mechanism, the dose of the mixture components, and the exposure scenario. PBPK models represent unique tools that facilitate the simulation of the magnitude and threshold of pharmacokinetic interactions, on the basis of interaction mechanisms and animal physiology. Finally, analysis of mixture toxicity data on the basis of pharmacokinetic interaction mechanisms would only represent part of the puzzle, the remainder of which can be resolved by taking into consideration the available information on the mode of action of the components and pharmacodynamic interactions (see Chapters 5 and 22).

Acknowledgments

This work was supported through the USEPA funded Environmental Bioinformatics and Computational Toxicology Center (GAD R 832721-010), USEPA funded Center for Exposure and Risk Modeling (CERM - EPAR827033), and the NIEHS sponsored UMDNJ Center for Environmental Exposures and Disease (Grant #: NIEHS P30ES005022). KK's work in this area is supported by the Natural Sciences and Engineering Research Council of Canada under partnership with Imperial Oil (CRDPJ 335163-05).

References

1 Krishnan, K. and Andersen, M.E. (1994) Physiologically based pharmacokinetic modeling in toxicology, in *Principles and Methods of Toxicology* (ed. A.W. Hayes), Raven Press, New York, pp. 149–188.
2 Calabrese, E.J. (1991) *Multiple Chemical Interactions*, Lewis Publishers, Chelsea, MI.
3 Krishnan, K. and Brodeur, J. (1991) Toxicological consequences of combined exposure to environmental pollutants. *Arch. Complex Environ. Stud.*, **3**, 1–106.
4 Krishnan, K. and Brodeur, J. (1994) Toxic interactions among environmental pollutants: corroborating laboratory observations with human experience. *Environ. Health Perspect.*, **102** (Suppl. 9), 11–17.
5 Krishnan, K., Clewell, H.J. III, and Andersen, M.E. (1994) Physiologically based pharmacokinetic analyses of simple mixtures. *Environ. Health Perspect.*, **102** (Suppl. 9), 151–155.
6 Krishnan, K., Haddad, S., Beliveau, M., and Tardif, R. (2002) Physiological modeling and extrapolation of pharmacokinetic interactions from binary to more complex chemical mixtures. *Environ. Health Perspect.*, **110** (Suppl. 6), 989–994.
7 Haddad, S., Beliveau, M., Tardif, R., and Krishnan, K. (2001) A PBPK modeling-based approach to account for interactions in the health risk assessment of chemical mixtures. *Toxicol. Sci.*, **63**, 125–131.
8 Yang, H.T., Chou, H.J., Han, B.C., and Huang, S.Y. (2007) Lifelong inorganic arsenic compounds consumption affected blood pressure in rats. *Food Chem. Toxicol.*, **45**, 2479–2487.
9 Haddad, S., Charest-Tardif, G., Tardif, R., and Krishnan, K. (2000) Validation of a physiological modeling framework for simulating the toxicokinetics of chemicals in mixtures. *Toxicol. Appl. Pharmacol.*, **167**, 199–209.
10 Dennison, J.E., Andersen, M.E., Clewell, H.J. III, and Yang, R.S. (2004) Development of a physiologically based pharmacokinetic model for volatile fractions of gasoline using chemical lumping analysis. *Environ. Sci. Technol.*, **38**, 5674–5681.
11 Poulin, P. and Krishnan, K. (1996) A mechanistic algorithm for predicting blood:air partition coefficients of organic chemicals with the consideration of reversible binding in hemoglobin. *Toxicol. Appl. Pharmacol.*, **136**, 131–137.
12 Beliveau, M., Tardif, R., and Krishnan, K. (2003) Quantitative structure–property relationships for physiologically based pharmacokinetic modeling of volatile organic chemicals in rats. *Toxicol. Appl. Pharmacol.*, **189**, 221–232.
13 Dreisbach, R.H. (1977) Cyanides, sulfides and cobalt, in *Handbook of Poisoning*, Lange Medical Publishing Co., Los Altos, CA, p. 241.
14 Hayes, W.J. and Pearce, G.W. (1953) Pesticide formulation: relation to safety in use. *J. Agric. Food Chem.*, 466–469.
15 Jacobs, S.W., Bischel, M., and Herschler, R.J. (1964) Dimethylsulfoxide: effects on the permeability of biologic membranes (preliminary report). *Curr. Ther. Res. Clin. Exp.*, **6**, 193–198.
16 Keil, H.L. (1965) DMSO shows great promise as a carrier of agricultural toxicants. *Agric. Chem.*, **20**, 23–24.
17 O'Brien, R.D. and Dannelley, C.E. (1965) Penetration of insecticides through rat skin. *J. Agric. Food Chem.*, **13**, 245–247.
18 Shellenberger, T.E., Newell, G.W., Okamoto, S.S., and Sarros, A. (1965) Response of rabbit whole-blood cholinesterase *in vivo* after continuous intravenous infusion and percutaneous application of dimethyl organophosphate inhibitors. *Biochem. Pharmacol.*, **14**, 943–952.
19 Hatanaka, T., Shimoyama, M., Sugibayashi, K., and Morimoto, Y. (1990) Influence of ethanol and polyethylene glycol upon the skin permeation of drugs. *J. Pharmacobiodyn.*, **13**, s-82.
20 Baynes, R.E. and Riviere, J.E. (1998) Influence of inert ingredients in pesticide formulations on dermal absorption of carbaryl. *Am. J. Vet. Res.*, **59**, 168–175.

21 Deichmann, W.B., Brown, P., and Downing, C. (1952) Unusual protective action of a new emulsifier for the handling of organic phosphates. *Science*, **116**, 221.

22 Weinberg, M.S., Morgareidge, K., and Osler, B.L. (1966) The influence of dispersing agents on the acute oral toxicity of pesticides. Abst. Meeting of the American Chemical Society, January 17–20, 1966, Phoenix, AZ.

23 Riihimaki, V. (1979) Percutaneous absorption of *m*-xylene from a mixture of *m*-xylene and isobutyl alcohol in man. *Scand. J. Work Environ. Health*, **5**, 143–150.

24 Raabe, O.G., Al-Bayati, M.A., Shulec, P.D., Gielow, F., Uyeminami, D., and Shimasaki, N. (1992) Dermal absorption of methanol and methanol/gasoline mixtures. Final report. Institute of Toxicology and Environmental Health, School of Veterinary Medicine University of California, Sacramento, California, California Environmental Protection Agency, Air Resources Board, Research Division, A933-186.

25 Marjukka Suhonen, T., Bouwstra, J.A., and Urtti, A. (1999) Chemical enhancement of percutaneous absorption in relation to stratum corneum structural alterations. *J. Control Release*, **59**, 149–161.

26 Baynes, R.E., Yeatts, J.L., Brooks, J.D., and Riviere, J.E. (2005) Pre-treatment effects of trichloroethylene on the dermal absorption of the biocide, triazine. *Toxicol. Lett.*, **159**, 252–260.

27 Ross, J.S. and Shah, J.C. (2000) Reduction in skin permeation of *N,N*-diethyl-*m*-toluamide (DEET) by altering the skin/vehicle partition coefficient. *J. Control Release*, **67**, 211–221.

28 Turkall, R.M., Skowronski, G.A., Suh, D.H., and bdel-Rahman, M.S. (2003) Effect of a chemical mixture on dermal penetration of arsenic and nickel in male pig *in vitro*. *J. Toxicol. Environ. Health A*, **66**, 647–655.

29 Rowland, M. and Tozer, T.N. (1995) *Clinical Pharmacokinetics: Concepts and Applications*, Williams & Wilkins, Baltimore.

30 Chowdhury, B.A. and Chandra, R.K. (1987) Biological and health implications of toxic heavy metal and essential trace element interactions. *Prog. Food Nutr. Sci.*, **11**, 55–113.

31 Telisman, S. (1995) Interactions of essential and/or toxic metals and metalloid regarding interindividual differences in susceptibility to various toxicants and chronic diseases in man. *Arh. Hig. Rada Toksikol.*, **46**, 459–476.

32 Goyer, R.A. (1997) Toxic and essential metal interactions. *Annu. Rev. Nutr.*, **17**, 37–50.

33 Peraza, M.A., yala-Fierro, F., Barber, D.S., Casarez, E., and Rael, L.T. (1998) Effects of micronutrients on metal toxicity. *Environ Health Perspect.*, **106** (Suppl. 1), 203–216.

34 Reeves, P.G. and Chaney, R.L. (2004) Marginal nutritional status of zinc, iron, and calcium increases cadmium retention in the duodenum and other organs of rats fed rice-based diets. *Environ. Res.*, **96**, 311–322.

35 Kakkar, P. and Jaffery, F.N. (2005) Biological markers for metal toxicity. *Environ. Toxicol. Pharmacol.*, **19**, 335–349.

36 Gunshin, H., Mackenzie, B., Berger, U.V., Gunshin, Y., Romero, M.F., Boron, W.F., Nussberger, S., Gollan, J.L., and Hediger, M.A. (1997) Cloning and characterization of a mammalian proton-coupled metal-ion transporter. *Nature*, **388**, 482–488.

37 Picard, V., Govoni, G., Jabado, N., and Gros, P. (2000) Nramp 2 (DCT1/DMT1) expressed at the plasma membrane transports iron and other divalent cations into a calcein-accessible cytoplasmic pool. *J. Biol. Chem.*, **275**, 35738–35745.

38 Rolfs, A. and Hediger, M.A. (2001) Intestinal metal ion absorption: an update. *Curr. Opin. Gastroenterol.*, **17**, 177–183.

39 Garrick, M.D., Dolan, K.G., Horbinski, C., Ghio, A.J., Higgins, D., Porubcin, M., Moore, E.G., Hainsworth, L.N., Umbreit, J.N., Conrad, M.E., Feng, L., Lis, A., Roth, J.A., Singleton, S., and Garrick, L.M. (2003) DMT1: a mammalian transporter for multiple metals. *Biometals*, **16**, 41–54.

40 Bressler, J.P., Olivi, L., Cheong, J.H., Kim, Y., and Bannona, D. (2004) Divalent metal transporter 1 in lead and cadmium transport. *Ann. N. Y. Acad. Sci.*, **1012**, 142–152.

41 Bressler, J.P., Olivi, L., Cheong, J.H., Kim, Y., Maerten, A., and Bannon, D. (2007) Metal transporters in intestine and brain: their involvement in metal-associated neurotoxicities. *Hum. Exp. Toxicol.*, **26**, 221–229.

42 Yokel, R.A., Lasley, S.M., and Dorman, D.C. (2006) The speciation of metals in mammals influences their toxicokinetics and toxicodynamics and therefore human health risk assessment. *J. Toxicol. Environ. Health B Crit. Rev.*, **9**, 63–85.

43 Madden, E.F. (2003) The role of combined metal interactions in metal carcinogenesis: a review. *Rev. Environ. Health*, **18**, 91–109.

44 Kwong, W.T., Friello, P., and Semba, R.D. (2004) Interactions between iron deficiency and lead poisoning: epidemiology and pathogenesis. *Sci. Total Environ.*, **330**, 21–37.

45 Cory-Slechta, D.A. (2005) Studying toxicants as single chemicals: does this strategy adequately identify neurotoxic risk? *Neurotoxicology*, **26**, 491–510.

46 Schumann, K. and Elsenhans, B. (2002) The impact of food contaminants on the bioavailability of trace metals. *J. Trace Elem. Med. Biol.*, **16**, 139–144.

47 Raymond, L.J. and Ralston, N.V.C. (2004) Mercury:selenium interactions and health implications. *SMFJ*, **7**, 72–77.

48 Brocker, E.R. and Schlatter, C. (1979) Influence of some cations on the intestinal absorption of Maneb. *J. Agric. Food Chem.*, **27**, 303–306.

49 Bridges, C.C. and Zalups, R.K. (2005) Molecular and ionic mimicry and the transport of toxic metals. *Toxicol. Appl. Pharmacol.*, **204**, 274–308.

50 Tandon, S.K., Hashmi, N.S., and Kachru, D.N. (1990) The lead-chelating effects of substituted dithiocarbamates. *Biomed. Environ. Sci.*, **3**, 299–305.

51 Weiss, B., Cory-Slechta, D.A., and Cox, C. (1990) Modification of lead distribution by diethyldithiocarbamate. *Fundam. Appl. Toxicol.*, **15**, 791–799.

52 Wieczorek, H. and Oberdorster, G. (1989) Effects of chelating on organ distribution and excretion of manganese after inhalation exposure to 54MnCl2. II: inhalation of chelating agents. *Pol. J. Occup. Med.*, **2**, 389–396.

53 Brzoska, M.M. and Moniuszko-Jakoniuk, J. (2001) Interactions between cadmium and zinc in the organism. *Food Chem. Toxicol.*, **39**, 967–980.

54 Bobillier-Chaumont, S., Maupoil, V., and Berthelot, A. (2006) Metallothionein induction in the liver, kidney, heart and aorta of cadmium and isoproterenol treated rats. *J. Appl. Toxicol.*, **26**, 47–55.

55 Segel, I.H. (1975) *Enzyme Kinetics: Behavior and Analysis of Rapid Equilibrium and Steady State Enzyme Systems*, John Wiley & Sons, Inc., New York.

56 Kuby, S.A. (1990) *A Study of Enzymes*, CRC Press, Boca Raton, Florida.

57 Sato, A. and Nakajima, T. (1979) Dose-dependent metabolic interaction between benzene and toluene *in vivo* and *in vitro*. *Toxicol. Appl. Pharmacol.*, **48**, 249–256.

58 Baud, F.J., Borron, S.W., and Bismuth, C. (1995) Modifying toxicokinetics with antidotes. *Toxicol. Lett.*, **82–83**, 785–793.

59 Uaki, H., Kawai, T., Mizunuma, K., Moon, C.S., Zhang, Z.W., Inui, S., Takada, S., and Ikeda, M. (1995) Dose-dependent suppression of toluene metabolism by isopropyl alcohol and methyl ethyl ketone after experimental exposure of rats. *Toxicol. Lett.*, **81**, 229–234.

60 Morel, G., Lambert, A.M., Rieger, B., and Subra, I. (1996) Interactive effect of combined exposure to glycol ethers and alcohols on toxicodynamic and toxicokinetic parameters. *Arch. Toxicol.*, **70**, 519–525.

61 Tardif, R., Charest-Tardif, G., Brodeur, J., and Krishnan, K. (1997) Physiologically based pharmacokinetic modeling of a ternary mixture of alkyl benzenes in rats and humans. *Toxicol. Appl. Pharmacol.*, **144**, 120–134.

62 Timchalk, C., Poet, T.S., Hinman, M.N., Busby, A.L., and Kousba, A.A. (2005) Pharmacokinetic and pharmacodynamic interaction for a binary mixture of

chlorpyrifos and diazinon in the rat. *Toxicol. Appl. Pharmacol.*, **205**, 31–42.

63 Buratti, F.M., D'Aniello, A., Volpe, M.T., Meneguz, A., and Testai, E. (2005) Malathion bioactivation in the human liver: the contribution of different cytochrome P450 isoforms. *Drug Metab. Dispos.*, **33**, 295–302.

64 Purcell, K.J., Cason, G.H., Gargas, M.L., Andersen, M.E., and Travis, C.C. (1990) In *vivo* metabolic interactions of benzene and toluene. *Toxicol. Lett.*, **52**, 141–152.

65 Jeong, E.J., liu, X., Jia, X., Chen, J., and Hu, M. (2005) Coupling of conjugating enzymes and efflux transporters: impact on bioavailability and drug interactions. *Curr. Drug Metab.*, **6**, 455–468.

66 Ito, K., Hallifax, D., Obach, S.R., and Houston, B.J. (2005) Impact of parallel pathways of drug elimination and multiple cytochrome P450 involvement on drug–drug interactions: CYP2D6 paradigm. *Drug Metab. Dispos.*, **33**, 837–844.

67 Moore, M.R. (2004) A commentary on the impacts of metals and metalloids in the environment upon the metabolism of drugs and chemicals. *Toxicol. Lett.*, **148**, 153–158.

68 Kadiiska, M. and Stoytchev, T. (1980) Effect of acute intoxication with some heavy metals on drug metabolism. *Arch. Toxicol.*, (Suppl. 4), **4**, 363–365.

69 Ke, Q., Yang, Y., Ratner, M., Zeind, J., Jiang, C., Forrest, J.N. Jr., and Xiao, Y.F. (2002) Intracellular accumulation of mercury enhances P450 CYP1A1 expression and Cl-currents in cultured shark rectal gland cells. *Life Sci.*, **70**, 2547–2566.

70 Alexidis, A.N., Rekka, E.A., and Kourounakis, P.N. (1994) Influence of mercury and cadmium intoxication on hepatic microsomal CYP2E and CYP3A subfamilies. *Res. Commun. Mol. Pathol. Pharmacol.*, **85**, 67–72.

71 Albores, A., Sinal, C.J., Cherian, M.G., and Bend, J.R. (1995) Selective increase of rat lung cytochrome P450 1A1 dependent monooxygenase activity after acute sodium arsenite administration. *Can. J. Physiol. Pharmacol.*, **73**, 153–158.

72 Phillips, J.C., Lake, B.G., Gangolli, S.D., Grasso, P., and Lloyd, A.G. (1977) Effects of pyrazole and 3-amino-1,2,4-triazole on the metabolism and toxicity of dimethylnitrosamine in the rat. *J. Natl. Cancer Inst.*, **58**, 629–633.

73 Garro, A.J., Seitz, H.K., and Lieber, C.S. (1981) Enhancement of dimethylnitrosamine metabolism and activation to a mutagen following chronic ethanol consumption. *Cancer Res.*, **41**, 120–124.

74 Pelkonen, O. and Sotaniemi, E. (1982) Drug metabolism in alcoholics. *Pharmacol. Ther.*, **16**, 261–268.

75 Peng, R., Tu, Y.Y., and Yang, C.S. (1982) The induction and competitive inhibition of a high affinity microsomal nitrosodimethylamine demethylase by ethanol. *Carcinogenesis*, **3**, 1457–1461.

76 Miller, K.W. and Yang, C.S. (1984) Studies on the mechanisms of induction of *N*-nitrosodimethylamine demethylase by fasting, acetone, and ethanol. *Arch. Biochem. Biophys.*, **229**, 483–491.

77 Swann, P.F., Coe, A.M., and Mace, R. (1984) Ethanol and dimethylnitrosamine and diethylnitrosamine metabolism and disposition in the rat. Possible relevance to the influence of ethanol on human cancer incidence. *Carcinogenesis*, **5**, 1337–1343.

78 Erill, S. (1987) Discussion on the paper "Enzyme inhibition by environmental agents" by EE Ohnhaus, in *Interaction between Drugs and Chemicals in Industrial Societies* (eds S. Erill, P. du Souich, and G.L. Plaa), Elsevier, Amsterdam, p. 69.

79 Stevens, J.T., Stitzel, R.E., and McPhillips, J.J. (1972) Effects of anticholinesterase insecticides on hepatic microsomal metabolism. *J. Pharmacol. Exp. Ther.*, **181**, 576–583.

80 Stevens, J.T., Stitzel, R.E., and McPhillips, J.J. (1972) The effects of subacute administration of anticholinesterase insecticides on hepatic microsomal metabolism. *Life Sci.*, **11**, 423–431.

81 Stevens, J.T., Greene, F.E., and Passanati, G.T. (1973) Binding of malathion, parathion and their oxygenated analogs to cytochrome P-450 in the rat. *Fed. Proc.*, **32**, 761.

82 Poulin, P. and Krishnan, K. (1999) Molecular structure-based prediction of the toxicokinetics of inhaled vapors in humans. *Int. J. Toxicol.*, **18**, 7–18.

83 Emond, C., Charbonneau, M., and Krishnan, K. (2005) Physiologically based modeling of the accumulation in plasma and tissue lipids of a mixture of PCB congeners in female Sprague-Dawley rats. *J. Toxicol. Environ. Health A*, **68**, 1393–1412.

84 Andersen, M.E., Gargas, M.L., Clewell, H.J. III, and Severyn, K.M. (1987) Quantitative evaluation of the metabolic interactions between trichloroethylene and 1,1-dichloroethylene *in vivo* using gas uptake methods. *Toxicol. Appl. Pharmacol.*, **89**, 149–157.

85 Kudsk, F.N. (1965) The influence of ethyl alcohol on the absorption of mercury vapour from the lungs in man. *Acta Pharmacol. Toxicol.*, **23**, 263–274.

86 Reynolds, K.E., Whitford, G.M., and Pashley, D.H. (1978) Acute fluoride toxicity: the influence of acid–base status. *Toxicol. Appl. Pharmacol.*, **45**, 415–427.

87 Zeng, H., Uthus, E.O., and Combs, G.F. Jr., (2005) Mechanistic aspects of the interaction between selenium and arsenic. *J. Inorg. Biochem.*, **99**, 1269–1274.

88 Elsenhans, B., Forth, W., and Richter, E. (1991) Increased copper concentrations in rat tissues after acute intoxication with 2,3,7,8-tetrachlorodibenzo-*p*-dioxin. *Arch. Toxicol.*, **65**, 429–432.

89 Kamstra, L.D. and Bonhorst, C.W. (1953) Effect of arsenic on the expiration of volatile selenium compounds by rats. *Proc. S. Dak. Acad. Sci.*, **32**, 72–75.

90 Ganther, H.E. and Baumann, C.A. (1962) Selenium metabolism. I. Effects of diet, arsenic and cadmium. *J. Nutr.*, **77**, 210–216.

91 Guiney, P.D., Yang, K.H., Seymour, J.L., and Peterson, R.E. (1978) Effects of 2,3,7,8-tetrachlorodibenzo-*p*-dioxin on the distribution and biliary excretion of polychlorinated biphenyls in rats. *Toxicol. Appl. Pharmacol.*, **45**, 403–414.

92 Ashby, S.L., King, L.J., and Parke, D.V. (1981) The effect of cadmium administration on the biliary excretion of copper and zinc and tissue disposition of these metals. *Environ. Res.*, **26**, 95–104.

93 Bernard, A.M. and Lauwerys, R.R. (1981) The effects of sodium chromate and carbon tetrachloride on the urinary excretion and tissue distribution of cadmium in cadmium-pretreated rats. *Toxicol. Appl. Pharmacol.*, **57**, 30–38.

94 Komsta-Szumska, E., Reuhl, K.R., and Miller, D.R. (1983) Effect of selenium on distribution, demethylation, and excretion of methylmercury by the guinea pig. *J. Toxicol. Environ. Health*, **12**, 775–785.

95 Scheufler, E. and Rozman, K. (1984) Effect of hexadecane on the pharmacokinetics of hexachlorobenzene. *Toxicol. Appl. Pharmacol.*, **75**, 190–197.

96 Jasim, S. and Tjalve, H. (1986) Effect of zinc pyridinethione on the tissue disposition of nickel and cadmium in mice. *Acta Pharmacol. Toxicol.*, **59**, 204–208.

97 Tamashiro, H., Arakaki, M., Akagi, H., Murao, K., Hirayama, K., and Smolensky, M.H. (1986) Effects of ethanol on methyl mercury toxicity in rats. *J. Toxicol. Environ. Health*, **18**, 595–605.

98 Jones, M.M., Cherian, M.G., Singh, P.K., Basinger, M.A., and Jones, S.G. (1991) A comparative study of the influence of vicinal dithiols and a dithiocarbamate on the biliary excretion of cadmium in rat. *Toxicol. Appl. Pharmacol.*, **110**, 241–250.

99 Zalups, R.K. and Barfuss, D.W. (2002) Simultaneous coexposure to inorganic mercury and cadmium: a study of the renal and hepatic disposition of mercury and cadmium. *J. Toxicol. Environ. Health A*, **65**, 1471–1490.

100 Simmons, J.E. (1996) Application of physiologically based pharmacokinetic modelling to combination toxicology. *Food Chem. Toxicol.*, **34**, 1067–1073.

101 Nagata, O., Murata, M., Kato, H., Terasaki, T., Sato, H., and Tsuji, A. (1990) Physiological pharmacokinetics of a new muscle-relaxant, inaperisone, combined with its pharmacological effect on blood flow rate. *Drug Metab. Dispos.*, **18**, 902–910.

102 El-Masri, H.A., Tessari, J.D., and Yang, R.S. (1996) Exploration of an interaction

threshold for the joint toxicity of trichloroethylene and 1,1-dichloroethylene: utilization of a PBPK model. *Arch. Toxicol.*, **70**, 527–539.

103 Dobrev, I.D., Andersen, M.E., and Yang, R.S. (2002) *In silico* toxicology: simulating interaction thresholds for human exposure to mixtures of trichloroethylene, tetrachloroethylene, and 1,1,1-trichloroethane. *Environ. Health Perspect.*, **110**, 1031–1039.

104 Tardif, R., Lapare, S., Charest-Tardif, G., Brodeur, J., and Krishnan, K. (1995) Physiologically-based pharmacokinetic modeling of a mixture of toluene and xylene in humans. *Risk Anal.*, **15**, 335–342.

105 Tardif, R., Lapare, S., Krishnan, K., and Brodeur, J. (1993) Physiologically based modeling of the toxicokinetic interaction between toluene and *m*-xylene in the rat. *Toxicol. Appl. Pharmacol.*, **120**, 266–273.

106 Diamond, G.L., Goodrum, P.E., Felter, S.P., and Ruoff, W.L. (1998) Gastrointestinal absorption of metals. *Drug Chem. Toxicol.*, **21**, 223–251.

107 Satarug, S., Nishijo, M., Lasker, J.M., Edwards, R.J., and Moore, M.R. (2006) Kidney dysfunction and hypertension: role for cadmium, P450 and heme oxygenases? *Tohoku J. Exp. Med.*, **208**, 179–202.

108 Degawa, M., Arai, H., Miura, S., and Hashimoto, Y. (1993) Preferential inhibitions of hepatic P450IA2 expression and induction by lead nitrate in the rat. *Carcinogenesis*, **14**, 1091–1094.

109 Degawa, M., Arai, H., Kubota, M., and Hashimoto, Y. (1994) Ionic lead, a unique metal ion as an inhibitor for cytochrome P450IA2 (CYP1A2) expression in the rat liver. *Biochem. Biophys. Res. Commun.*, **200**, 1086–1092.

110 Satarug, S., Nishijo, M., Ujjin, P., Vanavanitkun, Y., Baker, J.R., and Moore, M.R. (2004) Evidence for concurrent effects of exposure to environmental cadmium and lead on hepatic CYP2A6 phenotype and renal function biomarkers in nonsmokers. *Environ. Health Perspect.*, **112**, 1512–1518.

111 Satarug, S., Ujjin, P., Vanavanitkun, Y., Nishijo, M., Baker, J.R., and Moore, M.R. (2004) Effects of cigarette smoking and exposure to cadmium and lead on phenotypic variability of hepatic CYP2A6 and renal function biomarkers in men. *Toxicology*, **204**, 161–173.

112 Seubert, J.M., Sinal, C.J., and Bend, J.R. (2002) Acute sodium arsenite administration induces pulmonary CYP1A1 mRNA, protein and activity in the rat. *J. Biochem. Mol. Toxicol.*, **16**, 84–95.

113 Maier, A., Dalton, T.P., and Puga, A. (2000) Disruption of dioxin-inducible phase I and phase II gene expression patterns by cadmium, chromium, and arsenic. *Mol. Carcinog.*, **28**, 225–235.

114 Maier, A., Schumann, B.L., Chang, X., Talaska, G., and Puga, A. (2002) Arsenic co-exposure potentiates benzo[*a*] pyrene genotoxicity. *Mutat. Res.*, **517**, 101–111.

115 Vakharia, D.D., Liu, N., Pause, R., Fasco, M., Bessette, E., Zhang, Q.Y., and Kaminsky, L.S. (2001) Polycyclic aromatic hydrocarbon/metal mixtures: effect on PAH induction of CYP1A1 in human HEPG2 cells. *Drug Metab. Dispos.*, **29**, 999–1006.

116 Vakharia, D.D., Liu, N., Pause, R., Fasco, M., Bessette, E., Zhang, Q.Y., and Kaminsky, L.S. (2001) Effect of metals on polycyclic aromatic hydrocarbon induction of CYP1A1 and CYP1A2 in human hepatocyte cultures. *Toxicol. Appl. Pharmacol.*, **170**, 93–103.

117 Sugita, O., Sawada, Y., Sugiyama, Y., Iga, T., and Hanano, M. (1984) Kinetic analysis of tolubutamide-sulfonamide interaction in rabbits based on clearance concept. *Drug Metab. Dispos.*, **12**, 131–138.

10
Chemical Mixtures and Cumulative Risk Assessment

John C. Lipscomb, Jason C. Lambert, and Linda K. Teuschler

10.1
Introduction

Humans are voluntarily and involuntarily exposed daily to chemicals, including early exposures *in utero* and in breast milk. Chemicals are important for everyday life and can have both positive and negative impacts, such as the amelioration of symptoms of sickness and the curing of diseases and the induction of toxicity and pathological conditions, respectively. For chemical mixture exposures, the type of joint toxic action that may be observed is a function of exposure, and sometimes the type and/or severity of the effect is modulated by the coexposure to other chemicals. Because the result depends on the interaction between the chemical and the biological/molecular target at the cellular level, interactions between chemicals that alter the disposition of the chemical (toxicokinetics, TK) may alter its molecular interactions. In addition, chemical exposure that alters the molecular receptor or molecular or cellular processes that are related to its biochemical function (toxicodynamics, TD) may alter the response. These scenarios may result in a combined chemical mixture risk from joint toxic action that may be recognized as resulting from additivity (i.e., dose additive or response additive), or from interaction effects (i.e., greater than additive, referred to as synergism, or less than additive, referred to as antagonism). The choices to be carefully considered are (1) whether or not it is important to undertake a mixtures risk assessment (MRA) and (2) among the MRA approaches available, which method is most suitable for the risk assessment problem at hand.

There are "real-world" issues that need to be acknowledged in decisions concerning whether to conduct a mixtures risk assessment. Contaminated sites offer an example through which to present some of these issues. When chemical concentrations and/or exposures are very low (under anticipated no-effect levels in humans), performing an MRA may represent a rather conservative approach, inasmuch as joint toxic action is generally expected to occur at higher, rather than at lower, concentrations/exposures. In such cases, when an assessment is designed to screen for hazards, the application of MRA methods can identify a worst-case scenario. When this results in a prediction of no or negligible risk, public confidence is maintained,

Principles and Practice of Mixtures Toxicology. Edited by Moiz Mumtaz

ISBN: 978-3-527-31992-3

and this information can be factored into risk management (e.g., remediation) decisions. Alternatively, an assessment may be conducted to give a more representative (i.e., central tendency) measure of risk. In this case, a mixtures approach is chosen to produce the most accurate risk estimate for the exposure, resulting in estimates that are health protective on average.

A more complex form of addressing multiple chemical exposures is to conduct a cumulative risk assessment (CRA). CRAs are akin to MRAs, but may be differentiated on the basis of their coverage of population variability as applied to aggregate and multiroute exposures to multiple chemicals from multiple sources of pollutants. MRA and CRA efforts can consume appreciable resources and should be undertaken judiciously. The International Life Science Institute's Risk Science Institute conducted a workshop in which a framework for CRA was developed [1]. Importantly, this group addressed the question, "When should a CRA be done?" That report indicates that several conditions, which also apply to the conduct of an MRA, should be evaluated, namely,

- whether multiple chemicals share the same mechanism of toxicity;
- whether multiple chemicals share common biological and/or molecular events;
- whether multiple chemicals cause toxicity by acute or chronic exposures;
- the comparative potency of multiple chemicals;
- whether there is exposure to multiple chemicals;
- the timing of exposures to multiple chemicals;
- the routes of exposure for multiple chemicals; and
- the spatial scale of exposure to multiple chemicals.

The US EPA has provided leadership in this area by conducting a considerable amount of research on CRA and developing documents that provide methods, data sources, and guidance. Its 2003 *Framework for Cumulative Risk Assessment* and earlier reports on the initial planning and scoping phase needed to conduct a CRA laid a broad foundation for continued development of CRA approaches [2–4]. The 2003 Framework describes some basic considerations for conducting a CRA and outlines four areas of population vulnerability, specifically susceptibility or sensitivity, differential exposure (e.g., living in close proximity to pollutant sources), differential preparedness (e.g., lack of disease immunizations), and differential ability to recover from exposures (e.g., poor nutrition). The US EPA has published technical documents and guidelines that address MRAs [5, 6], contaminated site assessments [7], and multiple pathway exposures [8], which can be used to support a CRA, and has conducted additional research to address MRAs in combination with multiple exposures, effects, and exposure durations [9]. In addition, under Food Quality Protection Act (FQPA) of 1996, the EPAs Office of Pesticide Programs has developed methods, guidance, and assessments for the cumulative risk of pesticides that act by a common mechanism of toxicity [10–13].

In addition to the scientific issues that drive a decision to undertake an MRA or CRA (diversity and magnitude of chemical exposure, similarity of health effects from mixture components, relationships of modes of action among components, etc.), there are political and legal drivers as well. Important laws have been established on an international scale, and the actions mandated are to be undertaken independent of the type or extent of scientific information.

- **Comprehensive Environmental Response, Compensation, and Liability Act (CERCLA):** More than two decades ago, the US Congress [14] passed this law defining pollutants or contaminants as any substance, compound, or mixture that, when released to the environment, humans may be exposed to either directly or through the ingestion of contaminated food.
- **Canada's 1992 Environmental Assessment Act:** This law considers cumulative environmental effects and the effects of changes to the biophysical environment on health and socioeconomic conditions, physical and cultural heritage, and other environmental effects.
- **Food Quality Protection Act:** Passed in 1996 [15], this law requires the US EPA to consider the available evidence on pesticides and residues (chemicals) in foods in the assessment of the cumulative risks from multiple route exposures to these substances in combination with other substances that "have a common mechanism of toxicity." A critical component of the regulated activities also covered developing nonoccupational exposure estimates for the population, especially groups (e.g., children) that may be susceptible to toxicities.
- **Safe Drinking Water Act Amendments of 1996:** Congress [16] passed a series of amendments to the Safe Drinking Water Act that required the US EPA to develop new approaches to study the effects of complex mixtures of chemicals such as those found in drinking water. Key in the verbiage is the requirement to evaluate the likelihood of synergistic or antagonistic interactions among mixture components that may alter the dose-response relationship from what may have been expected from additive effects predicted from data on single-chemical exposures. These requirements also call for a population-based approach, emphasizing the evaluation of population subgroups deemed potentially susceptible.
- **Registration, Evaluation, Authorization, and Restriction of Chemical Substances (REACH) :** [17]: A European Community regulation was passed on chemicals and their safe use. This law prompted the formation of research efforts on CRA under the name, NoMiracle (Novel Methods for Integrated Risk Assessment of Cumulative Stressors in Europe).
- **Consumer Product Safety Improvement Act of 2008:** This US law mandated consideration of the cumulative effect of total exposure to phthalates, both from children's products and from other sources, such as personal care products.

Once a decision to proceed has been documented, the choice of MRA methods becomes the next step. There are two general types of approaches to MRA (Figure 10.1), whole mixture and component-based methods. Whole mixtures approaches treat the mixture as a single chemical entity and require data on the mixture of concern or a mixture that is sufficiently similar to the mixture of concern. Once appropriate whole mixture dose-response data are identified, they are evaluated and the results extrapolated to produce a reference value and/or a cancer risk estimate (slope factor and unit risk). These values are combined with human exposure

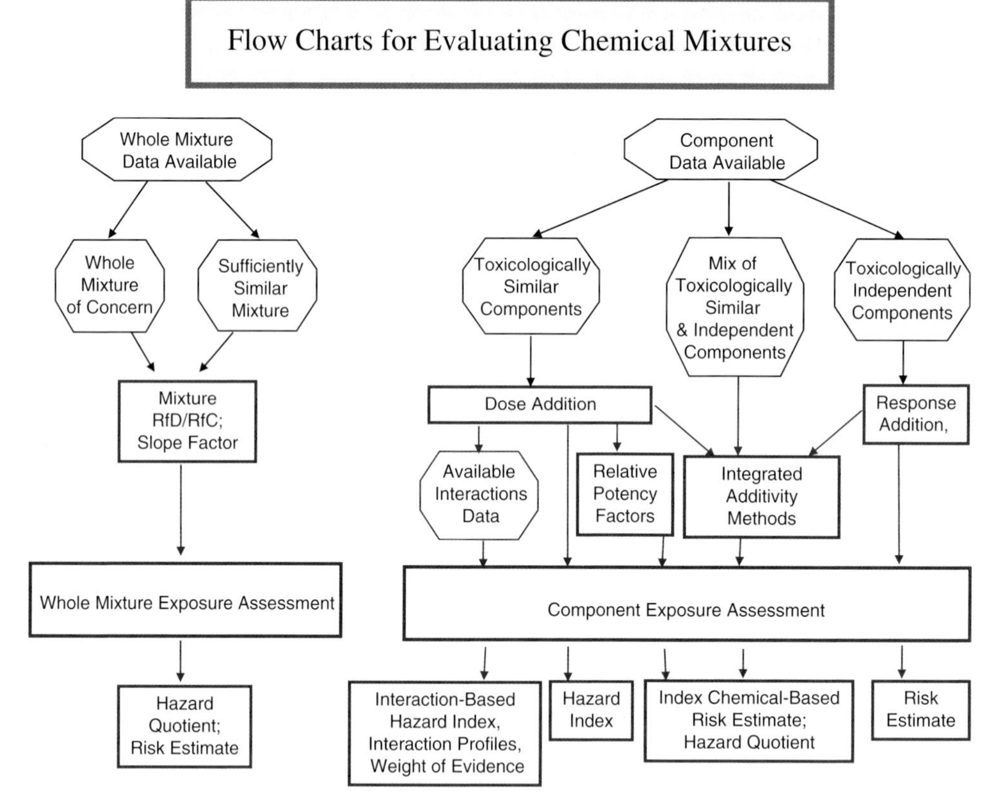

Figure 10.1 Availability of data drives the selection of the best approach to mixture toxicity.

estimates to characterize the risk. Although these procedures appear to be straightforward, an important area of uncertainty is the chemical composition of the complex mixture to which humans are exposed. Chemical mixtures often degrade or transform as they travel through environmental media and may be quite different in chemical composition from the original mixture released into the environment. Thus, when a reference value or cancer risk estimate based on data from a laboratory bioassay is applied to an environmental mixture, an effort should be made to evaluate the chemical similarity of the two mixtures and provide an evaluation of the confidence in the assessment.

There are three general component-based approaches, and the choice among them depends on the extent to which the toxic modes of action among components can be described as similar or independent. Within a mixture, compounds may act through similar or independent modes of action; in a given mixture, components may act through the same mode of action (MOA), through independent modes of action, or there may be a mix of components acting through a similar MOA and component(s) that act through unique (independent) modes of action. While this may sound complex, the three approaches may be easily followed.

For components acting through a similar mode of action, dose additive approaches are employed. Under dose additivity, data describing interactions (i.e., departures from dose additivity) among the components are evaluated to ascertain whether dose additive assumptions are justified. If so, then the relative potency of individual components is determined; response to the mixture (or group) is determined by summing the potency-adjusted doses of components. The approach to determining the degree of similarity of mode of action, and a complete description of the dose additivity approach, is covered later in this chapter.

The second component-based approach is that of response addition (RA). This approach is used to combine the likelihood of responses for components that cause the same general health outcome but act through independent modes of action. Here, the responses for the individual components are determined and the response for the mixture is estimated by summing the responses estimated for each of the components. This approach is also covered in more details later.

The third type of the component-based approach is characterized as integrated additivity. This may be considered a hybrid additivity approach that applies to mixtures whose components cause the same health outcome but where some components act via a similar mechanism and other components act via independent modes of action. In this case, the chemicals may be organized into common mode of action subgroups of one or more chemicals where the subgroups represent modes of action distinctly different from each other. Dose addition (DA) is then applied within the common mode of action subgroups to estimate risk and then these subgroup risks are added under RA because the subgroups represent independent modes of action.

These approaches are selected on the basis of several factors. While whole mixtures approaches may be preferred, data may limit their applicability. Seldom are data available for the mixture of concern, necessitating evaluation of surrogate mixtures. For a mixture to be selected as a surrogate, it must be deemed to be sufficiently similar – toxicologically and chemically indistinguishable from the mixture of interest. However, to make that justification, some data must be available to (1) estimate the chemistry and potential toxicity of the mixture of concern and/or of its major components, and (2) provide a reliable characterization of the toxicity of the potential surrogate mixture and/or its major components, most of which should overlap with those of the mixture of concern. Identifying a sufficiently similar mixture is an important issue in selecting an MRA approach.

Failing the identification of applicable whole mixtures data, component based approaches must be considered. When mixtures contain a large number of components, identifying appropriate qualitative toxicity information and dose-response data for each component becomes an issue. For components, mode of action data should be evaluated not only to determine the choice between dose or response (or integrated) approaches but also to serve as the basis for chemical grouping. The number of components, and the potentially limiting data for those components, is an important point – when the mixture contains a large enough number of components and/or the data for components is limited, uncertainty in the risk estimated from additivity approaches may limit success. When the uncertainty becomes excessive, this defines the demarcation between simple mixtures and complex mixtures. This condition clouds the justification of approaches, may warrant risk estimation via several methods, and may invoke policy-based decisions (e.g., health conservatism) to identify the most appropriate MRA approach. Finally, dose-dependent transitions in toxic effects and modes of action [18] should also be considered. The toxicities seen at lower or higher doses than those on which regulatory levels are based are accounted for in the target organ toxicity dose (TTD) method, which is a type of hazard index (HI) approach (discussed later). In addition, the number of modes of action may be increased at higher doses, whereas humans may be expected to be exposed to lower (more environmentally relevant) doses. The likelihood of toxicological interactions is recognized to be lower at lower doses, and some investigators have determined thresholds for chemical interactions based largely on competition for metabolism [19].

10.2
Toxicology Basis for Mixtures and Cumulative Risk Assessment

A major source of variability and uncertainty in CRA and MRA is the lack of information relating exposure doses to internal doses and the internal dose metric to toxicity, in particular toxic mode or mechanism of action data. This is primarily due to the limited number of studies that explore mixtures toxicity, thus forcing component-based approaches that require predictions based on single chemical data. However, mixtures of chemicals (and biota) do not occur in a vacuum; therefore, the potential risk of adverse health effects from cumulative exposures in humans is difficult to predict. As such, the importance of characterizing biological activity (i.e., mode or mechanism of action) in organisms exposed to myriad xenobiotics cannot be overstated, and indeed is not new to risk or safety assessment.

A persistent limiting factor in risk assessment in general, and in CRA in particular, is defining a common language for the qualitative interpretation and subsequent integration of mode of action data in quantitative assessment and uncertainty analysis. In particular, government agencies, industry, and academic institutes use the terms "mode" and "mechanism" interchangeably. While this may not be so much a concern in basic research or didactic environments, it is critical for those applying the National Research Council's risk assessment paradigm [20]. Mode and mechanism of toxic action are mutually supportive concepts that commonly differ only in the level of biological detail used as descriptors regarding the processes involved in toxicity. A toxic mode of action, as defined by the US EPA, is a series of "key" events and processes starting with interaction of an agent with a cell, and proceeding through operational and anatomical changes leading to disease, whereas a toxic mechanism of action implies a more detailed description and understanding of biological events, often at the molecular level, than is presented in a MOA [6, 21]. Thus, if the complete mechanism of action is known for a given environmental hazard, then the complete MOA is known; however, the reverse cannot be said. A postulated MOA may entail as few as one or two key events up to a fully characterized mechanistic pathway critical in the toxic phenotype.

Prior to the -omics generation of exploratory research, identification of critical molecular and cellular events involved in some toxic phenotype was often limited in scope and output. That is, the dose-response and temporal aspects of signal transduction pathways involved in a given toxic effect were either unknown or poorly characterized. With the advent of more high-throughput platforms such as toxicogenomics, proteomics, and metabolomics, the prototypical "blackbox" of toxic mode or mechanism of action in risk assessment of environmental contaminants is becoming increasingly less cryptic. However, as more information becomes available from these high-throughput assays, identification of key events will be crucial as this aspect of data interpretation will ultimately influence MOA grouping and subsequent quantitative cumulative mixtures risk predictions. Only the obligatory steps requisite for a toxicity of concern would be further considered in a postulated MOA grouping, in particular those key events that are related concordantly with dose-response and/or temporal behavior of the biological system. It should be noted that focusing on key events does not suggest disregard of those biological details that do not merit inclusion into a particular postulated MOA. Rather, in CRA considering all data, mechanistic and otherwise, is necessary such that potential overlap of the effects caused by two or more chemicals may be evaluated more effectively. Specifically, potential commonalities among mechanistic pathways of mixture components, in addition to key obligatory events, may lead to identification of biological interactions not anticipated from an evaluation of traditional MOA descriptions developed from data on single-chemical exposures.

Mixture components are typically grouped according to whether their modes of action are similar or independent. This segregation defines the additivity approaches used in combining the individual risks from cumulative mixture components, and is perhaps the least objective part of the process.

10.2.1 Mode of Action Characterization

The International Program on Chemical Safety (IPCS) has developed a framework for evaluating mode of action for single chemicals, but it also contains general principles useful for MRA. The framework is primarily directed at MOA analysis for cancer risk assessment and is intended to foster harmonization [22]. The conceptual framework presents an analytical approach that culminates in a form of weighing the evidence for carcinogenic modes of action. The framework is not a checklist and it was not intended to give an absolute answer to what is sufficient information, in that minimum requirements will vary depending on the situation. The 10-step process centers on the identification of key molecular/biochemical events for tumorigenesis, but it has been somewhat broadened here.

- **Introduction:** provides a description of the health end point of concern.
- **Postulated Mode of Action:** comprises a brief description of the sequence of events for the postulated mode of action.

- **Key Events:** presents the body of literature describing measurable events (key events) that are critical to the induction of the health effect.
- **Dose-Response Relationships:** the extent to which the dose-response for key events parallels the dose-response relationship for the health effect should be presented.
- **Temporal Association:** the temporal relationships between the key events and the appearance of the health effect should be presented.
- **Strength, Consistency, and Specificity of Association of Response with Key Events:** a weight of evidence description should be developed that links key events, precursor lesions, and the health effect.
- **Biological Plausibility and Coherence:** consideration must be given to whether the postulated mode of action does conflict with what is known about the health effect or about other effects for the chemical of concern. Given an often limited amount of data, knowledge gaps in this area should not be viewed as success-limiting.
- **Other Modes of Action:** when alternative modes of action become evident, they should also be given consideration.
- **Association of Postulated Modes of Action:** the outcome of the analysis should be presented along with a statement of the level of confidence placed in the results.
- **Uncertainties, Inconsistencies, and Data Gaps:** uncertainties in the data for the health effect and the chemical of concern should be identified; areas representing data gaps, and for which the development of data would be most beneficial should be identified.

The US EPA [21] has also addressed mode of action characterized in its guidelines for carcinogen risk assessment; the guidelines are generally similar to those presented by IPCS [23], above. Here, emphasis is also placed on understanding both the health effect (the toxicity) itself and the description of the toxicity produced by the chemical. The fit between what is known about the biology of the response and the specific nature of chemical insult provide a level of confidence in the developed mode of action. Under analyses conducted for carcinogen risk assessment, the US EPA [21] notes that multiple modes of action may be involved and that modes of action may be dose-dependent, and indicates that the dose-response relationship for each of the modes of action should be considered. The US EPA [21] and the IPCS [24] have each addressed the need to ascertain the relevance of mode of action in animal species to the human for carcinogen risk assessment. The IPCS strategy for determining human relevance comprises the following questions:

1) Is the weight of evidence sufficient to establish a mode of action in animals?
2) Can human relevance of the MOA be reasonably excluded on the basis of fundamental, qualitative differences in key events between animals and humans?
3) Can human relevance of the MOA be reasonably excluded on the basis of quantitative differences in either kinetic or dynamic factors between animals and humans?

Those strategies are essentially an extension of the strategy of IPCS to determine the mode of action [22]. Extending confidence in the mode of action observed in animals to the human increases confidence in the risk assessment for mixtures of chemicals.

10.2.2
Chemical Grouping Strategy

One common misconception is that CRA is restricted to chemicals producing the same toxic end point via a common mechanism of toxicity.[1] This is perhaps due to the manner in which Congress has mandated the US EPA's Office of Pesticide Programs to approach pesticide risk assessment under FQPA. More broadly, CRA can include chemicals that act both through similar and through independent modes of action. Thus, the grouping of chemicals according to toxicological similarity can heavily influence the outcome of CRA.

The grouping of chemicals is a key component of MRA and CRA, and is determined by the degree of similarity or independence in the manner in which toxicity is produced. The type and depth of information may limit the certainty of grouping of chemicals for MRA or CRA. Risk assessors may first think of data in the quantitative sense (e.g., dose-response data). However, qualitative data, such as those describing mode or mechanism, are also critical determinants.

These qualitative data determine the groupings of chemicals and the mathematical models used to combine and extrapolate estimates of response. While there is to date no government-endorsed objective method by which to document or defend similarity or independence of mode of action, it has been recommended [25] that at minimum, the chemicals share a common mechanism if they

- cause the same critical effect;
- act on the same molecular target at the same target tissue; and
- act by the same biochemical mechanism of action or share a common toxic intermediate.

This definition has been codified and used as the basis for a CRA by the US EPA [3], and addresses chemical grouping: "Common Mechanism of Toxicity pertains to two or more pesticide chemicals or other substances that cause a common toxic effect(s) by the same, or essentially the same, sequence of major biochemical events (i.e., interpreted as mode of action)."

Subsequently, a cumulative risk assessment was published to assess 24 organophosphate (OP) insecticides based on dose addition using a relative potency factor (RPF) approach [11]. The OP assessment quotes the ILSI work, "OPs exert their neurotoxicity by binding to and phosphorylating of the enzyme acetylcholinesterase in both the central (brain) and the peripheral nervous systems [23]." The US EPA's Office of Pesticide Programs ultimately conducted its cumulative risk assessment using female rat brain cholinesterase inhibition data. These data represented a direct

1) Cumulative risk has been defined as the risk of a common toxic effect associated with concurrent exposure by all relevant pathways and routes of exposure to a group of chemicals that share a common mechanism of toxicity. http://www.epa.gov/pesticides/glossary/index.html#cumulative.

measure of the common mechanism of toxicity and were similar in toxic potency to red blood cell cholinesterase inhibition for oral, dermal, and inhalation exposures [11].

To assist in the development of CRA undertaken for pesticides in response to the Food Quality Protection Act of 1996, the US EPA has developed some guidance for grouping pesticides and other chemicals according to similarity of mechanism of action [10, 13]. This approach was subsequently adopted by Health Canada [26]. There are at least three key concepts in this approach – common toxic effect, mechanism of toxicity, and common mechanism of toxicity.

- **Common Toxic Effect:** A pesticide and another substance that are known to cause the same toxic effect in or at the same anatomical or physiological site or locus (e.g., same organ or tissue) are said to cause a common toxic effect. Thus, a toxic effect observed in studies involving animals or humans exposed to a pesticide chemical is considered common with a toxic effect caused by another chemical if there is concordance with both site and nature of the effect [10].
- **Mechanism of Toxicity:** Mechanism of toxicity is defined as the major step leading to a toxic effect following interaction of a pesticide with biological targets. All steps leading to an effect do not need to be specifically understood. Rather, it is the identification of the crucial events, following chemical interaction, that are required in order to describe a mechanism of toxicity. In general, the more that is understood about the various steps in the pathway leading to an adverse effect, the more confident one is about the mechanism of toxicity. For instance, a mechanism of toxicity may be described by knowing the cascade of effects such as the following: a chemical binds to a given biological target *in vitro* and causes the receptor-related molecular response; *in vivo* it also leads to the molecular response and causes a number of intervening biological and morphological steps that result in an adverse effect. Other processes may describe a mechanism of toxicity in other cases [25].
- **Common Mechanism of Toxicity:** Common mechanism of toxicity pertains to two or more pesticide chemicals or other substances that cause a common toxic effect to human health by the same, or essentially the same, sequence of major biochemical events. Hence, the underlying basis of the toxicity is the same, or essentially the same, for each chemical [10].

The five-step process (Figure 10.2) begins with two broadly inclusive steps. This is done to avoid overlooking any chemical that should be evaluated. The grouping of chemicals according to the first two steps is not sufficient for application in a CRA; additional exercises described in steps 3–5 must be undertaken. Step 1 is a preliminary grouping of chemicals, which can be based on known similarities in (a) mechanism of pesticidal action, (b) pharmacokinetics, (c) general mechanisms of mammalian toxicity, or (d) a particular toxic effect. Toxic effects that may have a general or ill-defined basis, such as death or decreased body weight, are not appropriate for chemical grouping. Substances that are metabolic precursors to chemicals preliminarily grouped together should also be included in the group.

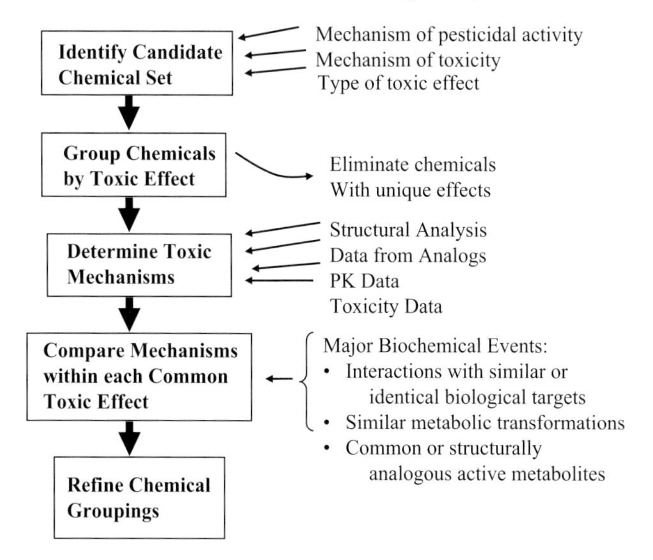

Figure 10.2 The process developed by the US EPA [10] to evaluate commonality of mechanism of action for pesticides evaluated under the Food Quality Protection Act of 1996.

Step 2 is the definitive identification of chemicals causing a common toxic effect. Because chemicals can produce a spectrum of effects depending on dose, all effects caused by the chemical, regardless of dose should be identified. When more than one toxic effect is observed, the chemical will be placed into multiple groups according to the toxic effects produced.

Step 3 is the determination of the mechanism of action. The US EPA [10] describes mechanism of action as the most crucial biochemical interactions leading to toxicity. Postulated or partially described mechanisms, as well as more well-developed descriptions of mechanisms, should be examined. This examination should include (a) a structural analysis and consideration of toxicity data for structural analogues, (b) available pharmacokinetic and metabolic data, and (c) toxicity data. This information is compiled and a comparison is made with the information available for each chemical (step 4). With the identification of commonalities in mechanisms of action, chemicals with a common mechanism of action are retained in the group; chemicals for which the mechanisms are not common are eliminated (step 5). With the process completed, chemicals in the common mechanism group (CMG) are forwarded for inclusion in the CRA.

Regardless of their definitions, delineating similarities or independence among the biological activities of mixtures components is a key process. Lambert and Lipscomb [27] have treated the issue by expanding the approach to include assessing the degree to which mixture components affect biological functions, regardless of whether the biological functions comprise toxic mode or mechanism of action. Under examples presented, effects that may prolong the biological longevity of biomolecules important in the toxicity of another component may logically be viewed as constituting a similarity – clearly not an "independence." When data are insufficient to defend a particular position, the components are typically assumed to act via independent modes of action.

Chemical and/or biological interactions between xenobiotics may lead to biological responses nonconcordant with the dose of a given component or the whole mixture estimated on the basis of additivity models. That is, when in mixture, chemicals may influence the relative biological activity of one another within a given target tissue. When the net response of a mixture is more or less than that anticipated using a specified assumption of additivity (e.g., response addition or dose addition), an interaction is indicated. These joint toxic actions may be characterized either as additive or as interactions (i.e., synergistic or antagonistic). Additive effects are those that can be adequately explained by the addition of either doses or responses. Synergistic responses are those exceeding that predicted by simple addition; this includes interactions that may also be characterized as potentiation. Antagonistic interactions are those in which the response is less than predicted by additivity. The nature of interactions of mixture components is exceedingly difficult to characterize but may be attributable to alterations in toxicokinetics and/or toxicodynamics. Often, interactions are related to changes in some aspect of the ADME of mixture components, which is more a liability of the available data than the true nature of the interaction(s). Toxicokinetic interactions between components of mixtures, when those components compete for metabolism, have shown dose thresholds, below which competition and toxicokinetic interactions do not occur [28, 29]. In contrast, the toxicodynamic events induced by chemicals in target organs are often so poorly characterized that no mode or mechanism of toxic action can be confidently postulated for a given mixture. Outside of a fully complimented biologically based dose-response (BBDR) model, there is no clear remedy or simplified alternative available to address mixtures' interactions. Because additive models are the models used for mixtures risk assessment in the absence of data that contraindicate their application, they will be discussed in some detail in subsequent sections.

10.3
Mixtures and Cumulative Risk Assessment Methods

MRA is the field of study that examines and predicts the responses following exposure to multiple chemicals. The composition of the mixture may be known or uncharacterized, and the mixture may be relatively simple (containing few components) or complex. MRA may be conducted for a given exposed population, some predefined segment of the population, or the population at large. Typically, MRA addresses the risks for a given mixture emanating from a specified source, though it can address exposure by multiple routes, across the population, and the exposure time frame may be extended, depending on the nature of exposure and the toxic responses. If a component produces an effect that is long-lived, the result of the exposure (but not the exposure itself) may overlap with exposures to and effects from other chemicals. It is thus important to characterize not only the dose-response data but also the data describing the toxicity itself.

10.3.1 Experimental Mixtures Data

Data may be available for combinations of chemicals, and may be used under whole mixtures approaches, or data may be available for single chemicals that are components of the mixture and may be applied using component-based approaches (Figure 10.1). Optimally, data are available for the mixture of concern. However, when those data are not available, then data may be used on (1) a sufficiently similar mixture of chemicals or (2) components of the mixture of interest. When there are data identifying interactions between chemicals (when coexposures result in responses other than those predicted by additivity), those interactions must also be considered and incorporated into the assessment.

10.3.2 Whole Mixture Approaches

Environmentally encountered mixtures may have components from the scores to hundreds, differentiating simple mixtures from complex mixtures. Complex mixtures are those mixtures whose components are numerous so that attempts to address their risks via component-based approaches would be fraught with excessive uncertainty. This is a point for the application of judgment and justification, rather than a point addressed by quantitative guidance.

Examples of complex environmental mixtures include exhausts, drinking water disinfection by-products, mixtures of polycyclic aromatic hydrocarbons, and fuels. Because these mixtures contain many toxicologically uncharacterized (even unidentified) components, a component-based approach to assessing the risk from exposures to these mixtures is severely hindered.

When adequate exposure and dose-response data are available on the mixture of concern, then toxicity values for use in risk assessment can be derived using methods analogous to those for single chemicals: the mixture can be treated as if it were a single chemical entity. Three environmental mixtures have been toxicologically characterized to the point that the US EPA has treated them as a single entity: Arochlor 1016, Arochlor 1254, and coke oven emissions. Arochlor 1016 comprises congeners that have been well characterized. A common toxicity is developmental toxicity, and it appears to be well justified as the critical effect. Dose-response data for Arochlor 1016 were used to identify a point of departure (a NOAEL value) to which uncertainty factors were applied to develop the reference value. While environmental mixtures of PCBs may not match the pattern of congeners in Arochlor 1016, these data are generally applicable, though the confidence placed in the derived acceptable exposure values may be somewhat reduced. In this case, a discussion regarding uncertainties in the chemical composition of Arochlor 1016 due to the degraded environmental mixture of concern is warranted.

The US EPA [6] has developed a procedure for deriving reference values for environmental mixtures, as follows:

1) Data on components, on mixtures of similar composition, and on the whole mixture of interest should be collected and evaluated. This includes both epidemiological data and laboratory data.
2) The constituency of the mixture should be examined to determine whether and to what extent it varies; care should be taken to address the constancy of chemical constituents.
3) When evaluating a potential surrogate mixture, evaluate the number of common components and their proportions, and whether the two mixtures are emitted from a common source or are derived from similar industrial processes.
4) Using the procedures for single chemical, a dose-response assessment should be conducted.
5) Finally, the uncertainties and the assumptions made in the assessment should be made clear. The relevance of the health effects data for the surrogate mixture to the health effects identified for the mixture of concern should be addressed. Additional attention should focus on the extent to which both mixtures remain stable during environmental fate and transport.

When whole mixtures data are not available, a component-based approach may be undertaken, but the uncertainty involved in a component-based approach may raise a level of concern such that additional whole mixture testing becomes a priority. Optimally, dose-response and exposure data should be available from the mixture of concern, but this is seldom the case. Another valid approach is to identify another mixture whose components are similar to the mixture of concern, and revise and apply data from the similar mixture. In other cases, data may be available for some individual components or subsets of components. Regardless of the approach to whole mixtures risk assessment, chemical identification and toxicological characterization are important.

10.3.3 Sufficient Similarity

While the preferred whole mixtures approach is to use toxicity (dose-response) data from the mixture of concern, such data are seldom available, necessitating a search for available and applicable data sets. The application of toxicity data developed for other chemical mixtures should be considered, and consideration must be given to the degree to which the mixtures are similar [6]. Under the tenets of sufficient similarity for whole mixtures risk assessment, two principles have emerged: (1) the proposed surrogate mixture should comprise similar components and (2) these components should account for similar proportions of the compared mixtures. The goal is to identify a surrogate mixture for which there are small differences in component composition and proportion between it and the mixture of interest and whose chemical and toxicological properties are indistinguishable from the mixture of concern. Additional attention should be paid to characterizing differences in the environmental fate and transport, bioavailability, and pharmacokinetics of the components of each mixture.

In making the judgment regarding similarity, the compositions of the mixture of concern and the tested mixture are examined for similar components and their proportions in the two mixtures. Similarity of dose-response can also be evaluated among a common group of components using statistical procedures to identify common components that can be characterized with the same dose-response slope [30] and to define sufficient similarity in dose-response using statistical equivalence testing for mixtures of many chemicals containing the same components with different ratios [31]. Furthermore, it is not unusual for a large portion of complex mixtures to be chemically uncharacterized, resulting in an unidentified fraction whose contribution to toxicity may be unknown but must be considered when evaluating similarity. In this case, "summary measures" are needed that represent and integrate the toxicity and chemistry of the complex mixture, including the unidentified fraction. *In vitro* data (e.g., from mutagenicity or cytotoxicity studies), *in vivo* toxicological data where animals are directly exposed to the complex mixture or concentrates of the mixture, and epidemiological results offer insights into whole mixture toxicity. Chemical composition can be evaluated using GC-MS (gas chromatography–mass spectrometry) chromatograms, other analytical methods, and summary measures of chemical characteristics of a mixture. A series of papers on the similarity of drinking water disinfection by-products shows important summary measures for those mixtures include factors such as mutagenic potency, pH level, percent total organic bromine relative to total organic halide, percent bromination of the trihalomethanes, or total organic halide, among others (see [32]). An example of an integrated protocol for evaluating similarity can be found in Cizmas *et al.* [33] using *in vitro* bioassays (*Salmonella*/microsome, *Escherichia coli* prophage induction, chick embryo toxicity screening assays) and chemical analysis (GC-MS) for fractionated mixtures of wood preserving wastes containing polycyclic aromatic hydrocarbons and pentachlorophenol. Although these particular mixtures appeared chemically similar, their biological responses were dissimilar. In addition, Eide *et al.* [34] developed a method to predict the mutagenicity of sufficiently similar mixtures of organic extracts of exhaust particles using a combination of pattern recognition techniques and multivariate regression modeling. They constructed a predictive mutagenicity model by regressing the peaks observed in GC-MS chromatograms of these mixtures on the mixtures' mutagenicity measures. Thus, the mutagenicity of a new mixture of exhaust particles can be estimated by the model using input from its GC-MS chromatogram.

10.3.4
Component-Based Approaches

When data are available, they more often than not describe results from single-chemical exposures from studies with experimental animals. Occasionally, data may be available for a mixture of interest or a mixture similar to that of interest. Data from the former may be applicable in component-based mixtures approaches, including

RA and DA models. In the latter instance, these data may be applicable under whole mixtures approaches, depending on the degree to which the evaluated mixture is deemed toxicologically similar to the mixture of interest.

10.3.5 Additivity Models and Single-Chemical Data

At very low doses, interactions between components of chemical mixtures are most often absent, allowing the application of additivity models. This represents a default position, to be discarded in favor of more applicable models, based on the available data. There are two fundamental types of additivity models, dose addition and response addition. The choice between these models depends on data that demonstrate whether mixture components act via similar and independent modes of action, characterized by Feron and Groten [35] as simple similar action and simple dissimilar action, respectively.

In the case of DA, components are essentially toxicological "clones" of one another such that the relative proportions of each in a mixture are treated as dilutions of one another [6, 36]. There are two DA models that are frequently used in MRA: relative potency factors and hazard index. Although some exceptions are noted, several studies have demonstrated that the addition of doses or concentrations provides adequate estimates of responses of a variety of chemicals that have similar and dissimilar types of effects [37–42]. Lambert and Lipscomb [27] have explored the possibility of including commonality of effect on significant biological processes not related to mode of action as a basis for similarity. Dose additivity seems to be a reasonable default when compounds affect the same target organ, but should be closely examined when the compounds have dissimilar (independent) modes of action. While the requirements for data sufficient to serve as the basis for selecting dose- or response additive models may be stringent, the application of these default additivity models seems warranted; the choice between DA and RA should be carefully considered in interpreting the results of MRA [27].

Once compounds are grouped into similar MOA groupings, their relative potencies should be determined. This is differentially accomplished under the RPF and the HI approaches. In the case of RPF, relative potency is determined in one of the two ways: (1) as the ratio of effective dose levels (producing the same level of response) of an index chemical (IC) to another chemical in the mixture or (2) as the ratio of the slope factors of a chemical in the mixture to a designated index chemical. MRA methodology indicates that the calculation of RPF should be made in the low-dose region and specifies that dose-response curves should be similarly shaped in this region (in contrast, response additive models do not require similarly shaped *D-R* curves). This is a key decision (whether the curves are similar shaped), and when the dose-response curves for two chemicals produce RPF values that are a function of response level, the point along the *D-R* curve at which the RPF is calculated will impact the RPF value (Figure 10.3). Under the RPF approach, an index chemical is selected and doses of other mixture components are scaled to doses of the index chemical, based on relative potency.

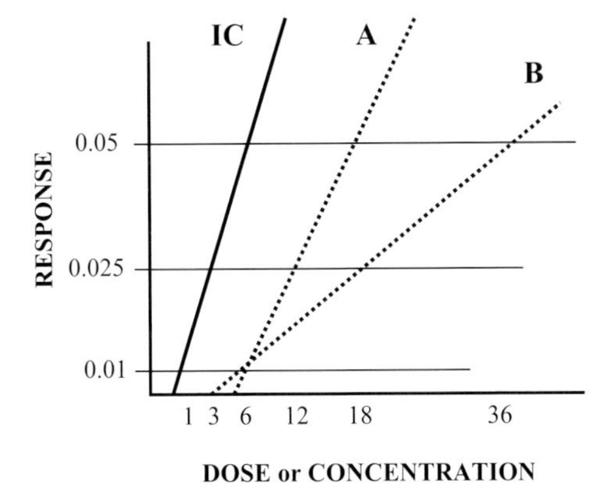

Figure 10.3 *D-R* curves for two chemicals. The degree of similarity in the response region where RPF is calculated may impact the value of the RPF.

Toxicity data for the components must be sufficient to support their grouping according to a common mode or mechanism of toxicity. This determination may be best supported relative to the adverse health outcome, rather than to isolated biochemical events evaluated individually. For mixture components that are determined to cause the same general health outcome acting via independent modes of action, the net risk estimated for the mixture is equivalent to the sum of the individual component risks [6]. The assumption is that for a mixture whose components cause a similar health outcome (e.g., cancer), the type or degree of adverse effect caused by one component (e.g., thyroid cancer) has no direct impact on the type or degree of adverse effect caused by a second component (e.g., skin cancer); in this case, the probability of observing cancer is the sum of the individual component cancer risks. Response addition requires dose-response data for all components of a mixture sufficient to define the slope of the dose-response relationship with a degree of certainty sufficient to support predictions of risk at uncharacterized, environmentally relevant doses. This is somewhat in contrast with the data requirements for dose addition [27].

10.3.5.1 Hazard Index Approaches

Hazard index approaches have been employed by the US EPA's Superfund Program to screen for the potential health risks associated with chemical exposures at specific locations (i.e., National Priority List sites) [7]. The initial step is the identification of mixture components, their exposure concentrations, and development of chemical-specific hazard quotients (HQ). HQ is calculated as E/AL, where E represents the exposure and AL represents the acceptable limit, which may be an oral reference dose (RfD), inhalation reference concentration TTD or MRL. Because E and AL are in the same units of exposure, the HQ value is unitless. The HI for the mixture is calculated as the sum of HQ values for the mixture components (Table 10.1).

Because acceptable limits of exposure are often and have typically been developed only on the basis of the single most sensitive end point, HI values address the

Table 10.1 Exposure levels, acceptable levels, and screening level hazard index calculation.

Chemical	Exposure[a)]	Affected organ/system[a)]				Screening hazard quotient
		Neuro.	Hemat.	Nephro.	Develop.	
A	6E-4	5E-2[b)]	NA	8E-2	6E-2	1.2E-2
B	1.5E-3	4E-2	2E-2[b)]	4E-2	NA	7.5E-2
C	1.2E-2	NA	4E-2	2E-2[b)]	6E-2	6E-1
D	6E-5	8E-2	5E-2	9E-3[b)]	1E-2	6.7E-3
E	8E-4	2E-2[b)]	NA	NA	NA	4E-2
F	4E-5	1E-3	4E-3	6E-4[b)]	8E-4	6.7E-2
G	6E-4	2.5E-1	4E-2[b)]	8E-2	6E-2	1.5E-2
H	8.4E-4	NA	8E-3[b)]	NA	2E-2	1.1E-1
Hazard index						9.2E-1

a) Exposure and acceptable level for affected organ or system are each expressed in units of mg/kg/day.

b) Only HQ values for the critical effect is used to calculate screening level hazard index. For example, the screening level HI for chemical A would be based on neurological effects, with an acceptable exposure limit of 5E-2 mg/kg/day. Dividing the exposure of 6E-4 mg/kg/day by the acceptable limit gives a hazard quotient of 1.2E-2.NA: Not affected.

likelihood of a response by the critical (most sensitive) end point. When HQ values rise above a value of 1, concern increases. For screening purposes, the HI is calculated without taking into account affected organ or organ system (e.g., liver, kidney, developmental toxicity). For a group of n chemicals in a mixture and using the RfD as an allowable level, the HI for oral exposure is calculated according to Eq. (10.1):

$$\mathrm{HI} = \sum_{i=1}^{n} \frac{E_i}{\mathrm{RfD}_i} \tag{10.1}$$

where E_i is the exposure level of the ith chemical and RfD_i denotes the reference dose of the ith chemical

To be more consistent with the assumption of a common mode of action across components, the HI is calculated by segregating components on the basis of affected organ or organ system (Table 10.2). Here, HI values are typically reduced compared to values developed under the screening approach. This occurs because the RfD is based on the critical effect (i.e., the adverse effect observed at the lowest does) for that chemical and because of the likelihood that the multiple mixture components will affect a spectrum of organs or tissues. The third HI approach uses target organ toxicity (TTD) values [6, 43] (see Table 10.2). Under the TTD approach, ALs are developed for tissues in addition to those representing the critical effect, and HQ values are developed for those. Naturally, for a given mixture component, the E value will be the same, but the AL values will represent several responding tissues. Because the AL values represent allowable exposure values, and because they represent multiple less sensitive tissues, regulatory values may not be available. In this case, uncertainty factor values will have to be developed and applied to the point of

Table 10.2 Organ-specific hazard quotients, hazard index calculation, and target organ toxicity dose calculation.

Chemical	Affected organ/system			
	Neuro.	Hemat.	Nephro.	Develop.
A	0.012[a]	0	0.08	0.01
B	0.038	0.075[a]	0.038	0
C	0	0.30	0.60[a]	0.20
D	0.001	0.001	0.007[a]	0.006
E	0.04[a]	0	0	0
F	0.04	0.01	0.067[a]	0.050
G	0.02	0.015[a]	0.008	0.010
H	0	0.105[a]	0	0.042
Hazard index[a]	0.052	0.195	0.674	0
TTD	0.133	0.506	0.726	0.318

a) Only HQ values for critical effects used to calculate hazard index. HQ values for critical and secondary effects used to calculate TTD values.

departure to develop AL values for these noncritical (secondary) effects. The uncertainty factors for each effect may differ, resulting in different AL values, even when the same point of departure dose is determined from animal studies. The TTD approach develops HQ values for affected organs and tissues and segregates HQ values for mixture components on the basis of affected tissues. Thus, under the TTD approach it is possible that a higher level of concern may be demonstrated due to the inclusion of secondary effects.

An important assumption in the HI approach is that mixture components affecting the same tissue will have a similar mode of action. Compounds affecting the same organ or tissue might or might not share a mode of action. Another issue is data quality. Some studies might not support more stringent limits of confidence in estimating acceptable exposure (e.g., dose spacing may result in inappropriately low NOAEL values). Finally, HI or TTD values may frequently include AL values that are derived from NOAEL or LOAEL values. This approach to determining points of departure takes less amount of the dose-response data into account than the application of benchmark dose (BMD) modeling. BMD modeling takes all data into account and estimates doses (and confidence bounds on doses), producing a predetermined response level; representing an advantage over the traditional NOAEL-based approach. These and other points are considered in developing the weight of evidence narrative, which characterizes the assumptions made in the assessment and provides a measure of strength to the conclusions reached.

10.3.5.2 Relative Potency Factors

Another approach to assessing the risks of groups of similar components is the relative potency factor approach. Here, *D-R* data must be sufficient to fully characterize the *D-R* function of the index chemical; some data must be available to estimate effective dose levels for the other chemicals in the mixture, but a full dose response curve is not required. In the best-case scenario, well-characterized toxic responses will be available for all of the component chemicals. Because of the extent to which the toxicity of dioxin-like compounds has been demonstrated to rely on activation of the AH receptor, the toxicity of these compounds has been evaluated [4, 7, 44, 45] using the toxicity equivalence factor (TEF) approach; this is a special case of the RPF approach where TEFs are considered valid across all toxic end points, exposure routes, and timeframes. Under the RPF approach, an index chemical is selected on the basis of several points including the extent to which its toxicity represents the toxicity of other components and the strength at which its *D-R* function has been characterized. Again, the shape of the *D-R* curves for the index chemical and the other components is assumed to be similar in the low-dose region. Because some professionals espouse that the *D-R* curves must be parallel in the low-dose region, we have inserted an example here that demonstrates what is intended – that relative potency estimates must be consistent in the data available. Figure 10.3 and Table 10.3 demonstrate that parallel curves will result in different RPF estimates depending on the response level chosen for comparison (e.g., chemical A). While the *D-R* data for chemical A have the same slope (0.05) as for the index chemical, the RPF developed depends on the response level chosen for quantitation. Results for chemical B demonstrate a markedly nonparallel *D-R* curve and a slope (0.008) markedly different from that of the index chemical. Regardless, equivalent RPF estimates can be developed for several points along the *D-R* curve. Note, however, that with this same slope, but shifted to the left of the index chemical (i.e., with RPF values above 1.0), similar RPF values will not be developed at multiple points along the curve. This example is intended to highlight the importance of evaluating multiple data points when calculating RPF values.

With the above in mind, a scaling factor, or proportionality constant, is derived for each component chemical relative to the index chemical, with that for the index chemical assigned a value of 1.0. The component-specific constants are called the RPF values. The value of the RPF for respective components will vary, and components may have RPF values lesser or greater than 1.0. In the example below, the doses producing a response of 0.05 are 1 and 6 for the index chemical and chemical A, respectively. Since it takes a higher dose of chemical A to produce this level of

Table 10.3 Impact of shape of D-R curve on RPF estimates.

Response level	Index chemical	Chemical A		Chemical B	
	Effective dose	Effective dose	RPF	Effective dose	RPF
0.05	1	6	0.17	6	0.17
0.15	3	8	0.33	18	0.17
0.30	6	11	0.50	36	0.17

a) These data are graphically presented in Figure 10.3.

Table 10.4 Example calculation of summed index chemical equivalent dose.

Component	RPF	Dose	ICED
Index chemical	1	15	15
Chemical A	0.2	46	0.9
Chemical B	1.5	22	33
		Summed ICED:	49

response, it is less potent than the index chemical, and the RPF will be lower than 1. Mathematically, $1/6 = 0.17$.

While an optimal data set will describe a large portion of the *D-R* relationship, seldom will data be complete enough for the same measured end point to develop a dose-response understanding sufficient to determine the degree of similarity of slope of the *D-R* relationship in the low-dose region. Here, knowledge about the relationship between the dose(s) studied and the anticipated exposures, as well as information about the mode of action for toxicants, can be combined with the limited *D-R* data to serve as the justification for an estimated RPF.

After assigning RPF values, exposures are taken into account. The expected response for the mixture R_m is determined by summing the index chemical equivalent dose (ICED) values for each component (including the index chemical) and reading the anticipated response from the *D-R* curve of the index chemical. ICEDs are summed to express the mixture exposure in terms of an equivalent exposure to the index chemical, as shown for *n* chemicals in Eq. (10.2) (see Table 10.4):

$$R_m = f_1\left(\sum_{i=1}^{n}(\mathrm{RPF}_i \times D_i)\right) \qquad (10.2)$$

where R_m is the risk posed by chemical mixture, $f_1(\cdot)$ denotes dose-response function of index chemical 1, D_i is the dose of the *i*th mixture component ($i = 1, \ldots, n$), and RPF_i is the toxicity proportionality constant relative to index chemical for the *i*th mixture component ($i = 1, \ldots, n$).

10.3.6 Response Addition

RA is applied when the observed effects cannot be explained using a common mode of action assumption; its application is warranted on the basis of information or policy regarding independent modes of action among components or groups of components. Under RA, data for individual components must be sufficient to characterize the dose-response, including (where relevant) the identification of a level below which a biological response cannot be detected (the threshold). Exposure to a given component is characterized, and the response from that exposure (dose) is determined from the dose-response data for that component. If the dose results in a response below the threshold, then that component is assigned a response value of zero. If the exposure corresponds to a dose producing a response above the threshold, then the response value (e.g., 0.15) is captured from the single-chemical data. Regarding the threshold concept, it is applied at the single-chemical level. The mathematical form of the model for an n-component mixture is presented in Eq. (10.3).

$$R_m(\mathrm{RA}) = 1 - \prod_{i=1}^{n}(1 - r_i) \tag{10.3}$$

where $R_\mathrm{m}(\mathrm{RA})$ is the response predicted for the mixture under response addition, and the risk for each individual ith component is depicted as r_i. When the individual risks are very small (e.g., in the range of 10E-4), the equation can be simplified to equal the sum of the individual risks.

10.3.7 Integrated Additivity

Integrated additivity is applied when a mixture contains components that cause the same adverse health outcome and act through more than one mode of action that may be classified as independent of each other. The necessity of applying this method increases with the complexity of the mixture. Mixtures may contain components that act through similar modes of action and components that act through independent modes of action. Under the integrated additivity approach, components that act through a similar mode of action are grouped together and the response for the group is estimated by dose additivity. When components of the mixture can be grouped into multiple mode-of-action subgroups, the responses for each subgroup of mode of action is separately estimated by dose additivity. For individual components acting through unique (independent) modes of action, their responses are estimated via response addition. Once the response for each component and/or mode of action group is estimated, these are combined through response additivity [4, 46].

10.3.8 Chemical Interactions

When organisms are coexposed to multiple chemicals, the response is called a joint toxic response. It does not necessarily mean that chemical interactions have occurred. In technical terms, interaction is defined as a mixture response that cannot be explained by an additivity model. Several terms have been used to describe interactions, including antagonism, inhibition, synergism, potentiation, and masking. While antagonism and inhibition may sound similar, they differ in an important distinction. Antagonism results when the response from two chemicals results in a less than additive response. Both chemicals are toxic, but the coexposure to the two results in a less than anticipated response. In contrast, inhibition results from the coexposure to two compounds, only one of which is toxic. The second can be thought of as being an antitoxin. Given that the mode of action of a given component may change with (increasing) dose [18], interactions among mixture components may also be a function of dose.

The characterization of chemical interactions is resource intensive. ATSDR has characterized the interactions of 11 sets of chemicals, evaluated as binary mixtures. A binary weight of evidence (BINWOE) has been developed for each pair, and these data are useful in characterizing the likelihood of interactions in mixtures containing these components. ATSR recommends that a qualitative weight of evidence narrative be developed for mixtures when the HI value for the mixture increases above 0.1 [47]. This is accomplished to document the likelihood of encountering responses not taken into account by additivity models.

10.3.9 Interaction-Based Hazard Index

To quantitatively handle toxicological interactions, the US EPA recommends using the interaction-based hazard index (HI_{INT}), which employs a numerical modification to HI values determined for mixtures containing components known to interact in a nonadditive manner [6, 48]. This method uses binary interaction data, assuming that these data can adequately account for responses due to higher order interactions. HI_{INT} is based on early work by Mumtaz and Durkin [49]. Noting that the first summation shown is the additive HI and the second summation shown is the modification for interactions, the formula for the HI_{INT} is

$$HI_{INT} = \sum_{j=1}^{n} HQ_j \sum_{k \neq j}^{n} f_{jk} M_{jk}^{B_{jk} g_{jk}} \qquad (10.4)$$

where HI_{INT} is HI modified by binary interactions data, HQ_j is hazard quotient for chemical j (e.g., daily intake/acceptable level), and f_{jk} denotes toxic hazard of the kth chemical relative to the total hazard from all chemicals potentially interacting with chemical j (thus k cannot equal j). To calculate, the formula is

$$f_{jk} = \frac{HQ_k}{\left[\sum_{j-1}^{n} HQ_j\right] - HQ_j} \qquad (10.5)$$

where M_{jk} is interaction magnitude, the influence of chemical k on the toxicity of chemical j. To calculate, estimate from binary data or use default value = 5. B_{jk} is the score for the weight of evidence that chemical k will influence the toxicity of chemical j (see Ref. [6] for numeric values). g_{jk} is the degree to which chemicals k and j are present in equitoxic amounts. To calculate, the formula is

$$g_{jk} = \frac{\sqrt{HQ_j{}^{*}HQ_k}}{(HQ_j + HQ_k)/2} \qquad (10.6)$$

Hertzberg and Teuschler [48] argue that Eq. (10.4) is well structured in that it performs as one might expect when parameter values are at their extremes. For example, they show that when all information on the chemical pairs indicates that each pair is dose additive (i.e., no data show toxicological interactions), then the value of the parameter in Eq. (10.5), f_{jk}, is consistent with Eq. (10.4) being equal to the dose additive HI.

10.4 Future Directions

Firmly rooted in regulation, policy, and practice, the fields of mixtures and cumulative risk assessment have matured over the past decade or so and now provide viable tools for community and site-based risk assessment. Cumulative risk assessment is a markedly interdisciplinary approach, and the several pertinent areas continue to mature. Several important advancements seem poised to further improve the certainty and accuracy of CRA in the near term. Among them are physiologically based pharmacokinetic (PBPK) modeling, evaluations of alterations beyond those in the mode of action, improved experimental design, and refined whole mixtures approaches by identifying sufficiently similar mixtures.

The PBPK modeling is more frequently being cited as a tool to refine dose-response approaches and intra- and interspecies extrapolation in health risk assessments. These models have been used to determine the likelihood of chemical interactions that are based on competition for metabolism. Inasmuch as toxicity is mediated by the interaction between the toxicologically active chemical species (parent or a metabolite) and the biological receptor, this tool offers the ability to predict time-dependent changes in tissue concentrations of toxicants and metabolites in experimental animals and humans. Despite providing a framework through which to integrate biological and biochemical understanding of chemical metabolism and disposition, often attempts to apply these models to situations beyond their capacity or intended purpose results in skepticism. The WHO's International Program on Chemical Safety has embarked on a project to document the best practices for PBPK modeling as applied to health risk assessment [50]. By addressing the toxicity, target organ/tissue, and toxic chemical species and knowledge about mode of action early in the process, the developed PBPK model will be more purpose-oriented and readily accepted. With the development and inclusion of an uncertainty analysis, identifying key determinants of dosimetry and documenting confidence in measured or estimated values for those parameters, an increased level of confidence in model application will follow.

The concept of similarity in chemicals' effects might be broadened to include those events that are not part of the mode of action. Chemicals are not "neat" in the adverse effects that they induce; rather they produce an array of alterations, some of which consist in the mode of action for the observed toxicity, others not. The development of an adverse health effect is a function of the biochemical and cellular events that consist in the mode of (toxic) action. The organization of these events into a mode of action serves as the basis for deciding whether chemicals act through common or dissimilar modes of action. While chemicals with similar modes of action are subjected to risk analysis via dose additive models, those with independent modes of action are relegated to analysis via response additivity. This choice is based on comparison of events within the mode of action for toxicity, and at present only seldom includes consideration of alterations beyond key steps in the mode of action, ignoring the potential that such alterations may be shared with the key steps in the mode of action for other mixture components. Increased awareness of alterations of biochemical and physiological processes that are not part of the toxic mode of action is necessary to more fully understand the likelihood of chemical interactions. Certain alterations may not be part of the mode of action for one chemical but may be key processes in the mode of action for another component of the same mixture.

A revised approach to experimental design is warranted. An increased awareness of the dose-response relationship would benefit health risk assessment in general, and mixtures and cumulative risk assessments, in particular. Often, constraints on experimental design and/or analytical sensitivity force exposures of experimental animals to doses of chemicals that far exceed the anticipated human exposure. With an increasing dose, two events occur – an increase in the frequency of responding individuals and an increase in the severity of the response. The first condition is often used as the basis for higher doses on the ground that it may overcome analytical insensitivity. Often, the second condition is underappreciated or overlooked altogether. Shifts in toxicity and mode of action with changes in dose may result in shifts in the pattern of interactions predicted. An increased attention to low-dose responses would reduce the uncertainty in extrapolating high-dose events to lower doses.

Toxicity data on whole mixtures, in general, are limited. An updated approach to identify surrogate mixtures has been launched under the auspices of "sufficiently similar mixtures." Two criteria have been established to guide the identification of toxicity data from a mixture that may be sufficiently similar to serve as a surrogate for the mixture in question: (1) the mixture components must be qualitatively similar and (2) the mixture components must occur in roughly the same proportions. Often, large fractions of complex mixtures may be unidentified; in these cases, additional data such as the types of toxicities observed for each mixture can be informative. The continued advance of this area by developing case studies where surrogate mixtures are identified and their dose-response data are applied mixtures and cumulative risk assessments will be valuable.

The interdisciplinary approach to cumulative risk assessment forces the inclusion of scientists whose focus may be quite narrow. It remains important to foster cross-disciplinary discussions that may result in additional, fruitful cumulative risk efforts. This will also ensure that discipline-specific advances can be brought to bear on cumulative risk efforts. Incorporation of these advances will further bolster confidence in and acceptance of mixtures and cumulative risk assessments.

References

1 ILSI (1999) A Framework for Cumulative Risk Assessment. International Life Sciences Institute, Risk Sciences Institute, Washington, DC. Available at http://rsi.ilsi.org/Publications/RAFramework.htm.

2 U.S. EPA (1997) Guidance on Cumulative Risk Assessment, Part 1. Planning and Scoping. U.S. Environmental Protection Agency, Science Policy Council, Washington, DC. Attachment to memo dated July 3, 1997 from the Administrator, Carol Browner, and Deputy Administrator, Fred Hansen, titled "Cumulative Risk Assessment Guidance – Phase I Planning and Scoping." Available at http://www.epa.gov/OSA/spc/2cumrisk.htm.

3 U.S. EPA (2002) Lessons Learned on Planning and Scoping of Environmental Risk Assessment. Memorandum from Science Policy Council. Available at http://www.epa.gov/osp/spc/llmemo.htm.

4 U.S. EPA (2003) Framework for Cumulative Risk Assessment. U.S. Environmental Protection Agency, Office of Research and Development, National Center for Environmental Assessment, Washington, DC. EPA/600/P-02/001F. Available at http://cfpub.epa.gov/ncea/cfm/recordisplay.cfm?deid=54944.

5 U.S. EPA (1986) Guidelines for the Health Risk Assessment of Chemical Mixtures. U.S. Environmental Protection Agency, Office of Research and Development, Washington, DC. EPA/630/R-98/002.

6 U.S. EPA (2000) Supplementary Guidance for Conducting Health Risk Assessment of Chemical Mixtures. U.S. Environmental Protection Agency, Risk Assessment Forum, Washington, DC. EPA/630/R-00/002. Available at http://www.epa.gov/ncea/raf/pdfs/chem_mix/chem_mix_08_2001.pdf.

7 U.S. EPA (1989) Risk Assessment Guidance for Superfund: Volume 1, Human Health Evaluation Manual (Part A). U.S. Environmental Protection Agency, Office of Emergency and Remedial Response, Washington, DC. EPA/540/1-89/002. (Also see Parts B–D.)

8 U.S. EPA (1998) Methodology for Assessing Health Risks Associated with Multiple Pathways of Exposure to Combustor Emissions. U.S. Environmental Protection Agency, Office of Research and Development, National Center for Environmental Assessment, Cincinnati, OH. EPA/600/R-98/137.

9 U.S. EPA (2007) Concepts, Methods, and Data Sources for Health Risk Assessment of Multiple Chemicals, Exposures, and Effects. EPA/600/R-06/013A.

10 U.S. EPA (2002) Guidance on Cumulative Risk Assessment of Pesticide Chemicals That Have a Common Mechanism of Toxicity. U.S. Environmental Protection Agency, Office of Pesticide Programs, Washington, DC. Available at http://www.epa.gov/oppfead1/trac/science/cumulative_guidance.pdf.

11 U.S. EPA (2006) Organophosphorus Cumulative Risk Assessment 2006 Update. U.S. Environmental Protection Agency, Office of Pesticide Programs, Washington, DC. Available at http://www.epa.gov/oppsrrd1/cumulative/2006-op/op_cra_main.pdf.

12 U.S. EPA (2006) Cumulative Risk from Triazine Pesticides. U.S. Environmental Protection Agency, Office of Pesticide Programs, Washington, DC. Available at http://www.epa.gov/oppsrrd1/REDs/triazine_cumulative_risk.pdf.

13 U.S. EPA (2007) Revised N-Methyl Carbamate Cumulative Risk Assessment. U.S. Environmental Protection Agency, Office of Pesticide Programs, Washington, DC. Available at http://www.epa.gov/oppsrrd1/REDs/nmc_revised_cra.pdf.

14 U.S. Congress (1980) Title 42 – The Public Health and Welfare Chapter 103. The Comprehensive Environmental Response, Compensation, and Liability Act (CERCLA). Available at http://www.epa.gov/epahome/laws.htm.

15 U.S. Congress (1996) The Food Quality Protection Act. Public Law 104-170-Aug. 3, 1996. 104th Congress. Washington, DC.

Available at http://www.epa.gov/oppfead1/fqpa/gpogate.pdf.

16 U.S. Congress (1996) 1996 Amendments to the Safe Drinking Water Act – Public Law 104-182 104th Congress. Washington, DC. Available at http://www.epa.gov/safewater/sdwa/text.html.

17 European Union (EU) (2006) Regulation (EC) No 1907/2006 of the European Parliament and of the Council of 18 December 2006 concerning the Registration, Evaluation, Authorisation and Restriction of Chemicals (REACH), establishing a European Chemicals Agency. Available at http://reach-compliance.eu/english/legislation/docs/launchers/launch-2006-1907-EC-06.html.

18 Slikker, W. Jr., Andersen, M.E., Bogdanffy, M.S., Bus, J.S., Cohen, S.D., Conolly, R.B., David, R.M., Doerrer, N.G., Dorman, D.C., Gaylor, D.W., Hattis, D., Rogers, J.M., Setzer, R.W., Swenberg, J.A., and Wallace, K. (2004) Dose-dependent transitions in mechanisms of toxicity: case studies. *Toxicol. Appl. Pharmacol.*, **201**, 226–294.

19 Yang, R.S. and Dennison, J.E. (2007) Initial analyses of the relationship between "Thresholds" of toxicity for individual chemicals and "Interaction Thresholds" for chemical mixtures. *Toxicol. Appl. Pharmacol.*, **223**, 133–138.

20 NRC (National Research Council) (1983) *Risk Assessment in the Federal Government: Managing the Process*, Committee on the Institutional Means for Assessment of Risks to Public Health, Commission on Life Sciences, NRC, National Academy Press, Washington, DC.

21 U.S. EPA (2005) Guidelines for Carcinogen Risk Assessment. U.S. Environmental Protection Agency, Washington, DC, EPA/630/P-03/001F. Available at http://cfpub.epa.gov/ncea/cfm/recordisplay.cfm?deid=116283.

22 Sonich-Mullin, C., Fielder, R., Wiltse, J., Baetcke, K., Dempsey, J., Fenner-Crisp, P., Grant, D., Hartley, M., Knaap, A., Kroese, D., Mangelsdorf, I., Meek, E., Rice, J.M., and Younes, M. (2001) IPCS conceptual framework for evaluating a mode of action for chemical carcinogenesis. *Regul. Toxicol. Pharmacol.*, **34**, 146–152.

23 Mileson, B.E., Chambers, J.E., and Chen, W.L. (1998) Common mechanism of toxicity: a case study of organophosphorous pesticides. *Toxicol. Sci.*, **41**, 8–10.

24 Boobis, A.R., Cohen, S.M., Dellarco, V., McGregor, D., Meek, M.E., Vickers, C., Willcocks, D., and Farland, W. (2006) IPCS framework for analyzing the relevance of a cancer mode of action for humans. *Crit. Rev. Toxicol.*, **36**, 781–792.

25 U.S. EPA (1999) Guidance for Identifying Pesticide Chemicals and Other Substances That have a Common Mechanism of Toxicity. January 29, 1999. Office of Pesticide Programs, Office of Prevention, Pesticides and Toxic Substances, Washington, DC. Available at http://www.epa.gov/fedrgstr/EPA-PEST/1999/February/Day-05.

26 Health Canada (2001) Guidance for Identifying Pesticide Chemicals and Other Substances that have a Common Mechanism of Toxicity. Science Policy Note, January 25, 2001. Pest Management regulatory Agency, Health Canada, Ottawa, Ontario, Canada. Available at http://www.hc-sc.gc.ca/pmra-arla/.

27 Lambert, J.C. and Lipscomb, J.C. (2007) Mode of action as a determining factor in human health risk assessment of chemical mixtures. *Regul. Toxicol. Pharmacol.*, **49**, 183–194.

28 Dobrev, I.D., Andersen, M.E., and Yang, R.S. (2001) Assessing interaction thresholds for trichloroethylene in combination with tetrachloroethylene and 1,1,1-trichloroethane using gas uptake studies and PBPK modeling. *Arch. Toxicol.*, **75**, 134–144.

29 el-Masri, H.A., Constan, A.A., Ramsdell, H.S., and Yang, R.S. (1996) Physiologically based pharmacodynamic modeling of an interaction threshold between trichloroethylene and 1,1-dichloroethylene in Fischer 344 rats. *Toxicol. Appl. Pharmacol.*, **141**, 124–132.

30 Chen, J.J., Chen, Y.J., Teuschler, L.K., Rice, G., Hamernik, K., Protzel, A., and Kodell, R.L. (2003) Cumulative risk assessment for quantitative response data. *Environmetrics*, **14**, 339–353.

31 Stork, L.G., Gennings, C., Carter, W.H. Jr., Teuschler, L.K., and Carney, E.W. (2008) Empirical evaluation of sufficient similarity in dose-response for environmental risk assessment of chemical mixtures. *J. Agric. Biol. Environ. Stat.*, **13**, 313–333.

32 Rice, G.E., Teuschler, L.K., Bull, R.J., Simmons, J.E., and Feder, P.I. (2009) Evaluating the similarity of complex drinking water disinfection by-product mixtures: overview of the issues. *J. Toxicol. Environ. Health A*, **72**, 429–436.

33 Cizmas, L., McDonald, T.J., Phillips, Traci D., Gillespie, A.M., Lingenfelter, R.A., Kubena, L.F., Phillips, Timothy D., and Donnely, K.C. (2004) Toxicity characterization of complex mixtures using biological and chemical analysis in preparation for assessment of mixture similarity. *Environ. Sci. Technol.*, **38**, 5127–5133.

34 Eide, I., Neverdal, G., Thorvaldsen, B., Grung, B., and Kvalheim, O.M. (2002) Toxicological evaluation of complex mixtures by pattern recognition: correlating chemical fingerprints to mutagenicity. *Environ. Health Perspect.*, **110**, 985–988.

35 Feron, V.J. and Groten, J.P. (2002) Toxicological evaluation of chemical mixtures. *Food Chem. Toxicol.*, **40**, 825–839.

36 U.S. EPA (2003) The Feasibility of Performing Cumulative Risk Assessments for Mixtures of Disinfection By-Products in Drinking Water. ORD/ NCEA Cincinnati, OH, EPA/600/R-03/ 051. Available at http://cfpub.epa.gov/ ncea/cfm/recordisplay.cfm?deid=56834.

37 Pozzani, U.C., Weil, C.S., and Carpenter, C.P. (1959) The toxicological basis of threshold values: 5. The experimental inhalation of vapor mixtures by rats, with notes upon the relationship between single dose inhalation and single dose oral data. *Am. Ind. Hyg. Assoc. J.*, **20**, 364–369.

38 Smyth, H.F., Weil, C.S., West, J.S., and Carpenter, C.P. (1969) An exploration of joint toxic action. I. Twenty-seven industrial chemicals intubated in rats in all possible pairs. *Toxicol. Appl. Pharmacol.*, **14**, 340–347.

39 Smyth, H.F., Weil, C.S., West, J.S., and Carpenter, C.P. (1970) An exploration of joint toxic action. II. Equitoxic versus equivolume mixtures. *Toxicol. Appl. Pharmacol.*, **17**, 498–503.

40 Murphy, S.D. (1980) Assessment of the potential for toxic interactions among environmental pollutions, in *The Principles and Methods in Modern Toxicology* (eds C.L. Galli, S.D. Murphy, and R. Paoletti), Elsevier/North-Holland Biomedical Press, Amsterdam.

41 Ikeda, M. (1988) Multiple exposure to chemicals. *Regul. Toxicol. Pharmacol.*, **8**, 414–421.

42 Feron, V.J., Groten, J.P., Jonker, D., Cassess, F.R., and van Bladeren, P.J. (1995) Toxicology of chemical mixtures: challenges for today and the future. *Toxicology*, **105**, 415–427.

43 Mumtaz, M.M., Poirier, K.A., and Coleman, J.T. (1997) Risk assessment for chemical mixtures: fine-tuning the hazard index approach. *J. Clean Technol. Environ. Toxicol. Occup. Med.*, **6**, 189–204.

44 Van den Berg, M., Birnbaum, L., Bosveld, A.T.C., Brunström, B., Cook, P., Feeley, M., Giesy, J.P., Hanberg, A., Hasegawa, R., Kennedy, S.W., Kubiak, T., Larsen, J.C., van Leeuwen, F.X.R., Liem, A.K.D., Nolt, C., Peterson, R.E., Poellinger, L., Safe, S., Schrenk, D., Tillitt, D., Tysklind, M., Younes, M., Wærn, F., and Zacharewski, T. (1998) Toxic equivalency factors (TEFs) for PCBs, PCDDs, PCDFs for humans and wildlife. *Environ. Health Perspect.*, **106**, 775–792.

45 Van den Berg, M., Birnbaum, L.S., Denison, M., De Vito, M., Farland, W., Feeley, M., Fiedler, H., Hakansson, H., Hanberg, A., Haws, L., Rose, M., Safe, S., Schrenk, D., Tohyama, C., Tritscher, A., Tuomisto, J., Tysklind, M., Walker, N., and Peterson, R.E. (2006) The 2005 World Health Organization reevaluation of human and mammalian toxic equivalency factors for dioxins and dioxin-like compounds. *Toxicol. Sci.*, **93**, 223–241.

46 Teuschler, L.K., Rice, G.E., Wilkes, C.R., Lipscomb, J.C., and Power, F.W. (2004) A feasibility study of cumulative risk

assessment methods for drinking water disinfection by-product mixtures. *J. Toxicol. Environ. Health A*, **67**, 755–777.

47 Pohl, H.R., Roney, N., Wilbur, S., Hansen, H., and De Rosa, C.T. (2003) Six interaction profiles for simple mixtures. *Chemosphere*, **53**, 183–197.

48 Hertzberg, R.C. and Teuschler, L.K. (2002) Evaluating quantitative formulas for dose-response assessment of chemical mixtures. *Environ. Health Perspect.*, **110**, 965–970.

49 Mumtaz, M.M. and Durkin, P.R. (1992) A weight-of-evidence scheme for assessing interactions in chemical mixtures. *Toxicol. Ind. Health*, **8**, 377–406.

50 IPCS (2008) Draft Guidance on Principles of Characterizing and Applying PBPK Models in Risk Assessment. World Health Organization, International Programme on Chemical Safety, Geneva. Available at http://www.who.int/ipcs/methods/harmonization/areas/pbpk_guidance/en/index.html.

11
Application of ATSDR's Mixtures Guidance for the Toxicity Assessment of Hazardous Waste Sites[1]

David Mellard, Mark Johnson, and Moiz Mumtaz

11.1
Introduction

The major goal of the risk assessment process is to identify, characterize, and quantify risk so as to protect the public and the environment. To have an informed environmental decision making, it is crucial that the ever-evolving science be correctly used to seek solutions to contemporary issues encountered in our lives. This is possible only through a thorough information analysis, development of a strategic action plan, and prioritization of research undertakings on the basis of anticipated impacts and outcomes. The decision-making process can be vastly improved by a better understanding of the mechanism or mode of action of chemicals and their interactions. The better is the understanding of the relationship between environmental levels and health effects, the better are the risk assessment conclusions and recommendations.

Critical to assessing the environmental and public health impact of individual chemicals and chemical mixtures is interaction with the public. The Agency for Toxic Substances and Disease Registry (ATSDR) has a long history of involving the public in the agency's work at hazardous waste sites. This involvement is so important that the agency has community involvement staff devoted solely to fostering and enabling interactions between the staff scientists and the public. One model that has worked particularly well is to have an initial public availability session in which ATSDR's scientist(s), health educators, and community involvement specialists meet residents who are concerned about hazardous waste in their community. From this initial meeting, staff gathers health concerns that will be addressed as part of the agency's work at the site. Quite often, staff establishes a health team consisting of community representatives and environmental organizations, as well as other federal, state, and local health and environmental agencies. At very complicated sites, this team will develop a plan to identify various public health activities that should take place as part of ATSDR's health assessment or health consultation process. ATSDR staff members work with this team to address and evaluate hazardous waste issues in a community,

1) The findings and conclusions in this chapter are those of the authors and do not necessarily represent the views of the Agency for Toxic Substances and Disease Registry.

Principles and Practice of Mixtures Toxicology. Edited by Moiz Mumtaz

ISBN: 978-3-527-31992-3

constantly getting the team's input into the agency's activities and assisting the agency in planning public meetings to present the agency's findings. This interaction is crucial to understanding the community's concerns, addressing those concerns, and partnering with the community in making public health decisions. This approach is consistent with phase I, problem formulation and scoping, in the National Research Council's recommendations called "Science and Decisions: Advancing Risk Assessment" and the book by the same name [1].

The process of environmental health assessment starts with exposure assessment that entails multiple steps, including identification of chemicals present in various environmental media, evaluation of exposure pathways, and determination of integrated exposure(s). Establishment of a completed exposure pathway (CEP) is the key to determination of integrated exposures at waste sites [2]. A CEP exists when there is direct evidence or a strong likelihood that people have come in contact with site-related contaminants, in the past or at present. For a route to be categorized as CEP, the source of contamination, environmental fate and transport, exposure point, and exposed population all must be identified. If a CEP is identified, a plan to mitigate exposures that could often compromise human or environmental health is developed and implemented. Of the 1706 waste sites assessed by ATSDR, 743 had CEPs. Of these, mixtures of chemicals were present at 588. At the sites with chemical mixtures, 64% had at least three chemicals [3]. These data underscore the pervasiveness of unintended exposure to mixtures of chemicals.

Environmental monitoring, biomonitoring, surveillance, and population surveys help put exposure assessment results into perspective. It is determined if the exposure levels are significant and constitute a potential health hazard. Environmental data provide information regarding the nature and extent of contamination and the magnitude of potential exposures. Depending on the availability of data, one of the appropriate methods described in ATSDR's Guidance Manual for the Assessment of Joint Toxic Action of Chemical Mixtures is used to estimate the potential toxicity of the mixture (http://www.atsdr.cdc.gov/interactionprofiles/ipga.html) [4].

The estimation of joint toxicity, thus obtained, is compared with epidemiological findings of health outcomes or disease prevalence. This comparison of toxicity assessment with the epidemiological facts leads to appropriate public health actions. Thus, the final recommendations are arrived at through pragmatic and realistic environmental toxicity assessments conducted by considering issues beyond single chemicals within the risk assessment paradigm of exposure assessment, hazard identification, dose response assessment, and risk characterization.

11.2
ATSDR's Process for Evaluating Chemical Mixtures

Assessing multiple chemical exposures involves a critical evaluation of environmental and health effect data, including toxicological, experimental laboratory, occupational, and epidemiological information. Three different, alternative approaches are available for toxicity assessment of chemical mixtures [4–6]. The most direct and accurate form of risk assessment with the least uncertainty is the evaluation of available toxicologic or epidemiologic data about the mixture of concern. When such data are available and reviewed, a criterion or regulatory standard is derived for occupational exposures and environmental mixtures. With the second approach, if data are unavailable for the mixture of concern but are available for a "similar mixture," these data are used to derive a criterion. Before using such data, however, the qualitative and quantitative aspects of the mixture's composition are carefully considered. This second approach is used on a case-by-case basis or if one is grouping chemicals [7] since no criteria exist for determining a similar mixture. In both the above approaches, the mixture under consideration is treated as a single chemical because data are available on the whole mixture. If data are not available on the whole mixture but are available for individual components of the mixture, then a third approach, the "hazard index," is used. This most often used approach is based on the concept of potency-weighted dose or response additivity.

ATSDR adopted these three methods in its guidance for assessment of chemical mixtures ([4]; http://www.atsdr.cdc.gov/interactionprofiles/ipga.html). The assessment process starts with an interaction profile (http://www.atsdr.cdc.gov/interactionprofiles/) for the specific mixture of concern [4]. Interaction profiles are documents that evaluate mixtures that are of special interest to environmental and public health. Such profiles provide the risk assessor's information about the types of interactions that can occur among the chemical components of a mixture. These documents also recommend methods for incorporating concerns about interactions into the public health assessment of a waste site. Based on published experimental and theoretical studies, these recommendations provide an assessment of interactions and generalized rules that might be used inferentially for other related exposure scenarios. These documents also provide the relevance of interaction data and their use in public health assessments. Thus, health assessors can combine information from interaction profiles with site-specific exposure data in health assessments and consultations (step 1). If an interaction profile is not available, then a toxicological profile or a whole-mixture study is consulted for guidance (Figure 11.1).

The toxicological profiles (http://www.atsdr.cdc.gov/toxpro2.html) are summaries of ATSDR's evaluations concerning exposure to chemicals and their consequences, including the establishment of adverse and no-adverse effects levels. The profiles include information about exposure and environmental fate, helping to determine the significance of levels found in the environment. These documents are for the informed public and health professionals who need succinct interpretations of toxicological data. Thus, these documents provide interpretations of data, a function that distinguishes them from other toxicological reviews.

Interpreting data often requires judgment and implicit assumptions that are more a matter of policy than objective science. Specifically, the profiles incorporate ATSDR's evaluations of the validity of particular studies and the inferences that can be made from them. A section of these documents subtitled "Interactions with Other Substances" provides mechanisms of the interaction, if known, including the influence of pharmaceuticals and other substances on the toxicity of the profiled substance. The limitations of the human and animal studies are addressed with

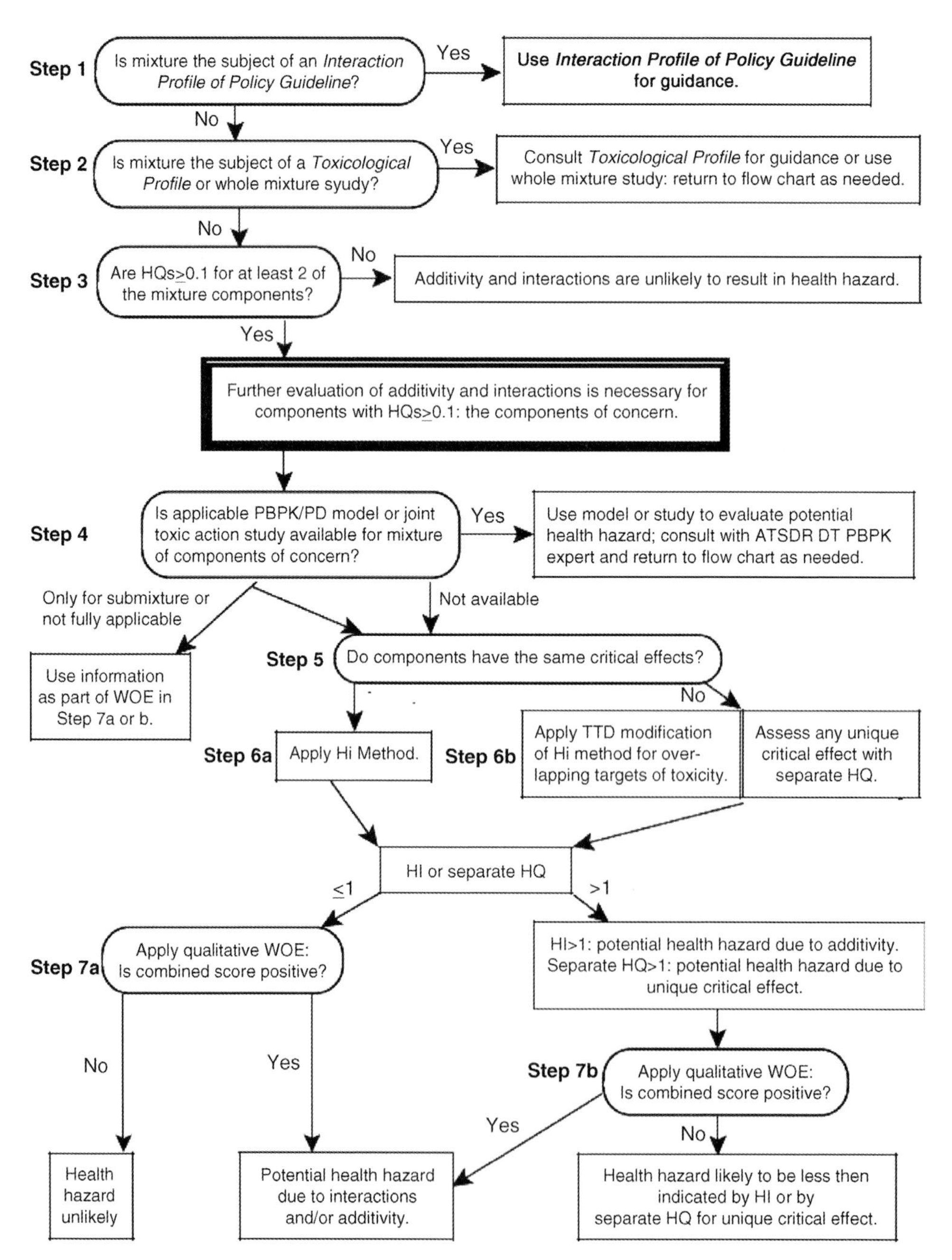

Figure 11.1 ATSDR's flow chart showing the steps involved in assessing the joint toxicity of chemical mixtures.

specific discussions regarding interactions, if interactions have been demonstrated only in animals or *in vivo*, and the studies speculate, providing the basis as to whether such interactions are likely to occur in humans. The following types of interactions are described in these documents:

- Direct interaction among chemicals, causing a chemical change in one or more of the chemicals.
- The effect of one chemical on the pharmacokinetics/metabolism of another, such that the quantity of the biologically active moiety reaching the target organ is altered.
- The capability of one chemical to modify the biological actions of the second by exerting biological effects that enhance or counteract the actions of the second one.
- The capacity of one chemical to cause alterations in the target or receptor(s) for the second.

As mentioned previously, when information about a mixture or a similar mixture is not available, the process of using individual chemical data and the potency-weighted dose or response additivity models is employed. A detailed analysis (steps 2–7) is performed as described in the Guidance Manual for the Joint Toxicity Assessment of Chemical Mixtures [4] (see Chapter 10).

Step 1: Use interaction profile or policy guideline if available for the mixture of concern

The interaction profiles recommend specific approaches to be used with site-specific exposure data to assess potential health hazard for joint toxicity assessment of mixtures. The recommended approaches may include the use of whole-mixture data, assessment of components singly, physiologically based pharmacokinetic/pharmacodynamic (PBPK/PD) models, toxicity equivalent factor (TEF), hazard index (HI), target toxicity dose (TTD), weight of evidence (WOE), indicator chemical, or other approaches. Such information as available, including TEF, binary weight of evidence (BINWOE), and TTD values, is provided in these documents. If the assessments therein offer only partial coverage of the mixture, the information is used as appropriate, and the flow chart shown in Figure 11.1 is followed for additional guidance. For example, an interaction profile may cover some of the component chemicals in the mixture but not others. The flow chart in Figure 11.1 can be used to further define the components of concern before deciding whether the mixture "matches" the mixture in the interaction profile, and to account for components of concern not covered by the profile. If no ATSDR documents are available and relevant information is available from another agency's documents, suitable information from the other agency can be used if deemed appropriate. Otherwise, one should return to the flow chart and proceed to step 2.

Step 2: Consult toxicological profile or use whole-mixture study, if available, for the mixture of concern

A number of toxicological profiles deal with intentional and generated mixtures and are posted on the ATSDR web site (http://www.atsdr.cdc.gov/toxpro2.html).

These mixtures include fuels [8]; polychlorinated biphenyls (PCBs) [9]; chlorinated dibenzodioxins (CDDs) [10]; polycyclic aromatic hydrocarbons (PAHs) [11]; pesticides, such as toxaphene [12]; and total petroleum hydrocarbons [13]. Some mixtures are assessed as whole mixtures (e.g., certain fuels, pesticides, and PCBs); some mixtures are assessed with minimal risk levels (MRLs) for individual components or by using a fraction approach (e.g., PAHs); and some mixtures are assessed by using dose-additivity of the components (chlorinated dibenzodioxins and chlorinated dibenzofurans – CDD and CDF). For complex mixtures of petroleum hydrocarbons, similar components are lumped into fractions for exposure and health effect assessment, and the use of MRLs for the fractions is recommended on the basis of a single representative (surrogate) component or a similar mixture.

For some fractions, an indicator chemical approach is used. ATSDR has considered some mixtures, such as gasoline and Stoddard Solvent, too variable in composition for MRL derivation [14, 15]. In this indicator chemical approach, when appropriate, the most toxic (known) component of the mixture is selected as the indicator of toxicity; this selection assumes that the indicator chemical would drive the risk assessment. For example, benzene is used as a marker or an indicator chemical for environmental exposure to automotive gasoline [16]. Alternatively, a fraction approach for complex mixtures of petroleum hydrocarbons [13], in conjunction with a components approach for the nonhydrocarbon components (such as methyl-*t*-butyl ether), may be useful for gasoline. If the toxicological profile does not provide MRLs or recommendations for health assessment approaches, and if relevant documentation from other agencies is not available or is not suitable, the literature can be searched for studies on the mixture of concern (whole mixture), and any available study can be evaluated for possible use as the basis for an MRL or to identify potential health effects of concern from exposure to the mixture. Studies of wildlife or companion animals exposed to site-related chemicals may be useful in identifying that a hazard exists at environmental levels of exposure, if such studies are evaluated for relevance to potential effects on human health. MRLs are derived in accordance with ATSDR guidance, which is provided in the appendix to each toxicological profile. Additional guidance regarding implementation of a "whole mixture" approach is provided by the agency [4]. If information sufficient to assess a mixture is not identified, return to the flow chart in Figure 11.1 and proceed to step 3.

Step 3: If no ATSDR document is available for the mixture of concern, select components of concern

If steps 1 and 2 cannot be followed, a component-based approach is employed. This approach focuses on components that are likely contributors to a health hazard, either because their individual exposure levels exceed health guidelines or because their joint toxicity with other components, including additivity or synergy, may pose a health hazard. Chemicals are considered unlikely to pose a health hazard due to interactions if the hazard quotient (HQ) is less than 0.1 with the use of noncancer health guidelines, such as MRLs or reference doses (RfDs) [4]. Unless there are a relatively large number of chemicals that act similarly, the mixture of chemicals is not likely to pose an increased hazard due to additivity. Such chemicals are eliminated from further consideration in step 3. The value 0.1 is chosen as a reasonable point of departure for simple mixtures consisting of approximately 10 components or fewer. If all the chemicals have HQs < 0.1, additivity and interactions among the chemicals are unlikely to result in a hazard to public health and further assessment of the mixture is not necessary. If two or more components have HQs > 0.1, these chemicals require further evaluation to determine if joint toxicity could occur. Proceed with the evaluation of additivity and interactions in steps 4–7 for these chemicals of concern. Judgment should be used, however, in applying this value. With a mixture of more than 10 components that act similarly, or with several components with HQs just slightly below 0.1 and other HQs above 0.1, a slightly lower point of departure may be appropriate.

Step 4: Evaluate and use PBPK/PD model or joint toxic action studies, if available and appropriate

If a PBPK model, a PBPK/PD model, or a mixture study is available for the complete mixture of components of concern, one should evaluate its relevance to human exposure by the anticipated route(s) and duration, as well as the noncancerous health effects of concern for the chemicals. Follow-up studies are commonly performed to validate such models. The effects of concern will include the critical effects and any relatively sensitive effects in common among two or more of the mixture components. The critical effect is the effect that forms the basis for the MRL (or RfD or RfC). Evaluation of the model should also include whether the models for the individual chemicals have been linked in a reasonable manner, based on the chemicals' toxicokinetics and mechanisms of action and the extent of validation of the model. There are some models or studies of joint action that may be directly useful, for example, if they report apparent threshold exposures for interactions relevant to human exposure or state that the chemicals will not interact.

This information can be used during the WOE evaluation (step 7) and as part of the rationale for the mixtures approach. The availability of linked PBPK and PBPK/PD models for mixtures is very limited, but research in this area is being highly actively pursued. Therefore, an appropriate database, such as TOXLINE, should be searched to identify pertinent PBPK or PBPK/PD models. For some mixtures, models may be available only for pairs of components of the mixture. In such a case, the hazard index method (step 6a) or the TTD modification of the hazard index or a separate HQ (step 6b) can be chosen, as appropriate, and reported results of the modeling for pairs of components can be used as part of a WOE approach [17]. If no suitable models are available or if the models are to be used as part of the WOE evaluation, proceed to step 5.

Step 5: Evaluate whether chemicals have the same or different critical effects

Assess whether the chemicals that contribute to a particular exposure pathway of concern appear to affect primarily the same end points, particularly in terms of critical effects or critical target organs. If so, apply the hazard index method (step 6a), using each chemical's individual HQs derived from the health guidelines, such as the MRL, RfD, or RfC. If the components appear to have a variety of critical effects, apply the TTD modification to the hazard index (step 6b) so that individual HQs can be developed through the use of the same target organ or critical effect. If most, but not all, of the mixture's chemicals have the same critical effect, the hazard index method can be chosen on the grounds of practicality, thus allowing the use of already established MRLs, RfDs, or RfCs to derive the individual HQs that make up the HI for the mixture.

Step 6a: Apply hazard index method to components with similar critical effects

The hazard index method for multiple chemicals is based on EPA's HQ approach for single chemicals. The HQ for an individual chemical requires estimation of a site-specific dose for residents who might be exposed to chemicals via an identified pathway. The formulas for deriving the HQ of a single chemical via ingestion and inhalation follow:

$$\text{Oral HQ}_{\text{individual chemical}} = \text{estimated oral dose in people/oral health guideline},$$

$$\text{Inhalation HQ}_{\text{individual chemical}} = \text{concentration in air/inhalation health guideline}.$$

Once the individual HQs are determined for a specific route and group of chemicals, the HI for the chemical mixture can be determined. The formula for determining the HI for a mixture containing three chemicals follows:

$$\text{Oral HI}_{\text{mixture}} = \text{oral HQ}_{\text{chemical one}} + \text{oral HQ}_{\text{chemical two}} + \text{oral HQ}_{\text{chemical three}},$$

$$\text{Inhalation HI}_{\text{mixture}} = \text{inhalation HQ}_{\text{chemical one}} + \text{inhalation HQ}_{\text{chemical two}} + \text{inhalation HQ}_{\text{chemical three}}.$$

A hazard index is estimated for a specific receptor population (e.g., preschool children or adults), for the duration (e.g., acute, intermediate, and chronic) and pathway (e.g., oral, inhalation) of concern. The exposure units should be the same as the units for the health guideline (e.g., mg/kg/day for oral exposure pathways, when using oral MRLs or RfDs). When the pathway is air, the units for air concentration should be the same as the units used for the inhalation MRL or RfC.

If the resulting hazard index exceeds 1, a potential health hazard may exist due to chemical interactions (e.g., additivity or synergy). Further evaluation of the interactions is needed to gauge whether a hazard truly exists and the extent of that hazard (step 7). If the resulting hazard index is less than or equal to 1, further evaluation of interactions is required to assess the potential for synergistic or other interactive effects that would be harmful in the exposed population (step 7).

Step 6b: Apply TTD modification of hazard index method for components with different critical effects

The target toxicity dose modification to the hazard index method [18] allows to estimate separate hazard indices for each major end point or target organ affected by two or more components of the mixture (i.e., the overlapping targets of toxicity). For example, to evaluate liver toxicity for a mixture of chemicals, one will need to determine a liver target toxicity dose for those chemicals in instances in which the health guideline is based on another target organ. TTDs do not have to be developed if the health guideline (e.g., MRL) is already based on liver toxicity. Once specific TTDs are developed for a mixture, they are used as health guidelines to develop the HI for the mixture. Using this approach makes it possible to develop organ-specific HIs (e.g., liver HI, kidney HI, etc.).

TTD-HI Method Example

Four-Component Mixture
Three components affect liver (#1, 2, 4)
Three components affect kidney (#1, 3, 4)
Three components affect neurological system (#1, 2, 3)
All four components affect fetal development
Critical targets: liver for #1 and 2; kidney for #3; development for #4

$$\mathrm{HI}_{\mathrm{hepatic}} = \frac{E_1}{\mathrm{MRL}_1} + \frac{E_2}{\mathrm{MRL}_2} + \frac{E_4}{\mathrm{TTD}_{4\ \mathrm{hepatic}}}$$

$$\mathrm{HI}_{\mathrm{renal}} = \frac{E_1}{\mathrm{MRL}_{1\ \mathrm{renal}}} + \frac{E_3}{\mathrm{MRL}_3} + \frac{E_4}{\mathrm{TTD}_{4\ \mathrm{renal}}}$$

$$\mathrm{HI}_{\mathrm{neuro}} = \frac{E_1}{\mathrm{MRL}_{1\ \mathrm{neuro}}} + \frac{E_2}{\mathrm{MRL}_{2\ \mathrm{neuro}}} + \frac{E_3}{\mathrm{TTD}_{3\ \mathrm{neuro}}}$$

$$\mathrm{HI}_{\mathrm{dev}} = \frac{E_1}{\mathrm{MRL}_{1\ \mathrm{dev}}} + \frac{E_2}{\mathrm{MRL}_{2\ \mathrm{dev}}} + \frac{E_3}{\mathrm{TTD}_{3\ \mathrm{dev}}} + \frac{E_4}{\mathrm{MRL}_4}$$

If any of the end point-specific HIs exceed 1, a *potential* health hazard may exist due to additivity for the mixture. Further evaluation of interactions is needed to gauge the extent of the hazard (step 7). If all the end point-specific hazard indexes are 1 or less than 1, further evaluation is required to assess the potential for synergistic interactions to increase the hazard (step 7). In addition, if any chemical in the mixture has a unique critical effect (an effect not produced by any of the other chemicals), this effect should be addressed by assessing whether the individual HQ exceeds unity, in which case it could be a *potential* health hazard. The qualitative WOE method should also be applied (step 7) to gauge whether any of the other mixture components may influence the toxicity of a chemical with regard to this critical effect.

Step 7 (a and b): Apply qualitative WOE

The qualitative WOE methodology is used to predict the joint toxicity when the data are not sufficient (as is usually the case) to allow the use of more quantitative means [17]. In this approach, a mixture is broken down into binary pairs and the interaction potential of each pair of chemicals is determined. This determination is the binary weight of evidence. More specifically, a BINWOE is a qualitative judgment, based on empirical observations and mechanistic considerations, that categorizes the most plausible nature of any potential influence of one compound on the toxicity of another compound for a given exposure scenario. Thus, this approach allows the integration and evaluation of the available information on the metabolism, health effects, and other pertinent data available in the literature on the chemicals. Each BINWOE gives the direction of an interaction (greater than additive, less than additive, or additive) and a classification that is based on the mechanistic understanding and toxicological significance of the interaction.

The qualitative WOE also helps to add professional judgment in deciding the true risk of harmful effects from a mixture. As part of the WOE, the BINWOE determinations for two chemicals are used to make judgments whether the health hazard may be greater or less than that would be predicted on the basis of the hazard index alone. BINWOEs need to be route, duration, and end point or target organ specific. This specificity may be accommodated within a single BINWOE determination for two chemicals or through separate BINWOE determinations for various combinations of two chemicals within a multichemical mixture. Before using a BINWOE, you should make sure it is applicable to the route(s), duration(s), and effect(s) of concern for the particular assessment.

Step 7a: This part of step 7 describes the application of the qualitative WOE to mixture HIs that are less than or equal to 1. If the BINWOE alphanumeric scores indicate less than additivity or additivity, or the combined numerical score is negative or very close to zero, the mixture is unlikely to be a health hazard at the hazardous waste site at the estimated doses. Thus, the HI for the mixture probably overestimates the risk because the health hazard is likely to be less than the calculated HI score. In contrast, if the BINWOE alphanumeric scores indicate greater than additivity, or if the combined BINWOE numerical score is positive and significantly greater than zero, exposure to the mixture may be underestimated and a potential health hazard could exist.

Step 7b: This part of step 7 describes the application of the qualitative WOE when the HI exceeds unity ($HI > 1$). If the BINWOE alphanumeric scores indicate greater than additivity or additivity, or if the combined BINWOE numerical score is positive, the estimated doses for the mixture may be a *potential* health hazard due to interactions and/or additivity.

After identifying the HI and BINWOE scores, the health assessor then compares the estimated doses to the no-observed adverse effects levels (NOAELs) and the lowest observed adverse effect levels (LOAELs) and uses professional judgment to decide if harmful effects might be possible. Further explanation of these methods can be found in ATSDR's mixture guidance [4] and the ATSDR Public Health Assessment Guidance Manual [2].

11.3
Case Studies

Two hazardous waste sites are used to illustrate the application of ATSDR's mixtures risk assessment guidance. At the Village of Endicott, New York, residents used municipal water for drinking and bathing; this water contained very low levels of chlorinated solvents. At the Conrail Rail Yard Site, residents used water from their private wells; this water was contaminated with carbon tetrachloride and trichloroethylene (TCE). These examples were chosen to show how to evaluate the joint toxicity of chemical mixtures and how to communicate scientific findings to residents.

The Health Consultation for the Endicott Area Investigation and the Public Health Assessment for the Conrail Rail Yard Site are available at ATSDR's web site: http://www.atsdr.cdc.gov [19, 20].

11.3.1 The Endicott Area Case Study

11.3.1.1 Site Description and History

The Village of Endicott is a mixed residential, commercial, and industrial community in the Town of Union, Broome County, New York. Endicott has a rich industrial heritage that included large manufacturing operations at the Endicott Johnson Tannery and the International Business Machines (IBM) facility. Many former and current businesses within the Village of Endicott also used solvents containing volatile organic compounds (VOCs). Such businesses include, but are not limited to, Endicott Johnson, IBM, automotive repair facilities, print shops, and dry cleaners. As a result of leaks and spills associated with these operations and runoff from local landfills, groundwater in the Endicott area is contaminated with VOCs.

Several municipal water supply wells provide drinking and household water for 48 000 customers in Endicott. The major source of public water is and has been the Ranney well, which supplies about 92% of the public water. In the early 1980s, routine monitoring of this main public water supply well detected VOCs at levels above New York State drinking water guidelines. Several measures were taken from 1983 to 1992 to reduce VOC levels in groundwater from the Ranney well. These efforts have largely eliminated VOC contamination in the Ranney well. At present, VOCs are seldom detected, and when they are found to be present, almost all levels are below the maximum levels permitted by federal and state guidelines. The South Street and Park wells are intermittently used to supply water to the public water system and have been found to contain low levels of several VOCs. The South Street/Park wells are used to supply about 6% of the public water and might have supplied a higher percentage of water in the past.

The Endicott municipal water supply operates on a grid-water system, meaning that neighborhoods closest to the wells are usually supplied at a greater rate from nearby wells than from wells farther away. For instance, those persons living close to the South Street/Park wells are likely served with greater amounts of water from the South Street/Park wells than are persons living farther away. However, due to system pressure variations, it is not possible to define strictly the areas of Endicott served by the South Street/Park wells, as opposed to those areas served by the Ranney well.

Members of a community advisory panel are concerned that VOCs in the South Street/Park wells, although individually below New York State drinking water standards, could produce an increased level of risk because the chemicals occur as a mixture. These concerns prompted ATSDR to evaluate the public health significance of a possible mixture effect when VOCs are below drinking water standards. Because our evaluation was prompted by members of the community advisory panel, we worked closely with those members by explaining how we reached our decisions about the risk of harmful effects from the mixture of VOCs.

11.3.1.2 Groundwater Contaminant Levels

Water samples from the South Street and Park wells were collected and analyzed for VOCs between January 1980 and July 1981, respectively. Table 11.1 contains summary statistics for the South Street and Park wells for the following VOCs: 1,1-dichloroethane (1,1-DCA), *cis*-1,2-dichloroethene (*cis*-1,2-DCE), 1,1,1-trichloroethane (TCA), vinyl chloride (VC), tetrachloroethene (PCE), and *trans*-1,2-dichloroethene (*trans*-1,2-DCE). Table 11.1 provides a summary of the findings regarding total VOC levels.

For VOCs, the detection limits for the South Street wells were about 1.0 part per billion (ppb). To address residents' concerns, we used the detection limit to express nondetectable concentrations. In this case, however, using the detection limit, ½ the detection limit, or even zero had no effect on the final outcomes.

We chose to use the arithmetic mean to estimate doses because the distribution of concentrations was normally distributed for most chemicals. In addition, using the arithmetic mean was a more conservative approach because the arithmetic means were always higher than or equal to geometric means in this data set. VOC concentrations from the South Street/Park wells were used to estimate dose because (1) the levels were higher in these wells and (2) members of the community advisory panel wanted to know the risk for people who were supplied by these wells.

Table 11.1 Summary statistics of VOC levels in the South Street and Park wells for the Village of Endicott.

Well #	Statistical parameter	1,1-DCA	*cis*-1,2-DCE	TCA	TCE	VC	*trans*-1,2-DCE
South Street	Geometric mean	0.71	1.48	0.71	0.29	0.28	0.27
South Street	Arithmetic mean	0.79	1.69	0.92	0.33	0.29	0.28
South Street	Maximum	1.76	4.00	3.00	2.00	0.90	0.50
South Street	Number of detections	44	52	42	3	1	0
South Street	Frequency of detection (%)	74.58	88.10	71.20	5.08	1.69	0
Park	Geometric mean	0.67	1.56	0.31	0.28	0.30	0.34
Park	Arithmetic mean	0.76	1.73	0.39	0.29	0.34	0.58
Park	Maximum	1.71	3.00	2.50	0.50	2.00	5.00
Park	Number of detections	41	50	5	0	4	6
Park	Frequency of detection (%)	70.69	86.20	8.62	0	6.9	10.34
NYS standard		5	5	5	5	2	5

We explained that residents who depended upon the entire water supply were not likely exposed to water solely from the South Street wells, given that the amount of water from the South Street wells represents only a small portion of the entire water supply. Due to the water distribution system, residents geographically closer to the South Street wells likely received a greater portion of their water supply from these wells. Therefore, these residents would likely be the group exposed to the higher VOC levels.

11.3.1.3 Brief Review of Exposure Pathways

We were asked to evaluate the possibility of harmful effects when people are exposed to VOCs from drinking public water. To estimate the oral dose, we used the arithmetic mean of each VOC from either the South Street or the Park wells, whichever was higher.

The average tap water intake for adults is 1.4 l/day, while the average tap water intake for children 1–3 years is 0.6 l/day. Table 11.2 shows the average daily tap water intake for various age groups in the United States [21]. From these intake rates, it is possible to estimate the dose (expressed in mg/kg/day) of chemicals for people who drank public water. The first step in evaluating exposure to each chemical is to compare the estimated doses to health guidelines, such as ATSDR's MRL. An MRL is an estimate of the daily human exposure to a chemical that is likely to be without an appreciable risk of adverse noncancerous effects over a specified duration of exposure. In other words, if the estimated dose for a person is below the MRL, then the person is not likely to experience (noncancerous) harmful effects from drinking the water. When oral MRLs are not available, doses can be compared to US EPA's reference dose or other health guidelines if they are available.

If the oral MRL is exceeded, further toxicological evaluation is required to determine if noncancerous harmful effects might be possible. This additional evaluation involves comparing the estimated dose in people exposed to a contaminant to doses in human and animal studies that showed harmful effects. The estimated dose can also be compared to doses in human and animal studies that did not show harmful effects. Certain sections in ATSDR's toxicological profile for a chemical provide additional insight that allows a more informed judgment in

Table 11.2 Average tap water intake rate and body weight for various age groups.

Age group	Number of 8 ounce glasses per day	Amount of tap water drank per day (l)	Body weight (kg)
Preschool children (1–3 years)	2–3	0.6	10
Preschool children (3.1–5 years)	3–4	0.87	16
Elementary school children	3	0.7	35
Teenagers	4	0.97	55
Adult women	6	1.4	60
Adult men	6	1.4	70

deciding possible health effects. The sections on toxicokinetics, mechanism of action, children's susceptibility, interactions with other chemicals, and populations that are unusually susceptible provide much in-depth scientific information. From this comparison of estimated dose to human and animal studies and review of the toxicological literature, it is possible to decide if harmful effects might be possible in people who are exposed to a contaminant. Table 11.2 provides data on average tap water intake and body weight for various age groups of people who used the water supply.

The equation for determining an oral dose for people who drank public water is as follows:

$$\text{Oral dose} = (\text{chemical concentration in water} \times \text{tap water intake/day})/\text{body weight}.$$

Because children and adults drink different amounts of tap water each day, the chronic oral dose of each chemical from drinking tap water will slightly vary, depending on the age and weight of the person. Table 11.3 shows the chronic oral dose of six VOCs for different age groups, together with the chronic oral MRL or the RfD for each chemical. For example, the estimated dose of 1,1-DCA for children 1–3 years old is 0.000048 mg/kg/day. This dose can be compared with US EPA's RfD of 0.1 mg/kg/day. This comparison shows that the estimated dose of 1,1-DCA in children 1–3 years of age is below the health guideline of 0.1 mg/kg/day. This means that children 1–3 years of age who drank tap water are not likely to experience noncancerous harmful effects from 1,1-DCA in tap water. Even if the tap water intake is doubled to include the group of children with higher-than-

Table 11.3 Oral doses of individual chemicals for various age groups from drinking public water.

Age group	1,1-DCA (dose in mg/kg/day)	*cis*-1,2-DCE (dose in mg/kg/day)	TCA (dose in mg/kg/day)	VC (dose in mg/kg/day)	*t*1,2-DCE (dose in mg/kg/day)	TCE (dose in mg/kg/day)
Children 1–3 years	0.0 000 482	0.0 001 031	0.0 000 561	0.0 000 207	0.0 000 354	0.0 000 201
Children 3.1–5 years	0.000 043	0.0 000 919	0.00 005	0.0 000 185	0.0 000 315	0.0 000 179
Children 6–12 years	0.0 000 158	0.0 000 338	0.0 000 184	0.0 000 068	0.0 000 116	0.0 000 066
Teenagers 13–17 year	0.0 000 139	0.0 000 298	0.0 000 162	0.000 006	0.0 000 102	0.0 000 058
Adult women 18 years and older	0.0 000 186	0.0 000 397	0.0 000 216	0.000 008	0.0 000 136	0.0 000 078
Adult men 18 years and older	0.0 000 159	0.0 000 340	0.0 000 185	0.0 000 068	0.0 000 117	0.0 000 066
Oral MRL	None	None	None	0.003	None	None
Oral RfD	0.1	0.01	0.28	0.003	0.02	0.00 146[a)]

a) Health Canada developed this reference dose for TCE.

Table 11.4 Oral hazard quotient for individual chemicals in different age groups from drinking public water.

Age group	1,1-DCA	*cis*-1,2-DCE	TCA	VC	t1,2-DCE	TCE
Children 1–3 years	0.0005	0.01	0.0002	0.007	0.002	0.01
Children 3.1–5 years	0.0004	0.009	0.0002	0.006	0.002	0.01
Children 6–12 years	0.0002	0.003	0.00 007	0.002	0.0006	0.004
Teenagers 13–17 years	0.0001	0.003	0.00 006	0.002	0.0005	0.004
Adult women (18 years and older)	0.0002	0.004	0.00 008	0.003	0.0007	0.005
Adult men (18 years and older)	0.0002	0.003	0.00 007	0.002	0.0006	0.005

a) The HQ is rounded off to one significant figure.

average intake, the estimated dose is still well below the health guideline established by the RfD. The estimated dose of the other VOCs is also below each chemical's health guideline; therefore, noncancerous harmful effects are unlikely when one is evaluating the VOCs individually.

Table 11.4 shows the oral HQ for each chemical. The oral HQ is determined by dividing the estimated dose by the chronic health guideline (i.e., the chronic oral MRL or the oral RfD). For example, the derivation of the oral HQ for 1,1-DCA in children aged 1–3 years follows:

$$\text{Oral HQ}_{\text{children aged 1-3 years}} = \text{estimated dose for children 1-3 years old/oral RfD},$$

$$\text{Oral HQ}_{\text{children aged 1-3 years}} = 0.000048\ \text{mg/kg/day}/0.1\ \text{mg/kg/day},$$

$$\text{Oral HQ}_{\text{children aged 1-3 years}} = 0.00048.$$

The oral HQ of 0.00048 is rounded off to 0.0005. Because the oral HQ for chronic exposure is below 1, noncancerous harmful effects are not likely for children 1–3 years old who drank tap water. The advantage of calculating an oral HQ is that one number shows whether the estimated oral dose in each age group is above (i.e., the oral HQ is greater than 1) or below (i.e., the oral HQ is less than 1) a chronic health guideline. Using the HQ allows the scientist to quickly review a set of numbers and determine if further evaluation is needed to assess the possibility of harmful effects.

As you can see from Table 11.4, none of the HQs for the chlorinated solvents exceeds 1; therefore, individually, none of the chemicals is likely to cause noncancerous harmful effects in children or adults.

One more step, however, is needed to determine if a mixture's effect might cause noncancerous harmful effects.

11.3.1.4 **Evaluating Oral Exposure to the Mixture of Chemicals in Endicott's Water**

Once the oral HQ for each chemical is determined, the next step is to evaluate the mixture of chemicals in Endicott's water. This portion of the mixture's evaluation determines possible interactions among the VOCs in Endicott's water.

The health scientist first reviews the individual oral HQs for each chemical to decide if an oral hazard index is needed for the mixture of chemicals. ATSDR's mixture guidance states that if all the oral HQs for each chemical are less than 0.1, then interactions among the chemicals in the mixture are unlikely. Stated another way, the chemical mixture will not have any significant interactions (either additive, synergistic, or antagonistic) if each of the individual oral HQs is less than 0.1. ATSDR's mixtures guidance also states that if only one HQ exceeds 0.1, then interactions between that chemical and other chemicals in the mixture also are unlikely. A review of the HQs in Table 11.4 shows that the HQ for each age group and for each chlorinated solvent is well below 0.1; therefore, interaction effects are not likely to be significant.

11.3.1.5 Conclusions for the Mixture of Chemicals and Drinking Endicott's Water

Children and adults are unlikely to experience noncancerous harmful effects from the mixture of chemicals in Endicott's water. Noncancerous harmful effects are unlikely because the individual chemicals in the mixture will not interact in any way that might be harmful.

11.3.1.6 Presentation to Residents on the Community Advisory Board

An integral part of public health practice requires talking with residents who live near hazardous waste sites and involving them in reaching decisions about possible health effects. The greater the involvement of residents in the decision-making process about exposure and health effects, the greater the acceptance the final message will receive. The goal of this interaction is not only to answer their questions about chemical contamination of the environment but also to convince them that our conclusions accurately describe their risk. For the Endicott area investigation, many residents on the community advisory board were already convinced that groundwater contamination with VOCs, even at levels below the drinking water standards, was a public health hazard because of the mixture of chemicals.

One approach that allows some residents to accept conclusions from public health officials is to explain the rather complicated steps used in arriving at the conclusions. For the Endicott area investigation, we sat down with the community advisory board and explained the steps previously described and showed them graphically where the individual doses (i.e., the HQs) were in relation to health guidelines, as well as to levels known to cause harmful effects (i.e., LOAELs) and levels known not to cause harmful effects (i.e., NOAELs).

This discussion requires explaining the concept of dose, using tap water concentrations of VOCs and how much tap water someone drinks. A brief explanation is needed to explain how the dose and health guideline are used to calculate an HQ. Figure 11.2 is a graphic representation of the individual HQs for each VOC; it can be compared easily with the health guideline and with HQs based on NOAELs and LOAELs. HQs were also provided for children and adults because residents were concerned about their children.

Members of the community advisory panel could easily see that the HQs calculated for water coming from the South Street/Park wells were far below the health guidelines for each chemical. Residents raised questions about people who drink

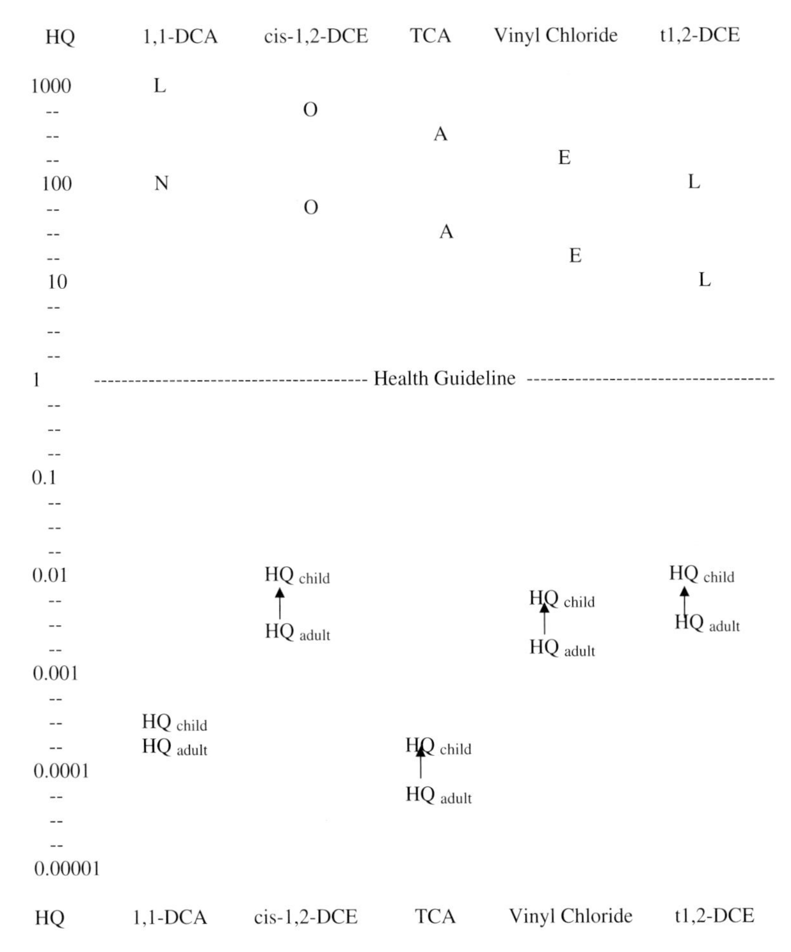

Figure 11.2 Noncancer mixtures evaluation for drinking Endicott's public water. The placement of LOAELs and NOAELs is provided to give a general sense of how high the HQ needs to be before one might expect to see a no-effect level compared to an effect level. The specific LOAELs and NOAELs for each chemical not identified. NOAEL: no-observed adverse effects level; LOAEL: lowest observed adverse effect level.

excessive amounts of tap water (e.g., people who exercise regularly). We were able to show that doubling or tripling the tap water intake provides a slightly higher HQ, but the HQ is still well below the health guideline. Therefore, even people with significantly higher tap water intakes were not at risk of harmful effects. This conclusion was the same for the individual HQs and for the mixture of chemicals. We explained further that because the HQs were all below 0.1 – that is, 10 times below the health guideline – that no interaction effects that would lead to the occurrence of harmful health effects were expected.

Many residents believe that if a health guideline is exceeded, then they will get sick. This graphical presentation also allowed residents to see that even if a health guideline is exceeded (e.g., HQ 3), the exposure is still far below the levels that are known to cause harmful effects. Figure 11.2 became a powerful tool in helping residents visualize this complicated science of dose, health guideline, and HQs. Using this figure allowed us to field questions such as those about people who might drink two or three times more tap water and easily and quickly answer about the risk of harmful effects.

11.3.2 Conrail Rail Yard Case Study

11.3.2.1 Site Description and History

Conrail Rail Yard is a 675-acre facility about 1 mile southwest of Elkhart, Indiana (see Figure 11.3). The rail yard began operating in 1956, serving as one of the largest freight car classification and switching yards in the country. In addition, rail cars are repaired and engines are cleaned there, and a diesel refueling station operated at the facility.

Over the years, many chemical spills have been documented at the facility, including caustic soda solution, hydrochloric acid, grain alcohol, a hydrofluoric gas leak, and diesel fuel. In 1986, Elkhart County Health Department received information that chemical wastes from track cleaning and equipment degreasing operations had been buried on the site. In response to this information, the Elkhart County Health Department tested the drinking water at the Conrail facility. Small amounts of toluene and xylenes were found in the water, at levels below those considered to cause harm. However, testing of private wells near the facility showed levels of trichloro-

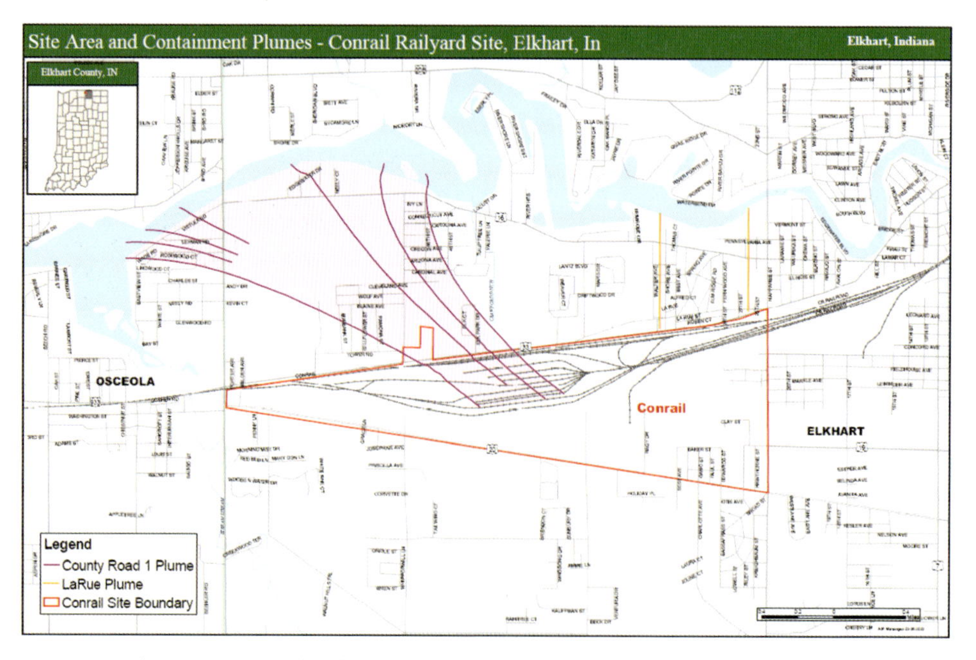

Figure 11.3 Conrail Rail Yard and surrounding area.

ethylene and carbon tetrachloride (CCl_4) that greatly exceeded drinking water standards established by the US EPA for public water supplies (referred to as Maximum Contaminant Levels, MCLs).

11.3.2.2 **Investigation of Groundwater Contamination**

Elkhart County Health Department immediately requested that EPA confirm its findings and provide assistance to characterize the extent of groundwater contamination. In June 1986, EPA collected and analyzed water from two private wells. One well contained 800 ppb of TCE and 485 ppb of CCl_4, while the other well contained 75.6 ppb TCE and 26.5 ppb CCl_4. The MCL for both these chemicals is 5 ppb. The EPA started an emergency action that included testing more private wells in the area and providing alternative drinking water for those residents whose wells were contaminated. Sampling was performed for over 500 private wells in the area. Homes with impacted wells were provided either a point-of-use activated carbon filter unit or a whole-house filter unit.

Figure 11.3 shows the Conrail Rail Yard, surrounding areas, and the general direction of groundwater flow. Three groundwater zones were tested to determine their general direction of flow from the site. All groundwater zones generally flow northwest from the site. The soil in the area is mostly sandy. Sandy soil promotes faster groundwater movement, both vertically and horizontally, than does clay soil.

Additional groundwater sampling was conducted by a private development company, which installed six monitoring wells. The shallow wells that were less than 30 feet deep were not found to contain contamination. However, wells that collected water from deeper than 110 feet were all contaminated. The maximum level for TCE was 2495 ppb, and for CCl_4 it was 388 ppb [22].

Figure 11.4 displays the concentration of CCl_4 and TCE levels for each of the individual private wells. A wide range of concentrations for both chemicals were detected, with many wells having concentrations that greatly exceeded the drinking water standards (5 ppb). The figure also shows that virtually all wells were cocontaminated with both CCl_4 and TCE. In addition to TCE and CCl_4, well sampling also detected dichloroethylene and low levels of tetrachloroethylene (2.4 ppb), chloroform (0.8 ppb), and 1,1,1-trichloroethane (19 ppb).

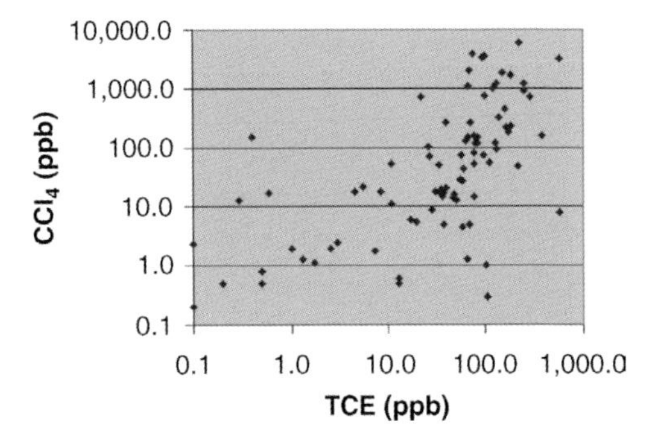

Figure 11.4 Scatter plot of TCE and CCl_4 concentrations for each well.

11.3.2.3 Exposure Pathway Analysis

Residents whose drinking water wells drew groundwater from contaminated aquifers were exposed to those chemicals through ingestion of the water. Evaporation of chemicals from contaminated water while bathing and showering also resulted in exposure through inhalation of TCE and CCl_4 vapors. This inhalation exposure was most significant while a resident was showering and while the resident remained in the bathroom shortly after the shower. Some TCE and CCl_4 also passed through the skin during showering or bathing. Due to the volatility of groundwater contaminants, residents were also exposed through the migration and accumulation of TCE and CCl_4 vapors into the interior air within their homes. The multiple exposure pathway analysis is summarized in Figure 11.5.

Exposure Assessment On the basis of well testing, it was estimated that over 600 people may have been exposed to TCE and CCl_4 as a result of using contaminated well water. Exposure doses are estimated for specific ranges of TCE concentration in water, from less than 5 μg/l (the federal drinking water standard) to greater than 30 000 μg/l. The total dose from all completed routes of exposure (ingestion of the water, dermal contact during showering or bathing, and inhalation of TCE vapors released during water use) was derived by summing the individual doses. This total dose served as the basis for comparison to the toxicological studies that were included in the review. The assumption was that people were exposed to CCl_4 in their drinking water within about a year of a reported tank car spill – from 1968 until safe water was provided, beginning in 1986. Indoor air exposures continued until 1999, when vapor extraction systems were installed on homes and businesses following discovery of the vapor intrusion problem.

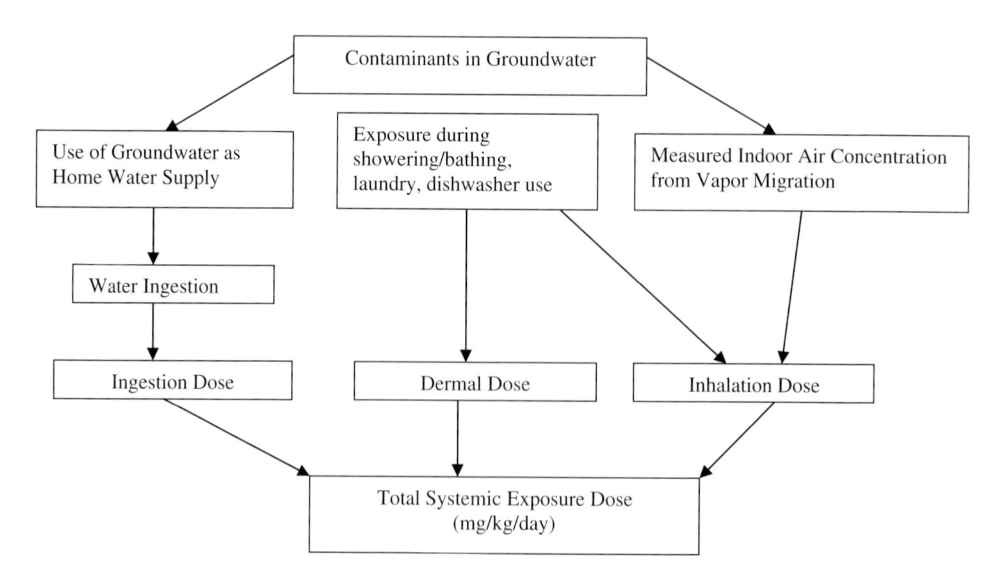

Figure 11.5 Conceptual model for multipathway exposure to groundwater contaminants.

11.3.2.4 Toxicity Assessment for Individual Chemicals

TCE Exposure *Human studies* Most of the information about the effects of TCE in humans is based on occupational studies of cases in which TCE is used as a solvent and degreasing agent. The Department of Health and Human Services/National Toxicology Program has designated TCE as reasonably anticipated to be a carcinogen, and the International Agency for Research on Cancer classifies TCE as a probable human carcinogen. The EPA's cancer designation for TCE is under review. The occupational studies have generally evaluated the effects of inhalation of high concentrations of TCE vapors, with evidence of associations with increased incidence of kidney cancer [23] and non-Hodgkin's lymphoma (NHL) [24]. Other studies have shown weaker associations with cancer risk [25].

There is a much greater level of uncertainty about estimating potential harmful effects associated with exposure to *low* levels of TCE. There are several studies of communities that have been exposed to TCE in their municipal water supply. One study was initiated as an investigation of a cancer cluster study of childhood leukemia cases in Woburn, Massachusetts, in 1986. The area with the reported leukemia cases corresponded to a part of the city where TCE and other solvents had been detected in two of the eight municipal drinking water wells, dating from 1979. This study is of interest because the levels of TCE found in the Woburn wells (maximum TCE detection of 267 μg/l) were within the range of concentrations detected in the private wells affected by Conrail. Results of the epidemiologic analysis of these cases identified a weak association between the potential for exposure to contaminated water during maternal pregnancy and leukemia diagnosis in the child. However, a child's potential for exposure from birth to diagnosis showed no association with leukemia risk.

Another study of the health effects of exposure to TCE-contaminated drinking water was conducted in New Jersey, where 75 towns were evaluated from 1979 to 1987. Investigators compared towns without detectable TCE in drinking water with towns with TCE levels greater than 5 μg/l in their drinking water. The comparison revealed an increase in the incidence of total leukemia among females, particularly for acute lymphocytic leukemia in females under 20 years of age. The study also noted an elevated incidence of chronic myelogenous leukemia among females, chronic lymphocytic leukemia among males and females, non-Hodgkin's lymphoma in females, diffuse large-cell NHL in females and males, and non-Burkitt's high-grade NHL among females and males. The results suggest a link between TCE and leukemia/NHL incidence. However, the conclusions are limited by lack of information about the long-term exposure levels to TCE and the confounding influence of other chemicals found in the drinking water. The levels of TCE found in the New Jersey study are relatively low (maximum detection = 67 μg/l) compared to the levels found in some private wells affected by the Conrail site.

Health effects other than cancer have also been examined. A study of people in Arizona exposed to TCE in their drinking water identified an association with congenital heart malformations [26]. This observation has been confirmed by an animal study described in the next section [27]. These health effects are summarized in Table 11.5.

Table 11.5 Summary of health effects associated with specific levels of exposure to TCE.

TCE exposure conc. (ppb)	Number of wells	Chronic exposure dose (mg/kg/day)	Possible health effects from chronic TCE exposure				Acute inhalation exposure (ppm)	Possible health effects from acute TCE exposure	
			Human studies		Animal studies			Human studies	Animal studies
			Cancer	Other effects	Cancer	Other effects			
	None	500		Woburn exposure group: cardiac, respiratory, immune, dermal	Renal and kidney tumors	Liver, kidney, neurological, reproductive, developmental	4128		Immune, respiratory, renal
		250			No information	Liver, kidney, neurological, reproductive, developmental	2064		
		100					830		
		10					83	Neurological, eye	Immune, respiratory
>30 000		3.27					27	No demonstrated effect	Immune
>10 000		1.09					9		
>3000	2	0.34				Developmental	2.7		No demonstrated effect
>1000	5	0.11				Fetal heart abnormalities at 0.18 mg/kg/day	0.9		
>300	9	0.03	Woburn exposure group: increased acute lymphocytic leukemias			Fetal heart abnormalities at 0.05 mg/kg/day	0.27		
>100	30	0.01	Increased risk of leukemia and non-Hodgkin's lymphoma in drinking water study (23–67 μg/l; [28])	No demonstrated effect		No demonstrated effect	0.09		
>30	76	0.0033					0.03		
>5	100	0.0005	No demonstrated effect				0.004		
<5									

Animal studies The effects of TCE have been more extensively studied in experimental animals. TCE is associated with the development of liver and kidney tumors in animals but only at relatively high doses. Heart defects have been detected in newborn rats that were exposed to TCE during embryo development [27–30]. However, other animal studies have not demonstrated these effects [31]. Effects in the liver, kidney, and neurological systems occur at higher doses. Table 11.5 summarizes the health effects at various air concentrations and oral doses, as well as corresponding drinking water concentrations.

In Table 11.6, the exposure dose for each level of TCE is compared to several health-based criteria, including the acute minimal risk level for neurological effects and the intermediate/chronic LOAELs for developmental (intermediate) and kidney (chronic) effects. The purpose of these comparisons is both to better define the levels of exposure where there is confidence that no-adverse health effects are likely and to define the levels at which health effects have been observed.

Conclusions about TCE exposure It should be acknowledged that there is a significant amount of uncertainty in attempting to characterize the magnitude of the health risk associated with exposure to TCE, particularly at low levels. There is conflicting information regarding the health effects of TCE in both human and animal studies. What is clear is that the magnitude of the hazard is directly proportional to the concentration of TCE in the drinking water and the duration of time that people may have been exposed to the contaminated water. On the basis of a weight-of-evidence approach to evaluating the human and animal studies, the conclusion of this assessment is that individuals who used TCE-contaminated well water above 300 μg/l have estimated doses similar to identified LOAELs and may be at significant risk for cancer and possibly birth defects.

The 300 μg/l concentration is not intended to be a threshold level that defines a safe level, but rather a level that is likely to be associated with possible health effects. The risk of harmful effects is uncertain for residents who used well water containing TCE at levels between 5 and 300 μg/l. It is reasonable to assume that as the TCE levels increase above the drinking water standard of 5 μg/l toward 300 μg/l, the risk of harmful effects increases. This uncertainty is indicated by the light gray in Figure 11.6.

According to the well sampling records conducted since the mid-1980s, nine wells were contaminated by TCE at levels greater than 300 ppb. It is possible that a larger number of wells could have been affected but were either not sampled at any time or were sampled after the peak levels had passed through the well field.

CCl_4 Exposure

Human studies The Department of Health and Human Services/National Toxicology Program classified CCl_4 to be "reasonably anticipated to be a human carcinogen, based on evidence of liver cancer in experimental animals when CCl_4 is administered by ingestion." Table 11.7 summarizes the exposure doses of CCl_4 that are associated with specific health effects. The exposure doses are estimated for

Table 11.6 Comparison of total absorbed TCE dose to specific toxicity end points.

TCE conc. in water (μg/l) (lower end of range)	Ingestion dose (mg/kg/day)	Showering dermal dose (mg/kg/day)	Showering inhalation dose (mg/kg/day)	Total absorbed dose (mg/kg/day)	Comparison to MRL		Other comparisons to total absorbed dose		
					Developmental and neurological effects		Neurological	Developmental cardiac effects	Kidney effects
					Acute MRL (mg/kg/day)	Acute hazard quotient*	Acute LOAEL (mg/kg/day)	Intermediate LOAEL (mg/kg/day)	Chronic LOAEL (mg/kg/day)
30 000	0.86	1.4	1.0	3.3	0.2	16.6	50	0.05	50
10 000	0.29	0.5	0.3	1.1	0.2	5.5	50	0.05	50
3000	0.09	0.1	0.1	0.33	0.2	1.7	50	0.05	50
1000	0.03	0.05	0.03	0.11	0.2	0.6	50	0.05	50
500	0.01	0.02	0.02	0.06	0.2	0.3	50	0.05	50
300	0.009	0.01	0.010	0.03	0.2	0.2	50	0.05	50
100	0.003	0.005	0.003	0.01	0.2	0.06	50	0.05	50
30	0.0009	0.001	0.0010	0.003	0.2	0.02	50	0.05	50
5	0.0001	0.0002	0.0002	0.0006	0.2	0.003	50	0.05	50

Doses are in units of mg/kg/day (ingestion assumes 100% absorption, dermal dose is based on an absorption model). Exposure durations: acute (up to 14 days); intermediate (14 days to 1 year); chronic (longer than 1 year). See ATSDR's toxicological profile for TCE for toxicity values [31].

*ratio of total absorbed dose to MRL.

specific ranges of CCl_4 concentration in water, from less than 5 ppb (the US federal drinking water standard) to greater than 30 000 ppb. Occupational studies are generally limited to high levels of exposure through inhalation of CCl_4, with reports of gastrointestinal, liver, and neurological effects. However, studies of the effects of human exposure to relatively low doses of CCl_4 are very limited. In fact, there is essentially only one study that has examined health effects in the range of exposures that are likely to have occurred in the communities affected by Conrail. An epidemiologic study was conducted using birth outcome and drinking water exposure databases from a four-county area in northern New Jersey [32–34]. Estimated CCl_4 concentrations in the drinking water of greater than 1 part per billion were associated with a statistically significant finding of smaller babies (decrease in full-term birth weight) and an increased incidence of neural tube defects, with weaker associations with central nervous system defects and cleft-lip or cleft-palate. A limitation of this study is the lack of defined exposure levels and the possible complication of other contaminants in the drinking water. Therefore, the 1 ppb CCl_4 in drinking water cannot be used as a threshold for adverse effects.

Animal studies As summarized in Table 11.7, animal studies have shown effects on the liver, fetal weight gain, immune function, neurological function, and the kidney at relatively high doses. The doses for these effects are shown in Table 11.7 along with the estimated concentration in drinking that would yield similar doses in people. Table 11.8 shows the total estimated exposure dose for Endicott residents from ingestion, dermal, and inhalation routes for specific intervals of CCl_4 concentrations in water. These levels are then compared to several health-based criteria from studies of CCl_4 ingestion, including the chronic minimal risk level and the acute and intermediate LOAEL values for developmental [35], liver [36], and kidney effects [37].

As described earlier, the HQ is a ratio of the exposure dose to the health-based guideline – in this case, the oral chronic MRL. The exposures to CCl_4 exceed the HQ of 1.0 at concentrations in water greater than 140 ppb, indicating that exposures above this level were evaluated further. Also, as with TCE, we compared the estimated exposure doses with the levels that have been found to cause adverse health effects on various organ systems. The most sensitive effects of CCl_4 exposure are on liver function. CCl_4 is also known to cause adverse health effects on various organ systems, including the respiratory, kidney, and neurological systems, as well as developmental effects. At the doses estimated in residents exposed at the Conrail site, levels of exposure to CCl_4 in water that are associated with liver toxicity may occur at concentrations greater than 300 ppb.

Conclusions about CCl_4 exposure This analysis is intended to characterize the magnitude of the risk that residents may have experienced as a result of their exposure to the contaminants found in their drinking water wells. After evaluating human and animal studies, we concluded that individuals who used CCl_4-contaminated well water at levels above 300 ppb have an increased risk for liver and kidney damage. That level should not be considered a threshold for health effects below which there is no concern. The magnitude of that risk is directly related to the

Table 11.7 Summary of health effects associated with specific levels of exposure to CCl_4 and corresponding concentrations of CCl_4 in water [38].

CCl_4 conc. in water (ppb)	No. of Wells	Total chronic exposure dose (mg/kg/day)	Possible health effects from chronic CCl_4 exposure				Acute inhalation exposure (ppm)	Possible health effects from acute CCl_4 exposure	
			Human studies		Animal studies			Human studies	Animal studies
			Cancer	Other effects	Cancer	Other effects			
		1200	No information			Mild kidney effects			
		250		Neurological (acute); hepatic effects		Neurological effects (intermediate)	500		Developmental effects (body weight); hematological effects
		50		Nausea and vomiting (acute); serious hepatic effects (acute)	Heptatocellular carcinomas at 47 mg/kg/day	Reduced fetal weight gain for gestational days 6–8 (acute); decreased immune function (acute)	200	Serious respiratory, renal, hepatic effects	Serious neurological
		10		Developmental impacts at drinking water concentrations >1 ppb (low birth weight, CNS defects, neural tube defects, cleft-lip, and cleft-palate [32, 33]	Hepatoma at 20 mg/kg/day (intermediate)		100	No demonstrated effects	Hepatic effects (7 h/day)
>30 000	0	3.27				Hepatic effects at 10 and 5 mg/kg/day	22.5		
>10 000	1	1.09				No demonstrated effects	7.5		
>3000	6	0.34			10^{-2} cancer risk at 0.1		2.25		No demonstrated effects
>1000	14	0.11					0.8		
>300	20	0.03			10^{-3} cancer risk at 0.01		0.23		
>100	37	0.01					0.08		
>30	47	0.0033			10^{-4} cancer risk at 0.001		0.02		
>5	75	0.0005					0.004		
<5	88								

Table 11.8 Estimation of total absorbed CCl_4 dose from ingestion, dermal, and inhalation exposure to water.

CCl_4 conc. in water (μg/l)	Ingestion dose (mg/kg/day)	Showering inhalation dose (mg/kg/day)	Showering dermal dose (mg/kg/day)	Total absorbed dose (mg/kg/day)	Chronic oral exposures		Acute and intermediate oral exposures		
					Liver effects		Developmental effects	Kidney effects	Liver cancer
					CCl_4 chronic oral MRL (mg/kg/day)	Chronic oral hazard quotient	LOAEL (mg/kg/day) acute	LOAEL (mg/kg/day) acute	LOAEL (mg/kg/day) intermediate
30 000	0.9	1.0	1.4	3.31	0.02	165.7	50	180	20
10 000	0.3	0.3	0.5	1.10	0.02	55.2	50	180	20
3000	0.1	0.1	0.1	0.33	0.02	16.6	50	180	20
1000	0.03	0.03	0.05	0.11	0.02	5.5	50	180	20
300	0.01	0.01	0.01	0.03	0.02	1.7	50	180	20
100	0.003	0.003	0.005	0.01	0.02	0.6	50	180	20
30	0.001	0.001	0.001	0.003	0.02	0.2	50	180	20
5	0.0001	0.0002	0.0002	0.0006	0.02	0.03	50	180	20

Doses are in units of mg/kg/day (ingestion assumes 100% absorption, dermal dose is based on an absorption model). Exposure durations: acute (up to 14 days); intermediate (14 days to 1 year); chronic (longer than 1 year). See ATSDR's toxicological profile for carbon tetrachloride for toxicity values [38].

duration of their use of water contaminated at that level. According to the records that were examined for well water sampling conducted since the mid-1980s, 20 wells, serving an estimated 80 people, were contaminated with CCl_4 at that level or higher. It is possible that a larger number of wells could have been affected but were either not sampled or were sampled after the CCl_4 peak levels had passed through groundwater.

11.3.2.5 Hazard Assessment for the Mixture of TCE and CCl_4

Hazard Index As described earlier in this chapter, the initial evaluation of exposure to mixtures of TCE and CCl_4 consisted of summing the hazard quotient for each chemical and for each route of exposure to calculate the hazard index for the mixture as follows:

$$\text{Oral HI}_{\text{mixture}} = \text{oral HQ}_{\text{TCE}} + \text{oral HQ}_{\text{CCl}_4},$$

$$\text{Inhalation HI}_{\text{mixture}} = \text{inhalation HQ}_{\text{TCE}} + \text{inhalation HQ}_{\text{CCl}_4}.$$

Target Organ Toxicity As already described in this chapter, the next step was to evaluate the combined effects of TCE and CCl_4 on specific target organs. The sum of organ-specific HQs for each chemical is referred to as a target toxicity dose for each chemical. As with calculating an MRL, the resulting TTD for TCE and CCl_4 is then used to develop an organ- or system-specific HI. Again, the HIs based on organ-specific target toxicity doses are calculated for each route of exposure as follows:

$$\text{Oral HI-TTD}_{\text{specific organ/system}} = \text{oral TCE HQ}_{\text{specific organ/system}} + \text{oral CCl}_4\ \text{HQ}_{\text{specific organ/system}},$$

$$\text{Inhalation HI-TTD} = \text{inhalation TCE HQ}_{\text{specific organ/system}} + \text{inhalation CCl}_4\ \text{HQ}_{\text{specific organ/system}}.$$

Similar to evaluating possible effects of single-chemical exposure, when the HI-TTD for a specific organ exceeds 1.0, a comparison of the combined estimated exposure doses for the chemicals to NOAELs and to LOAELs for both chemicals provides a better understanding of which exposures might pose a greater risk of resulting in adverse health effects. For the Conrail site, we calculated HI-TTDs for several target organs and specific health end points. Because of the levels of exposure, some of the values exceeded 1.0, requiring a final comparison of the estimated exposure dose to doses that are known to cause harmful effects in animals and humans.

Of the 257 private wells that contained either TCE or CCl_4, 152 wells contained both chemicals. The objective of this analysis was to determine if synergistic effects might have resulted from combined exposures to TCE and CCl_4.

Binary Weight of Evidence In addition to the dose comparisons described previously, another important step to consider is the interactions that chemicals might have to cause toxicity. Chemicals can interact in the body, resulting in effects that might be additive, greater than additive, or less than additive. If additive, the dose of each chemical would have an equal weight in its ability to cause harmful effects. In that case, the combined HI for the two chemicals is an indication of the degree to which possible harmful effects could occur in people. When the chemicals act in a greater-than-additive manner, known as synergism, one chemical is enhancing the effect of the other chemical. In that case, the combined HI for the two chemicals underestimates the potential toxicity of the mixture of them. For chemicals that act in a less than additive manner, known as an antagonistic effect, the combined HI overestimates the potential toxicity of the mixture of the chemicals. In other words, one chemical might be thought of as protecting against adverse effects from the other chemical. In that case, the HI for exposure to that mixture is less than simply adding the individual HQs for each chemical.

To evaluate whether a mixture of chemicals could be acting additively, synergistically, or antagonistically, ATSDR developed an approach known as the binary weight of evidence analysis. The results of the BINWOE analysis for TCE and CCl_4 at the Conrail site are as follows:

TCE impact on CCl_4 toxicity	
Hepatic effects	+ 1.0 (synergistic)
Developmental effects	0.5 (more than additive)
CCl_4 impact on TCE toxicity	
Hepatic effects	0 (additive)
Developmental effects	0 (additive)

From the BINWOE analysis, we know that the combined HI for TCE and CCl_4 will underestimate the possibility of developmental and liver toxicity. It will be important to evaluate the presence of TCE in the well water and its contribution to a household's HI score. For example, the concern is greater for a household with an HI of 10 if TCE is present along with CCl_4 because TCE is likely to synergistically enhance the effects of CCl_4.

Experimental Evidence for Greater than Additive Effects of TCE and CCl_4 Exposure
Several studies show that TCE will enhance the toxic effects of CCl_4, particularly toxic effects on the liver. Thus, TCE acts in a greater-than-additive manner to enhance the toxic effects of CCl_4. Over a wide range of nontoxic doses (injected into the body cavity of rats), TCE was shown to lower the threshold for liver toxicity from CCl_4, demonstrating the potentiation of hepatotoxicity by TCE [39]. The same study showed that CCl_4 did not increase the toxic effects of TCE. Because the Pessayre study exposed rats by injecting them with TCE and CCl_4, it is difficult to determine precisely how much TCE or CCl_4 in water would be needed to cause these synergistic effects. However, a study by another investigator showed that toxic effects to the liver from a mixture of TCE and CCl_4 were similar by either injection into the body cavity or by oral administration [40]. Steup showed a rather complex relationship between TCE dosing and CCl_4. Relatively low doses of CCl_4 (25 mg/kg) required higher doses of TCE (790 mg/kg) to cause liver damage, while relatively high doses of CCl_4 (51 mg/kg) resulted in liver damage from lower doses of TCE (79 mg/kg). The interplay

between TCE and CCl_4 doses makes it difficult to pinpoint the lowest levels of each chemical that might interact to cause harmful effects to the liver. It is also important to remember that Steup's investigations evaluated very short periods of exposure (usually a day), making the doses he identified as critical not as applicable to longer exposure periods that are typical for residents at the Conrail site. Nevertheless, the principle is established that TCE at some level will synergistically increase the toxic effects on the liver from exposure to some levels of CCl_4.

Another study that administered TCE and CCl_4 orally to rats supports the findings from Steup [41]. Borzelleca exposed rats to relatively large amounts of CCl_4 and TCE (ranging from 100 to 400 milligrams of chemical per kilogram body weight (mg/kg)). The results clearly show that TCE synergistically increases the toxic effects of CCl_4.

Another series of studies reported the occurrence of hepatitis in people who sniffed solvents containing predominantly TCE and small amounts (7–20%) of CCl_4 [42–44]. Because these were human exposures, the concentration of TCE and CCl_4 in the air that these individuals breathed could not be measured. Therefore, it is not possible to know the concentrations of TCE needed to enhance the toxic liver effects of the small amounts of CCl_4 in the solvent. Because of animal studies, however, it is reasonable to assume that TCE most likely enhanced the toxic liver effects of CCl_4 in the solvent.

Sensitive Groups Studies in rodents have shown that other factors might increase the risk of harmful effects from exposure to CCl_4. Diet, diabetes, and stress have been shown to increase the harmful effects of CCl_4 in rodents [45–48]. For instance, nontoxic doses of CCl_4 caused liver damage when stress was induced in rats from shock treatment [48]. Similarly, when diabetes was induced in rats by treating the animals with alloxan, previously nontoxic doses of CCl_4 caused liver damage in rats. Evidence from a dietary study suggests that a low-protein diet might protect against the harmful effects of CCl_4 [45]. In addition, alcohol has been shown to potentiate the harmful effects of CCl_4 in mice [49]. Therefore, it is reasonable to assume that people who drink alcohol might be at increased risk of harmful effects should their drinking water contain CCl_4.

Possible Health Effects from Exposure to the Mixture TCE and CCl_4 Table 11.9 shows the HQ, HQ-TTD, and the HI for well water containing 300 ppb CCl_4 and 7350 ppb

Table 11.9 Comparison of summaries for HI for inhalation exposure route and TTD methods for a specific CCl_4 and TCE mixture, based on intermediate duration exposure.

Chemical concentration	HQ-inhalation (based on MRL)	HQ-TTD (based on target organs)		
		Neurologic	Hepatic	Body weight
CCl_4: 300 ppb	15.7	0.1	15.7	4.7
TCE: 7350 ppb	134.9	134.9	36.5	13.5
HI total	151	135	52	18

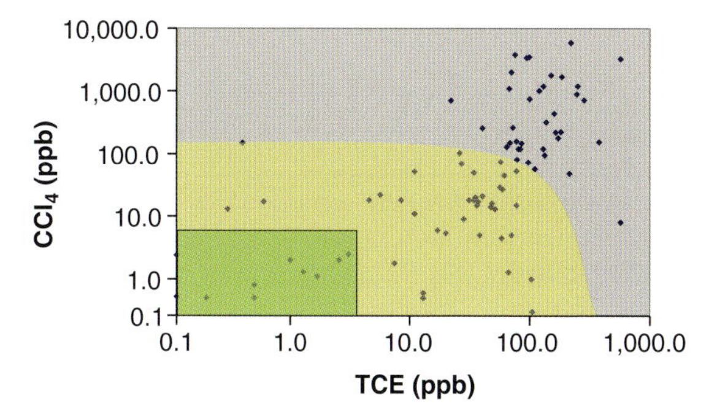

Figure 11.6 Scatter plot of detectable levels of TCE and CCl_4 from well sampling (◆, individual wells).

TCE. Because the HI score of 151 is greater than 1, it is necessary to determine the HQ-TTD scores for sensitive end points (i.e., neurologic, hepatic, and body weight). These sensitive end points show the neurologic system to have the highest HQ-TTD (i.e., 135), thus indicating that the neurologic system is at greatest risk of harmful effects. Some risk for harmful effects also exists for the liver (HQ-TTD = 52) and for body weight (HQ-TTD = 18).

Because the HI and HQ is a complex evaluation not easily explained to the public, a graph of CCl_4 and TCE concentration in well water was developed to show the public which combination of CCl_4 and TCE might be a concern. Figure 11.6 graphically shows the various concentrations of CCl_4 and TCE in well water. When private well water contains 5 ppb or less of TCE and CCl_4 (the drinking water standard), no appreciable risk is apparent from exposure to a mixture of the two chemicals (shown in gray box in Figure 11.6). As TCE and CCl_4 levels increase above the drinking water standard, the risk of harmful effects increases (shown in light gray in Figure 11.6). While it is difficult to be precise in determining the specific levels at which harmful effects might begin for someone who was exposed to a mixture containing CCl_4 and TCE, the risk is greater for people whose private wells had CCl_4 levels greater than 100 ppb and TCE levels greater than 500 ppb (shown in dark gray in Figure 11.6). Table 11.10 shows the harmful effects that might be possible from exposure to a mixture of CCl_4 and TCE.

11.4 Overall Conclusions from the Case Studies

For the Endicott area case study, ATSDR's mixtures guidance provided a standardized framework that was used to calculate individual HQs. The calculation of a HI for various mixtures was not necessary because the HQ for each chemical was below 0.1. According to ATSDR's mixtures guidance, when the HQ for two chemicals is below 0.1, it is unlikely that they will interact in an

Table 11.10 Organ systems and possible health end points resulting from simultaneous exposure to CCl_4 and TCE.

Organ system	Possible harmful effects
Liver	Mild to severe damage to liver cells; changes in enzyme levels indicating injury, cirrhosis, or harmful scarring of liver tissue
Neurological	Sleepiness, depression
Respiratory	Fluid in the lungs
Developmental	Lower fetal weight (small babies), decrease in fetal weight gain (small babies), maternal weight loss, decreased length in fetuses
Kidney	Decreased urine output, altered kidney function indicating injury; protein in urine

additive or synergistic manner to be a health concern for noncancerous health effects. Using graphs to show the HQs in relation to health guidelines, NOAELs, and LOAELs (see Figure 11.2) is a useful approach to visually and easily explain to the public the findings that the mixture of chemicals is not harmful to them.

For the Conrail Rail Yard case study, a more complex evaluation was needed because the HI for the individual HQs for TCE and CCl_4 exceeded 1. Exceeding an HQ of 1 required a focused toxicological evaluation involving the calculation of HQs for specific target organs and health end points (referred to as HQ-TTDs). In addition, a weight-of-evidence evaluation of TCE and CCl_4 studies showed that TCE synergistically enhanced the effects of CCl_4, thus complicating the interpretation of the HI scores for the combined HQ-TTDs for TCE and CCl_4. Studies show that TCE is strongly synergistic for CCl_4 effects on the liver. In addition, people with diabetes or added stress and people who drink alcohol may be particularly sensitive to CCl_4, and this sensitivity may be exacerbated by coexposure to TCE. Residents who used water containing TCE and CCl_4 whose levels exceeded the US drinking water standard of 5 μg/l may have an increased risk of experiencing harmful effects on the liver and kidney, as well as the neurological, respiratory, and developmental systems. This risk is greatest for residents who used well water containing TCE greater than 300 μg/l and CCl_4 greater than 500 μg/l.

At the Conrail Rail Yard site, the chemical-specific and mixtures evaluation resulted in numerous actions to protect public health. Safe water was provided to residents whose private wells were contaminated with TCE and CCl_4. In addition, vapor mitigation systems were installed in buildings that contained CCl_4 levels greater than 3 ppbv in air. Health education about exposures to and harmful effects from TCE and CCl_4 was provided to residents and health care providers. At the Endicott site, residents were reassured that their municipal water supply was of high quality and was suitable for drinking and bathing, which allowed them to focus on other environmental issues.

Evaluating the potential for the biological interaction of chemicals in water once they are absorbed by humans requires a complex approach of estimating the total dose from various routes of exposures. The evaluation is further complicated by applying mathematical tools to combine doses in addition to incorporating a weight-of-evidence that allows professional judgment to interpret the numerical hazard index of combined chemicals. The goal is either to assure the public that no harm is expected or to inform the public that harmful effects might be expected.

References

1 National Research Council (U.S.) (2009) Committee on improving risk analysis approaches used by the U.S. EPA, in *Science and Decisions: Advancing Risk Assessment*, National Academies Press, Washington DC.

2 Agency for Toxic Substances and Disease Registry (2005) Public Health Assessment Guidance Manual (Update), US Department of Health and Human Services, Atlanta.

3 Mumtaz, M.M., Poirier, K.A., and Colman, J.T. (1997) Risk assessment for chemical mixtures: fine-tuning the hazard index approach. *J. Clean Technol. Environ. Toxicol. Occup. Med.*, **6** (2), 189–204.

4 Agency for Toxic Substances and Disease Registry (2004) Guidance Manual for the Assessment of Joint Toxic Action of Chemical Mixtures, US Department of Health and Human Services, Atlanta.

5 U.S. Environmental Protection Agency (1986) Guidelines for the health risk assessment of chemical mixtures. *Fed. Reg.*, **51** (185), 34014–34025.

6 U.S. Environmental Protection Agency (2000) Supplementary Guidance for Conducting Health Risk Assessment of Chemical Mixtures, EPA/630/R-00/002, U.S. Environmental Protection Agency, Washington, DC.

7 Durkin, P., Hertzberg, R., Stiteler, W., and Mumtaz, M.M. (1995) The identification and testing of interaction patterns. *Toxicol. Lett.*, **79**, 251–264.

8 Agency for Toxic Substances and Disease Registry (1998) Toxicological Profile for Jet Fuels (JP5 and JP8), U.S. Department of Health and Human Services, Atlanta.

9 Agency for Toxic Substances and Disease Registry (2000) Toxicological Profile for Polychlorinated Biphenyls (Update), U.S. Department of Health and Human Services, Atlanta.

10 Agency for Toxic Substances and Disease Registry (1998) Toxicological Profile for Chlorinated Dibenzo-*p*-dioxins (Update), U.S. Department of Health and Human Services, Atlanta.

11 Agency for Toxic Substances and Disease Registry (1995) Toxicological Profile for Polycyclic Aromatic Hydrocarbons (PAHs) (Update), U.S. Department of Health and Human Services, Atlanta.

12 Agency for Toxic Substances and Disease Registry (1996) Toxicological Profile for Toxaphene (Update), U.S. Department of Health and Human Services, Atlanta.

13 Agency for Toxic Substances and Disease Registry (1999) Toxicological Profile for Total Petroleum Hydrocarbons (Update), U.S. Department of Health and Human Services, Atlanta.

14 Agency for Toxic Substances and Disease Registry (1995) Toxicological Profile for Automotive Gasoline, U.S. Department of Health and Human Services, Atlanta.

15 Agency for Toxic Substances and Disease Registry (1995) Toxicological Profile for Stoddard Solvent, U.S. Department of Health and Human Services, Atlanta.

16 Pohl, H.R., Roney, N., Fay, M., Chou, C.-H., Wilbur, S., and Holler, J. (1999) Site-specific consultation for a chemical mixture. *Toxicol. Ind. Health*, **15**, 470–479.

17 Mumtaz, M.M. and Durkin, P.R. (1992) A weight-of-evidence approach for assessing interactions in chemical mixtures. *Toxicol. Ind. Health*, **8**, 377–406.

18 Mumtaz, M.M., De Rosa, C.T., and Durkin, P.R. (1994) Approaches and challenges in risk assessments of chemical mixtures, in *Toxicology of Chemical Mixtures: Case Studies, Mechanisms, and Novel Approaches* (ed. R.S.H. Yang), Academic Press, New York, pp. 565–597.

19 Agency for Toxic Substances and Disease Registry (2006) Health Consultation, Public Health Implications of Exposures to Low-Level Volatile Organic Compounds in Public Drinking Water, Endicott Area Investigation, Broome County, New York, U.S. Department of Health and Human Services, Atlanta, November 30. Available at http://www.atsdr.cdc.gov/HAC/pha/EndicottAreaInvestigation113006/EndicottAreaInvestigationHC113006.pdf.

20 Agency for Toxic Substances and Disease Registry (2005) Public Health Assessment for Conrail Rail Yard, Elkhart, Elkhart County, Indiana, U.S. Department of Health and Human Service, Atlanta, August 11. Available at http://www.atsdr.cdc.gov/HAC/pha/ConrailRailYd/ConrailRailYardPHA081105.pdf.

21 U.S. Environmental Protection Agency. (1997). Exposure factors handbook; August.

22 Ecology and environment; inc. 1994 – remedial investigation/feasibility study Corail site, Elkhart, Indiana, volume 1; work assignment number 01-5L7Y; and volume 2; work assignment number 01-5L7Y.

23 Henschler, D., Vamvakas, S., Lammert, M., Delamt, W., Kraus, B., Thomas, B., and Ulm, K. (1995) Increased incidence of renal cell tumors in a cohort of cardboard workers exposed to trichloroethene. *Arch. Toxicol.*, **69**, 291–299.

24 Anttila, A., Pukkala, E., Sallmen, M., Hernberg, S., and Hemminki, K. (1995) Cancer incidence among Finnish workers exposed to halogenated hydrocarbons. *J. Environ. Occup. Med.*, **37**, 797–806.

25 U.S. Environmental Protection Agency (2001) Trichloroethylene Health Risk Assessment, NCEA-W-1041, Washington, DC, April.

26 Goldberg, S.J., Lebowitz, M.D., and Graver, E.J. (1990) An association of human congenital cardiac malformations and drinking water contaminants. *J. Am. Coll. Cardiol.*, **16**, 155–164.

27 Dawson, B.V., Johnson, P.D., Goldberg, S.J., and Ulreich, J.B. (1993) Cardiac teratogenesis of halogenated hydrocarbon-contaminated drinking water. *J. Am. Coll. Cardiol.*, **21**, 1466–1472.

28 Cohn, P., Klotz, J., Bove, F., et al., Drinking Water contamination and the incidence of Leukemia and Non-Hodkins lymphoma. Environ Health Perspect 1994, 102: 556–561.

29 Johnson, P.D., Dawson, B.V., and Goldberg, S.J. (1998) A review: trichloroethylene metabolites: potential cardiac teratogens. *Environ. Health Perspect.*, **106** (Suppl. 4), 995–999.

30 Johnson, P.D., Goldberg, S.J., Mays, M.Z., and Dawson, B.V. (2003) Threshold of trichloroethylene contamination in maternal drinking waters affecting fetal heart development in the rat. *Environ. Health Perspect.*, **111** (3), 289–292.

31 Agency for Toxic Substances and Disease Registry (1997) Toxicological Profile for Trichloroethylene (Update), U.S. Department of Health and Human Services, Atlanta.

32 Bove, F.J., Fulcomer, M.C., and Klotz, J.B. (1992) Population-based surveillance and etiological research of adverse reproductive outcomes and toxic wastes report on phase IV-A, public drinking water contamination and birthweight, fetal deaths, and birth defects, a cross-sectional study, New Jersey Department of Health, Trenton, New Jersey.

33 Bove, F.J., Fulcomer, M.C., and Klotz, J.B. (1992) Population-based surveillance and etiological research of adverse reproductive outcomes and toxic wastes report on phase IV-B, public drinking water contamination and birthweight, fetal deaths, and birth defects, a cross-sectional study, New Jersey Department of Health, Trenton, New Jersey.

34 Bove, F.J., Fulcomer, M.C., Klotz, J.B., Esmart, J., Dufficy, E.M., and Savrin, J.E. (1995) Public drinking water contamination and birth outcomes. *Am. J. Epidemiol.*, **141** (9), 850–862.

35 Narotsky, M.G. and Kavlock, R.J. (1995) A multidisciplinary approach to toxicological screening: II. Developmental toxicity. *J. Toxicol. Environ. Health*, **45**, 145–171.

36 Eschenbrenner, A.B. and Miller, E. (1946) Liver necrosis and the induction of carbon tetrachloride hepatomas in strain A mice. *J. Natl. Cancer Inst.*, **6**, 325–341.

37 Docherty, J.F. and Burgess, E. (1922) The action of carbon tetrachloride on the liver. *Br. Med. J.*, **2**, 907–908.

38 Agency for Toxic Substances and Disease Registry. Toxicological profile for carbon tetrachloride draft for public comment. Atlanta: US Department of Health and Human Services; 2003 September.

39 Pessayre, D., Cobert, B., and Descatoire, V. (1982) Hepatotoxicity of trichloroethylene–carbon tetrachloride mixtures in rats. *Gastroenterology*, **83**, 761–772.

40 Steup, D.R., Wiersma, D., McMillan, D.A., and Sipes, I.G. (1991) Pretreatment with drinking water solutions containing trichloroethylene or chloroform enhances the hapatotoxicity of carbon tetrachloride in Fischer 344 rats. *Fundam. Appl. Toxicol.*, **16**, 798–809.

41 Borzelleca, J.F., O'Hara, T.M., Gennings, C., Granger, R.H., Sheppard, M.A., and Condie, L.W, Jr. (1990) Interactions of water contaminants. I. Plasma enzyme activity and response surface methodology following gavage administration of CCl4 and CHCl3 or TCE singly and in combination in the rat. *Fundam. Appl. Toxicol.*, **14**, 477–490.

42 Conso, F., Efthymiou, M.L., Fournier, E., and Garnier, R. (1980) Interet de la toxicovigilance: exemple do certains trihcloroethylenes du commerce. *Arch. Mal. Prof.*, **41**, 198–200.

43 Conso, F., Fournier, E., and Riboulet, G. (1980) Acute hepatonephritis in so-called trichloroethylene inhalation, in *Toxicological Aspects* (ed. A.N. Kovatsis), Tecknika Editions, Thessaloniki.

44 Bouygues, M., Danne, O., and Bouvry, M. (1980) Hepatite au trichloroethylene, actions synergique du tetrachlorure de carbone. *Nouv. Press. Med.*, **9**, 3277.

45 McLean, A.E.M. and McLean, E.K. (1966) The effect of diet and 1,1,1-trichloro-2,2-bis (*p*-chlorophenyl)ethane (DDT) on microsomal hydroxylating enzymes and on sensitivity of rats to carbon tetrachloride poisoning. *Biochem. J.*, **100**, 564–571.

46 Hanasono, G.K., Cote, M.G., and Plaa, G.I. (1975) Potentiation of carbon tetrachloride-induced hepatotoxicity in alloxan- or streptozotocin-diabetic rats. *J. Pharmacol. Exp. Ther.*, **192**, 592–604.

47 Hanasono, G.K., Witschi, H., and Plaa, G.L. (1975) Potentiation of the hepatotoxic responses to chemicals in alloxan-diabetic rats. *Proc. Soc. Exp. Biol. Med.*, **149**, 903–907.

48 Iwai, M., Saheki, S., Ohta, Y., and Shimazu, T. (1986) Footshock stress accelerates carbon tetrachloride-induced liver injury in rats; implication of the sympathetic nervous system. *Biomed. Res.*, **7**, 145–154.

49 Weber, L.W.D., Boll, M., and Stampfl, A. (2003) Hepatotoxicity and mechanism of action of haloalkanes: carbon tetrachloride as a toxicological model. *Crit. Rev. Toxicol.*, **33** (2), 105–136.

12
Application of Mixture Methodology for Workplace Exposures

Frank J. Hearl

12.1
Introduction

Modern workplaces are complex environments in which workers are often exposed to multiple stressors. Workers from agriculture, construction, mining, and other industries are commonly exposed to combinations of chemical, biological, and physical agents. Yet the knowledge of the potential health effects of mixed exposures to a combination of these stressors is limited. Additional nonwork-related exposures, such as the consumption of alcohol or tobacco, the use of insect repellents, cosmetics, or other chemicals, and individual susceptibility add to the complexity of exposure assessment and the resulting biological responses. Simplified risk assessment and risk management strategies are available to control mixed exposures in the workplace. Various equations for combining risks have been applied to the occupational environment for several decades. Most occupational risk assessment and risk management approaches used to date have been based on the additivity principle, although there has been a recognition of the possibility of nonadditive combined effects.

In the workplace, there are several known cases of interactions involving chemical and nonchemical stressors. Two examples of known or perceived health risks with toxicological end points and consequences from mixed exposures are (1) hearing loss due to the combined exposure to excessive noise and certain ototoxic chemicals; and (2) synergistic carcinogenesis of asbestos and smoking. Mixtures of chemicals and physical agents not only interact within the human system but can also undergo chemical transformations in the environment. For example, chlorinated hydrocarbons can be converted into toxic phosgene in the presence of ultraviolet light. Another example is the enhanced transport of radionuclides into the lungs when adsorbed on respirable particulate.

The US labor force continues to grow both as a percentage of the overall population and in total number. In 1998, 138 million people (67% of the population) made up the American workforce that is expected to grow to 158 million in 2010. Over the past 20 years, the topic of mixture research has witnessed a transition from outright

Principles and Practice of Mixtures Toxicology. Edited by Moiz Mumtaz

ISBN: 978-3-527-31992-3

avoidance to carrying out simple, descriptive studies of binary mixtures to planning and carrying out sophisticated studies using new technologies in biological and computational sciences. The complete elucidation of the human genome, the related development of nanotechnologies in genomics and proteomics, and the exponential growth of computational technologies provide essential opportunities to deal with the effects of mixture exposures on complex biological systems. The current "state of science" is right for addressing research in the complex area of multiple stressors [1].

12.2
Occupational Exposure Limits

Occupational exposure limits (OELs) were developed largely during the mid-twentieth century. In the United States, many OELs were originally developed by a private professional standard-setting organization known as the American Conference of Governmental Industrial Hygienists, Inc. (ACGIH) that was founded in 1938. The ACGIH first published a list of maximum allowable concentrations (MACs) for 63 substances in 1942 [2]. The ACGIH's list of maximum allowable concentrations later became known as Threshold Limit Values® or TLVs®. The TLV list is now published annually by the ACGIH, and includes updates and additions to the list of chemical substances and physical agents with a TLV, as well as biological exposure indices (BEIs®) for substances that can be monitored through collection of biological specimen [3].

The ACGIH states that TLVs are not consensus standards, but are developed by committees of experts who after reviewing available peer-reviewed scientific literature and using their professional judgment recommend health-based exposure limits. TLVs do not take into account economic or technical feasibility and are set at levels the ACGIH believes that "nearly all workers may be repeatedly exposed without adverse health effects." Furthermore, the ACGIH states that TLVs are neither fine lines for determining between safe and unsafe conditions nor are they relative measures of toxicity between substances, but should be used and interpreted by trained industrial hygienists in consultation with background documentation [3].

An OEL, such as a TLV, is an important component of a risk management system. Without an OEL, measurement of a contaminant concentration has no reference scale to give it meaning [4]. When a measured concentration is scaled using an OEL, the normalized concentration in occupational settings is often referred to as the "severity" index, calculated as

$$\text{Severity}(i) = C(i)/\text{OEL}(i),$$

where $C(i)$ is the measured concentration of chemical species i, and OEL(i) is its occupational exposure limit.

In 1963, at the 25th annual meeting of the ACGIH, a new appendix was added to the TLV list, describing an approach for applying TLVs for mixed exposures [5]. The new mixtures approach was to apply when two or more hazardous substances are present. In the absence of information to the contrary, the effects of exposures were recommended to be considered "additive." To determine if the TLV of the mixture is exceeded, one would sum the severity fractions to compute the mixture severity, for example,

$$\text{Severity(mixture)} = C(1)/\text{TLV}(1) + C(2)/\text{TLV}(2) + \cdots + C(n)/\text{TLV}(n).$$

If the severity computed by this formula exceeds 1.0, then the TLV of the mixture is exceeded.

This rule was to apply to all simple mixtures, defined as mixtures in which the components are individually identifiable, when there is no information that would contradict using the additivity assumption.

The 1963 appendix provided some guidance on exceptions to the additivity rule above, when the chief effects of the harmful substances in question have local effects on different organ systems. Under those circumstances, control of the environment may reduce to examining each component's exposure to its own exposure limit. Hence, the mixture would be in control if the following condition were true:

$$C(1) < \text{OEL}(1); \quad C(2) < \text{OEL}(2); \ldots; \quad C(n) < \text{OEL}(n).$$

Or, in other words, if the severity index of all the components taken individually is less than 1.0, then the mixture is considered to be in compliance with the mixture limit. An example of such a situation where independent effects are clearly warranted would be the combined exposure to lead (CNS, blood, and kidney) and sulfuric acid (lung function).

The appendix also described the possibility that substances could have potentiating or antagonistic effects that might merit deviations from the additivity formula. However, the TLV Booklet provides no particular insight into when such exceptions could be made.

In the most recent guidance from the ACGIH, the individual exposure limits are now accompanied by a "TLV Basis" column that lists the physiological effect on which the TLV is based. For example, acetylene's basis is "asphyxia," benzene's basis is "leukemia," and tetramethyl lead's basis is "CNS impair." According to current guidance, when two or more hazardous substances have a *similar toxicological effect* on the same target organ or system, their combined effect should be considered [3]. As with the 1963 guidance, additivity should be assumed without information to the contrary.

Workplace mixtures also occur as complex mixtures composed of many related species and where quantification of all components is not feasible. Examples of complex mixtures include diesel exhaust emissions, welding fume, oil mist, and petroleum distillates. For many complex mixtures, the approach that has been taken is to develop a surrogate measure that is applied to the mixture regardless of variability in the composition of the mixture.

An example of such a complex mixture is coal tar pitch volatiles (CTPVs that include the fused polycyclic hydrocarbons that volatilize from the distillation residues of coal, petroleum (excluding asphalt), wood, and other organic matter. CTPVs describe the types of substances that are emitted as coke oven emissions. They generally include varying concentrations of multiring aromatic hydrocarbons including but not limited to phenanthrene, anthracene, pyrene, chrysene, and benzo(*a*) pyrene. The exposure limit for CTPVs is based on the measurement of the total mass of the benzene soluble mass of the mixture. The CTPV aerosol is collected on a glass fiber filter by drawing the aerosol-laden air through the sampler at a known rate for a measured time. The collected material is extracted using the solvent, benzene. After evaporating the benzene solvent from the extracted liquid, the residues are redeposited, weighed, and reported as "benzene soluble." The present TLV for CTPV is 0.2 mg/m^3 of benzene soluble aerosol, which can be compared with that measured by this process.

To measure CTPV, the benzene soluble mass as a surrogate index of exposure does not vary when composition of the mixture changes. The varying carcinogenic potential that may result from variations in composition is not considered. However, quantifying each component of the mixture for routine risk management would be prohibitively expensive. It would also be a frustrating experience because there are no accepted algorithms for combining the exposure measurements and estimating the likely risk from each unique mixed exposure. As a result, the approach using the surrogate measure, however imperfect, remains a practical alternative for practitioners.

Additional examples of complex mixtures that use a surrogate measure of exposure include aliphatic hydrocarbon gases (liquefied petroleum gas), asphalt fume, cotton dust, coal dust, diesel fuel, grain dust, gasoline, graphite, kerosene, natural rubber latex (as total proteins), oil mist (mineral), particles not otherwise classified (PNOC), Portland cement, Stoddard solvent, synthetic vitreous fibers, and wood dusts.

12.3
Regulating Mixed Exposures in the United States

OELs are used by enforcement agencies to ascertain compliance with legal exposure limits. As such they do provide a target for workplace exposure controls. The occupational hygienist can compute single-agent severities, mixture severities when appropriate, or can consider other safety factors in order to recommend appropriate control strategies to keep the severity indices below 1. The severity computed from an OEL is used directly in the selection of respiratory protective equipment. The worst-case expected severity is estimated, and then a respirator class is selected that has an "assigned protection factor" greater than the expected worst-case severity [6, 7].

In the United States, exposure limits for air contaminants in general industry are found in Title 29, US Code of Federal Regulations (CFRs), Part 1910.1000. The limits found in this part, known as Permissible Exposure Limits (PELs), were adopted shortly after the enactment of the Occupational Safety and Health Act of 1970, hereafter referred to as the OSHAct [8]. Section 6(a) of the OSHAct gave the Department of Labor 2 years after the date of passage to promulgate rules adopting any national consensus standard or established federal standard. Prior to the OAHAct, the federal government established occupational safety and health standards for all government contracts work in excess of $10,000. These federal occupational safety and health standards were authorized by the Walsh-Healey Public Contracts Act of 1936. In 1969, the Department of Labor incorporated a list of about 400 workplace exposure limits, using the 1968 Threshold Limit Value set by the ACGIH. Thus, the ACGIH TLV list became part of the federal standards covered by the Walsh-Healey Act [9].

When the OSHAct was passed in 1970, the Department of Labor's Occupational Safety and Health Administration (OSHA) promulgated rules under Section 6(a) of the OSHAct incorporating the 1968 TLVs that were then existing federal standards under the Walsh-Healey Act. Although the list of TLVs were incorporated into the new list of OSHA PELs, it is not clear in the record if the appendices from the ACGIH TLV booklet, such as the appendix on mixed exposures, were also included. The TLV list, now incorporated into the Code of Federal Regulations exists in Title 30, Part 1910.1000, also known as "supbpart Z," and the list of chemicals with PELs is sometimes referred to as the "Z-tables." This subpart, in addition to the PEL values for the individual substances, includes the following language related to mixtures:

1910.1000(d)(2)(i)

In case of a mixture of air contaminants, an employer shall compute the equivalent exposure as follows:

$$E_{\mathrm{m}} = C_1/L_1 + C_2/L_2 + \cdots + C_n/L_n,$$

where E_{m} is the equivalent exposure for the mixture, C is the concentration of a particular contaminant, and L is the exposure limit for that substance specified in subpart Z of 29 CFR Part 1910. The value of E_{m} shall not exceed unity (1).

1910.1000(d)(2)(ii)

To illustrate the formula prescribed in paragraph (d)(2)(i) of this section, consider the following exposures:

Substance	Actual concentration of 8 h exposure (ppm)	8 h TWA PEL (ppm)
B	500	1000
C	45	200
D	40	200

Substituting in the formula, we have

$$E_{\mathrm{m}} = 500/1000 + 45/200 + 40/200,$$

$$E_{\mathrm{m}} = 0.500 + 0.225 + 0.200,$$

$$E_{\mathrm{m}} = 0.925.$$

Since E_m is less than unity (1), the exposure combination is within acceptable limits.

The regulation, though it was incorporated from the ACGIH 1968 booklet that included the mixture appendix, provides no insights into when the additivity formula should be applied and when it should not.

However, OSHA does provide its field inspectorate with guidance from its Field Inspection Reference Manual (FIRM) that says:

C. 3. d. Additive and Synergistic Effects.

1) Substances which have a known additive effect and, therefore, result in a greater probability/severity of risk when found in combination shall be evaluated using the formula found in 29 CFR 1910.1000(d)(2). The use of this formula requires that the exposures have an additive effect on the same body organ or system.
2) If the CSHO (Compliance Safety and Health Officer) suspects that synergistic effects are possible, it shall be brought to the attention of the supervisor, who shall refer the question to the Regional Administrator. If it is decided that there is a synergistic effect of the substances found together, the violations shall be grouped, when appropriate, for purposes of increasing the violation classification severity and/or the penalty [10].

This guidance is far from clear on when to apply the additivity formula. The FIRM does incorporate the concepts from the original ACGIH appendix that the agent should affect the same body organ or system. The second part of the guidance requires the OSHA inspector, the CSHO, to refer questions of synergism or potentiation to the OSHA Regional Administrator. In case of potential synergism, the guidance only refers to modification of the violation classification and modification of assigned penalties rather than some guidance on multiplication factors for assessing compliance based on toxicological differences.

Due to the cumbersome processes that evolved from the Act (US Government 1970, Section 6(b)), by regulation, and as a result of several landmark court decisions, changing existing PELs or establishing limits for new substances, the PELs have become stagnant for 37 years, with most currently enforced standards dating back to the ACGIH's 1968 TLV list. Furthermore, with the introduction of new materials into the work environment, and with the recognition of the hazardous properties of previously unregulated substances, the updating of old PELs and the creation of new PELs have proved challenging [11]. As a result, there has been no change in the application of the additive mixture formulas since these were first incorporated into federal regulations as described above.

There is one special case, the exposure limit for respirable silica dust exposure, in which additivity was "built in" to the exposure limit for the combined exposure of silica and respirable dust (i.e., particles not otherwise classified). The silica PEL is described by a formula whereby one analyzes the composition of the whole dust to determine the percentage of silica concentration. Using the percentage of silica based on this analysis, the exposure limit of whole respirable dust is computed as

$$\mathrm{PEL} = 10/[\%\mathrm{Quartz} + 2].$$

So, for example, if the crystalline silica concentration is 18%, the PEL for respirable dust would be PEL = 10 mg/m^3/[18 + 2] = 0.5 mg/m^3. On the extremes, if there is no crystalline silica in the sample (%Quartz = 0), the formula reduces to a respirable dust limit of 5 mg/m^3, and if the sample is 100% silica, the PEL is approximately 0.1 mg/m^3. If one assumes that respirable crystalline silica (quartz) has a PEL of 0.1 mg/m^3 and a measured concentration of C_q, and that respirable dust has a PEL of 5 mg/m^3 and a measured concentration of C_{rd}, then using the additivity formula at the mixture exposure limit:

$$1 = (C_{rd}/5) + (C_q/0.1).$$

Multiplying each term by $(10/C_{rd})$ gives

$$10/C_{rd} = 2 + 100[C_q/C_{rd}].$$

Substituting the identity into the equation above, 100 $[C_q/C_{rd}]$ = %Quartz

$$10/C_{rd} = 2 + \%\text{Quartz}.$$

Rearranging this expression generates the PEL formula for respirable dust containing silica,

$$C_{rd} = 10/[2 + \%\text{Quartz}].$$

So the existing PEL, which still uses the formula for regulation of crystalline silica, automatically incorporates the additivity assumption.

In addition to the simple mixtures that might be covered by the additivity formula and the singular case of silica and respirable dust described above, OSHA has PELs for the following complex mixtures: CTPV, cotton dust, coal dust, grain dust, liquefied petroleum gas, oil mist (mineral), PNOC, PCBs, petroleum distillates, Portland cement, and Stoddard solvent.

12.4 Hazard Communications

OSHA's hazard communications standard was first adopted in 1983 for the manufacturing sector of industry and expanded to cover all sectors in 1987 [12]. This standard requires employers to establish a comprehensive written hazard communications program that includes lists of hazardous chemicals present, labeling of containers of chemicals in the workplace, preparation and distribution of material safety data sheets (MSDS) for employees and downstream employers, and training in protective measures and hazard control for employees who may contact hazardous substances in their workplace.

The hazard communications standard excludes certain chemicals and mixtures from these requirements such as chemicals covered by the Federal Insecticide, Fungicide, and Rodenticide Act; any substance or mixture subject to labeling requirements under the Toxic Substance Control Act; any food, food additive, color additive, drug, and cosmetic or medical or veterinary device or product; distilled spirits including wine and malt beverages; consumer products subject to consumer product safety standards; and a variety of other specific exceptions. The standard does include any chemical or substance with a PEL as listed in OSHA's "Z-tables," any substance with a current ACGIH TLV, and substances listed as carcinogens by the National Toxicology Program (NTP), or the International Agency for Research on Cancer (IARC).

When the substance is a mixture, the hazard communications standard provides a logic for dealing with the mixture. The first level of evaluation would be if the mixture has been tested as a whole, then the results of those tests would be used to assess the toxicity or hazards associated with the mixture. If the mixture has not been tested as a whole, then the mixture is presumed to have the hazard associated with its components that may be present at a concentration of 1% or greater by weight or by volume. If any carcinogenic component is present at or greater than 0.1% by weight or volume, then the mixture is presumed to be a carcinogen. In evaluating the physical hazards of mixture, employers are to use whatever scientifically valid data are available to assess the hazard associated with the mixture.

Interestingly, the hazard communication rule also requires employers to assess the possibility that a substance present at less than 1% or a carcinogen present at less than 0.1% composition in the mixture could be released into the environment at levels above the PEL or the ACGIH TLV of that substance. In those cases, the mixture would be presumed to have the hazards associated with those components.

12.5 Emerging Approaches

Recently, a group of researchers in Canada have developed a web-based tool to assist practioners in identifying and evaluating mixed exposures [13]. The tool was developed in several phases. The first phase involved gathering the basic toxicological data for the 695 chemical substances that are regulated in Quebec. For each, information on the target organ, the principle toxicity mechanism, and the resulting toxic effects were catalogued. Only those effects were considered that might occur at realistic exposure concentrations. Using these basic data, an algorithm can be addressed to apply the additivity formula where appropriate.

The second phase of their effort was to assess the available research literature for various binary mixtures that are known to exist in industry. In this case, a job-exposure matrix developed by the National Institute for Occupational Safety and Health (NIOSH) was used to identify observed or likely mixtures. Using this as their starting point, they evaluated 209 binary mixtures involving 114 substances and found 24 mixtures with "supra-additivity" or synergism, and 3 mixtures with infra-additivity or antagonism. Overall, the supporting data for most of the binary mixture scenarios were sparse. For a number of mixtures, supporting documentation was prepared by the group and is available along with the evaluation tool on the web site: http://www.irsst.qc.ca/en/_outil_100037.html

Despite the limited data available on specific mixtures, tools such as this hold promise that practioners may apply appropriate combined effect calculations for their workplace risk assessments.

12.6 Summary

Workplace environments are often complex settings with exposure to multiple stressors including mixtures of chemicals, noise, heat, and other physical and mental stressors. Although some simple procedures are available to use to assess exposures in combination, essentially reducing single-agent exposure limits to account for the multiple exposures, the techniques are not often applied by practioners. Recent advances in technology and access to more detailed information via the Internet hold promise for improving the use of available knowledge for worker protection and risk reduction.

References

1 NIOSH (2004) Mixed Exposures Research Agenda: A Report by the NORA Mixed Exposures Team, DHHS (NIOSH) Publication No. 2005-106. Available at www.cdc.gov/niosh.

2 Paustenbach, D. (1998) Occupational exposure limits, Chapter 30, in *Encyclopedia of Occupational Health and Safety*, 4th edn (ed. J.M. Stellman), International Labour Organization, Geneva, Switzerland, pp. 27–34.

3 ACGIH (2007) TLVs® and BEIs® Based on the Documentation of the Threshold Limit Values for Chemical Substances and Physical Agents & Biological Exposure Indices. ACGIH Worldwide, Cincinnati, Ohio.

4 Ettinger, H.J. (2005) Henry Smyth Jr. Award lecture – Occupational exposure limits: do we need them? Who is responsible? How do we get them? *J. Occup. Environ. Hyg.*, **2**, D25–D30.

5 ACGIH (1963) Transactions of the Twenty-Fifth Annual Meeting of the American Conference of Governmental Industrial Hygienists, May 6–10, 1963, Cincinnati, Ohio.

6 NIOSH (2004) NIOSH Respirator Selection Logic 2004, DHHS (NIOSH) Publication No. 2005-100. Available at www.cdc.gov/niosh.

7 OSHA (2006) Assigned Protection Factors. Final Rule, *Fed. Reg.*, **71** (164), 50121–50192, August 24. Available at www.osha.gov.

8 OSHA (1970) The Occupational Safety and Health Act of 1970, P.L. 91-596. Available at www.osha.gov.

9 Ashford, N.A. (1976) *Crisis in the Workplace: Occupational Disease and Injury. A Report to the Ford Foundation*, The MIT Press, Cambridge, Massachusetts.

10 OSHA (2008) Field Inspection Reference Manual. Occupational Safety and Health Administration (OSHA). Available at http://www.osha.gov/Firm_osha_data/100007.html.

11 Howard, J. (2005) Setting Occupational exposure limits: are we living in a post-OEL world? University of Pennsylvania. *J. Labor Employ. Law*, **7** (3), 513–528.

12 OSHA (2006) Hazard communication: advanced notice of proposed rulemaking. *Fed. Reg.*, **71**, 53617–53627, September 12. Available at www.osha.gov.

13 Vyskocil, A., Drolet, D., Viau, C., Lemay, F., Lapointe, G., Tardif, R., Truchon, G., Baril, M., Gagnon, N., Gagnon, F., Bégin, D.,and Gérin, M.J. (2007) *Occup. Environ. Hyg.*, **4**, 281–287.

13
Assessing Risk of Drug Combinations

Christopher J. Borgert and Alexander A. Constan

Therapeutic strategies for certain medical conditions often involve simultaneous administration of two or more drugs, and patients are often prescribed several medications simultaneously when there is a need to treat more than a single medical condition. Thus, the risk and safety assessment of drug combinations is required both in the design of products containing two or more prescription drugs and in the assessment of adverse outcomes in patients taking more than a single medication or supplement, whether prescription or over the counter. This chapter is divided into two main sections. Section 13.1 addresses considerations for development of drug combination products and Section 13.2 addresses issues involved in the assessment of adverse drug interactions (see also Chapter 25) and patient outcomes.

13.1
Safety Considerations for Drug Combination Products

In the past several years, there has been an increase in the development of drug combinations. Disease indications in therapeutic areas such as oncology, HIV, dyslipidemia, and hypertension, are often treated by more than one drug from different classes (i.e., different mechanisms of action). In other instances, the clinical effectiveness of one drug can be increased by concomitant administration of another drug. Combination drugs in which the active ingredients have been combined into a single pill have the advantage of reducing the number of pills a patient must take on a daily basis, thereby increasing compliance. Regardless of the clinical reason, the development of most drug combinations will require some level of nonclinical safety assessment to support clinical trials and product approval. The goal of such an assessment is to evaluate the potential additivity, synergy, potentiation, and/or antagonism when two or more drugs are concurrently administered.

Principles and Practice of Mixtures Toxicology. Edited by Moiz Mumtaz

ISBN: 978-3-527-31992-3

13.1.1
Categories of Drug Combinations

The development stage of a drug is important when considering further development as part of a drug combination. Marketed (approved) drugs often have abundant clinical safety data in addition to a robust, if not complete, nonclinical safety assessment package. A new chemical entity (NCE)[1)] that is considered for combination development is usually in early or late stage clinical trials with the nonclinical safety package not yet completed. For each compound being considered in a drug combination product, the amount of clinical safety data and the stage of the nonclinical safety assessment package are key to outlining the appropriate combination development program.

For regulatory toxicology, there are three main categories of drug combinations: a combination of a marketed drug (MD) with one or more marketed drugs; a combination of one or more MD plus one or more NCE; or an NCE with one or more NCEs. Although the future of drug development will likely see an increase in triple or quadruple (or more) combination therapies, for ease of discussion this section will focus on binary drug combinations. The development strategy is basically the same regardless of the number of individual components and will be discussed as MD/MD, MD/NCE, or NCE/NCE development programs.

In addition to the development stage of the combination components, the final product form of the combination is another important variable. The two main forms of drug combinations are fixed dose combination (FDC) products and copackaged products. An FDC product has two or more of the drug components (active pharmaceutical ingredients) combined in a single dosage form (e.g., tablets). Because of the single dosage form, the concentration of each ingredient provides fixed doses for the patient. Copackaged drug products, also referred to as a free combination, contain two or more drugs in their separate final dosage forms, packaged together (e.g., blister packs). An attractive feature for FDC drug products is patient compliance (less pills); however, there are more inherent chemistry, manufacturing, and control (CMC) development issues such as compound compatibility, stability, and bioequivalence of the final form. Overall, the nonclinical safety assessment for FDC and copackage drug combinations adheres to the same principles.

Drug combinations, whether an FDC or copackaged drug products, should not be confused with the phrase "combination products." Combination products are composed of any combination of a drug or biological product with a device (i.e., drug/device, biologic/device, or drug/device/biologic). An example of a combination product would be drug-coated stents. In the United States, combination products are regulated by the FDA Office of Combination Products [1], whereas drug combinations are regulated by the FDA Center for Drug Evaluation and Research (CDER).

1) Also known as a new molecular entity (NME).

13.1.2
Regulatory Guidance

Until recently, there was limited regulatory guidance on the nonclinical safety evaluation for development of drug combinations, both in the United States and globally. In 2004, the FDA issued draft guidance on FDC and copackaged drug products for the treatment of HIV [2]. Because concomitant therapy is common for antiviral compounds, and many times needed to suppress viral load and prevent resistance, this document was primarily written for combinations of individual antiviral drugs already approved by the FDA and on which there is adequate evidence of clinical safety and efficacy when concomitantly administered.

In the European Union (EU), the EMEA provided draft guidance in 2002 on the nonclinical requirements for combinations of medicinal products with well-established use [3]. As with the FDA draft guidance for antiviral treatment of HIV, the EU draft guidance indicated that nonclinical studies would generally not be required when there is sufficient clinical safety data for each compound.

In the past few years, three key documents became available that provide insight into the nonclinical development expectations for drug combinations by US and European regulatory authorities [4–6]. More importantly, these documents outline the basic nonclinical study types needed to support drug combinations that do not have robust clinical safety data – those containing NCEs or MDs with limited safety experience. The intention of these publications is to both provide a nonclinical framework to evaluate potential additivity, synergy, potentiation, and/or antagonism of the known effects of individual compounds when used together for a drug combination, and identify any unexpected interaction and finding.

In March 2006, the FDA released final guidance on the nonclinical safety evaluation of drug combinations [4]. This guidance harmonizes the expectations of different divisions within the FDA regarding nonclinical development of drug or biologic combinations and establishes fundamental criteria for the three general types of combinations: MD/MD, MD/NCE, and NCE/NCE.

Similarly, a detailed European guidance on the nonclinical development of fixed drug combinations was adopted by the EMEA and its scientific committee (CHMP). The EU guideline on the topic is quite similar to that from the US FDA. A publication by Liminga and Lima [5] summarized the CHMP's views over the previous years on nonclinical development of drug combinations and the thinking behind the EMEA guidance document. Like the FDA guidance, the nonclinical recommendations are separated into categories based on the development stage of the individual compounds, which are (1) individual drugs already approved for concomitant use, (2) well-known/approved drugs not approved for concomitant use, (3) one or more NCEs with one or more approved drugs, and (4) a combination of two or more NCEs.

When reading both regulatory perspectives, it is apparent that the development strategy for each combination program will be unique based on the development stage, extent of existing clinical data, known safety issues of the components, and the intended indication/patient population. Because of this, there is no simple "one size fits all" approach to designing a nonclinical safety evaluation program. Although basic guidance is provided for the main categories of drug combinations, it is best to consult with the appropriate regulatory agency regarding the proposed nonclinical development program prior to embarking on clinical trials or submitting a clinical trial application.

The basic nonclinical framework provided in the FDA and EMEA guidances is provided below, with differences noted. First, the sponsor should consider the development stage of each compound. Both regulatory agencies indicate that for approved compounds with significant patient safety history of concomitant use and known pharmacokinetic interactions (if any), additional nonclinical safety studies may not be needed to support development of the drug combination. Of course, there are exceptions to this based on the safety profile for each compound, and the guidance documents should be reviewed to determine the appropriate nonclinical program.

Second, the sponsor should assess each nonclinical safety study for each compound in the proposed combination for compliance with current International Conference on Harmonization of Technical Requirements for Registration of Pharmaceuticals for Human Use (ICH) guidance. For instance, the sponsor should consider the robustness of the genetic toxicology package, reproductive toxicology studies, and safety pharmacology studies, especially for drugs approved several years ago that may need additional studies to meet current regulatory expectations. Both regulatory views suggest that if the genetic toxicology studies for the individual agents have been tested to current standards (ICH) and have not identified any genotoxic potential, then genetic toxicity studies of the combination may not be necessary. Similarly, if the carcinogenicity studies have been (or will be) completed for each of the individual agents, then carcinogenicity studies of the combination may not be necessary. In the situation in which an NCE/NCE combination is being developed without conducting individual ICH nonclinical safety assessment programs for each component, then a nonclinical ICH program for the combination may be required for regulatory approval, including a carcinogenicity study.

General toxicology studies, which will likely be needed to support MD/NCE and NCE/NCE combinations, should be conducted prior to a long-term administration in clinical trials. Both regulatory perspectives suggest that a general toxicology bridging study of the combination with repeated dosing up to 3 months should be conducted and used to compare with results from monotherapy administration. Any unexpected finding or interaction identified in this study may trigger subsequent nonclinical safety evaluation.

One general difference in the US and EU view on drug combinations is the need for reproductive and developmental toxicity studies. The FDA guidance recommends that embryo/fetal developmental studies of the combination should be conducted, in addition to a complete reproductive and developmental toxicology assessment for each compound. An exception to this would be if one compound is known to have significant risk for developmental toxicity (e.g., finding that requires category X labeling[2]), hence

2) FDA pregnancy category "X" indicates that studies in animals or humans have demonstrated fetal abnormalities and/or risk, therefore, the use of the product is contraindicated in women who are or may become pregnant.

the health risk in the label for that compound would carry over to the combination label. The EU regulatory perspective suggests that if the individual components have adequate reproductive and developmental toxicology packages and have sufficiently characterized the toxicity profile, then a reproductive toxicity study of the combination may not be necessary. Again, the actual decision will depend on the safety profile for each of the individual compounds and their potential interactions.

However, neither regulatory perspective provides sufficient details on when nonclinical drug combination studies should be conducted to support either single-dose clinical studies or longer-term multidose studies (e.g., phase 3) with the combination. The timing will likely depend on the nonclinical safety profile for the individual components, any available clinical safety data, and expected pharmacokinetic interactions, among other aspects. As mentioned earlier, the simple message is that the sponsor company should review these aspects and their proposed nonclinical safety assessment plan with the regulatory agency prior to starting the clinical program with the combination.

13.1.3 Considerations for Nonclinical Development of Drug Combinations

There are several compound-related aspects to consider prior to starting nonclinical studies of a drug combination. The following discussion is by no means exhaustive but rather is intended to help the investigator think about and address key study design concerns. As mentioned above, each drug combination program is unique, and there may be more than one safety assessment strategy available to support a clinical plan and approval. The primary purpose of the nonclinical safety assessment program is to evaluate the additivity, synergy, potentiation, and/or antagonism of known effects exhibited by the individual compounds when administered together, as well as identify possible new adverse effects and interactions. When conducting a toxicology study with a drug combination, remember that the study is intended to "bridge" results with that of monotherapy nonclinical data. Hence, assessing relevant findings identified in the toxicology studies conducted with the individual compounds is done in the combination study. The following questions should be answered when designing nonclinical drug combination studies:

- Are there any known clinical drug–drug interactions between the individual components?
- Are the nonclinical safety assessment packages complete for the individual components?
- Were one or more of the combination components approved long ago, so that the safety assessment portions (e.g., genetic toxicology) need to be contemporized to meet current ICH criteria?
- What are the target organs identified in nonclinical toxicology studies with each compound? Is there a single test species in which the relevant target organs for all combination components have been observed? Are there overlapping target organs in a single species?

- Is the combination dose ratio and dose range of the individual components known for the final drug combination? This information is essential to help determine the nonclinical doses needed to provide a therapeutic index.

13.1.3.1 **Nonclinical Study Formulations**

One of the first things to consider when contemplating how to dose animals with more than one drug is the route of administration and dosage formulation(s). In most instances, oral gavage dosing is the preferred route of administration, especially if the individual compounds have been evaluated in nonclinical studies using this route of administration. Although oral gavage administration may be transferred to the combination study, a significant amount of formulation work will still be needed prior to conducting the GLP combination bridging study.

Physicochemical characteristics of each component should be assessed to determine if they are compatible when concomitantly administered. The current regulatory perspectives [4, 5] do not recommend whether nonclinical studies need to use a single dosage formulation in the combination study or whether the individual components can be administered separately. Regardless of the sponsor's decision, compatibility of the dosage formulation(s) should be evaluated because the individual compounds will ultimately be mixed together in the gut, under variable pH conditions. For oral gavage formulations that are solutions, precipitation of the active ingredient could occur with concomitant administration of other compounds. Furthermore, stability assessment of the individual components should be conducted for combined dosing formulations.

13.1.3.2 **Species Selection**

For general toxicology studies of the combination, the choice of testing species should made considering the following points: the species used in the nonclinical safety assessment program for each component, most sensitive species to assess target organ toxicities and clinically relevant effects, and species that demonstrate pharmacological activity for compounds used in the combination. However, the "appropriate" test model for the combination may not demonstrate all these qualities. The easiest decision may be to continue using the same test species as used in the individual toxicology studies, but sometimes the components may have used different test species in their nonclinical programs (e.g., dog versus monkey as a nonrodent test species). Under these situations, exploratory monotherapy toxicology studies may be needed to provide background data in the desired test species or guide dose selection.

With these considerations in mind, both regulatory agencies suggest using an appropriate test species to conduct a bridging study from which the combination dose groups can be compared with monotherapy dose groups. In many instances, the rat will be an appropriate test model to conduct the safety evaluation of the combination. However, in some instances the rat may not be the best model to assess certain toxicities (e.g., relevant target organ toxicity is observed in the dog) and in these cases a nonrodent model, or both rodent and nonrodent models, should be used. As stated before, each combination will present unique development aspects that will dictate the nonclinical assessment strategy.

13.1.3.3 Pharmacokinetic (PK) Considerations

In addition to any potential clinical drug–drug interaction, consideration of non-clinical interactions is critical to designing the bridging toxicology studies for the combination. The metabolism pathway of each compound should be known for the combination test species. Specifically, will the compounds compete for the same hepatic cytochrome P450 (CYP) enzyme system? Do any of the compounds induce or inhibit CYP enzymes in the test species? CYP3A induction by pharmaceuticals is not uncommon in rats and many pharmaceutical compounds are substrates for CYP3A, sometimes producing autoinduction and decreased systemic exposure of the test compound following repeated administration. For example, hepatic enzyme induction by compound A increases the metabolism of the coadministered compound B, thereby decreasing systemic exposure of compound B, and a decrease in clinical safety margins previously established during monotherapy toxicology studies. Hence, PK exposures for the same dose of a compound may not be similar between monotherapy administration and combination administration. Evaluation of potential PK interactions can usually be made in short-term combination range finding studies, prior to selecting doses for the definitive bridging toxicology study of the combination.

Another PK aspect to consider in nonclinical combination studies is the analysis of metabolites for each of the components. Known major metabolites for each compound will likely need to be followed in the combination toxicology study, in addition to parent compounds, to assess potential PK interactions and clinical safety margins. The difficulty comes when trying to determine the appropriate PK sampling time points, especially if the PK profile (T_{max}, $T_{1/2}$, etc.) differs significantly between the combination components. Simply put, if compound A reaches T_{max} in less than 3 h and compound B reaches T_{max} after 8 h postdose, then multiple PK sampling time points may be needed to adequately assess the PK profile for each compound (or metabolite).

Furthermore, the PK sample volume needed at each time point could increase in a combination study compared to that needed in monotherapy studies. Most likely, analytical methods will not be available that can measure two different parent compounds and their major metabolites in a single assay. Because sample volume collection from rodents is limited, additional PK animals may be needed in the study design to allow adequate PK assessment.

13.1.3.4 Dose Selection

Selecting doses for the combination toxicology studies require some basic knowledge from the clinical program, toxicology studies, and the expected final product form, such as

- anticipated final product dose levels (some components may have multiple approved doses);

- anticipated dose ratios of the combination in the final product and in the clinical trials;
- dose response for known target organs;
- maximum tolerated dose for each of the components in the appropriate test species.

The doses selected for the combination study should be able to adequately assess the potential for interaction of known toxicities with each compound and also the potential for new toxicity findings. For example, it is important to set dose levels so as to be able to assess whether a given toxicity is enhanced or attenuated by coadministration. The study should also assess the expected dose ratios of the final product (if known). Sometimes it is difficult to thoroughly address both objectives in one study while keeping the number of dose groups manageable. In addition, monotherapy groups should be incorporated into the study design to adequately compare effects observed in the combination with those observed with the individual components. Relying on historical data from earlier monotherapy toxicology studies may not be in the best interest of the sponsor because a number of variables can change from study to study, especially over long periods of time. As mentioned above, dose selection may best be narrowed down after initial PK interaction and/or short-term (e.g., 2 weeks) nonclinical range finding studies. Dose groups that should be considered in a nonclinical combination toxicology study are as follows:

- Vehicle control group, including any significant excipients to be used in the final product.
- Monotherapy dose group for each component, same as the highest dose used in a combination dose group.
- Combination dose group of the high dose for each component. In general, these doses should be lower than the identified MTD but high enough to assess the potential additivity, synergy, potentiation, and antagonism of known toxicities.
- Combination dose groups of a low dose for each component. Sometimes this group will be at doses that produce a 1*x* clinical safety margin above anticipated human exposure for each component of the combination.
- Mixed dose groups (e.g., high/low, low/high, mid/high, etc.) to support various dose ratios and provide possible dose-response assessment for effects observed in the high-high dose group.
- An alternative consideration is to maintain a constant dose level for one component while increasing the dose level of the other component. This works well when the final commercial product has a similar fixed dose profile.

The number of variables to consider when designing a nonclinical program to support development of a drug combination is significant. In a way, development of a drug combination can easily produce a synergistic effect on the nonclinical workload. Each nonclinical program must be developed on a case-by-case basis, of which the existing nonclinical and clinical safety data of the individual components will drive the development strategy. Global regulatory guidance on nonclinical safety evaluation for drug combinations allows flexibility in the design of studies and overall development strategy. Once the nonclinical program to support a drug combination development program has been drafted, it should be discussed with the appropriate regulatory agencies prior to submitting clinical trial applications.

13.2
Evaluating Adverse Drug Interactions and Patient Outcomes

13.2.1
Magnitude of the Problem

The task of monitoring drug therapy in patients is becoming increasingly complicated for physicians and pharmacists. The inherent complexity of monitoring drug therapy in patients to ensure efficacy and to avoid adverse events is compounded when patients are taking multiple drugs. The government's recently published *Annual Report on Americans' Health* cites that 44% of Americans are taking at least one prescription medication and over 16% are taking at least three – and this represents an increase of 39 and 12%, respectively, over similar statistics reported 10 years ago [7]. According to a US study, 12% of women aged at least 65 years take at least 10 medications, and 23% take at least 5 prescription drugs [8]. Even for younger people, the norm tends toward an increasing use of prescription and over-the-counter medications, often in combination with various vitamin, nutritional, and dietary supplement products. Not only are more Americans taking multiple prescription medications, their use of dietary supplements has increased during the past decade. The third National Health and Nutrition Examination Survey (NHANES III) found, "40% of Americans are taking at least one dietary supplement; a substantial proportion of the U.S. population takes vitamins, minerals, and/or other dietary supplements" [9]. These numbers suggest that when physicians and pharmacists monitor a patient's drug therapy, it is increasingly likely they will be monitoring a combination of prescription medications and dietary supplements. Practitioners must therefore routinely consider the potential for drug–supplement interactions and anticipate the impact that combination drug–supplement therapy will have on their patients. Although there are regulatory programs to oversee almost every type of product that might be ingested and to help consumers manage potential risks, evaluating the potential risks of taking combinations of several medications and other products is still problematic both from a scientific and from a regulatory perspective.

Although clinical outcomes of drug therapy *may* be affected when prescription medications are combined, with or without dietary supplements, the recent proliferation of published reports of significant interactions between prescription medications and dietary supplements gives the mistaken impression that these interactions are *known* to occur and that they *usually* manifest a clinically significant outcome. Neither premise can be generally substantiated on the basis of data, although a few exceptions exist. In this chapter, we review some fundamental concepts underlying the demonstration of interactions, discuss the reasons why biologically significant interactions between drugs and supplements appear to be rare and interactions with significant therapeutic impact rarer still. Unfortunately, the publication of adverse interactions has become something of a growth industry in clinical pharmacy, pharmacology, and medicine. There are web sites devoted to drug and dietary supplement interactions, clinical textbooks include drug and supplement interactions, and the primary literature is full of reports of similar information. While the number of reports of interactions is exploding, it is difficult to interpret the value of the literature due to the lack of a standard methodology for studying drug–supplement interactions (or drug–drug interactions) and because the language used to describe the nature of the interactions is unclear.

Literature reports of interactions fail to establish terms that accurately reflect the nature of the interactions; much of the data reported comes from sources remote from the patient, for example, from animal or *in vitro* experiments; studies are often undertaken in healthy young humans after a single dose or at nontherapeutic doses. Other studies attribute changes in clinical parameters to drug–supplement interactions without considering the contribution of various disease states. In other words, the drug–supplement literature consists of poorly validated anecdotal reports without sound appraisal of all the relevant pharmacological and clinical factors. To avoid the accumulation of more and more "interactions" of doubtful relevance, investigators and perhaps editors of journals have an obligation to demand some confirmation of their clinical validity. Simply put, the literature and even the reference resources encourage clinicians attempting to monitor combination drug or drug–supplement therapy to draw conclusions about interactions based on the application of subjective and imprecise methods to woefully inadequate data.

Two common problems involve clinical diagnostic probability scales and adverse event reporting data. A number of probability scales and algorithms have been developed to inform clinical judgments about the likelihood that a specific patient outcome was due to a particular drug or combination of drugs [10–12]. These scales provide a guideline for drawing subjective conclusions on the basis of information about the patient, the timing of drug administration, prior reports of an association between the drug and the observed patient outcome, and other parameters. Drug-induced liver injury (DILI) was one of the first adverse outcomes for which such a probability scale was developed. Over the past two decades, several modifications of the scale have been published. These scales are of considerable interest because, except for a very few drugs (e.g., acetaminophen), there is no objective test that can distinguish liver injury caused by drugs from that caused by other factors, or for distinguishing among several candidate drug causes. In fact, a "high degree of suspicion" on the part of the examining physician is a key requirement of the determination [13, 14]. The scales can be validated only against the consensus of other experts, and unlike objective clinical tests amenable to determination of statistical precision, agreement typically decreases with the number of experts asked to test the conclusion [15].

Despite their subjective basis, diagnostic probability scales may be useful in some circumstances for treating physicians who must make decisions regardless of the availability of objective data. However, such scales can also be misused as if they were proper surrogates for objective determinants of cause. Shapiro and Lewis [15] note

> The lack of a specific biomarker or characteristic histologic feature to identify a drug as causal further hampers this effort and has fostered reliance on clinical assessment techniques that are based largely on medical judgment and expert opinion rather than on a truly objective means of assessing causality accurately. As a result, depending on the knowledge and experience of the clinician performing the diagnostic evaluation, the final assessment may lack precision and seems at times to reduce the process to one of making a diagnosis of exclusion based on circumstantial evidence. The consequences of erroneously attributing the cause of hepatic injury to a drug can be dire for patients and health care providers and for the pharmaceutical industry and regulatory bodies as well.

Recently, Horn *et al.* [12] proposed a drug interaction probability scale to be used in lieu of the Naranjo probability scale for adverse drug reactions, which the authors contend is inapplicable to drug–drug interactions because some of the questions to be answered in the Naranjo scale are relevant only to single drugs. Horn and coauthors also discuss the power of dechallenge/rechallenge scenarios for evaluating causation. While carefully controlled dechallenge/rechallenge scenarios could enhance the objectivity of the assessment, they must be interpreted with caution for concluding the cause of an adverse patient outcome. Because drug–drug interactions are not simply predictable additive pharmacological or toxicological effects, but are phenomena that fundamentally alter dose-response characteristics of a drug in the presence of other drugs, it is difficult to envision an unambiguous determination of a drug–drug interaction without information from more than a single-dose level of the drugs. Some type of dose-response information would be necessary to determine whether an interaction was responsible for a particular outcome as opposed to the combination of drugs additively achieving the threshold of effect in a particular patient.

Gaikwad *et al.* [16] evaluated the accuracy of drug interaction alerts triggered by electronic medical record systems commonly used to safeguard against prescribing errors and adverse drug–drug interactions in the elderly. The elderly might be particularly affected by drug–drug interactions because this patient population receives a disproportionate number of prescription medications, and are more sensitive to the effects of many drugs than younger patients. The systems tested showed only limited potential to identify clinically significant, severe drug–drug interactions but considerable probability for triggering spurious alerts.

Much of the difficulty in designing effective alerting systems and unambiguous assessment procedures surely stems from the confused notions about what constitutes a drug–drug interaction as well as the subjective means of assessing them commonly applied in clinical practice. Regardless how well tailored a clinical assessment scale might be for drug–drug interactions, these carry the inherent weakness of subjective judgments, devoid of objective measurements whose precision and accuracy can be empirically tested. In the parlance of sports, evaluating drug–drug interactions is a judged sport like diving or figure skating, whose outcome is subject to the individual characteristics and biases of judges, which often become the object of public controversy. These stand in stark contrast to sports like swimming and speed skating, where outcomes are determined dispassionately by electronic timing devices that remove the variable influence of judges, and whose results are consequently less exposed to debate and controversy. Advancements in the science of drug–drug interactions are sorely needed to remove clinical practice from the realm of subjective determinations made by panels of judges to a more evidence-based foundation reliant on scientific measurement.

Misusing subjective interaction probability scales to evaluate drug–drug and drug–herbal interactions portends similar consequences. Yet, for drug–drug interactions, objective means of evaluation are available that avoid the risk of circular reasoning inherent in the use of subjective probability scales. Given the increasing number of individuals taking several medications, the exponential number of drug combinations that might be evaluated erroneously, and the potential impact of those errors on patients, health care providers, and the pharmaceutical industry, it seems worthwhile to at least differentiate between conclusions about drug interactions based on objective science versus those based on subjective judgments.

13.2.2
Evidence that Interactions Occur is a Scientific Question, Requiring Data

The minimum elements of scientific data are measurement verification, control of extraneous influences on the measurement, and replicability [17–19]. Verifying that an interaction occurs, therefore, requires a measurement of the interaction. Before describing the means of measuring interactions, it is useful to clarify the basic terminology. Pharmacological and toxicological interactions are of two main types, often called synergism and antagonism. Synergism, including a subcategory called potentiation, is an interaction in which the response to a combination of drugs is greater than predicted based on the dose-response characteristics of the individual drugs. In other words, synergism occurs when a particular level of response can be achieved by lower doses of drugs given together than would be predicted based on their effects when given separately. Antagonism is the converse; a lesser response is produced, or more drug is required to produce a given level of response than would be predicted based on the dose-response characteristics of the drugs given separately. It is critical to appreciate that drug interactions are meaningful only for specified responses, whether pharmacologic and toxicologic. In almost all cases, drug interactions alter the dose-response characteristics of drugs given in combination, but do not alter the spectrum of therapeutic or adverse effects that can be produced by the drugs administered separately at a sufficient dose. Interactions are not intensive properties of drugs, as are physical characteristics such as molecular weight or boiling point, but instead are extensive properties wholly dependent upon the particular response under consideration, the doses of the drugs administered, and how one defines the response expected from the combination of two or more drugs.

In other words, interactions are not qualitative properties, but quantitative relationships, and hence, the need for measurement.

Interactions are the difference between what occurs and what was expected to occur, and hence, requires defining the expected, or "zero interaction," in quantitative terms. Zero interaction, usually called noninteraction, is essentially the null hypothesis in an interaction experiment. Two classical models of noninteraction are commonly accepted. "Dose addition" was described by Loewe and Muischnek in 1926 [20] and states that no interaction has occurred if the response produced by dose A plus the response produced by dose B is equal to the response produced by A and B given together, in other words, $R(\mathrm{a}) + R(\mathrm{b}) = R(\mathrm{a} + \mathrm{b})$. For a single drug, this relationship will always hold [21]: for example, two 325 mg doses of aspirin, by definition, produce the same response as one 650 mg dose. Dose addition deems this the expected relationship, and thus, noninteraction. Different drugs that exhibit dose addition simply behave as dilutions of one another, with the expected effect of their combination being the sum of their doses and relative potencies according to the relationship given by Eq. (13.1).

$$\frac{D_\mathrm{a}}{D_\mathrm{A}} + \frac{D_\mathrm{b}}{D_\mathrm{B}} = 1, \tag{13.1}$$

where D_A and D_B are doses of drugs A and B, respectively, that produce a specific level of response when administered separately, and D_a and D_b are doses of the drugs that produce the same level of response when given in combination. The other commonly accepted model of noninteraction was given by Bliss in 1939 [22] and is often called either "independence" or "response addition." Response addition is based on statistical models of probabalistic independence, meaning that the effect of each drug is manifest as if the other drugs were not present. Response addition (independence) requires that the response produced by each agent is summed according to Eq. (13.2), and the interaction term becomes zero when two or more drugs in combination are noninteractive under the model.

$$E_{\mathrm{A+B}} = E_\mathrm{A} + E_\mathrm{B} - (E_\mathrm{A} \times E_\mathrm{B}), \tag{13.2}$$

where E is the effect produced by drugs A and B.

The distinction between these models may seem esoteric, but the differences are critical. Dose addition predicts that a combination of three drugs, each present at a dose one-half its threshold dose for a particular effect, will produce a suprathreshold response, that is, $0.5 + 0.5 + 0.5 = 1.5$; whereas response addition predicts that the same combination of drugs and doses will produce zero response because each individual response is zero. Thus, these two models of noninteraction produce quite different predictions that can be critically important in certain circumstances, for example, when predicting whether otherwise nontoxic doses of various medications will be toxic when given together to the same patient. Only under special constraints, such as when the drugs under consideration have linear and parallel dose-response curves for the effect in question would dose addition and response addition yield the same noninteraction prediction.

13.2.3 Criteria for Evaluating Interactions

Recently, Borgert *et al.* [23] published a critical evaluation of the literature on drug and chemical interactions wherein they proposed five criteria for evaluating interaction studies. In that paper, they concluded that much of the published literature appears to ignore the foundations of interaction study design and interpretation. Borgert was not the first to point out this deficiency; in 1989, Berenbaum listed eight common approaches for analyzing interactions, noting that the most commonly used is one he labeled the "no method approach" in which authors claim to have demonstrated synergy without specifying any method or criterion at all, apparently assuming that the conclusion is self-evident, or possibly being unaware that there is a problem of selection [24]. The criteria proposed by Borgert *et al.* [23] for evaluating interaction studies incorporate the fundamental requirements of interaction analysis developed by Berenbaum and others, and address the most common mistakes that appear in the literature. The five criteria can be summarized as follows. Criterion 1 addresses the fact that it is usually necessary to define the dose-response curves for the individual drugs in a mixture, including parameters such as the maxima, the minima, the inflection points and areas of linearity. Criterion 2 states that a no interaction hypothesis should be explicitly stated and used as the basis for assessing synergy and antagonism. Indeed, the very conclusion that synergism or antagonism has been produced depends upon the model of no interaction against which the data for a drug combination are compared. Criterion 3 states that an adequate number of combinations should be tested across a sufficient dose range to meet the goals of the study. Criterion 4 specifies that formal statistical tests must be applied to distinguish the combined response from that predicted by the noninteraction model. It is not sufficient to simply compare treated with control or to compare the response produced by a combination of drugs with the responses produced by the drugs individually. Finally, criterion five states that interactions should be assessed at the relevant level or levels of biological organization.

Criterion 5 is not readily amenable to an explanatory figure, so will be clarified with the following example from the literature (Ref. [23] and citations therein). Patients on propranolol for control of blood pressure have sometimes exhibited a malignant hypertensive response when treated with epinephrine to maintain blood pressure during emergency surgical procedures. This response seems paradoxical considering that the drugs combined have the opposing pharmacological activities of an antihypertensive and a vasopressor. If one were to investigate this apparent drug interaction in the laboratory using receptor-based transactivation systems, a system comprised of beta-adrenergic receptors would yield antagonism because propranolol is a nonspecific beta receptor blocker, and epinephrine, a beta receptor agonist. In contrast, a system comprised of alpha-adrenergic receptors would yield zero interaction because propranolol has no effect at alpha-adrenergic receptors. None of these results explains the clinical observation, however, because the experiments are not conducted at a level of biological organization appropriate for testing the interaction.

The malignant hypertension observed clinically is due to the fact that in the presence of propranolol, the alpha agonist activity of epinephrine is unopposed by the normal compensatory physiological effects of its beta-receptor activation. This conclusion cannot be reached by experiments at the molecular or cellular level because the interaction requires the intact, integrated physiological system. On the other hand, it would be difficult to explain the clinical response in patients without knowledge of both the physiological and the molecular pharmacological action of these drugs. Criterion 5, therefore, implies more than a simple need to corroborate *in vitro* with *in vivo* results, and requires considering at what level, or levels, of biological organization two drugs might interact to produce a particular type of response [23]. A set of interaction data from the molecular, cellular, physiological, or whole organism level may be required.

The need to fulfill the first two criteria can be illustrated with a theoretical interaction experiment using two doses of the same drug. Performing a theoretical interaction experiment with a single drug will simplify the concepts and underscore the importance of these criteria. Figure 13.1a shows a hypothetical dose-response plot with data points E1 and E2 measured for doses D1 and D2 of the same drug, and zero response is measured at zero dose. The predicted no interaction result for an experiment in which D1 and D2 are combined, and D2 and D2 are combined, would be responses EZ3 and EZ4, based on the line (Z) extrapolated from the dose-response data collected for points E1 and E2. This is an instance in which Loewe

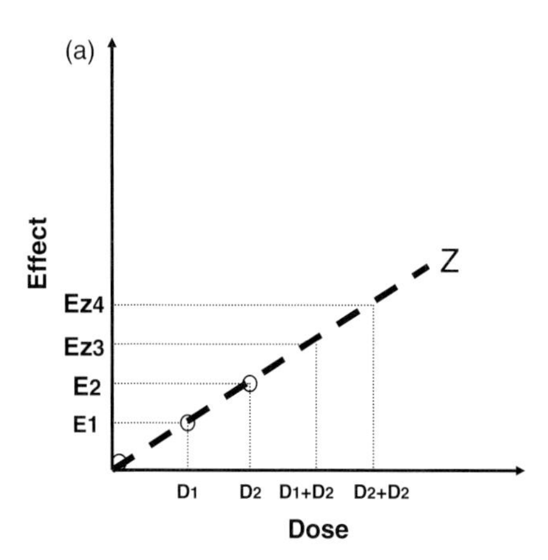

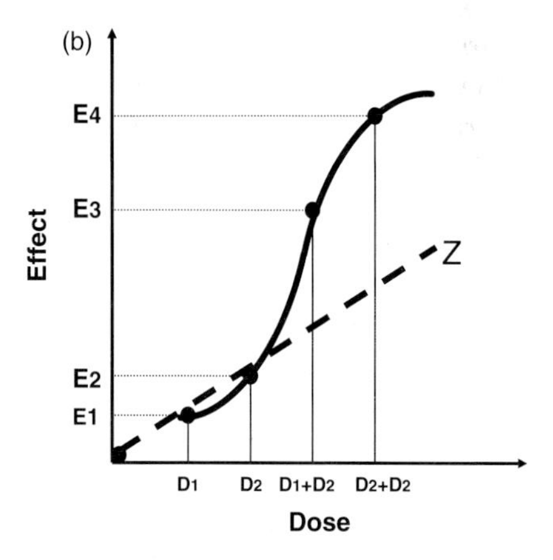

Figure 13.1 a) A hypothetical dose response plot with data points E-1 and E-2 measured for doses D-1 and D-2 of the same drug, and zero response is measured at zero dose. The predicted no-interaction result for an experiment in which D-1 and D-2 are combined, and D-2 and D-2 are combined, would be responses EZ3 and EZ4, based on the line (Z) extrapolated from the dose response data collected for points E-1 and E-2. If the experiment is actually performed and instead yields E3 for D1 plus D2, and E4 for D2 plus D2, one might conclude that these drug combinations are synergistic, which implies that the drug synergizes with itself. b) If the dose response curve for the drug is actually sigmoidal, the result would simply be dose additive (zero interaction), and the previous conclusion of synergism would be revealed as a misinterpretation due to inadequate characterization of the dose response curve. [30]

additivity (dose addition) and Bliss independence (response addition) would give the same prediction because the limited dose-response data collected appear to define a linear response, which must be parallel with itself since the experiment employs doses of the same drug. If the experiment is actually performed and instead yields E3 for D1 + D2, and E4 for D2 + D2, one might conclude that these drug combinations are synergistic, which implies that the drug synergizes with itself.

If, however, the dose-response curve for the drug is actually sigmoidal, as shown in Figure 13.1b, the result would simply be dose additive (zero interaction), and the previous conclusion of synergism would be revealed as a misinterpretation due to inadequate characterization of the dose-response curve. The predicted no interaction response under Bliss independence would change as well, but would be different from the one prediction under Loewe additivity, and raises the possibility for results to be interpreted as synergistic under one model but antagonistic under the other [25]. Besides illustrating why it is important to adequately characterize dose-response curves (criterion 1), this example also illustrates why it is necessary to define a specific no interaction hypothesis (criterion 2). A corollary point raised by this example is that antagonism can result in an increased pharmacological effect, even though the magnitude of the effect is less than predicted on the basis of dose addition. Hence, antagonism does not always imply that a combination of drugs has lesser effects than the drugs administered singly. For this reason, synergism and antagonism cannot be equated with therapeutic advantage or disadvantage, with respect to either efficacy or toxicity [26].

The necessity of the third and fourth criteria can be illustrated by a graphical technique based on the null hypothesis of dose addition, called the isobologram. Figure 13.2a shows that an isobologram is a Cartesian dose by dose plot for two agents, in this case, agents T and R, that produce a specific, measurable pharmacological or toxicological effect or clinical response, for example, the EC_{50}. The dose of T increases along the ordinate and R along the abscissa, so that every other point in the coordinate region represents a mixture of T and R. Doses of T and R that individually produce a specific level of response – in this case, the EC_{50} – are plotted on the axes. The equation for Loewe additivity (dose addition) defines a line that connects those equally effective doses of T and R administered individually. Points other than those on the axes define dose combinations of T and R that produce the EC_{50} when administered together. Synergistic combinations are those that produce the EC_{50} with lesser drug than expected, and those fall below the line of additivity. Antagonistic combinations fall above the line of additivity, representing the converse, that is, that more drug is required to produce the EC_{50} than predicted by dose addition. In practice, such experiments often produce data points that fall both above and below the line of additivity, indicating that some dose combinations of two agents may be synergistic and others antagonistic. When this situation occurs, there will necessarily be some dose combination that falls on the line of additivity between the synergistic and antagonistic points. This illustrates the extensive quality of pharmacological interactions in that they strongly depend on the doses and ratios of the drugs administered rather than on some intensive property of the drug molecules themselves.

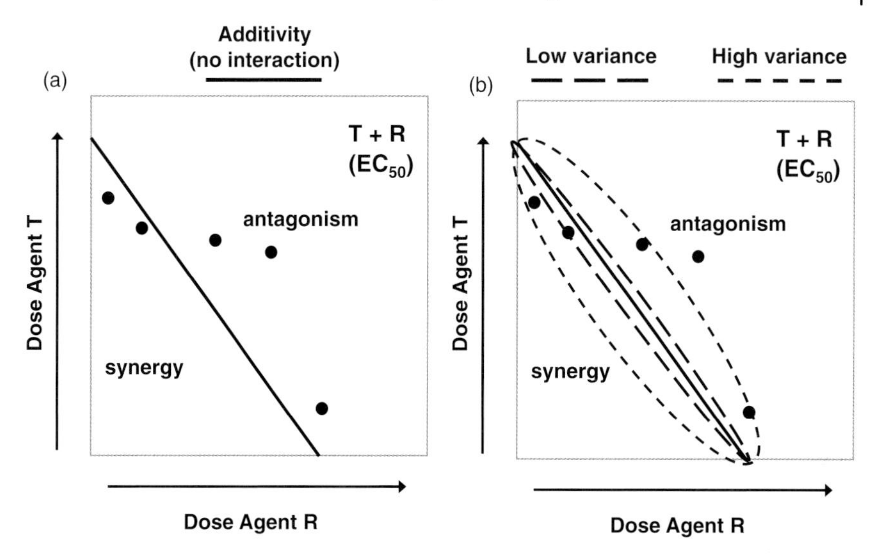

Figure 13.2 a) An isobologram is a Cartesian dose by dose plot for two agents, in this case, agents T and R, that produce a specific, measurable pharmacological or toxicological effect or clinical response, for example, the EC_{50}. The dose of T increases along the ordinate and R along the abscissa, so that every other point in the shaded region represents a mixture of T and R. Doses of R and T that individually produce a specific level of response – in this case, the EC_{50} - are plotted on the axes. The equation for Loewe additivity (dose addition) defines a line that connects those equally effective doses of agent R and T administered individually. Points other than those on the axes, define dose combinations of agent T and R that produce the EC_{50} when administered together. Synergistic combinations are those that produce the EC_{50} with lesser drug than expected, and these fall below the line of additivity. Antagonistic combinations fall above the line of additivity, representing the converse, i.e., that more drug is required to produce the EC_{50} than predicted by dose addition. b) Envelopes of statistical significance about the line of additivity for an endpoint measurement with low variance (long-dashed oval) and higher variance (short-dashed oval). The former requires a lesser departure from additivity for a dose combination to be statistically significant than for an endpoint measurement with greater variance. [30]

Rarely can a single interaction type be used to describe the interaction for all combinations of two drugs, which is why the number of combinations tested must coincide with the goals of the research. If the interest is in defining interactions for a very narrow range of doses, then fewer combinations must be tested than if the goal is to characterize the interactions for a broad range of doses. In order to make a global statement about how two drugs interact, combinations spanning the entire dose-response range would need to be tested, which is an onerous task, particularly for whole animal and clinical studies. Borgert *et al.* [23] were also not the first to point out the lack of appropriate statistical treatment of interaction data in much of the published literature (criterion 4); Berenbaum [24] and the US EPA [27] raised this issue earlier with respect to the literature on drugs and chemical interactions. The isobologram can also be used to illustrate the importance of statistical analysis for interaction data. The envelope of statistical significance about the line of additivity is much wider for an end point that is more highly variable than for one that is less variable (Figure 13.2b). If the end point has a relatively low variance, a number of the data points on the isobologram might be statistically significantly different from the null hypothesis of dose addition, and an interaction conclusion would be justified. When the assay variance is higher, however, fewer data points will represent a statistically significant departure from dose addition. Statistical methods and study designs that address biological variability in interaction analysis have been published (for example, see Refs [23, 28, 29]). These concepts are frequently ignored in the published literature that claims to evaluate interactions in general, and so this criterion is essential in assessing the clinical applicability of literature on interactions between drugs and herbal supplements.

13.2.4 Scoring Algorithm

Any particular study can be evaluated with respect to these five criteria, and a simple algorithm can be used to compare studies or to assess a broader body of literature [30]. If a study fails a particular criterion, a value of zero is assigned for that criterion; if a study partially satisfies the criterion, a score of 0.5 is assigned; and, if a study fully satisfies the criterion, a score of 1 is assigned. A composite score can be calculated for any particular study by dividing the sum of the scores for all the criteria by 5; hence, a composite score of zero is the lowest score that can be attained, and one the highest. This algorithm provides no weighting of the criteria, although it might be argued that some of the criteria are more important than others and should be assigned greater weight. The criteria and the algorithm do not directly evaluate the clinical relevance of a study, although we would argue that a consideration of these criteria is important in evaluating the clinical relevance of study results. The criteria must be applied cautiously to mechanistic and pharmacokinetic studies whose goals may not have been to fully define or characterize an interaction. Nonetheless, the criteria can help define the limits of interpretation for a particular set of data, regardless of the goals of the study that produced them. It can be challenging to incorporate all these criteria into whole animal or clinical studies simply because of the numbers of treatment groups that might be required.

Data that fulfill these simple criteria from human trials is rarely available. Even if the design of a human trial were sufficiently detailed to measure drug interactions, controlling or eliminating extraneous factors that could affect the measurement are often quite difficult due to human variability in genetics, diet, habitat, behaviors, and so on. In the absence of direct human data demonstrating an interaction for a particular drug effect, indirect evidence from an animal model of the effect would be sought. Animal models exist for both pharmacologic and toxicologic effects of drugs, but few have been validated to ensure that these models reliably predict qualitatively and quantitatively drug interaction in humans. The physiological and biochemical pathways underlying interactions would need to be quite similar between the animal model and the human, but where feasible, controlling or eliminating extraneous factors that could influence the interaction is much more tractable in an experimental animal model than in human clinical trials.

Establishing that an interaction occurs between two or more drugs and determining that a drug–drug interaction caused an adverse patient outcome requires scientific data and the application of established scientific principles. Making clinical judgments about medication therapy for a particular patient may be done with the best available information, which could involve the application of nonscientific clinical diagnostic probability scales. The literature on drug–drug and drug–supplement interactions often fails to make this critical distinction, but physicians, regulatory agencies, and the pharmaceutical industry would all be better served if it did. Editors and peer-reviewers should be aware of this problem and take appropriate steps to improve the quality of published literature on drug and supplement interactions.

References

1 US FDA (2004) Current good manufacturing practice for combination products: draft guidance for industry and the Food and Drug Administration. *Fed. Reg.*, 69 FR 59239.
2 US FDA (2004) Fixed dose combination and co-packaged drug products for treatment of HIV: draft guidance for industry. *Fed. Reg.*, 69 FR 28931.
3 EMEA (2002) Note for Guidance on the Non-Clinical Documentation of Medicinal Products With Well Established Use (Draft), CPMP/SWP/799/95/draft, November.
4 US FDA (2006) Center for Drug Evaluation and Research (CDER): Nonclinical Safety Evaluation of Drug or Biologic Combinations, March.
5 Liminga, U. and Lima, B.A. (2004) Non-clinical development of fixed combinations: a European regulatory perspective. *Int. J. Pharm. Med.*, **18** (3), 135–138.
6 European Medicines Agency (EMEA) (2008) Guideline on the Non-Clinical Development of Fixed Combinations of Medicinal Products, Document reference EMEA/CHMP/SWP/258498/2005, January 24.
7 US Department of Health and Human Services Centers for Disease Control and Prevention, National Center for Health Statistics. Health, United States, 2006, Table 93.
8 Kaufman, D.W., Kelly, J.P., Rosenberg, L., Anderson, T.D., and Mitchell, A.A. (2002) Recent patterns of medication use in the ambulatory adult population of the United States. *JAMA*, **287** (3), 337–344.
9 Ervin, R.B., Wright, J.D., and Kennedy-Stephenson, J. (1999) Use of dietary supplements in the United States, 1988–1994. *Vital Health Stat.*, **11** (244) i–iii, 1–14.
10 Naranjo, C.A., Busto, U., Sellers, E.M., Sandor, P., Ruiz, I., Roberts, E.A., Janecek, E., Domecq, C., and Greenblatt, D.J. (1981) A method for estimating the probability of adverse drug reactions. *Clin. Pharmacol. Ther.*, **30** (2), 239–245.
11 Wittkowsky, A.K. (2001) Drug interactions update: drugs, herbs, and oral anticoagulation. *J. Thromb. Thrombolysis*, **12** (1), 67–71.
12 Horn, J.R., Hansten, P.D., and Chan, L.N. (2007) Proposal for a new tool to evaluate drug interaction cases. *Ann. Pharmacother.*, **41** (4), 674–680.
13 Lucena, M.I., Camargo, R., Andrade, R.J., Perez-Sanchez, C.J., and Sanchez De La Cuesta, F. (2001) Comparison of two clinical scales for causality assessment in hepatotoxicity. *Hepatology*, **33** (1), 123–130.
14 Andrade, R.J., Camargo, R., Lucena, M.I., and Gonzalez-Grande, R. (2004) Causality assessment in drug-induced hepatotoxicity. *Exp. Opin. Drug Saf.*, **3** (4), 329–344.

15 Shapiro, M.A. and Lewis, J.H. (2007) Causality assessment of drug-induced hepatotoxicity: promises and pitfalls. *Clin. Liver Dis.*, **11** (3), 477–505.

16 Gaikwad, R., Sketris, I., Shepherd, M., and Duffy, J. (2007) Evaluation of accuracy of drug interaction alerts triggered by two electronic medical record systems in primary healthcare. *Health Inform. J.*, **13** (3), 163–177.

17 Gori, G.B. (1999) The EPA and the courts: inching toward a showdown. *Regul. Toxicol. Pharmacol.*, **30** (3), 167–168.

18 Gori, G.B. (2001) The costly illusion of regulating unknowable risks. *Regul. Toxicol. Pharmacol.*, **34** (3), 205–212.

19 Gori, G.B. (2002) Considerations on guidelines of epidemiologic practice. *Ann. Epidemiol.*, **12** (2), 73–78.

20 Loewe, S. and Muischnek, H. (1926) Effect of combinations: mathematical basis of problem. *Arch. Exp. Pathol. Pharmacol.*, **114**, 313–326.

21 Berenbaum, M.C. (1981) Criteria for analyzing interactions between biologically active agents. *Adv Cancer Res.*, **35**, 269–335.

22 Bliss, C.I. (1939) The toxicity of poisons applied jointly. *Ann. Appl. Biol.*, **26**, 585–615.

23 Borgert, C.J., Price, B., Wells, C., and Simon, G.S. (2001) Evaluating chemical interaction studies for mixture risk assessment. *Hum. Ecol. Risk Assess.*, **7**, 259–306.

24 Berenbaum, M.C. (1989) What is synergy? *Pharmacol. Rev.*, **41**, 93–141 [published erratum appears in *Pharmacol. Rev.* 1990, 41 (3), 422].

25 Greco, W.R., Bravo, G., and Parsons, J.C. (1995) The search for synergy: a critical review from a response surface perspective. *Pharmacol. Rev.*, **47**, 331–385.

26 Berenbaum, M.C. (1988) Synergy and antagonism are not synonymous with therapeutic advantage and disadvantage [letter]. *J. Antimicrob. Chemother.*, **21**, 497–500.

27 US EPA (US Environmental Protection Agency) (1990) Technical Support Document on Risk Assessment of Chemical Mixtures (Dated November 1988). EPA/600/8-90/064. Office of Health and Environmental Assessment, Cincinnati, OH.

28 Gennings, C., Carter, W.H.J., Carney, E.W., Charles, G.D., Gollapudi, B.B., and Carchman, R.A. (2004) A novel flexible approach for evaluating fixed ratio mixtures of full and partial agonists. *Toxicol. Sci.*, **80**, 134–150.

29 Price, B., Borgert, C.J., Wells, C.S., and Simon, G.S. (2002) Assessing toxicity of mixtures: the search for economical study designs. *Hum. Ecol. Risk Assess.*, **8**, 305–326.

30 Borgert, C.J., Borgert, S.A., and Findley, K.C. (2005) Synergism, antagonism, or additivity of dietary supplements: application of theory to case studies. *Thromb. Res.*, **117**, 123–132.

14
Dermal Chemical Mixtures

Ronald E. Baynes and Jim E. Riviere

14.1
Introduction

Drug and chemical dermal absorption typically involves experiments conducted using single chemicals making the mechanisms of absorption of individual chemicals extensively studied. Similarly, most risk assessment profiles and mathematical models are based on the behavior of single chemicals. A primary route of occupational and environmental exposure to toxic chemicals is through the skin; however, such exposures are often to complex chemical mixtures. In fact, the effects of coadministered chemicals on the rate and extent of absorption of a topically applied systemic toxicant may determine whether toxicity is ever realized. This dichotomy between the availability of data and the absorption models based on single chemicals with field exposure scenarios dominated by complex mixtures deserves further attention.

It is axiomatic that for a topically applied chemical to exert systemic toxicity, absorption across the dermal barrier is required. For a topically applied compound to be absorbed into the skin, it must first pass through the stratum corneum, continue through the epidermal layers and penetrate the dermis where absorption into the dermal microcirculation becomes the portal for systemic exposure. For a chemical with direct toxicity to the skin, systemic absorption is not required as the target cells could be any of those comprising the epidermis or dermis.

The application of risk assessment to dermal absorption by US regulatory agencies (EPA, OSHA, and ATSDR) is varied and highly dependent upon available data [25, 26, 44]. A similar concern over the lack of data exists for overall risk assessment of chemical mixtures [16, 23, 24, 45]. A Congressional Commission on Risk Assessment and Risk Management [20] recommended going beyond individual chemical assessments and to focus on the broader issue of toxicity of chemical mixtures. With complex mixtures of hundreds of components, current approaches border on the impossible. Importantly, interactions involving the modulation of dermal absorption, and thus systemic bioavailability, have not been addressed.

It is impossible to assess all potential combinations of chemicals to determine which has the greatest potential to modulate absorption of a known toxic entity

Principles and Practice of Mixtures Toxicology. Edited by Moiz Mumtaz

ISBN: 978-3-527-31992-3

topically exposed to a chemical mixture. The present state of knowledge in this area is particularly weak since the significance of specific interactions has not been quantitated, let alone in many cases even identified. In many ways, this same concern continues to define the very nature of chemical mixture toxicology. The appreciation of the importance of chemical mixture interactions to effect chemical and drug disposition, biotransformation, pharmacokinetics, and activity has been well recognized for many years and is extensively reviewed elsewhere [15, 16, 28–30, 45]. A large body of literature exists in the field of drug–drug interactions and a body of work is developing in the area of food additive interactions on drug absorption. Despite this widespread knowledge base of the importance of drug–drug interactions and formulation effects in pharmaceutical science, the potential role of chemical interactions in systemic toxicology has not garnered significant attention. This chapter provides an overview of the potential mechanisms operative in topical chemical mixtures and presents some of our laboratory's efforts to quantitate these effects.

In cases where the potential toxicity of a specific mixture is of concern (e.g., at a specific toxic waste site), the complete mixture is often tested [35]. However, how does one quantitate the absorption of a mixture consisting of 50 chemicals? How are markers selected? How are these data expressed? Unfortunately, even after a complete toxicological profile of a specific mixture (e.g., "standard" mixture of 50 environmentally relevant compounds, surrogate jet fuels, etc.) is defined using all the techniques modern toxicology and toxicogenomics have to offer, one cannot define the links between the absorption and the effects seen. Was the observed toxicity exerted because a specific toxicant was in the mixture or because two synergistic toxicants were absorbed *or* was it exerted simply by the presence of a mixture component (e.g., alcohol, surfactant, fatty acid) that enhanced the absorption of a normally minimally absorbed toxicant? We have demonstrated [9, 52] such an interaction with the putative toxins involved in the Gulf War Syndrome [1] where systemic pyridostigmine bromide or coexposure to jet fuel was shown to greatly enhance the dermal absorption of topical permethrin. Would other pesticides have this effect? How does one take into account such critical interactions so that a proper risk assessment may be conducted? An inclusive approach to this problem is to define chemicals on the basis of how they would interact both with other components of a mixture and with the barrier components of the skin. What are the physical–chemical properties that would significantly modify absorption and potentiate systemic exposure to a toxicant? What are the properties of molecules susceptible to such modulation?

One recently reported approach to address this problem assesses potential interactions in dermal absorption by fractionating the effects of a vehicle on drug penetration into two primary parameters describing permeation according to Fick's law: partitioning (PC) and diffusivity (D) (see below; permeability (K_p) = $D \cdot$PC/membrane thickness) [56]. Although this study only reported on four compounds, one (diazepam) was not predictable using this approach as its physiochemical properties were already optimal for absorption, and only absorption enhancers were investigated. This study illustrates the difficulty of making broad generalizations across compounds solely on physical–chemical properties.

The problem of dermal mixture absorption is conceptually similar to that of dermatological formulations in the pharmaceutical arena. The primary difference is that most pharmaceutical formulation components are added for a specific purpose relative to the delivery, stability, or activity of the active ingredient. In the environmental and occupational scenarios, additives are a function on either their natural occurrence or presence in a mixture for a purpose related to uses of that mixture (e.g., a fuel performance additive, stability), and not for their effects on absorption or toxicity of the potential toxicant. Unlike pharmaceutical formulation additives in a dermal medication, chemical components of a mixture are not classified on how they could modulate percutaneous absorption of simultaneously exposed topical chemicals. They are present *functionally* for specific purposes (e.g., performance additives, lubricants, and modulators of some biological activity), *sequentially* because they were applied to the skin independently at different times for unrelated purposes (cosmetics followed by topical insect repellent, sunscreens), *accidentally* because they were simultaneously disposed of as waste, or *coincidentally* associated as part of a complex occupational or environmental exposure.

14.2 Mechanisms of Interactions

Chemical interactions that may modulate dermal absorption can be conveniently classified according to physical location where an interaction may occur. The advantage of this approach is that potential interactions may be defined both on the basis of specific mechanisms of action involved and by the biological complexity of the experimental model required to detect it.

Surface of Skin:

Chemical–chemical (binding, ion-pair formation, etc.)

Altered physical–chemical and solvatochromatic properties (e.g., solubility, volatility, critical micelle concentration (CMC))

Altered rates of surface evaporation

Occlusive behavior

Binding or interaction with adnexial structures or their products (e.g., hair, sweat, sebum).

Stratum Corneum:

Altered permeability through lipid pathway (e.g., enhancer)

Altered partitioning into stratum corneum

Extraction of intercellular lipids

Epidermis:

Altered biotransformation

Induction of and/or modulation of inflammatory mediators

Dermis:

Altered vascular function (direct or secondary to mediator release)

The first and most widely studied area of chemical–chemical interactions is the surface of the skin. The types of phenomena that could occur are governed by the laws of solution chemistry and include factors such as altered solubility, precipitation, supersaturation, solvation, or volatility, as well as physical–chemical effects such as altered surface tension from the presence of surfactants, changed solution viscosity, and micelle formation [4, 32, 40, 61]. For some of these so-called solvatochromatic effects, chemicals act independent of one another. However, for many the presence of other component chemicals may modulate the effect seen.

Chemical interactions may further be modulated by interaction with adnexial structures or their products such as hair, sebum, or sweat secretions. The result is that when a marker chemical is dosed on the skin as a component of a chemical mixture, the amount freely available for subsequent absorption may be significantly affected. The primary driving force for chemical absorption in skin is passive diffusion that requires a concentration gradient of thermodynamically active (free) chemical.

A second level of potential interaction is of those involving the marker and/or component chemicals with the constituents of the stratum corneum. These include the classic enhancers such as oleic acid, Azone®, or ethanol, widely reviewed elsewhere [61]. These chemicals alter a compound's permeability within the intercellular lipids of the stratum corneum. Organic vehicles persisting on the surface of the skin may extract stratum corneum lipids that would alter permeability to the marker chemical [38, 48]. Compounds may also bind to stratum corneum constituents forming a depot.

Another level of interaction would be with the viable epidermis. The most obvious point of potential interaction would be with a compound that undergoes biotransformation [18, 43]. A penetrating chemical and mixture component could interact in a number of ways, including competitive or noncompetitive inhibition for occupancy at the enzyme's active site, or induction or inhibition of drug-metabolizing enzymes. Other structural and functional enzymes could also be affected (e.g., lipid synthesis enzymes) that would modify barrier function [22]. A chemical could also induce keratinocytes to release cytokines or other inflammatory mediators [3, 33, 37] that could ultimately alter barrier function in the stratum corneum or vascular function in the dermis. Alternatively, cytokines may modulate biotransformation enzyme activities [39].

The last level of potential interaction is in the dermis where a component chemical may directly or indirectly (e.g., via cytokine release in the epidermis) modulate vascular uptake of the penetrated toxicant [49, 62]. In addition to modulating transdermal flux of chemicals, such vascular modulation could also affect the depth and extent of toxicant penetration into underlying tissues.

The optimal experimental approach to define such interactions is to conduct studies *in vivo* since all potential mechanisms of interactions are present. However, this approach is cost-prohibitive and is not amenable to determination of specific permeability parameters that would elucidate basic principles that could then be used to predict interactions in the future. Although the biologically intact skin seen in the *in vivo* setting might seem to be ideal, in reality it is difficult to dissect out interactions that occur in different phases of absorption. This is due to the confounding influence of multiple biological and surface chemical factors, as well as systemic effects, being present. This scenario is further aggravated by the high level of interindividual variability typical of *in vivo* studies that mask the detection of important interactions that might be exerted in other mixtures of slightly different composition. Most importantly, acute studies or those using extremely toxic chemicals could never be conducted in humans for ethical reasons or even in intact animals due to humane considerations, making an alternative approach such as outlined here a necessity.

14.3 Mixture Interactions in Skin

Our research program has focused on the effects of chemical mixture components on dermal absorption of select "marker" chemicals in the mixture [8, 47, 53, 54]. As varied as these potential interactions may be, experimentally isolating them is difficult due to confounding effects from simultaneously occurring multiple interactions (e.g., enhanced partitioning coupled with decreased diffusivity, solubility, or vascular uptake). To dissect out these processes, our laboratory's approach has been to use different model systems with increasing levels of biological complexity.

Silicone Membrane Flow-Through Diffusion Cells (SMFT): Sensitive to solvatochromatic processes
Porcine Skin Flow-Through Diffusion Cells (PSFT): Also sensitive to changes in lipid permeability
Isolated Perfused Porcine Skin Flaps (IPPSF): Also sensitive to irritation and vascular events

This hierarchy of experimental models allows interactions to be independently studied using efficient and humane *in vitro* model systems. These systems, as well as the basic principles of percutaneous absorption of single chemicals, have been extensively described in the literature [17, 50].

In order to investigate the nature of mixture interactions on chemical absorption, a series of studies were conducted on 12 diverse chemicals representing three chemical classes [58, 59]:

Substituted Phenols: nonylphenol, pentachlorophenol (PCP), phenol, *p*-nitrophenol (PNP)

Organophosphate Pesticides: chlorpyrifos, ethylparathion, fenthion, methyl parathion
Triazines: atrazine, propazine, simazine, and triazine

These compounds have molecular weights ranging from 94 to 350 and log octanol–water partition coefficients (log $K_{o/w}$) ranging from −1 to 5. Compounds were studied using a complete factorial experimental design. There were three vehicles and three binary-vehicle combinations: water, ethanol, propylene glycol (PG), water/ethanol, water/PG, and ethanol/PG. For each vehicle combination, a control (no additive) or one or both of two mixture components, the surfactant sodium lauryl sulfate (SLS) or vasodilator methyl nicotinate (MNA), were added. This resulted in 24 mixture combinations per chemical, each conducted in two flow-through diffusion cell systems: silastic membrane (SMFT) or porcine skin (PSFT). Porcine stratum corneum partition coefficient (log $K_{SC/MIX}$) across all vehicles was determined. Parameters determined in the diffusion cell studies included permeability and diffusivity. A restricted number of specific vehicles and compound combinations were then selected for study in the IPPSF. Data were initially analyzed using compass plots to determine mixture interactions. Compass plots have been useful to visually examine data to probe complex mixture interactions [6, 19, 41, 47]. Analysis of means (ANOM) is used to visualize statistical significance of effects seen [19].

The underlying hypothesis in this study was that permeability parameters are a function of log $K_{o/w}$ and molecular weight, which may be affected when a compound is administered in a chemical mixture. This forms the basis of recent EPA dermal absorption guidelines based on the work of Potts and Guy [46]. Comparisons where this relationship held confirm that diffusion processes remain rate limiting. However, when such relations break down, a significant interaction may be present that is not diffusion limited, and thus not directly related to log $K_{o/w}$. This would have obvious impact on risk assessment procedures as it would affect the mathematical form of any quantitative structure permeability relationship (QSPR) equation linking log $K_{o/w}$ to dermal absorption. These occurred when extrapolating across solvent systems (e.g., ethanol/water), when surfactant SLS was added to different systems, and occasionally when individual compounds deviated from their predicted fluxes.

14.3.1 Compound Susceptibility to Solvent Interactions

A consistent finding across all mixtures was that as expected, some compounds behave differently in specific solvent systems. This interaction tended to be consistent across SMFT and PSFT systems, a finding supporting our hypothesis that the interaction is solvatochromatic and does not involve interaction with other constituents of the stratum corneum or epidermis. Figure 14.1 is a series of compass plots illustrating permeability in PSFT of all 12 study compounds in 12 propylene glycol mixtures. Compounds are arranged clockwise around the plot in order of descending

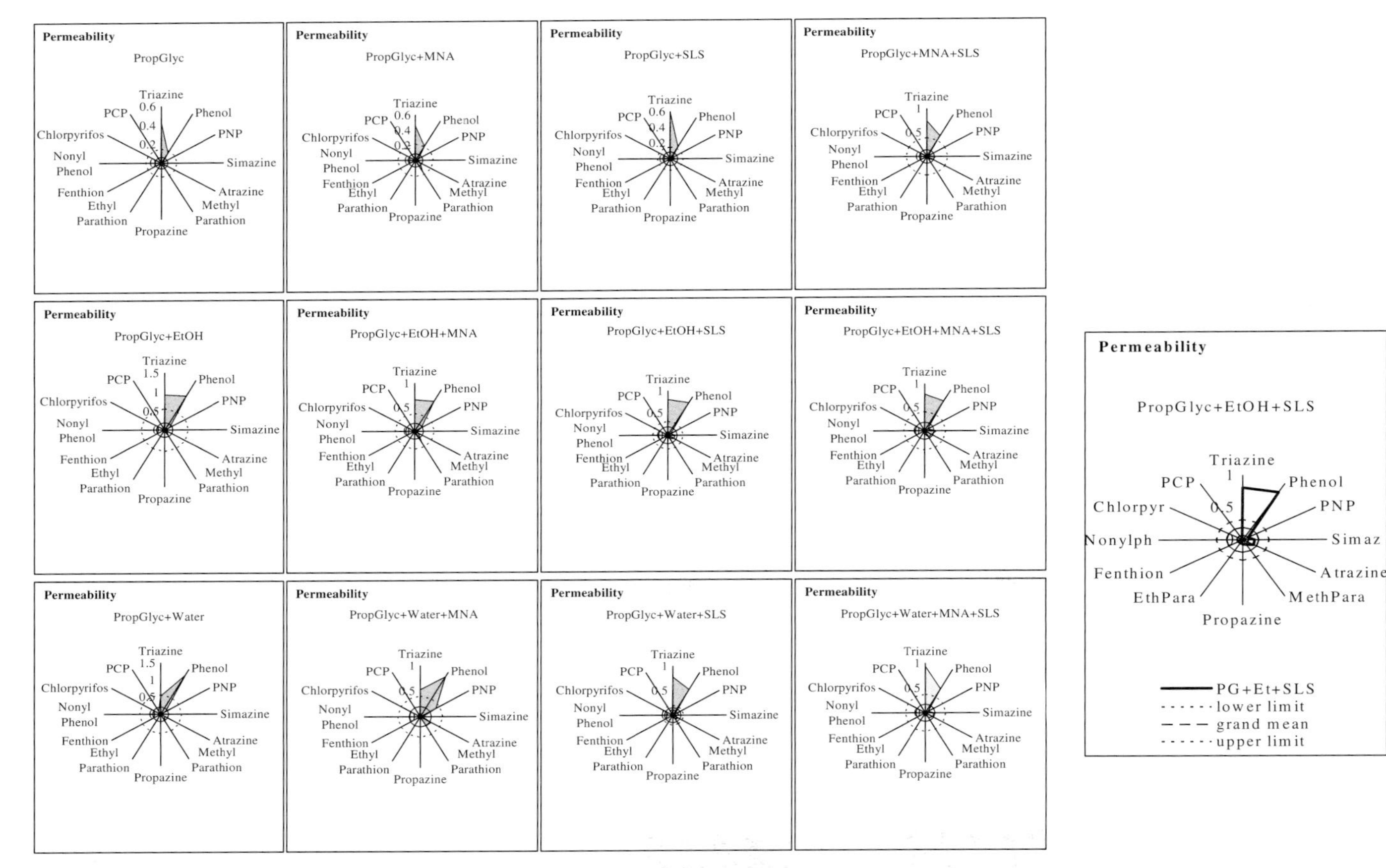

Figure 14.1 Compass plots for penetrants in propylene glycol mixtures in PSFT. The interactions noted in this plot at left for triazine and phenol were positive and outside of the upper confidence interval for significant interactions ($p < 0.05$). One cell is expanded above to illustrate upper and lower bounds.

log $K_{o/w}$ with the lowest (triazine, −0.9) at 12 o'clock and the highest (PCP, 5.12) at 11 o'clock. Permeability from flow-through diffusion cells are plotted on the radii that is further demarcated by mean, upper, and lower ANOM confidence intervals. These plots illustrate that in this system, compounds behave similarly except for triazine and phenol, two compounds with the lowest $K_{o/w}$. Exposure to these in PG would not be predicted from a log $K_{o/w}$ relationship. Ethanol modulates these fluxes, especially for phenol. Compass plots across SMFTs (data not shown) and PSFTs were comparable. Absorption in PG was low compared to the other solvents studied (except for low log $K_{o/w}$ compounds), and mixture interactions detected were opposite of what was seen in other systems (e.g., water/ethanol). These complex interactions make simple extrapolations difficult.

14.3.2
Mixture Interactions Across Model Systems

An analysis of compass plots detected a series of statistically significant ($p < 0.05$) interactions in some solvent systems that did not clearly extrapolate across model systems and demonstrated individual compound specificity. Those identified are illustrated in Figure 14.2 that depicts absorption flux (% dose/h) versus time profiles for four compounds in two chemical classes (PNP, PCP, atrazine, and propazine) exposed in water or water + SLS, studied in SMFT or PSFT models as well as IPPSFs. It must be noted that the shape of the absorption profiles across the three model systems is always different due to differences in model system structure. Flux through the SMFT is an order of magnitude greater than that of the PSFT making the time of peak flux to occur earlier in this system. It is the *correlation* of permeability between these systems, and not the absolute magnitude, that is important for assessing mixture effects on absorption. The delayed absorption seen in the PSFT compared to the SMFT reflects the greater lag time, and hence reduced diffusivity, for transport across porcine skin. Similarly, the shorter lag time generally seen in IPPSFs reflects the shorter diffusional distance seen when a model is perfused by dermal capillaries. A characteristic of these models, compound flux in the PSFT exceeds those in IPPSF that possesses a more complex membrane.

The first comparison is a relatively consistent SLS effect across compounds and models. The next comparison involved the substituted phenols PCP and PNP in the same solvent systems. In the SMFT, PCP flux was greater than PNP as predicted from its greater log $K_{o/w}$ of 5.1 versus 1.9. However, the fluxes were of a similar order of magnitude in the PSFT experiments, which carried forward into the IPPSF studies where the major difference was a slightly earlier peaking of PCP. These compounds have significantly different molecular properties (solubility, H-bonding indices) that could explain these differences both on the basis of stratum corneum penetration and subsequent dermal disposition. log $K_{o/w}$ alone, used in risk assessment models, would not predict this IPPSF response. Again, in all systems, SLS reduced flux compared to controls. This suggests that the SLS surfactant effect is detectable in the simplest model system; however, the actual shape of the transdermal flux profile is not directly predictable from the simpler systems. In

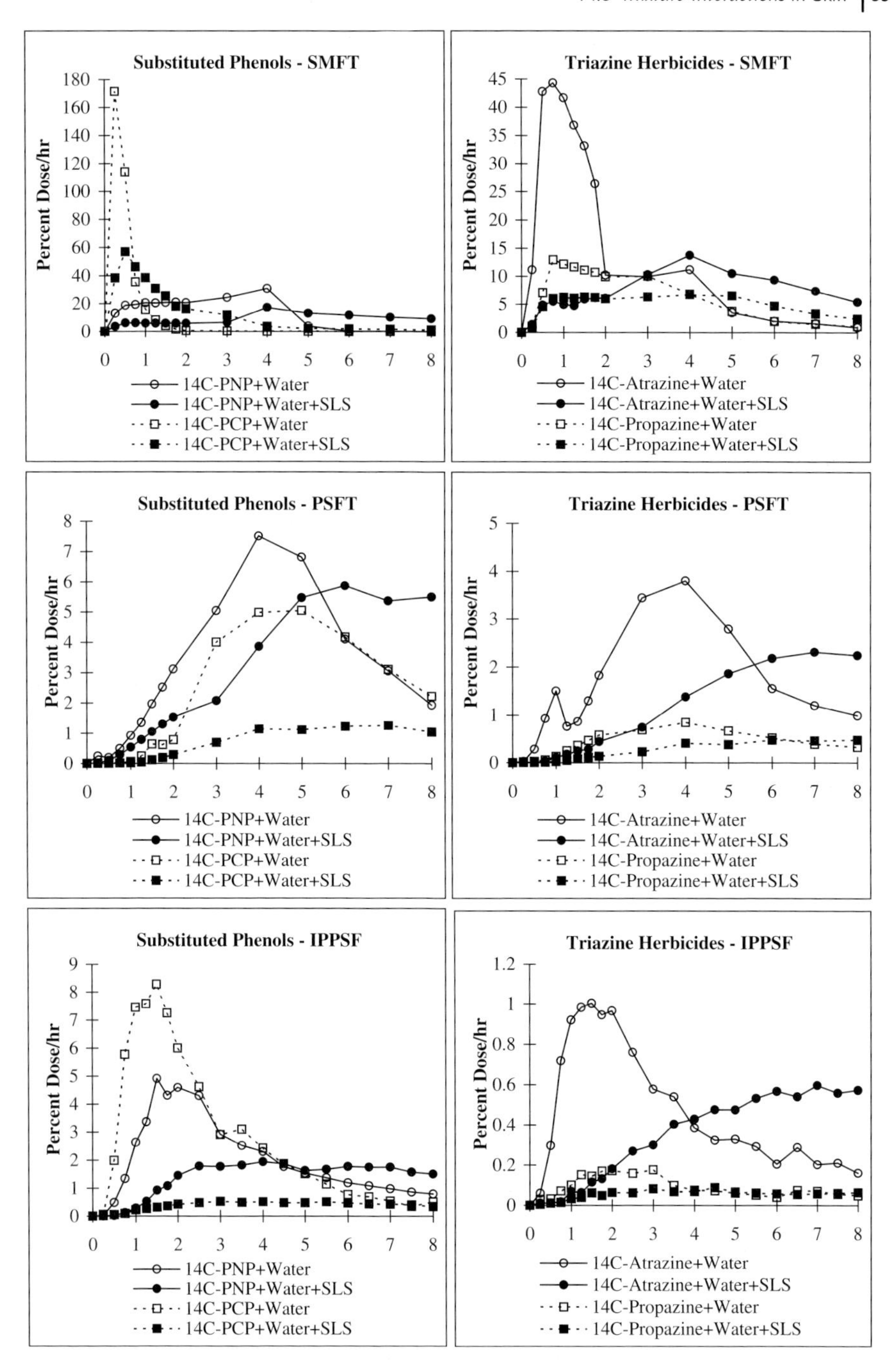

Figure 14.2 Effects of sodium lauryl sulfate on dermal flux profiles of four model penetrants (PNP, PCP, atrazine, and propazine) in SMFT, PSFT, and IPPSF model systems. Open symbols are control and filled symbols are SLS treatments.

fact, the broadening of the SLS absorption profiles may reflect modified dermal deposition within the stratum corneum that would be consistent with previously reported enhancing effects of SLS. We would expect the shape of the absorption profile to be a function of the kinetic model parameters describing these flux profiles. Second, it suggests that the SLS effect dominates over any log $K_{o/w}$ QSPR relationship for permeability of a single-compound absorption. This key finding provided the insight into developing the modified linear free energy relationship (LFER) approaches discussed below.

The mechanism most likely responsible for the flux reduction from SLS at this concentration (10%) under these aqueous solvent conditions may be via reducing the CMC in solution [7], an effect that reduces availability of compound for absorption independent of its initial permeability. We have shown that mixture interactions modulate CMC, a phenomenon detectable across all model systems. These SLS effects are fully documented in Refs [58, 59]. This effect would be present across all model systems. Once the compound is absorbed into the stratum corneum, SLS could affect its structure and further modulate absorption reflected in the altered shape profiles. It should be noted that in acetone or DMSO vehicles, parathion flux had been shown to increase with SLS [47], reflecting a differential response of this mixture additive depending upon the solvent system in which it is exposed. This dependency of mixture additive effects on the dosing vehicle was a major finding of this work.

Effects of these mixture interactions on IPPSF kinetic parameters are listed in Table 14.1 as they illustrate the most interesting mixture-specific data. SLS in water results in a decreased absorption rate (smaller *b*, *d*) compared to water alone. Ethanol in water reduces "*b*" for all compounds. Ethanol significantly reduces "*d*" except for propazine and simazine. These independent changes in rate parameters are consistent with different shape profiles seen under different mixture conditions.

14.3.3
Can Partition Coefficient Predict Mixture Behavior?

If log *K* is a primary determinant of diffusion, our hypothesis was that disposition from a mixture could be predicted from stratum corneum/mixture partition coefficient ($K_{SC/MIX}$). Validation of this relationship would offer strong support for using LFER equations to predict mixture absorption. This was addressed by

Table 14.1 Effects of SLS or ethanol on *b*/*d* mean IPPSF kinetic parameters in water.

	Atrazine	Chlorpyrifos	Ethyl-parathion	Methyl-parathion	PCP	PNP	Propazine	Simazine
Water	0.45/0.68	0.13/0.29	0.46/0.64	0.57/0.74	0.65/0.77	0.47/0.58	0.39/0.62	0.36/0.69
Water/SLS	0.06/0.11	0.02/0.10	0.11/0.26	0.18/0.31	0.17/0.33	0.16/0.22	0.18/0.38	0.14/0.46
Water/EtOH	0.23/0.46	0.09/0.22	0.15/0.46	0.28/0.65	0.26/0.34	0.36/0.50	0.14/0.72	0.28/1.28

comparing log $K_{SC/MIX}$ to permeability across SMFT and PSFT models, as well as to total IPPSF penetration (absorption + skin deposition). Figure 14.3 is a histogram of these parameters across all systems using PCP as a model penetrant to illustrate this effect. Chemical mixtures are ordered by decreasing log $K_{SC/MIX}$ (top panel). Permeability in SMFT and PSFT models, as well as IPPSF absorption, generally follows the order of descending log $K_{SC/MIX}$ with a few clear exceptions. SMFT permeability mirrored log $K_{SC/MIX}$ ($R^2 = 0.83$). A unique mixture was ethanol/water/MNA in PSFTs where permeability was less than predicted from log $K_{SC/MIX}$, an effect that carried into the IPPSF suggesting a potential interaction with epidermal cells or dermal components; the only consistent factors different between isolated stratum corneum and SMFT compared to PSFT and IPPSFs. This interaction was also seen with other compounds. For PCP, stratum corneum partitioning appears to be the dominant factor. These findings support the hypothesis that a mixture component effect (e.g., SLS) in a specific solvent system will reduce permeability across penetrants (independent of the compound-specific QSPR relation to log $K_{o/w}$) and can be estimated by partition coefficients in simpler system.

14.3.4 Solvent–Water Interactions

A significant interaction detected was the different effects seen for absorption across mixtures consisting of water, water/ethanol, and ethanol. Although log $K_{SC/MIX}$ correlated highly with log $K_{o/w}$ in water (system in which most dermal absorption QSPR analyzes are defined), there was no clear correlation between log $K_{o/w}$ across these other relatively simple solvents. This mechanism was explored in more detail using log $K_{SC/MIX}$ [58, 59]. Figure 14.4 depicts both log $K_{SC/MIX}$ for all 12 compounds ordered by log $K_{o/w}$ and the regression analyses for log $K_{SC/MIX}$ versus log $K_{o/w}$ in the three separate solvent mixtures. As expected, there is a reasonable correlation between these parameters in water. However, the correlation significantly weakened when water/ethanol and ethanol system were analyzed. Viewed from another perspective, rank order of log $K_{SC/MIX}$ was expected to be water > water/ethanol > ethanol, which held for compounds with log $K_{o/w}$ at the extremes. However for compounds with *mid-range* PCs (e.g., atrazine–ethylparathion), this clear-cut order was lost, suggesting molecular properties not predicted by log $K_{o/w}$ may modulate these relationships. Other researchers [56] showed that diazepam with mid-range log $K_{o/w}$ did not respond to chemical enhancement. It is evident that a mixture containing ethanol has a narrower range of altered log $K_{SC/MIX}$ than does a pure water system, potentially reflecting stronger molecular interactions seen with this solvent. Significantly, a 50/50 mixture could not be interpolated from data in pure solvents. Finally, one must mention that molecular weight was also an important covariate in these analyses; however, they did not add a significant ability to discriminate these interactions or predict absorption over that which partition coefficient already provided.

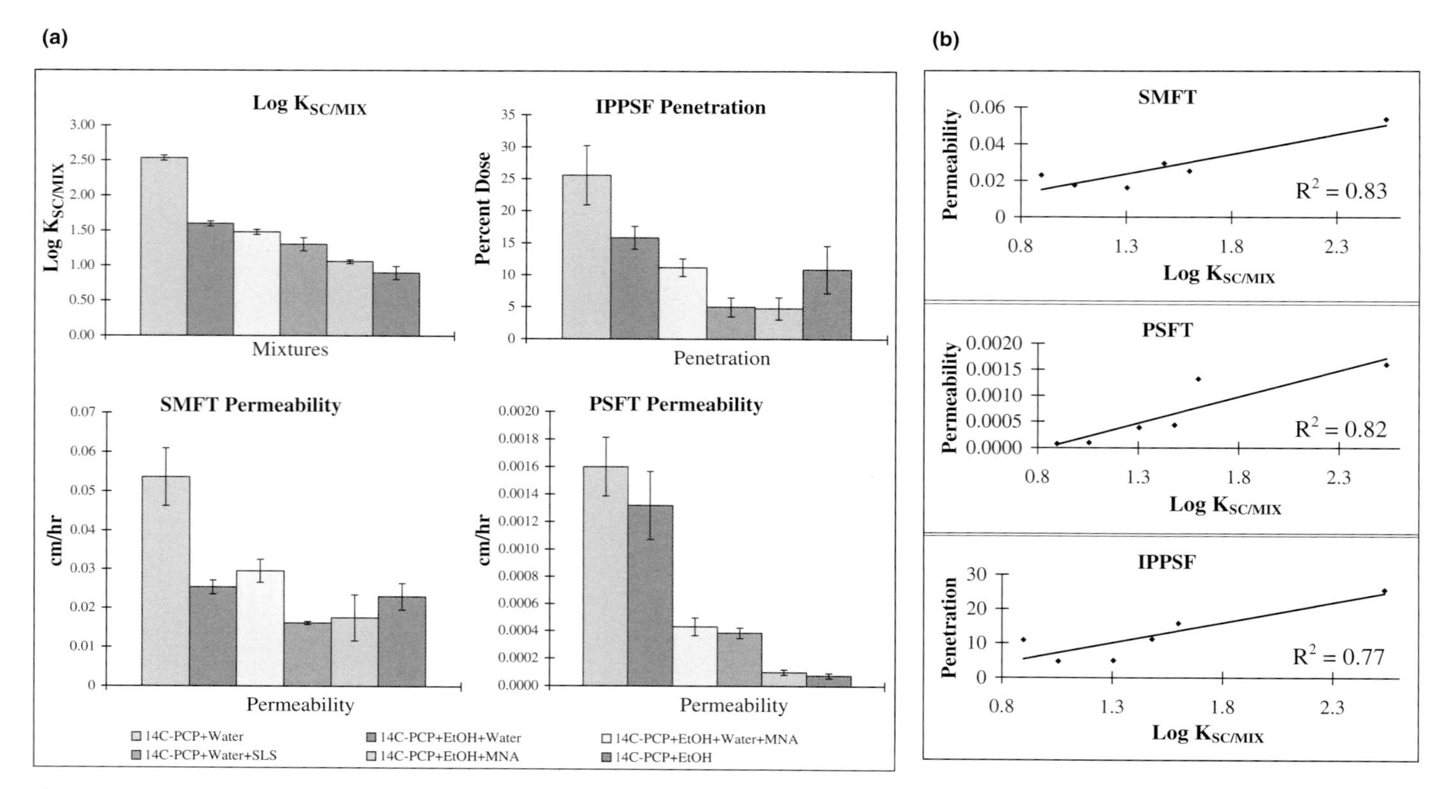

Figure 14.3 Comparative PCP partitioning (log $K_{SC/MIX}$) and permeability from six mixtures across all three model systems, SMFT, PSFT, and IPPSF (a) and correlation between PCP permeability and penetration from the six mixtures and log $K_{SC/MIX}$ (b).

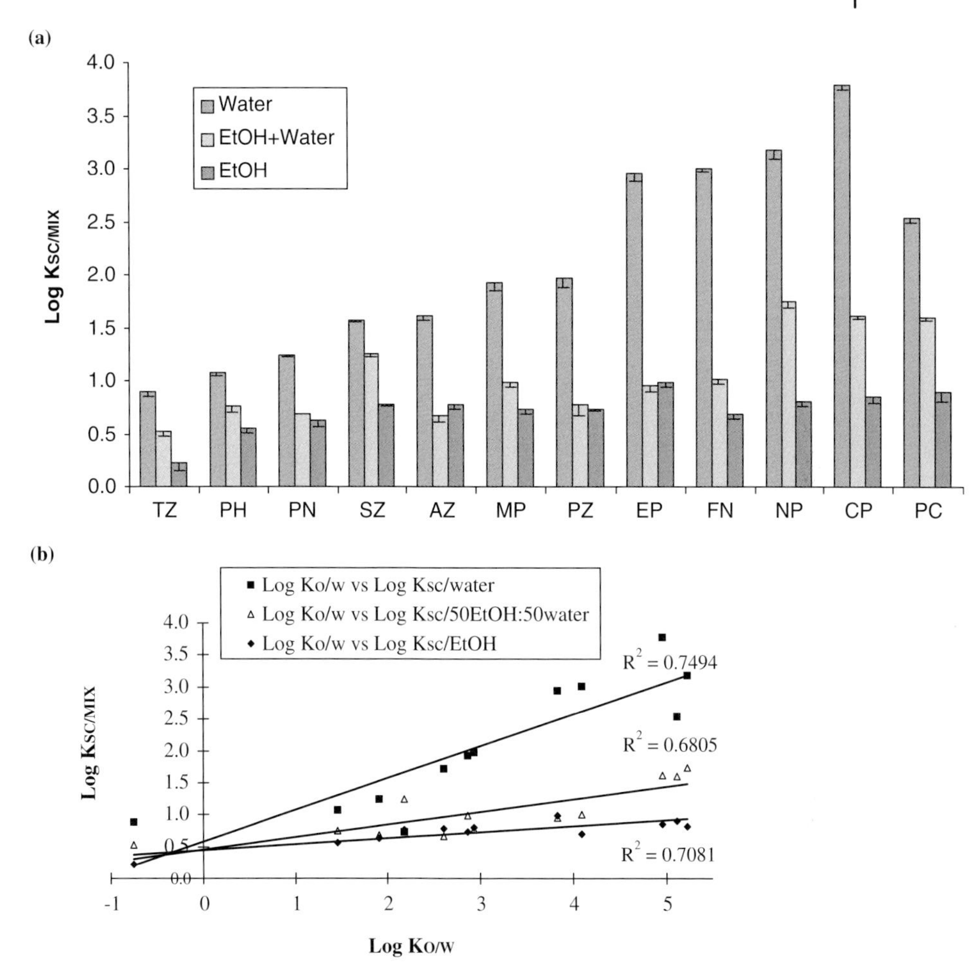

Figure 14.4 log $K_{SC/MIX}$ for 12 pesticides in water, 50% ethanol/water, and ethanol mixtures (a) and the correlation between these log $K_{SC/MIX}$ experimental values and literature values for log $K_{o/w}$ (b).

14.3.5
Modified QSPR Equations that Predict Chemical Absorption from a Mixture

The above analysis suggested that mixture interactions (e.g., vehicles, surfactants) modify the normal correlation seen between a parameter such as log $K_{o/w}$ and absorption. Such correlations are often embedded in the QSPR based on linear free energy relationship. For individual chemicals, there is often a strong correlation with parameters such as log $K_{o/w}$ and molecular weight [46] or with multiple molecular descriptors that allow a wider range of chemicals to be accounted for [2]. As can be seen from the above data, the addition of SLS or ethanol modified the correlation between absorption and log $K_{SC/MIX}$ or log $K_{o/w}$, and did not destroy this relationship. Such "mixture effects" were often constant across all model systems studied as was demonstrated with the compass plot analyses. Finally, the fact that mixture effects seen in the IPPSF somewhat reflected changes in log $K_{SC/MIX}$ suggested that an LFER approach incorporating both molecular descriptors of the penetrants of interest (classical single chemical approach) and mixture effects described by physical descriptors of the mixture might hold promise.

We elected to use Abraham's LFER model as our base equation since it is representative of the dermal QSPR approaches presently available [27]. Preliminary analyses applying 16 different LFER equations reviewed by Geinoz *et al.* [27] to the entire PSFT (288 treatment combinations) and IPPSF data set (32 treatment combinations) demonstrated a superior fit of our data set to the Abraham equation compared to most other models reviewed. It must be stressed that the purpose of this research was neither to identify the *optimal* LFER for predicting dermal permeation nor to validate that this model is predictive of dermal absorption. Rather, we selected this model since it best described the data generated in this specific research and is widely accepted by the scientific community.

$$\log k_p = c + a\Sigma\alpha_2^H + b\Sigma\beta_2^H + s\pi_2^H + rR_2 + vV_x,$$

where k_p is the permeability constant for the PSFT experiments, $\Sigma\alpha_2^H$ is the hydrogen-bond donor acidity, $\Sigma\beta_2^H$ is the hydrogen-bond acceptor basicity, π_2^H is the dipolarity/polarizability, R_2 represents the excess molar refractivity, and V_x is the McGowan volume. Molecular descriptor values for all these parameters were calculated for the 12 penetrants studied with ABSOLV® Solute Property Prediction Software (Sirius Analytical Instruments, Ltd, East Sussex, UK). The parameters a, b, s, r, and v are *strength coefficients* coupling the molecular descriptors to skin permeability in the specific experimental system (e.g., PSFT or IPPSF).

To incorporate mixture effects, another term called the mixture factor (MF) is added, yielding

$$\log k_p = c + mMF + a\Sigma\alpha_2^H + b\Sigma\beta_2^H + s\pi_2^H + rR_2 + vV_x.$$

This concept allows to define an LFER equation across data collected from different mixtures. Hostynek and Magee [31] had used indicator variables embedded in LFER equations to allow analysis across exposures consisting of different vehicles or occlusive conditions. Unlike our approach, these indicator variables did not contain any information concerning the vehicles, but were a statistical regression tool to allow the base LFER model to be applied to penetrants dosed under different experimental conditions.

Figure 14.5 depicts the predicted versus observed permeability constants (log K_p) for all 288 treatment combinations studied without taking into account the specific mixtures these chemicals were dosed. The residuals of this model showed no further correlation with penetrant properties. However, when vehicle/mixture component properties were analyzed, trends in residuals became evident. An excellent single parameter explaining some variability of this residual pattern (R^2 of 0.44) was log(1/Henry Constant) (1/HC). Figure 14.6 depicts the modified LFER model including MF = log(1/HC).

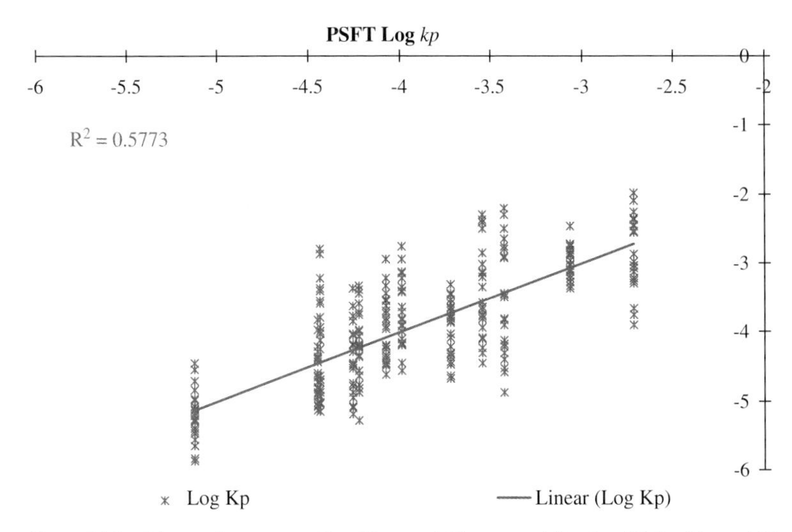

Figure 14.5 Observed versus predicted for pesticide permeability (log K_p) for 12 pesticides from various chemical mixtures. Data obtained from PSFT experiments and regression analysis performed without inclusion of a mixture factor.

The improvement on the prediction of PSFT permeability across all treatment is clearly evident. It must be stressed that an MF related to 1/HC is not the final form of this analysis but was the first property suggestive that this approach might work. Other physical parameters of the mixture similar to the molecular descriptors of the penetrant were also correlated. A similar approach was used for the smaller data sets

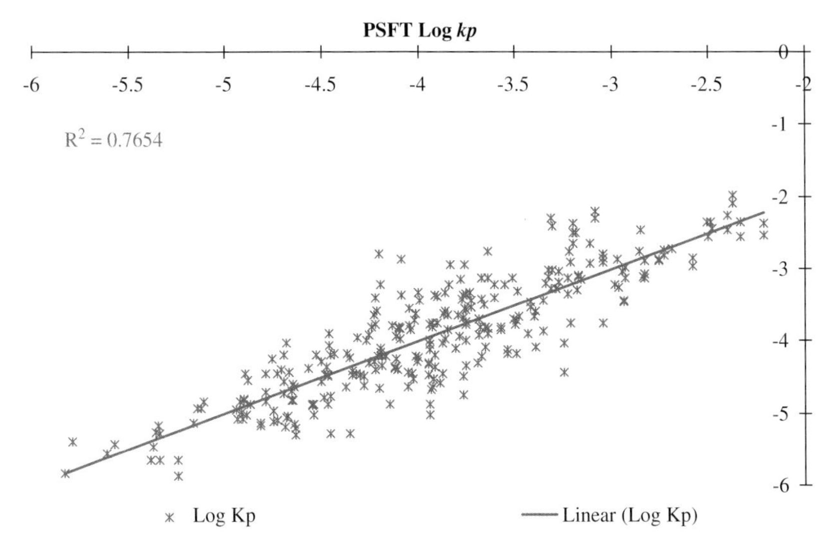

Figure 14.6 Observed versus predicted for pesticide permeability (log K_p) for 12 pesticides from various chemical mixtures. Data obtained from PSFT experiments and regression analysis performed with inclusion of a mixture factor.

available in the IPPSF mixture exposures, and inclusion of a mixture factor improved predictability. The final form of such a modified mixture LFER must await further research. This includes defining which physical–chemical property of the mixture best describes the mixture factor. However, the finding that an MF can be used to extrapolate chemical dermal absorption data across different vehicles and mixture combinations is significant that we believe would have an impact on the occupational risk assessment process. It would allow single-chemical data collected in multiple research environments, and resulting single-chemical LFER determined in different studies, to be modified on the basis of mixture component or vehicle properties.

14.3.6
Novel MCF Approach to Calibrate Dermal Absorption of Mixtures

The membrane-coated fiber (MCF) has been used as a solid-phase microextraction technique for chemical analysis of drug and environment samples [60, 64]. In brief, the MCF (Figure 14.7a) is immersed in the sample solution containing the analyte of interest, and after a given permeation time to allow solute equilibrium between the solution and the MCF, the MCF is injected directly into an LC or GC injector for dissolution or desorption and MS analysis to determine phase distribution of

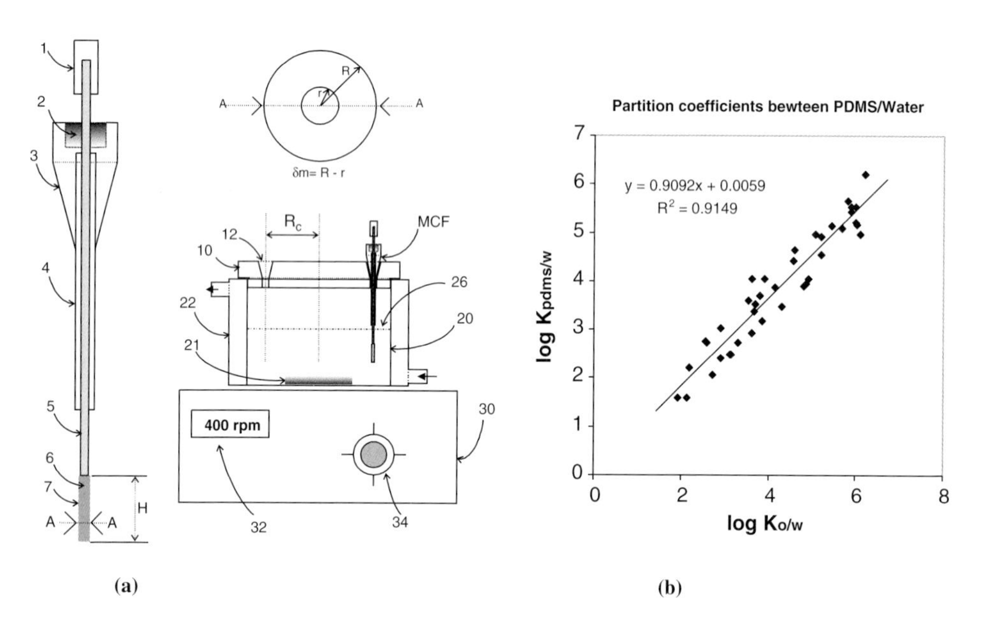

Figure 14.7 (a) MCF fiber system (left side) and experimental apparatus (right side) and (b) correlation of log $K_{PDMS/w}$ versus log $K_{o/w}$. (1) Holding tip, (2): sealing septum, (3) needle base, (4) piercing needle, (5) fiber attachment tubing, (6) inert fiber, (7) PDMS membrane, (10) needle holding cap, (12) MCF positioning holes, (20) solution container, (21) magnetic stirring bar, (22) water jacket, (26) donor solution, (30) magnetic stirrer, (32) tachometer, (34) stirring speed control. r: radius of the inert fiber; R: radius of the membrane, δ_m: thickness of the membrane, and R_c: hole radius on the cap.

analyte between the solution and the MCF membrane. The membrane coatings of commercially available MCF assemblies can range from 7 to 100 μm thick coated onto a fused silica or metal alloy fiber. There are at least seven commercially available coatings, of which only three MCFs display absorption characteristics (polydimethylsiloxane, PDMS; polyacrylate, PA; and carbowax, WAX) and the others display mixed absorption and adsorption characteristics (e.g., carboxen, divinylbenzene). PDMS, PA, and WAX have been demonstrated to display an equilibrium process based on Fick's first law of diffusion. Our laboratory was the first to capitalize on the unique physicochemical characteristics of these various three MCF membranes to develop mathematical models [65, 66] that predict solute partitioning and initial solute permeation rates following MCF exposure to a diverse series of 39 solutes (Figure 14.7b).

These MCFs were able to characterize partitioning behavior of the predominant aromatic components of jet fuels into human stratum corneum [13]. The purpose of these experiments was to demonstrate the reproducibility of the MCFs to predict octanol–water coefficients and by extension its versatility to assess chemical partitioning into skin from chemical mixtures. The latter is worth noting in that traditional approaches of assessing octanol–water partition coefficients are limited to single chemicals and not chemical mixtures. Solute partitioning into bilayers (e.g., skin, MCFs) differs from that into bulk isotrophic liquid hydrocarbons such as octanol [34, 36]. Furthermore, the octanol–water partition coefficient has sometimes been characterized as a poor descriptor of solute partitioning into the stratum corneum [2, 36] as the latter consists of a heterogenous matrix of diverse lipids and proteins that cannot be simply replaced with octanol. Thus, the use of multiple fibers (MCF array) with diverse physicochemical properties can theoretically better characterize solute partitioning into skin than octanol. The jet fuel data generated from our partitioning studies demonstrated that the more hydrophobic the jet fuel component is, the more likely it is able to partition into the MCFs. However, this partitioning ability was reduced in the presence of a solvent that in reality mimics partitioning behavior of these components from a jet fuel mixture into skin. The two MCFs, PDMS and PA, optimally predicted jet fuel component partitioning into skin ($R^2 = 0.86$–0.93) and demonstrated that the presence of a solvent in the mixture decreased component partitioning by a factor greater than 2. This finding is consistent with previous partitioning studies involving the use of stratum corneum and chemical mixtures [8, 58, 59] and can be explained by the fact that the presence of some solvents in some mixtures can increase solute solubility in the chemical mixture thereby reducing the solute thermodynamic activity in the membrane [21, 57].

Our experiments also utilized the 3-MCF array approach to predict skin permeability with three MCFs and was based on a multiple linear regression analysis of the permeability ($\log K_p$) data sets generated in our lab with porcine skin and the three $\log K_{MCF}$ data sets [67]. The mathematical model describing this relationship is

$$\log K_p = -2.34 - 0.124 \log K_{PDMS} + 1.91 \log K_{PA} - 1.17 \log K_{WAX}$$
$$(n = 25, R^2 = 0.93).$$

The 1-MCF or any combination of 2-MCF arrays performed poorly in predicting skin permeability. While the 3-MCF array approach resulted in an excellent correlation, the chemical space was limited to 25 solutes used in the calibration of the three MCFs and these solutes were limited to $K_{o/w}$ values ranging from 2 to 6. Future research efforts should be focused on exploring the utility of these regression models using the extremes of the physicochemical properties of chemicals, that is, chemicals outside of the defined chemical space and of specific occupational and environmental concern.

With this in mind, it was hypothesized that MCF array exposure to chemical mixtures containing either a solvent or a surfactant should result in changes in solute absorption into the MCF that are correlated with changes in skin permeability. When the MCF array was exposed to chemical mixtures of either 50% ethanol or 1% SLS (Figure 14.8), there was a decrease in absorption into all three fibers that is consistent with observations from previous dermal permeability studies [8, 53, 55]. Data from the Riviere *et al.* study [55] were used to develop a predictive model for the skin permeability of chemicals from 50% ethanol (E50) established by using multiple regression analysis of the matrix [log $k_{\mathrm{Skin/E50}}$: log $K_{\mathrm{PDMS/E50}}$, log $K_{\mathrm{PA/E50}}$, log $K_{\mathrm{WAX/E50}}$];

$$\log k_{\mathrm{Skin/E50}} = -1.18 + 0.36 \log K_{\mathrm{PDMS/E50}} + 0.80 \log K_{\mathrm{PA/E50}} - 1.32 \log K_{\mathrm{WAX/E50}}$$

$$n = 25,\ R^2 = 0.91,\ s = 0.133,\ F = 75.$$

This analysis demonstrated the feasibility of the MCF array to predict solute permeability in skin in the presence of 50% ethanol. The next step was to ascertain whether the chemical-induced mixture interactions in the MCF array are predictive of similar solvatochromatic interactions in skin. This first required correlating the log $K_{\mathrm{MCF/mixture}}$ or the log $K_{\mathrm{Skin/mixture}}$ experimental values with the five solvatochromatic

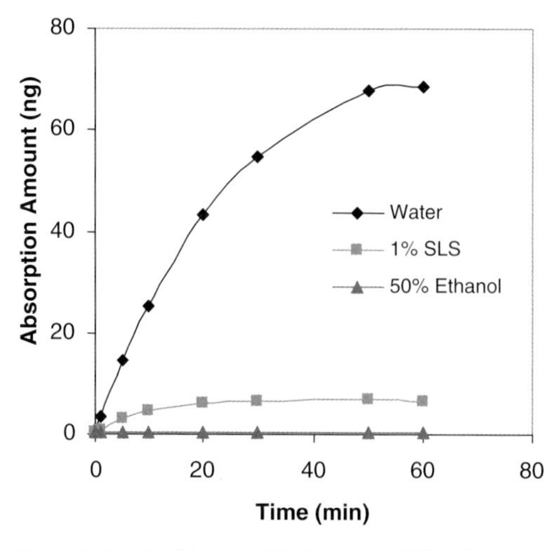

Figure 14.8 Partitioning of biphenyl into WAX fiber in water, 1% SLS and 50% ethanol solutions.

descriptors already described in this chapter (E, S, A, B, and V) to generate corresponding system or strength coefficients for skin and for each MCF exposed to water and either of the solvent or surfactant mixtures. The solvatochromatic relationships for each of the MCF fibers can be described by the following equations:

$$\log K_{\mathrm{MCF/mixture}} = c + a\Sigma\alpha_2^{\mathrm{H}} + b\Sigma\beta_2^{\mathrm{H}} + s\pi_2^{\mathrm{H}} + rR_2 + V_x$$

$$\log K_{\mathrm{MCF/water}} = c + a\Sigma\alpha_2^{\mathrm{H}} + b\Sigma\beta_2^{\mathrm{H}} + s\pi_2^{\mathrm{H}} + rR_2 + vV_x$$

and the solvatochromatic relationship for the skin can be described by the following equations:

$$\log K_{\mathrm{Skin/mixture}} = c + a\Sigma\alpha_2^{\mathrm{H}} + b\Sigma\beta_2^{\mathrm{H}} + s\pi_2^{\mathrm{H}} + rR_2 + vV_x$$

$$\log K_{\mathrm{Skin/water}} = c + a\Sigma\alpha_2^{\mathrm{H}} + b\Sigma\beta_2^{\mathrm{H}} + s\pi_2^{\mathrm{H}} + rR_2 + vV_x$$

The difference in system coefficients ($\Delta = \text{mixture} - \text{water}$) for corresponding molecular descriptors following skin or MCF exposure to either 1% SLS or 50% ethanol can be calculated by utilizing the same data set as described above [55, 67, 68].

$$[\Delta r\ \Delta p\ \Delta a\ \Delta b\ \Delta v] = [rpabv]_x - [rpabv]_o = [r_x - r_o\ p_x - p_o\ a_x - a_o\ b_x - b_o\ v_x - v_o],$$

where $[rpabv]_o$ are the system coefficients of the water mixture in either skin or MCF; $[rpabv]_x$ are the system coefficients after the change of a major component in the chemical mixture; and $[\Delta r\ \Delta p\ \Delta a\ \Delta b\ \Delta v]$ are the changes of the system coefficients.

This generated a matrix of changes in system coefficients ($[\Delta r\ \Delta p\ \Delta a\ \Delta b\ \Delta v]$ for both the 3-MCF array and the skin permeability and these delta changes were correlated within a multilinear regression analysis framework as shown below [14].

$$\begin{pmatrix}\Delta r\\ \Delta p\\ \Delta a\\ \Delta b\\ \Delta v\end{pmatrix}_{\mathrm{Skin}} = a_0 + a_1\begin{pmatrix}\Delta r\\ \Delta p\\ \Delta a\\ \Delta b\\ \Delta v\end{pmatrix}_{\mathrm{WAX}} + a_2\begin{pmatrix}\Delta r\\ \Delta p\\ \Delta a\\ \Delta b\\ \Delta v\end{pmatrix}_{\mathrm{PDMS}} + a_3\begin{pmatrix}\Delta r\\ \Delta p\\ \Delta a\\ \Delta b\\ \Delta v\end{pmatrix}_{\mathrm{PA}}$$

The WAX, PDMS, and PA in the above matrix refer to the carbowax, polydimethylsilaxane, and polyacrylate fibers, respectively, which make up the 3-MCF array system.

Multiple linear regression analysis relating the 3-MCF array to skin permeability for mixtures containing 1% SLS resulted in the following regression equation:

$$y = 0.548 - 0.931x_1 + 3.81x_2 - 2.66x_3$$

with x_1, x_2, and x_3 representing delta (Δ) values for WAX, PDMS, and PA, respectively. An R^2 of 0.9445 demonstrated a strong correlation between changes

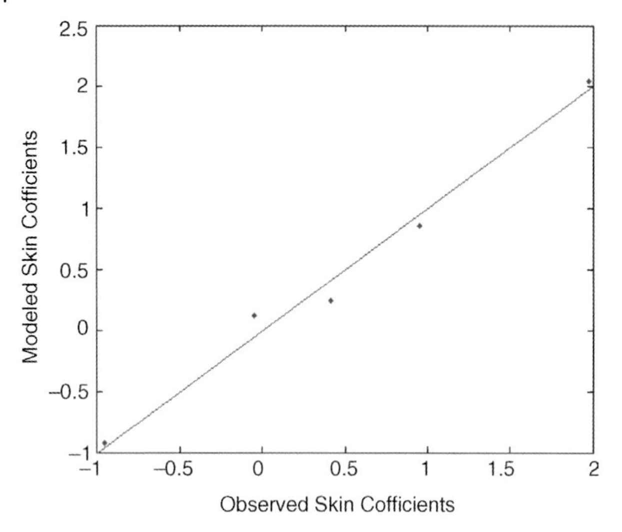

Figure 14.9 The predicted versus observed Δ system coefficients for skin for the full 3-MCF array exposed to 1% SLS.

in system coefficients across skin and the 3-MCF array. Very weak correlations were observed when Δ values from a single or pairs of MCFs were correlated with Δ values from skin permeability. Note that the above model accounts for over 94% of the variability in observations. Figure 14.9 demonstrates the strong relationship (R^2 0.984) between the five observed changes in system coefficients (Δr Δp Δa Δb Δv) and those predicted Δ values from the polynomial model used in the previously described matrix.

The MCF array therefore has the unique potential as a tool to not only simulate dermal permeability of individual chemicals but also assess how chemical mixtures can modulate solute permeability in skin. The above experiments demonstrated that the chemical mixture interactions can be quantified within a solvatochromatic framework. The solvatochromatic approach compares the physicochemical interactions between a solute and a diffusion barrier (e.g., skin or MCF) for one chemical mixture scenario with another mixture scenario. The presence of chemical mixtures can result in significant changes in physicochemical properties as described above for solvents and surfactants and where the MCF array can simulate these physicochemical interactions. For some occupational and environmental exposures, these changes may be significant as described above but there is the very likelihood that some exposure scenarios may result in little or no change in physicochemical properties. The MCF array may not be sensitive to such small or subtle changes and will require further calibration for these exposure scenarios and exposure scenarios involving higher levels of mixture interactions such as more complex solvent + surfactant mixtures. These interactions will also need to be assessed across different classes of solvents and surfactants to be of practical use.

14.4
Potential Impact of Multiple Interactions

The complexity occurs when one considers that the above interactions are all independent making the observed effect *in vivo* a vectorial sum of all interactions. This allows the so-called "emergent properties" of complex systems [5] to become evident when the individual interactions are finally combined in the intact system, in our case *in vivo* skin. For example, assume that mixture component A decreases the absorption of a chemical across skin due to increased binding to skin components. In contrast, mixture component B increases its absorption due to an enhancer effect on stratum corneum lipids. When A and B are administered together, the transdermal flux of the chemical under study may not differ from control. This is illustrated by the effect of two different jet fuel performance additives, *N,N*-disalyclidene-1,2-propanediamine (MDA) and butylated hydroxytoluene (BHT), on dermal absorption of naphthalene administered from the base fuel JP-8 not containing these additives, or in combination as often is the case with JP-8(100) jet fuel (Figure 14.10). BHT functions as an antioxidant additive that prevents formation of soluble gums and insoluble particulate deposits in jet fuel systems produced by oxidation, and MDA functions as a metal deactivator additive that suppresses the catalytic effect that some metals in fuels induce on the surfaces of fuel systems and tanks. In this case, we hypothesize that MDA increases surface retention of naphthalene thereby decreasing

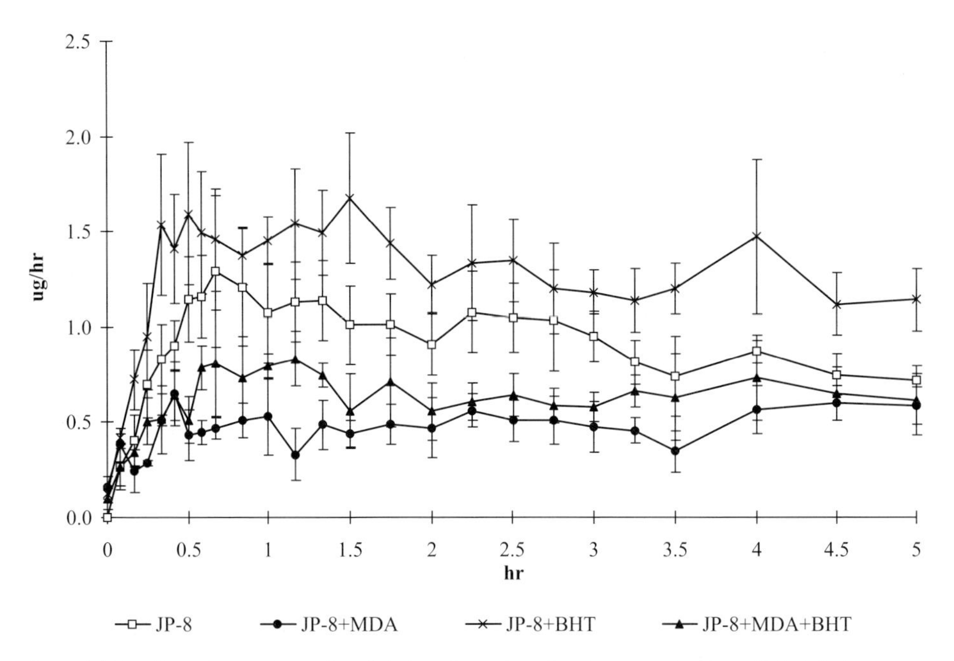

Figure 14.10 Dermal absorption of naphthalene through perfused porcine skin from JP-8 jet fuel administered with additives BHT and MDA alone and in combination.

its absorption, while BHT functions more like a penetration enhancer within skin [41]. When both are present, flux returns to base levels. We have previously seen similar effects with other combinations of additives on absorption of jet fuel hydrocarbons [6, 51].

It may be a mistake to assume that these opposite effects simply cancel one another out and that the flux of chemical is now equivalent to it being applied alone. The mechanisms behind the similarity in fluxes are different. Fick's first law of diffusion can be used to illustrate this. In the base situation (∅), compound flux would

$$\text{Flux}_{\varnothing} = (K_p)(\Delta C),$$

where K_p is the permeability coefficient and ΔC is the concentration gradient driving the absorption process. We will consider ΔC the effective dermal dose since increasing concentration on the surface of skin effectively increases ΔC. In the presence of additives, we had two scenarios where additive A decreased absorption by retaining chemical on the surface, effectively reducing ΔC:

$$\downarrow\text{Flux}_{A} = (K_p)\,(\downarrow\Delta C)$$

and scenario B where flux increased due to an increased K_p:

$$\uparrow\text{Flux}_{B} = (\uparrow K_p)\,(\Delta C).$$

When both A and B are present, the flux is now back to base levels, but is governed by a fundamentally different set of diffusion parameters:

$$\text{Flux}_{A+B} \cong \text{Flux}_{\varnothing} = (\downarrow K_p)(\downarrow\Delta C).$$

One can appreciate how different factors that would interact with these altered parameters could change dermal disposition patterns within skin compared to the baseline scenario.

In occupational exposure to metal working fluids (MWFs), workers are often exposed to complex chemical mixtures of performance additives such as biocides (e.g., triazine), surfactants (e.g., linear alkylbenzene sulfonate, LAS), anticorrosive agents (triethanolamine, TEA), and lubricants (e.g., sulfated ricinoleic acid, SRA) that can result in the previously described interactions. The dermal absorption of commonly used MWF biocide, triazine, is limited to 2.41–3.89% dose in PSFT and 12.61–18.63% dose in SMFT [10]. In a synthetic MWF formulation neither LAS nor SRA had an individual effect on triazine absorption, but TEA alone, SRA + TEA, and the complete surrogate MWF formulation significantly increased triazine permeability (Figure 14.11a). This trend was also demonstrated in a soluble oil-based MWF formulation where the physicochemical interactions are expected to be significantly different from those in a synthetic MWF. It should, however, be noted that triazine deposition in stratum corneum after an 8 h exposure was significantly reduced by the presence of additives in soluble oil-based MWF, but the trend was reversed in synthetic MWF. The presence of MWF additives can also significantly reduce SRA permeability in both types of MWFs (Figure 14.11b) with LAS having one level of the

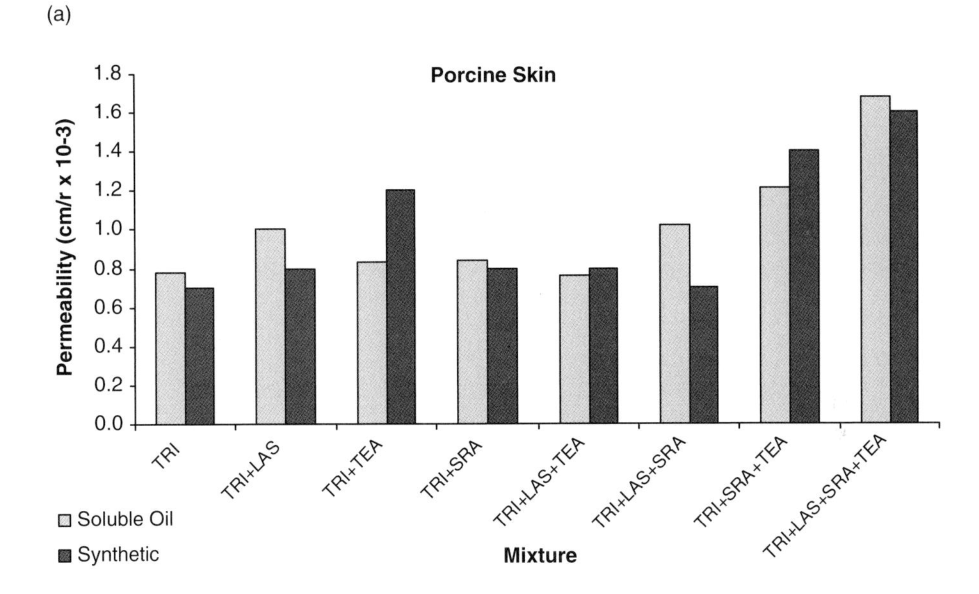

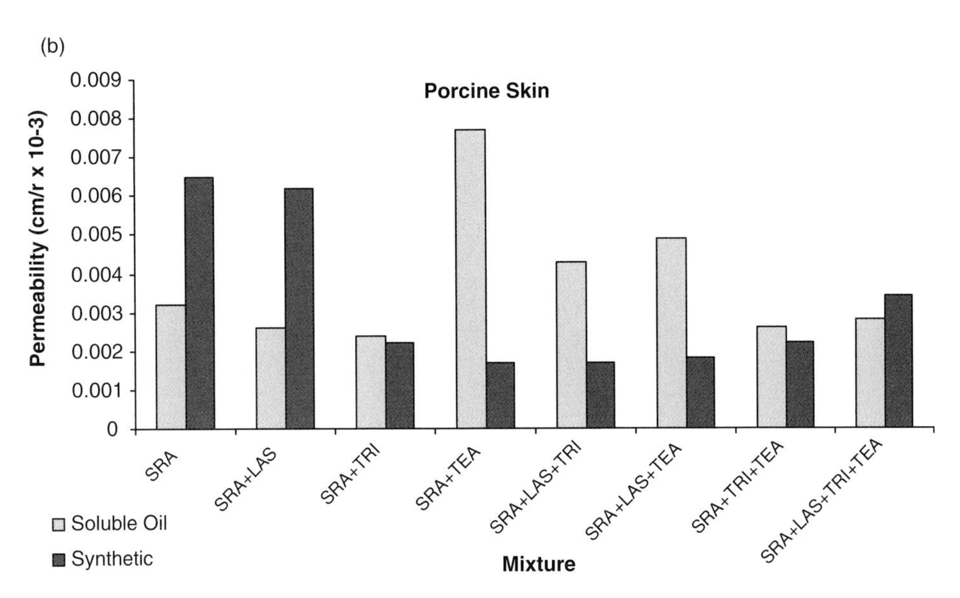

Figure 14.11 Permeability of (a) triazine and (b) sulfated ricinoleic acid in PSFT system in mixtures containing linear alkylbenzene sulfonate, triethanolamine, and/or SRA in either soluble oil mixtures (gray histograms) or synthetic fluid mixtures (black histograms).

inhibitory effect while other additives and combination of additives providing a more significant second-level inhibitory effect [11]. These inhibitory effects were apparently more evident in synthetic MWF (3.8-fold) than in soluble oil-based MWF (1.2-fold), and this was consistent with the less preferential partitioning of SRA into the stratum corneum from the synthetic MWF than from a soluble oil-based MWF. Not surprisingly, TEA by itself significantly increased SRA permeability in PSFT exposed to soluble oil-based formulations.

Our laboratory for the most part was able to reproduce these trends in a synthetic membrane system such as the SMFT where permeability for either triazine or SRA was approximately 6–20-fold greater than in the PSFT. Although these performance additives can potentially modulate the chemical–biological interactions in skin during solute permeability, these PSFT versus SMFT comparisons demonstrated that chemical–chemical interactions may play a more significant role in predicting mixture interactions that modulate dermal absorption. The dermal enhancer effects of alkanolamines such as TEA and fatty acids such as SRA are well documented in the literature [63] as it applies to transdermal delivery of cosmetics and pharmaceutics; however, these series of studies were the first to demonstrate how chemical interactions during an occupational scenario such as an exposure to MWF can influence the eventual delivery of potential toxicants into the systemic circulation. Data from our physicochemical evaluation of these soluble oil-based and synthetic MWF support this hypothesis as we have seen mixture-induced changes in viscosity, stratum corneum/mixture partition coefficients, changes in critical micelle concentration in the presence of SRA, and changes in additive solubility [7, 10, 11].

The above scenario becomes a bit more complicated when the worker's skin is repeatedly exposed to chemical mixtures that actually modify the skin biology. This is especially true for repeated exposures to industrial solvents that are known to either remove intercellular lipids or alter their structure and orientation in the epidermis [38]. *In vivo* pretreatment of porcine skin by the industrial solvent cleanser, TCE, for 4 days almost doubled triazine permeability and was more important in synthetic MWF than soluble oil-based MWF [12]. Simultaneous exposure to TCE did not, however, influence dermal permeability of triazine. Conversely, both classes of MWF caused a 4–10-fold increase in TCE absorption, which is significantly less than the effect the MWF additives had on triazine absorption. The differences here could be attributed to the relative polar nature of the MWF compared to a penetrant that is polar such as triazine and a penetrant that is nonpolar such as TCE. In similar simulations in our laboratory, we demonstrated that repeated occupational exposure to jet fuels can significantly increase dermal absorption of aromatic and aliphatic hydrocarbon components in jet fuel mixtures [42]. A twofold increase in absorption was observed after a 1-day pre-exposure and a fourfold increase was observed after a 4-day pre-exposure. These *in vitro* simulations highlight the fact that the skin biology is altered after repeated occupational exposure and this in addition to the complex chemical–chemical interactions previously described should be considered in any dermal absorption assessment for workers in these scenarios.

14.5 Summary

This chapter has reviewed the potential effects that topical exposure to a chemical mixture might produce compared to exposure to a single neat chemical. In fact, most single-chemical exposures are not to neat compound but rather to a chemical in a defined vehicle such as water, ethanol, or acetone in many toxicological studies. It is clear that there are significant vehicle effects that may overshadow differences between absorption of chemicals with different properties. Such vehicle effects have been previously noted. However, what is just as significant is the phenomenon that chemical interactions may also be vehicle specific depending upon the physiochemical properties of both the penetrating chemical and the vehicle/mixture in which it is dosed. The ability to dramatically improve the prediction of permeability for 12 diverse chemicals across multiple mixtures when the physical–chemical properties of the solvent/mixture are also taken into account presents an approach that may have future utility once fully validated. The MCF array approach provides a novel experimental tool to quantify these mixture interactions using solvatochromatic parameters. Even in situations where chemical permeability is similar across mixtures, it is possible that opposing mixture interactions are at work that would alter the toxicological interpretation of this permeability compared to control values where no mixture additives are present.

Acknowledgments

The authors would like to acknowledge all of the staff and students in the NCSU/CCTRP for their continued efforts, as well as NIOSH grants R01-OH-07555 and RO1-OH-03669 and AFOSR grant F49620-01-1-0080 for supporting this mixture research.

References

1 Abou-Donia, M.B., Wilmarth, K.R., Jensen, K.F., Oehme, K.W., and Kurt, T.L. (1996) Neurotoxicity resulting from coexposure to pyridostigmine bromide, DEET and permethrin. *J. Toxicol. Environ. Health*, **48**, 35–56.

2 Abraham, M.H. and Martins, F. (2004) Human skin permeation and partition: general linear free-energy relationship analyse. *J. Pharm. Sci.*, **93**, 1508–1523.

3 Allen, D.G., Riviere, J.E., and Monteiro-Riviere, N.A. (2000) Induction of early biomarkers of inflammation produced by keratinocytes exposed to jet fuels Jet-A, JP-8, and JP-8(100). *J. Biochem. Mol. Toxicol.*, **14**, 231–237.

4 Barry, B.W. (2001) Novel mechanisms and devices to enable successful transdermal drug delivery. *Eur. J. Pharm. Sci.*, **14**, 101–114.

5 Bar-Yum, Y. (1997) *Dynamic of Complex Systems*, Addison-Wesley, Reading, MA.

6 Baynes, R.E., Brooks, J.D., Budsaba, K., Smith, C.E., and Riviere, J.E. (2001) Mixture Effects of JP-8 additives on the dermal disposition of jet fuel components. *Toxicol. Appl. Pharmacol.*, **175**, 269–281.

7 Baynes, R.E., Brooks, J.D., Barlow, B.M., and Riviere, J.E. (2002) Physiochemical determinants of linear alkylbenzene sulfonate (LAS) disposition in skin exposed to aqueous cutting fluid mixtures. *Toxicol. Ind. Health*, **18**, 237–248.

8 Baynes, R.E., Brooks, J.D., Mumtaz, M., and Riviere, J.E. (2002) Effects of chemical interactions in pentachlorophenol mixtures on skin and membrane transport. *Toxicol. Sci.*, **69**, 295–305.

9 Baynes, R.E., Monteiro-Riviere, N.A., and Riviere, J.E. (2002) Pyridostigmine bromide modulates the dermal disposition of [^{14}C] permethrin. *Toxicol. Appl. Pharmacol.*, **181**, 164–173.

10 Baynes, R.E., Barlow, B., and Riviere, J.E. (2003) Dermal disposition of triazine in cutting fluid mixtures. *J. Toxicol. Cut. Ocular Toxicol.*, **22** (4), 215–229.

11 Baynes, R.E. and Riviere, J.E. (2004) Mixture additives inhibit the dermal permeation of the fatty acid, ricinoleic acid. *Toxicol. Lett.*, **147** (1), 15–26.

12 Baynes, R.E., Yeatts, J.L., Brooks, J.D., and Riviere, J.E. (2005) Pretreatment effects of trichloroethylene on the dermal absorption of the biocide, triazine. *Toxicol. Lett.*, **159** (3), 252–260.

13 Baynes, R.E., Xia, X.R., Barlow, B., and Riviere, J.E. (2007) Partitioning behavior of aromatic components in jet fuel into diverse membrane coated fibers. *J. Toxicol. Environ. Health*, **70** (22), 1879–1887.

14 Baynes, R.E., Xia, X.R., Imran, M., and Riviere, J.E. (2008) Quantification of chemical mixture interactions modulating dermal absorption using a multiple membrane fiber array. *Chem. Res. Toxicol.*, **21**, 591–599.

15 Bliss, C.I. (1939) The toxicity of poisons applied jointly. *Ann. Appl. Biol.*, **26**, 585–615.

16 Bogert, C.J., Price, B., Wells, C.S., and Simon, G.S. (2001) Evaluating chemical interaction studies for mixture risk assessment. *Hum. Ecol. Risk Assess.*, **7**, 259–306.

17 Bronaugh, R.L. and Stewart, R.F. (1985) Methods for in vitro percutaneous absorption studies IV: the flow-though diffusion cell. *J. Pharm. Sci.*, **74**, 64–67.

18 Bronaugh, R.L., Stewart, R.F., and Strom, J.E. (1989) Extent of cutaneous metabolism during percutaneous absorption of xenobiotics. *Toxicol. Appl. Pharmacol.*, **99**, 534–543.

19 Budsaba, K., Smith, C.E., and Riviere, J.E. (2000) Compass plots: a combination of star plot and analysis of means (ANOM) to visualize significant interactions in complex toxicology studies. *Toxicol. Methods*, **10**, 313–332.

20 CRARM (Commission on Risk Assessment and Risk Management) (1997) U.S. Congress, Washington, DC.

21 Cross, S.E., Pugh, W.J., Hadgraft, J., and Roberts, M.S. (2001) Probing the effect of vehicles on topical delivery: understanding the basic relationship between solvent and solute penetration using silicone membranes. *Pharm. Res.*, **18**, 999–1005.

22 Elias, P.M. and Feingold, K.R. (1992) Lipids and the epidermal water barrier: metabolism, regulation, and pathophysiology. *Semin. Dermatol.*, **11**, 176–182.

23 EPA (1986) Guidelines for the health risk assessment of chemical mixtures. *Fed. Reg.*, **51**, 34014–34025.

24 EPA (1988) Technical Support Document on Risk Assessment of Chemical Mixtures, EPA/600/8-90/064, November.

25 EPA (1995) Dermal Exposure Assessment: Principles and Applications, EPA/600/8-91/011B, March.

26 EPA (2000) Options for Revising CEB's Method for Screen-Level Estimates of Dermal Exposure, Chemical Engineering Branch, June.

27 Geinoz, S., Guy, R.H., Testa, B., and Carrupt, P.A. (2004) Quantitative structure-permeation relationships (QSPeRs) to predict skin permeation: a critical evaluation. *Pharm. Res.*, **21**, 83–92.

28 Groton, J.P., Feron, V.J., and Suhnel, J. (2001) Toxicology of simple and complex mixtures. *Trends Pharmacol. Sci.*, **22**, 316–322.

29 Haddad, S., Charest-Tardif, G., Tardif, R., and Krishnan, K. (2000) Validation of a physiological modeling framework for simulating the toxicokinetics of chemicals in mixtures. *Toxicol. Appl. Pharmacol.*, **167**, 199–209.

30 Haddad, S., Beliveau, M., Tardif, R., and Krishnan, K. (2001) A PBPK modeling-based approach to account for interactions in the health risk assessment of chemical mixtures. *Toxicol. Sci.*, **63**, 125–131.

31 Hostynek, J.J. and Magee, P.S. (1997) Modeling *in vivo* human skin absorption. *Quant. Struct. Act. Relat.*, **16**, 473–479.

32 Idson, B. (1983) Vehicle effects in percutaneous absorption. *Drug Metab. Rev.*, **14**, 207–222.

33 Luger, T.A. and Schwarz, T. (1990) Evidence for an epidermal cytokine network. *J. Invest. Dermatol.*, **95**, 104–110.

34 Marqusee, J.A. and Dill, K.A. (1986) Solute partitioning into chain molecule interphases: monolayers, bilayer membranes, and micelles. *J. Chem. Phys.*, **85** (1), 434–444.

35 McDougal, J.N. and Robinson, P.J. (2002) Assessment of dermal absorption and penetration of components of a fuel mixture (JP-8). *Sci. Total Environ.*, **288**, 23–30.

36 Mitragotri, S. (2003) Modeling skin permeability to hydrophilic and hydrophobic solutes based on four permeation pathways. *J. Control. Release*, **86**, 69–92.

37 Monteiro-Riviere, N.A., Baynes, R.E., and Riviere, J.E. (2003) Pyridostigmine bromide modulates topical irritant-induced cytokine release from human epidermal keratinocytes and isolated perfused porcine skin. *Toxicology*, **183**, 15–28.

38 Monteiro-Riviere, N.A., Inman, A.O., Mak, V., Wertz, P., and Riviere, J.E. (2001) Effects of selective lipid extraction from different body regions on epidermal barrier function. *Pharm. Res.*, **18**, 992–998.

39 Morgan, E.T. (2001) Regulation of cytochrome P450 by inflammatory mediators: why and how? *Drug Metab. Dispos.*, **29**, 207–212.

40 Moser, K., Kriwet, K., Kalia, Y.N., and Guy, R.H. (2001) Enhanced skin permeation of a lipophilic drug using supersaturated formulations. *J. Control. Release*, **73**, 245–253.

41 Muhhamad, F., Brooks, J.D., and Riviere, J.E. (2004) Comparative mixture effects of JP-8(100) additives on the dermal absorption and disposition of jet fuel hydrocarbons in different membrane systems. *Toxicol. Lett.*, **150**, 351–365.

42 Muhammad, F., Monteiro-Riviere, N.A., Baynes, R.E., and Riviere, J.E. (2005) Effect of *in vivo* jet fuel exposure on subsequent *in vitro* dermal absorption of individual aromatic and aliphatic hydrocarbon fuel constituents. *J. Toxicol. Environ. Health Part A*, **68**, 719–737.

43 Mukhtar, H. (1992) *Pharmacology of the Skin*, CRC Press, Boca Raton, FL.

44 Poet, T.S. and McDougal, J.N. (2002) Skin absorption and human risk assessment. *Chem. Biol. Interact.*, **140**, 19–34.

45 Pohl, H.R., Hansen, H., and Chou, C.H.S.J. (1997) Public health guidance values for chemical mixtures: Current practice and future directions. *Reg. Toxicol. Pharmacol.*, **26**, 322–329.

46 Potts, R.O. and Guy, R.H. (1992) Predicting skin permeability. *Pharm. Res.*, **9**, 663–669.

47 Qiao, G.L., Brooks, J.D., Baynes, R.E., Monteiro-Riviere, N.A., Williams, P.L., and Riviere, J.E. (1996) The use of mechanistically defined chemical mixtures (MDCM) to assess component effects on the percutaneous absorption and cutaneous disposition of topically exposed chemicals. I. Studies with parathion mixtures in isolated perfused porcine skin. *Toxicol. Appl. Pharmacol.*, **141**, 473–486.

48 Rastogi, S.K. and Singh, J. (2001) Lipid extraction and transport of hydrophilic solutes through porcine epidermis. *Int. J. Pharm.*, **225**, 75–82.

49 Riviere, J.E. and Williams, P.L. (1992) Pharmacokinetic implications of changing blood flow to the skin. *J. Pharm. Sci.*, **81**, 601–602.

50 Riviere, J.E., Monteiro-Riviere, N.A., and Williams, P.L. (1995) Isolated perfused porcine skin flap as an *in vitro* model for predicting transdermal pharmacokinetics. *Eur. J. Pharm. Biopharm.*, **41**, 152–162.

51 Riviere, J.E., Monteiro-Riviere, N.A., Brooks, J.D., Budsaba, K., and Smith, C.E. (1999) Dermal absorption and distribution of topically dosed jet fuels Jet A, JP-8, and JP-8(100). *Toxicol. Appl. Pharmacol.*, **160**, 60–75.

52 Riviere, J.E., Monteiro-Riviere, N.A., and Baynes, R.E. (2002) Gulf War Illness-related exposure factors influencing topical absorption of ^{14}C-permethrin. *Toxicol. Lett.*, **135**, 61–71.

53 Riviere, J.E., Qiao, G., Baynes, R.E., Brooks, J.D., and Mumtaz, M. (2001) Mixture component effects on the in vitro dermal absorption of pentachlorophenol. *Arch. Toxicol.*, **75**, 329–334.

54 Riviere, J.E., Baynes, R.E., Brooks, J.D., Yeatts, J.L., and Monteiro-Riviere, N.A. (2003) Percutaneous absorption of topical diethyl-*m*-toluamide (DEET): effects of exposure variables and coadministered toxicants. *J. Toxicol. Environ. Health Part A*, **66**, 133–151.

55 Riviere, J.E., Baynes, R.E., and Xia, X.R. (2007) Membrane-coated fiber array approach for predicting skin permeability of chemical mixtures from different vehicles. *Toxicol. Sci.*, **99** (1), 153–161.

56 Rosado, C., Cross, S.E., Pugh, W.J., Roberts, M.S., and Hadgraft, J. (2003) Effect of vehicle pretreatment on the flux, retention, and diffusion of topically applied penetrants *in vitro*. *Pharm. Res.*, **20**, 1502–1507.

57 Sloan, K.B., Koch, S.A.M., Siver, K.G., and Flowers, F.P. (1986) Use of solubility parameters of drug and vehicle to predict flux through skin. *J. Invest. Dermatol.*, **87**, 244–252.

58 van der Merwe, D. and Riviere, J.E. (2005) Effect of vehicles and surfactants on xenobiotic permeability and stratum corneum partitioning in porcine skin. *Toxicology*, **206**, 325–335.

59 van der Merwe, D. and Riviere, J.E. (2005) Comparative studies on the effects of ethanol and water/ethanol mixtures on chemical partitioning into porcine stratum corneum and silastic membrane. *Toxicol. In Vitro*, **19** (1), 69–77.

60 Wang, Z., Xiao, C., Wu, C., and Han, H. (2000) High-performance polyethylene glycol-coated solid-phase microextraction fibers using sol-gel technology. *J. Chromatogr. A*, **893**, 157–168.

61 Williams, A.C. and Barry, B.W. (1998) Chemical penetration enhancement, in *Dermal Absorption and Toxicity Assessment* (eds M.S. Roberts and K.A. Walters), Marcel Dekker, New York, pp. 297–312.

62 Williams, P.L. and Riviere, J.E. (1993) Model describing transdermal iontophoretic delivery of lidocaine incorporating consideration of cutaneous microvascular state. *J. Pharm. Sci.*, **82**, 1080–1084.

63 Woodford, R. and Barry, B. (1986) Penetration enhancers and the percutaneous absorption of drugs: an update. *J. Toxicol. Cut. Ocular Toxicol.*, **5**, 167–177.

64 Xia, X.R. and Leidy, R.B. (2001) Preparation and characterization of porous silica-coated multifibers for solid phase microextraction. *Anal. Chem.*, **73**, 2041–2047.

65 Xia, X.R., Baynes, R.E., Monteiro-Riviere, N.A., Leidy, R.B., Shea, D., and Riviere, J.E. (2003) A novel *in vitro* technique for studying percutaneous permeation with a membrane-coated fiber and gas chromatography/mass spectrometry: Part I. Performances of the technique and determination of the permeation rates and partition coefficients of chemical mixtures. *Pharm. Res.*, **20**, 272–279.

66 Xia, X.R., Baynes, R.E., Monteiro-Riviere, N.A., and Riviere, J.E. (2004) A compartment model for the membrane-coated fiber technique used for determining the absorption parameters of chemicals into lipophilic membranes. *Pharm. Res.*, **21** (8), 1345–1352.

67 Xia, X.R., Baynes, R.E., Monteiro-Riviere, N.A., and Riviere, J.E. (2007) An experimentally based approach for predicting skin permeability of chemicals and drugs using membrane-coated fiber array. *Toxicol. Appl. Pharmacol.*, **221** (3), 320–328.

68 Xia, X.R., Baynes, R.E., Monteiro-Riviere, N.A., and Riviere, J.E. (2007) A system coefficient approach for quantitative assessment of the solvent effects on membrane absorption from chemical mixtures. *SAR QSAR Environ. Res.*, **18** (5), 579–593.

15
Synergy: A Risk Management Perspective

Paul S. Price

15.1
Introduction

Any discussion of the risks from exposure to mixtures quickly comes round to the concept of synergy and the role that synergy plays in the toxicity of mixtures. All too often, at this point, the conversation ceases to be hopeful and the tone becomes one of quiet frustration [1]. There is a recitation of examples where synergy has been shown to occur and then a litany of reasons why it is difficult to address the issue. One might hear statements similar to the following: "Well the currently available *in vivo* tools are not amenable to investigating synergy," "There are an infinite number of combinations we can't test them all," "We acknowledge the potential for synergy, but do not address it quantitatively," and "There are no agreed upon methodologies for accounting for synergy." At the end, one is left with the impression that whether or not synergistic effects actually pose a risk to public health, the understanding of the risks posed by synergy is still a work in progress.

This chapter investigates the issue of synergy using an approach that has been touched upon in prior publications [2], but has not been fully explored. Instead of asking how risk management can account for the occurrence of synergy, this chapter asks what characteristics would be required in order for synergy cause risk management processes (that do not account for synergy), to underestimate the risks posed by mixtures of chemicals? This chapter begins with a review of synergy and the current practice of risk management of chemical mixtures. The estimates of the safe doses of mixtures and their components for test animals and humans permitted under current models of mixture toxicity are examined and the implications of the models for exposure to components of the mixture are discussed. A graphic approach is then presented for integrating findings of synergy into the risk management frameworks established by current models of mixture toxicity. An illustration of the approach is provided. Finally, we review the current literature on findings for low-dose interactions and recommendations for extrapolation of synergy observed at high doses to lower doses.

Principles and Practice of Mixtures Toxicology. Edited by Moiz Mumtaz

ISBN: 978-3-527-31992-3

This chapter focuses on risk management of chronic noncancer effects in humans from exposure to discrete mixtures. Elevated toxicity due to synergistic interactions is an issue that also needs to be considered for acute noncancer and for cancer end points and for effects on environmental receptors (algae, fish, birds, etc.); however, these issues are not addressed in this chapter. Exposure to mixtures occurs in a variety of ways, such as exposures to discrete mixtures, cumulative exposures from multiple sources, coexposure to parent compounds and metabolites, and so on. This chapter focuses on exposure to discrete mixtures; however the conclusions in this chapter may be relevant to other types of mixtures.

15.2 Synergy

The term synergy has been applied to a number of different phenomena [2, 3]. In this chapter, synergy is defined as a supra-additive dose response as a result of concurrent exposure to two or more chemicals [4]. Synergy has been observed to occur in a number of instances [5] and there are a number of mechanisms by which synergy occurs [5–7].

Synergy poses a number of challenges to assessing risks from mixtures. It appears as an emergent property of mixtures and is difficult to predict from the results of the limited *in vivo* studies that are typically available for chemicals [1]. The concern is that like a "wild card" synergy could potentially turn any combination of low-toxicity components into a toxic mixture of concern. Because synergy can elevate toxicity and is not readily predictable, it hangs like a cloud over mixture risk assessments.

The occurrence of synergy and other interactions is favored by the laws of mathematics. Exposure to a mixture of two compounds (A and B) produces one opportunity for synergy since there is one pair of chemicals (AB). Mixtures of three compounds (A, B, and C) have three opportunities for synergy since there are three unique combinations (AB, BC, and CA). For a mixture of 30 compounds there are 435 pairs and for a mixture of 100 compounds, 4950 pairs. By this argument, if synergy occurs in only one pair of chemicals in 10 000, then at least one instance of synergy would occur in half the mixtures containing 100 compounds.

Weighing against this prediction of enhanced toxicity in complex mixtures is the observation that most of the objects and media we encounter in life and all foods we consume are complex mixtures containing hundreds or thousands of compounds. In the case of food, many of these compounds are known to have considerable toxicity [8, 9]. A full discussion of risk and benefits of the complex mixtures that form our foods is beyond the scope of this chapter, but it is clear that synergy must be either extremely rare or not result in significant increases in toxicity in foods that humans consume (see chapter 23 for additional discussion of mixtures in our diets).

Strategies to account for the occurrence of synergy in mixtures include

- performing testing of whole mixtures;
- incorporating dose-response data from the joint testing of binary components; or
- modeling the mechanism of dose interaction.

The potential for synergy to affect a mixture's toxicity is most directly addressed by testing the toxicity of the entire mixture. This approach fully captures the interactions between components of the mixture. Testing individual mixtures; however is generally not feasible because of the vast number of mixtures and the expense of performing *in vivo* testing. This may change in the future with the rise of *in vitro* approaches to toxicity testing [7] (also see chapter 7).

The second option is to evaluate the interactions between individual components in mixtures and use these data to predict the behavior of the entire mixtures. While synergy has been reported to occur in combinations of three or more compounds, characterizing the mechanisms of synergy has often been based on understanding the interaction between pairs of compounds [10–13]. The interactions in more complex mixtures are characterized by building on the understanding of binary interactions. Techniques to use data on binary mixtures include both PBPK-based approaches [10] and conceptual approaches (weight of evidence) [14–16] (see chapter 5 for additional discussion on this approach).

Finally, if the mechanism of action of each component in the mixture for the critical effect is known, then mechanistic models of the mixtures can be developed [7, 17]. All of these approaches require compound-specific data that may not be available for every component of mixtures of interest.

15.3 Risk Management and Synergy

Risk management of chemicals has been divided into two basic approaches. One approach is to assume no threshold and use a response addition model. This approach has been applied to cancer and genotoxic effects [15]. For other effects, the focus of risk management has been on estimating doses of chemicals that are unlikely to cause adverse effects in sensitive humans [18]. Under this approach, a battery of *in vivo* toxicity tests are used to identify a chemical's critical effect. The critical effect is the effect that occurs at the lowest dose as a result of either chronic or acute exposure. The dose response for this effect is used to establish a point of departure (POD). The POD could be a lowest observed effect level (LOAEL), a no-observed adverse effect level (NOAEL), no-observed effect level (NOEL), or benchmark dose (EPA noncancer policy [19]). The POD is then used to estimate a dose that is protective of sensitive individuals using a series of adjustment factors. These adjustment factors and POD produce conservative estimates of safe doses that reflect the toxicity of the chemicals and the quality of the toxicity data available for a particular chemical [20]. These estimates of safe doses have been given a variety of different names by different standard setting organizations including reference dose (RfD), tolerable daily intake, allowable daily intake, or the derived no-adverse effect level.

Setting safe levels of a chronic exposure to a mixture is a two-step process. The first step is to develop an estimate of the toxicity of the mixture, the POD, based on *in vivo* study data of the mixture's components, the mixture itself, or the related mixtures [15]. The second step is the extrapolation of mixture's toxicity data in the test

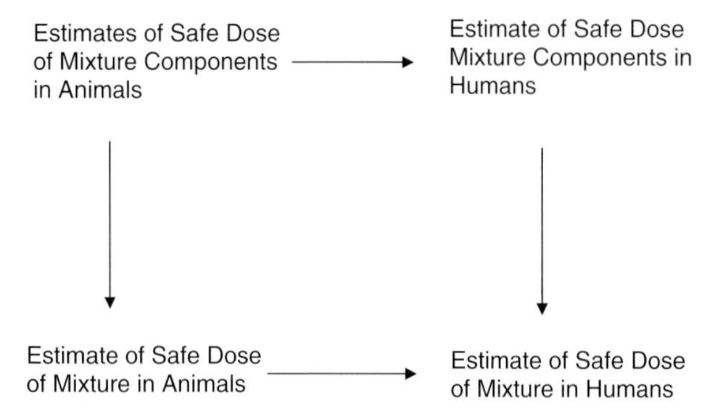

Figure 15.1 Two approaches for setting safe doses for mixtures.

animal to an estimate of the safe dose for sensitive humans. Interestingly, different mixture risk assessments have performed these two steps in different orders, Figure 15.1. For example, the toxicity of exposure to mixtures of organophosphorous pesticides in diet performed by EPA under the FQPA used relative potency factors (RPF) to estimate the animal toxicity of the mixture and then extrapolated the estimates to a safe dose for sensitive humans [21]. In contrast, the hazard index (HI) method for evaluating mixtures first converts the animal data on the components of the mixture to estimates of safe dose in humans (RfDs) and then estimates the safe dose of the mixture for humans [15].

Not all findings of synergy have a direct impact on the risk management of mixtures. Figure 15.2 presents results from two hypothetical examples of experimental data demonstrating synergy in mixtures containing two compounds A and B. In both examples, a dose of B (that does not cause an adverse effect) coadministered with a range of doses of A increases the response to A. In Example 1, B changes the threshold of A. This type of synergy is very relevant since it changes the POD of the mixture. Example 2 shows that B increases the response rate of A above the threshold but does not change A's threshold. This type of synergy does not affect the current system of

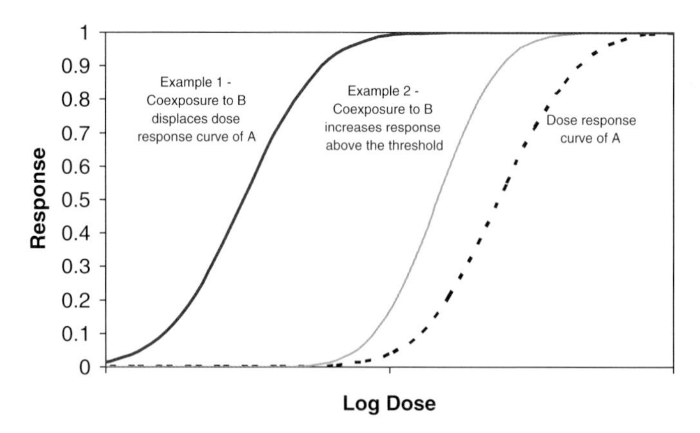

Figure 15.2 Examples of synergy that changes risk management decisions for mixture and synergy that has minimal impact.

noncancer risk assessment since the threshold dose does not change. For this reason, findings about synergy that occur when one or both chemicals are below the NOAELs of the individual chemicals should receive additional attention. Such findings demonstrated that synergy both occurs and changes the mixture's threshold dose.

Mechanisms of chemical interaction can provide plausible justification for both types of synergy. If B facilitates the uptake of A, then it would shift the entire curve by increasing the effective dose at all dose levels of A. If B inhibits the recovery from the effects of A, then it would change the slope of the response but not the threshold dose.

There are also risk based issues why certain findings of synergy may not be relevant to the management of chronic effects. Demonstrating synergy is a resource-intensive process since testing of both the mixture and the mixture components is required. As a result, many studies of mixtures rely on acute and subchronic end points [22–24]. The organ or mechanism for the end points measured in the studies may not be the same as the critical end point used to set the chronic RfDs. If the end point is not related, then a finding of synergy may be of little importance for the management of the risk of chronic effects.

More importantly, acute end points generally require much higher doses than chronic end points (see Section 15.6). At these higher doses, the chemical components of the mixture cause a number of effects that are not adverse over the duration of the acute tests and appear as NOAELs. These effects are known to occur because over longer durations of exposure they lead to chronic toxicity. These effects can form the bases for observations of synergy when one or both chemicals may be at or below the acute NOAEL. This suggests that findings of synergy involving doses below an acute NOAEL would not necessarily imply that synergy would occur below chronic NOAELs, and again may not be relevant to setting chronic standards.

15.4 Models of Mixture Toxicity

A number of recent publications review models that can be used to estimate the safe doses of mixtures based on toxicity data on mixture components [6, 15, 25–27]. In the case of chronic noncancer effects where no data on interactions are available, the toxicity of mixtures is addressed using additive or independence models. The additive model has been described as being neutral with respect to synergy [15]. This model assumes that the components will neither suppress nor enhance each other's responses or that any deviations from additivity will cancel each other out. Neither the additive nor the independence models of mixture toxicity, however, directly account for the possibility that synergy will increase a mixture's toxicity.

15.4.1 Additive Models

Additive models of mixture toxicity assume that each chemical produces the same effects but at different doses. Adjusting for these differences using relative potency factors (RPFs) yields equivalent doses that can be summed to determine the mixture's response [6] (This approach is also discussed in chapters 3, 4, 10 and 18.)

Under additive models, the safe dose of a mixture can be determined from the prediction of the composition of the mixture and information on the toxicity of each component. The safety of mixtures is often expressed in terms of the hazard index (HI). Intakes of mixtures that result in HI values of less than 1 are considered to be safe. (In this approach, the RPF for a chemical is the inverse of each chemical's RfD). The value of the HI from exposure to one or more sources of multiple chemicals is given by Eq. (15.1).

$$\mathrm{HI} = \sum^{i} \frac{D_i}{\mathrm{RfD}_i}, \tag{15.1}$$

where D_i is the dose of the ith component of the mixture (mg/kg/day) and RfD_i is the reference dose of the ith component (mg/kg/day). The ratio of the dose to the RfD has traditionally been called the hazard quotient (HQ) [15].

Additive models of mixture risk can be used to derive an estimate of the safe dose of any discrete mixture where the composition of the mixture is known and RfDs are available for each component of the mixture. This mixture RfD is calculated using Eq. (15.2) [15]

$$\text{Mixture RfD} = \frac{1}{\sum^{i} \frac{F_i}{\mathrm{RfD}_i}}, \tag{15.2}$$

where F_i is the fraction of the mass of the mixture for the ith component of the mixture.

An alternative additive model that has been used for pesticides is based on a benchmark chemical and RPFs. Ratios of consistent measurements of response such as ED_{10} have been used to set the RPF values. Under this approach, the toxicity of a mixture is

$$\text{Mixture RfD} = \frac{\mathrm{RfD_B}}{\sum^{i} \frac{F_i}{\mathrm{RPF}_i}}, \tag{15.3}$$

where $\mathrm{RfD_B}$ is the reference dose for the benchmark chemical, and RPF_i is the RPF for the ith chemical. Other variations on the additive models also exist [15, 26].

15.4.2 Independence Models

The independence models make the explicit assumption that none of the mixture's components interact. While there is no interaction, the responses from each component are still added [6]. This response addition could influence the determination of the POD if the POD is based on a benchmark dose [15]. (chapter 10 provides additional discussion on the independence model.)

Setting a safe level for mixtures using the independence assumption involves setting a separate assessment for each mixture component. In these component-specific estimates, the safe dose of the mixture is determined on the basis of the toxicity of the one component and assuming that the other contaminants merely dilute the component. A dose of a mixture that would allow a safe dose of the ith component of the mixture where the mass fraction of the component in the mixture (F_i) is given by

$$\text{Safe dose from } i\text{th component} = \frac{\text{RfD}_i}{F_i}. \tag{15.4}$$

The mixture reference dose determined by an independence model would be the minimum value of *the safe dose from the ith component* for all of the mixture's components.

$$\text{Mixture RfD} = \min\left(\frac{\text{RfD}_i}{F_i}\right) \tag{15.5}$$

15.4.3
Exposure to the Components Permitted Under Additive and Independence Models

The dose of each chemical component of a mixture that will occur when an individual receives a safe dose of a mixture (e.g., a dose less than the safe dose estimated by an additive or independence model) is a function of the dose of the mixture and the values of F_i for the individual components. The doses of the components permitted under the additive and independence models are

$$\text{Dose}_i = \frac{F_i}{\sum_i \frac{F_i}{\text{RfD}_i}}, \tag{15.6}$$

$$\text{Dose}_i = F_i \times \min\left(\frac{\text{RfD}_i}{F_i}\right). \tag{15.7}$$

Under the additive model, individuals exposed to the maximum dose of the mixture will not receive a dose of any mixture component greater than the safe dose (RfD) of the component. Under the independence model, individuals exposed to the maximum dose of the mixture will receive a dose of one mixture component that is equal to the RfD of that component. Doses of remaining components will be at or below their respective RfDs.

15.5
Placing Doses Used in Studies Demonstrating Synergy into a Risk Management Framework

A graphic approach is used to place the doses of mixtures used in synergy studies and the doses allowed under different risk models into a common framework. This approach builds on the isobologram format used in many mixtures studies. The presentation of this approach takes into account the two-step process for setting safe levels of mixtures for sensitive humans (see Figure 15.1). The first step of the process is to characterize the doses permitted *to test animals* that would be allowed under additive and independent models using the Eqs. (15.7) and (15.8). The second step is to examine how the RfD-setting process affects the doses permitted for humans under the use of RfD-based models of mixture risks. The first step is presented in this section and the second step in Section 15.7.

We begin the presentation of the approach with mixtures containing two compounds then expands the approach to more complex mixtures.

Figure 15.3 presents the range of possible doses of two chemicals A and B (D_A and D_B) normalized to the chemicals' chronic NOAELs. The chronic NOAELs are defined in terms of the absence of all adverse effects of the chemical in a test animal. The normalized doses appear as two axes in the graph. The doses of A and B received as a result of exposure to a specific mixture appear as a point on this graph. The single point in Figure 15.3 indicates a study where an exposure to a mixture resulted in a dose that is three times the chronic NOAELs of chemicals A and B.

The graph can be divided into different regions. Region 1 consists of the combinations of doses where both the doses of both compounds exceed their respective chronic NOAELs. Region 2 consists of the combinations of doses where one of the chemical's doses exceeds the chemical's NOAEL. Under risk management

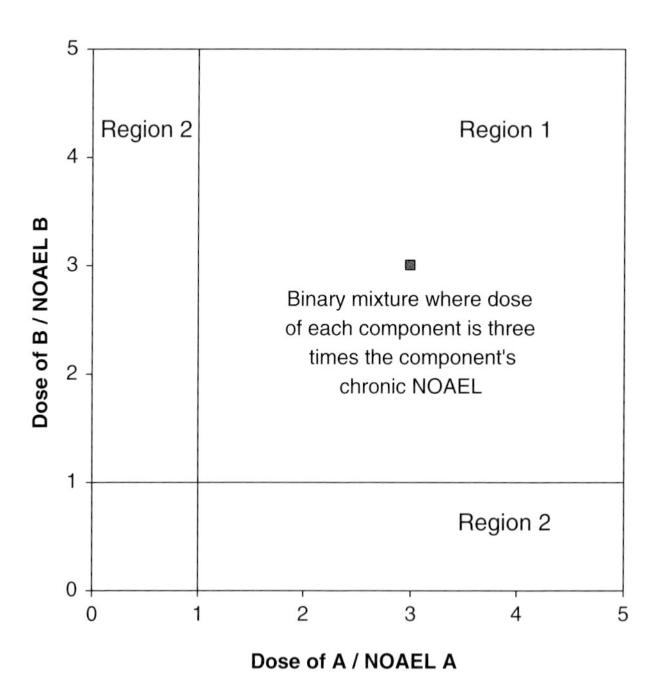

Figure 15.3 Plotting the normalized doses and identifying regions of unacceptable chronic doses of binary mixtures.

decision making, chronic doses of mixtures that fall into regions 1 and 2 are not permitted, not because of concerns for synergetic effects but because of the toxicity of either one or both mixture components.

In the case of a mixture of chemicals A and B, the maximum doses of chemicals A and B that are permitted from safe level of exposure to the mixture for the test animals are limited. For the additive model these Levels are given by Eq. (15.8).

$$1 \leq \frac{D_A}{NOAEL_A} + \frac{D_B}{NOAEL_B} \qquad (15.8)$$

Under independence models of mixture risk, Eq. (15.5), the permitted doses are also restricted to doses below the NOAELs; however, the permitted doses are independent of the concentration of the other component. The doses of the components are given by Eq. (13.9).

$$1 \leq \frac{D_A}{NOAEL} \quad \text{and} \quad 1 \leq \frac{D_B}{NOAEL} \qquad (15.9)$$

The independence model (13.9) allows dose combinations that fill the entire remaining region in Figure 15.4. The additive model (13.8) allows the dose combinations that fill a triangular area close to the origin. Region 3 is defined as dose combinations that are not allowed under an additive model but are allowed under

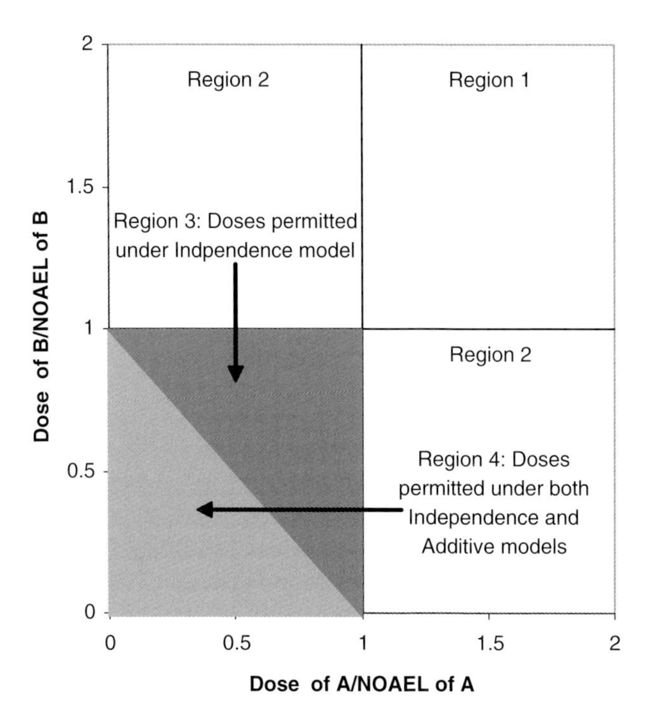

Figure 15.4 Identifying regions of acceptable chronic doses under additive and independence models of mixture toxicity.

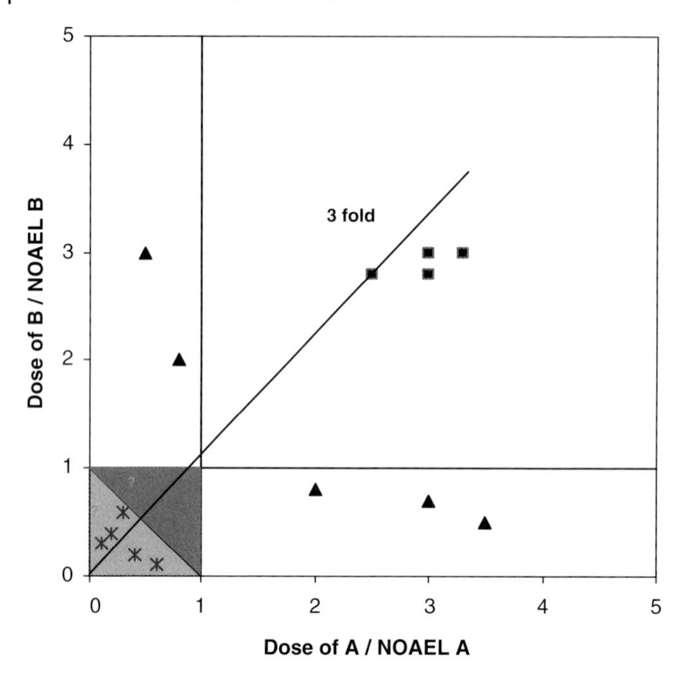

Figure 15.5 Placing doses from synergy studies into the four regions and describing extrapolation of synergy findings from region 1 to regions 3 and 4.

an independence model. Region 4 is defined as doses that are allowed under both models.

With this framework, it is possible to place data on synergy in binary mixtures into a risk management context. The doses of the binary components in studies that demonstrate synergy can be assigned a location in one of these four regions (Figure 15.5). Studies that have demonstrated synergy for dose combinations that fall in regions 1 and 2 are not directly relevant to setting safe doses of mixtures for chronic exposures. The additive and independence mixture models already have determined that such mixtures are unsafe because of the concern for the toxicity of the mixture components. Thus the finding of synergy does not change the risk management decision for the mixture. Observations of synergy in mixtures that fall into region 3 are relevant for setting safe doses of mixtures under an independent model since such doses would be permitted under such a model. The doses that fall into region 4 would be relevant for setting safe doses under either independence or additive models.

Data on synergy from dose combinations that fall into regions 1 and 2 could still suggest an indirect concern if the available data imply that the synergy observed at higher doses persists at lower doses. (These lower doses appear as a ray in Figure 15.5.) Thus, if data suggest that a threefold increase in response above additivity observed at doses of A and B that are three times the test animal's chronic NOAELs, and the data suggest that some portion of the increase also occurs at doses at the animal's NOAELs, then the finding will be relevant for independence models of mixture toxicity. If the data imply that an increase also occurs at doses less than 50% of the animal's chronic NOAELs (region 4), it also would be relevant to doses permitted under additive models.

15.6 Extending the Approach to Mixtures of Three or More Chemicals

The above approach for displaying the doses of mixture components can be extended to more complex mixtures. As discussed above, mixture of three or more compounds will have three or more unique pairs of chemicals. Each pair from a mixture that has been shown to cause synergistic effects can be plotted in one graph. For consistency sake, the larger of the two normalized doses is plotted on the *x*-axis and the smaller of the two normalized doses are plotted on the *y*-axis. This allows the display of how each pair of the chemicals in a mixture compares to the chronic NOAELS of the chemicals.

An illustration of how a more complex mixture can be plotted is provided in Figure 15.6. This figure presents the normalized doses of components in a 10 mg/kg dose of a mixture of five organophosphorous pesticides tested by Moser *et al.* [23].

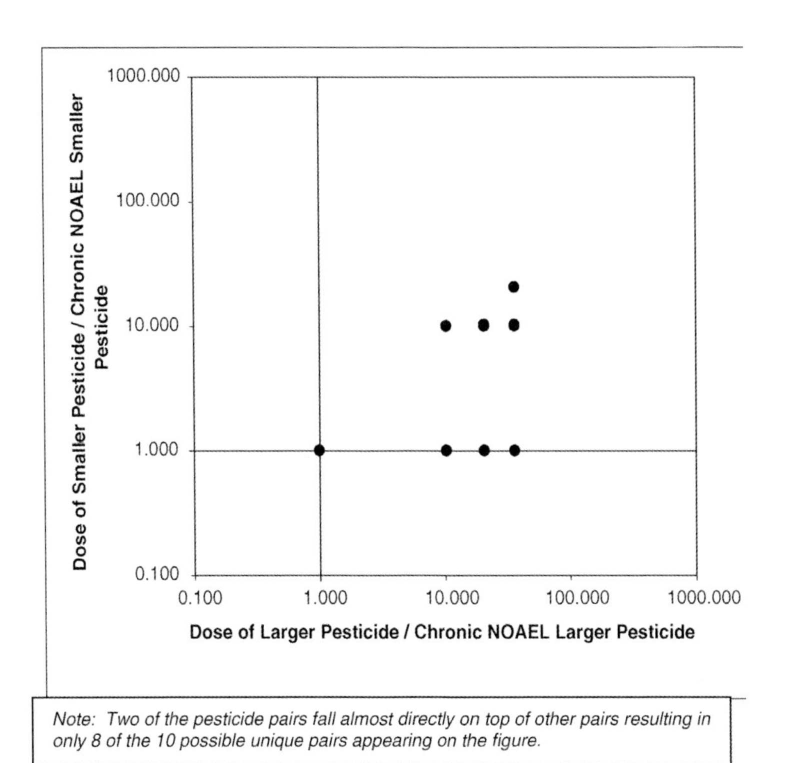

Note: Two of the pesticide pairs fall almost directly on top of other pairs resulting in only 8 of the 10 possible unique pairs appearing on the figure.

Figure 15.6 Plotting doses from synergy studies of more complex mixtures.

Since there are 5 pesticides, there are 10 unique pairs of pesticides that appear as 10 points on the graph. All doses of the mixture components in the 10/mg dose groups are at or above the no-observed effect level used to set the chronic standards of the components. The data point for the largest pair of the pesticides (based on a fraction in the mixture and toxicity) occurs between Malathion (36 times its (No Observed Effect Level (NOEL)) and Dimethoate (20 times its NOEL).

15.7
Using the Graphic Framework to Place Data on Synergy into a Risk Management Context

The Moser *et al.* study [23] investigated a mixture of five pesticides determined to cooccur at specific relative ratios in U.S. diets. The mixture was studied in a ray design where the composition of the mixture is held constant but the doses of the mixture are varied. The individual pesticides were studied separately and an additive model of the mixture's toxicity was developed and used to predict the response of the mixture at the different doses. The actual responses were higher than those predicted by the additive model indicating a synergic response. Table 15.1 presents the fraction of each pesticide in the mixture, the doses administered in the three lowest doses in the study (10, 20, and 40 mg/kg of the mixture), the degree of synergy (ratio of the value from a response curve fitted to the observed data to the value from the response curve predicted by the additive model), and the NOEL used in setting the compounds' chronic standards. The magnitude of synergy at the lowest dose estimated was 2.5 times higher than the additive model predicted and increased to 20 times higher at 40 mg/kg.

Because the composition of the mixture is known and the NOELs for each compound are available, it is possible to estimate the maximum doses of the mixture and its components that would be allowed under additive and independence models (Table 15.2). Figure 15.7 presents the data points for top pair of pesticides in the three test doses and in the doses permitted under additive and independence models. In order to display the results of the study, the dose/NOEL ratios are presented on a log scale. In addition, the size of the synergistic factor associated with each of the doses is given above the data point for each dose group. Note, in order to easily place the data on the figure, the ratios of dose to NOAEL are placed on a log scale. As a result, the boundary between regions 3 and 4 now appears as a curve.

Table 15.1 Data on components of organophosphorous pesticide mixture.

Mixture	Chlorpyrifos	Acephate	Diazinon	Dimethoate	Malathion
NOAEL[a)]	0.03	0.04	0.02	0.05	0.23
Fraction of mixture[b)]	0.031	0.04	0.002	0.102	0.825

a) As reported for the pesticide's RfDs [46] or in the Reregistration Decision [47].
b) Moser *et al.* [23].

Table 15.2 Dose and synergy measurements for mixtures of organophosphorous pesticide mixtures.

	Chlorpyrifos	Acephate	Diazinon	Dimethoate	Malathion	Total mixture	Synergy observed[a)]
Doses used in three lowest dose groups (mg/kg/day)							
Dose 3	1.2	1.6	0.080	4.1	33	40	20
2	0.62	0.80	0.040	2.0	17	20	5.6
1	0.31	0.40	0.020	1.0	8.3	10	2.5
Doses permitted under additive and independence models (mg/kg/day)							
Independence model	0.0086	0.011	0.00 056	0.028	0.23	0.28	—
Additive model	0.0040	0.0052	0.00 026	0.013	0.11	0.13	—

a) Synergy is defined as the rate of response observed divided by the rate predicted by the author's additive model at each of the dose levels as presented in Figure 6 of Ref. [23].

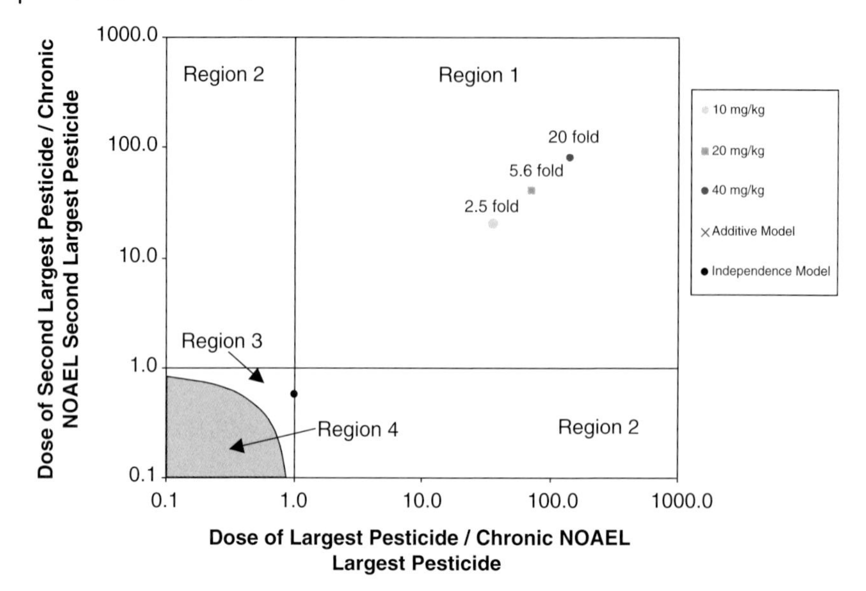

Figure 15.7 Plot of the highest pair of components of the 10, 20, 40 mg/kg/day dose groups from Moser *et al.* [23] and the corresponding maximum doses permitted under independence and additive models.

As Table 15.2 and Figure 15.7 show, the smallest of the administered doses, 10 mg/kg, in the Moser *et al.* study is 78 times larger than the doses that would have been permitted under an additive model and 36 times the doses that would be allowed under an independence model of animal toxicity. This finding is not surprising since a single dose was used in the study; however this indicates that the finding of synergy in this study falls into region 1 and is not directly relevant to the management of mixtures of organophosphorous pesticides.

The data from the study also provide insight into whether the study's findings could provide indirect evidence of synergy at lower doses. The level of the synergy observed was highly dose dependent. The synergy declined at lower doses dropping from 20- to 2.5-fold over a 4-fold change in dose. Given this finding, it is unlikely that synergy will persist over an additional 36–78-fold drop in dose. Thus, while the finding of synergy in this study may be useful in understanding the mechanism by which various organophosphorous pesticides interact at high doses, it does not provide compelling evidence that an additive model will underestimate chronic risks from the organophosphorous pesticides.

This illustration also demonstrates that chemicals that have lower toxicity and occur at lower concentrations will be restricted to smaller doses under both models of mixture toxicity. As shown in Table 15.2, exposure to one of the five pesticides was limited to 1/80 of the chemical's NOEL and two others to 1/8 of their NOELs. The impact of this is a reduction of the number of pairs of chemicals where synergy is likely to occur. While there are ten unique pairs of pesticides in the mixture, there is only one pair of pesticides, Malathion and Dimethoate, where the doses of both chemicals are greater than 25% of the chronic NOAELs.

15.8 Doses of Mixture Components Permitted Under Current Models of Mixture Risks for Humans

Discussions in Sections 15.5 through 15.7 only address the issue of synergy in determining safe doses of mixtures in animals. The ultimate goal of mixture risk assessment is to identify safe doses that protect sensitive humans. As a result, assessments of the risks posed to humans must address from animal data to a dose that is protective of a sensitive individual. This extrapolation is performed using a series of uncertainty factors that convert the POD to RfDs.

15.8.1 Recent Findings on the Relationship Between Chronic Toxicity in Sensitive Humans and Reference Doses

In the case of a two chemical mixture, the doses of chemicals A and B that are permitted for humans under additive models are given by Eq. (15.10) and independence models by Eq. (15.11).

$$1 \leq \frac{D_A}{RfD_A} + \frac{D_B}{RfD_B} \tag{15.10}$$

$$1 \leq \frac{D_A}{RfD_A} \quad \text{and} \quad 1 \leq \frac{D_B}{RfD_B} \tag{15.11}$$

The relationship between the RfD and the chronic sensitive human NOAEL (CSHN) has been the subject of considerable debate. Hertzberg and Teuschler [16] acknowledged the issue but concluded that the RfD could be considered to be "equally uncertain and equally biased" estimates of the sensitive human NOAEL. Support for this position is implicit in the EPA definition of the RfD.

> An estimate (with uncertainty spanning perhaps an order of magnitude) of a daily oral exposure to the human population (including sensitive subgroups) that is likely to be without an appreciable risk of deleterious effects during a lifetime. (EPA [18])

This finding implies that the same regions can be established for sensitive humans using RfDs instead of the chronic animal NOAEL (Figure 15.8).

This position has been challenged by the work of a number of researchers in the United States and European Union who have investigated the uncertainty in the CSHN and its relationship with the RfD [28–34]. These studies consistently find that RfDs are in fact a very biased estimate of the chronic sensitive human NOAEL (CSHN). The majority of the published RfDs (>90%) established using the current system of safety factors have values that are at or below the chemical's actual CSHN. The typical RfD set with two 10-fold safety factors is likely to be an order of magnitude below the actual CSHN for the chemical. When additional safety factors are used, the typical value of the RfD is even lower compared to the chemical's actual CSHN.

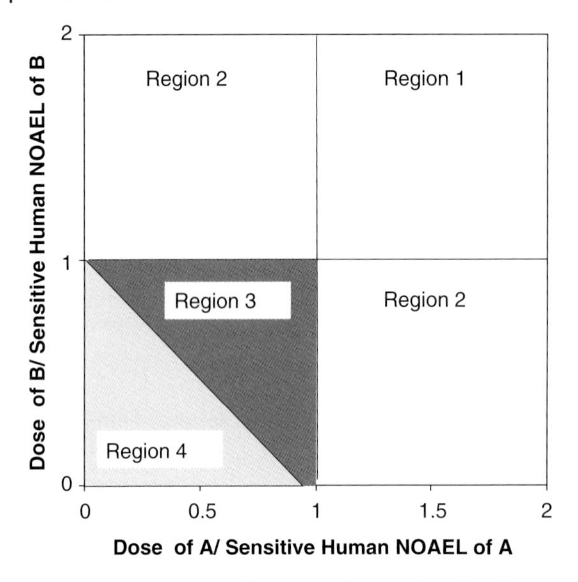

Figure 15.8 Regions 1–4 for sensitive humans based on the assumption that the RfD is an accurate and unbiased estimate of the chronic NOAEL for the sensitive human.

The impact of this systematic underestimation of the CSHN by the RfD results in a reduction in the fraction of the sensitive human NOAEL that is permitted in the doses of the components in the mixture risk models. When additive models or independence models are used with RfDs that there are lower than the CSHN the areas covered by regions 3 and 4 are reduced. Figure 15.9 presents the size of the regions when the RfD of mixtures A and B are both 25%[1)] of the CSHN.

In order for synergy to be relevant under this situation it must occur at doses that are a small fraction of the CSHN.

15.8.2
Impact of the Current System of Safety Factors on the Moser *et al.* [23] Data

The chronic standards for the organophosphorous pesticides are established with various numbers of safety factors and the total safety factors applied to the pesticides' NOAELs range from 10 to 3000. The impact of the use of the safety factors in extrapolating from animal to sensitive humans can be shown in Table 15.3.

The permitted doses of each pesticide for the test animals can be calculated using Eqs. (15.6) and (15.7). Table 15.3 presents these doses normalized to the animal NOAELs. The results show that under the additive model the compound with the highest fraction of its NOAEL is Malathion (0.46). Under the independence model,

1) The value of 25% was selected as a reasonable amount of over estimation (4 fold). The cited references suggest that this degree of over estimation of toxicity would occur for the majority of chemicals established with the current system of uncertainty factors.

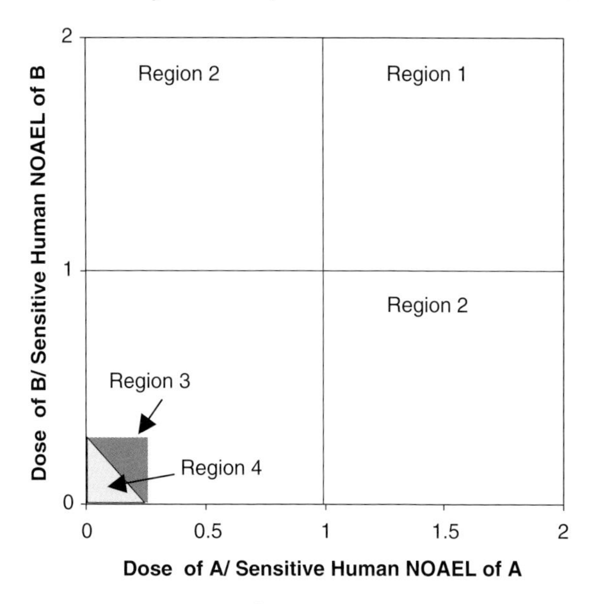

Figure 15.9 Regions 1–4 for sensitive humans based on the assumption that the RfD is an overestimate of the chronic NOAEL for the sensitive human by a factor of 4.

the highest fraction is again Malathion (1.0). If we assume that the RfD is 25% of the CSHN the equivalent values for humans would be 0.12 and 0.25.

These findings indicate that it is likely that the additive and independence model will keep human exposure to small fractions of the CSHN. This finding has profound implications for synergy. It requires that synergy must affect humans at doses that are significantly less than the CSHN. In the case of the additive model, the pair of components Dimethoate and Malathion with the highest ratios (0.088 and 0.15). All other pairs of pesticides involve exposures that occur at even smaller fractions of the CSHN.

Table 15.3 Doses to animals and humans permitted under additive and independence models.

	Maximum dose permitted to animals/animal NOEAL		Maximum dose permitted to humans/actual chronic sensitive human NOAEL	
	Additive model	**Independence model**	**Additive model**	**Independence model**
Chlorpyrifos	0.13	0.29	0.034	0.072
Acephate	0.13	0.28	0.033	0.072
Diazinon	0.013	0.028	0.0032	0.0070
Dimethoate	0.26	0.57	0.066	0.14
Malathion	0.46	1.0	0.12	0.25

15.9
Relationship between Toxicity and Synergistic Potential

The above discussion has demonstrated that in order for synergy to be relevant to the characterization of chronic effects from discrete mixtures, it must occur at doses well below CSHN. The existence of synergy at levels well below NOAELs is not a new question [2]. This section presents a brief summary of the literature on this topic.

15.9.1
What are "Low Doses?"

Publications on chemical interactions have defined "low doses" in a number of ways. These multiple definitions have led to a great confusion in the relevance of findings to the risk assessment process. The term "low dose" has been used in multiple ways.

1) Doses causing frank effects but below acutely toxic doses (LD_{50}) in single dose studies. Synergy has often been reported to occur on the basis of observations in acute toxicity (single doses) of mixtures.
2) Repeated doses but over a small number of days (subacute studies) that cause frank effects but not lethality. In many instances, the end point where synergy is observed may not be a measure of the critical effect for chronic durations.
3) Doses near or below the NOAEL seen in chronic studies [36].
4) Doses below the chronic health standards [2].

As discussed above, from a risk management perspective, the third use is the correct one. As demonstrated in Section 15.6, Eqs. (15.4) and (15.6) provide a direct method to determine the maximum chronic doses of a mixture, and its components, that are permitted under additive and independence models. These estimates can be used to check the dose levels of any mixture where synergy has been identified.

15.9.2
Does Synergy Occur at Low Doses?

In the 1980s, TNO began a program to look at the occurrence of chemical interaction at low doses [37]. The results of this work were published in a series of papers [38–43] that form the largest set of data mixture interaction for subchronic effects. These papers are the largest group of studies that look at doses closest to the chronic NOAELs of chemicals. The general conclusion reached from this body of work was that at these doses additive models appeared to be appropriate for groups of chemicals that have a common end point. Mixtures of chemicals that operated by different mechanisms and affect different organs were best predicted by an independence model [37]. Little or no evidence of synergy was observed. Konemann and Pieters [2] argue similar findings based on data from aquatic toxicology.

These findings are based on a relatively small number of studies (<20). As a result, there is a need for additional research to determine whether these findings will be sustained by additional testing and the authors acknowledged that there might be exceptions to these findings [41].

15.9.3 Can Findings of Synergy at High Doses be Used to Characterize the Potential for Synergy at Low Doses?

The literature presents a surprisingly consistent answer to this question. Findings on interactions at high doses do not predict the existence of interactions at lower doses [24]. Empirical findings from ray studies such as Gennings [22], Moser [23], and Crofton [35] have shown that synergy is a function of dose and that as doses decline the intensity of synergistic effects declines. Other studies have shown even more complex dose responses including changes from antagonism at high doses to synergy and additivity at lower doses [44]. Because of these observations, Groten [23], Konemann and Pieters [2], and Feron and Groten [36] have strongly argued that there is no evidence that synergy observed at high doses can be automatically applied to lower doses. Gennings *et al.* [45] has termed the dose dependent nature of synergy as "interaction threshold." (The reader is referred to the extensive discussion on the measurement of the interaction threshold in chapters 4 and 7.)

Konemann and Pieters further argue that the mechanisms by which synergy occurs imply that synergy is a dose-dependent phenomena. They state that there is "little doubt that combined dose action of compounds is a dose-dependent phenomenon. At low doses, physiochemical interactions are of relatively low importance and toxicokinetic, and toxicodynamic may also be very rare" [2].

Finally, the doses of mixture components that occur as a result of exposure to discrete mixtures also favor dose dependency in synergistic effects. In discrete mixtures, the exposures to any two chemicals are fixed by their relative fractions in a mixture. As the dose of one chemical declines so will the second. Consider Example 1 from Figure 15.2 where a dose of chemical B enhanced the toxicity of chemical A at all doses of A. In the case of a discrete mixture, the dose of B declines with A. If the effect of B on A occurs by any mechanism that is dose dependent, then reducing the exposure to the mixture will result in a reduction in synergy because the dose of B declines, so does the dose of A.

15.10 Discussion

This chapter has outlined a series of arguments on the potential occurrence of synergy in humans exposed to the maximum doses permitted under different models currently used by risk assessors to manage exposure to mixtures. Current risk management practices limit chronic exposures to mixture components to levels that are below chronic NOAELs for the chemicals in test animals. In simple binary mixtures, the permitted exposures could approach the component chemical's NOAELs, especially when independence models are used. However, if additive models are used and if the mixtures include larger numbers of components, the mixtures' components will be restricted to doses much lower than the component's animal NOAELs. Current practices for extrapolating from animal data to sensitive humans have the net effect of further reducing the ratio of permitted doses to the mixture components' actual CSHNs. As seen in the Moser *et al.* example, only one of 10 possible pairs of chemicals had appreciable to both chemicals (Table 13.3). This pattern is expected to occur in all mixtures with 3 or more components.

A review of the literature suggests that synergy is least likely to occur at such low doses. Existing studies that have demonstrated synergy have been performed at higher doses (region 1). Extrapolation of synergistic effects at higher doses to lower doses does not appear to be justified on either an empirical or a theoretical basis. As a result, the existing empirical findings on synergy do not appear to provide evidence that the estimates of safe doses produced by additive and independence models in chronic noncancer risk assessments are not protective. The existing findings of synergy may be more relevant to the establishment of acute toxicity standards, but the assessment of such standards is beyond the scope of this chapter.

These findings present a conundrum for mixture toxicologists who are trying to empirically confirm the safety of the doses of mixtures predicted by the models. The costs of long-term animal studies are so large that it is impossible to investigate the occurrence of chemical interactions at doses below the chronic NOAELs used in assessing mixtures. Short-term studies require higher doses that fall in region 1 and thus are not directly relevant to noncancer assessments. Finally, the existing literature strongly warns against extrapolating findings of synergy from high doses to low doses.

The solution to this problem will not be found by continuing the current practice of testing mixtures at high doses. Rather, new techniques and short-term assays are required to detect and evaluate the low-level molecular and cellular changes that are precursors to chronic toxicity [7, 17]. Techniques, such as toxicogenomics, proteomics, and signaling pathway assessments could be used to investigate chemical interactions.

15.11
Summary and Conclusions

A complete review of the literature on synergy and a determination of the relevance of each published study for the assessment of the toxicity of mixtures are beyond the scope of a single book chapter. The key finding of this effort is that synergy is a phenomenon that occurs only at doses where one or both chemicals pose unacceptable risk to humans and thus has no direct impact on chronic risk management decisions. In this chapter, we have outlined an approach for determining the relevance of existing and future findings of synergy that can be applied to any discrete mixture where the mixture's composition is defined and the chronic toxicity of the components are known.

Acknowledgment

The author would like to thank Dr Ed Carney of The Dow Chemical Company for his critical review of this chapter.

References

1 Borgert, C.J. (2004) Chemical mixtures: an unsolvable riddle? *Hum. Ecol. Risk Assess.*, **10**, 619–629.

2 Konemann, W.H. and Pieters, M.N. (1996) Confusion of concepts in mixture toxicology. *Food Chem. Toxicol.*, **34**, 1025–1031.

3 Groten, J.P. (2000) Mixtures and interactions. *Food Chem. Toxicol.*, **38**, S65–S71.

4 Gennings, C., Carter, W.H., Jr., Carchman, R.A., Teuschler, L.K., Simmons, J.E., and Carney, E.W. (2005) A unifying concept for assessing toxicological interactions: changes in slope. *Toxicol. Sci.*, **88**, 287–297.

5 Greco, W.R., Bravo, G., and Parsons, J.C. (1995) The search for synergy: a critical review from a response surface perspective. *Pharmacol. Rev.*, **47**, 331–385.

6 IGHRC (2009) Chemical Mixtures: A Framework for Assessing Risks.

7 Yang, R. (2004) Chemical mixture toxicity: from descriptive to mechanistic and going on to *in silico* toxicology. *Environ. Tox. Pharmacol.*, **18**, 65–81.

8 Culliney (1992) Pesticides and natural toxicalns in foods. *Agric. Ecosyst. Environ.*, **41**, 297–320.

9 Mattsson, J.L. (2007) Mixtures in the real world: the importance of plant self-defense toxicants, mycotoxins, and the human diet. *Toxicol. Appl. Pharmacol.*, **223**, 125–132.

10 Haddad, S. and Krishnan, K. (1998) Physiological modeling of toxicokinetic interactions: implications for mixture risk assessment. *Environ. Health Perspect.*, **106** (Suppl. 6), 1377–1384.

11 Haddad, S., Tardif, R., Charest-Tardif, G., and Krishnan, K. (1999) Physiological modeling of the toxicokinetic interactions in a quaternary mixture of aromatic hydrocarbons. *Toxicol. Appl. Pharmacol.*, **161**, 249–257.

12 Haddad, S., Charest-Tardif, G., Tardif, R., and Krishnan, K. (2000) Validation of a physiological modeling framework for simulating the toxicokinetics of chemicals in mixtures. *Toxicol. Appl. Pharmacol.*, **167**, 199–209.

13 Haddad, S., Charest-Tardif, G., and Krishnan, K. (2000) Physiologically based modeling of the maximal effect of metabolic interactions on the kinetics of components of complex chemical mixtures. *J. Toxicol. Environ. Health A*, **61**, 209–223.

14 Mumtaz, M.M. and Durkin, P.R. (1992) A weight-of-evidence approach for assessing interactions in chemical mixtures. *Toxicol. Ind. Health*, **8**, 377–406.

15 US EPA (2000) Supplementary Guidance for Conducting Health Risk Assessment of Chemical Mixtures. US Environmental Protection Agency, Office of Research and Development, Washington, DC.

16 Hertzberg, R.C. and Teuschler, L.K. (2002) Evaluating quantitative formulas for dose-response assessment of chemical mixtures. *Environ. Health Perspect.*, **110** (Suppl. 6), 965–970.

17 Teuschler, L., Klaunig, J., Carney, E., Chambers, J., Conolly, R., Gennings, C., Giesy, J., Hertzberg, R., Klaassen, C., Kodell, R., Paustenbach, D., and Yang, R. (2002) Support of science-based decisions concerning the evaluation of the toxicology of mixtures: a new beginning. *Regul. Toxicol. Pharmacol.*, **36**, 34–39.

18 US EPA (2008) Reference Dose (RfD): Description and Use in Health Risk Assessments: Background Document 1A, December 1. http://www.epa.gov/IRIS/rfd.htm.

19 US EPA (2008) EPA Guidance for Noncancer Risk Assessments, March 3. http://www.epa.gov/ttn/atw/toxsource/noncarcinogens.html.

20 Dourson, M.L. (1994) Methodology for establishing oral reference doses (RfDs), in *Risk Assessment of Essential Elements*, ISLI Press.

21 US EPA (2002) Organophosphate Pesticides: Revised OP Cumulative Risk Assessment; Washington, DC.

22 Gennings, C., Carter, J.W.H., Casey, M., Moser, V., Carchman, R., and Simmons, J.E. (2004) Analysis of functional effects of a mixture of five pesticides using a ray design. *Environ. Tox. Pharmacol.*, **18**, 115–125.

23 Moser, V.C., Simmons, J.E., and Gennings, C. (2006) Neurotoxicological interactions of a five-pesticide mixture in preweanling rats. *Toxicol. Sci.*, **92**, 235–245.

24 Groten, J.P. (2000) Mixtures and interactions. *Food Chem. Toxicol.*, **38**, S65–S71.

25 Agency for Toxic Substances and Disease Registry (ATSDR) (2002) Guidance Manual for the Assessment of Joint Toxic Actions of Chemical Mixtures. US Department of Health and Human Services, Public Health Service, Atlanta, GA.

26 Mumtaz, M.M., Poirier, K.A., and Colman, J.T. (1997) Risk assessment for chemical mixtures: fine-tuning the hazard index approach. *J. Clean. Technol., Environ. Toxicol. Occup. Med.*, **6**, 189–204.

27 Teuschler, L.K. (2007) Deciding which chemical mixtures risk assessment methods work best for what mixtures. *Toxicol. Appl. Pharmacol.*, **223**, 139–147.

28 Baird, S.J.S., Cohen, J.T., Graham, J.D., Shlyakter, A.I., and Evans, J.S. (1996) Noncancer risk assessment: a probabilistic alternative to current practice. *Hum. Ecol. Risk Assess.*, **2**, 79–102.

29 Swartout, J.C., Price, P.S., Dourson, M.L., Carlson-Lynch, H.L., and Keenan, R.E. (1998) A probabilistic framework for the reference dose (probabilistic RfD). *Risk Anal.*, **18**, 271–282.

30 Slob, W. and Pieters, M.N. (1998) A probabilistic approach for deriving acceptable human intake limits and human health risks from toxicological studies: general framework. *Risk Anal.*, **18**, 787–798.

31 Bosgra, S., Bos, PM., Vermeire, TG., Luit, RJ., Slob, W. (2005) Probabilistic risk characterization: an example with di(2-ethylhexyl) phthalate. *Regul Toxicol Pharmacol* **43** (1): 104–113.

32 Gaylor, D.W. and Kodell, R.L. (2000) Percentiles of the product of uncertainty factors for establishing probabilistic reference doses. *Risk Anal.*, **20**, 245–250.

33 Kalberlah, F., Schneider, K., and Schuhmacher-Wolz, U. (2003) Uncertainty in toxicological risk assessment for non-carcinogenic health effects. *Regul. Toxicol. Pharmacol.*, **37**, 92–104.

34 Schneider, K., Oltmanns, J., and Hassauer, M. (2004) Allometric principles for interspecies extrapolation in toxicological risk assessment–empirical investigations. *Regul. Toxicol. Pharmacol.*, **39**, 334–347.

35 Crofton, K.M., Craft, E.S., Hedge, J.M., Gennings, C., Simmons, J.E., Carchman, R.A., Carter, W.H., Jr., and DeVito, M.J. (2005) Thyroid-hormone-disrupting chemicals: evidence for dose-dependent additivity or synergism. *Environ. Health Perspect.*, **113**, 1549–1554.

36 Hendriksen, P.J., Freidig, A.P., Jonker, D., Thissen, U., Bogaards, J.J., Mumtaz, M.M., Groten, J.P., and Stierum, R.H. (2007) Transcriptomics analysis of interactive effects of benzene, trichloroethylene and methyl mercury within binary and ternary mixtures on the liver and kidney following subchronic exposure in the rat. *Toxicol. Appl. Pharmacol.*, **225**, 171–188.

37 Feron, V.J. and Groten, J.P. (2002) Toxicological evaluation of chemical mixtures. *Food Chem. Toxicol.*, **40**, 825–839.

38 Jonker, D., Woutersen, R.A., van Bladeren, P.J., Til, H.P., and Feron, V.J. (1990) 4-Week oral toxicity study of a combination of eight chemicals in rats: comparison with the toxicity of the individual compounds. *Food Chem. Toxicol.*, **28**, 623–631.

39 Jonker, D., Woutersen, R.A., van Bladeren, P.J., Til, H.P., and Feron, V.J. (1993) Subacute (4-week) oral toxicity of a combination of four nephrotoxins in rats: comparison with the toxicity of the individual compounds. *Food Chem. Toxicol.*, **31**, 125–136.

40 Jonker, D., Woutersen, R.A., and Feron, V.J. (1996) Toxicity of mixtures of nephrotoxicants with similar or dissimilar mode of action. *Food Chem. Toxicol.*, **34**, 1075–1082.

41 Cassee, F.R., Groten, J.P., van Bladeren, P.J., and Feron, V.J. (1998) Toxicological evaluation and risk assessment of chemical mixtures. *Crit. Rev. Toxicol.*, **28**, 73–101.

42 Groten, J.P., Schoen, E.D., van Bladeren, P.J., Kuper, C.F., van Zorge, J.A., and Feron, V.J. (1997) Subacute toxicity of a mixture of nine chemicals in rats: detecting interactive effects with a fractionated two-level factorial design. *Fundam. Appl. Toxicol.*, **36**, 15–29.

43 Feron, V.J., Cassee, F.R., Groten, J.P., van Vliet, P.W., and van Zorge, J.A. (2001) International issues on human health effects of exposure to chemical mixtures. Conference on Application of Technology to Chemical Mixtures Research, Fort Collins, CO, January 9–11, 2001, Fort Collins, CO, Abstract S-5.

44 Gennings, C., Carter Jr., W.H., Campain, J.A., Bae, D., Yang, R.S.H. (2002). Statistical analysis of interactive cytotoxicity in human epidermal heratinocytes following exposure to a mixture of four metals. *J. Agric. Biol. Environ. Stat.* **7**, 58–73.

45 Gennings, C., Carter, W.H., Carchman, R.A., DeVito, M.J., Simmons, J.E., and Crofton, K.M. (2007) The impact of exposure to a mixture of eighteen polyhalogenated aromatic hydrocarbons on thyroid function: estimation of an interaction threshold. *JABES*, **12** (1), 96.

46 Integrated Risk Information System (IRIS). Accessed March 1, 2008.

47 US EPA (2006) Finalization of interim reregistration eligibility decisions (IREDs) and interim tolerance reassessment and risk management decisions (TREDs) for the organophosphate pesticides, and completion of the tolerance reassessment and reregistration eligibility process for the organophosphate pesticides.

16
Chemistry, Toxicity, and Health Risk Assessment of Drinking Water Disinfection By-Products[1)]

Jane Ellen Simmons and Linda K. Teuschler

16.1
Introduction

Chemical disinfection of drinking water has resulted in significant gains in public health protection in developed countries, effectively decreasing morbidity and mortality caused by microbial exposures responsible for waterborne diseases such as gastrointestinal illness, cholera, and typhoid [1]. During water treatment, however, a wide variety and number of chemical disinfection by-products (DBPs) are formed. Oxidants such as chlorine, ozone, chlorine dioxide, and chloramines, used as disinfectants, react with naturally occurring organic and inorganic matter in the source water, resulting in chronic, low-level exposure to DBP mixtures that is ubiquitous across all segments of the population. Source-water characteristics, such as concentration and type of organic matter, bromide concentration, temperature, and pH, and water treatment scenarios, such as the disinfectant employed, its concentration and where in the process it is used, are factors that influence the number, chemical types, and concentrations of DBPs formed [2–5].

Although more than 600 DBPs have been identified [6], the known and quantified DBPs account for approximately less than 50% of the mass of total organic halide (TOX) formed as a result of chlorination [7, 8]. Quantitative concentration data are readily available for relatively few compounds, including the regulated trihalomethanes (THMs), chlorinated and brominated haloacetic acids (HAAs), haloacetonitriles, haloketones, aldehydes, bromate, chloral hydrate, and chloropicrin, among others [8–10]. Historically, most other DBPs have only been identified by qualitative analyses; thus, the size of the unknown fraction is large. The Nationwide DBP Occurrence Study [11–14] has advanced the state of knowledge in this area by focusing on quantitative analysis of approximately 50 "high-priority" DBPs whose concentrations are not routinely measured in drinking

1) This article has been reviewed by the National Health and Environmental Effects Research Laboratory of the United States Protection Agency and approved for publication. Approval does not signify that the contents necessarily reflect the views and policies of the Agency, nor does the mention of trade names or commercial products constitute endorsement or recommendation for use.

Principles and Practice of Mixtures Toxicology. Edited by Moiz Mumtaz

ISBN: 978-3-527-31992-3

water (e.g., iodinated THMs, nitrosamines), as well as collecting measurement data on the routinely monitored DBPs. Selection of the high-priority DBPs was made from among those identified as being present in water, but not routinely quantified, by using predictions of the likelihood of being capable of causing adverse health effects [15].

Data collected on DBP health effects can be classified in three ways: studies of individual DBPs; defined mixtures of a small number of DBPs; or, complex mixtures, including epidemiological studies and toxicological testing of either concentrated water (extracts) or finished drinking water. Toxicological research has been focused primarily on individual DBPs, with fewer studies conducted on DBP mixtures and little effort expended on understanding the interactions among simple, defined mixtures of DBPs. Within class interactions among the THMs and the HAAs have been the general focus of research with defined mixtures of DBPs [16]. Early research with complex mixtures of DBPs was conducted primarily with XAD-resin extracts prepared from drinking water samples, with an emphasis on the detection of mutagenicity [16]. More recent research has utilized reverse osmosis technology to prepare water concentrates [17–19] for coordinated chemical and toxicological evaluation with an emphasis on reproductive and developmental effects [17, 18, 20]. Finally, epidemiological studies inherently investigate the whole DBP mixture and real human exposure scenarios. An important goal of such whole-mixture studies is to provide data to elucidate the relative contribution of the known DBPs and the unidentified DBP fraction to the toxicity of the whole mixture.

16.2
Regulation of DBPs in the United States

The US EPA's Office of Water (OW) is responsible for the regulation of DBPs. Drinking water regulations have been promulgated (based on a negotiated rule-making process) and others have been proposed with the goal of controlling levels of DBPs in the drinking water [21–24]. As these rules have gone into effect, alternative drinking water treatment technologies have been developed and employed to help utilities meet these new standards. OW conducts its regulatory assessments by setting maximum contaminant levels, defined by the US EPA [25] as

Maximum Contaminant Level Goal (MCLG): The level of a contaminant in drinking water below that there is no known or expected risk to health. MCLGs allow a margin of safety and are nonenforceable public health goals.
Maximum Contaminant Level (MCL): The highest level of a contaminant that is allowed in drinking water. MCLs are set as close to MCLGs as feasible using the best available treatment technology and taking cost into consideration. MCLs are enforceable standards.

MCLs and MCLGs are derived for single chemicals, with a few notable exceptions [24]. The four regulated trihalomethane and five haloacetic acid are drinking water DBPs that are each regulated as a group, with MCLs of 80 and 60 μg/l for each chemical mixture, respectively. The Safe Drinking Water Act Amendments of 1996 [26] charge the US EPA to "develop new approaches to the study of complex mixtures, such as mixtures found in drinking water . . .."

16.3 DBP Mixture Health Effects Data Collection and Related Risk Assessment Approaches

Guidelines for and guidance on conducting health risk assessment of chemical mixtures have been articulated in publications by the US EPA [27, 28] and the Agency for Toxic Substances and Disease Registry [29]. In these documents, three approaches to quantify health risk are recommended depending on the type of data available to the risk assessor: data on the complex mixture of concern; data on a sufficiently similar mixture; or data on the individual components of the mixture or on their interactions.

Figure 16.1 (adapted from Ref. [30]) illustrates that health effects studies provide input data to these three risk assessment approaches whose results can be used to yield information on DBP toxicity and health risk assessment. In the first approach (top 2 rows of Figure 16.1), epidemiological data or toxicological data are available on the complex mixture of concern, in this case, real-world drinking water samples or extracts, and the quantitative risk assessment is done directly from these data. Epidemiological analyses include the development of odds ratios and relative risk

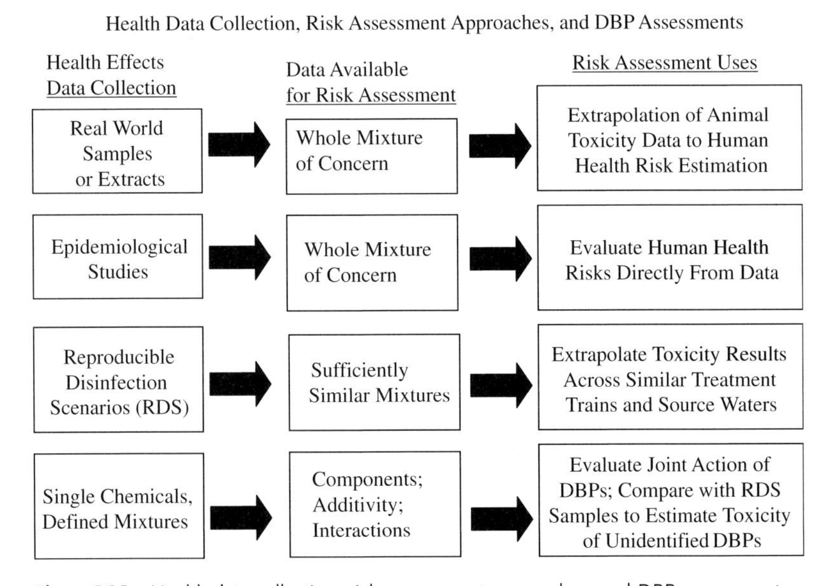

Figure 16.1 Health data collection, risk assessment approaches, and DBP assessments.

estimates of health effects, such as bladder cancer; toxicological analyses may include mixture dose-response modeling to estimate potential human health risks or to derive a reference value (RfV)[2] for the mixture.

In the second approach (third row of Figure 16.1), health effects data are produced on a "sufficiently similar" mixture. When two or more mixtures are thought to be sufficiently similar for use in risk assessment, the differences in their major chemical components and their relative proportions are small [28, 32]; thus, when similarity criteria are met, the toxicity data from a tested DBP mixture may be used to estimate the expected toxicity of another DBP mixture or group of DBP mixtures. This concept is particularly important in risk assessment both when considering the generalization of epidemiologic associations involving exposures to environmental mixtures and when evaluating the relevance of dose-response data for a mixture that has been tested toxicologically to those of environmental mixtures.

The concept of *sufficient similarity* for mixtures can be evaluated from at least two perspectives: toxicological and compositional. Sufficient similarity entails the assumption that the toxicological consequences of exposure to the two mixtures (i.e., the environmental mixture and the mixture on which data are available) will be identical or at least indistinguishable from one another at the level of organization evaluated (i.e., whole organism, target organ, target tissue, cellular, and subcellular). Evaluations of sufficient similarity from the perspective of mixtures composition entail assumptions or measures regarding the identity and quantity of mixture component(s) as well as the relative proportions of the concentrations of various components. Analyses of compositional similarity include evaluations of the consistency of mixtures generated through various processes. For DBPs, this may include consideration of factors such as the natural organic matter in source water, type of treatment and treatment conditions, and pH.

In the third mixture risk assessment approach (bottom row of Figure 16.1), toxicological data are available on individual DBPs or simple DBP mixtures, including data that support either additivity assumptions or the presence of toxicological interactions. Additive risk assessment methods to be applied may include dose addition, in the form of the hazard index or relative potency factors, and response addition [28]. Dose addition is a chemical mixture risk assessment method in which doses are summed (after scaling for relative potency) across chemicals that have a similar toxic mode of action; risk is then estimated using the combined total dose. Response addition is a chemical mixture risk assessment method applied to chemicals whose modes of action are independent of each other (i.e., the presence of one chemical in the body does not influence the effects caused by another chemical); the risk of a whole-body effect (e.g., nonspecific cancer) is then estimated by summing the risks (e.g., skin cancer, liver cancer) of the individual chemicals [28]. Interactions are defined generically as observing toxicity greater than or less than

2) Reference Value (RfV) is an estimate of an exposure for a given duration to the human population (including susceptible subgroups) that is likely to be without an appreciable risk of adverse health effects over a lifetime [31].

expected under a specified definition of additivity, such as dose addition or response addition [28].

When a DBP mixture contains component chemicals that have differing toxic modes of action, dose addition and response addition may be integrated into one method. The US EPA illustrated the use of this method to address the assessment of multiple route exposures to defined DBP mixtures [33–35]. These particular methods assume that at low concentration levels interaction effects either do not occur at all or are small enough to be insignificant to the risk characterization. If interactions are expected to occur, then the quantitative interaction-based hazard index [28, 36] or the qualitative interactions weight-of-evidence approaches [29] may be used to articulate potential health concerns.

As is typical of many chemical classes, the vast majority of laboratory-based toxicological studies on DBPs have been conducted with individual chemicals. Outstanding risk assessment issues of concern for DBPs can be articulated and used to inform new data collection efforts. These include (1) improved understanding of the relationship(s) between human exposure to DBPs and the health risk(s) associated with these exposures; (2) evaluations of DBP exposures and potential human health risks for various source waters and alternative water disinfection treatments; and (3) reducing any real or potential health risks from DBP exposures without increasing the risks of waterborne microbial diseases. Two important aspects of DBP mixture exposures must be considered. One aspect is that disinfected drinking water contains highly complex DBP mixtures whose composition depends on both source-water and treatment scenario characteristics. Another aspect is that humans are exposed to very low levels of DBPs, particularly when compared with levels that produce adverse health effects in experimental animals. Given the complexity of the problem, toxicological experiments must be carefully designed to answer specific questions regarding DBP toxicity and health risk, using cross-disciplinary expertise in areas such as toxicology, statistics, risk assessment, water treatment engineering, DBP formation/occurrence/exposure, analytical chemistry, and epidemiology.

16.4 Health Effects Data on DBP Mixtures

16.4.1 Epidemiological Data

The human health effects of complex DBP mixtures are inherently evaluated by epidemiological studies. Although epidemiological results have been inconsistent across different studies, positive associations, albeit weak, have been observed between human consumption of chlorinated drinking water and increased spontaneous abortions, increased duration of menstrual cycles; stillbirth, low birth weight, small-for-gestational-age infancy, birth defects, bladder cancer, and rectal cancer [37–50]. Several of these studies [39, 44, 47] examined the relationship between adverse reproductive outcome and individual THMs and noted correlation of effects with BDCM.

16.4.2 Toxicological Data on Individual DBPs and Simple Defined DBP Mixtures

The most frequently available data are on individual DBPs, with some associated toxicological data on simple, defined mixtures of 2 to approximately 12 component DBPs of interest. At high dose levels of individual DBPs, numerous studies have been performed on experimental animals that provide toxicological evidence of carcinogenicity, reproductive effects, developmental effects, and other systemic effects [51–60]. A variety of reproductive and developmental effects have been observed, including decreased fertility, adverse sperm effects, full-litter resorption, decreased prenatal or neonatal survival, reduced fetal or pup weight, and cardiac malformations [52, 60–69]. A relatively small number of toxicological studies have been conducted with defined mixtures of DBPs (for a review, see Ref. [16]). On the basis of the available literature to date, the nature of the interactive effects among DBPs appears to be dose-dependent. While both greater than additive and less than additive toxicities have been reported at high dose levels, dose additive toxicity appears to be more commonly observed in lower portions of the mixture dose-response curve than nonadditivity. The nature of the interactions among DBPs, as for other chemicals, may depend on whether the multiple chemical exposure is concurrent or temporally separated, and for temporally separated exposures, the order in which the chemical exposure occurs and the length of time separating the exposures. In addition, as illustrated below for the haloacetic acids, the mixing ratio of the component chemicals contained in the mixture may influence toxic outcome.

A limited number of experiments have been conducted that examined the joint toxic action of mixtures of the regulated DBPs. Sistrunk and DeAngleo [70] evaluated the formation of aberrant crypt foci (ACF), putative preneoplastic precursors of colon neoplasia. Other work from this lab has demonstrated that exposure to individual THMs resulted in ACF formation, with the number of ACF formed related to THM bromination [71]. Male F344 rats were exposed chronically, by drinking water, to a mixture of the four regulated THMs reflective of a source water extremely high in bromine (bromoform was the predominant THM) at 10×, 100×, 1000×, and 10 000× ambient levels. After 26 weeks of exposure, significantly increased numbers of ACF were observed in rats exposed to 100× and higher concentrations. While toxicity at these low dose levels is striking, it is important to keep in mind that such highly brominated drinking water is not typical of tap waters in the United States.

More recently, Narotsky *et al.* [72] assessed, in a screening-level reproductive/developmental bioassay, the combined toxicity of mixtures of the four regulated THMs (THM4), the five regulated HAAs (HAA5), and mixtures of nine DBPs (DBP9) comprised of THM4 and HAA5. Chemical proportions reflected those in chlorinated tap water, with chloroform accounting for approximately 50% of THM4. Mixtures, prepared in 10% Alkamuls® EL-620, were administered daily to F344 rats by gavage on gestation days 6–20. All three mixtures, THM4, HAA5, and DBP9, caused dose-dependent pregnancy loss. For the HAA5 mixture, dose-dependent increases in eye malformations (anophthalmia or microphthalmia) were observed following exposure to HAA5, but not THM4 or DBP9. Thus, the presence of THMs in the full mixture reduced the incidence of HAA-induced eye defects.

As outlined in the next section, experiments have been conducted to evaluate the hepatotoxic interactions of the six binary combinations of the four THMs regulated together under the Stage1 D/DBP (disinfectant/disinfectant by-product) rule at a maximum contaminant level of 80 μg/l. These data are being used to examine the reasonableness of assuming dose addition for the hepatotoxic effects of THMs and to estimate hepatotoxic risk by the interaction-based hazard index approach.

16.4.3 Multiple-Purpose Design Approach

Toxicologists, risk assessors, and statisticians have developed a multimethod research project, the Multiple-Purpose Design Approach Project, to investigate the hepatotoxicity of simple, defined mixtures of the four regulated THMs [73, 74]. These THMs, namely, chloroform ($CHCl_3$), bromodichloromethane (BDCM), chlorodibromomethane (CDBM), and bromoform ($CHBr_3$), are among the most abundant DBPs formed as a result of chemical disinfection of water. Exposure of experimental animals to individual THMs at levels markedly higher than those found in disinfected water has been demonstrated to result in adverse health effects [75], including cancer [55–58], aberrant colonic crypts [71], hepatotoxicity [76, 77], nephrotoxicity [77], and pregnancy loss [60]. Hepatotoxicity was selected as the health end point for this project because liver effects were determined to be the most sensitive end point for three ($CHCl_3$, CDBM, and $CHBr_3$) of the four THMs during the development of oral RfD values [78–81].

The primary objectives of the multiple-purpose design approach [73, 82, 83] are to provide improved methods for risk assessment of mixtures and data relevant to the assessment of the health risk of exposure to DBP mixtures. While the data being used to develop and evaluate the risk assessment methods are specific to DBP mixtures, the methods themselves should be broadly applicable to defined mixtures where component data are available. To this end, both project-level and data-level goals have been established and addressed. One project-level goal is development of efficient experimental designs with associated analysis methodology. Another project-level goal is development, refinement, and evaluation of three quantitative statistical and risk assessment methods: (1) detection of departure from additivity [84], (2) the interaction-based hazard index [85], and (3) proportional response addition, a methodology in development [86]. Because the DBPs formed during disinfection by alternative disinfection scenarios differ both in actual DBP concentration (e.g., μg/l) and in relative DBP proportions (e.g., their mixing ratios), a data-level goal is comparison, based on disinfection of the same source water, of the toxicity of DBP mixing ratios resulting from chlorination alone and ozonation followed by chlorination. Another data-level goal is development of an understanding of the hepatotoxicity, including both potency and the nature of the interactions, of the four

THMs. The experimental regimen calls for female CD-1 mice to be exposed by gavage in an aqueous vehicle daily for 14 days, with hepatotoxicity assessed on the 15th day. Aqueous dosing was used to avoid the confounding influence of corn oil vehicles [77].

Given the many goals that must be achieved with the same data sets to meet the objectives of the project, careful attention had to be paid to the number of dose groups, the mixing ratios, and the placement of dose groups along the dose-response curve. Experiments were designed to study the individual THMs, several four-THM mixtures, and the six possible binary combinations of the four THMs. Dose levels for the individual THMs ranged from 0.1 to 3.0 mmol/kg/day. The binary combinations of the THMs were studied in separate experiments, each containing 12 dose groups: a vehicle control group; 3 dose levels of DBP "A" alone; 3 dose levels of DBP "B" alone; 3 mixture groups at a 1 : 1 mixing ratio of A:B and dose levels of 0.1, 1.0, and 3.0 mmol/kg/day; and 2 mixture groups with the mixing ratio based on the relative proportions of the two DBPs reported by Krasner *et al.* [87] at dose levels of 1.0 and 3.0 mmol/kg/day. Dose levels were selected on the basis of examination of the dose-response curves of the individual THMs and the data needs of the three quantitative statistical and risk assessment methods considered in this project. Placement of the doses below, near, and well above the potential threshold for hepatotoxic response allows to (a) estimate slope and threshold parameters to test for departure from additivity; (b) assess interactive effects higher on the dose-response curve to increase the likelihood that nonadditive interactions would be observed that could then be used to develop parameters for the interaction-based hazard index method; and (c) assess interactive effects lower on the dose-response curve to evaluate the likelihood of nonadditive interactions at environmentally more relevant dose levels. As described by Teuschler *et al.* [82], placement of the highest mixture dose level required consideration that greater than additive effects (synergy) are difficult to detect when the additive response of the chemicals is near or at the maximal response level of the system and that, conversely, less than additive effects (antagonism) are difficult to detect when the additive response of the chemicals is near or at the background response level of the system. Among the overall conclusions from evaluation of the data for departure from dose additivity are that in the low dose region, additivity was the predominant response and that at environmentally relevant mixing ratios, additivity appeared to prevail. Nonadditivity, when detected was not frequent. Where detected, nonadditivity tended to be a function of higher doses and the 1 : 1 mixing ratio. The direction of nonadditivity tended to be less than additive.

16.4.4
In Vitro Toxicological Data on Defined and Complex DBP Mixtures

Other than the evaluation of the mutagenicity of XAD resin extracts of DBPs from finished drinking water, there is a paucity of experimental effort to examine the target organ toxicity of either extracts of DBPs or drinking water concentrates. This is probably due to the limited sample volumes that can be obtained by XAD resin extraction techniques compared to those needed for *in vivo* animal experimentation.

The extraction procedure causes loss of the volatiles, so that the concentrates tested consist of the semivolatiles and nonvolatiles [16]. In addition, XAD resin extraction produces concentrates in an organic matrix that is not suitable for oral exposure. Matrix exchange is typically into dimethylsulfoxide, also an unsuitable solvent for oral exposure. Furthermore, removal of the organic solvent and resuspension into an aqueous medium is considered difficult as the solubility of the extracted material becomes an issue.

In one of the larger *in vitro* efforts to date, an *in vitro* Chinese hamster ovary cytotoxicity assay [87] was used to examine the influence of mixing ratio on the toxicity of mixtures of the five regulated HAAs, the nine common HAAs (the five regulated HAAs plus tribromoacetic acid, bromodichloroacetic acid, dichlorobromoacetic acid, and bromochloracetic acid), and the nine common HAAs plus iodoacetic acid. Seven different mixtures were evaluated, including three mixtures comprised of equimolar mixing ratios of the five, nine, and ten HAAs, two mixtures (a five HAA, a nine HAA) based on the relative proportions of HAAs in water disinfected by postchlorination, and two mixtures (a five HAA, a nine HAA) based on the relative proportions of the HAAs in water disinfected by ozonation with postchlorination (ozone/chlorine). Looking across the mixtures, mixing ratio had a clear influence on the dose-dependent relationship between total mixture dose and decreased cell density. Based on the concentration that resulted in a 50% decrease in cell density (%C1/2 value), the 10 HAA equimolar mixture was the most toxic and the 9 HAA chlorine mixture was the least toxic; there was an approximately 7.5-fold difference in toxicity between these two mixtures. Considering the environmentally relevant mixing ratios, the five HAA ozone/chlorine mixture was more toxic than the five HAA chlorine, the nine HAA ozone/chlorine, or the nine HAA chlorine mixtures [88]. These data, with the cautions that must be employed when extrapolating from *in vitro* to *in vivo*, indicate that the relationship between dose and response will depend on the relative proportions (the mixing ratio) of the DBPs present in disinfected water. Understanding the influence of mixing ratio on dose-response relationships for key health end points is needed to improve assessments of the change in risk associated with a change in disinfection scenario because different disinfection scenarios result in both different absolute concentrations and different relative proportions of DBPs. In addition, understanding the influence of mixing ratio on toxic outcome will aid in understanding when two DBP mixtures may be judged sufficiently similar.

16.4.5
In Vivo Toxicological Data on Complex DBP Mixtures

A few *in vivo* studies have been conducted in experimental animals to assess the effects of complex mixtures of DBPs on reproduction and development, as shown in Table 16.1 [20, 89–92]. These known studies are screening level only. The sole study that incorporated power calculations found a significant increase in the incidence of dams with one or more resorptions when the group receiving tap water was compared with the group of dams receiving deionized water but not when the tap water group was compared with the bottled water group [92].

Table 16.1 Developmental toxicity studies on complex mixtures of tap water.

Citation	Study Details	Species & Strain	Sample Sizes & Study Power	Tap Water Concentration	Results
Narotsky et al. [20]	Developmental toxicity screen; Dosing Gestation Days 1–14	Sprague-Dawley Rats	Tap water: 19–20 Control: 19–20 No power calculations noted	~130×	Marginal or subtle effects of chlorination and ozonation/postchlorination tap water on rodent development.
Uriu-Hare et al. [92]	Developmental toxicity screen; Dosing $\geq$ 2 weeks prior to Gestation and for Gestation Days 0–20	Sprague-Dawley Rats	Tap water: 173 Bottled water: 77 Deionized water: 79 Power calculations were conducted.	1×	Significant increase in the incidence of dams with one or more resorptions for dams drinking tap water compared to dams drinking deionized water. Note: tap water supply disruption due to earthquake decreased sample sizes.
Keen et al., [91]	Developmental toxicity screen; Dosing Gestation Days 0–20	Fischer 344 Rats	Tap water: 60 Bottled water: 60 Liquid chromatography grade water: 60 No power calculations noted	1×	Mostly negative results. Rats drinking unchlorinated tap water from California households receiving untreated ground water had a somewhat higher resorption frequency (resorptions per litter) compared to rats drinking bottled water.
Chernoff et al. [89]	Developmental toxicity screen; Dosing Gestation Days 1–14	CD-1 Mice	Control: 304 Tap water: 284 No power calculations noted	1×	Mostly negative results. Fetuses of mice drinking Durham, NC, tap water had a significantly increased incidence of supernumerary ribs compared to mice drinking purified water.
Kavlock et al. [90]	Developmental toxicity screens over 8 months; Dosing Gestation Days 1–14/18	CD-1 Mice	Tap water: 11–18 Control: 72 Vehicle control: 69 No power calculations noted	300× 1000 3000×	No adverse developmental effects in offspring of pregnant mice treated by gavage to concentrated tap waters (in dimethyl sulfoxide) from Miami, Seattle, New Orleans, Philadelphia, and Ottumwa, IO.

Scientists from the US EPA's Office of Research and Development (ORD) have joined hands to address concerns related to the potential adverse health effects of DBP exposure that cannot be addressed by investigation of either individual DBPs or simple, defined DBP mixtures. The resulting research effort, "Integrated Disinfection ByProducts Mixtures Research: Toxicological and Chemical Evaluation of Alternative Disinfection Treatment Scenarios" [17, 75], was motivated by several key facts: (a) significant amounts of the material that make up the total organic halide formed during water disinfection have not been identified; (b) epidemiologic data, while not conclusive, are suggestive of possible reproductive/developmental or carcinogenic effects in humans exposed to DBPs; and (c) because of concerns over the potential health effects of THMs and other DBPs formed when water is disinfected with chlorination for both pre- and posttreatment, water treatment plants in the United States are switching to alternative disinfection scenarios. It was also noted that although several hundred DBPs have been identified, fewer than 20 have been subjected to toxicity assessments suitable for risk assessment [32]. It would be extremely difficult to screen this many individual chemicals for adverse health effects in experimental animals. Given the low probability that such a massive effort could be undertaken and completed and that both the possible interactions among the DBPs and the toxicity of the unknown fraction would remain as outstanding questions, a mixture-as-a-whole approach was taken. Although the traditional whole-mixture approach does not separately identify the toxicity of the unknown DBP fraction, its influence is included in the estimate of the toxicity of the whole mixture. In addition, all interactions (additive, greater than additive and less than additive) between and among the known and unknown chemicals are captured when the toxicity of the whole mixture is assessed.

The multidisciplinary ORD research team identified a critical data gap in DBP mixtures research: toxicological evaluation in experimental animals of those end points identified as of concern from epidemiological studies. These end points are reproductive, developmental, and cancer. A peer-reviewed research plan, described in detail by Simmons *et al.* [17], was formulated to provide data to improve the estimation of the potential human health risks associated with exposure to the mixtures of DBPs formed during disinfection of drinking water. Because the studies proposed by the research plan involved steps that were logical and necessary, but difficult to implement due to a number of challenging technical issues, a phased series of experiments was planned and implemented. The ability to implement the planned multigenerational reproductive developmental bioassay depended on, among other things, the successful outcome of experiments to determine that representative brominated and iodinated species were formed during water disinfection [10, 93]; development of methods for concentration of water using reverse osmosis membrane technology [19]; and assessment of the feasibility of concentration of large volumes of water [10]. Because the route of exposure is oral, with the water concentrates provided to the rats as their source of drinking water, it was necessary to develop water concentrates palatable to experimental animals and to assess the chemical stability of the water concentrates during both storage and placement in the animal cages.

The third stage of experimentation proposed by Simmons *et al.* [17] has recently been completed and the data are being analyzed. The results of the first and second phases were used to inform and revise the proposed plan for the third phase. The first and second phase experiments involved a number of scientists within and outside EPA in five physical locations around the United States, as well as an experiment designed to assess the potential of artifact formation. Disinfection both by chlorination and by ozonation/chlorination was examined. The results of these experiments are described in detail elsewhere [10, 19, 20, 93–96]. In brief, a method was developed for the preparation of water concentrates by concentration through reverse osmosis membranes. The experiments also demonstrated that representative brominated and iodinated species were formed during water disinfection; water concentrates were acceptable to experimental animals, so that confounding of results by voluntary water restriction will not be a problem; larger volumes of concentrate can be prepared by reverse osmosis methods than by XAD resin extraction, although the available amounts of concentrate will remain rate-limiting until further technological advances are achieved; and despite some technical difficulties with analytical chemistry, the concentrates were sufficiently stable for further experimentation to proceed.

16.4.6
Reproducible Disinfection Scenarios

Reproducible disinfection scenario (RDS) mixtures (Figure 16.1) have been proposed as an appropriate approach for developing related sets of mixtures that can be subsequently evaluated toxicologically and used in risk assessments under the sufficiently similar mixture approach [30, 97]. The strategy is to produce a matrix of RDS samples that represent specific source-water characteristics and water treatment/disinfection scenarios. Important source-water characteristics such as pH and the concentrations of natural organic matter, bromide, and iodide would be carefully controlled. Intake waters would then be disinfected by alternative disinfection scenarios and quantitative and qualitative analytical chemistry would be used to document the DBP mixture composition. For use in sufficient similarity assessments, it is important to not only document the chemistry of the individual DBPs but also to provide "summary measures" that reflect the composition of the unidentified material in the mixture. Example summary measures would include total organic carbon (TOC) in the source water and total organic halide in the finished water. In addition, measures that further distinguish the chemical nature of the complex mixture are also useful, such as measuring the percentage of the TOX that is brominated versus chlorinated, as these types of chemicals may produce different toxicological effects.

Part of the RDS strategy might be to evaluate source waters that differ in factors thought to be relevant to adverse human health effects and to compare the resulting DBP mixtures across treatment scenarios. For example, the levels of bromide and iodide can be varied in a source-water sample, while keeping the other characteristics of the water the same to evaluate the importance of brominated and iodinated DBPs. Finally, to aid in selection of disinfection schemes that lower adverse health risks, RDS samples could be evaluated following different disinfection processes.

16.5 Summary and Conclusions

Component-based and whole-mixture techniques are currently available to evaluate the toxicity of defined mixtures of DBPs and highly complex, environmentally realistic mixtures of DBPs. Evaluation of defined mixtures is useful for risk assessment, but fails to take into account the toxicity of the unknown DBP fraction. Mixture-as-a-whole *in vivo* experiments with environmentally realistic mixtures of DBPs are valuable but technically difficult and relatively few such mixtures can be tested. The advantage of mixture-as-a-whole experiments is that they account for the toxicity of the unknown DBP fraction and any interaction that might occur between the known and the unknown DBPs. Research is in progress on methods that combine component-based and whole-mixture approaches that may let us estimate that portion of observed toxicity that is caused by the unknown fraction. Such experiments require assessment of the complex mixture in concert with assessment of a defined mixture where the known chemicals have the same relative proportions in both the whole complex mixture and the defined chemical subcomponent mixture. An improved understanding of the potential human health risks from exposure to DBPs will be gained through integration of knowledge on the toxicity of individual DBPs; simple, defined DBP mixtures; partially chemically characterized complex, environmentally realistic mixtures of DBPs; and the unidentified fraction.

References

1 Centers for Disease Control and Prevention (CDC) (1999) *Morb. Mortal. Wkly. Rep.*, **48** (29), 621.

2 Fair, P.S. (1995) Influence of water quality on formation of chlorination by-products, in *Drinking Water: Critical Issues in Health Effects Research: Workshop Report*, ILSI Press, Washington, DC, pp. 14–17.

3 Singer, P.C. (1995) Disinfection by-products: from source to tap, in *Drinking Water: Critical Issues in Health Effects Research: Workshop Report*, ILSI Press, Washington, DC, pp. 7–8.

4 Krasner, S.W. (2001) Chemistry and occurrence of disinfection by-products, in *Microbial Pathogens and Disinfection By-Products in Drinking Water* (eds G.F. Craun, F.S. Hauchman, and D.E. Robinson), ILSI Press, Washington, DC, pp. 197–209.

5 International Program on Chemical Safety (IPCS) (2000) Chemistry of disinfectants and disinfectant byproducts, in *Disinfectants and Disinfectant By-Products. Environmental Health Criteria 216*, World Health Organization, Geneva, pp. 27–80.

6 Richardson, S.D., Plewa, M.J., Wagner, E.D., Schoeny, R., DeMarini, D.M. (2007) Occurrence, genoloxicity, and carcinogenicity of regulated and

emerging disinfection by-products in drinking water: a review and roadmap for research. *Mutation Research*, **636**, pp. 178–242.

7 Stevens, A.A., Moore, L.A., Slocum, C.J., Smith, B.L., Seeger, D.R., and Ireland, J.C. (1989) *Disinfection By-Products: Current Perspectives*, American Water Works Association, Denver, CO.

8 Richardson, S.D., Simmons, J.E., and Rice, G. (2002) DBPs: the next generation. *Environ. Sci. Technol.*, **36**, 197A–205A.

9 Krasner, S.W., McGuire, M.J., Jacangelo, J.G., Patania, N.L., Reagan, K.M., and Aieta, E.M. (1989) The occurrence of disinfection byproducts in U.S. drinking water. *J. Am. Water Works Assoc.*, **81**, 41–53.

10 Miltner, R.J., Speth, T.F., Richardson, S.D., Krasner, S.W., Weinberg, H.S., and Simmons, J.E. (2008) Integrated disinfection byproducts mixtures research: disinfection of drinking waters by chlorination and ozonation/postchlorination treatment scenarios. *J. Toxicol. Environ. Health Part A*, **71** (17), 1133–1148.

11 Krasner, S.W., Weinberg, H.S., Richardson, S.D., Pastor, S., Chinn, R., Sclimenti, M.J., Onstad, G.D., and Thruston, A.D., Jr. (2006) Occurrence of a new generation of DBPs. *Environ. Sci. Technol.*, **40**, 7175–7185.

12 Onstad, G.D., Weinberg, H.S., Krasner, S.W., and Richardson, S.D. (2001) Evolution of analytical methods for halogenated furanones in drinking water. Proceedings of the American Water Works Association Water Quality Technology Conference, 6–10 November, Nashville, TN, American Water Works Association. Denver, CO.

13 Weinberg, H.S., Krasner, S.W., Richardson, S.D., and Thurston, S.D., Jr (2001) The occurrence of disinfection by-products of health concern in drinking water. Proceedings of the American Water Works Association Water Quality Technology Conference, American Water Works Association, Denver, CO, 2001.

14 Weinberg, H.S., Krasner, S.W., Richardson, S.D., and Thurston, S.D., Jr (2002) The Occurrence of Disinfection By-products of Health Concern in Drinking Water: Results of a Nationwide DBP Occurrence Study. National Exposure Research Laboratory, US Environmental Protection Agency, Athens, GA. EPA/600/R-02/068. www.epa.gov/athens/publications/EPA_600-R02_068.pdf.

15 Woo, Y.-T., Lai, D., McLain, J.L., Manibusan, M.K., and Dellarco, V. (2002) Use of mechanism-based structure–activity relationships analysis in carcinogenic potential ranking for drinking water disinfection by-products. *Environ. Health Perspect.*, **110** (Suppl. 1), 75–87.

16 Simmons, J.E., Teuschler, L.K., and Gennings, C. (2001) The toxicology of disinfection by-products: methods for multi-chemical assessment, present research efforts and future research needs, in *Microbial Pathogens and Disinfection By-Products in Drinking Water: Heath Effects and Management of Risks* (eds G.F. Craun, F.S. Hauchman, and D.E. Robinson), ILSI Press, Washington, pp. 325–340.

17 Simmons, J.E., Richardson, S.D., Speth, T.F., Miltner, R.J., Rice, G., Schenck, K.M., Hunter, S.E., III, and Teuschler, L.K. (2002) Development of a research strategy for integrated technology-based toxicological and chemical evaluation of complex mixtures of drinking water disinfection byproducts. *Environ. Health Perspect.*, **110** (Suppl. 6), 1013–1024.

18 Simmons, J.E., Richardson, S.D., Speth, T.F., Miltner, R.J., Rice, G., Schenck, K.M., Hunter, S.E., III, and Teuschler, L.K. (2008) Integrated disinfection byproducts mixtures research: issues underlying the integrated chemical and toxicological assessment of complex mixtures. *J. Toxicol. Environ. Health Part A*, **71** (17), 1125–1132.

19 Speth, T.F., Miltner, R.J., Richardson, S.D., and Simmons, J.E. (2008) Integrated disinfection byproducts (DBPs) mixtures research:

concentration by reverse osmosis membrane techniques of DBPs from water disinfected by chlorination and ozonation/postchlorination. *J. Toxicol. Environ. Health Part A*, **71** (17), 1149–1164.

20 Narotsky, M., Best, D., Rogers, E., McDonald, A., Sey, Y., and Simmons, J.E. (2008) Integrated disinfection byproducts mixtures research: assessment of developmental toxicity in Sprague-Dawley rats exposed to concentrates of water disinfected by postchlorination and preozonation/ postchlorination. *J. Toxicol. Environ. Health Part A*, **71** (17), 1216–1221.

21 U.S. Environmental Protection Agency (1979) National interim primary drinking water regulations: trihalomethanes. *Fed. Reg.*, **44** (231), 68624.

22 U.S. Environmental Protection Agency (1996) National primary drinking water regulations: monitoring requirements for public drinking water supplies; final rule. *Fed. Reg.*, **61** (94), 24354–24388.

23 U.S. Environmental Protection Agency (1998) National primary drinking water regulations: disinfectants and disinfection byproducts; final rule. *Fed. Reg.*, **63** (241), 69389–69476.

24 U.S. Environmental Protection Agency (2006) National primary drinking water regulations: stage 2 disinfectants and disinfection byproducts rule. *Fed. Reg.*, **71** (2), 387–493.

25 U.S. Environmental Protection Agency (2008) Setting Standards for Safe Drinking Water. Available at http://www.epa.gov/safewater/standard/setting/html (accessed March 5, 2008).

26 U.S. Congress (1996) Amendments to the Safe Drinking Water Act: Public Law 104–182 104th Congress, Washington, DC. Available at http://www.epa.gov/safewater/sdwa/text.html.

27 U.S. EPA (1986) Guidelines for the health risk assessment of chemical mixtures. *Fed. Reg.*, **51** (185), 34014–34025.

28 U.S. EPA (2000a) Supplementary Guidance for Conducting Health Risk Assessment of Chemical Mixtures. EPA/630/R-00/002. U.S. Environmental Protection Agency, Washington DC.

29 ATSDR (Agency for Toxic Substances and Disease Registry) (2004) Guidance Manual for the Assessment of Joint Toxic Action of Chemicals. U.S. Department of Health and Human Services, Atlanta, GA.

30 Teuschler, L.K. and Simmons, J.E. (2003) Approaching disinfection byproduct (DBP) toxicity as a mixtures problem. *Am. Water Works Assoc. J.*, **95**, 131–138.

31 U.S. EPA (2008) Integrated Risk Information System (IRIS). Glossary.

32 U.S. EPA (2000) Conducting a Risk Assessment for Mixtures of Disinfection Byproducts in Drinking Water Treatment Systems. EPA/600/R-03/040. U.S Environmental Protection Agency, Office of Research and Development/National Center for Environmental Assessment, Cincinnati, OH.

33 Teuschler, L.K., Rice, G.E., Wilkes, C.R., Lipscomb, J.C., and Power, F.W. (2004) A feasibility study of cumulative risk assessment methods for drinking water disinfection by-product mixtures. *J. Toxicol. Environ. Health A*, **67** (8–10), 755–777.

34 U.S. EPA (2003) The Feasibility of Performing Cumulative Risk Assessments for Mixtures of Disinfection By-products in Drinking Water. EPA/600/R-03/051. U.S. Environmental Protection Agency, Office of Research and Development, National Center for Environmental Assessment, Cincinnati, OH.

35 U.S. EPA (2006) Exposures and Internal Doses of Trihalomethanes in Humans: Multi-Route Contributions from Drinking Water. EPA/600/R-06/087. 2006 ORD/NCEA, Cincinnati, OH.

36 Hertzberg, R.C. and Teuschler, L.K. (2002) Evaluating quantitative formulas for dose-response assessment of chemical mixtures. *Environ. Health Perspect.*, **110** (Suppl. 6), 965–970.

37 Bove, F., Fulcomer, M., Klotz, J., Esmarat, J., Duffey, E.M., and Savrin, J.E.

(1995) Public drinking water contamination and birth outcomes. *Am. J. Epidemiol.*, **141**, 850–862.

38 Cantor, K.P., Lynch, C.F., Hildesheim, M.E., Dosemeci, M., Lubin, J., Alavanja, M., and Craun, G. (1998) Drinking water source and chlorination byproducts. I. Risk of bladder cancer. *Epidemiology*, **9**, 21–28.

39 Dodds, L. and King, W.D. (2001) Relation between trihalomethane compounds and birth defects. *Occup. Environ. Med.*, **58**, 443–446.

40 Dodds, L., King, W., Allen, A.C., Armson, B.A., Fell, D.B., and Nimrod, C. (2004) Trihalomethanes in public water supplies and risk of stillbirth. *Epidemiology*, **15** (2), 179–186.

41 Gallagher, M.D., Nuckols, J.R., Stallones, L., and Savitz, D.A. (1998) Exposure to trihalomethanes and adverse pregnancy outcomes. *Epidemiology*, **9** (5), 484–489.

42 Hildesheim, M.E., Cantor, K.P., Lynch, C.F., Dosemeci, M., Lubin, J., Alavanja, M., and Craun, G. (1998) Drinking water source and chlorination byproducts. II. Risk of colon and rectal cancers. *Epidemiology*, **9**, 29–35.

43 Klotz, J.B. and Pyrch, L.A. (1999) Neural tube defects and drinking water disinfection by-products. *Epidemiology*, **10** (4), 383–390.

44 King, W.D., Dodds, L., and Allen, A.C. (2000) Relation between stillbirth and specific chlorination byproducts in public water supplies. *Environ. Health Perspect.*, **108**, 883–886.

45 King, W.D. and Marett, L.D. (1996) Case–control study of bladder cancer and chlorination byproducts in treated water (Ontario, Canada). *Cancer Causes Control*, **7**, 596–604.

46 Toledano, M.B., Nieuwenhuijsen, M.J., Best, N., Whitaker, H., Hambly, P., de Hoogh, C., Fawell, J., Jarup, L., and Elliott, P. (2005) Relation of trihalomethane concentrations in public water supplies to stillbirth and birth weight in three water regions in England. *Environ. Health Perspect.*, **113** (2), 225–232.

47 Waller, K., Swan, S.H., DeLorenze, G., and Hopkins, B. (1998) Trihalomethanes in drinking water and spontaneous abortion. *Epidemiology*, **9** (2), 134–140.

48 Waller, K., Swan, S.H., Windham, G.C., and Fenster, L. (2001) Influence of exposure assessment methods on risk estimates in an epidemiologic study of total trihalomethane exposure and spontaneous abortion. *J. Exposure Anal. Environ. Epidemiol.*, **11**, 522–531.

49 Windham, G.C., Waller, K., Anderson, M., Fenster, L., Mendola, P., and Swan, S. (2003) Chlorination by-products in drinking water and menstrual cycle function. *Environ. Health Perspect.*, **111** (7), 935–941.

50 Wright, J.M., Schwartz, J., and Dockery, D.W. (2004) The effect of disinfection by-products and mutagenic activity on birth weight and gestational duration. *Environ. Health Perspect.*, **112** (8), 920–925.

51 Bull, R.J. and Kopfler, F.C. (1991) *Health Effects of Disinfectants and Disinfection By-Products*, AWWA Research Foundation.

52 Christ, S., Christ, S.A., Read, E.J., Stober, J.A., and Smith, M.K. (1995) The developmental toxicity of bromochloroacetonitrile in pregnant Long-Evans rats. *Int. J. Environ. Health Res.*, **5**, 175–188.

53 DeAngelo, A.B., George, M.H., Kilburn, S.R., Moore, T.M., and Wolf, D.C. (1998) Carcinogenicity of potassium bromate administered in the drinking water to male B6C3F1 mice and F344/N rats. *Toxicol. Pathol.*, **26**, 587–594.

54 George, M.H., Moore, T., Kilburn, S., Olson, G.R., and DeAngelo, A.B. (2000) Carcinogenicity of chloral hydrate administered in drinking water to the male F344/N rat and male B6C3F1 mouse. *Toxicol. Pathol.*, **28** (4), 610–618.

55 National Cancer Institute (NCI) (1976) Report on the Carcinogenesis Bioassay of Chloroform. NTIS PB-264-018. National Cancer Institute, Bethesda, MD.

56 NTP (National Toxicology Program) (1985) Toxicology and Carcinogenesis Studies of Chlorodibromomethane in

F344/N Rats and B6C3F Mice (Gavage Studies). NTP TR282.

57 NTP (National Toxicology Program) (1986) Toxicology and Carcinogenesis Studies of Bromodichloromethane in F344/N Rats and B6C3F Mice (Gavage Studies). NTP Technical Report, Ser. No. 321, NIH Publ. No. 87-2537.

58 NTP (National Toxicology Program) (1989) Toxicology and Carcinogenesis Studies of Tribromomethane and Bromoform in F344/N Rats and B6C3F Mice (Gavage Studies). NTP-350.

59 Keegan, T.E., Simmons, J.E., and Pegram, R.A. (1998) NOAEL and LOAEL determinations of acute hepatotoxicity for chloroform and bromodichloromethane delivered in an aqueous vehicle to F-344 rats. *J. Toxicol. Environ. Health*, **55**, 65–75.

60 Narotsky, M.G., Pegram, R.A., and Kavlock, R.J. (1997) Effect of dosing vehicle on the developmental toxicity of bromodichloromethane and carbon tetrachloride toxicity in rats. *Fundam. Appl. Toxicol.*, **40**, 30–36.

61 Klinefelter, G.R., Hunter, E.S., III, and Narotsky, M.G. (2001) Reproductive and developmental toxicity associated with disinfection by-products in drinking water, in *Microbial Pathogens and Disinfection By-Products in Drinking Water: Heath Effects and Management of Risks* (eds G.F. Craun, F.S. Hauchman, and D.E. Robinson), ILSI Press, Washington, DC, pp. 309–324.

62 Klinefelter, G.R., Strade, L.F., Suarez, J.D., Roberts, N.L., Goldman, J.M., and Murr, A.S. (2004) Continuous exposure to bibromoacetic acid delays pubertal development and compromises sperm quality in the rat. *Toxicol. Sci.*, **81**, 419–429.

63 U.S. EPA (2000) Toxicological review of chlorine dioxide and chlorite (CAS nos. 10049–04–4 and 7757-19-2) in support of summary information on the Integrated Risk Information System (IRIS). U.S. Environmental Protection Agency, Washington, DC. EPA/635/R–00/007. Available at http://www.epa.gov/iris/index.html.

64 Bielmeier, S.R., Best, D.S., Guidici, D.L., and Narotsky, M.G. (2001) Pregnancy loss in the rat caused by bromodichloromethane. *Toxicol. Sci.*, **59** (2), 309–315.

65 Bielmeier, S.R., Best, D.S., and Narotsky, M.G. (2004) Serum hormone characterization and exogeneous hormone rescue of bromodichloromethane-induced pregnancy loss in the F344 rat. *Toxicol. Sci.*, **77**, 101–108.

66 Smith, M.K., Randall, J.L., Read, E.F. *et al.* (1992) Developmental toxicity of dichloroacetate in the rat. *Teratology*, **46** (3), 217–223.

67 Epstein, D.L., Nolen, G.A., Randall, J.L., Christ, S.A., Read, E.J., Stober, J.A., and Smith, M.K. (1992) Cardiopathic effects of dichloroacetate in the fetal Long-Evans rat. *Teratology*, **46** (3), 225–235.

68 Johnson, P.D., Dawson, B.V., and Goldberg, S.J. (1998) Cardiac teratogenicity of trichloroethylene metabolites. *J. Am. Coll. Cardiol.*, **32** (2), 540–545.

69 Christ, S.A., Read, E.J., Stober, J.A., and Smith, K. (1996) Developmental effects of trichloroacetonitrile administered in corn oil to pregnant Long–Evans rats. *J. Toxicol. Environ. Health*, **47** (3), 233–247.

70 Sistrunk, S. and DeAngelo, A.B. (2002) The induction of colorectal neoplasia in male F344/N rats exposed to a mixture high in brominated trihalomethanes (THMs) in the drinking water. *Int. J. Toxicol.*, **20**, 408–409.

71 DeAngelo, A.B., Geter, D.R., Rosenberg, D.W., Crary, C.K., and George, M.H. (2002) The induction of aberrant crypt foci (ACF) in the colons of rats by trihalomethanes administered in the drinking water. *Cancer Lett.*, **187**, 25–31.

72 Narotsky, M.G., Best, D.S., McDonald, A., Myers, E.A., Hunter, E.S., and Simmons, J.E. (2010) Effects of trihalomethane and haloacetic acid mixtures on pregnancy maintenance and eye development in F344 rats, submitted.

73 Teuschler, L.K., Gennings, C., Hartley, W.R., Stitler, W.M., Colman, J.T.,

Hertzberg, R.C., Thiyagarajah, A., Lipscomb, J.C., and Simmons, J.E. (2000) A multiple-purpose design approach to the evaluation of risks from mixtures of disinfection by-products. *Drug Chem. Toxicol.*, **23**, 307–321.

74 Simmons, J.E., Teuscher, L.K., Gennings, C., Speth, T.F., Richardson, S.D., Miltner, R.J., Narotsky, M.G., Schenck, K.D., Hunter, E.S., III, Hertzberg, R.C., and Rice, G. (2004) Component-based and whole-mixture techniques for addressing the toxicity of drinking-water disinfection-byproduct mixtures. *J. Toxicol. Environ. Health Part A*, **67**, 741–754.

75 International Program on Chemical Safety (IPCS) (2000) Toxicology of disinfectant byproducts, in *Disinfectants and Disinfectant Byproducts. Environmental Health Criteria 216*, World Health Organization, Geneva, pp. 110–276.

76 Keegan, T.E., Simmons, J.E., and Pegram, R.A. (1998) NOAEL and LOAEL determinations of acute hepatotoxicity for chloroform and bromodichloromethane delivered in an aqueous vehicle to F-344 rats. *J. Toxicol. Environ. Health*, **55**, 65–75.

77 Lilly, P.D., Simmons, J.E., and Pegram, R.A. (1994) Dose-dependent vehicle differences in the acute toxicity of bromodichloromethane. *Fundam. Appl. Toxicol.*, **23**, 132–140.

78 U.S. EPA (2003) Integrated Risk Information System (IRIS). Bromoform. www.epa.gov/iris.

79 U.S. EPA (2003) Integrated Risk Information System (IRIS). Chlorodibromomethane. www.epa.gov/iris.

80 U.S. EPA (2003) Integrated Risk Information System (IRIS). Chloroform. www.epa.gov/iris.

81 U.S. EPA (2003) Integrated Risk Information System (IRIS). Dichlorobromomethane. www.epa.gov/iris.

82 Teuschler, L.K., Hertzberg, R.C., Rice, G.E., and Simmons, J.E. (2004) EPA project-level research strategies for chemical mixtures: targeted research for meaningful results. *Environ. Toxicol. Pharmacol.*, **18** (3), 193–199.

83 US EPA (1999) Research Report on the Risk Assessment of Mixtures of Disinfection By-products (DBPs) in Drinking Water. EPA/600/R-03/039. US Environmental Protection Agency, Office of Research and Development/ National Center for Environmental Assessment, Cincinnati, OH.

84 Gennings, C., Schwartz, P., Carter, H., and Simmons, J.E. (1997) An efficient approach for detecting departure from additivity in mixtures of many chemicals with a threshold additivity model. *J. Agric. Biol. Environ. Stat.*, **2**, 198–211; Erratum 2000, 5: 257–259.

85 Hertzberg, R.C. and Teuschler, L.K. (2002) Evaluating quantitative formulas for dose–response assessment of chemical mixtures. *Environ. Health Perspect.*, **110**, 965–970.

86 Simmons, J.E., Stiteler, W., Hertzberg, R.C., McDonald, A., Sey, Y.M., Teuschler, L.K., Rice, G., Colman, J., and Durkin, P. (2007) Additivity assessment of trihalomethane mixtures by proportional response addition. Abstract 1603. *Toxicologist CD*, **102** (S-1), March 2008.

87 Plewa, M.J., Wagner, E.D., Jazwierska, P., Richardson, S.D., Chen, P.H., and McKague, A.B. (2004) Halonitromethane drinking water disinfection byproducts: chemical characterization and mammalian cell cytotoxicity and genotoxicity. *Environ. Sci. Technol.*, **38**, 62–68.

88 Simmons, J.E., Wagner, E.D., and Plewa, J.E. (2006) Mixing ratio influences haloacetic acid (HAA) mixture toxicity. Abstract. *Toxicologist CD*, **101** (S-1), March 2008.

89 Chernoff, N., Rogers, E., Carver, B., Kavlock, R., and Gray, E. (1979) The fetotoxic potential of municipal drinking water in the mouse. *Teratology*, **19**, 165–169.

90 Kavlock, R., Chernoff, N., Carver, B., and Kopfler, F. (1979) Teratology studies in mice exposed to municipal drinking-water concentrates during organogenesis. *Food Cosmet Toxicol.*, **17**, 343–347.

91 Keen, C.L., Uriu-Hare, J.Y., Swan, S.H., and Neutra, R.R. (1992) The effects of water source on reproductive outcome in Fischer-344 rats. *Epidemiology*, **3** (2), 130–133.

92 Uriu-Hare, J.Y., Swan, S.H., Bui, L.M., Neutra, R.R., and Keen, C.L. (1995) Drinking water source and reproductive outcomes in Sprague-Dawley rats. *Reprod. Toxicol.*, **9**, 549–561.

93 Richardson, S.D., Thruston, A.D., Krasner, S.W., Weinberg, H.S., Miltner, R.J., Schenck, K.M., Narotsky, M.G., McKague, A.B., and Simmons, J.E. (2008) Integrated disinfection byproducts mixtures research: comprehensive characterization of water concentrates prepared from chlorinated and ozonated/postchlorinated drinking water. *J. Toxicol. Environ. Health Part A*, **71** (17), 1165–1186.

94 Claxton, L.D., Pegram, R.A., Schenck, K.M., Simmons, J.E., and Warren, S.H. (2008) Integrated disinfection byproducts mixtures research: *Salmonella* mutagenicity of water concentrates disinfected by chlorination and ozonation/postchlorination. *J. Toxicol. Environ. Health Part A*, **71** (17), 1187–1194.

95 Crosby, L.M., Ward, W.O., Moore, T.M., Simmons, J.E., and DeAngelo, A.B. (2008) Integrated disinfection byproducts mixtures research: gene expression alterations in primary rat hepatocytes cultures exposed to DBP mixtures formed by chlorination and ozonation/postchlorination. *J. Toxicol. Environ. Health Part A*, **71** (17), 1195–1215.

96 Rice, G., Teuschler, L.K., Speth, T.F., Richardson, S.D., Miltner, R.J., Schenck, K.M., Gennings, C., Hunter, E.S., Narotsky, M.G., and Simmons, J.E. (2008) Integrated disinfection byproducts mixtures research: assessing reproductive and developmental risks posed by complex disinfection byproduct mixtures. *J. Toxicol. Environ. Health Part A*, **71** (17), 1222–1234.

97 International Life Sciences Institute (ILSI) (1998) Assessing Toxicity of Exposure to Mixtures of Disinfection Byproducts: Research Recommendations. ILSI Risk Science Institute, Washington, DC.

17
Endocrine Active Chemicals

Ed Carney, Kent Woodburn, and Craig Rowlands

17.1
Introduction

Launch your Internet browser, punch in the search terms "endocrine disruptor" and within minutes you are likely to be immersed in a captivating story mixed with alarm for your personal well being, allegations of injustice, and a short lesson on receptor biology thrown in for good measure. Catch phrases such as "complex chemical cocktail" or "sea of estrogens" draw attention to the fact that people are typically exposed to multiple rather than single chemicals and that some of these chemical mixtures contain compounds possessing the ability to modulate hormonal processes. In the toxicology and risk assessment vernacular, this is referred to as cumulative exposure to hormonally active or endocrine active chemicals (EACs). The public has become increasingly aware of such exposures, and many have raised concerns that these combined doses of EACs will lead to disease and other adverse health outcomes that would not have been predicted on the basis of individual chemical risk assessments.

The general public, as well as policy makers, look to toxicologists and risk assessment professionals to provide advice and perspective on this emerging area of health science. Although not all of the answers are readily agreed upon by the scientific community, there is nonetheless a large and growing body of scientific information from which to derive perspective on this issue. These data range from the vast amount of knowledge on the molecular action of steroid hormones to human experience with hormonally active pharmaceutical agents.

The ability of natural and synthetic chemicals to interact with endogenous hormone receptors has been known since the 1930s [1–3]. However, interest in EAC mixtures was suddenly galvanized by a 1996 report claiming fantastic synergy among a mixture of four weakly estrogenic agents [4]. Attention to this issue persisted, despite the fact that the results could not be reproduced by others [5, 6] and subsequent retraction of the original report [7]. It was around the same time that the United States Environmental Protection Agency initiated development of an extensive program to screen EACs as mandated in the 1996 amendments to the Food Quality Protection and Safe Drinking Water Acts. Since then, there have been

Principles and Practice of Mixtures Toxicology. Edited by Moiz Mumtaz

ISBN: 978-3-527-31992-3

numerous investigations into the effects of multiple EACs, as well as assessments of exposure to EACs from consumption of groundwater, use of commercial products, and diet.

In this chapter, we will review and assimilate the currently available information on EAC mixtures, keeping in mind those pragmatic concerns of the public and policy makers. The approach taken in this review reflects the principles espoused in a recent paper emanating from a multistakeholder workshop entitled "New Beginnings in Chemical Mixture Toxicology" [8]. The authors emphasized the need to understand the composition of real-world mixtures in terms of typical exposure levels, numbers of chemicals, variability in mechanism of action, and to design experiments and interpret data in the light of these characteristics. The chapter starts with a characterization of some common "real-world" mixtures containing EACs. We then review experimental data on the effects of EAC mixtures on wildlife and mammalian species *in vivo*. Based mainly on the availability of data, the focus will primarily be on estrogenic and antiestrogenic compounds, that is, chemicals with the ability to mimic or interfere with the action of estrogens and/or estrogen receptors (ER) in humans or wildlife. However, the general principles discussed herein should apply to other compounds whose toxicity is mediated directly or indirectly through any number of steroid hormone receptors, such as the androgen receptor (AR), retinoid receptors, or aryl hydrocarbon (AH) receptor.

17.2 Common Characteristics of EAC Mixtures

Estrogenic and antiestrogenic compounds have drawn considerable attention from a chemical mixtures perspective mainly because the ER has over 100 different ligands to which it will bind [9]. Synthetic chemicals with estrogenic or antiestrogen activity, often referred to as "xenoestrogens," include bisphenol A, used in polycarbonate, can coatings, and dental sealants; alkylphenol ethoxylates used in making detergents; phthalates found in certain plastics; parabens in body lotions; benzophenones in sun block; and environmental contaminants such as DDT, methoxychlor, some PCBs, dioxins, and furans (Table 17.1). There are also many pharmaceuticals with (intentional) hormonal activity, such as diethylstilbestrol (DES), contraceptive agents, and the selective estrogen receptor modulating agents (SERMs) used in the treatment of diseases such as osteoporosis. In addition, there are abundant natural sources of estrogenic compounds found in plants. These phytoestrogens include lignans, such as enterolactone, found in high concentrations in flaxseed products, sunflower seed, green tea, coffee, and broccoli; isoflavones, such as genistein, daidzein, naringenin, genistein, daidzin, glycitein, and puerarin [10]; and polyphenols, such as resveratrol, a constituent of red wine [11]. Finally, exposure to synthetic and plant-derived estrogens occurs against a background of endogenous estrogens, such as 17β-estradiol (E2), estrone (E1), and estriol (E3), the levels of which in blood and tissues can dramatically vary depending on gender, age, and physiological state (e.g., pregnancy, pre-, or postmenopause).

Table 17.1 Examples of estrogenic compounds in foods.

Mixture	EAC	Concentration in mixture	Reference
Soy formula, powdered	GEN	16–23 μg/g	[43]
	Daidzein	8–19 μg/g	
	Glycitein	3–4 μg/g	
	Bisphenol A	0.1–13.2 ng/g (US); <0.002 ng/g (UK); 44–113 ng/g (Taiwan)	[46]
	DEHP	0.12 μg/g	[101]
	Dieldrin	0.05–0.08 ng/g	
Soy products	Genistein	0.2–11 μg/g	[102]
	Daidzein	0.2–9 μg/g	
	Lignans	9 μg/g	
Canned food			[103]
Soft drinks	Bisphenol A	Not detected (<10 ng/g)	
Most foods	Bisphenol A	<10–29 ng/g	
Tuna, coconut cream, corned beef	Bisphenol A	98–191 ng/g	
Vegetables	Lignans	1–4 μg/g	[102]
	Genistein	10–30 μg/g	
	Daidzein	10–30 μg/g	[44, 46]
	DEHP	48 ng/g	
Liquids from canned vegetables	Bisphenol A	12–76 ng/g	

17.2.1
Rivers and Aquatic Environments

The presence of natural hormones in wastewater treatment outfalls in the United States was noted in 1965 [12] and this work was expanded in 1970 to include synthetic estrogens used as birth control agents [13]. By the mid-1970s, scientists began to detect other pharmaceutical agents near wastewater outlets in the United States [14, 15]. The presence of estrogenic substances and pharmaceuticals in wastewater effluents was largely ignored until the 1990s when reports of fish deformities in the United Kingdom were linked to estrogenic compounds in wastewater effluents [16–18]. The relationship between estrogenicity of wastewater effluents and the presence of estrogenic compounds, both natural and synthetic, has now been well established around the globe [19–22]. The measured concentrations of potent natural and synthetic estrogens in domestic wastewater effluents range from <1 to 48 ng/l for 17β-estradiol (E2), from <1 to 87 ng/l for estrone (E1), and from <1 to 7 ng/l for 17α-ethinylestradiol (EE2) [23–27]. Estrogen mimics of lower endocrine potency that may be found in wastewater effluents include phytoestrogens such as the isoflavones (e.g., genistein) and coumestranes (e.g., coumestrol), and industrial chemicals such as bisphenol A and phthalates [28, 29]. Average concentrations of genistein and coumestrol have been reported at 12 and 10 ng/l, respectively, in surface water downstream from Italian wastewater discharges [30], while in North America, median concentrations of BPA and diethylphthalate of 140 and 200 ng/l, respectively, have been reported downstream of wastewater treatment plant effluents [31]. The concentrations of these compounds and other EACs in downstream river water and their potential impact on fish and other aquatic wildlife are dictated by the dilution of the wastewater by the receiving water. Over time and distance, additional dilution from other streams, biodegration, and sorptive processes may act to further lower concentrations of the EACs [32–34].

While wastewater effluents have received a bulk of the attention regarding endocrine activity and the potential for impact on downstream aquatic species, discharge from pulp/paper mills and runoff of agricultural waste may also contain EACs. The levels of EACs in paper/pulp mill effluents vary with the specific process performed in a mill [35]. Reductions in circulating sex steroids, gonad size, and expression of secondary sex characteristics have been observed in fish exposed to some mill effluents [36, 37]. The androgenicity of fish downstream of some paper/pulp mills has been primarily attributed to the processing of plant material resulting in the release of the natural steroid, stigmasterol [38, 39]. Animal manure may contain both endogenous hormones and hormones used to treat the animals prophylactically. Studies have identified E2 in poultry litter [40], and the use of poultry litter as fertilizer has resulted in concentrations as high as 1280 ng/l in runoff water entering adjacent streams [41]. Irwin *et al.* [42] detected E2 in farm ponds receiving runoff from beef cattle pastures at levels of 0.5–1.8 ng/l [42].

17.2.2 Human Exposure

The most fundamental and important chemical mixture that humans are exposed to is food. EACs are ubiquitous in plant-derived foods (Table 17.1) due to the presence of flavonoids or other phytoestrogens in a multitude of plant products. One food that is among the richest and best-studied sources of phytoestrogens is infant soy formula. It also is of great toxicological interest because human exposure to soy formula occurs during early developmental stages, and for many infants it constitutes their entire diet for the first several months of life. A US Department of Agriculture survey of infant formulas reported genistein and daidzein concentrations ranging from 16 to 23 and 8 to 19 mg/kg, respectively [43]. Cumulative exposure estimates for genistein in the United States range from a high end of 1000–8000 μg/kg bw/day for infants fed soy formula to 140 μg/kg bw/day for vegetarians and 100 μg/kg bw/day for omnivores [43]. In Japanese adults, average genistein intake is estimated at 200–400 μg/kg bw/day.

As a general rule, exposures to manmade xenoestrogens tend to be at least a few orders of magnitude lower than exposure to phytoestrogens. Exposures to bisphenol A and di(2-ethylhexyl)phthalate (DEHP), which are among the most widely studied EACs, illustrate this point. Aggregate exposure to bisphenol A from food, air, dust, and soil, as well as from specialized applications (e.g., dental sealants), have been extensively evaluated (reviewed in Ref. [44]). Aggregate exposure estimates for bisphenol A in infants, children, and adults range from 0.02 to 8, 0.005 to 15, and 0.002 to 1.4 μg/kg bw/day, respectively, with 99% of the bisphenol A exposures coming from food. Many of the latter estimates are based on food packaging migration studies, with the variability in reported values stemming from differences in temperatures, duration, the type of food stimulant, and so on. In fact, the highest values for bisphenol A have been considered to be of uncertain reliability [45], and are likely to be overly conservative estimates. For DEHP, aggregate exposure estimates for infants, children, and adults were 5.0–7.3, 10–25.8, and 8.2 μg/kg bw/day, respectively (reviewed in Ref. [46]). The majority (>90%) of DEHP exposures in children and adults came from food, whereas in infants, 44–60% of the DEHP exposure was from food, with the remainder attributed to ingestion of dust. Figure 17.1 provides an example of cumulative exposure to some common EACs in infants (0–11 months of age), which are often assumed to be among the most sensitive and highly exposed subpopulation. Again, the data show phytoestrogen exposures that are several orders of magnitude greater than those of the synthetic EACs.

In addition to measured concentrations of EACs in foods, data on EAC levels in human blood and other body fluids serve as composite indicators of EAC exposure, in that they reflect the contributions of a wide range of potential sources that occur by multiple routes of exposure. Importantly, these measures also reflect pharmacokinetic differences between different EACs. For example, phytoestrogens such as genistein, and many xenoestrogens such as bisphenol A, are eliminated much more rapidly (elimination half lives in humans of 2–7 and 4–5.4 h, respectively; [43, 44]) than some persistent environmental estrogens, such as DDT or its metabolite, DDE. Nonetheless, internal levels of phytoestrogens tend to be much higher than those of

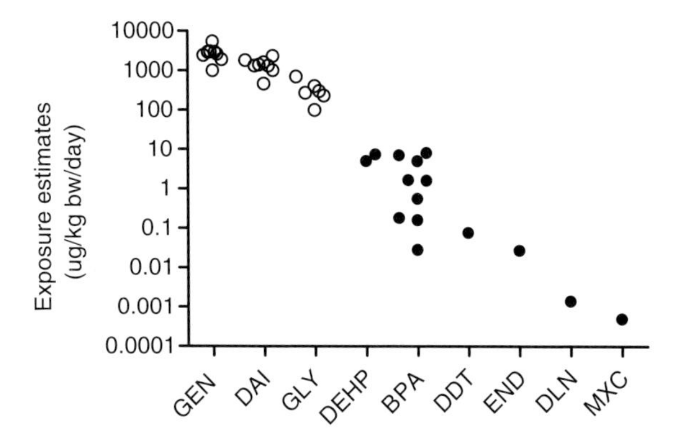

Figure 17.1 Exposure estimates of various EACs in human infants (0–11 months of age). Open circles denote phytoestrogens; closed circles denote synthetic EACs. GEN, genistein/genistein; DAI, daidzein/daidzin; GLY, glycitein/glycitin; DEHP, di-(2-ethylhexyl)phthalate; BPA, bisphenol A; DDT, 1,1,1-trichloro-2,2-bis (*p*-chlorophenyl)ethane; ESF, endosulfan; DLN, dieldrin; MXC, methoxychlor [43, 44, 46, 100].

synthetic EACs presumably due to their abundance in food and the constant loading that occurs through the diet, relative to environmental xenoestrogen exposures that are characterized by lower and more sporadic exposures (reviewed in Refs [43, 44, 46, 47]. Considering both exposure levels and estrogenic potency, Safe [47] argued that xenoestrogens contribute imperceptibly to the overall estrogenic activity as compared to that attributable to endogenous phytoestrogens, a premise supported by the aforementioned exposure data, much of which have been collected in more recent years.

17.3
Toxicity of EAC Mixtures

Many kinds of assays have been developed for the assessment of an individual compound's estrogenicity or antiestrogenicity, ranging from ER binding, transcriptional activation, and cell-based assays, to whole animal studies in fish, laboratory mammals, and other organisms (reviewed by Safe [48]). The receptor-based and cell-based assays serve mainly as first pass screens to flag compounds with EAC activity, but their focus on a very restricted set of biological activities limits their utility for assessing risk. Conversely, the whole animal tests are more apical in nature and serve to detect adverse effects but are less specific for mechanism of action. When applying various assays to study EAC mixtures, it becomes especially important to appreciate the inherent limitations of each assay. For example, a reporter gene assay system involves only a single cell type and one target gene. As such, interactions involving pharmacokinetic factors (e.g., enhanced absorption or clearance), cross talk with other cell types, or feedback with other hormonal systems cannot be detected. A whole animal test like the uterotrophic assay is responsive to a wider array of potential interactions, although it still evaluates only a single tissue, and may not predict interactions involving other organ systems. Thus, it is critical that mixtures be evaluated in systems utilizing appropriate levels of biological organization, and often this would ideally be an *in vivo* system. To make such systems practical for the study of mixtures, economical study designs such as the fixed ratio ray design [49] can be used to greatly reduce the number of treatment groups relative to designs such as the full factorial.

In addition, mixture studies require consideration of some additional experimental design principles in order to be useful for risk assessment. These include the need for well-characterized dose-response data for the individual mixture components, an explicit no-interaction hypothesis stated a priori, testing of a sufficient range of mixture combinations, and a formal statistical test of the no interaction hypothesis [50]. There are far too many examples in the scientific literature of simple 2×2 factorial designs using high doses of each chemical. Interactions at high doses and at a single mixing ratio are of little predictive value for effects of real-world mixtures, which are far more complex, and tend to involve low doses of individual mixture components [8].

17.3.1 Mixtures of EACs in Aquatic Organisms

Measured effects of EACs with wildlife species have been demonstrated in controlled laboratory studies with fish, reptilian, amphibian, mammalian, avian, and invertebrate models following exposure to a variety of synthetic and natural chemicals [51]. The degree to which natural populations of wildlife species have been affected by EACs is less clear, though there is some evidence of effects such as gonadal alterations, reproductive activity, delayed sexual maturity, and reduced gonadal development to have been observed [26, 52–55].

There is a considerable difference in the potency of chemicals possessing endocrine activity, with the best-known example being the estrogenic potency of the natural estrogen (E1, E2) and synthetic estrogen (EE2) exceeding the xenoestrogens by 3 to 4 orders of magnitude [56]. Potency is allied with environmental concentration, so a weakly estrogenic chemical can cause or add to effects if it is present in the environment at suitably high concentrations [17].

Early mixture studies that concluded synergistic responses observed from combinations of estrogenic compounds in cell lines [57, 58], fish [59], and reptiles [60] have been hampered by an experimental approach where single agents and the mixture were tested at only one dose level, and the combined effects of multiple compounds are assumed to be equal to the sum of the effects of the constituents; deviation from this expectation is identified either as synergism or as antagonism. However, it has been noted that using this method with compounds that exhibit the common sigmoidal dose-response curve, rather than the linear dose-response curves needed to support such a conclusion, will lead to logical inconsistencies and incorrect conclusions concerning synergistic interactions [61–63]. Mixture studies predicated on developing a full dose-response concentration range of individual components report additive, not synergistic, behavior as a worst-case description of mixture endocrine response [26, 64–66]. Thorpe *et al.* [26] noted that the *in vitro* recombinant yeast estrogen screen overestimated presumed additive endocrine activity by an average of 50% compared to a fish *in vivo*vitellogenin (VTG) assay, while Brian *et al.* [66] determined that the combined effect of an EAC mixture did not deviate from an additivity assumption.

Xie *et al.* [67] have claimed "greater than additive" effects of mixtures of aquatic herbicides and surfactants using a trout VTG assay as an endocrine end point. However, the authors failed to present objective numerical data supporting a statistically significant difference in the endocrine mixture VTG response. The authors note that the herbicide triclopyr produced "greater than additive estrogenic responses under laboratory conditions and in a field setting," but the underlying data for triclopyr tested alone are not presented in the paper and analytical measurements for triclopyr in a "field" setting failed to detect triclopyr in the water samples tested. It is insufficient to suggest that because the field pond was previously treated with a triclopyr-containing herbicide, this chemical elicited the observed effects in their tests. Finally, the use of a mixed gender population of juvenile trout and limited replication ($N = 6$) suggests that the observed VTG responses cannot be attributed with confidence to the effects of the treatment(s). The gender of each individual trout used in the studies was never identified, and juvenile female rainbow trout may exhibit blood VTG levels up to 1000 times that of male rainbow trout [68]. Because the sex of the juvenile fish was apparently not known or controlled as part of the experimental design, it is plausible that the results observed were solely due to the presence of more females in any particular treatment compared to a lesser number of females in the controls. In addition, the VTG results as presented are further clouded and obscured by the normalization of the VTG levels in the blood to the total protein in the blood, for example, nanograms VTG/milligram protein. This normalization procedure deviates from the commonly accepted practice within the aquatic toxicology community of presenting VTG results simply as the concentration in the blood, for example, micrograms or nanograms VTG/milliliter plasma [59–72].

17.3.2 Mixtures of EACs *In Vitro*

As mentioned earlier, the study of EAC mixtures was greatly stimulated by an experiment in which binary combinations of the weak xenoestrogens, dieldrin, endosulfan, or toxaphene reportedly caused up to a 1000-fold increase in ER-mediated gene transcription relative to the individual response to any of these chemicals [4]. A number of other laboratories attempted to replicate these provocative findings but were unable to do so, whether the test system was the yeast estrogen system as used by Arnold *et al.*, or a battery of yeast-based, human cell-based, and mammalian *in vivo* assays [5, 6]. This is not to say that synergy is not theoretically possible. It is well established that endogenous signaling molecules (e.g., IGF-1, EGF) can interact with estradiol in a synergistic fashion [73–75] as part of normal endocrine signaling.

Other attempts to address the potential for synergy within mixtures of estrogenic compounds includes a thorough assessment in which many different ternary mixtures were tested in an MCF-7 cell-based ER-alpha reporter gene assay using a full factorial design in which four concentrations of each chemical were assessed in all possible combinations (64 treatment groups) [76]. The mixtures included various combinations of organochlorine pesticides (methoxychlor, *o,p*-DDT, dieldrin), polyaromatic hydrocarbons (benzo[*a*]pyrene, 1,2-benzanthracene, chrysene), a phytoestrogen (genistein), and the endogenous estrogen, 17β-estradiol. In addition, mixtures were tested in both a low range (concentrations near the individual chemical response thresholds) and a high range (approximately 2–10× higher) experiment. The majority of mixtures yielded responses that were consistent with additivity, and in no case was a synergistic interaction observed. The mixture of estradiol, genistein, and DDT exhibited antagonistic interactions at both the low- and high-concentration ranges.

As the search for synergy largely has come up dry, efforts have shifted to exploration of additivity among ER ligands and its potential implications for human and environmental effects [63]. A series of studies from the laboratory of Andreas Kortenkamp have explored this hypothesis by combining multiple xenoestrogens at levels below their individual effect thresholds, and examining the combined responses in the yeast estrogen reporter gene system [64, 65, 77, 78]. These studies were notable in their acquisition of full dose-response data for all chemicals, the formation of an explicit no-interaction hypothesis and the application of formal statistical tests to assess their hypothesis, thus satisfying several of the criteria outlined by Borgert *et al.* [50]. The key hypothesis addressed in this body of work was that the combined exposures to multiple xenoestrogens can yield a biological effect even when each component of the mixture is at a concentration below its individual threshold.

In one of these studies [65], full dose-response data for seven xenoestrogens and one phytoestrogen (genistein) were obtained and EC_{01} concentrations, that is, the concentration inducing 1% of the maximal response induced by 17β-estradiol, were determined. The eight compounds were then combined in a fixed ratio, equipotent mixture design using a range of test concentrations for the mixture. Interestingly, when the eight chemicals were present at a level near or just below their EC_{01}, a statistically significant response for the mixture was detected. This response was not due to synergy, but rather, fit with an expected response predicted under the assumption of additivity. A similar approach was used by Rajapakse *et al.* [78], using an equipotent mixture of 11 xenoestrogens. This mixture of 11 xenoestrogens was then combined with 17β-estradiol in mixing ratios of 1 : 25 000, 1 : 50 000, or 1 : 100 000 (17β-estradiol:xenoestrogen pool) to address the question of whether or not mixtures of weak xenoestrogens can incrementally contribute to the activity above and beyond that attributable to a potent endogenous estrogen. Indeed, the xenoestrogen mixture shifted the dose-response curve to the left relative to the curve for estradiol alone. However, the degree of shift was relatively slight and was consistent with additivity. As in the study by Silva *et al.* [65], when all 11 xenoestrogens were combined at concentrations just below their individual thresholds, a response from the mixture was clearly detectable.

Based on these and similar data, the authors coined the phrase "something from nothing," attracting interest to the supposition that the effects of xenoestrogens cannot be dismissed on the basis of their low individual potency or that synergy must be invoked for such risks to be plausible. However, there are several caveats to this series of studies that might temper the implications suggested. One of these is the inability of the yeast estrogen system to detect antiestrogenic effects. In fact, all of the chemicals tested behaved as full agonists in the assay, whereas most weak xenoestrogens behave as partial agonists or even antagonists, depending on cell context. Also, it is also unlikely that environmental mixtures would contain a sufficiently large number of ER agonists at concentrations at or just below their threshold for induction of biological activity. Similarly, the experiments in which xenoestrogens were mixed with estradiol at ratios up to 1 : 100 000 do not seem to align with the ratio exposure data reviewed previously.

17.3.3 Mixtures of EACs in Mammalian Animal Models

In vivo assessment affords the greatest opportunity to detect synergistic or antagonistic interactions and represents a level of biological organization that is most relevant to human exposure. However, pragmatic constraints have limited the number of studies assessing interactions among estrogenic chemicals to just a few, and these have mainly utilized short-term, limited end point screens, such as the rodent uterotrophic assay. This assay involves dosing of immature rats or mice for 3 days, followed by measurement of uterine weight and sometimes ancillary markers of estrogen stimulation.

Wade *et al.* [79] examined potential interactions between dietary isoflavones and either a strong (EE) or a weak (bisphenol A) estrogenic compound in the rodent uterotrophic assay. Dose responses for each of the test compounds were determined in rats fed diets supplemented with increasing levels of a soy isoflavone mix (Novasoy) or an isoflavone-free diet. Uterine weights for the two test compounds were not affected by the isoflavone content of the diet, even at dietary isoflavone concentrations as high as 1250 mg/kg. There also was no effect of the isoflavones on uterine epithelial cell height, although there was a significant increase in uterine peroxidase activity at the highest isoflavone dose level. Charles *et al.* [80] conducted a similar study on potential interactions between phytoestrogens and xenoestrogens, but instead of feeding the isoflavones, they dosed the rats with varying concentrations of a genistein/daidzein preparation. They then posed the question as to whether or not low levels of a six-chemical xenoestrogen mixture could measurably increase the uterotrophic response over that induced by phytoestrogens alone. At low doses of the xenoestrogen mixture, uterine weights were identical to those of the genistein/daidzein alone. However, high doses of the xenoestrogen mixture, that is, at doses in which each xenoestrogen was near or at its individual threshold for induction of uterotrophic activity, led to an incremental increase in uterine weight over that caused by the phytoestrogens alone. In the latter case, statistical analyses indicated that the xenoestrogens and phytoestrogens acted additively. In another *in vivo* study, Naciff *et al.* [81] coupled the uterine weight and epithelial cell height responses with an assessment of gene expression to compare responses to EE in rats fed a casein-based diet (low phytoestrogen) versus standard rodent diet (higher phytoestrogen content). Although there were some genes whose expression differed in animals fed the different diets, the genes that differed were not associated with estrogen stimulation. Overall, there was no impact of diet on uterine weight, epithelial cell height or the expression of estrogen-regulated genes. Finally, Tinwell and Ashby [82] looked at mixtures containing up to seven estrogenic chemicals with widely differing potencies (range of 6 orders of magnitude). Consistent with the Charles *et al.* [80] and Silva *et al.* [65] studies, combining multiple estrogenic chemicals at doses slightly below or at their minimal effect levels led to a measurable response of the mixture. However, responses in the seven-chemical mixture were lower than those predicted using simple addition calculations [82]. This was likely due to the action of weaker estrogens competing with more potent estrogens, a phenomenon that was reported over 40 years ago [83].

An alternative approach to component-based interaction studies is the whole mixture experiment. Such experiments can serve to truth the predictions derived from interaction studies. Wade *et al.* [84] examined the effects of subchronic oral exposure to a complex mixture containing 2,3,7,8-tetrachlorodibenzo-*p*-dioxin, polychlorinated biphenyls, DDE, DDT, dieldrin, endosulfan, methoxychlor, hexachlorobenzene, other chlorinated benzenes, hexachlorocyclohexane, mirex, heptachlor, lead, and cadmium. Male rats were dosed for 70 days, with the majority of chemicals at their regulatory limit (minimal risk level) or at doses 10, 100, or 1000 times higher. Apart from some liver and kidney effects at high doses, effects of the mixture on reproductive parameters were limited to minor changes in epididymal weight at doses in which all chemicals were present at $\geq 10\times$ their regulated level.

17.4 Is the Concept of "Common Mechanism" Relevant for EAC Mixtures?

Regulatory guidance documents issued by government agencies usually discuss the concept of common mechanism of action, asserting that chemicals that share a common mechanism would be expected to act in an additive manner [85]. This assumption is a key scientific pillar of the toxic equivalency factor (TEF) approach used to regulate agents that bind to the AH receptor, and is commonly invoked for other types of receptor-mediated toxicity. While the concept seems reasonable at first glance, in practice, it can be quite challenging to define, and predictions based on such assumptions are fraught with problems [86].

Let us use ER ligands as a case to illustrate the inherent challenges of applying the common mechanism of toxicity concept to mixtures of chemicals that bind to steroid hormone receptors. Here, the assumption will be that any xenoestrogen, phytoestrogen, endogenous estrogen, or pharmaceutical preparation would potentially be included in this category on the basis of its common property of binding to the ER. Such an approach would be consistent with the notion of "something from nothing" in which concerns were raised about multiple ER ligands acting additively to induce adverse health effects.

17.4.1 Receptor, Cell, and Tissue Specificity

Using this train of logic, we are immediately confronted with the fact that there are two ERs, α and β, not to mention additional subtypes. The distribution of these two receptors is variable, with tissues such as bone having predominantly the β isoform, uterus containing mainly ERα, but many tissues expressing both α and β. Furthermore, α and β often regulate opposing functions. In addition, the biological effects elicited by specific ER ligands are highly cell- and tissue specific. One of the best-known examples of this phenomenon is the drug tamoxifen, which has been used for treatment of early-stage breast cancer and as a chemoprevention agent for women at high risk for this disease [87]. Tamoxifen is a SERM that is an ER antagonist in breast cancer, but an ER agonist in the bone and uterus [88]. Another example is the phytoestrogen resveratrol, which shows more activity in cells that uniquely express ERβ or that express higher levels of ERβ than ERα, and also appears to exhibit E2 antagonist activity for ERα in certain promoter contexts [89].

Another factor complicating common mechanism approaches is that only a small percentage of ER ligands are pure estrogen agonists or antagonists. Most phytoestrogens and xenoestrogens are partial agonists, meaning that they are unable to induce a maximal response matching that of the endogenous ligand, even at very high concentrations [90]. The latter compounds tend to act as agonists under some conditions, but act as antagonists when their concentration is very high relative to more potent ligands. Thus, mixing ratio becomes a key determinant of the response to mixtures containing partial ER agonists. Finally, the mixture approaches based on common mechanism require assignment of a relative potency factor.

17.4.2 Signaling Pathways

If receptor isoform-, cell-, and tissue-specificity were not enough, things get really difficult for the common mechanism concept when we consider signaling events that occur downstream of receptor binding. Steroid hormones such as estrogens and androgens are essentially endocrine signaling molecules that act by binding to and activating nuclear hormone receptors, for example, ERα, ERβ, and androgen receptor [91]. These nuclear receptors are ligand-activated transcription factors that modulate target gene expression, and with the completion of the human genome sequence, it is now known that there are some 48 nuclear receptors expressed in humans, of which only half have identified ligands such as steroid hormone receptors that constitute only a minority of this large superfamily of receptors [91]. These nuclear receptor proteins are typically divided into three major domains with the amino terminus (A/B region) being the transactivation domain where coregulator proteins (coactivators, corepressors) and other transcription factors interact with the receptor, the central DNA binding domain (DBD), and the carboxy terminus ligand binding domain (LBD) that also contains an activation function (AF2) with coregulatory protein interactions [91]. The general scheme for steroid hormone receptor activation (Figure 17.2) involves ligand binding to the receptor in the cytoplasm that initiates nuclear translocation whereby the receptor homodimerizes with itself and binds to inverted repeat DNA half sites, called hormone response elements, or HREs in regulated genes [91]. Other nonsteroid hormone receptors such as thyroid hormone receptors (TRs), retinoic acid receptors (RARs), and vitamin D receptors (VDRs) are localized in the nucleus in the unliganded state and following ligand activation usually bind as heterodimers with a retinoid X receptor (RXRα, RXRβ, RXRγ) to direct repeats in the regulated genes [91]. Prior to DNA binding and continuing afterward, nuclear localized hormone activated steroid receptors recruit numerous coregulatory proteins to assemble and stabilize the transcriptional machinery needed for gene transcription or alternatively for gene repression [91–95].

Like the nuclear receptor superfamily, coregulators are themselves a large and mixed family of proteins and possess a varied array of enzymatic activities necessary for chromosomal maintenance and gene regulation [94]. They can be categorized

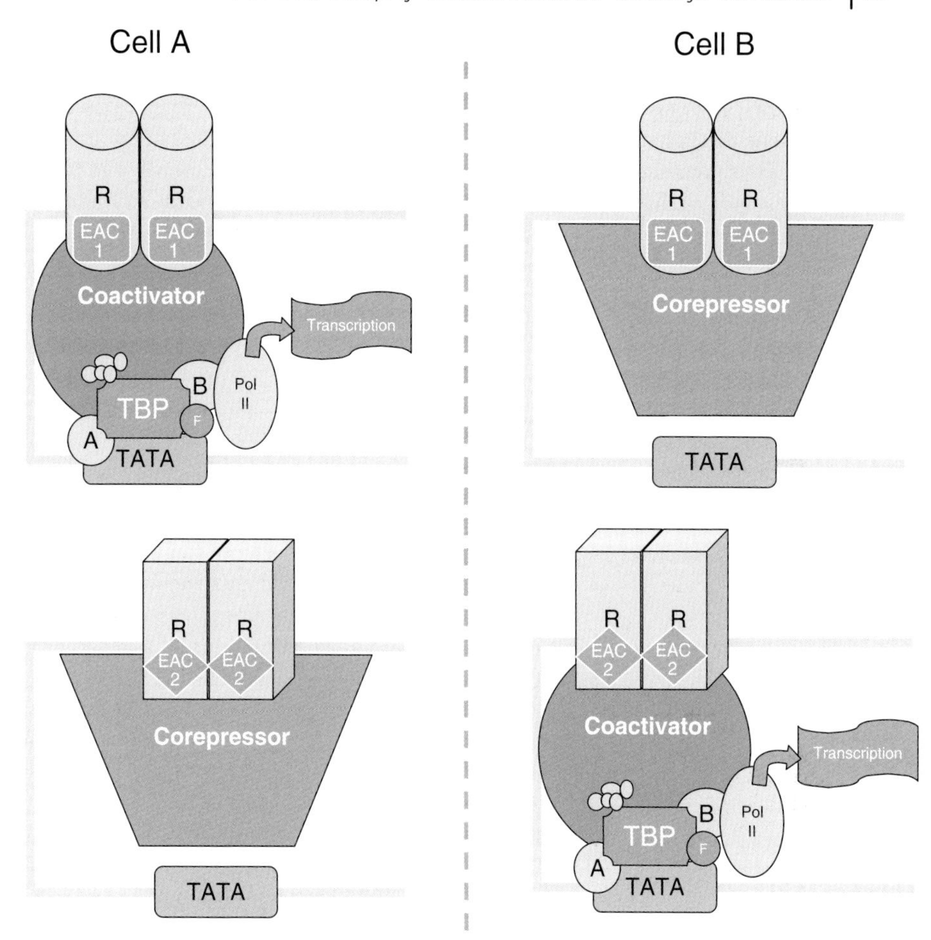

Figure 17.2 Actions of hormone receptors depend both on ligand and on cell context. Hormone receptors (R) form a different tertiary structure depending on the ligand that activates them (EAC-1 or EAC-2). This in turn causes the receptors to recruit different coregulators, for example, coactivators or corepressors, that lead to gene induction or repression, respectively. An EAC can act as an agonist or antagonist depending on the coregulators present in different cell environments (Cell A, Cell B). Although not shown, the gene architecture influences hormone receptor/coregulator recruitment as well.

based on their wide-ranging functional properties including acetyltransferases, ubiquitin ligases, ATP-coupled chromatin remodeling complexes, protein methylases, cell cycle regulators, RNA helicases, and bridging proteins that facilitate direct contact with components of the basal transcription machinery [93]. At present, there are 285 coregulators identified and many are expressed differentially by cell type, tissue type, gender, developmental stage, and species [92, 95]. Moreover, coregulators can be differentially recruited to the same receptor by ligand-induced conformational changes in the hormone receptor. Differences in the tissue-specific receptor agonist and antagonist activities have long been recognized for the steroid hormone receptors and form the basis for the development of drugs for treatment of hormone-dependent diseases. Tamoxifen's tissue-specific activity is now thought to be due to recruitment of different coregulators to the ER by tamoxifen in breast tissue versus uterine and bone tissue [91]. Thus, coregulators extend the actions of the same hormone receptor by producing cell type-specific positive or negative stimuli from the same hormone ligand or different receptor ligands depending on the cell-specific coregulator availability and/or ligand-specific induced receptor conformations with differential coregulator affinities.

17.4.3
Evidence from Toxicogenomics Data

The complexity of biological effects produced by hormones through their receptors has become even more evident with the advent of genomic technologies (e.g., genomics). The ability to measure the induced mRNA expression changes from every gene in a cell or tissue simultaneously, the so-called gene expression fingerprint, now allows a direct comparison of fingerprints from several suspected endocrine active chemicals to see if they match the fingerprint of a well-known hormone such as estrogen. Examples where genomics has to be used for characterization of endocrine active chemicals in toxicology (toxicogenomics) exist for mammals and fish [96, 97]. By focusing on gene expression changes in the premature rat uterus and ovary treated with 17α-ethynylestradiol, bisphenol A and genistein, Daston and coworkers [96] identified approximately 66 genes that were consistently and significantly regulated in the same direction, but with different magnitudes, and many of these genes were well-characterized estrogen-regulated genes. In addition, many novel genes were discovered and there were genes not in common with each other chemical. While not as mature as mammalian toxicogenomics, gene expression profiling in fish is also being applied as an environmental monitoring strategy for endocrine active chemicals. The feasibility of this approach was provided from a study in juvenile rainbow trout where 49 genes were identified to be responsive to 17β-estradiol including well-characterized estrogen responsive genes associated with vitellogenesis (e.g., vitellogenesis; VTG1). In addition, tamoxifen induced VTG and antagonized many E2-induced genes consistent with tamoxifen's known properties as both an ER agonist and an antagonist [87]. These investigators concluded that a set of estrogen-responsive genes has been identified that can serve as a biomarker of environmental exposure to xenoestrogens [97]. Collectively, these rat and fish studies indicate that transcription profiling for endocrine active chemicals could identify the similarities and specificities for each individual chemical at the gene expression level and therefore have the propensity to provide useful initial screens for individual chemicals that may possess endocrine activity *in vivo*. Whether this approach proves to be predictive and practical for EAC mixtures that can operate through a multitude of different mechanisms as described above remains to be seen.

17.4.4
Implications for Risk Assessment of EAC Mixtures

In order to simplify risk assessment for complex mixtures, many regulatory agencies are developing relative potency and toxic equivalency methods. The most widely used method applies to dioxins and related halogenated aromatic hydrocarbons and is referred to as the toxic equivalency quotient (TEQ) approach [98]. This method requires that a relative potency for a given chemical called the toxic equivalency factor for dioxins is determined relative to a standard chemical (e.g., TCDD). The total TEQ for a mixture is a sum of the TEQ for each individual chemical, which is its TEF-adjusted value (e.g., concentration, gram weight). A similar approach has been suggested for ER ligands, an estrogen equivalency quotient (EEQ), based on the assumption that they act through the same hormone receptor-mediated pathways in exactly the same way and that their differences are only due to relative potencies. Some of the many assumptions for EEQ approach include

1) all EACs are full agonists, the only difference is their potency and there is no allowance for partial agonist or antagonists;
2) the hormone receptor "activated" in exactly the same way by all EACs, no allowance for differences in induced receptor conformation;
3) for all species in which the EEQ applies, the same mechanism of action for the hormone receptor occurs in all cell types and with the same kinetics;
4) the relative potency value for a given EAC predicts the same magnitude of endocrine effects in all cells and tissues in the relevant species.

Clearly, few, if any, of these assumptions are likely to be valid for environmental estrogens. There is now extensive evidence suggesting that individual EACs, such as estrogens, are very different even though they are receptor agonists, are actually selective receptor modulators [91]. Moreover, as described above, many differences in cell types and environments will markedly affect the activity of EACs, as will ligand-dependent differences in the conformation of the hormone receptor that determines the types and affinities with various coregulators. In addition, assignment of a relative potency value would be extremely problematic, as relative potency varies widely according to the assay in which it is assessed. In fact, this remains a critical problem undermining the TEQ approach for AH receptor ligands [99]. Overall, there appears to be overwhelming evidence contraindicating the utilization of an EEQ approach for assessing the risk of cumulative exposure to EACs.

17.5
Summary and Conclusions

The study of EAC mixtures, particularly mixtures of compounds with estrogenic or antiestrogenic activity, has received a great deal of attention in the past 10 years.

In contrast to initial (nonreproducible) findings of synergy among weak xenoestrogens that invoked a considerable degree of concern, the body of knowledge accumulated since then overwhelmingly indicates that ER ligands act in either an additive or less than additive fashion. While additivity has been clearly observed under certain conditions, often the assay conditions favor such a result. This is either because of the high dose levels evaluated, the use of single cell or tissue types restricting the range of potential interacting factors, or the selection of compounds that are much more homogeneous than is typical of human exposure.

While some have maintained that additivity among low-level exposures to multiple weak estrogens could pose a health threat [65], it is difficult to envision numerous human exposure scenarios where this would be the case. Under the assumption of additivity, a large number of ER agonists would need to be present in the mixture, all or most of them at levels very near their individual thresholds. The human exposure data reviewed in this chapter do not mesh with that type of scenario. Furthermore, the basic biology of ligand-specific steroid hormone signaling, as well as the SERM-like behavior of xenoestrogens and phytoestrogens, all make simple additivity an unlikely proposition. In aquatic species, a wide variety of reproductive and developmental problems have been attributed to the additive effects of EACs present in wastewater effluents, though several authors have fractionated wastewater effluents for endocrine activity and determined that the potent steroid estrogens (i.e., E1, E2, and EE2) may be present at sufficient concentrations to account for the observed effects in UK waters [18].

Given the almost infinite array of possible EAC mixture combinations and exposure levels, it is not practical to pursue each of them experimentally. Instead, it is imperative that toxicologists and risk assessment professionals identify the mixtures of greatest concern apart from those unlikely to pose a significant risk, so that finite resources can be applied in a manner that will be of greatest benefit to public health. That "sea of estrogens" alluded to at the outset of this chapter mainly comprises endogenous hormones and phytoestrogens that have been part of our environment since the beginning of time. As maintained by Safe in 1995 [47], exposure to manmade EACs contributes almost imperceptibly to the overall estrogenic exposure. Since 1996, concerns about synergy among EACs have largely faded, and the increasing knowledge of both receptor-mediated signaling and human exposure characteristics has tended to provide reassurances about the safety of typical xenoestrogen exposures. Increasing certainty about such potential risks is expected to come from further study of mechanisms of action for single EACs, characterization of their behavior when combined, and anchoring of cell and molecular responses to phenotypic changes.

Acknowledgments

The authors thank Dr. Reza Rasoulpour, Sarah Bell, and Tori Ordway for assistance in preparing the manuscript, and Paul Price, M.S., for critical review.

References

1 Walker, B.S. and Janney, J.C. (1930) Estrogenic substances. II. Analysis of plant sources. *Endocrinology*, **14**, 389–392.
2 Schueler, F.W. (1946) Sex hormonal action and chemical constitution. *Science*, **103**, 221–223.
3 Sluczewski, A. and Roth, P. (1948) Effects of androgenic and estrogenic compounds on the experimental metamorphoses of amphibians. *Gynecol. Obstet.*, **47**, 164–176.
4 Arnold, S.F., Klotz, D.M., Collins, B.M., Vonier, P.M., Guillette, L.J. Jr., and McLachlan, J.A. (1996) Synergistic activation of estrogen receptor with combinations of environmental chemicals. *Science*, **272**, 1489–1492.
5 Ramamoorthy, K., Wang, F., Chen, I.C., Norris, J.D., McDonnell, D.P., Leonard, L.S., Gaido, K.W., Bocchinfuso, W.P., Korach, K.S., and Safe, S. (1997) Estrogenic activity of a dieldrin/toxaphene mixture in the mouse uterus, MCF-7 human breast cancer cells, and yeast-based estrogen receptor assays: no apparent synergism. *Endocrinology*, **138**, 1520–1527.
6 Ashby, J., Lefevre, P.A., Odum, J., Harris, C.A., Routledge, E.J., and Sumpter, J.P. (1997) Synergy between synthetic oestrogens? *Nature*, **385**, 494.
7 McLachlan, J.A. (1997) Synergistic effect of environmental estrogens: report withdrawn. *Science*, **277**, 462–463.
8 Teuschler, L., Klaunig, J., Carney, E., Chambers, J., Conolly, R., Gennings, C., Giesy, J., Hertzberg, R., Klaassen, C., Kodell, R., Paustenbach, D., and Yang, R. (2002) Support of science-based decisions concerning the evaluation of the toxicology of mixtures: a new beginning. *Regul. Toxicol. Pharmacol.*, **36**, 34–39.
9 Blair, R.M., Fang, H., Branham, W.S., Hass, B.S., Dial, S.L., Moland, C.L., Tong, W., Shi, L., Perkins, R., and Sheehan, D.M. (2000) The estrogen receptor relative binding affinities of 188 natural and xenochemicals: structural diversity of ligands. *Toxicol. Sci.*, **54**, 138–153.
10 Boue, S.M., Wiese, T.E., Nehls, S., Burow, M.E., Elliott, S., Carter-Wientjes, C.H., Shih, B.Y., McLachlan, J.A., and Cleveland, T.E. (2003) Evaluation of the estrogenic effects of legume extracts containing phytoestrogens. *J. Agric. Food Chem.*, **51**, 2193–2199.
11 Klinge, C.M., Risinger, K.E., Watts, M.B., Beck, V., Eder, R., and Jungbauer, A. (2003) Estrogenic activity in white and red wine extracts. *J. Agric. Food Chem.*, **51**, 1850–1857.
12 Stumm-Zollinger, E. and Fair, G.M. (1965) Biodegradation of steroid hormones. *J. Water Pollut. Control Fed.*, **37**, 1506–1510.
13 Tabak, H.H. and Bunch, R.L. (1970) Steroid hormones as water pollutants. I. Metabolism of natural and synthetic ovulation-inhibiting hormones by microorganisms of activated sludge and primary settled sewage. *Dev. Ind. Microbiol.*, **11**, 367–376.
14 Garrison, A.W., Pope, J.D., and Allen, F.R. (1975) GC/MS analysis of organic compounds in domestic wastewaters, in *Chemical Congress of the North American Continent*, Ann Arbor Science.
15 Hignite, C. and Azarnoff, D.L. (1977) Drugs and drug metabolites as environmental contaminants: chlorophenoxyisobutyrate and salicyclic acid in sewage water effluent. *Life Sci.*, **20**, 337–341.
16 Purdom, C.E., Hardiman, P.A., Bye, V.J., Eno, N.C., Tyler, C.R., and Sumpter, J.P. (1994) Estrogenic effects of effluents from sewage treatment works. *Chem. Ecol.*, **8**, 275–285.
17 Desbrow, C., Routledge, E.J., Brighty, G.C., Sumpter, J.P., and Waldock, M. (1998) Identification of estrogenic chemicals in STW effluent. I. Chemical fractionation and *in vitro* biological screening. *Environ. Sci. Technol.*, **32**, 1549–1558.
18 Routledge, E.J., Sheahan, D., Desbrow, C., Brighty, G.C., Waldock, M., and Sumpter, J.P. (1998) Identification of estrogenic chemicals in STW effluent. 2.

In vivo responses in trout and roach. *Environ. Sci. Technol.*, **32**, 1559–1565.

19 Snyder, S.A., Keith, T.L., Verbrugge, D.A., Snyder, E.M., Gross, T.S., Kannan, K., and Giesy, J.P. (1999) Analytical methods for detection of selected estrogenic compounds in aqueous mixtures. *Environ. Sci. Technol.*, **33**, 2814–2820.

20 Snyder, S.A., Villeneuve, D.L., Snyder, E.M., and Giesy, J.P. (2001) Identification and quantification of estrogen receptor agonists in wastewater effluents. *Environ. Sci. Technol.*, **35**, 3620–3625.

21 Ternes, T.A., Stumpf, M., Mueller, J., Haberer, K., Wilken, R.D., and Servos, M. (1999) Behavior and occurrence of estrogens in municipal sewage treatment plants. I. Investigations in Germany, Canada and Brazil. *Sci. Total Environ.*, **225**, 81–90.

22 Daughton, C.G. and Ternes, T.A. (1999) Pharmaceuticals and personal care products in the environment: agents of subtle change? *Environ. Health Perspect.*, **107** (Suppl. 6), 907–938.

23 Johnson, A.C., Belfroid, A., and Di Corcia, A.D. (2000) Estimating steroid oestrogen inputs into activated sludge treatment works and observations on their removal from the effluent. *Sci. Total Environ.*, **256**, 163–173.

24 Komori, K., Tanaka, H., Okayasu, Y., Yasojima, M., and Sato, C. (2004) Analysis and occurrence of estrogen in wastewater in Japan. *Water Sci. Technol.*, **50**, 93–100.

25 Niven, S.J., Snape, J., Hetheridge, M., Evans, M., McEvoy, J., Sutton, P.G., and Rowland, S.J. (2001) Investigations of the origins of estrogenic A-ring aromatic steroids in UK sewage treatment works effluents. *Analyst*, **126**, 285–287.

26 Thorpe, K.L., Gross-Sorokin, M., Johnson, I., Brighty, G., and Tyler, C.R. (2006) An assessment of the model of concentration addition for predicting the estrogenic activity of chemical mixtures in wastewater treatment works effluents. *Environ. Health Perspect.*, **114** (Suppl. 1), 90–97.

27 Larsson, D.G.J., Adolfsson-Erici, M., Parkkonen, J., Pettersson, M., Berg, A.H., Olsson, P.E., and Forlin, L. (1999) Ethinyloestradiol: an undesired fish contraceptive? *Aquat. Toxicol.*, **45**, 91–97.

28 Rutishauser, B.V., Pesonen, M., Escher, B.I., Ackermann, G.E., Aerni, H.R., Suter, M.J., and Eggen, R.I. (2004) Comparative analysis of estrogenic activity in sewage treatment plant effluents involving three *in vitro* assays and chemical analysis of steroids. *Environ. Toxicol. Chem.*, **23**, 857–864.

29 Liney, K.E., Jobling, S., Shears, J.A., Simpson, P., and Tyler, C.R. (2005) Assessing the sensitivity of different life stages for sexual disruption in roach (*Rutilus rutilus*) exposed to effluents from wastewater treatment works. *Environ. Health Perspect.*, **113**, 1299–1307.

30 Bacaloni, A., Cavaliere, C., Faberi, A., Foglia, P., Samperi, R., and Lagana, A. (2005) Determination of isoflavones and coumestrol in river water and domestic wastewater sewage treatment plants. *Anal. Chim. Acta*, **531**, 229–237.

31 Kolpin, D.W., Furlong, E.T., Meyer, M.T., Thurman, E.M., Zaugg, S.D., Barber, L.B., and Buxton, H.T. (2000) Pharmaceuticals, hormones and other organic wastewater contaminants in U.S. streams, 1999-2000: a national reconnaissance. *Environ. Sci. Technol.*, **36**, 1202–1211.

32 Williams, R.J., Jurgens, M.D., and Johnson, A.C. (1999) Initial predictions of the concentrations and distribution of 17 beta-oestradiol, oestrone, and ethinyl oestradiol in three English waters. *Water Res.*, **33**, 1663–1671.

33 Jurgens, M.D., Holthaus, K.I., Johnson, A.C., Smith, J.L., Hetheridge, M., and Williams, R.J. (2002) The potential for estradiol and ethinylestradiol degradation in English rivers. *Environ. Toxicol. Chem.*, **21**, 480–488.

34 Williams, R.J., Johnson, A.C., Smith, J.J., and Kanda, R. (2003) Steroid estrogens profiles along river stretches arising from sewage treatment works discharges. *Environ. Sci. Technol.*, **37**, 1744–1750.

35 Oakes, K.D., Tremblay, L.A., and van der Kraak, G.J. (2005) Short-term lab exposures of immature rainbow trout (*Oncorhynchus mykiss*) to sulfite and kraft pulp-mill effluents: effects on oxidative

stress and circulating sex steroids. *Environ. Toxicol. Chem.*, **24**, 1451–1461.

36 Oakes, K.D., McMaster, M.E., and van der Kraak, G.J. (2004) Oxidative stress responses in longnose sucker (*Catostomus catostomus*) exposed to pulp and paper mill and municipal sewage effluents. *Aquat. Toxicol.*, **67**, 255–271.

37 Kovacs, T.G., Voss, R.H., Megraw, S.R., and Martel, P.H. (1997) Perspectives on Canadian field studies examining the potential of pulp and paper mill effluent to affect fish reproduction. *J. Toxicol. Environ. Health*, **51**, 305–352.

38 Howell, W.M. and Denton, T.E. (1989) Gonopodial morphogenesis in female mosquitofish, *Gambusia affinis affinis*, masculinized by exposure to degradation products from plant sterols. *Environ. Biol. Fish.*, **24**, 43–51.

39 Parks, L.G., Lambright, C.S., Orlando, E.F., Guillette, L.J., Jr., Ankley, G.T., and Gray, L.E., Jr. (2001) Masculinization of female mosquitofish in kraft mill effluent-contaminated Fenholloway River water is associated with androgen receptor agonist activity. *Toxicol. Sci.*, **63**, 257–267.

40 Shore, L.S., Gurevitz, M., and Shermesh, M. (1993) Estrogen as an environmental pollutant. *Bull. Environ. Contam. Toxicol.*, **51**, 361–366.

41 Nichols, D.J., Daniel, T.C., Moore, P.A., Edwards, D.R., and Pote, D.H. (1997) Runoff of estrogen hormone 17 (β-estradiol from poultry litter applied to pasture. *J. Environ. Qual.*, **26**, 1002–1006.

42 Irwin, L.K., Gray, S., and Oberdorster, E. (2001) Vitellogenin induction in painted turtle, *Chrysemys picta*, as a biomarker of exposure to environmental levels of estradiol. *Aquat. Toxicol.*, **55**, 49–60.

43 Rozman, K.K., Bhatia, J., Calafat, A.M., Chambers, C., Culty, M., Etzel, R.A., Flaws, J.A., Hansen, D.K., Hoyer, P.B., Jeffery, E.H., Kesner, J.S., Marty, S., Thomas, J.A., and Umbach, D. (2006) NTP-CERHR expert panel report on the reproductive and developmental toxicity of soy formula. *Birth Defects Res. B Dev. Reprod. Toxicol.*, **77**, 280–397.

44 CERHR (2007) Draft monograph on bisphenol A. Accessed November 26, 2007, in National Toxicology Program, Center for the Evaluation of Risks to Human Reproduction http://cerhr., niehs., nih., gov/chemicals/bisphenol/ bisphenol-eval., html.

45 ECB (2003) European Chemicals Bureau, European Union Risk Assessment Report: 4,4′-Isopropylidenediphenol (Bisphenol A) Risk Assessment.

46 Kavlock, R., Barr, D., Boekelheide, K., Breslin, W., Breysse, P., Chapin, R., Gaido, K., Hodgson, E., Marcus, M., Shea, K., and Williams, P. (2006) NTP-CERHR expert panel update on the reproductive and developmental toxicity of di(2-ethylhexyl)phthalate. *Reprod. Toxicol.*, **22**, 291–399.

47 Safe, S.H. (1995) Environmental and dietary estrogens and human health: is there a problem? *Environ. Health Perspect.*, **103**, 346–351.

48 Safe, S., Connor, K., and Gaido, K. (1998) Methods for xenoestrogen testing. *Toxicol. Lett.*, **102–103**, 665–670.

49 Gennings, C., Carter, W.H., Jr., Carney, E.W., Charles, G.D., Gollapudi, B.B., and Carchman, R.A. (2004) A novel flexible approach for evaluating fixed ratio mixtures of full and partial agonists. *Toxicol. Sci.*, **80**, 134–150.

50 Borgert, C.J., Price, B., Wells, C.S., and Simon, G.S. (2001) Evaluating chemical interaction studies for mixture risk assessment. *Hum. Ecol. Risk Assess.*, **7**, 259–306.

51 Ankley, G.T., Mihich, E., Stahl, R., Tillitt, D., Coborn, T., McMaster, S., Miller, R., Bantle, J., Campbell, P., Denslow, N., Dickerson, R., Folmar, L., Fry, M., Giesy, J.P., Gray, E., Guiney, P., Hutchinson, T.H., Kennedy, S., Kramer, V., LeBlanc, G., Mayes, M., Nimrod, A., Patino, R., Peterson, R., Purdy, R., Ringer, R., Thomas, P., Touart, L., van der Kraak, G.J., and Zacharewski, T. (1998) Overview of a workshop on screening methods for detecting potential (anti-) estrogenic/androgenic chemicals in wildlife. *Environ. Toxicol. Chem.*, **17**, 68–87.

52 Ankley, G.T. and Giesy, J.P. (1998) Endocrine disruptors in wildlife:

a weight-of-evidence perspective, in *Principles and Processes for Assessing Endocrine Disruption in Wildlife* (eds R. Kendall, R. Dickerson, W. Suk, and J.P. Giesy), SETAC Press, Pensacola, FL.

53 Edwards, T.M., Miller, H.D., and Guillette, L.J., Jr. (2006) Water quality influences reproduction in female mosquitofish (*Gambusia holbrooki*) from eight Florida springs. *Environ. Health Perspect.*, **114** (Suppl. 1), 69–75.

54 Liney, K.E., Hagger, J.A., Tyler, C.R., Depledge, M.H., Galloway, T.S., and Jobling, S. (2006) Health effects in fish of long-term exposure to effluents from wastewater treatment works. *Environ. Health Perspect.*, **114** (Suppl. 1), 81–89.

55 Matthiessen, P. (2003) Historical perspective on endocrine disruption in wildlife. *Pure Appl. Chem.*, **75**, 2249–2262.

56 Sumpter, J.P. and Johnson, A.C. (2005) Lessons from endocrine disruption and their application to other issues concerning trace organics in the aquatic environment. *Environ. Sci. Technol.*, **39**, 4321–4332.

57 Soto, A.M., Chung, K.L., and Sonnenschein, C. (1994) The pesticides endosulfan, toxaphene, and dieldrin have estrogenic effects on human estrogen-sensitive cells. *Environ. Health Perspect.*, **102**, 380–383.

58 Arnold, S.F., Vonier, P.M., Collins, B.M., Klotz, D.M., Guillette, L.J., Jr., and McLachlan, J.A. (1997) *In vitro* synergistic interaction of alligator and human estrogen receptors with combinations of environmental chemicals. *Environ. Health Perspect.*, **105** (Suppl. 3), 615–618.

59 Sumpter, J.P. and Jobling, S. (1995) Vitellogenesis as a biomarker for estrogenic contamination of the aquatic environment. *Environ. Health Perspect.*, **103** (Suppl. 7), 173–178.

60 Vonier, P.M., Crain, D.A., McLachlan, J.A., Guillette, L.J., Jr., and Arnold, S.F. (1996) Interaction of environmental chemicals with the estrogen and progesterone receptors from the oviduct of the American alligator. *Environ. Health Perspect.*, **104**, 1318–1322.

61 Yang, R.S. (1998) Some critical issues and concerns related to research advances on toxicology of chemical mixtures. *Environ. Health Perspect.*, **106** (Suppl. 4), 1059–1063.

62 Kortenkamp, A. and Altenburger, R. (1998) Synergisms with mixtures of xenoestrogens: a reevaluation using the method of isoboles. *Sci. Total Environ.*, **221**, 59–73.

63 Kortenkamp, A. and Altenburger, R. (1999) Approaches to assessing combination effects of oestrogenic environmental pollutants. *Sci. Total Environ.*, **233**, 131–140.

64 Payne, J., Rajapakse, N., Wilkins, M., and Kortenkamp, A. (2000) Prediction and assessment of the effects of mixtures of four xenoestrogens. *Environ. Health Perspect.*, **108**, 983–987.

65 Silva, E., Rajapakse, N., and Kortenkamp, A. (2002) Something from "nothing": eight weak estrogenic chemicals combined at concentrations below NOECs produce significant mixture effects. *Environ. Sci. Technol.*, **36**, 1751–1756.

66 Brian, J.V., Harris, C.A., Scholze, M., Backhaus, T., Booy, P., Lamoree, M., Pojana, G., Jonkers, N., Runnalls, T., Bonfa, A., Marcomini, A., and Sumpter, J.P. (2005) Accurate prediction of the response of freshwater fish to a mixture of estrogenic chemicals. *Environ. Health Perspect.*, **113**, 721–728.

67 Xie, L., Thrippleton, K., Irwin, M.A., Siemering, G.S., Mekebri, A., Crane, D., Berry, K., and Schlenk, D. (2005) Evaluation of estrogenic activities of aquatic herbicides and surfactants using an rainbow trout vitellogenin assay. *Toxicol. Sci.*, **87**, 391–398.

68 Bon, E., Barbe, U., Nunez Rodriguez, J., Cuisset, B., Pelissero, C., Sumpter, J.P., and Le Menn, F. (1997) Plasma vitellogenin levels during the annual reproductive cycle of the female rainbow trout (*Oncorhynchus mykiss*): establishment and validation of an ELISA. *Comp. Biochem. Physiol. B Biochem. Mol. Biol.*, **117**, 75–84.

69 Tyler, C.R., van Aerle, R., Hutchinson, T.H., Maddix, S., and Trip, H. (1999)

An *in vivo* testing system for endocrine disruptors in fish early life stages using induction of vitellogenin. *Environ. Toxicol. Chem.*, **18**, 337–347.

70 Tyler, C.R., van Aerle, R., Nilsen, M.V., Blackwell, R., Maddix, S., Nilsen, B.M., Berg, K., Hutchinson, T.H., and Goksoyr, A. (2002) Monoclonal antibody enzyme-linked immunosorbent assay to quantify vitellogenin for studies on environmental estrogens in the rainbow trout (*Oncorhynchus mykiss*). *Environ. Toxicol. Chem.*, **21**, 47–54.

71 Nichols, K.M., Snyder, E.M., Snyder, S.A., Pierens, S.L., Miles-Richardson, S.R., and Giesy, J.P. (2001) Effects of nonylphenol ethoxylate exposure on reproductive output and bioindicators of environmental estrogen exposure in fathead minnows *Pimephales promelas*. *Environ. Toxicol. Chem.*, **20**, 510–522.

72 Zhang, F., Bartels, M.J., Brodeur, J.C., and Woodburn, K.B. (2004) Quantitative measurement of fathead minnow vitellogenin by liquid chromatography combined with tandem mass spectrometry using a signature peptide of vitellogenin. *Environ. Toxicol. Chem.*, **23**, 1408–1415.

73 Aronica, S.M. and Katzenellenbogen, B.S. (1993) Stimulation of estrogen receptor-mediated transcription and alteration in the phosphorylation state of the rat uterine estrogen receptor by estrogen, cyclic adenosine monophosphate, and insulin-like growth factor-I. *Mol. Endocrinol.*, **7**, 743–752.

74 Dupont, J., Karas, M., and LeRoith, D. (2000) The potentiation of estrogen on insulin-like growth factor I action in MCF-7 human breast cancer cells includes cell cycle components. *J. Biol. Chem.*, **275**, 35893–35901.

75 Charles, G.D., Gennings, C., Zacharewski, T.R., Gollapudi, B.B., and Carney, E.W. (2002) An approach for assessing estrogen receptor-mediated interactions in mixtures of three chemicals: a pilot study. *Toxicol. Sci.*, **68**, 349–360.

76 Charles, G.D., Gennings, C., Zacharewski, T.R., Gollapudi, B.B., and Carney, E.W. (2002) Assessment of interactions of diverse ternary mixtures in an estrogen receptor-alpha reporter assay. *Toxicol. Appl. Pharmacol.*, **180**, 11–21.

77 Payne, J., Scholze, M., and Kortenkamp, A. (2001) Mixtures of four organochlorines enhance human breast cancer cell proliferation. *Environ. Health Perspect.*, **109**, 391–397.

78 Rajapakse, N., Silva, E., and Kortenkamp, A. (2002) Combining xenoestrogens at levels below individual no-observed-effect concentrations dramatically enhances steroid hormone action. *Environ. Health Perspect.*, **110**, 917–921.

79 Wade, M.G., Lee, A., McMahon, A., Cooke, G., and Curran, I. (2003) The influence of dietary isoflavone on the uterotrophic response in juvenile rats. *Food Chem. Toxicol.*, **41**, 1517–1525.

80 Charles, G.D., Gennings, C., Tornesi, B., Kan, H.L., Zacharewski, T.R., Bhaskar Gollapudi, B., and Carney, EW. (2007) Analysis of the interaction of phytoestrogens and synthetic chemicals: an *in vitro/in vivo* comparison. *Toxicol. Appl. Pharmacol.*, **218**, 280–288.

81 Naciff, J.M., Overmann, G.J., Torontali, S.M., Carr, G.J., Tiesman, J.P., and Daston, G.P. (2004) Impact of the phytoestrogen content of laboratory animal feed on the gene expression profile of the reproductive system in the immature female rat. *Environ. Health Perspect.*, **112**, 1519–1526.

82 Tinwell, H. and Ashby, J. (2004) Sensitivity of the immature rat uterotrophic assay to mixtures of estrogens. *Environ. Health Perspect.*, **112**, 575–582.

83 Edgren, R.A. and Calhoun, D.W. (1960) Oestrogen antagonisms: the effects of various steroids on the uterine growth produced in mice by oestriol. *Experientia*, **16**, 188–189.

84 Wade, M.G., Foster, W.G., Younglai, E.V., McMahon, A., Leingartner, K., Yagminas, A., Blakey, D., Fournier, M., Desaulniers, D., and Hughes, C.L. (2002) Effects of subchronic exposure to a complex mixture of persistent contaminants in male rats: systemic, immune, and

reproductive effects. *Toxicol. Sci.*, **67**, 131–143.

85 EPA US (2002) Guidance on Cumulative Risk Assessment of Pesticide Chemicals That Have a Common Mechanism of Toxicity. States Environmental Protection Agency, Office of Pesticide Programs 2002: http://www.epa.gov/pesticides/trac/science/cumulative_guidance.pdf.

86 Borgert, C.J., Quill, T.F., McCarty, L.S., and Mason, A.M. (2004) Can mode of action predict mixture toxicity for risk assessment? *Toxicol. Appl. Pharmacol.*, **201**, 85–96.

87 Jordan, V.C. (2003) Antiestrogens and selective estrogen receptor modulators as multifunctional medicines. 1. Receptor interactions. *J. Med. Chem.*, **46**, 883–908.

88 Fong, C.J., Burgoon, L.D., Williams, K.J., Forgacs, A.L., and Zacharewski, T.R. (2007) Comparative temporal and dose-dependent morphological and transcriptional uterine effects elicited by tamoxifen and ethynylestradiol in immature, ovariectomized mice. *BMC Genom.*, **8**, 151.

89 Bowers, J.L., Tyulmenkov, V.V., Jernigan, S.C., and Klinge, C.M. (2000) Resveratrol acts as a mixed agonist/antagonist for estrogen receptors alpha and beta. *Endocrinology*, **141**, 3657–3667.

90 Zhu, B.T. (2005) Mechanistic explanation for the unique pharmacologic properties of receptor partial agonists. *Biomed. Pharmacother.*, **59**, 76–89.

91 Gronemeyer, H., Gustafsson, J.A., and Laudet, V. (2004) Principles for modulation of the nuclear receptor superfamily. *Nat. Rev. Drug Discov.*, **3**, 950–964.

92 Lonard, D.M., Lanz, R.B., and O'Malley, B.W. (2007) Nuclear receptor coregulators and human disease. *Endocr. Rev.*, **28**, 575–587.

93 McKenna, N.M. and O'Malley, B.W. (2003) Nuclear receptor signaling: concepts and models, In *NURSA: The Nuclear Receptor Signaling Atlas*. www.nursa.org/flash/gene/nuclearreceptor/start.html.

94 Roeder, R.G. (2005) Transcriptional regulation and the role of diverse coactivators in animal cells. *FEBS Lett.*, **579**, 909–915.

95 Smith, C.L. and O'Malley, B.W. (2004) Coregulator function: a key to understanding tissue specificity of selective receptor modulators. *Endocr. Rev.*, **25**, 45–71.

96 Naciff, J.M. and Daston, G.P. (2004) Toxicogenomic approach to endocrine disrupters: identification of a transcript profile characteristic of chemicals with estrogenic activity. *Toxicol. Pathol.*, **32** (Suppl. 2), 59–70.

97 Benninghoff, A.D. and Williams, D.E. (2008) Identification of a transcriptional fingerprint of estrogen exposure in rainbow trout liver. *Toxicol. Sci.*, **101**, 65–80.

98 Van den Berg, M., Birnbaum, L.S., Denison, M., De Vito, M., Farland, W., Feeley, M., Fiedler, H., Hakansson, H., Hanberg, A., Haws, L., Rose, M., Safe, S., Schrenk, D., Tohyama, C., Tritscher, A., Tuomisto, J., Tysklind, M., Walker, N., and Peterson, R.E. (2006) The 2005 World Health Organization reevaluation of human and mammalian toxic equivalency factors for dioxins and dioxin-like compounds. *Toxicol. Sci.*, **93**, 223–241.

99 Poland, A. (1997) Reflections on risk assessment of receptor-acting xenobiotics. *Regul. Toxicol. Pharmacol.*, **26**, 41–43.

100 Winter, C.K. (1992) Dietary pesticide risk assessment. *Rev. Environ. Contam. Toxicol.*, **127**, 23–67.

101 Cressey, P.J. and Vannoort, R.W. (2003) Pesticide content of infant formulae and weaning foods available in New Zealand. *Food Addit. Contam.*, **20**, 57–64.

102 Whitten, P.L. and Patisaul, H.B. (2001) Cross-species and interassay comparisons of phytoestrogen action. *Environ. Health Perspect.*, **109** (Suppl. 1), 5–20.

103 Thomson, B.M. and Grounds, P.R. (2005) Bisphenol A in canned foods in New Zealand: an exposure assessment. *Food. Addit. Contam.*, **22**, 65–72.

18
Evaluation of Interactions in Chemical Mixtures

Hana R. Pohl, Moiz Mumtaz, Mike Fay, and Christopher T. De Rosa

18.1
Introduction

A large number of chemicals are found in our environment, and humans are likely to be exposed to many of them during their lifetime. Such exposures may constitute a human health hazard, depending on the duration, route, and level of exposure. To protect human health and the environment, chemical releases are monitored and permissible levels are defined. Chemicals of concern are identified through priority setting processes by various branches of government, affected communities, and the private sector. For some of these chemicals, the existing toxicological database pertinent to human and environmental impact, including epidemiological, occupational, animal, and *in vitro* data, is thoroughly reviewed and an acceptable exposure level is derived and documented for each individual chemical of concern. However, sufficient information is available only for a small subset of chemicals in use in industry throughout the world.

For decades, the development of health criteria and chemical regulation has focused on single chemicals. However, human exposures most often involve combinations of chemicals. The National Toxicology Program's 11th report on carcinogens lists 11 mixtures suspected of causing cancer, such as environmental tobacco smoke, coal tar, diesel exhaust particulates, mineral oils, polychlorinated biphenyls (PCBs), polybrominated biphenyls (PBBs), and polycyclic aromatic hydrocarbons (PAHs) [1]. In some instances, exposures entail complex mixtures for which the exact composition of the mixture is poorly characterized (e.g., coke oven emissions, diesel exhaust, asphalt fumes, or welding fumes). Conversely, exposures of concern at hazardous waste sites (HWS) typically entail only a handful of chemicals, although multiple chemical exposures are still the rule. It is clear that human exposure to chemicals in the environment often involves multiple chemicals, from a wide spectrum of potential sources including uncontrolled hazardous waste sites, permitted industrial releases, accidental spills, and consumer products ranging from food to cosmetics.

Principles and Practice of Mixtures Toxicology. Edited by Moiz Mumtaz

ISBN: 978-3-527-31992-3

In this chapter, we present some of the methods used for risk assessment of chemical mixtures and their application in evaluation of exposures to [1] background levels of chemical mixtures through media including air, water, and food, and [2] exposure to mixtures of systemic toxicants and radionuclides at hazardous waste sites. (See also chapter 11 for practical applications of these methods.)

18.2
Methodology for Identification of Priority Mixtures

Exposures to multiple chemicals in the environment may occur in acute, episodic, sequential, and/or chronic scenarios. In populations living in the vicinity of a hazardous waste site, an industrial chemical facility, or an incinerator, potential long-term exposures may occur through a range of environmental media (air, water, and soil). Some HWSs are the source of significant exposures and hence have been placed on the US Environmental Protection Agency's National Priorities List (NPL) [2]. Such sites typically exhibit completed exposure pathways (CEPs) through which the public was exposed [3]. A CEP is defined as a pathway for which there is either clear evidence or a reasonable likelihood that people are, or have been, in contact with site-related contaminants. More specifically, four elements must be identified to categorize a pathway as a CEP: a source of contamination, environmental fate and transport, exposure point, and exposed population all have to be identified [3]. If a CEP is identified, a plan to mitigate exposures that could compromise human or environmental health is implemented. Of the 1706 HWSs ATSDR has assessed, 743 have been found to have one or more CEPs [4]. Of these, 588 sites (79%) have at least two chemicals in a CEP and 475 sites (64%) have at least three. This data underscores the public health hazards posed by exposure to mixtures at HWSs.

The potential impact of exposure to a mixture of chemicals evaluated by analogous methods irrespective of whether the exposure arises from the environment, a work setting, an emergency spill or release, contaminated food, or other scenarios is described below.

18.3
Methodology for the Joint Toxicity Assessment of Mixtures

Several approaches are available for the risk assessment of chemical mixtures; these have been summarized in previous reviews [5–9]. Toxicological data are often not available for exposure to the whole mixture. In the absence of such data, risk assessments are performed using a potency-weighted dose addition based on the toxicity of the individual components of the mixture. This approach, called hazard index (HI) or component-based approach, attempts to estimate the toxicity of a mixture based on the information available on individual components of the mixture and relies heavily on some form of additivity or potency-weighted dose (or response) additivity. The HI is the sum of the individual hazard quotients (HQ) for chemicals in the mixture. The HQ is the ratio between the exposure level (numerator) and an acceptable exposure level (denominator) for each component. The goal of the HI approach is to construct a toxicity index that is most plausible in lieu of the one based on testing of the mixture itself. Two major sources of uncertainty need to be carefully considered when using this approach.

In general, the HI method usually focuses on the critical effects of chemicals in the mixture; acceptable exposure levels for other major effects are often not covered. To address this, the target-organ toxicity dose (TTD) method is used. It allows fine-tuning of the HI approach and estimation of separate HIs for all major end points or target organs. TTDs are derived the same way as other health-based guidance values such as minimal risk levels (MRLs) [7, 10]. A combined WOE score is computed for each target organ/effect of concern. For example,

$$\text{(a) } \mathrm{HI}_{\mathrm{hepatic}} = \frac{\mathrm{E}_1}{\mathrm{MRL}_1} + \frac{\mathrm{E}_2}{\mathrm{MRL}_2} + \frac{\mathrm{E}_3}{\mathrm{TTD}_{4\ \mathrm{hepatic}}},$$

$$\text{(b) } \mathrm{HI}_{\mathrm{renal}} = \frac{\mathrm{E}_1}{\mathrm{TTD}_{1\ \mathrm{renal}}} + \frac{\mathrm{E}_3}{\mathrm{MRL}_3} + \frac{\mathrm{E}_4}{\mathrm{TTD}_{4\ \mathrm{renal}}},$$

$$\text{(c) } \mathrm{HI}_{\mathrm{neuro}} = \frac{\mathrm{E}_1}{\mathrm{TTD}_{1\ \mathrm{neuro}}} + \frac{\mathrm{E}_2}{\mathrm{TTD}_{2\ \mathrm{neuro}}} + \frac{E_3}{\mathrm{TTD}_{3\ \mathrm{neuro}}},$$

$$\text{(d) } \mathrm{HI}_{\mathrm{dev}} = \frac{\mathrm{E}_1}{\mathrm{TTD}_{1\ \mathrm{dev}}} + \frac{\mathrm{E}_2}{\mathrm{TTD}_{2\ \mathrm{dev}}} + \frac{\mathrm{E}_3}{\mathrm{TTD}_{3\ \mathrm{dev}}} + \frac{\mathrm{E}_4}{\mathrm{MRL}_4},$$

where E is the exposure level, HI denotes hazard index, MRL is the minimal risk level, and TTD is the target-organ toxicity dose. (See also chapter 10.)

If the HIs for any end points of potential concern are found to be above unity [1] as a result of the TTD analysis, those end points are then examined more closely.

The other source of uncertainty with the HI approach is the potential of any interaction among the chemical-specific components of the mixture. While most toxicologists believe there is ample evidence – both empirical and mechanistic – that chemicals interact, there is no clear agreement, and hence limited guidance, on using such information for risk assessment. Most of the information on compound interactions cannot be used to quantify interactions. The weight-of-evidence (WOE) scheme proposed by Mumtaz and Durkin [11] provides a framework for systematically assessing the qualitative likelihood of interactions (i.e., whether the mixture is likely to be more or less toxic than anticipated, versus the default assumption of additivity). The approach also suggests ways to consider the magnitude of the interaction and to quantitatively adjust risk assessments using dose-response or dose-severity data (Table 18.1). Nonetheless, due to a considerable degree of uncertainty, in most cases, it is rarely used for *quantitative* risk assessment adjustments in the field. Therefore, a WOE evaluation is a qualitative judgment. The methodology characterizes the plausibility of joint toxicity of pairs of toxicants. It produces an alphanumeric result that takes into consideration important factors such as the

Table 18.1 BINWOE classification.

Direction of interaction
= Additivity
> More than additivity
< Less than additivity
Mechanistic understanding
I. Direct and unambiguous mechanistic data: The mechanism(s) by which the interactions could occur has been well characterized and leads to an unambiguous interpretation of the direction of the interaction
II. Mechanistic data on related compounds: The mechanism(s) by which the interactions could occur has not been well characterized for the chemicals of concern, but structure–activity relationships, either quantitative or informal, can be used to infer the likely mechanisms(s) and the direction of the interaction
III. Inadequate or ambiguous mechanistic data: The mechanism(s) by which the interactions could occur has not been well characterized or information on the mechanism(s) does not clearly indicate the direction that the interaction will have
Toxicological significance
A. The toxicological significance of the interaction has been directly demonstrated
B. The toxicological significance of the interaction can be inferred or has been demonstrated for related chemicals
C. The toxicological significance of the interaction is unclear

quality of the data, its mechanistic understanding, its toxicological significance, and the route and duration of exposure. All of these elements can play a critical role in the expression of the integrated joint toxicity of a mixture. For an example of a WOE evaluation of eight chemicals found in combination at a hazardous waste site, see Ref. [12].

18.4 Evaluations of Mixtures Related to Background Exposures

The ambient environment is characterized by low-level background concentrations for many chemicals. However, as a result of human activities (industry, urban development, mining, and natural disasters), large areas of surface land, air, or water can come to have elevated levels of chemicals. In this way, natural and manmade chemicals, singly or in aggregate, may have ubiquitous background levels that can approach concentrations of concern, sometimes across multiple media.

18.4.1 Air Pathway

Air pollutants can pose a mixture of potentially harmful chemicals for exposed populations. Environmental exposures vary locally depending on emission sources in local geographical areas and can also be impacted by global events (e.g., acid rain or the Chernobyl disaster) that may serve to amplify exposures that already occur in the ambient environment.

In addition to outdoor air pollution, indoor-air quality is also of concern. Levels of some pollutants in household air are often higher than in outdoor air. The pollutants include chemicals released from carpet, pressurized wood, foam, paint, and other solvents. Pollutants are also generated by everyday activities such cooking, heating, and cleaning. The US Environmental Protection Agency reported that carbon monoxide levels ranged from 5 to 15 ppm in homes with properly adjusted gas stoves and could be up to 30 ppm in homes with poorly adjusted stoves [13]. Similarly, indoor levels of nitrogen dioxide often exceed outdoor levels, with increased levels associated with gas stoves, kerosene use, and gas heaters. In addition, volatile organic compounds (VOCs) are typically two to three times higher inside homes than outside them. Activities such as bringing clothes home from a dry cleaner, painting, or paint stripping, can further substantially increase VOC concentrations. Based on these indoor-air findings of EPA, ATSDR evaluated a mixture of chemicals that included carbon monoxide, formaldehyde, methylene chloride, nitrogen dioxide, and tetrachloroethylene [14].

Before evaluating the relevance of joint toxic action data for these chemicals, an assessment of end points of concern for inhalation exposure to the mixture is conducted. While no one end point is a common toxicity target for all five components, there are common end points of concern for many of them as follows: *Cardiovascular:* carbon monoxide and tetrachloroethylene; *hematological:* carbon monoxide, methylene chloride, and tetrachloroethylene; *hepatic:* methylene chloride and tetrachloroethylene; *neurological:* carbon monoxide, methylene chloride, and tetrachloroethylene; *respiratory:* formaldehyde, methylene chloride, and nitrogen dioxide; *cancer:* formaldehyde and methylene chloride.

Due to the fact that there are no toxicity data for the entire five-component mixture, its toxicity was evaluated following the HI approach augmented with WOE analysis (Table 18.2). The scientific evidence for greater than additive or less than additive interactions among the components of the mixture is limited, with a majority of existing data suggesting additive interactions.

For example, carbon monoxide and formaldehyde are created in the body during the metabolism of methylene chloride. The mechanism by which methylene chloride induces effects on the hematological and nervous systems at least partially involves the metabolism of methylene chloride to carbon monoxide by cytochrome P4502E1. A study comparing the neurological effects of methylene chloride with those of equivalent concentrations of carbon monoxide relative to blood carboxyhemoglobin (COHb) levels found a more pronounced performance deficit for methylene chloride [15]. Thus, the neurological effects of methylene chloride are only partially mediated by carbon monoxide formation, suggesting an additive response for coexposures of these two chemicals (=IIB). Similarly, the carboxyhemoglobin-related effects are expected to be additive, based on total blood COHb levels formed by the two compounds. This has been verified for acute exposures by Kurppa *et al.* [16], who found additive COHb levels following joint exposures to methylene chloride and carbon monoxide in rats.

Table 18.2 Matrix of BINWOE determinations for a mixture of indoor-air chemicals.

		On toxicity of				
		Carbon monoxide	**Formaldehyde**	**Chloroform**	**Nitrogen dioxide**	**Trichloroethylene**
Effect of	Carbon monoxide		?	= IA hepatic neuro	?	?
	Formaldehyde	?		= IB cancer; = IIC renal	= IIIC renal	?
	Chloroform	= IA hepatic; = IIB neuro	= IB renal, cancer		?	?
	Nitrogen dioxide	?	= IIIC renal	?		?
	Trichloroethylene	?	?	?	?	

Neuro: neurological. See Table 18.1 for definition of classifications and modifiers.

The hematological effects of methylene chloride generally involve the formation of COHb. COHb-related effects are therefore expected to be additive, based on total blood COHb levels formed by these two chemicals. This prediction is supported by Kurppa *et al.* [16], who found additive COHb levels following joint acute exposures to methylene chloride and carbon monoxide in rats (=IA).

Data are inadequate to characterize the possible joint action on other pertinent health end points, as denoted with question marks in Table 18.2. In summary, ATSDR concluded that additivity should be assumed as a protective public health measure in exposure-based assessments of health hazards from mixtures of these indoor-air contaminants. The HI in conjunction with expert judgment characterizes the biological plausibility of joint toxic action of the whole mixture.

18.4.2
Water Pathway

As part of the National Water-Quality Assessment Program of the US Geological Survey, untreated groundwater samples were collected from 1255 domestic (rural) wells and 242 public water-supply wells, and analyzed for 60 VOCs, 83 pesticides, and nitrates [17]. VOCs were found in 44%, pesticides in 38%, and nitrates in 28% of the samples. The most frequently occurring four-chemical mixture in these groundwater samples consisted of two triazine herbicides (atrazine and simazine), the metabolite deethylatrazine, and nitrate. Also, diazinon was the most frequently detected organophosphorus insecticide. On the basis of this study, ATSDR performed a risk assessment on the four-component mixture of atrazine, simazine, diazinon, and nitrate [18]. Since toxicity data were not available on the whole mixture, a component-based approach was implemented. The results are presented in Table 18.3.

Studies of the joint action of these triazine herbicides (atrazine, deethylatrazine, and simazine) indicate that they have a common mechanism of toxicity with regard to

Table 18.3 Matrix of BINWOE determinations for a mixture of chemicals found in drinking water.

		On toxicity of			
		Atrazine	**Simazine**	**Diazinon**	**Nitrate**
Effect of	Atrazine		=IA repro	>IIB neuro	>IIB geno
	Simazine	=IA repro		>IIB neuro	>IIB geno
	Diazinon	?	?		?
	Nitrate	>IIB geno	>IIB geno	?	

Geno: genotoxic; neuro: neurological; repro: reproductive. See Table 18.1 for definition of classifications and modifiers.

attenuation of the luteinizing hormone surge in female and male rats, alteration of the estrous cycle, delayed pubertal development in both sexes of rats, and altered pregnancy maintenance [19]. This mechanism involves neuroendocrine disruption of hypothalamic–pituitary–gonadal function, and it is predicted to be dose additive (=IA). Importantly, this disruption is expected to be relevant to humans based on our knowledge of the analogue systems in both species.

In contrast, based on the potentiation of diazinon neurotoxicity by atrazine in the midge [20], potentiation of diazinon lethality and acetylcholinesterase inhibition in amphipods [21], metabolic activation of a similar organophosphorus insecticide (chlorpyrifos) by atrazine [20], and similar mechanisms of neurotoxicity in invertebrates and humans, the interaction is expected to be greater than additive for neurotoxicity (>IIB). The WOE for this diazinon–atrazine pair is a good example of the inferred mechanistic information (see the WOE classification in Table 18.1). Furthermore, a similar mechanism (induction of metabolic activation) can also be inferred for atrazine's potentiation of diazinon, based on the similarity in structure and mechanism of action of diazinon and chlorpyrifos.

Interaction of atrazine (or simazine) with nitrate is a purely chemical reaction. Atrazine and nitrite (the metabolite of nitrate) react at acidic pH to form *N*-nitrosoatrazine [22–25]. The formation of *N*-nitrosoatrazine from atrazine and nitrite has been seen in human gastric juice [26] and in mice [23]. In laboratory testing, the clastogenicity of *N*-nitrosoatrazine in cultured human lymphocytes is much greater than that of atrazine or nitrate, and *N*-nitrosoatrazine is mitogenic, although atrazine is not [27]. *N*-Nitrosoatrazine causes chromosomal aberrations in a Chinese hamster fibroblast-derived cell line at concentrations much lower than those of atrazine; atrazine alone gives negative results in the same study [28, 29]. Therefore, the WOE analysis predicts greater than additive genotoxic action for atrazine on nitrate (>IIB).

This scenario with well-water pollutants illustrates the interactive process beginning with the HI, augmenting with WOE, and reaching the conclusion that a greater than additive interaction is possible. In this situation, the health assessor should carefully evaluate information on potential health effects, the extent of combined exposure, and available public health outcome data.

18.4.3 Food Pathway

Another major exposure route for humans is via food, often fish. For example, in 1985, the US and Canadian International Joint Commission (IJC) identified 11 of 362 contaminants as "critical Great Lakes pollutants" due to their persistence and pervasiveness: polychlorinated biphenyl, dichlorodiphenyl trichloroethane (DDT), dieldrin, toxaphene, mirex, methylmercury, benzo[*a*]pyrene, hexachlorobenzene (HCB), chlorinated dibenzo-*p*-dioxins (CDDs), chlorinated dibenzo-*p*-furans (CDFs), and alkylated lead [30]. Eight of these bioaccumulate in organisms, biomagnify in the food chain, and persist at elevated levels in some areas of the Great Lakes. This was evidenced by, for example, a group of Great Lakes fish consumers who averaged 49 meals of Great Lakes sportfish per year for an average of 33 years and were found to have higher mean blood levels of CDDs, CDFs, PCBs, *p,p′*-DDE, and mercury than reference groups [31].

Epidemiological studies suggest an association between consumption of these contaminated fish and human neurodevelopmental and neurological effects. For example, a prospective study of children whose mothers consumed Lake Michigan fish before and during pregnancy found small, but statistically significant, changes in neurological end points (including cognitive functions) at several stages of development compared to children of mothers who did not consume such fish [32–37]. The association between fish consumption from two lakes of the Upper St. Lawrence River and performance in a neurofunctional test battery in a group of 63 matched pairs of fisheating and nonfisheating adult residents of Southwest Quebec was examined in another study [38]. The fish eaters had small performance deficits in neurological tests involving "higher levels of information processing" such as auditory recall and word naming.

ATSDR evaluated possible joint toxic action of CDDs, hexachlorobenzene, *p,p′*-DDE, methylmercury, and PCBs found in fish (Table 18.4; [39]). A common target for these chemicals is the neurological system, so there is certainly the potential for interaction. No information was available for the whole mixture, so a component-based approach was used. Only a limited amount of evidence is available for greater than additive or less than additive interactions between pairs: (1) hexachlorobenzene's potentiation of tetrachlorodibenzo-*p*-dioxin (TCDD) reducing body and thymus weights (>IIIA greater than additive interaction) [40]; (2) PCB's antagonism of TCDD immunotoxicity (<IIIB less than additive interaction) [41, 42]; (3) PCB's antagonism of TCDD developmental toxicity (<IIIC less than additive interaction) [43]; and (4) synergism between PCB and methylmercury in altering brain levels of dopamine, including during development (>IIC greater than additive interaction) [44].

WOE evaluations reveal that the scientific evidence for these interactions is limited and is inadequate to fully characterize the possible modes of joint action for the whole mixture. For the remaining pairs, additive joint action at shared targets of toxicity is either supported by data or is recommended as an assumption to protect public health. This is not only due to a lack of interaction data but also due to conflicting

Table 18.4 Matrix of BINWOE determinations for a mixture of chemicals found in fish.

		On toxicity of				
		TCDD	**Hexachlorobenzene**	**DDE**	**Methylmercury**	**PCBs**
Effect of	TCDD		?	= IIIC repro	= IIIB immuno	= IIIC weight
	Hexachlorobenzene	>IIIA weight		?	?	?
	DDE	= IIIC repro	?		?	?
	Methylmercury	?	= IIIC liver	?		>IIC neuro
	PCBs	<IIIB immuno	?	?	>IIC neuro	

Immuno: immunological: neuro: neurological; repro: reproductive.See Table 18.1 for definition of classifications and modifiers.

interaction data or the lack of a mechanistic understanding to reliably presume nonadditive interactions. In conclusion, the HI approach with qualitative adjustments can be used to assess the whole mixture of persistent chemicals in fish.

18.5 Evaluation of Mixtures Related To Hazardous Waste Sites

ATSDR is mandated to evaluate the potential health effects of chemicals found at hazardous waste sites, as well as episodic and emergency releases to the environment. Often, more than one contaminant is found in the environmental media in and near such sites. ATSDR's congressional mandate specifically directs it to address potential human health effects due to exposure to mixtures of chemicals.

18.5.1 Air and Water Pathways

As mentioned previously, a USGS study reported that VOCs were found in 44% of well-water samples [17]. ATSDR found that VOCs are the most common type of contaminants in both air and water completed exposure pathways at hazardous waste sites; indeed, 8 of the 10 most frequent chemicals for both of these media were VOCs [4]. Tables 18.5 and 18.6 show the top 10 mixtures in air and water CEPs, respectively.

A quarter (182) of the 743 HWSs with CEPs had one or more chemicals in air CEPs; these were primarily VOCs (Table 18.5). The top-most binary pairs found in air CEPs consisted of combinations such as toluene and benzene (at 28 sites; 3.8% of 743), trichloroethylene (TCE) and tetrachloroethylene at 19 sites (2.6%), and trichloroeth-

Table 18.5 Top 10 mixtures in air completed exposure pathways at hazardous waste sites.

Single substances				Combinations of two					Combinations of three						Combinations of four						
Rank	Sites	CEPs	Name	Rank	Sites	CEPs	Name 1	Name 2	Rank	Sites	CEPs	Name 1	Name 2	Name 3	Rank	Sites	CEPs	Name 1	Name 2	Name 3	Name 4
Air pathways																					
1	43	57	Benzene	1	21	26	Benzene	Toluene	1	15	17	TCE	PCE	Benzene	1	9	11	TCE	PCE	Benzene	1,1,1-TCA
2	38	61	VOCs N.O.S.	2	19	22	TCE	Benzene	2	14	17	Benzene	Toluene	Ethylbenzene	1	9	10	TCE	PCE	Benzene	Ethylbenzene
3	28	36	Lead	2	19	22	TCE	PCE	3	11	13	TCE	Benzene	1,1,1-TCA	1	9	10	TCE	PCE	Benzene	Toluene
4	26	32	TCE	4	16	22	Benzene	Ethylbenzene	3	11	13	TCE	Benzene	Toluene	4	8	10	TCE	PCE	Benzene	MeCl
5	25	30	Toluene	4	16	18	PCE	Benzene	5	10	12	TCE	Benzene	MeCl	4	8	9	TCE	Benzene	1,1,1-TCA	Toluene
6	22	30	PCE	6	15	18	Toluene	Ethylbenzene	5	10	12	TCE	PCE	1,1,1-TCA	4	8	9	TCE	PCE	1,1,1-TCA	Ethylbenzene
7	21	30	MeCl	7	14	17	Benzene	MeCl	5	10	11	Benzene	Toluene	Total xylenes	4	8	9	TCE	PCE	1,1,1-TCA	Toluene
8	19	26	Ethylbenzene	8	13	16	Benzene	1,1,1-TCA	5	10	11	TCE	1,1,1-TCA	Toluene	4	8	9	TCE	PCE	Toluene	Ethylbenzene
9	18	26	Arsenic	8	13	15	TCE	1,1,1-TCA	5	10	11	TCE	PCE	Ethylbenzene	9	7		(16 combinations appear at seven sites and tie for ninth place)			
10	15	18	1,1,1-TCA	8	13	15	TCE	Toluene	5	10	11	TCE	PCE	Toluene							

"Rank" is based on decreasing site count. The rank is repeated if the site count is the same.TCE: trichloroethylene; PCE: perchloroethylene (tetrachloroethylene); MeCl: methylene chloride; PAH: polycyclic aromatic hydrocarbons; VOCs: volatile organic compounds; PCB: polychlorinated biphenyls; N.O.S.: not otherwise specified. In general, "CE": chloroeth*E*ne and "CA": chloroeth*A*ne.

ylene and benzene at 19 sites (2.6%). All of the top 10 most-frequent combinations of 2, 3, or 4 chemicals in air exposure pathways consist solely of VOCs.

Table 18.6 shows similar information for water pathways. Many HWSs with CEPs had at least one chemical in a water exposure pathway (70.4%; 523 of the 743). As can be seen, these chemicals were frequently VOCs: trichloroethylene was in a water CEP at 34.6% of the 743 sites, tetrachloroethylene at 24.6%, 1,1,1-trichloroethane at 11.3%, and 1,1-dichloroethane at 9.6% of the sites. The most frequent binary pair in water CEPs was, by far, trichloroethylene and tetrachloroethylene. This pair was detected in 184 water CEPs at 136 sites (24.8% of 743). The most frequent quaternary mixture found in water CEPs was 1,1,1-trichloroethane, 1,1-dichloroethane, trichloroethylene, and tetrachloroethylene, in 31 CEPs at 26 sites (3.5% of sites).

For the mixture risk assessment of these VOCs that are most often found in combination (Table 18.7) [45], no toxicological data were available for the most frequent quaternary mixture itself (1,1,1-trichloroethane, 1,1-dichloroethane, trichloroethylene, and tetrachloroethylene). The central nervous system is the primary target of toxicity for these chemicals, but most of them have low potency because they are poorly metabolized [46–49]. Only trichloroethylene is metabolized to a considerable degree [46, 50]. It is believed that effects of trichloroethylene on the central nervous system may involve not only the parent chemical but also metabolites such as trichloroethanol. Mild liver or kidney effects are also observed in rodents, caused by reactive metabolic intermediates formed via CYP 2B1/2 or 2E1 catalysis.

Interaction studies are scarce. Tetrachloroethylene inhibited the rates of urinary excretion of trichloroethanol, a 1,1,1-trichloroethane metabolite, in rats exposed to an inhaled mixture of 350 ppm 1,1,1-trichloroethane and 100 ppm tetrachloroethylene [51]. Tetrachloroethylene inhibits the metabolism of trichloroethylene at low exposure levels (<20 ppm) in studies of urinary metabolites in workers exposed to trichloroethylene alone, tetrachloroethylene alone, or mixtures of trichloroethylene and tetrachloroethylene [52]. PBPK simulations of VOC interactions indicate that competitive metabolic interactions occur only at high concentrations [53]. Results of the WOE determinations for low exposures are presented in Table 18.7. A component-based hazard index approach that assumes additive joint toxic action is recommended for the whole mixture. The HI can be qualitatively adjusted to reflect the predicted less than additive effects for the interaction between trichloroethylene and tetrachloroethylene (see Table 18.7). However, the model does not have enough data to state the level of exposure at which the interaction occurs.

Additional common mixtures of VOCs at HWSs are formed from the chemicals chloroform, 1,1-dichloroethylene, trichloroethylene, and vinyl chloride. Trichloroethylene and 1,1-dichloroethylene were found in 82 CEPs at 62 sites (8.3% of the 743 HWSs with CEPs), trichloroethylene and chloroform were found in 60 CEPs at 58 sites (7.8%), trichloroethylene and vinyl chloride were found in 55 CEPs at 46 sites (6.2%), and 1,1-dichloroethylene and chloroform were found in 34 CEPs at 28 sites (3.8% of sites).

All four of these chemicals are known to have effects on the liver, kidney, and developing organism. The immunological system is a common target of three of the chemicals (chloroform, trichloroethylene, and vinyl chloride), and the nervous

Table 18.6 Top 10 mixtures in water completed exposure pathways at hazardous waste sites.

Single substances				Combinations of two					Combinations of three						Combinations of four						
Rank	Sites	CEPs	Name	Rank	Sites	CEPs	Name 1	Name 2	Rank	Sites	CEPs	Name 1	Name 2	Name 3	Rank	Sites	CEPs	Name 1	Name 2	Name 3	Name 4
Water pathways																					
1	257	391	TCE	1	136	184	TCE	PCE	1	50	60	TCE	PCE	1,1,1-TCA	1	26	31	TCE	PCE	1,1,1-TCA	1,1-DCA
2	183	253	PCE	2	67	85	TCE	1,1,1-TCA	2	46	59	TCE	PCE	1,1-DCE	2	25	31	TCE	PCE	1,1,1-TCA	1,1-DCE
3	141	181	Lead	3	62	82	TCE	1,1-DCE	3	36	48	TCE	1,1,1-TCA	1,1-DCE	3	23	31	TCE	1,1,1-TCA	1,1-DCE	1,1-DCA
4	94	126	Arsenic	4	59	68	Lead	Arsenic	3	36	41	TCE	PCE	Benzene	3	23	29	TCE	PCE	1,1-DCE	1,1-DCA
4	94	117	Benzene	5	56	66	TCE	Benzene	5	35	45	TCE	PCE	1,1-DCA	5	21	26	LEAD	Arsenic	Cadmium	Manganese
6	84	105	1,1,1-TCA	6	55	65	PCE	1,1,1-TCA	5	35	44	TCE	1,1,1-TCA	1,1-DCA	6	20	22	TCE	PCE	1,1,1-TCA	*trans*-1,2-DCE
7	81	107	1,1-DCE	7	53	68	PCE	1,1-DCE	7	34	39	LEAD	Arsenic	Cadmium	7	18	20	TCE	PCE	1,1,1-TCA	Chloroform
8	78	95	Vinyl chloride	8	52	60	TCE	Chloroform	8	32	36	TCE	PCE	Chloroform	8	17	22	LEAD	Cadmium	Chromium	Manganese
9	77	95	Chloroform	9	50	64	TCE	1,1-DCA	9	31	41	TCE	1,1-DCE	1,1-DCA	8	17	21	PCE	1,1,1-TCA	1,1-DCE	1,1-DCA
10	76	106	VOCs N.O.S.	9	50	57	Lead	TCE	10	30	32	TCE	PCE	*trans*-1,2-DCE	8	17	21	TCE	1,1,1-TCA	1,1-DCA	*trans*-1,2-DCE
				9	50	56	Lead	Cadmium							8	17	19	TCE	PCE	1,1,1-TCA	MeCl
															8	17	19	TCE	PCE	1,1-DCA	*trans*-1,2-DCE
															8	17	18	TCE	PCE	Benzene	1,1-DCE

"Rank" is based on decreasing site count. The rank is repeated if the site count is the same.
TCE: trichloroethylene; PCE: perchloroethylene (tetrachloroethylene); MeCl: methylene chloride;
PAH: polycyclic aromatic hydrocarbons; VOCs: volatile organic compounds; PCB: polychlorinated biphenyls; N.O.S.: not otherwise specified. In general, "CE": chloroeth*Ene* and "CA": chloroeth*Ane*.

Table 18.7 Matrix of BINWOE determinations for a mixture of VOCs found in water and air.

	On toxicity of			
	1,1,1-Trichloroethane	**1,1-Dichloroethane**	**Trichloroethylene**	**Tetrachloroethylene**
Effect of 1,1,1-Trichloroethane		= IIC neuro, hepato, renal	= IIC neuro; = IIB hepato, renal	= IIC neuro; = IIB hepato, renal
1,1-Dichloroethane	=IIC neuro, hepato, renal		= IIC neuro, hepato, renal	= IIC neuro, hepato, renal
Trichloroethylene	= IIC neuro; = IIB hepato, renal	= IIC neuro, hepato, renal		= IIC neuro; = IIB hepato, renal
Tetrachloroethylene	= IIC neuro; = IIB hepato, renal	= IIC neuro, hepato, renal	= IIC neuro; <IIB hepato, renal	

Hepato: hepatological; neuro: neurological. See Table 18.1 for definition of classifications and modifiers.

system of two (chloroform and trichloroethylene). Carcinogenicity is an end point of concern for three of the chemicals (chloroform, trichloroethylene, and vinyl chloride).

ATSDR's assessment used binary weight-of-evidence evaluations in lieu of data on the whole mixture [54]. Most of the effects of chemicals in this mixture are believed to be the result of metabolites that react with target tissues to cause toxicity. The exceptions are neurological effects that are caused by chloroform and trichloroethylene. An acute coexposure to high inhaled concentrations of 1,1-dichloroethylene and vinyl chloride in fasted rats that had been depleted of glutathione resulted in elimination of the hepatotoxicity seen with 1,1-dichloroethylene alone [55]. Rat inhalation studies have also shown that at high doses, trichloroethylene can compete with 1,1-dichloroethylene for CYP2E1 active sites, resulting in a less than additive metabolic interaction [56, 57]. Similarly, a rat study using acute intraperitoneal injections of chloroform and trichloroethylene demonstrated less liver toxicity when administered together than from either chemical by itself [58]. These results are consistent with our mechanistic understanding that predicts competitive inhibition of CYP2E1 metabolism for these chemicals at high-dose exposures. PBPK models for some of the binary combinations clarified the question of what represents the "high" exposure by predicting thresholds of interaction levels in rats (Table 18.8). However, no data or models are available regarding interaction levels in humans. The hazard index (additivity) is recommended for the whole mixture with the stipulation that the index may overestimate toxicity at higher exposure levels.

Another VOC mixture evaluated by ATSDR comprised of benzene, ethylbenzene, toluene, and xylenes (BTEX), a common indicator of petroleum contamination. This assessment used PBPK modeling [59]. The model predicted toxicokinetic interactions in the quaternary mixture following inhalation exposure, as based on venous blood levels of chemicals, by using information on binary interactions among the component chemicals [60]. The information on blood levels was based on results

Table 18.8 PBPK modeling predictions of interaction thresholds.

Binary mixture	Thresholds in rats	Reference
1,1-Dichloroethylene Trichloroethylene	>100 ppm for each chemical	[57]
Trichloroethylene Vinyl chloride	>30 ppm for each chemical	[53]

from rats [60, 61]. All four chemicals have similar structures and are known substrates for the same cytochrome P450 isozyme (CYP2E1). PBPK analyses of blood kinetic data from binary exposure studies strongly suggest that competitive inhibition of hepatic metabolism is the most plausible mechanism of interaction at high exposures [60]. The study concluded that, with increasing mixture complexity, the blood level of a chemical is increased according to the potency and number of inhibitors, rather than by modification of the metabolic inhibition constant (K_i) for binary interactions. Later the PBPK model for BTEX [60] was linked to a PBPK model for dichloromethane to construct a quinary model for DBTEX in rats [62]. The model was scaled from rats to humans by changing rat physiological and physiochemical parameters to human values and by keeping the biochemical parameters species-invariant (except for the K_m of dichloromethane) [63]. The assumption that the metabolic interaction constants are species-invariant was based on previous findings in a PBPK study of the ternary mixture toluene/ethylbenzene/xylene, where rat to human extrapolation was validated with human experimental data [61]. Results of PBPK simulations and experimental exposures with BTEX and ternary and quinary mixtures of its components strongly suggest that joint neurotoxic action is expected to be additive at concentrations below approximately 20 ppm for each component. Exposures above 20 ppm of each chemical are expected to increase the potential for neurotoxicity and decrease the potential for hemato-toxicity/carcinogenicity due to competitive metabolic interactions among the mixture components. This evaluation demonstrates that it is often impossible to make quantitative interaction predictions when using PBPK in the mixture assessment. Here, it allowed for estimating the concentrations at which the mixture components begin to interact.

18.5.2 Soil Pathway

Soil exposures represent a common medium by which the public might come in contact with chemicals at HWSs; 320 of the 743 sites with CEPs (43%) had one or more soil exposure pathways. Metals are the predominant contaminant, as can be seen in Table 18.9. The primary routes of exposure are oral and dermal.

Table 18.9 Top 10 Mixtures in soil completed exposure pathways at hazardous waste sites.

Single substances				Combinations of two					Combinations of three						Combinations of four						
Rank	Sites	CEPs	Name	Rank	Sites	CEPs	Name 1	Name 2	Rank	Sites	CEPs	Name 1	Name 2	Name 3	Rank	Sites	CEPs	Name 1	Name 2	Name 3	Name 4
Soil pathways																					
1	162	318	Lead	1	102	189	Lead	Arsenic	1	64	110	Lead	Arsenic	Cadmium	1	39	56	Lead	Arsenic	Cadmium	Chromium
2	141	263	Arsenic	2	75	128	Lead	Cadmium	2	54	86	Lead	Arsenic	Chromium	2	34	60	Lead	Arsenic	Cadmium	Zinc
3	90	151	Chromium	3	70	122	Arsenic	Cadmium	3	44	63	Arsenic	Cadmium	Chromium	3	31	54	Lead	Arsenic	Cadmium	Manganese
4	87	150	Cadmium	4	67	112	Arsenic	Chromium	4	43	61	Lead	Cadmium	Chromium	4	30	53	Lead	Arsenic	Cadmium	Copper
5	80	168	PCBs	5	66	106	Lead	Chromium	5	40	69	Lead	Cadmium	Zinc	4	30	43	Lead	Arsenic	Cadmium	Antimony
6	72	147	PAHs N.O.S.	6	53	73	Cadmium	Chromium	6	38	68	Lead	Arsenic	Zinc	6	29	50	Lead	Cadmium	Zinc	Copper
7	59	102	Zinc	7	50	88	Lead	Zinc	7	36	62	Arsenic	Cadmium	Zinc	7	28	49	Lead	Arsenic	Cadmium	Mercury
8	56	88	Benzo[*a*]pyrene	8	45	78	Lead	Copper	7	36	62	Lead	Arsenic	Copper	7	28	49	Lead	Arsenic	Zinc	Copper
9	55	103	Copper	8	45	76	Lead	PCBs	7	36	55	Lead	Arsenic	Antimony	7	28	48	Lead	Arsenic	Cadmium	Barium
10	48	95	Metals N.O.S.	10	44	76	Cadmium	Zinc	10	35	62	Lead	Zinc	Copper	7	28	44	Lead	Arsenic	Cadmium	Nickel
10	48	82	Barium						10	35	61	Lead	Arsenic	Manganese	7	28	41	Lead	Cadmium	Chromium	Zinc
10	48	73	Nickel						10	35	60	Lead	Cadmium	Copper							

"Rank" is based on decreasing site count. The rank is repeated if the site count is the same.

TCE: trichloroethylene; PCE: perchloroethylene (tetrachloroethylene); MeCl: methylene chloride;

PAHs: polycyclic aromatic hydrocarbons; VOCs: volatile organic compounds; PCBs: polychlorinated biphenyls; N.O.S.: NOT otherwise specified. In general, "CE": chloroeth*Ene* and "CA": chloroeth*Ane*.

The most frequent quaternary mixture in soil CEPs consists of arsenic, cadmium, chromium, and lead [4]. This combination appeared in 56 CEPs at 39 sites (5.2% of the 743 HWSs with CEPs). Its component chemicals likewise comprise the chemicals most frequently found in binary pairs in soil CEPs: Pb and As appeared in 189 soil CEPs at 102 sites (13.7% of sites), Pb and Cd in 128 pathways at 75 sites (10.1%), and so on, as can be seen in the table.

No toxicological interaction data are available for all four chemicals together. Still, several sensitive effects can be identified, particularly neurological, hematological, and renal end points [64]. Some of these binary WOE determinations are supported by human or animal data (Table 18.10). For example, the predicted direction of joint toxic action for neurological effects is greater than additive (>IIIB) for the effect of lead on arsenic on the basis of effects of combined exposure on reading and spelling in children [65], and greater than additive for arsenic on lead (>IIB) and cadmium on lead (>IIIC) on the basis of a study of maladaptive classroom behavior in children [66]. Therefore, the potential health hazard might be underestimated on the basis of the end point-specific hazard index for neurological effects, particularly for waste sites with relatively high hazard quotients for lead and arsenic, and lower hazard quotients for the other chemical components.

In contrast, in this mixture the HI may overestimate the renal toxicity of the mixture. This conclusion was reached on the basis of data from animal studies and mechanistic information regarding antagonistic effects of these metals on renal end points. Less than additivity (<IIIB) was reported for lead on arsenic and vice versa [67], for arsenic on chromium(VI) and vice versa (<IIIB) [68], and for cadmium on lead (<IIA) [69, 70].

Table 18.10 Matrix of BINWOE determinations for a mixture of metals found in soil (first mixture).

		On toxicity of			
		Lead	**Arsenic**	**Cadmium**	**Chromium (VI)**
Effect of	Lead		>IIIB neuro; <IIIB hemato, renal	= IIC neuro; = IIA renal; >IIA testes	?
	Arsenic	>IIB neuro; <IIIB hemato, renal		<IIIB hemato; <IIIB testes	<IIIB renal
	Cadmium	>IIIC neuro; <IIIB hemato; <IIA renal	=IIB renal; <IIIB hemato		?
	Chromium (VI)	?	>IIIC dermal; <IIIB renal	?	

Hemato: hematological; neuro: neurological. See Table 18.1 for definition of classifications and modifiers.

Another set of metals frequently found in combination in soil pathways is of copper, lead, manganese, and zinc. For example, lead and zinc are found in 88 CEPs at 50 sites (6.7% of HWSs with CEPs), and lead and copper are found in 78 CEPs at 45 sites (6.1%).

As with the previous four metals, no toxicological study has been performed for the entire mixture. Binary combinations show a possible greater than additive effect of manganese on the toxicity of lead (>IC), while zinc (<IB) and copper (<IC) have less than additive effects on lead neurotoxicity (Table 18.11) [71]. The prediction of less than additivity for the effect of copper on the toxicity of lead is based on copper's protective influence on blood and tissue lead levels [72–74], protective influence on lead-inhibited heme biosynthesis, specifically ALAD activity [73], and copper's induction of metallothionein (which can sequester lead).

For both metal mixtures discussed above, the HI approach, augmented by WOE analysis, is an appropriate method for assessment of joint toxic action of chemicals. Although the current discussion focuses on HWSs, these assessments also apply to a wide range of other scenarios including household settings. For example, exposure of children to playground and decking wood that has been impregnated with chromated copper arsenate is quite common. Exposure to these combinations of metals occurs through the weathering of the treated wood (a soil/dust route) or through direct contact [75, 76].

18.5.3 Hazardous Waste Sites with Radioactive Chemicals

Hazardous waste sites contaminated by both radioactive and nonradioactive chemicals are a special challenge for risk assessors. Examples of such combinations, including stable or radioactive forms of cesium, cobalt, and strontium, have been found at several Department of Energy (DoE) sites. In addition, nonradioactive

Table 18.11 Matrix of BINWOE determinations for a mixture of chemicals found in soil (second mixture).

		On toxicity of			
		Lead	**Manganese**	**Zinc**	**Copper**
Effect of	Lead		= IIIC neuro	= IIB hemato	= IIIC hepatic
	Manganese	>IC neuro; >IIIB hemato		?	?
	Zinc	<IB neuro; <IA hemato	?		<IB hepatic
	Copper	<IC neuro; <1B hemato	?	<IIA hemato	

Hemato: hematological; neuro: neurological. See Table 18.1 for definition of classifications and modifiers.

Table 18.12 Matrix of BINWOE determinations for a mixture of radioactive and nonradioactive chemicals.

		On toxicity of				
		Strontium	**Cobalt**	**Cesium**	**Trichloroethylene**	**PCBs**
Effect of	Strontium		= IIC repro, immuno, devel, cancer	= IIC repro; = IIB immuno	?	?
	Cobalt	= IIC immuno, cancer? hemato		= IIB repro, immuno, devel, cancer	?	?
	Cesium	= IIB hemato, immuno, cancer	= IIB repro, immuno, devel, cancer		?	?
	Trichloroethylene	?	?	?		?
	PCBs	?	?	?	>IIB neuro, hepatic	

Devel: developmental; hemato: hematological; immuno: immunological; neuro: neurological; repro: reproductive. See Table 18.1 for definition of classifications and modifiers.

chemicals such as volatile organic compounds, semivolatile compounds, and heavy metals were also found at these sites. Frequently, the radioactive elements were reported together with TCE and PCBs. Therefore, ATSDR evaluated the mixture of cesium, cobalt, PCBs, strontium, and trichloroethylene [77] (Table 18.12).

No information was located on the toxicity of the whole mixture, so ATSDR assessed potential interactions among the components. Prior reports of chemical use and releases at the sites indicated that strontium, cobalt, and cesium radionuclides, rather than the stable forms of these metals, are of greatest concern for possible adverse health effects. This is because the radiation-related health effects of specific radionuclides will occur at lower exposure levels than those needed for the majority of chemical effects. Internalized cobalt distributes throughout the body and has a relatively short clearance half-life (<10 days), while cesium distributes throughout the body and has a longer half-life (∼70 days). Both cobalt and cesium emit beta and gamma radiation. The gamma radiation is believed to be responsible for most of the health effects caused by these radionuclides (e.g., testicular degeneration and hematological, immunological, and neurodevelopmental effects), mainly due to gamma radiation's higher penetration ability. Because the mechanism of action for the two radionuclides is quite similar, the default assumption of additivity should apply in this case. In contrast, strontium preferentially accumulates in bone with a long half-life (18 years), and it primarily emits beta radiation. Because radioactive strontium affects blood cells (decreased erythrocytes, lymphocytes), additivity of hematological and immunological effects is expected with cesium and cobalt.

Following evaluation of the mixture's components, the HI was applied according to ATSDR's mixtures guidance [7]. Although this guidance does not specifically address radioactive mixtures, nevertheless the HI approach as shown there is recommended for them. This is in accord with other agencies when assessing the risks posed by radionuclides. For example, the National Council on Radiation Protection and Measurements [78] has established screening levels for individual radionuclides that equate to a yearly radiation dose of 0.25 mSv/year. For multiple nuclides, the fraction (i.e., hazard quotient) from each radionuclide is added, just as it is with the hazard index. If the resulting sum is below unity (1), there is assumed to be no cause for concern. As can be seen, this approach also assumes additivity. The Nuclear Regulatory Commission [79] recommends a similar approach for dealing with exposure to radionuclides internally, wherein the sum of the ratios of doses from nuclides to their annual limit on intake (ALI) values should not exceed unity for the internal radiation dose, in order to conclude that there is no cause for concern.

The other chemicals in the DOE mixture were TCE and PCBs. PCB congeners were treated as a single component of concern in the whole mixture, according to the approach used to derive health-based guidance values for PCBs [80]. Information on PCBs and trichloroethylene interaction was obtained from Ref. [81]; for example, pretreatment of rats with Aroclor 1254 by daily gavage for 7 days resulted in a longer mean recovery time after TCE-induced anesthesia compared to rats exposed to TCE alone, with significant increases in numerous parameters: trichlorinated urinary metabolites, hepatic serum glutamic-oxaloacetic transaminase (SGOT) and cytochrome P450 activity, and hepatic sodium, potassium, and calcium levels. Anesthesia recovery time was positively correlated with mean SGOT levels. Animals pretreated with Aroclor 1254 showed necrotic bands of pyknotic hepatocytes, with calcium-rich necrotic cells in corresponding regions. The effect of PCB pretreatment on TCE toxicity is considered greater than additive; the increase in cytochrome P450 activity induced by Aroclor 1254 may be responsible for the increased hepatotoxicity and neurotoxicity of TCE.

In summary, additivity is assumed for the effects of radionuclides, and greater than additivity for the effect of PCBs on the toxicity of TCE, for the DOE mixture. However, no decisions were made on the joint toxic action of other binary combinations of chemicals in the mixture, due to insufficient information. The sites themselves are not likely to be a threat due to limited access to the public and groundwater restrictions. Nevertheless, some sites could be hazardous to on-site workers and those involved with environmental restoration and management.

18.6 Future Directions

Due to the enormity of the challenge posed by chemical mixtures, classical experimental toxicology will be hard-pressed to provide direct information on all the mixtures to which all environmental populations are exposed. Consequently, toxicity risk assessments do not typically address exposures to, or interactions with, other compounds. Even when a great deal of information is available on single chemicals, questions often still remain. How can the dose/response and dose/severity relationships be evaluated? Do all the toxic effects have thresholds? If so, can the uncertainty associated with these thresholds be estimated and the consequences of exceeding these thresholds measured? The answers to all such questions have direct and significant impacts on risk assessments for mixtures.

Pragmatic and realistic risk assessments for mixtures can only be done by considering issues beyond a single chemical within the risk assessment paradigm, for example, exposure assessment, hazard identification, dose-response assessment, and risk characterization. Data needs should be noted at every step of this paradigm. Environmental monitoring (including biomonitoring), surveillance, and population surveys are essential for an accurate exposure assessment, which is the fundamental basis of every risk assessment. The entirety of the exposure scenario – chemicals and environmental factors – must be evaluated to perform a total integrated exposure assessment through multiple routes and for multiple chemicals. (See also chapter 2.) Hazard identification and evaluation may take one of the several options available, with the choice driven by the quality and quantity of toxicity data. Understanding the mechanisms and mode of actions is essential for any advances in the joint toxicity assessment of mixtures. Only then can predictive models be developed for relevant complex exposures that occur in real life.

With respect to dose-response assessment, dose-dependent transitions in overt mixtures toxic effects may often mask more subtle but arguably more significant effects such as endocrine disruption. These effects are often realized before the onset of reproductive age and as such may modify the genome. This is particularly true in relation to genetic polymorphisms in regions populated by vulnerable communities by virtue of elevated exposure and/or increased physiological sensitivity at pivotal periods of development throughout life.

Body burden of the general populations, as reflected in the NHANES III report [82], may actually underestimate the vulnerability of underserved populations to mixtures exposure, particularly for individuals including the urban poor, sport and subsistence anglers and hunters, the immunologically compromised men and women of reproductive age, children, the fetus, and elderly.

The myriad of issues regarding mixtures risk assessment detailed in this chapter, clearly illustrate the need for alternative methods to complement more traditional epidemiologic and biologic tools. Computational toxicology tools can and will be increasingly relied upon to augment traditional approaches. Structure–activity analyses, physiologically based pharmacokinetic and pharmacodynamic models and low dose extrapolations must be used to interpolate, extrapolate, and predict chemical-specific toxicities.

References

1 NTP (2007) Report on Carcinogens, Eleventh Edition. US Department of Health and Human Services, Public Health Service, National Toxicology Program. http://ntp.niehs.nih.gov.

2 EPA (2007) National Priorities List (NPL). US Environmental Protection Agency. www.epa.gov/superfund/sites/npl.

3 ATSDR (2007) Public Health Assessment Guidance Manual. Agency for Toxic Substances and Disease Registry, US Department of Health and Human Services, Atlanta, GA. www.atsdr.cdc.gov/HAC/PHAManual.

4 Fay, M. Environmental pollutant mixtures in completed exposure pathways at hazardous waste sites. *Env. Tox. Pharm.*, (personal communication).

5 ACGIH (2006) TLV/BEI Resources. American Conference of Governmental and Industrial Hygienists. www.acgih.org/TLV.

6 ATSDR (2001) Guidance Manual for the Assessment of Joint Toxic Action of Chemical Mixtures (Draft for Public Comments). Agency for Toxic Substances and Disease Registry, US Department of Health and Human Services, Atlanta, GA.

7 ATSDR (2004) Guidance Manual for the Assessment of Joint Toxic Action of Chemical Mixtures (Final). Agency for Toxic Substances and Disease Registry, US Department of Health and Human Services, Atlanta, GA. www.atsdr.cdc.gov/interactionprofiles/ipga.html.

8 De Rosa, C.T., El-Masri, H.A., Pohl, H., Cibulas, W., and Mumtaz, M.M. (2004) Implications of chemical mixtures in public health practice. *J. Toxicol. Environ. Health, Part A*, **67**, 1–17.

9 EPA (2000) Supplementary Guidance for Conducting Health Risk Assessment of Chemical Mixtures. US Environmental Protection Agency, Washington, DC. EPA/630/R-00/002, August 2000. http://cfpub.epa.gov/ncea/raf/chem_mix.cfm.

10 Pohl, H. and Abadin, H. (1995) Utilizing uncertainty factors in minimal risk levels derivation. *Regul. Toxicol. Pharmacol.*, **22** (2), 180–188.

11 Mumtaz, M. and Durkin, P.R. (1992) A weight of the evidence approach for assessing interactions in chemical mixtures. *Toxicol. Ind. Health*, **8** (6), 377–406.

12 Pohl, H.R., Roney, N., Fay, M., Chou, C.-H., Wilbur, S., and Holler, J. (1999) Site-specific consultation for a chemical mixture. *Toxicol. Ind. Health*, **15**, 470–479.

13 EPA (2007) Indoor Air Quality. US Environmental Protection Agency. www.epa.gov/iaq.

14 ATSDR (2007) Interaction Profile for Carbon Monoxide, Formaldehyde, Methylene Chloride, Nitrogen Dioxide, and Tetrachloroethylene (Draft for Public Comment). Agency for Toxic Substances and Disease Registry, US Department of Health and Human Services, Atlanta, GA. www.atsdr.cdc.gov/interactionprofiles.

15 Winneke, G. (1981) The neurotoxicity of dichloromethane. *Neurobehav. Toxicol. Teratol.*, **3**, 391–395.

16 Kurppa, K., Kivisto, H., and Vainio, H. (1981) Dichloromethane and carbon monoxide inhalation: carboxyhemoglobin addition and drug metabolizing enzymes in rat. *Int. Arch. Occup. Environ. Health*, **49**, 83–87.

17 Squillace, P.J., Scott, J.C., Moran, M.J. *et al.* (2002) VOCs, pesticides, nitrate, and their mixtures in groundwater used for drinking water in the United States. *Environ. Sci. Technol.*, **36**, 1923–1930.

18 ATSDR (2006) Interaction Profile for Atrazine, Deethylatrazine, Diazinon, Nitrate, and Simazine. Agency for Toxic Substances and Disease Registry, U.S. Department of Health and Human Services, Atlanta, GA. www.atsdr.cdc.gov/interactionprofiles.

19 EPA (2002) The Grouping of a Series of Triazine) (Pesticides Based on a Common Mechanism of Toxicity. Office of Pesticides Programs, U.S. Environmental Protection Agency, Washington, DC. www.epa.gov/oppsrrd1/cumulative/triazines/triazinescommonmech.pdf.

20 Belden, J.B. and Lydy, M.J. (2000) Impact of atrazine on organophosphate insecticide toxicity. *Environ. Toxicol. Chem.*, **19** (9), 2266–2274.

21 Anderson, T.D. and Lydy, M.J. (2002) Increased toxicity to invertebrates associated with a mixture of atrazine and organophosphate insecticides. *Environ. Toxicol. Chem.*, **21**, 1507–1514.

22 Eisenbrand, G., Ungerer, O., and Preussmann, R. (1975) Formation of *N*-nitroso compounds from agricultural chemicals and nitrite, in *N-Nitroso Compounds in the Environment* (eds P. Bogovski, E.A. Walker, and W. Davis),

International Agency for Research on Cancer, Switzerland, pp. 71–74.
23 Krull, I.S., Mills, K., Hoffman, G. *et al.* (1980) The analysis of *N*-nitrosoatrazine and *N*-nitrosocarbaryl in whole mice. *J. Anal. Toxicol.*, **4** (5), 260–262.
24 Mirvish, S.S., Gannett, P., Babcook, D.M. *et al.* (1991) *N*-Nitrosoatrazine: synthesis, kinetics of formation, and NMR spectra and other properties. *J. Agric. Food Chem.*, **39** (7), 1205–1210.
25 Wolfe, N.L., Zepp, R.G., Gordon, J.A. *et al.* (1976) *N*-Nitrosamine formation from atrazine. *Bull. Environ. Contam. Toxicol.*, **15** (3), 342–347.
26 Cova, D., Nebuloni, C., Arnoldi, A. *et al.* (1996) *N*-Nitrosation of triazines in human gastric juice. *J. Agric. Food Chem.*, **44**, 2852–2855.
27 Meisner, L.F., Roloff, B.D., and Belluck, D.A. (1993) *In vitro* effects of *N*-nitrosoatrazine on chromosome breakage. *Arch. Environ. Contam. Toxicol.*, **24**, 108–112.
28 Ishidate, M., Sofuni, T., and Yoshikawa, K. (1981) Chromosomal aberration tests *in vitro* as a primary screening tool for environmental mutagens and/or carcinogens. *GANN Monogr. Cancer Res.*, **27**, 95–108.
29 Ishidate, M. (1983) Application of chromosomal aberration tests *in vitro* to the primary screening for chemicals with carcinogenic and/or genetic hazards, in *Tests courts de cancérogénèse: Quo vadis?* (ed. S. Garattini), Centre de Recherches Clin Midy, Montpellier, pp. 57–79.
30 IJC (2007) Persistent Organic Pollutants (POPs). International Joint Commission, Washington, DC, USA and Ottawa, Ontario, Canada. www.ijc.org; also see www.great-lakes.net/humanhealth/other/pops.html.
31 Anderson, H.A., Falk, C., Hanrahan, L. *et al.* (1998) Profiles of great lakes critical pollutants: a sentinel analysis of human blood and urine. *Environ. Health Perspect. Suppl.*, **106** (5), 279–289.
32 Fein, G.G., Jacobson, J.L., Jacobson, S.W. *et al.* (1984) Prenatal exposure to polychlorinated biphenyls: effects on birth size and gestational age. *J. Pediatr.*, **105** (2), 315–320.
33 Jacobson, J.L. and Jacobson, S.W. (1996) Intellectual impairment in children exposed to polychlorinated biphenyls *in utero. N. Engl. J. Med.*, **335** (11), 783–789.
34 Jacobson, J.L., Jacobson, S.W., Schwartz, P.M. *et al.* (1984) Prenatal exposure to an environmental toxin: a test of the multiple effects model. *Dev. Psychol.*, **20** (4), 523–532.
35 Jacobson, S.W., Fein, G.G., Jacobson, J.L. *et al.* (1985) The effect of intrauterine PCB exposure on visual recognition memory. *Child Dev.*, **56**, 853–860.
36 Jacobson, J.L., Jacobson, S.W., and Humphrey, H.E.B. (1990) Effects of *in utero* exposure to polychlorinated biphenyls and related contaminants on cognitive functioning in young children. *J. Pediatr.*, **116** (1), 38–45.
37 Jacobson, J.L., Jacobson, S.W., and Humphrey, H.E.B. (1990) Effects of exposure to PCBs and related compounds on growth and activity in children. *Neurotoxicol. Teratol.*, **12**, 319–326.
38 Mergler, D., Belanger, S., Larribe, F. *et al.* (1998) Preliminary evidence of neurotoxicity associated with eating fish from the upper St. Lawrence River Lakes. *Neurotoxicology*, **19** (4–5), 691–702.
39 ATSDR (2004) Interaction Profile for Persistent Chemicals Found in Fish (Chlorinated Dibenzo-p-dioxins, Hexachlorobenzene, p,p′-DDE, Methylmercury, and Polychlorinated biphenyls). Agency for Toxic Substances and Disease Registry, U.S. Department of Health and Human Services, Atlanta, GA. www.atsdr.cdc.gov/interactionprofiles.
40 Li, S.M.A., Denomme, M.A., Leece, B. *et al.* (1989) Hexachlorobenzene: biochemical effects and synergistic toxic interactions with 2,3,7,8-tetrachloro-dibenzo-*p*-dioxin. *Toxicol. Environ. Chem.*, **22**, 215–227.
41 Bannister, R., Davis, D., Zacherewski, T. *et al.* (1987) Aroclor 1254 as a 2,3,7,8-tetrachlorodibenzo-*p*-dioxin antagonist: effects on enzyme induction and immunotoxicity. *Toxicology*, **46**, 29–42.
42 Davis, D. and Safe, S. (1989) Dose-response immunotoxicities of commercial polychlorinated biphenyls

(PCBs) and their interaction with 2,3,7,8-tetrachlorodibenzo-*p*-dioxin. *Toxicol. Lett.*, **48**, 35–43.

43 Haake, J.M., Safe, S., Mayura, K. *et al.* (1987) Aroclar 1254 as an antagonist of the teratogenicity of 2,3,7,8-tetrachloro-dibenzo-*p*-dioxin. *Toxicol. Lett.*, **38**, 299–306.

44 Bemis, J.C. and Seegal, R.F. (1999) Polychlorinated biphenyls and methylmercury act synergistically to reduce rat brain dopamine content *in vitro*. *Environ. Health Perspect.*, **107**, 879–885.

45 ATSDR (2004) Interaction Profile for 1,1,1-Trichloroethane, 1,1-Dichloro-ethane, Trichloroethylene, and Tetrachloroethylene. Agency for Toxic Substances and Disease Registry, U.S. Department of Health and Human Services, Atlanta, GA. www.atsdr.cdc.gov/interactionprofiles.

46 Kaneko, T., Wang, P.-Y., and Sato, A. (1994) Enzymes induced by ethanol differently affect the pharmacokinetics of trichloroethylene and 1,1,1-trichloroethane. *Occup. Environ. Med.*, **51**, 113–119.

47 McCall, S.N., Jurgens, P., and Ivanetich, K.M. (1983) Hepatic microsomal metabolism of the dichloroethanes. *Biochem. Pharmacol.*, **32** (2), 207–213.

48 Mitoma, C., Steeger, T., Jackson, S.E. *et al.* (1985) Metabolic disposition study of chlorinated hydrocarbons in rats and mice. *Drug Chem. Toxicol.*, **8** (3), 183–194.

49 Nolan, R.J., Freshour, N.L., and Rick, D.L. (1984) Kinetics and metabolism of inhaled methyl chloroform (1,1,1-trichloroethane) in male volunteers. *Fundam. Appl. Toxicol.*, 4, 64–662.

50 Lash, L.H., Fisher, J.W., Lipscomb, J.C. *et al.* (2000) Metabolism of trichloro-ethylene. *Environ. Health Perspect.*, **108**, 177–200.

51 Koizumi, A., Kuami, M., and Ikeda, M. (1982) *In vivo* suppression of 1,1,1-trichloroethane metabolism by co-administered tetrachloroethylene: an inhalation study. *Bull. Environ. Contam. Toxicol.*, **29**, 196–199.

52 Seiji, K., Inoue, O., In, C. *et al.* (1989) Dose-excretion relationship in tetrachloroethylene-exposed workers and the effect of tetrachloroethylene co-exposure on trichloroethylene metabolism. *Am. J. Ind. Med.*, **16**, 675–684.

53 Barton, H.A., Creech, J.R., Godin, C.S. *et al.* (1995) Chloroethylene mixtures: pharmacokinetic modeling and *in vitro* metabolism of vinyl chloride, trichloroethylene, and *trans*-1,2-dichloroethylene in rat. *Toxicol. Appl. Pharmacol.*, **130**, 237–247.

54 ATSDR (2007) Interaction Profile for Chloroform, 1,1-Dichloroethylene, Trichloroethylene, and Vinyl chloride (Draft for Public Comment). Agency for Toxic Substances and Disease Registry, U.S. Department of Health and Human Services, Atlanta, GA. www.atsdr.cdc.gov/interactionprofiles.

55 Jaeger, R.J., Conolly, R.B., Reynolds, E.S., and Murphy, S.D. (1975) Biochemical toxicology of unsaturated halogenated monomers. *Environ. Health Perspect.*, **11**, 121–128.

56 Andersen, M.E., Gargas, M.L., Clewell, H.J. *et al.* (1987) Quantitative evaluation of the metabolic interactions between trichloroethylene and 1,1-dichloro-ethylene *in vivo* using gas uptake methods. *Toxicol. Appl. Pharmacol.*, **89** (2), 149–157.

57 El-Masri, H.A., Constan, A.A., Ramsdell, H.S. *et al.* (1996) Physiologically based pharmacodynamic modeling of an interaction threshold between trichloroethylene and 1,1-dichloroethylene in Fischer 344 rats. *Toxicol. Appl. Pharmacol.*, **141** (1), 124–132.

58 Anand, S.S., Mumtaz, M.M., and Mehendale, H.M. (2005) Dose-dependent liver tissue repair after chloroform plus trichloroethylene binary mixture. *Basic Clin. Pharmacol. Toxicol.*, **96** (6), 436–444.

59 ATSDR (2004) Interaction Profile for Benzene, Toluene, Ethylbenzene, and Xylenes (BTEX). Agency for Toxic Substances and Disease Registry, U.S. Department of Health and Human Services, Atlanta, GA. www.atsdr.cdc.gov/interactionprofiles.

60 Haddad, S., Tardif, R., Charest-Tardif, G. *et al.* (1999) Physiological modeling of the toxicokinetics interactions in a quaternary mixture of aromatic hydrocarbons. *Toxicol. Appl. Pharmacol.*, **161**, 249–257.

61 Tardif, R., Charest-Tarif, G., Brodeur, J. *et al.* (1997) Physiologically based pharmacokinetic modeling of a ternary mixture of alkyl benzenes in rats and humans. *Toxicol. Appl. Pharmacol.*, **144**, 120–134.

62 Haddad, S., Charest-Tardif, G., Tardif, R. *et al.* (2000) Validation of a physiological modeling framework for simulating the toxicokinetics of chemicals in mixtures. *Toxicol. Appl. Pharmacol.*, **167**, 199–209.

63 Haddad, S., Béliveau, M., Tardif, R. *et al.* (2001) A PBPK modeling-based approach to account for interactions in the health risk assessment of chemical mixtures. *Toxicol. Sci.*, **63**, 125–131.

64 ATSDR (2004) Interaction Profile for Arsenic, Cadmium, Chromium and Lead. Agency for Toxic Substances and Disease Registry, U.S. Department of Health and Human Services, Atlanta, GA. www.atsdr.cdc.gov/interactionprofiles.

65 Moon, C., Marlowe, M., Stellern, J. *et al.* (1985) Main and interaction effects of metallic pollutants on cognitive functioning. *J. Learn. Disabil.*, **18** (4), 217–221.

66 Marlowe, M., Cossairt, A., Moon, C. *et al.* (1985) Main and interaction effects of metallic toxins on classroom behavior. *J. Abnorm. Child Psychol.*, **13** (2), 185–198.

67 Fairhall, L.T. and Miller, J.W. (1941) A study of the relative toxicity of the molecular components of lead arsenate. *Public Health Rep.*, **56**, 1610–1625.

68 Mason, R.W. and Edwards, I.R. (1989) Acute toxicity of combinations of sodium dichromate, sodium arsenate and copper sulphate in the rat. *Comp. Biochem. Physiol.*, **93C** (1), 121–125.

69 Mahaffey, K.R. and Fowler, B.A. (1977) Effects pf concurrent administration of lead, cadmium, and arsenic in the rat. *Environ. Health Perspect.*, **19**, 165–171.

70 Mahaffey, K.R., Capar, S.G., Gladen, B.C. *et al.* (1981) Concurrent exposure to lead, cadmium, and arsenic: effects on toxicity and tissue metal concentrations in the rat. *J. Lab. Clin. Med.*, **98**, 463–481.

71 ATSDR (2004) Interaction Profile for Lead, Manganese, Zinc, and Copper. Agency for Toxic Substances and Disease Registry, U.S. Department of Health and Human Services, Atlanta, GA. www.atsdr.cdc.gov/interactionprofiles.

72 Flora, S.J.S., Coulombe, R.A., Sharma, R.P. *et al.* (1989) Influence of dietary protein deficiency on lead/copper interaction in rats. *Ecotoxicol. Environ. Saf.*, **18**, 75–82.

73 Flora, S.J.S., Jain, V.K., Behari, J.R. *et al.* (1982) Protective role of trace metals in lead intoxication. *Toxicol. Lett.*, **13**, 51–56.

74 Kies, C. and Ip, S.W. (1990) Lead bioavailability to humans from diets containing constant amounts of lead: impact of supplemental copper, zinc and iron. *Trace Subst. Environ. Health*, **24**, 177–184.

75 ATSDR (2005) Toxicological Profile for Arsenic. Agency for Toxic Substances and Disease Registry, U.S. Department of Health and Human Services, Atlanta, GA. www.atsdr.cdc.gov/toxpro2.html.

76 Hamula, C., Wang, Z., Zhang, H., Kwon, E., Li, X.-F., Gabos, S., and Le, X.C. (2006) Chromium on the hands of children after playing in playgrounds built from chromated copper arsenate (CCA)-treated wood. *Environ. Health Perspect.*, **114** (3), 460–465.

77 ATSDR (2004) Interaction Profile for Cesium, Cobalt, Polychlorinated Biphenyls, Strontium, and Trichloroethylene. Agency for Toxic Substances and Disease Registry, U.S. Department of Health and Human Services, Atlanta, GA. www.atsdr.cdc.gov/interactionprofiles.

78 NCRP (1999) Recommended Screening Limits for Contaminated Surface Soil and Review of Factors Relevant to Site-Specific Studies. National Council on Radiation Protection and Measurements, NCRPReport No. 129.

79 NRC (2001) Annual Limits on Intake (ALIs) and Derived Air Concentrations (DACs) on Radionuclides for Occupational Exposure. Effluent concentrations; Concentrations for Release to Sewerage. Code of Federal

Regulations. Chapter 1, Part 20, Appendix B. http://www.nrc.gov/NRC/CFR/PART020/part020-appb.html, May 23, NRC 2001.

80 ATSDR (2000) Toxicological Profile for Polychlorinated Biphenyls (PCBs). Agency for Toxic Substances and Disease Registry, U.S. Department of Health and Human Services, Atlanta, GA. www.atsdr.cdc.gov/toxpro2.html.

81 Moslen, M.T., Reynolds, E.S., and Szabo, S. (1977) Enhancement of the metabolism and hepatotoxicity of trichloroethylene and perchoroethylene. *Biochem. Pharmacol.*, **26**, 369–375.

82 CDC (Centers for Disease Control and Prevention) (2005) Third National Report on Human Exposure to Environmental Chemicals. US Department of Health and Human Services, Atlanta, GA.

19
Thyroid-Active Environmental Pollutants and Their Interactions on the Hypothalamic–Pituitary–Thyroid Axis

Eva D. McLanahan and Jeffrey W. Fisher

19.1
Thyroid-Active Environmental Pollutants

Many chemicals found in the environment have been shown to clearly alter thyroid function in laboratory animals and *in vitro* systems, and suggestive evidence exists for humans [1]. Thyroid hormones are indispensable for physiological body functions and play a crucial role in maintaining normal development of the fetus and infant. Therefore, public health concern exists for understanding the hazards that may be associated with population exposures to thyroid-active environmental pollutants. Thyroid-active chemicals, which mediate their toxicity by disrupting the hypothalamic–pituitary–thyroid (HPT) axis, can be viewed as indirect acting chemicals that first disturb the HPT axis and then, as a consequence of this, a myriad of body function deficits may occur. Several documented adverse effects in humans associated with the lack of thyroid hormone were elucidated from iodide-deficient populations around the world or from laboratory animal studies with iodide-restricted diets. Many chemicals have been identified as rodent thyroid-active chemicals, such as polychlorinated biphenyls (PCBs), phthalates, anions (e.g., nitrate, thiocyanate, chlorate, and perchlorate), dioxins, flame retardants, phenols, and pesticides [1]; however, interpreting the human relevance and dose-response characteristics of the findings for risk assessment purposes has been hindered by important differences in the HPT axis between humans and laboratory animals [2] and a lack of standardized methods for testing thyroid-active chemicals [3] for regulatory decisions.

A sufficient body of research exists to demonstrate modes of action through which several thyroid-active chemicals disturb the HPT axis. Figure 19.1 depicts a simplified view of the HPT axis with emphasis on processes where chemicals can disturb homeostasis. This is a conceptual framework useful for generating hypotheses and evaluating hypothesis-driven testing. There are numerous toxicity studies of thyroid-active chemicals; however, few are designed to ascertain the dose-response characteristics for HPT axis disturbances linked with relevant mode of action studies. Cunha and van Ravenzwaay [4] have reported on the dose-response properties and specific modes of action on the HPT axis for perchlorate, propylthiouracil, iodide,

Principles and Practice of Mixtures Toxicology. Edited by Moiz Mumtaz

ISBN: 978-3-527-31992-3

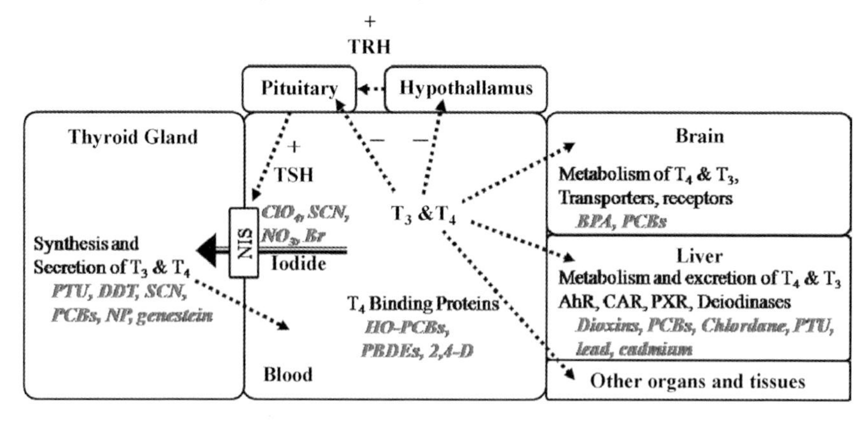

Figure 19.1 The hypothalamic–pituitary–thyroid axis is among the most complex endocrine systems. Iodide is actively sequestered by the thyroid via the sodium/iodide symporter (NIS) and incorporated into thyroglobulin to form thyroid hormones. Thyroid hormones are produced and secreted from the thyroid gland into systemic circulation, the majority of which is thyroxine (T_4), and 3,5,3′-triiodothyronine (T_3) is secreted from the normal thyroid in much smaller quantities. Once in the bloodstream, T_4 and T_3 bind to transport proteins. T_4 and T_3 are used by virtually every cell in the body as signals to maintain essential physiological processes. Within tissues, T_4 is metabolized to the active form of thyroid hormone (T_3) by deiodinase enzymes. Thyroid hormones can also be metabolized via several different phase II conjugation pathways within the liver. Furthermore, T_4 and T_3 provide feedback control on the hypothalamus and pituitary (negative feedback loops). They are the primary signaling molecules that regulate the production and secretion of thyroid releasing hormone (TRH) and thyroid stimulating hormone (TSH) within the hypothalamus and pituitary, respectively. TRH is secreted into portal circulation by the hypothalamus where it stimulates the pituitary to produce TSH that in turn upregulates many processes within the thyroid gland. Environmental pollutants and pharmacological drugs are able to disrupt HPT axis homeostasis by a variety of mechanisms, several of which are depicted in the figure. For example, several have been shown to affect the ability of the thyroid gland to synthesize and secrete thyroid hormones, while others affect the systemic transport or thyroid hormone metabolism in the liver. This is neither a comprehensive list of mechanisms for HPT axis disruption nor does it include all compounds with the ability to disrupt HPT axis homeostasis. However, the more commonly studied compounds with well-defined modes of action are shown.

pyrazole, minocycline, amiodrarone, 2,4-dichlorophenoxyacetic acid, cadmium, phenobarbital, temelastine, and SK&F 93479. Studies that tie these events to frank toxicity mediated by HPT disturbances remain a research need for risk assessment. Below is a brief introduction to the mode of action information for selected xenobiotics shown in Figure 19.1.

19.1.1
Inhibitors and Potential Inducers of Iodide Uptake into the Thyroid

A wide range of anions found in the environment inhibit or limit normal uptake of thyroidal iodide by the sodium/iodide symporter (NIS) protein, and in addition, these anions are also transported by the NIS protein into the thyroid gland. Blocking iodide uptake at the thyroid gland may cause hypothyroidism because of a decline in thyroid hormone production. Wolff [5] reports the potency of the following anions to inhibit radiolabeled iodide as follows: $ClO_4 > SCN > I > NO_3 > Br > Cl$. Thiocyanate (SCN) is found in excess in people who smoke tobacco, while nitrate (NO_3), perchlorate (ClO_4), bromide (Br), and chloride (Cl) can be found in water and food supplies. The NIS protein is a membrane-bound protein located in many extrathyroidal tissues (salivary gland, stomach, mammary gland, placenta, kidney, and testes) [5, 6] and plays a quantitative role in the transport of xenobiotics or iodide. Margret Hayes demonstrated the influence of the NIS in the stomach/GI tract on the serum kinetics of iodide in humans [7]. Perchlorate and bromide have been used to block NIS-mediated secretion of iodide into breast milk of rats [8, 9].

19.1.2 Alteration in Thyroid Hormone Synthesis/Secretion

DDT and the commercial PCB mixture Aroclor 1254 have been shown to interfere with adenylate cyclase (cAMP) activity and cAMP production *in vitro* suggesting interference with thyrotropin (TSH, thyroid stimulating hormone) signaling and thyroid hormone production [13]. Thyroperoxidase (TPO) is a heme-containing enzyme in the thyroid gland that, in the presence of hydrogen peroxide, organifies iodine. Organification refers to the covalent attachment of molecular iodine (I_2) to the thyroglobulin protein. Nonylphenol (NP), an industrial additive, inhibits TPO activity [14]. 6-Propyl-2-thiouracil (PTU) is a thioamide drug used extensively in both laboratory animals and humans for mechanistic studies. PTU reduces serum thyroid hormone and increases serum TSH levels. PTU inhibits thyroperoxidase [15] and type I 5′-deiodinase [16]. Soy-based chow formulations containing soy proteins, such as genistein, inhibit thyroperoxidase in the rat [17]; however, no change in serum thyroid hormones or TSH is observed.

19.1.3 Alteration in Thyroid Hormone Metabolism

Peripheral iodothyronine deiodinases convert thyroxine (T_4) to the active thyroid hormone, 3,5,3′-triiodothyronine (T_3), and also the conversion of T_3 to T_2, T_2 to T_1, and T_1 to T_0. This metabolic activity occurs in many organs in the body and controls the throughput or daily requirement for thyroid hormone production. Several persistent organic environmental chemicals, found in food and the environment, inhibit type I 5′-deiodinase in the liver such as 2,3,7,8-tetrachlorodibenzo-*p*-dioxin (TCDD) [10], organochlorines (e.g., chlordane), lead, and cadmium [11]. Thyroid hormones, in particular T_4, are metabolized by phase II conjugation. Coplanar PCBs and dioxins bind to and induce hepatic nuclear receptors (aryl hydrocarbon receptor (AhR)), while nonplanar *ortho*-substituted PCBs bind to and induce the orphan nuclear receptors, the pregnane X receptor (PXR), and constitutive androstane receptor (CAR). These hepatic nuclear receptors induce the phase II enzymes, UDP-glucuronosyl transferases (UDPGTs), sulfotransferases (SULTs), and glutathione *S*-transferases (GSTs). UDPGTs are induced by PCB congeners and result in increased biliary excretion of T_4 [12].

19.1.4
Alteration in Thyroid Hormone Transport Proteins or Receptor Proteins

About 99% of circulating thyroid hormones are bound to serum transport proteins. Some halogenated hydrocarbons structurally resemble thyroid hormones and have been shown to displace T_4 bound to serum proteins such as 2,4-D-dichlorophenoxyacetic acid, dinoseb, bromoxynil, and pentachlorophenol [18, 19]. These compounds are used as herbicides, and pentachlorophenol was also used as a wood preservative. Polybrominated diphenyl ethers (PBDEs), used as flame retardants in many household fabrics, also bind to the serum protein transthyretin and displace T_4 [20]. Hydroxylated metabolites of PCBs have been shown to bind to serum proteins and displace T_4 [21]. The consequences of this may be increased clearance of T_4 from the body. Aroclor 1254 has also been shown to displace T_3 from rat thyroid receptors [22]. Bisphenol A (BPA) has been shown to be a thyroid hormone receptor antagonist *in vitro* and via this mechanism is thought to disrupt thyroid hormone signaling pathways in the developing brain [23].

Many of these chemicals have been measured in blood or urine of people in the United States [24] documenting the fact that US population is exposed to thyroid-active chemicals. Unfortunately, the health consequences of these chemical exposures are not well understood.

19.2
Interaction of Chemical Mixtures on the HPT Axis

Several challenges exist in improving chemical risk assessment practices, particularly with the reliance of human health risk assessments on laboratory animal toxicology studies. DeVito *et al.* [25] suggested that measurement of serum T_4 and thyroid histology are important parameters in screening or testing chemicals for thyroid activity. Twenty-one chemicals on the US Environmental Protection Agency (EPA) Integrated Risk Information System (IRIS) were retrieved using thyroid toxicity as the critical effect [26] for reference dose (RfD) derivation. In these cases, other organ toxicity, perhaps mediated by hypothyroidism and/or serum thyroid hormone changes were used for risk assessment purposes.

An interest in our laboratory is to better understand the dose-response characteristics of thyroid-active chemical mixtures. As briefly discussed, many individual chemicals are known to disturb the HPT axis in rodents by fairly well-understood modes of action. Few studies on the effects of chemical mixtures on the HPT axis have been conducted; however, even fewer of the HPT axis toxicity studies have been designed specifically to evaluate the potential interactions from chemical mixtures. Assessing the interactions for chemical mixtures is important because people are exposed to mixtures of chemicals, not individual chemicals.

A few binary mixture studies have been conducted in rats that include perchlorate as one of the compounds. Khan *et al.* [27] reported synergistic interactions when rats ingested the binary combination of ammonium perchlorate (NH_4ClO_4) and sodium chlorate ($NaClO_3$) in drinking water for 7 days, as evidenced by greater decreases in serum T_4 levels than seen with individual chemicals. Interestingly, Khan *et al.* [27] also noted that male Fischer rats appear less sensitive than male Sprague–Dawley rats to NH_4ClO_4-induced alterations in the HPT axis. In a more recent study [28], perchlorate was administered alone or in combination with ethanol to female Myers' rats to examine the effects on plasma thyroid hormones and brain catecholamine concentrations. However, perchlorate doses administered (300 and 3000 μg/l; yielding average intake of 0.06 and 0.6 mg/kg bw) for 21 days were too low to cause statistically significant effects on serum T_4 and T_3 concentrations. Furthermore, when these doses of perchlorate were administered with 10% ethanol, no further reductions in total serum T_4 or T_3 were observed. Recently, research efforts have been expanded to better understand the contributions of other common pollutants that share the same mode of action as perchlorate on the thyroid gland, namely, nitrate and thiocyanate [29, 30]. These authors [30] report "response" additivity based on the affinity (Km) of the anion for the sodium/iodide symporter protein using a nonlinear Michaelis–Menten equation to describe competitive inhibition of thyroidal uptake of radiolabeled iodide.

In complex mixture studies with a different class of chemicals, namely, organochlorines, Desaulniers *et al.* [31] and Crofton *et al.* [32] administered rats cocktails of mixtures containing polychlorinated biphenyls, polychlorinated dibenzo-*p*-dioxins (PCDDs), and polychlorinated dibenzofurans (PCDFs) and evaluated serum T_4 concentrations. A primary mechanism of action by which these chemicals act on the HPT axis is through activation of the aryl hydrocarbon receptor and resulting in enhanced hepatic metabolism and clearance of T_4, although several other mechanisms are possible [32]. Additive decreases in serum T_4 concentrations were observed by Desaulniers *et al.* [31] after administration of mixtures of either 6 PCDDs, 3 non-*ortho*-PCBs, or 7 PCDFs to prepubertal female Sprague–Dawley rats. Crofton *et al.* [32] evaluated serum T_4 levels after 4 consecutive days of oral gavage dosing with a mixture of 18 polyhalogenated aromatic hydrocarbons, including PCB126. There was no deviation from additivity at the lowest mixture dose, but a greater than additive decrease in serum T_4 was observed at the three highest mixture doses.

Many studies have been conducted to determine the effect of complex mixtures of thyroid-active compounds; however, only the few mentioned above were conducted such that the contribution of each component of the mixture could be ascertained. For example, Zhou *et al.* [33] evaluated the effect of a complex mixture of polybrominated diphenyl ethers on the thyroid axis; however, the contribution of each component of the mixture was not analyzed, so understanding the mixture HPT axis effect relative to the individual compounds found in the mixture was not possible. It is not only important to determine the ability of complex mixtures to affect the HPT axis but also equally important to determine the contributions of the mixture components in order to determine if the mixture behaves in a subadditive, additive, or superadditive fashion.

19.3 Case Study with Binary Mixture of PCB126 and Perchlorate

In this case study, we discuss findings from our laboratory in which rats were pretreated with a wide range of single oral bolus doses of 3,3′,4,4′,5-pentachlorobiphenyl (PCB126) and then placed on drinking water containing perchlorate, also over a range of concentrations [12, 34]. PCB126 appears to act primarily on the HPT axis by AhR-mediated upregulation of phase II hepatic metabolism of T_4 [12], while perchlorate disrupts the axis primarily by competitive inhibition of NIS thyroidal iodide uptake [35]. This exposure scenario for PCB126 and perchlorate allowed for (1) the evaluation of perchlorate effects on the HPT axis in rats that were in various hypothyroid states when perchlorate treatment was initiated, (2) the evaluation of the dose-response characteristics for a binary mixture of chemicals that act on the HPT axis by two different modes of action, and (3) the collection of kinetic data sets for future development of a biologically based dose-response (BBDR) model to describe the interactions of a binary mixture on the HPT axis. The first two objectives are discussed in this chapter.

Both dose- and time-dependent perturbations of the HPT axis were characterized for each individual chemical and then compared with HPT axis perturbations caused by various binary combinations of PCB126 and perchlorate. PCB126 is long lived in the body of rats [12], while perchlorate is cleared more quickly [35]. Our working hypothesis for the binary mixture study was that for PCB126 pretreated rats that showed increased formation of hepatic T_4-glucuronide, along with decreased serum T_4 and elevated serum TSH concentrations, the blocking effects of perchlorate on thyroidal uptake of iodide would be diminished, resulting in subadditive perturbations in the HPT axis. TSH stimulates many thyroidal processes including the uptake of iodide into the thyroid gland and formation and the secretion of thyroid hormones. Thus, if perchlorate acts on the thyroid gland solely by competitive inhibition of uptake of thyroidal iodide, it would be less effective in the TSH-stimulated thyroid gland compared to the naïve condition. For conditions in which pretreatment with PCB126 did not alter serum T_4 and TSH concentrations, the observed effects of perchlorate on the HPT axis would be similar to a naïve animal [35].

19.3.1 Exposure Design

Dose combinations of PCB126 and perchlorate were selected to inflict moderate to minimal or no disturbances on the HPT axis in the adult male rat. Single oral bolus PCB126 doses of 7.5 and 75 μg/kg caused clear perturbations in the HPT axis [12], as well as ingestion rates of perchlorate in drinking water of 0.1 and 1.0 mg/kg/day [35]. In one set of mixture studies, on day 0 adult male Sprague–Dawley rats were administered single oral bolus doses of PCB126 dissolved in corn oil (0, 7.5, or 75 μg/kg). On day 9 after dosing with PCB126, rats were administered perchlorate (0, 0.01, 0.10, or 1.00 mg/kg/day) in their drinking water for an additional 14 days at which time the rats were killed and tissues harvested. In another set of mixture studies, rats were administered a single oral gavage dose of PCB126 (0, 0.075, 0.75, or 7.5 μg/kg) on day 0 and then placed on drinking water with perchlorate (0 or 0.01 mg/kg/day) on day 1 for 1–4 days at which time the rats were killed and tissues were harvested. In all cases, vehicle controls and corresponding single-chemical exposures were conducted in conjunction with the binary exposures.

When evaluating the temporal effects of PCB126 on the HPT axis [12], it was apparent that the timescale for the onset of perturbations was much longer than for perchlorate [35]. With single PCB126 at doses of 75 or 275 μg PCB126/kg, by day 5 postdosing, serum T_4 was decreased and rate of hepatic T_4-glucuronide formation increased, while a significant increase in serum TSH was not observed until day 9 postdosing [12]. When rats were administered an i.v. dose of 3.3 mg/kg of perchlorate, serum TSH was increased and serum T_4 decreased by 12 h after dosing, and for ingestion of perchlorate in drinking water (1, 3, or 10 mg/kg/day), a decrease in serum T_4 and increase in serum TSH occurred within 24 h of treatment [35].

The experimental mixture studies with PCB126 and perchlorate [34] were designed to provide information about their interactions under pharmacokinetic conditions (dose and time) where each chemical was known to disturb the HPT axis and also under pharmacokinetic conditions where each chemical produced minimal or no disturbances. In this case, PCB126 and perchlorate each act on the HPT axis by a different mode of action, leading to hypothyroidism. When a chemical mixture shares similar adverse outcomes (hypothyroidism) by disturbing the HPT axis, a nonindependent dose-response curve is expected. When this occurs, the HPT disturbance by the chemical mixture can be subadditive, additive, or superadditive. One public health concern is that a chemical mixture may act synergistically, or in a superadditive fashion, resulting in enhanced adverse outcomes, which may occur at exposure levels previously thought to be safe for individual chemicals. An additive response generally occurs when the dose-response curves to the individual chemicals are parallel and can be added to predict the combined response. A subadditive response occurs when the combined response is somewhat less than expected through simple addition but greater than the response expected from either chemical alone. The superadditive response occurs when the combined response is greater than simple addition of the individual responses occurs.

19.3.2 Results and Discussion

The design of this PCB126/perchlorate mixture study was to evaluate only one aspect of the binary exposure: when rats were pretreated with PCB126 and then placed on drinking water with perchlorate. Rats dosed with an HPT-responsive dose of PCB126 such as 75 μg/kg or placed on drinking water containing 1.0 mg/kg/day of perchlorate, both chemicals caused the serum-free and total T_4 levels to drop, the serum TSH levels to increase, and the hepatic T_4 phase II metabolic rates to increase (Figure 19.2). The combined exposure to these doses of PCB126 and perchlorate resulted in a masking of the effects of perchlorate on the HPT axis. That is, the PCB126 effects on the HPT axis dominated and effects on serum T_4 and TSH appeared similar to

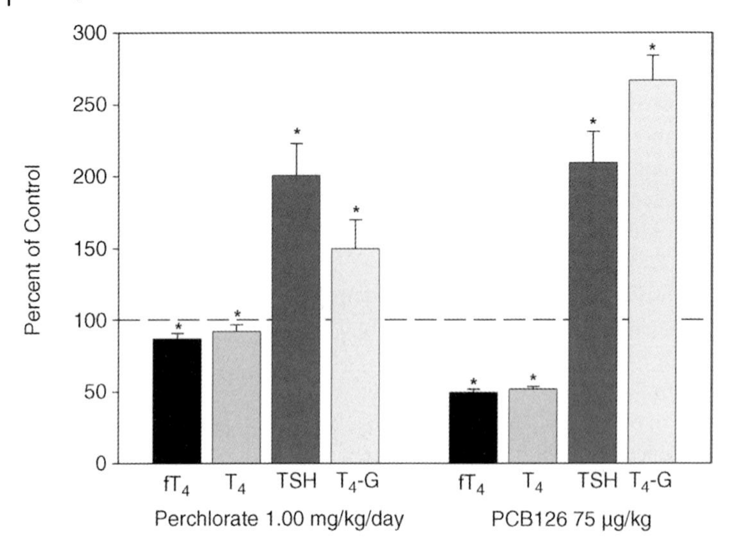

Figure 19.2 Changes in HPT axis indices (serum-free T_4 (fT_4), total T_4 (T_4), TSH, and rate of hepatic T_4-G formation (T_4-G)) expressed as percentage of control ± SEM. Bars on the left are results following 14-day exposure to 1.00 mg perchlorate/kg/day in drinking water and bars on the right correspond to animals administered a single oral bolus dose of 75 μg PCB126/kg on day 0 and measurements obtained 22 days postdosing. Results significantly different from control ($p \leq 0.05$) are indicated by an asterisk (*). Data adapted from McLanahan *et al.* [34].

PCB126 alone. The combined exposure response may be considered subadditive, where by definition, the response to the mixture was not greater than the response expected from either chemical alone (Figure 19.3) (still less than the sum of the two responses). One possible explanation for this observation is that this dose of PCB126 resulted in an increased loss of T_4 such that it triggered a stimulation of TSH production, which, in turn, stimulated the thyroidal NIS protein. This adaptive response by the HPT axis reduces the ability of this dose of perchlorate to block thyroidal uptake of iodide, and therefore the effective perchlorate dose to block thyroidal uptake of iodide is greater in the TSH-stimulated thyroid gland.

Lesson learned: *To understand and predict the adverse outcomes of chemical mixture insult on endocrine systems, such as the HPT axis, which has several adaptive mechanisms to retain homeostasis, the mode of action of the chemicals of interest and their pharmacokinetic behavior and pharmacodynamic influence on the HPT axis must be understood.*

The temporal relationship between administered dose, perturbations in the HPT axis, and frank toxicity is generally unknown for important end points such as developmental neurotoxicity.

Another exposure condition to evaluate for PCB126 and perchlorate was to administer doses of each chemical that caused minimal or no measured responses in the HPT axis. Will a binary exposure combination of PCB126 and perchlorate

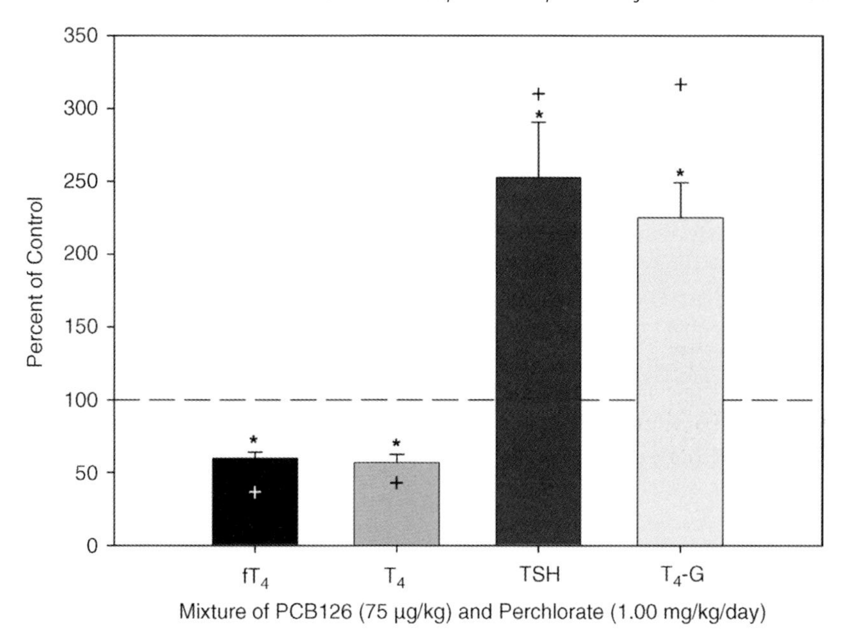

Figure 19.3 Changes in HPT axis indices (serum-free T_4 (fT_4), total T_4 (T_4), TSH, and rate of hepatic T_4-G formation (T_4-G)) expressed as percentage of control ± SEM. The bars represent data for coexposed rats, pretreated with 75 μg PCB126/kg on day 0, then placed on drinking water containing 1.00 mg perchlorate/kg/day on day 9 that continued until tissues were collected on day 22. The cross hairs (+) are the expected result based on the assumption of response additivity (determined from data in Figure 1). Results significantly different from control ($p \leq 0.05$) for coexposed animals are indicated by an asterisk (*). Data adapted from McLanahan *et al.* [34].

administered at concentrations considered to be below a no observed effect level result in a superadditive response, such that it will result in a measurable effect on the HPT axis?

There were no perturbations in serum TSH or T_4 concentrations or hepatic thyroxine phase II metabolic rates in rats administered a single dose of 0.075 μg/kg of PCB126 and evaluated at either 2 days or 5 days after dosing or in rats placed on 0.01 mg/kg/day perchlorate for either 1 day or 4 days. This was also true when rats were coexposed to 0.075 μg/kg of PCB126 and 0.01 mg/kg of perchlorate for up to 5 days. Thus, for these exposure conditions, considered to be low doses, an additive or superadditive effect was not observed. Interestingly, perchlorate appeared to increase hepatic T_4 phase II metabolism in a dose-dependent manner after administration of perchlorate in drinking water for 14 days [34]. When animals were exposed to PCB126 and then provided perchlorate in drinking water, perchlorate appeared to modestly suppress the effect of PCB126 on T_4-G formation. Although the significance of this finding is unknown, it suggests that perchlorate may be acting by more than one mechanism on the HPT axis and/or interfering with PCB126 metabolism.

19.4 Experimental Challenges

With the interest in understanding the public health implications of endocrine disrupting chemicals, there is a need to improve assay kits developed for the measurement of thyroid hormones and TSH to address subclinical or low-dose effects. The detection limit and variability in measured serum thyroid hormone and TSH concentrations are problematic for subtle disturbances in the HPT axis. Researchers have stated that statistically significant changes in serum TSH can be detected if adequate sample sizes are used [25]; however, determining the sample size to use is still difficult for weak acting compounds. Most studies referenced for this chapter used eight animals per treatment group [12, 34, 35]. Is this n sufficient and what is the variability between inbred rodent strains?

For thyroid hormone assays a coefficient of variation of 20% or less has been accepted [36], which may present problems for evaluating weak thyroid-active chemicals or low doses of chemicals. Typically, radioimmunoassay (RIA) methods commercially available (prepared commercial kits, for example, ELISA and clinical preparations) perform well, but the sensitivity of the same kit can substantially vary in different laboratories [37]. The use and validation of RIA kits for TSH and thyroid hormones have been recently reviewed [36]. Because of the variability and pulsatile nature of TSH, circadian rhythms, and inter- and intra-assay coefficients of variation, experiments should be timed to sample blood at the same time of the day. Assays should be carefully validated prior to use. Standardizing procedures across laboratories would help minimize differences between laboratories and allow a better comparison of studies when evaluating thyroid-active compounds. Alternative analytical methods such as reversed phase liquid chromatography–ICP–mass spectrometry (RPC-ICP-MS) [38] should become more widespread.

Traditionally, much of the published literature on the effects of thyroid-active chemicals on the HPT axis report on single time points following administration of a compound. To develop a better quantitative understanding of HPT axis disruption, time course studies are needed for HPT axis end points, similar to drug or chemical pharmacokinetic studies. This would help advance our interpretation of disturbances in the HPT axis by single chemical or chemical mixtures. The HPT axis is dynamic and has several adaptive mechanisms to overcome insult. A better quantitative and temporal understanding of the relationships between the administered dose of toxicant(s), internal dose linked to a mode of action, perturbation in the HPT axis, and ensuing frank toxicity due to hypothyroidism remains to be realized. This will require the measurement of thyroid hormones in target tissues, enzyme activity of thyroid metabolizing enzymes, and perhaps thyroid hormone receptor status.

In conclusion, our findings with a binary mixture of PCB126 and perchlorate suggest that for the HPT axis, when evaluating the consequences of chemical mixtures, it is important to understand the mode of action. In this case, the PCB126-induced perturbations in the HPT axis masked the perchlorate effect on the HPT axis. Qualitative and quantitative understanding of the HPT axis adaptive features is critical when interpreting results from mixture studies of thyroid-active chemicals.

19.5
Dose-Response Computational Modeling of Chemical Effects on the HPT Axis

Physiologically based pharmacokinetic (PBPK) models are useful tools to understand the fate of chemicals in biological systems. A quantitative description of chemical absorption, distribution, metabolism, and elimination in the body is needed to determine total tissue dose, which can then be linked to toxic outcomes. PBPK models can also be linked mathematically via mode of action to BBDR models to predict biological responses resulting from a given exposure. Model simulations of a combined PBPK/BBDR model can also provide quantitative insight into the chemical(s) mode of action.

For example, to describe the effects of perchlorate and PCB126 on the HPT axis in adult rats, the PBPK models for the individual chemicals should first be linked with the BBDR model of the HPT axis individually then together. The BBDR-HPT axis model has been described previously [39]. One mathematical description to allow quantitative evaluation of perchlorate's disruption of the HPT axis is based upon the well-known mode of action of perchlorate, competitive inhibition of the NIS. To describe this biological interaction quantitatively using the PBPK perchlorate model and BBDR-HPT axis model, the equation for NIS transport of iodide into the thyroid would also include competitive inhibition of iodide transport by perchlorate:

$$\mathrm{rTNIS_i} = \frac{\mathrm{Vmax\,T_i^{TSH}} \times \mathrm{Cvt_i}}{\mathrm{Cvt_i} + \mathrm{Km_i} \times \left(1 + \dfrac{\mathrm{Cvt_p}}{\mathrm{Ki_p}}\right)}, \tag{19.1}$$

where $\mathrm{VmaxT_i^{TSH}}$ is the TSH-regulated maximal rate of NIS iodide transport (nmol/h), $\mathrm{Cvt_i}$ is the free concentration of iodide in thyroid blood (nmol/l), $\mathrm{Km_i}$ is the affinity constant of iodide for the NIS (nmol/l), $\mathrm{Ki_p}$ is inhibition constant of perchlorate for NIS iodide transport (nmol/l), and $\mathrm{Cvt_p}$ is the concentration of perchlorate in the thyroid blood (nmol/l). Linking the models in this way allows evaluation of perchlorate's mode of action and its relationship with thyroidal iodide stores and decrease in circulating thyroid hormone concentrations.

Mathematically, the effects of PCB126 on the HPT axis can be explored with the BBDR-HPT axis model by mathematically describing an increase in the rate of T_4-G formation in the liver. Several approaches are available to describe an induction of the hepatic UDPGTs. One of the simplest approaches would be to describe the rate of T_4-G formation ($\mathrm{rUGT_{T4}}$) in relation to the PCB126 PBPK model-predicted blood concentration of PCB126 perfusing the liver such that

$$\mathrm{rUGT_{T4}} = \frac{\mathrm{Vmax_{T4}^{UGT}} * \mathrm{Cvl_{T4}^{free}}}{\mathrm{Km_{T4}^{UGT}} + \mathrm{Cvl_{T4}^{free}}}, \tag{19.2}$$

$$\mathrm{Vmax_{T4}^{UGT}} = \frac{\mathrm{Vmax1_{T4}^{UGT}} * \mathrm{Cvl_{PCB126}}}{\mathrm{Km_{PCB126}^{UGT}} + \mathrm{Cvl_{PCB126}}}, \tag{19.3}$$

where Cvl_{PCB126} (nmol/l) is the PCB126 PBPK model prediction of PCB126 hepatic venous blood, $Vmax1_{T4}^{UGT}$ (nmol/h) is the maximal rate of T_4-G formation in the presence of PCB126, Km_{PCB126}^{UGT} (nmol/l) is a constant that relates the concentration of PCB126 in the hepatic venous blood to the half-maximal induction of UGTs that catalyze formation of T_4-G, Km_{T4}^{UGT} (nmol/l) is the affinity constant of T4 for UDPGT, and Cvl_{T4}^{free} is the free T_4 in the hepatic venous blood (nmol/l) available for glucuronidation predicted by the BBDR-HPT axis model.

However, greater complexity is often needed to mathematically describe induction of proteins and enzymes in biological systems. In order to achieve changes in the rate of T_4-G formation described in the literature as a result of PCB126, a more complex approach, similar to what has been used to describe nuclear AhR binding, and the induction and binding of specific cytochrome P450s by TCDD [40, 41] could be employed. This would include more biological and physiological information such as binding of PCB126 to the AhR and subsequent increase in protein production and inclusion of intrinsic degradation rates of the enzyme.

The mathematical description of the interaction of PCB126 with the BBDR-HPT axis model depends partly on the amount of data available to determine model constants. The second approach requires many more model constants and is less desirable if the model constants have to be estimated rather than determined from literature, increasing model uncertainty.

Using a mode of action-based approach to link the PBPK models of PCB126 and perchlorate with the BBDR-HPT axis model provides a framework to quantitatively analyze the impact of the compounds on the HPT axis. This analysis can inform the quantitative impact of a chemical-specific mode of action on the disruption of the HPT axis. In addition, the mathematical models not only provide a basis to support the laboratory-determined mode of action but also inform and hypothesize additional modes of action, if necessary. Finally, once individual chemical perturbations of PCB126 and perchlorate are described with a PBPK/BBDR model, both PBPK models for perchlorate and PCB126 can be linked with the BBDR-HPT axis model to simulate chemical interactions on the HPT axis. The mixture model can also be used to explore how the dosing regimen can affect the HPT axis (e.g., chemicals administered concurrently, sequentially, etc.).

References

1 Boas, M., Feldt-Rasmussen, U., Skakkebaek, N.E., and Main, K.M. (2006) Environmental chemicals and thyroid function. *Eur. J. Endocrinol.*, **154**, 599–611.

2 Capen, C.C. (2001) Toxic responses of the endocrine system, in *Casarett and Doull's Toxicology, the Basic Science of Poisons*, 6th edn (ed. C.D. Klaassen), McGraw-Hill, New York.

3 Zoeller, R.T. and Tan, S.W. (2007) Implications of research on assays to characterize thyroid toxicants. *Crit. Rev. Toxicol.*, **37**, 195–210.

4 Cunha, G.C. and van Ravenzwayy, B. (2005) Evaluation of mechanisms inducing thyroid toxicity and the ability of the enhanced OECD Test Guideline 407 to detect these changes. *Arch. Toxicol.*, **79**, 390–405.

5 Wolff, J. (1998) Perchlorate and the thyroid gland. *Pharmacol. Rev.*, **50**, 89–105.

6 Riesco-Eizaguirre, G. and Santisteban, P. (2006) A perspective view of sodium iodide symporter research and its clinical implications. *Eur. J. Endocrinol.*, **155**, 495–512.

7 Hayes, M.T. and Solomon, D.H. (1965) Influence of the gastrointestinal iodide cycle on the early distribution of radioactive iodide in man. *J. Clin. Invest.*, **44**, 117–127.

8 Clewell, R.A., Merill, E.A., Yu, K.O., Mahle, D.A., Sterner, T.R., Fisher, J.W., and Gearhart, J.M. (2003) Predicting neonatal perchlorate dose and inhibition of iodide uptake in the rat during lactation using physiologically-based pharmacokinetic modeling. *Toxicol. Sci.*, **74**, 416–436.

9 Pavelka, S. (2004) Metabolism of bromide and its interference with the metabolism of iodine. *Physiol. Res.*, **53** (Suppl. 1), 81–90.

10 Viluksela, M., Raasmaja, A., Lebofsky, M., Stahl, B.U., and Rozman, K.K. (2004) Tissue-specific effects of 2,3,7,8-tetrachlorodibenzo-*p*-dioxin (TCDD) on the activity of 5′-deiodinases I and II in rats. *Toxicol. Lett.*, **147**, 133–142.

11 Wade, M.G., Parent, S., Finnson, K.W., Foster, W., Younglai, E., McMahon, A., Cyr, D.G., and Hughs, C. (2002) Thyroid toxicity due to subchronic exposure to a complex mixture of 16 organochlorines, lead, and cadmium. *Toxicol. Sci.*, **11**, 1075–1081.

12 Fisher, J.W., Campbell, J., Muralidhara, S., Bruckner, J.V., Ferguson, D., Mumtaz, M., Harmon, B., Hedge, J.M. *et al.* (2006) Effect of PCB126 on hepatic metabolism of thyroxine and perturbations in the hypothalamic–pituitary–thyroid axis in the rat. *Toxicol. Sci.*, **90**, 87–95.

13 Santini, F., Vitti, P., Ceccarini, G., Mammoli, C., Rosellini, V., Pelosini, C., Marsili, A., Tonacchera, M. *et al.* (2003) *In vitro* assay of thyroid disruptors affecting TSH-stimulated adenylate cyclase activity. *J. Endocrinol. Invest.*, **26**, 950–955.

14 Schmutzler, C., Hamann, I., Hofmann, P.J., Kovacs, G., Stemmler, L., Mentrup, B., Schomburg, L., Ambrugger, P. *et al.* (2004) Endocrine active compounds affect thyrotropin and thyroid hormone levels in serum as well as endpoints of thyroid action in liver, heart, and kidney. *Toxicology*, **205**, 95–102.

15 Engler, H., Taurog, A., and Nakashima, T. (1982) Mechanism of inactivation of thyroid peroxidase by thiourylene drugs. *Biochem. Pharmacol.*, **31**, 3801–3806.

16 Ortega, E., Osorio, A., and Ruiz, E. (1996) Inhibition of 5′DI and 5′DII L-tiroxine (T_4) monodeiodinases. Effect on the hypothalamo-pituitary ovarian axis in adult hypothyroid rats treated with T_4. *Biochem. Mol. Biol. Int.*, **39**, 853–860.

17 Doerge, D.R. and Sheehan, D.M. (2000) Goitrogenic and estrogenic activity of soy isoflavones. *Environ. Health Perspect.*, **110**, 349–353.

18 Van den Berg, K.J., van Raaij, J.A., Bragt, P.C., and Notten, W.R. (1991) Interactions of halogenated industrial chemicals with transthyretin and effects on thyroid hormone levels *in vivo*. *Arch. Toxicol.*, **65**, 15–19.

19 Kobal, S., Cebulji-Kadunc, N., and Cestnik, V. (2000) Serum T_3 and T_4 concentrations in the adult rats treated with herbicide 2,4-dichlorophenoxyacetic acid. *Pflugers Arch.*, **440** (Suppl. 5), R171–R172.

20 Hallgren, S. and Darnerud, P.O. (2002) Polybrominated diphenyl ethers (PBDEs), polychlorinated biphenyls (PCBs) and chlorinated paraffins (CPs) in rats – testing interactions and mechanisms for thyroid hormone effects. *Toxicology*, **177**, 227–243.

21 Darnerud, P.O., Törnwall, U., Morse, D., Klasson-Wehler, E., and Brouwer, A. (1996) Binding of a 3,3′,4,4′-tetrachlorobiphenyl (CB-77) metabolite to fetal transthyretin and effects on fetal thyroid hormone levels in mice. *Toxicology*, **106**, 105–116.

22 Bogazzi, F., Raggi, F., Ultimieri, F., Russo, D., Campomori, A., McKinney, J.D., Pinchera, A., Bartalena, L. *et al.* (2003) Effects of a mixture of polychlorinated biphenyls (Aroclor 1254) on the transcriptional activity of the thyroid

hormone receptor. *J. Endocrinol. Invest.*, **26**, 972–978.

23 Zoeller, R.T., Bansal, R., and Parris, C. (2005) Bisphenol-A, an environmental contaminant that acts as a thyroid hormone receptor antagonist *in vitro*, increases serum thyroxine, and alters RC3/neurogranin expression in the developing rat brain. *Endocrinology*, **146**, 607–612.

24 Centers for Disease Control and Prevention (2005) Third National Report on Human Exposure to Environmental Chemicals. Centers for Disease Control and Prevention, National Center for Environmental Health, Atlanta, GA.

25 DeVito, M., Biegel, L., Brouwer, A., Brown, S., Brucker-Davis, F., Cheek, O., Christensen, R., Colborn, T. *et al.* (1999) Screening methods for thyroid hormone disruptors. *Environ. Health Perspect.*, **107**, 407–415.

26 United States Environmental Protection Agency (US EPA) (2007). Integrated Risk Information System (IRIS). Available online at http://www.epa.gov/IRIS. Accessed October 2007.

27 Khan, M.A., Fenton, S.E., Swank, A.E., Hester, S.D., Williams, A., and Wolf, D.C. (2005) A mixture of ammonium perchlorate and sodium chlorate enhances alterations of the pituitary-thyroid axis caused by the individual chemicals in adult male F344 rats. *Toxicol. Pathol.*, **33**, 776–783.

28 James-Walke, N.L., Williams, H.L., Taylor, D.A., and McMillen, B.A. (2006) The effect of oral consumption of perchlorate, alone and in combination with ethanol, on plasma thyroid hormone and brain catecholamine concentrations in the rat. *Basic Clin. Pharmacol. Toxicol.*, **99**, 340–345.

29 Braverman, L.E., He, X., Pino, S., Cross, M., Magnani, B., Lamm, S.H., Kruse, M.B., Engel, A. *et al.* (2005) The effect of perchlorate, thiocyanate, and nitrate on thyroid function in workers exposed to perchlorate long-term. *J. Clin. Endocrinol. Metab.*, **90**, 700–706.

30 Tonacchera, M., Pinchera, A., Dimida, A., Ferrarini, E., Agretti, P., Vitti, P., Santini, F., Crump, K. *et al.* (2004) Relative potencies and additivity of perchlorate, thiocyanate, nitrate, and iodide on the inhibition of radioactive iodide uptake by the human sodium iodide symporter. *Thyroid*, **14**, 1012–1019.

31 Desaulniers, D., Leingartner, K., Musicki, B., Yagminas, A., Xiao, G.-H., Cole, J., Marro, L., Charbonneau, M. *et al.* (2003) Effects of postnatal exposure to mixtures of non-*ortho*-PCBs, PCDDs, and PCDFs in prepubertal female rats. *Toxicol. Sci.*, **75**, 468–480.

32 Crofton, K.M., Craft, E.S., Hedge, J.M., Gennings, C., Simmons, J.E., Carchman, R.A., Carter, W.H., Jr., and DeVito, M.J. (2005) Thyroid-hormone-disrupting chemicals: evidence for dose-dependent additivity or synergism. *Environ. Health. Perspect.*, **113**, 1549–1554.

33 Zhou, T., Taylor, M.M., DeVito, M.J., and Crofton, K.M. (2002) Developmental exposure to brominated diphenyl ethers results in thyroid hormone disruption. *Toxicol. Sci.*, **66**, 105–116.

34 McLanahan, E.D., Campbell, J.L., Jr., Ferguson, D.C., Harmon, B., Hedge, J.M., Crofton, K.M., Mattie, D.R., Braverman, L. *et al.* (2007) Low-dose effects of ammonium perchlorate on the hypothalamic–pituitary–thyroid axis of adult male rats pretreated with PCB126. *Toxicol. Sci.*, **97**, 308–317.

35 Yu, K.O., Narayanan, L., Mattie, D.R., Godfrey, R.J., Todd, P.N., Sterner, T.R., Mahle, D.A., Lumpkin, M.H. *et al.* (2002) The pharmacokinetics of perchlorate and its effect on the hypothalamus–pituitary–thyroid axis in the male rat. *Toxicol. Appl. Pharmacol.*, **182**, 148–159.

36 OECD (Organisation for Economic Co-operation and Development, Environment Directorate) (2006) Detailed Review Paper on Thyroid Hormone Disruption Assays. OECD Environment, Health and Safety Publications Series on Testing and Assessment No. 57. ENV/JM/MONO(2006)24. Available online at http://www.oecd.org/ehs.

37 Scanlon, M.R. and Toft, A.D. (2000) Regulation of thyrotropin secretion, in *Werner & Ingbar's The Thyroid, A Fundamental and Clinical Text*, 8th edn

(eds L.E. Braverman and R.D. Utiger), Lippincott Williams & Wilkins, New York.

38 Michalke, B., Schramel, P., and Witte, H. (2000) Method developments for iodine speciation by reversed-phase liquid chromatography-ICP-mass spectrometry. *Biol. Trace Elem. Res.*, **78**, 67–79.

39 McLanahan, E.D., Andersen, M.E., and Fisher, J.W. (2008) A biologically based dose-response model for dietary iodide and the hypothalamic–pituitary–thyroid axis in the adult rat: evaluation of iodide deficiency. *Toxicol. Sci.*, **102**, 241–253.

40 Andersen, M.E., Birnbaum, L.S., Barton, H.A., and Eklund, C.R. (1997) Regional hepatic CYP1A1 and CYP1A2 induction with 2,3,7,8-tetrachlorodibenzo-*p*-dioxin evaluated with a multicompartmental geometric model of hepatic zonation. *Toxicol. Appl. Pharmacol.*, **144**, 145–155.

41 Kohn, M.C., Sewall, C.H., Lucier, G.W., and Portier, C.J. (1996) A mechanistic model of effects of dioxin on thyroid hormones in the rat. *Toxicol. Appl. Pharmacol.*, **136**, 29–48.

20
Toxic and Genotoxic Effects of Mixtures of Polycyclic Aromatic Hydrocarbons

K.C. Donnelly and Ziad S. Naufal

20.1
Introduction

Exposure of human and ecological receptors to mixtures of polycyclic aromatic hydrocarbons (PAHs) is routine. Both natural (forest fires, volcanoes, etc.) and manmade (cooking, combustion of fossil fuels, cigarette smoking, etc.) sources of PAH mixtures exist. Approximately 250 years ago, small children exposed to PAH mixtures from soot from chimneys were observed to have an increased risk of scrotal cancer [1]. In 1930, Hieger [2] isolated a yellow powder from coal tar pitch that was found to induce cancer in rodents. More recently, occupational studies have observed excess rates of lung and bladder cancers in workers exposed to PAH mixtures [3–5].

Animal studies have been used to provide detailed information regarding the metabolism and dose-response relationship of individual PAH carcinogens. In addition, occupational studies have provided critical information regarding the potency of specific PAH mixtures. However, due to the broad range of chemical composition and concentrations in PAH mixtures, there remains a great deal of uncertainty regarding the toxicity and genotoxicity of PAH mixtures. This chapter provides a summary of the present state of knowledge regarding the toxic and genotoxic risk associated with PAH mixtures. Due to the rapid advance of new genomic tools, as well as cell culture and animal models and pharmacokinetic models, new information regarding the risk associated with these mixtures is being generated at a remarkable rate. Thus, this text is not intended to be exhaustive. Rather, the information provided in this chapter gives a brief history of these compounds, an overview of their metabolism and toxicity, and a summary of information needs that could reduce the uncertainty associated with exposure to PAH mixtures.

Complex chemical mixtures containing PAHs have been detected at almost half the 1609 hazardous waste sites listed as Superfund sites in the United States [6]. PAH mixtures are commonly found at wood preserving sites, coal gasification sites, refineries, petroleum production facilities, and other sites where petroleum products have been produced, stored, or disposed. In addition, PAH mixtures are common combustion by-products. Thus, sources of human exposure to PAH mixtures include

Principles and Practice of Mixtures Toxicology. Edited by Moiz Mumtaz

ISBN: 978-3-527-31992-3

not only the release of hazardous chemicals but also ingestion of cooked foods and inhalation of cigarette smoke or other combustion by-products.

20.2 Sources

The release of hydrocarbon mixtures into the environment may produce contamination of air, surface water, soil, sediment, and/or groundwater. Such mixtures generally persist in the environment and could pose a threat to human and/or ecological health by accumulating in the food chain. Humans are generally exposed to complex environmental mixtures capable of producing a broad range of biological effects [7, 8]. Environmental exposures are often repetitive, low-dose exposures involving multiple pathways including inhalation, ingestion, and dermal absorption. The severity of adverse health effects produced following such exposures varies greatly and depends not only on the dose and duration of exposure but also on intrinsic individual factors such as lifestyle exposures and genetic sensitivities.

PAHs are ubiquitous environmental contaminants. The incomplete combustion of virtually any type of organic material results in the production of these chemicals. PAHs share a similar chemical structure consisting of two or more fused benzene rings in linear, cluster, or angular arrangements [9]. PAHs are lipophilic nonpolar chemicals that can adsorb to particles in the air or water and generally persist in the environment for extended periods of time [10]. High concentrations of PAHs are present in crude oil, coal, and oil shale. These petroleum and petrochemical products are extensively used to produce fuels and synthetics (fibers and plastics) [11]. The widespread use of petroleum products has increased the level of PAHs in the environment. Coke oven emissions and the combustion of fossil fuels and refuse generate approximately 50% of the total emissions of the carcinogen, benzo[*a*]pyrene (BAP), in the United States. Vehicle emissions constitute another major source of PAHs, especially in urban areas where they generate approximately 35% of PAH emissions. Other sources of PAHs in the environment include fumes from manufacturing industries and tobacco smoking. Natural sources of PAHs include forest fires and volcanic eruptions. However, in most regions of the world, anthropogenic sources are generally assumed to be the major source of PAHs [11].

Levels of PAHs in urban atmospheres depend on the density, sources of local emissions, temperature, and local meteorological conditions among other factors. With respect to human health, the major concern is particles that are less than 5 μm in diameter as these are considered respirable. In rural areas, background levels of PAHs were reported to range between 0.02 and 1.2 ng/m^3, whereas in urban area levels ranged between 0.15 and 19 ng/m^3 [12]. Atmospheric PAH concentrations were also found to proportionally increase with population density, as described by Hafner *et al.* [13]. PAHs levels in air in developing countries were typically higher than those of developed countries that is assumed to be due to lack of regulation and technological innovation [13].

Table 20.1 Summary of median and range concentrations of BAP, carcinogenic (USEPA B2), and total PAH concentrations in different environmental media.

Environmental medium	Carcinogenic PAH	Total PAH	References
Outdoor air (ng/m^3)	12 (0.1–19 700)	20 (4–360)	[171–175]
Indoor air (ng/m^3)	2 (2–24)	290 (7–21 000)	[176–182]
Soil (mg/kg)	n/a	1 (0.01–2600)	[14, 183–186]
Sediment (mg/kg)	n/a	5 (0.03–17 280)	[184, 187–190]
Surface water (μg/l)	n/a	3 (0.01–340)	[15, 191–193]
Ground water (μg/l)	n/a	5200 (0.01–14 240)	[16, 17, 194, 195]

n/a = not available.

A brief summary of concentrations of PAHs detected in environmental media in several regions of the world is included in Table 20.1. Air concentrations of total and carcinogenic PAHs (cPAHs) appear to be affected primarily by proximity to high-density traffic areas and industrial facilities. In most studies, PAH concentrations were found to be higher in the winter than in other seasons. Median concentrations of cPAHs and total PAHs (tPAHs) in outdoor air were 12 and 20 ng/m^3, respectively, whereas, median concentrations of cPAHs in indoor air were 2 ng/m^3 and median concentrations of tPAHs in indoor air were 290 ng/m^3 (Table 20.1). Indoor air concentrations of PAHs may be higher than outdoor air because of the larger number of sources (tobacco smoke, cooking, heating, etc.) and because indoor air has a lower exchange rate than outdoor air.

In sediment and soil, PAH concentrations are often elevated in areas where coal, wood, gasoline, or other products have been burned. Total PAH concentrations in soil were found to range from 0.01 to 2600 mg/kg, while sediment PAH concentrations ranged from 0.03 to 17280 mg/kg (Table 20.1). The highest concentrations of PAHs in sediment were detected in an urbanized watershed. The most abundant PAHs included 4- and 5-ring PAHs that constituted around 75% of the PAH content in sediment samples. Significant levels of PAHs exist in soil in virtually all regions of the earth. PAH contamination tends to be higher in urban industrialized areas where levels of PAHs are usually 10–100 times more than those in undeveloped areas [11]. Depending on sampling methods and location, concentrations of PAHs are likely to vary significantly. Wang *et al.* [14] collected surface soil samples from four sampling sites in Dalian, a city located in Northeast China. The four sites included one close to traffic, another in a park or residential area, a suburban, and a rural site, and PAH concentrations were reported to range between 0.2 and 9 ppm. Total PAHs showed an urban–suburban–rural gradient with the traffic site having around 18 times higher concentrations of tPAHs compared to the rural site.

Atmospheric dispersion of particle-bound PAHs often results in the deposition of these chemicals into surface waters. Approximately, two-thirds of PAHs detected in surface water are particle-bound. Runoff of polluted ground sources or direct pollution of rivers and lakes by municipal and industrial effluents are also among the diverse sources of water contamination. Lower concentrations of PAHs may also be leached through soils into groundwater. PAH concentrations in surface waters were found to range from 0.01 to 340 μg/l (Table 20.1). In groundwater, PAH concentrations ranged from 0.01 to 14 240 μg/l. PAHs in water sampled near a bitumen field in Nigeria were found to range from 11 to 342 μg/l [15]. In a study on groundwater quality at Anoka Sand Plain Aquifer located in east central Minnesota, USA, a source of drinking water, benzo[*ghi*]perylene and indeno[1,2,3-*cd*]pyrene were the most elevated PAHs; however, their levels were in the ng/l range [16]. Groundwater sampled near a coal and oil gasification facility in Seattle, Washington, USA, which ceased operation in 1956 was found to contain PAH levels ranging from 230 to 14 240 μg/l. PAH levels were highest where groundwater was in contact with a nonaqueous phase liquid (NAPL) in the soil [17].

Direct contamination of food with PAHs has also been reported. Intake of PAH-rich foods was linked to cancer of the stomach and esophagus [18] as well as cancer of the colon and rectum in humans [19]. PAHs in food items can have plant or animal origins. Vegetables with large leaves accumulate PAHs on their surface and to a lesser extent in their internal tissue. Grazing cattle and poultry can accumulate PAHs in their tissues. PAH contamination was detected in leafy plants such as lettuce, spinach, tea, and tobacco, and in smoked meats and fish. PAHs present in plants are most likely due to atmospheric contamination. In fresh meats and seafood, PAHs exist due contamination of air, water, or animal feed. Cooking methods such as frying and charcoal broiling tend to increase the level of PAHs and other potentially carcinogenic chemicals such as heterocyclic amines in foods (HCAs). Depending on the method of food preparation, levels are highly variable but can reach parts per billion (ppb) levels [12]. Barbecued meat, for example, can have a PAH level as high as 10–20 ppb [20]. Epidemiological studies revealed a positive association between consumption of red meat cooked by deep-frying and risk of breast cancer [21].

Food may be considered a main source of exposure in human populations. Contamination of food by PAHs can be environmental or through cooking or processing. Average daily intake of PAHs in humans has been estimated to be 0.2 μg from air, 0.03 μg from water [12], and 2–3 μg from food. The average daily intake of PAHs from food compares to 2–5 μg PAHs per pack of cigarettes in a regular smoker [22]. Exposure from dietary sources can account for more than 70% of PAH exposure in nonsmokers and also significantly contributes to nonoccupational exposure to PAHs. Cereals, oils, and vegetables are some of the main sources of PAHs in the diet [20]. Food content was analyzed for potential carcinogens including PAHs. Levels of BAP and total PAHs in several food items are summarized in Table 20.2. For products from plant origin, the following median values were reported: vegetables 0.1 ppb BAP and 4.2 ppb total PAHs [23]; fruits 0.01 ppb BAP and 0.7 ppb total PAHs [24]; wheat grain 0.3 ppb BAP and 4 ppb total PAHs [25]; wheat flour 0.1 ppb BAP and 1.5 ppb total PAHs [26]; white bread 0.017 ppb BAP and 3.12 ppb total PAHs [27]; and coffee 0.9 ppb BAP and 25 ppb total PAHs [28]. Cooking often increases PAH concentrations in foods. Values reported for cooked foods (Table 20.2) include 50 ppb BAP and 800 ppb total PAHs for smoked fish [29], 1.5 ppb BAP and 45 ppb total PAHs for cooked meat [27], and 55 ppb BAP and 800 ppb total PAHs for grilled frankfurters [30].

Table 20.2 Summary of median and range concentrations of PAHs (in ppb) in different food items.

Source	BAP	Total PAH	References
Fruits, vegetables, and grains	0.1 (0.01–0.3)	3 (0.7–4.2)	[23–27]
Coffee	0.9	25	[28]
Vegetable oil	30 (0.1–110)	220 (25–750)	[196, 197]
Meat and fish	26 (nd–55)	90 (10–800)	[27, 29, 30, 198, 199]

nd = nondetect.

PAHs are a major component of tobacco smoke. About 150 different unsubstituted and methylated PAHs have been detected in tobacco smoke condensate. It has been estimated that approximately 10–50 μg of BAP is inhaled into the lungs from one cigarette [31, 32]. According to Harvey [11], more than 150 PAHs occur in gas phase and more than 2000 were identified in the particulate phase. Tobacco smoke is one of the most important sources of PAHs in indoor air.

Background exposures to PAHs are typically low (air, water, and diet) compared to occupational sources. PAH exposure can be significant during industrial processes such as coal tar production and aluminum smelting, and also in industries where petroleum and petroleum products are used. Exposure to PAHs in an aluminum production plant was reported at levels ranging between 1488 and 15 149 ng/m^3 depending on the job function [33], and as high as 200 000 ng/m^3 in coke ovens [34]. Monitoring of airborne PAHs in the work environment of an iron foundry and two steel plants revealed that the concentrations of PAHs ranged between 321 and 1331 ng/m^3 compared to 7 ng/m^3 in urban areas of the general environment [35]. Among other occupations, personal PAH monitoring revealed that traffic policemen are exposed to PAH levels ranging from 9 to 140 ng/m^3 in Budapest, Hungary [36], and around 1700 ng/m^3 in Beijing, China [37]. The total PAH exposure levels for tollbooth attendants were reported to range between 6300 and 15 000 ng/m^3 in Taipei, Taiwan [38].

20.3 Source Apportionment

On the basis of the ratio of various components, PAH mixtures can be classified as originating from petroleum or petrogenic sources, or from pyrolysis or pyrogenic sources. Specific marker compounds may help in the identification of the major sources of PAHs at a given location. In the case of petrogenic sources, low molecular weight compounds (two and three rings) tend to be predominant, whereas PAHs emitted by incomplete combustion of organic materials have a higher proportion of heavier compounds (four rings or larger). As an illustration, low molecular weight compounds contributed to 98% of total PAHs emitted by coke ovens, 76 and 73% of the emissions from diesel and petroleum engines, respectively, and 80% of the PAHs emitted from wood combustion [39]. Indeno[1,2,3-*cd*]pyrene (six rings) and benzo [*ghi*]perylene (five rings) were detected in PAH profiles associated with mobile sources. Benzo[*ghi*]perylene is known to be a marker of gasoline exhaust emissions [40]. The relative abundance of PAHs present in mixtures can therefore be used to elucidate sources and provide a source fingerprint [41]. Molecular indices based on ratios of selected PAH concentrations may differentiate between pyrolytic and petrogenic sources [42]. For example, a phenanthrene:anthracene ratio greater than 10 suggests a petrogenic source of PAHs. On the other hand, a fluoranthene:pyrene ratio greater than 1.0 indicates that PAH contamination is most likely due to combustion or pyrolytic processes.

20.4 Hazardous Effects of Polycyclic Aromatic Hydrocarbons

20.4.1 Ecological Effects

Wildlife may be affected by exposure to PAHs both through direct contact and through food chain poisoning. Laboratory and field studies on feral organisms have attempted to evaluate the toxic potential of PAHs. Increased tumor formation and other diseases have been reported in fish exposed to PAHs. However, field studies of PAH toxicity present the challenge of identifying biological responses caused specifically by PAHs due to the presence of common cocontaminants such as organochlorines and metals. Fish uptake and rapidly metabolize PAHs [43] and excrete them into the bile that is considered a major route of elimination [44]. Metabolism converts up to 99% of PAH to metabolites within 24 h of uptake, significantly reducing tissue concentrations of PAH parent compounds [44]. Thus, PAHs tend not to bioconcentrate and might cause adverse effects with little or no chemical trace. As a result, assessing exposure to PAHs in fish may yield misleading results. Determining the levels of PAH metabolites in bile is usually more significant since it reflects uptake, metabolism, and excretion [45]. The major adverse effects reported in feral fish and linked to PAHs can be categorized into biochemical, histopathologic, immunological, genetic, reproductive, developmental, and behavioral effects.

Developmental effects were observed from relatively low PAH concentrations in larval and juvenile fish. Moles and Norcross [46] reported reduced growth in juvenile flounder with chronic exposure to low levels of PAHs (around 1.6 ppm) in sediment. Among the behavioral changes that were potentially linked to PAH exposure are the alteration of feeding [47, 48] and swimming behaviors [49]. In lab feeding studies during which fish are exposed to higher PAH concentration than detected in the field, physiological changes were detected at PAH concentrations around 25 000 ng/g wet weight in juvenile chinook salmon whereas immunosuppression, CYP1A induction and DNA damage were observed in rainbow trout exposed to PAHs in diet at concentrations around 40 000 ng/g wet weight [50].

Effects of acute or chronic exposure to PAHs in wild mammals, reptiles, and birds are poorly defined. PAHs have been implicated as possible causal agents for the health problems of beluga whales found dead over a period of 8 years along the shores of St. Lawrence estuary in Canada [51]. This assumption was based on a finding of an elevated incidence of neoplasms and BAP-DNA adducts in whale tissues [52].

20.4.2 Human Cancer Risk

A primary concern associated with human exposures to complex mixtures of PAHs comes from the fact that many of these compounds have mutagenic and carcinogenic properties [53]. In addition, a typical PAH mixture includes hundreds of compounds that may act as enzyme inducers, carcinogens, or promoters. Thus, the interactions of the components of a specific PAH mixture are difficult to predict. Historically, human exposures to a variety of PAH complex mixtures have been associated with increased cancer rates [54–58].

As early as 1775, Sir Percivall Pott [1] linked scrotal cancer to exposure to PAHs in soot among British chimney sweeps. This study was the first to link an environmental exposure to a specific type of cancer. More than a century later, studies using rabbits observed the production of malignant epithelial tumors by repetitive application of coal tar to the ear skin [59]. Beginning in 1930 with the help of the British Gas, Light and Coke Company, Hieger [2] isolated approximately 7 g of a yellow powder from 2 tons of coal tar pitch by repetitive steps of fractional distillation, extraction, and crystallization. The yellow powder showed strong carcinogenic activity in rodents. Radioactively labeled PAHs that became available in the late the 1940s were found to bind to both proteins and DNA fractions in epidermal cells after administration to mice. DNA was then proposed to be the essential "cellular receptor for carcinogenesis."

The carcinogenic potencies of a series of PAHs and the extent to which these are bound to DNA *in vivo* were roughly correlated. It was further noted that this process was dependent on a series of additional cellular events and the presence of activating enzymes residing in the endoplasmic reticulum (ER). However, it was not until the early 1970s when Borgen *et al.* [60] reported that a metabolite of BAP, the 7,8-dihydrodiol, binds to a 10-fold greater extent to DNA *in vitro* than its parent compound. For this to occur, activating microsomal preparations were required. Sims *et al.* [61] proposed that a secondary metabolite, the 7,8-dihydrodiol-9,10-epoxide (diol epoxide), a derivative of BAP, is the chemical covalently interacting with DNA. Hence, the diol epoxide was considered to be the ultimate carcinogen.

Animal studies have demonstrated that individual PAHs and PAH mixtures may act as potent carcinogens in rodent skin models. These observations are in agreement with results from human studies indicating an increased incidence of skin cancer associated with the use of various coal tar preparations [62–69]. One of the first synthetic PAH congeners shown to be carcinogenic in laboratory animal studies was dibenz[*a*,*h*]anthracene [70]. BAP was later identified as an important carcinogenic PAH congener present in many carcinogenic PAH mixtures [70]. Several PAH congeners have been classified by the International Agency for Research on Cancer (IARC) as human carcinogens (either 2A-probable or 2B-possible) [71]. These include benz[*a*]anthracene, benzo[*b*]fluoranthene, benzo[*j*]fluoranthene, benzo[*k*] fluoranthene, BAP, dibenz[*a,h*]acridine, dibenz[*a,j*]acridine, dibenz[*a,h*]anthracene, dibenzo[*a,e*]pyrene, dibenzo[*a,h*]pyrene, dibenzo[*a,i*]pyrene, dibenzo[*a,l*]pyrene, indeno[1,2,3-*c,d*]pyrene, and 5-methylchrysene. PAHs categorized by the US Environmental Protection Agency (USEPA) as probable human carcinogens (B2) include benz[*a*]anthracene, chrysene, benzo[*b*]fluoranthene, benzo[*k*]fluoranthene, BAP, dibenz[*a,h*]anthracene, and indeno[1,2,3-*c,d*]pyrene [72].

Occupational studies are of particular importance in analyzing human health effects of PAH mixtures. They are usually considered to be the most complete studies in terms of exposure, dose-response, and adverse health effects. The IARC classified several PAH mixtures common in occupational settings and industrial processes as carcinogenic, probably carcinogenic (2A), or possibly carcinogenic (2B) to humans based on epidemiological and experimental evidence [73–78]. PAH mixtures evaluated for carcinogenicity by IARC include coal tars, diesel exhaust, soots, shale oils, and emissions from aluminum production, coal gasification, coke production, and iron or steel foundries.

Recently, Mastrangelo *et al.* [78] evaluated PAH carcinogenicity using a summary of results from occupational studies published between 1966 and 1996. Ten studies that investigated the link between PAH exposure and lung or bladder cancer were reviewed [79–88]. The study by Armstrong *et al.* [79] examined lung cancer in relation to BAP exposure in a cohort of aluminum workers. A smoking-adjusted lung cancer rate of 2.2 was observed and corresponded to a cumulative BAP exposure ranging from 100 to 199 $\mu g/m^3$ with a significant dose-response trend. Tremblay *et al.* [88] reviewed bladder cancer in the same cohort of aluminum workers. After adjusting for smoking, a risk of bladder cancer was reported to be 6.7 times higher in workers exposed to a cumulative level of BAP ranging between 200 and 299.9 $\mu g/m^3$ compared to nonexposed workers. However, none of the studies reviewed by Mastrangelo *et al.* [78] measured internal dose in workers as a biomarker of exposure to PAHs.

Boffetta *et al.* [73] published a more comprehensive review of cancer risk associated with occupational exposures to PAHs. Cancer incidence was reviewed in workers in aluminum production, coal gasification, coke production, iron and steel foundries, as well as workers exposed to diesel engine exhaust, coal tars and related products, carbon blacks, and other mixtures of PAHs. In a study on coal gasification workers in Germany, the relative risk (RR) of lung cancer was estimated to be 2.9 [89]. Laryngeal and renal cancers were not found to be significantly increased in these populations. Exposure misclassification and controlling for confounders such as tobacco smoking, however, remain an issue in these studies.

More recently, a review by Bosetti *et al.* [90] presented detailed results from cohort studies linking quantitative occupational PAH exposure to respiratory and urinary tract neoplasms. Epidemiological evidence from the reviewed studies confirms an elevation in lung/respiratory cancer risk among several PAH-related industrial processes, except for aluminum and carbon electrode manufacturers. The pooled relative risk by industry was 2.58 for coal gasification, 1.58 for coke production, 1.40 for iron and steel foundries, 1.51 for roofers, and 1.30 for carbon black production. Cancer of the bladder and the urinary system was less evident among the reviewed occupations, with the exception of aluminum production (pooled RR = 1.29), coal gasification (pooled RR = 2.39), and iron and steel foundries (pooled RR = 1.29) that were associated with a modest increase in risk.

Recent occupational studies are summarized in Table 20.3 Krishnadasan *et al.* [200] selected 362 prostate cancer cases and 1805 matched controls from a cohort of aerospace and radiation workers in the United States. PAHs were the most common chemical that workers were exposed to (39%). Crude risk was 1.3. This relation however was not significant after adjusting of exposure to other chemicals, physical activity and socioeconomic status. In a study on coal tar-derived substances and risk of bladder or lung cancer, Friesen *et al.* [4] selected a cohort of 6423 men who worked for three or more years at an aluminum smelter in Canada for a period extending from 1954 to 1997. Exposure in this study was evaluated using two different measures of PAH exposure, benzene-soluble materials (BSM) and BAP. These two exposure indices were found to be highly correlated ($r = 0.94$). The median BAP cumulative exposures were 20 and 18 $\mu g/m^3$/year for no lag and 20-year lag, respectively, and the maximum cumulative exposure was 300 $\mu g/m^3$/year. The BAP cumulative exposures and incidence of bladder cancer indicated a strong dose-response relationship with the highest relative risk (3.0) corresponding to the highest cumulative exposure. The relative risk of lung cancer was lower (1.8) for the highest cumulative exposure to BAP but the dose-response relationship was the same.

Hoshuyama *et al.* [5] monitored mortality in a cohort of more than 121 000 male iron-steel workers in China. Combined exposure to PAHs and two or more dusts increased the risks of lung cancer (SRR = 654, 95% CI 113–3780). Risks of all neoplasms evaluated in the study were also significantly increased with combined PAH exposure (SRR = 541, 95% CI 209–1395). The risk of bladder cancer was investigated in a cohort of asphalt workers by Burstyn *et al.* [3]. Although cumulative exposure was not associated with cancer risk, workers exposed to 1.0–1.8 $\mu g/m^3$ were found to have an estimated risk of 1.67 (95% CI 0.62–4.48). Rybicki *et al.* [91] estimated prostate cancer risk associated with dermal or inhalation exposure to PAHs. The prostate cancer risk estimate associated with PAH inhalation from coal was 1.29 (95% CI 0.73–2.3) and 1.48 (95% CI 0.68–3.2) through dermal exposure, although the associations were not significant. The primary objective of this study, however, was to determine if variation in the GSTP1 gene modifies the risk of prostate cancer following occupational exposures to PAHs. In prostate cancer cases, carriage of the *GSTP1 Val*^{105} allele was significantly increased in subjects in the highest quartile of PAH exposure, especially in cases with an earlier onset of disease (under age 60) when the risk was 4.52 (95% CI 1.96–10.41).

These studies confirm the increased incidence of different types of cancer, especially lung and bladder, following exposures to PAHs in occupational settings. A major flaw, however, in these studies is exposure assessment. No attempt was made in any of these studies to measure internal dose of chemicals that would help avoid exposure misclassification and would account for different exposure routes. However, it can be argued that exposure misclassification occurred nondifferentially leading to underestimation of risks. In addition, nonoccupational risks were seldom adjusted for and could have confounded results. Nonetheless, occupational studies of PAHs and cancer risk present many advantages of which are large sample sizes and relatively long follow-up periods needed to evaluate diseases with long latency such as cancer. In addition, assessment of dose-response trends enabled by these studies is of particular importance in determining "safe" doses of PAH exposure over long periods.

A number of factors can influence the carcinogenic potency of individual PAHs and PAH mixtures in animal studies. Factors such as species, age, sex, and strain of animal, route of administration, vehicle or solvent for the PAH(s), and presence or absence of exogenous and endogenous tumor promoters or inhibitors may significantly modify organ/tissue-specific genotoxic responses to PAHs. Differences in PAH sensitivity may result from differences in ease of penetration to target tissues, basal and/or inducible levels of drug-metabolizing enzyme activities in both target and nontarget tissue, and levels of various DNA repair enzyme systems. The difference in aryl hydrocarbon receptor (AhR) affinity among inbred mouse strains leads to difference in CYP enzymes inducibility that was associated with differences in PAH cancer risk [92]. Differences in expression, regulation, and catalytic activities of metabolic enzymes between rats, mice, and humans have also been previously documented [93]. Unidentified components in a mixture may contribute to its carcinogenic potential [94]. The carcinogenicity of BAP and other noncarcinogenic PAHs as well as their mixtures was investigated in male C3H/HeJ mouse skin [95]. A mixture of a noncarcinogenic dose of BAP in combination with five noncarcinogenic PAHs resulted in enhanced carcinogenic potency. Furthermore, ingestion of manufactured gas plant residue by female A/J mice exclusively induced lung tumors, whereas feeding or administering single intraperitoneal (i.p.) injection of BAP to the same mouse strain induced lung and forestomach tumors [96]. Route of administration of the chemical may therefore have implications for the tumor formation process.

20.4.3 Birth Defects

Birth defects are the leading cause of infant mortality worldwide. Birth defects can be categorized as structural, functional, metabolic, behavioral, or hereditary [97]. Exposure to genotoxic compounds, such as PAHs, is thought to be a contributing factor to the risk of birth defects. The linkage between coal combustion by-products and birth defects was investigated in a study conducted in Nova Scotia, Canada. The study found that the risk of congenital malformations (overall and specific categories) was increased in the offspring of residents near a coking operation compared to the overall risk in Nova Scotia [98]. Exposure to airborne PAHs has been associated with increased levels of DNA adducts and mutations in newborns as well as preterm birth and intrauterine growth restriction [99–101]. Cigarette smoking during pregnancy has been associated with several adverse birth outcomes including spontaneous abortions, delayed conception, and low birth weight [102]. Findings from a more recent study suggest that high levels of PAHs as measured by DNA adducts in cord blood, from the World Trade Center fires in 2001 in New York City may have contributed to reduced fetal growth in the offspring of exposed women [103]. In a review of the literature on ambient pollution and pregnancy outcomes by

Sram *et al.* [104], low birth weight, premature birth, and intrauterine growth retardation seemed to be the reproductive effects most often associated with air pollutants. Detection of DNA adducts in human placenta and cord blood confirms that PAHs may be transferred to the fetus and could predispose it for developing disease later in life [105]. Limited evidence is available from animal studies. However, a study by Wang and Yu [106] demonstrated relationship between exposures to PAH mixtures in cigarette smoke and neural tube defects in hamsters.

20.5 Pharmacokinetics

The absorption, distribution, metabolism, and excretion of individual PAHs and mixtures have significant implications for the toxicity of these compounds. Under most circumstances, PAH absorption through the skin is relatively slow. Exceptions would occur in the presence of more soluble (low molecular weight PAHs) and/or cosolvents. More rapid absorption of PAHs occurs through the respiratory or gastrointestinal tract. Once absorbed into systemic circulation, PAHs are usually modified through phase I (oxidation) and phase II (conjugation) to increase solubility and the rate of elimination. Phase I metabolism of PAHs is typically catalyzed by cytochrome P450 (CYP) enzymes. PAHs are also capable of inducing their own biotransformation. They can significantly increase expression of a variety of CYP enzymes in liver, lung, and most extrahepatic tissues by binding to the aryl hydrocarbon receptor.

Most PAHs are chemically inert. Transformation by phase I enzymes has been shown to convert nonreactive compounds to intermediate chemically and biologically highly reactive species. Thus, BAP as a model PAH can be converted by phase I enzymes (CYP, mEH, AKR, or peroxidases) to a large number of metabolites: arene oxide, phenols, *trans*-dihydrodiols, quinones, and diol-epoxide (Figure 20.1). Oxidative metabolism of PAHs therefore yields highly reactive diol-epoxide derivatives that may bind covalently to cellular macromolecules, including DNA, to form addition products referred to as adducts. Mutations that activate protooncogenes and deactivate tumor suppressor genes may occur and cause disruption of regulatory processes that might lead to the initiation of the tumor formation process [107]. Elevated levels of DNA adducts and p53 mutations have been associated with PAH exposures in human studies [108–110]. Cellular defense mechanisms exist especially through apoptosis and the global genome nucleotide excision repair pathway that provide an essential line of defense against the mutagenic and carcinogenic activity of PAH diol-epoxide metabolites.

BAP is often used as a model compound for PAH metabolism. CYP1A1 was demonstrated to have an essential role not only for PAH-mediated toxicity but also for detoxification of orally administered BAP [111, 112]. Mice models with knocked-out CYP1A1 gene were found to be protected against liver toxicity and death [112]. Conclusions drawn from this study indicated that the reported resistance of mice was due to a decrease in production of the normally large amounts of toxic metabolites.

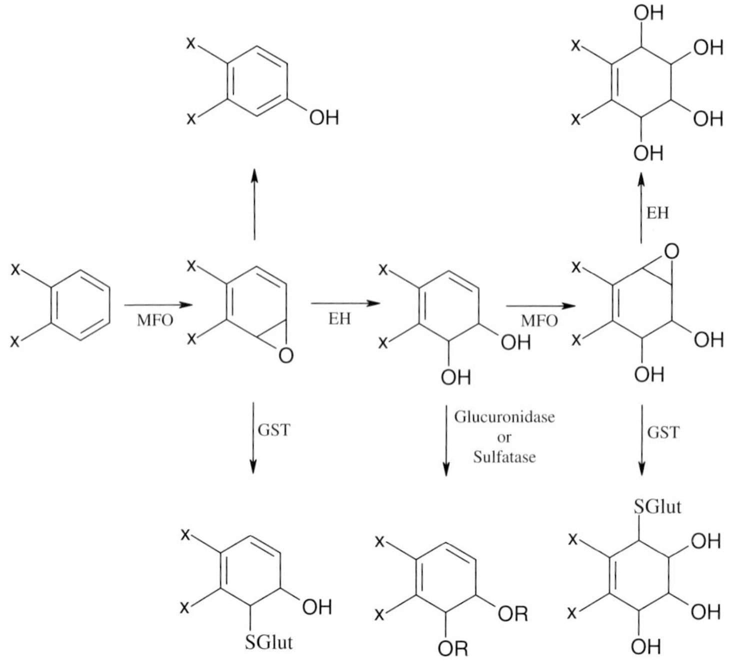

Figure 20.1 Metabolic pathways of PAH metabolism (adapted from Harvey [55]). MFO is mixed function oxidase, EH is epoxide hydrolase, GST is glutathione-S-transferase, R is glucuronate or sulfate.

Uno *et al.* [113] detected higher levels of DNA adducts in the *Cyp1a1*(−/−) knockout mice compared to the levels induced in *Cyp1a1*(+/+) wild-type mice. These results indicate that CYP1A1 is necessary for the detoxification of BAP.

As mentioned previously, USEPA classifies seven PAHs as probable human carcinogens. Numerous animal studies using several different routes of administration and numerous genotoxic assays established the carcinogenic potential of BAP. Species of animals demonstrating positive carcinogenic responses include rats, mice, hamsters, and guinea pigs. Routes of exposure indicated to be carcinogenic in animals include dietary, gavage, inhalation, intratracheal instillation, dermal application, intraperitoneal injection, subcutaneous injection, intravenous, transplacental, implantation in the stomach wall, lung, renal parenchyma, and brain, injection into the renal pelvis, and vaginal painting. Sites of tumor formation seen after oral administration of BAP include forestomach, squamous cell papillomas, and carcinomas.

The structure of BPDE, the most reactive metabolite of BAP, bound to the N-2 of the deoxyguanosine molecule is shown in Figure 20.2. BAP was administered at 0, 1, 10, 20, 30, 40, 45, 50, 100, and 250 ppm in the diet of male and female CFW-Swiss mice in a past study by Neal and Rigdon [114]. Forestomach tumors were detected in the 20+ ppm dose ranges. The incidence of tumors was also found to increase with dose. In a study by Brune *et al.* [115], Sprague–Dawley rats were fed BAP at

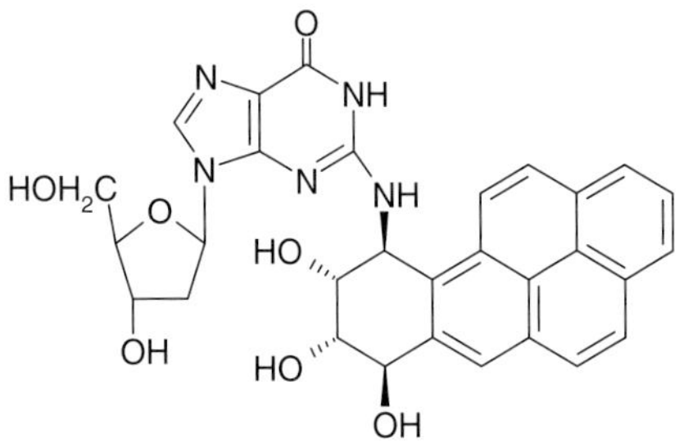

anti - BPDE-N-2-dG

Figure 20.2 Benzo[*a*]pyrene diol epoxide, *anti* configuration, adduct attached at the N-2 position of deoxyguanosine (adapted from Harvey [55].).

0.15 mg/kg. Dosing of the laboratory rats occurred every 9 days to five times a week until death, yielding an average dose of 6–39 mg/kg. Tumors were observed in the forestomach, esophagus, and larynx. The incidence of tumor followed a linear trend based on dose.

20.6 Genetic Sensitivities

Variations in genotype, or single-nucleotide polymorphisms (SNPs), in metabolizing and DNA repair enzymes are likely to exert a significant influence on the sensitivity of individuals toward carcinogenic PAHs. Increased risk of lung cancer has been associated with polymorphisms in CYP1A1 and GST enzymes according to several studies listed by Hecht *et al.* [116]. Hecht *et al.* investigated 11 polymorphisms in a group of 346 smokers. Female subjects and subjects with the GSTM1 null genotype exhibited a stronger correlation effect. It was also noted that the highest 10% of the parent PAH compound to metabolite ratios could not be predicted by any single polymorphism or by certain combinations. Coke oven workers were shown to have elevated BAP-DNA adduct levels in lymphocytes. The GSTM1 null genotype was found to be associated with significantly higher levels of adducts among workers (60 adducts per 10^9 nucleotides) compared to the GSTM1 active genotype (33 per 10^9 nucleotides) at the same exposure level [117].

Genetic polymorphisms of metabolic enzymes therefore have been shown to affect an individual's capacity to either activate or detoxify PAHs and their metabolites. The main CYPs in humans that participate in PAH metabolism are 1A1, 1A2, 1B1, 2C9, 3A4, and 3A5. Exposure to PAHs has been shown to induce expression of 1A1, 1A2, and 1B1 [118]. Three genetic polymorphisms were detected within the CYP1A1 gene. CYP1A1 MspI (CYP1A1*2A) and Ile/Val (CYP1A1*2B or *2C) are more prevalent in Asians than Caucasians [119]. GSTM1 and GSTP1 are major phase II enzymes that catalyze the conjugation reaction of BPDE. GSTT1 has both detoxification and activation properties and hence is difficult to predict the biological consequences of a null genotype [119]. Individuals with the GSTM1 null genotype have a slight increase in lung cancer risk and similarly modest increase in the risk of bladder cancer. Studies on the GSTM1 gene deletion are widely conducted to investigate the effects of GSTM1 enzyme deficiency that has been linked to lung cancers among cigarette smokers [120–123].

The assessment of a single polymorphic genotype may not provide a reliable estimate of individual susceptibility to PAH-induced cancers. Gene–gene interaction exists, for example GSTM1, may regulate the induction of other metabolizing enzymes such as CYP1A1 and CYP1A2 [121, 124]. GSTM1 deficiency not only leads to an increase in hepatic CYP1A2 activity in active smokers but also to significantly increased levels of bulky PAH-DNA adduct in lung tissues compared to individuals carrying wild-type GSTM1 [125, 126]. A commonly accepted concept is that individuals with a combination SNPs that increase PAH activation and decrease PAH detoxification would be at the greatest risk of developing PAH-induced cancers. However, data are not yet available from molecular epidemiology studies to confirm this assumption.

Polymorphisms in DNA repair genes are also likely to impact individual sensitivity to PAH-induced cancers. Some of the most common DNA repair genes investigated include ERCC1 (excision repair cross-complementing 1), ERCC2 or XPD (excision repair cross-complementing 2), XRCC1, and XRCC3 (X-ray repair cross-complementing groups 1 and 3). Matullo *et al.* [127] observed that nonsmokers with the XRCC1 399 Gln homozygote exhibited significantly higher levels of DNA adducts in their white blood cells (WBCs) compared to Gln/Arg heterozygotes. The presence of at least one variant allele in XPD exon 23 was associated with a significant (threefold) increase in lung cancer risk for lung cancer among younger nonsmoking individuals (<70 years) after adjusting for age, gender, and environmental tobacco smoke [128]. In addition, polymorphisms of XPD repair gene in exon 23 were found to be significant predictors for total DNA adduct levels whereas polymorphisms of XPD repair gene in exon 6 were related to the formation of BAP-"like" DNA adducts [129]. A recent study demonstrated a negative influence of exposure to PAHs from traffic emissions on DNA repair efficiency, and it also suggested that smoking might be a factor influencing that process. A significant decrease in repair efficiency due to exposure to PAHs was observed in the exposed individuals. In addition, the authors reported a negative influence of tobacco smoking on DNA repair efficiency [130].

20.7 Biomarkers of Exposure

Biomarkers represent a relatively new and potentially valuable tool for identifying sources of exposure to PAHs and preventing associated diseases. Biomarkers are generally observed in an exposed population prior to overt signs or symptoms of disease [131]. Concisely defined, biomarkers are biological particles that undergo detectable change when the individual is exposed to hazardous substances [132, 133]. An essential component of primary prevention of diseases induced by environmental contaminants is to rank various sources of exposure and identify methods that would effectively reduce these exposures.

Biomarkers are usually divided into three broad categories including biomarkers of exposure, effect, and susceptibility. Biomarkers of exposure usually consist of the unchanged compound or its metabolite measured in biological material such as blood, tissue, urine, feces, exfoliated cells, sweat, and nails [10]. One of the most common methods for monitoring PAH exposures in human populations is the determination of PAH metabolites in urine. Concentrations of the PAH unchanged parent compound in urine are generally low, and instead one or more metabolites of the predominant hydrocarbon are usually present in high enough concentrations to be determined [134, 135]. In mammals, 1-OHP represents the main metabolite of pyrene, a four-ringed PAH abundant in complex mixtures of PAHs [136]. PAH profiles may significantly vary from an exposure source to another; however, pyrene is a dominant compound in almost all PAH mixtures. The half-life of 1-OHP in urine is relatively long, lasting up to 48 h. Several studies have demonstrated that 1-hydroxypyrene is a good indicator of PAH exposure [137–139]. Sram and Binkova [140] concluded on the basis of their review of results from molecular epidemiology studies published between 1997 and 1999 that 1-OHP is a more effective biomarker in measuring occupational rather than environmental PAH exposures due to its reduced sensitivity to ambient PAH exposure levels in air. Selected data on urinary 1-hydroxypyrene levels are summarized in Table 20.4. These data indicate that a 100-fold difference in urinary concentrations of 1-hydroxypyrene may exist between an unexposed residential population (nonsmokers) and an occupationally exposed population. As discussed previously, these values are also likely to be influenced by genetic polymorphisms (Table 20.4).

A biomarker of effect is a biological measurement that indicates that the organism is responding to an exposure at some level. DNA adducts are one example of a biomarker of effect that can be used to measure early indicators of genotoxic effects that precede the onset of health effects such as adverse pregnancy outcomes or cancer. Since bulky DNA adducts reflect persistent genetic damage at a target site, they may not exhibit the same degree of variability as biomarkers that only reflect recent exposures. DNA adducts may also account for multiple routes of exposure and differences in toxicokinetics and repair amongst exposed subjects [141]. Phillips [142] recognizes that monitoring the formation of DNA adducts in lymphocytes as a surrogate tissue provides a valuable tool for investigating environmental exposure in healthy individuals. Other biomarkers of effects include protein adducts, levels of functionally critical protein such as p53 and p21, chromosome aberrations (CAs), and sister chromatid exchange (SCE) [143, 144].

The formation of DNA adducts is generally considered to be the earliest critical event that can be detected in the complex multistage process of chemically induced carcinogenesis caused by compounds such as PAHs [145–147]. Poirier and Beland [148] reported an overall linear relationship between levels of DNA adducts and the dose of a carcinogen administered to rodents. DNA adducts were also generally correlated with tumorigenesis. In human studies, a direct link was reported between bulky DNA adducts in WBCs and lung cancer risk in a prospective study

Table 20.3 Summary of occupational studies evaluating cancer risk associated with exposure to PAHs.

BAP/PAH exposure[a)]	Occupation (no. study subjects)	Type of study	Type of cancer	Risk estimate[b)] (95% CI)	References
Unexposed	Aerospace and radiation (392 case/ 1805 controls)	Nested case–control	Prostate	1.0	[200]
Low/moderate				1.0 (0.69, 1.6)	
High				1.3 (0.73, 2.5)	
108.3 μg BAP/m^3/year	Aluminum smelting (6423)	Retrospective cohort	Bladder	3.0	[4]
119.6 μg BAP/m^3/year			Lung	1.8	
n/a	Iron-steel foundry (121 846)	Retrospective cohort	Lung	6.54 (1.13–37.8)	[5]
0–0.4 μg BAP/m^3/year	Asphalt paving (7298)	Cohort	Bladder	1.0	[3]
0.4–1.0 μg BAP/m^3/year				1.13 (0.44–2.90)	
1.0–1.8 μg BAP/m^3/year				1.67 (0.62–4.48)	
1.8 + μg BAP/m^3/year				1.09 (0.30–3.99)	
N/A	Petroleum industry (637 case/244 control)	Case–control	Prostate	0.74–1.48	[91]

a) Cumulative exposure used when available.
b) Risk estimate corresponding to lagged exposure used when available.

within the Physician's Health Study [149]. Overall, *in vitro, in vivo,* and human studies indicate that as biomarkers of exposure, DNA adduct levels provide important information for predicting human cancer risk.

Among biomarkers of structural or functional alterations in humans are cytogenetic end points such as chromosome aberration, micronuclei (MN), and sister chromatid exchanges. These cytogenetic modifications are mainly due to errors of DNA replication, which can be caused by mutagens [150]. The frequency of chromosomal aberrations in peripheral blood lymphocytes was found to be a predictor of cancer risk in several human cohorts [151]. Previous studies have also confirmed the relationship between PAH exposure and p53 expression or chromosome damage [139, 152, 153]. With regard to the biomarkers of effect, data from previous studies have established the utility of measurements of protein levels and chromosome damage as an indicator of genotoxic effects [153–157]. A study population in an environmentally polluted part of Poland exhibited increased levels of various biomarkers such as DNA adducts, CA, SCE, and ras oncogene expression [158].

Table 20.4 Summary of urinary 1-hydroxypyrene (1-OHP) concentrations in µmol/mol creatinine in selected human populations.

Type of exposure	Number of individuals	1-OHP	References
General population (China)	70	0.5–1.6	[201]
General population nonsmoker	19	0.08	[202]
General population smoker	22	0.2	[202]
Children (living near a steel mill)	350	0.05	[203]
Cooking (females)	108	0.5	[204]
Highway toll station (preshift)	32	1	[38]
Highway toll station (postshift)		3	
Traffic police	89	0.14	[205]
Coal-electrodes production (preshift)	17	4	[206]
Coal-electrodes production (postshift)		10	
Coal tar distillation	4	4–12	[207]
Aluminum production	5	1.2–8.8	[33]
Coking plant	447	4.2–5.2	[208]

Chromosomal aberrations were shown to be an intermediate step in tumor formation pathway and act as indicators of both exposure and susceptibility [159, 160]. Coke oven workers exhibited an increased level of chromosomal aberrations and SCE compared to control subjects. The exposure among coke oven workers ranged from 0.6 to 550 $\mu g/m^3$ and 0.002 to 50 $\mu g/m^3$ for carcinogenic PAHs and BAP, respectively. The respective values in controls were 0.1–1.5 $\mu g/m^3$ and from 0.002 to 0.01 $\mu g/m^3$. The frequency of CA and SCE was found to be related to exposure to carcinogenic PAHs [161]. In Turkey, exposure to urban air pollution significantly increased the levels of CAs in traffic policemen and taxi drivers [162].

The tumor suppressor gene p53 has been reported to play a critical role in cell responses to genotoxic chemicals such as cell cycle arrest, DNA repair, and apoptosis [163]. The p53 gene, very frequently altered in human cancer cells, is found to mutate in around 50% of all human tumors [164]. Cells exposed to genotoxic agents exhibit increased levels of the p53 protein that in turn lead to an upregulation of the Cyclin-dependent kinase (Cdk) inhibitor, $p21^{WAF1/CIP1}$ protein [163]. The expression of both p53 and $p21^{WAF1/CIP1}$ proteins was found to be induced by PAHs *in vitro* [165, 166]. Thus, the presence of both proteins in blood serum has the potential to be used as a molecular marker of exposure to specific carcinogens in environmental monitoring and risk assessment studies. Previous studies have used p53 in blood serum as a marker of cancer [167, 168] and PAHs [138, 169], among other genotoxic agents. A more recent study by Rossner *et al.* [170], however, has found no correlation between p53 and $p21^{WAF1/CIP1}$ plasma levels, as well as a negative correlation between p53 levels and PAHs exposure. In addition, smoking was found to have no effect on the levels of either protein. The contradictory results presented in the study by Rossner *et al.* suggest that the use of p53 and $p21^{WAF1/CIP1}$ plasma levels as biomarkers of carcinogenic PAH effect might require further examination.

Finally, biomarkers of susceptibility are related to the genotype of an individual. The genetic makeup does not usually establish a disease condition but most likely identifies a certain sensitivity that leaves a person at higher risk for disease. SNPs in genes coding for drug metabolizing or DNA repair enzymes could result in a faster or slower metabolism and DNA repair efficiency in human tissues. Therefore, SNPs can modify the levels of biomarkers detected in tissues by altering retention and/or elimination of hazardous chemicals from the body.

20.8 Conclusions

Polycyclic aromatic hydrocarbon mixtures are one of the most common hazardous environmental chemicals. Human exposure to these compounds has clearly been linked to a variety of occupational cancers. However, the adverse health outcomes associated with ambient exposures are less clearly defined. PAHs are genotoxic, adversely impact the immune system, and are capable of altering the expression of a broad range of metabolic enzymes. Major research gaps at present fall into three categories including mixtures, biomarkers, and genetic impacts. Mixtures research is needed to improve analytical methods for characterization of mixtures, most importantly the nitroaromatics and high molecular weight compounds. In addition, more data are needed from cell culture and animal studies of mixtures and isolated fractions to identify the most toxic components of complex mixtures. Biomarker studies are needed to improve our understanding of the relationship between ambient exposures and adverse health effects. Perhaps, the most important is the impact of genetic polymorphisms on retention and elimination of PAH mixtures. This should include investigations as to the impact of circadian rhythms on PAH metabolism.

References

1 Pott, P. (2002) The chirurgical works of Percivall Pott, F.R.S., surgeon to St. Bartholomew's Hospital, a new edition, with his last corrections. 1808. *Clin. Orthop. Relat. Res.*, (398), 4–10.

2 Hieger, I. (1930) The spectra of cancer-producing tars and oils and of related substances. *Biochem. J.*, **24** (2), 505–511.

3 Burstyn, I. *et al.* (2007) Bladder cancer incidence and exposure to polycyclic aromatic hydrocarbons among asphalt pavers. *Occup. Environ. Med.*, **64** (8), 520–526.

4 Friesen, M.C. *et al.* (2007) Comparison of two indices of exposure to polycyclic aromatic hydrocarbons in a retrospective aluminium smelter cohort. *Occup. Environ. Med.*, **64** (4), 273–278.

5 Hoshuyama, T. *et al.* (2006) Mortality of iron-steel workers in Anshan, China: a retrospective cohort study. *Int. J. Occup. Environ. Health*, **12** (3), 193–202.

6 United States Environmental Protection Agency (USEPA) (2006) Superfund: New Report Projects Number, Cost and Nature of Contaminated Site Cleanups in the U.S. Over Next 30 Years. [cited; available at: http://www.epa.gov/superfund/accomp/news/30years.htm].

7 Gennings, C. (1995) An efficient experimental design for detecting departure from additivity in mixtures of many chemicals. *Toxicology*, **105** (2–3), 189–197.

8 Teuschler, L.K. and Hertzberg, R.C. (1995) Current and future risk assessment guidelines, policy, and methods development for chemical mixtures. *Toxicology*, **105** (2–3), 137–144.

9 Wilson, S.C. and Jones, K.C. (1993) Bioremediation of soil contaminated with polynuclear aromatic hydrocarbons (PAHs): a review. *Environ. Pollut.*, **81** (3), 229–249.

10 Brandt, H.C. and Watson, W.P. (2003) Monitoring human occupational and environmental exposures to polycyclic aromatic compounds. *Ann. Occup. Hyg.*, **47** (5), 349–378.

11 Harvey, R.G. (1997) *Polycyclic Aromatic Hydrocarbons*, John Wiley & Sons, Inc., New York, p. 667.

12 ATSDR (1995) Toxicological Profile for Polycyclic Aromatic Hydrocarbons. U.S. Department of Health and Human Services, Public Health Service, Atlanta, GA.

13 Hafner, W.D., Carlson, D.L., and Hites, R.A. (2005) Influence of local human population on atmospheric polycyclic aromatic hydrocarbon concentrations. *Environ. Sci. Technol.*, **39** (19), 7374–7379.

14 Wang, Z. *et al.* (2007) Distribution and sources of polycyclic aromatic hydrocarbons from urban to rural soils: a case study in Dalian, China. *Chemosphere*, **68** (5), 965–971.

15 Olajire, A.A. *et al.* (2007) Distribution of polycyclic aromatic hydrocarbons in surface soils and water from the vicinity of Agbabu bitumen field of Southwestern Nigeria. *J. Environ. Sci. Health A*, **42** (8), 1043–1049.

16 Trojan, M.D. *et al.* (2003) Effects of land use on ground water quality in the Anoka Sand Plain Aquifer of Minnesota. *Ground Water*, **41** (4), 482–492.

17 Tumey, G. and Goerlitz, D. (1990) Organic contamination of ground water at Gas Works Park, Seattle, Washington. *Ground Water Monitor. Rev.*, **19** (3), 187–198.

18 Ward, M.H. *et al.* (1997) Risk of adenocarcinoma of the stomach and esophagus with meat cooking method and doneness preference. *Int. J. Cancer*, **71** (1), 14–19.

19 Sinha, R. *et al.* (1999) Well-done, grilled red meat increases the risk of colorectal adenomas. *Cancer Res.*, **59** (17), 4320–4324.

20 Phillips, D.H. (1999) Polycyclic aromatic hydrocarbons in the diet. *Mutat. Res.*, **443** (1–2), 139–147.

21 Dai, Q. *et al.* (2002) Consumption of animal foods, cooking methods, and risk of breast cancer. *Cancer Epidemiol. Biomarkers Prev.*, **11** (9), 801–808.

22 Jagerstad, M. and Skog, K. (2005) Genotoxicity of heat-processed foods. *Mutat. Res.*, **574** (1–2), 156–172.

23 Tateno, T., Nagumo, Y., and Suenaga, S. (1990) Polycyclic aromatic hydrocarbons produced from grilled vegetables. *J. Food Hyg. Soc. Jpn.*, **31**, 271–276.

24 Falco, G. *et al.* (2003) Polycyclic aromatic hydrocarbons in foods: human exposure through the diet in Catalonia, Spain. *J. Food Prot.*, **66** (12), 2325–2331.

25 Jones, K.C. *et al.* (1989) Changes in the polynuclear aromatic hydrocarbon content of wheat grain and pasture grassland over the last century from one site in the U.K. *Sci. Total Environ.*, **78**, 117–130.

26 Dennis, M.J. *et al.* (1991) Factors affecting the polycyclic aromatic hydrocarbon content of cereals, fats and other food products. *Food Addit. Contam.*, **8** (4), 517–530.

27 Lodovici, M. *et al.* (1995) Polycyclic aromatic hydrocarbon contamination in the Italian diet. *Food Addit. Contam.*, **12** (5), 703–713.

28 Klein, H., Speer, K., and Schmidt, E.H.F. (1993) Polycyclic aromatic hydrocarbons in raw coffee and roasted coffee. *Bundesgesundheitsblatt*, **36**, 98–100.

29 Akpan, V., Lodovici, M., and Dolara, P. (1994) Polycyclic aromatic hydrocarbons in fresh and smoked fish samples from three Nigerian cities. *Bull. Environ. Contam. Toxicol.*, **53** (2), 246–253.

30 Larsson, B.K. *et al.* (1983) Polycyclic aromatic hydrocarbons in grilled food. *J. Agric. Food Chem.*, **31** (4), 867–873.

31 Luch, A. (2005) Polycyclic aromatic hydrocarbons-induced carcinogenesis: an introduction, in *The Carcinogenic Effects of PAHs* (ed. A. Luch), Imperial College Press, pp. 97–136.

32 Phillips, D.H. (1996) DNA adducts in human tissues: biomarkers of exposure to carcinogens in tobacco smoke. *Environ. Health Perspect.*, **104** (Suppl. 3), 453–458.

33 Vu Duc, T. and Lafontaine, M. (1996) 1-Hydroxypyrene in human urines as biomarker of exposure to PAH in work related processes in an aluminium production plant. *Polycycl. Aromat. Hydrocarbons*, **11**, 1–9.

34 Lewtas, J. *et al.* (1997) Air pollution exposure–DNA adduct dosimetry in humans and rodents: evidence for non-linearity at high doses. *Mutat. Res.*, **378** (1–2), 51–63.

35 Apostoli, P. *et al.* (2003) Influence of genetic polymorphisms of CYP1A1 and GSTM1 on the urinary levels of 1-hydroxypyrene. *Toxicol. Lett.*, **144** (1), 27–34.

36 Szaniszló, J. and Ungváry, G.R. (2001) Polycyclic aromatic hydrocarbon exposure and burden of outdoor workers in budapest. *J. Toxicol. Environ. Health A*, **62** (5), 297–306.

37 Liu, Y.N. *et al.* (2007) Exposure of traffic police to polycyclic aromatic hydrocarbons in Beijing, China. *Chemosphere*, **66** (10), 1922–1928.

38 Tsai, P.J. *et al.* (2004) Urinary 1-hydroxypyrene as an indicator for assessing the exposures of booth attendants of a highway toll station to polycyclic aromatic hydrocarbons. *Environ. Sci. Technol.*, **38** (1), 56–61.

39 Khalili, N.R., Scheff, P.A., and Holsen, T.M. (1995) PAH source fingerprints for coke ovens, diesel and, gasoline engines, highway tunnels, and wood combustion emissions. *Atmos. Environ.*, **29** (4), 533–542.

40 Eiguren-Fernandez, A. *et al.* (2004) Seasonal and spatial variation of polycyclic aromatic hydrocarbons in vapor-phase and PM 2.5 in southern California urban and rural communities. *Aerosol. Sci. Tech.*, **38** (5), 447–455.

41 Harrison, R.M., Smith, D.J.T., and Luhana, L. (1996) Source apportionment of atmospheric polycyclic aromatic hydrocarbons collected from an urban location in Birmingham, U.K. *Environ. Sci. Technol.*, **30** (3), 825–832.

42 Edgar, P.J. *et al.* (2006) Sediment influence on congener-specific PCB bioaccumulation by *Mytilus edulis*: a case study from an intertidal hot spot, Clyde Estuary, UK. *J. Environ. Monit.*, **8** (9), 887–896.

43 Lemaire, P. *et al.* (1990) The uptake metabolism and biological half-life of benzo[*a*]pyrene in different tissues of sea bass, *Dicentrarchus labrax*. *Ecotoxicol. Environ. Saf.*, **20** (3), 223–233.

44 Varanasi, U., Stein, J.E., and Nishimoto, M. (1989) Biotransformation and disposition of polycyclic aromatic hydrocarbons (PAH) in fish, in *Metabolism of Polycyclic Aromatic Hydrocarbons in the Aquatic Environment* (ed. U. Varanasi), CRC Press, Boca Raton, FL, pp. 94–149.

45 Meador, J.P. *et al.* (1995) Bioaccumulation of polycyclic aromatic hydrocarbons by marine organisms. *Rev. Environ. Contam. Toxicol.*, **143**, 79–165.

46 Moles, A. and Norcross, B.L. (1998) Effects of oil-laden sediments on growth and health of juvenile flatfishes. *Can. J. Fish Aquat. Sci.*, **55** (3), 605–610.

47 Gregg, J.C., Fleeger, J.W., and Carman, K.R. (1997) Effects of suspended, diesel-contaminated sediment on feeding rate in the darter goby, *Gobionellus boleosoma* (Teleostei: Gobiidae). *Mar. Pollut. Bull.*, **34** (4), 269–275.

48 Hinkle-Conn, C. *et al.* (1998) Effects of sediment-bound polycyclic aromatic hydrocarbons on feeding behavior in juvenile spot (*Leiostomus xanthurus* Lacepede: Pisces). *J. Exp. Mar. Biol. Ecol.*, **227** (1), 113–132.

49 Carls, M.G., Rice, S.D., and Hose, J.E. (1999) Sensitivity of fish embryos to weathered crude oil: Part I. Low-level exposure during incubation causes malformations, genetic damage, and mortality in larval pacific herring

(*Clupea pallasi*). *Environ. Toxicol. Chem.*, **18** (3), 481–493.

50 Johnson, L.L. *et al.* (2007) Contaminant exposure in outmigrant juvenile salmon from Pacific northwest estuaries of the United States. *Environ. Monit. Assess.*, **124** (1–3), 167–194.

51 Beland, P., DeGuise, S., Girard, C., Lagace, A., Martipeau, D., Michaud, R., Muir, D.C.G., Norstrom, R.J., Pelletier, E., Ray, S., and Shugart, L.R. (1993) Toxic compounds and health and reproductive effects in St. Lawrence beluga whales. *Great Lakes Res.*, **19**, 766–775.

52 Martineau, D. *et al.* (1994) Pathology and toxicology of beluga whales from the St. Lawrence Estuary, Quebec, Canada. Past, present and future. *Sci. Total Environ.*, **154** (2–3), 201–215.

53 Lijinsky, W. (1991) The formation and occurrence of polynuclear aromatic hydrocarbons associated with food. *Mutat. Res.*, **259** (3–4), 251–261.

54 Dipple, A., Moschel, R.C., and Bigger, C.A.H. (1984) Polynuclear aromatic carcinogens, in *Chemical Carcinogens, ACS Monograph*, pp. American Chemical Society Press, Washington, DC.

55 Harvey, R.G. (1991) *Polycyclic Aromatic Hydrocarbons, Chemistry and Carcinogenicity*, Cambridge University Press, Cambridge, UK, pp. 11–87.

56 International Agency for Research on Cancer (IARC) (1973) "Benzo[*b*] fluoranthene", in IARC Monographs on the Evaluation of Carcinogenic Risk of Chemicals to Man. Certain Polycyclic Aromatic Hydrocarbons and Heterocyclic Compounds, Lyon, France, pp. 69–81.

57 International Agency for Research on Cancer (IARC) (1983) "Benzo[*b*] fluoranthene", in IARC Monographs on the Evaluation of Carcinogenic Risk of Chemicals to Humans. Polycyclic Aromatic Compounds, Part 1, Chemical, Environmental and Experimental Data, Lyon, France, pp. 33–91, 147–153.

58 Warshawsky, D. (1992) Environmental sources, carcinogenicity, mutagenicity, metabolism and DNA-binding of nitrogen and sulfur heterocyclic aromatics. *Environ. Carcinog. Ecol. R.*, **10** (1), 1–71.

59 Yamagiwa, K. and Ichikawa, K. (1915) Experimentelle studie uber die pathogenese der epithelialgeschwulste. *Mitt. Med. Fak. Tokio*, **15**, 295–344.

60 Borgen, A. *et al.* (1973) Metabolic conversion of benzo(*a*)pyrene by Syrian hamster liver microsomes and binding of metabolites to deoxyribonucleic acid. *J. Med. Chem.*, **16** (5), 502–506.

61 Sims, P. *et al.* (1974) Metabolic activation of benzo(*a*)pyrene proceeds by a diol-epoxide. *Nature*, **252** (5481), 326–328.

62 Grimmer, G. *et al.* (1982) Quantification of the carcinogenic effect of polycyclic aromatic hydrocarbons in used engine oil by topical application onto the skin of mice. *Int. Arch. Occup. Environ. Health*, **50** (1), 95–100.

63 Lewis, S.C., King, R.W., Cragg, S.T., and Hillman, D.W. (1982) Skin carcinogenic potential of petroleum hydrocarbons 2. Carcinogenesis of crude oil, distillate fractions and chemical class subfractions, in *The Toxicology of Petroleum Hydrocarbons* (eds H.N. MacFarland, C.E. Holdsworth, J.A. MacGregor, R.W. Call and M.L. Kane), American Petroleum Institute, pp. 185–195.

64 Lewis, S.C. (1983) Crude petroleum and selected fractions. Skin cancer bioassays. *Prog. Exp. Tumor. Res.*, **26**, 68–84.

65 Mahlum, D.D. *et al.* (1984) Fractionation of skin tumor-initiating activity in coal liquids. *Cancer Res.*, **44** (11), 5176–5181.

66 Mukhtar, H. *et al.* (1982) Effect of topical application of defined constituents of coal tar on skin and liver aryl hydrocarbon hydroxylase and 7-ethoxycoumarin deethylase activities. *Toxicol. Appl. Pharmacol.*, **64** (3), 541–549.

67 Mumtaz, M.M. *et al.* (1996) ATSDR evaluation of health effects of chemicals. IV. Polycyclic aromatic hydrocarbons (PAHs): understanding a complex problem. *Toxicol. Ind. Health*, **12** (6), 742–971.

68 Robinson, M. *et al.* (1984) Comparative carcinogenic and mutagenic activity of coal tar and petroleum asphalt paints

used in potable water supply systems. *J. Appl. Toxicol.*, **4** (1), 49–56.

69 Wallcave, L. *et al.* (1971) Skin tumorigenesis in mice by petroleum asphalts and coal-tar pitches of known polynuclear aromatic hydrocarbon content. *Toxicol. Appl. Pharmacol.*, **18** (1), 41–52.

70 Cook, J.W., Hewett, C.L., and Hieger, I. (1933) The isolation of a cancer-producing hydrocarbon from coal tar. Parts I, II, and III. *J. Chem. Soc.*, 395–405.

71 International Agency for Research on Cancer (IARC) (2004) IARC Monographs on the Evaluation of Carcinogenic Risks to Humans. [cited; available at http://monographs.iarc.fr/].

72 United States Environmental Protection Agency (USEPA (2006) Integrated Risk Information System (IRIS).

73 Boffetta, P., Jourenkova, N., and Gustavsson, P. (1997) Cancer risk from occupational and environmental exposure to polycyclic aromatic hydrocarbons. *Cancer Causes Control*, **8** (3), 444–472.

74 International Agency for Research on Cancer (IARC) (1984) IARC Monographs on the Evaluation of the Carcinogenic Risk of Chemicals to Humans, in Polynuclear Aromatic Compounds, Part 2: Carbon Blacks, Mineral Oils (Lubricant Base Oils and Derived Products) and Some Nitroarenes, IARC, Lyon, France. p. 245.

75 International Agency for Research on Cancer (IARC) (1984) IARC Monographs on the Evaluation of the Carcinogenic Risk of Chemicals to Humans, in Polynuclear Aromatic Compounds, Part 3: Industrial Exposures in Aluminium Production, Coal Gasification, Coke Production, and Iron and Steel Founding, IARC, Lyon, France, p. 219.

76 International Agency for Research on Cancer (IARC) (1985) IARC Monographs on the Evaluation of the Carcinogenic Risk of Chemicals to Humans, in Polynuclear Aromatic Compounds, Part 4: Bitumens, Coal-Tars and Derived Products, Shale-Oils and Soots, IARC, Lyon, France, p. 271.

77 International Agency for Research on Cancer (IARC). (1989) IARC Monographs on the Evaluation of the Carcinogenic Risk of Chemicals to Humans, in Diesel and Gasoline Engine Exhausts and Some Nitroarenes, IARC, Lyon, France, p. 458.

78 Mastrangelo, G., Fadda, E., and Marzia, V. (1996) Polycyclic aromatic hydrocarbons and cancer in man. *Environ. Health Perspect.*, **104** (11), 1166–1170.

79 Armstrong, B. *et al.* (1994) Lung cancer mortality and polynuclear aromatic hydrocarbons: a case-cohort study of aluminum production workers in Arvida, Quebec, Canada. *Am. J. Epidemiol.*, **139** (3), 250–262.

80 Bonassi, S. *et al.* (1989) Bladder cancer and occupational exposure to polycyclic aromatic hydrocarbons. *Int. J. Cancer*, **44** (4), 648–651.

81 Clavel, J. *et al.* (1994) Occupational exposure to polycyclic aromatic hydrocarbons and the risk of bladder cancer: a French case-control study. *Int. J. Epidemiol.*, **23** (6), 1145–1153.

82 Costantino, J.P., Redmond, C.K., and Bearden, A. (1995) Occupationally related cancer risk among coke oven workers: 30 years of follow-up. *J. Occup. Environ. Med.*, **37** (5), 597–604.

83 Jockel, K.H. *et al.* (1992) Occupational and environmental hazards associated with lung cancer. *Int. J. Epidemiol.*, **21** (2), 202–213.

84 McLaughlin, J.K. *et al.* (1992) A nested case-control study of lung cancer among silica exposed workers in China. *Br. J. Ind. Med.*, **49** (3), 167–171.

85 Nadon, L. *et al.* (1995) Cancer risk due to occupational exposure to polycyclic aromatic hydrocarbons. *Am. J. Ind. Med.*, **28** (3), 303–324.

86 Spinelli, J.J. *et al.* (1991) Mortality and cancer incidence in aluminum reduction plant workers. *J. Occup. Med.*, **33** (11), 1150–1155.

87 Tola, S. *et al.* (1979) Lung cancer mortality among iron foundry workers. *J. Occup. Med.*, **21** (11), 753–759.

88 Tremblay, C. *et al.* (1995) Estimation of risk of developing bladder cancer among workers exposed to coal tar pitch volatiles

in the primary aluminum industry. *Am. J. Ind. Med.*, **27** (3), 335–348.

89 Berger, J. and Manz, A. (1992) Cancer of the stomach and the colon-rectum among workers in a coke gas plant. *Am. J. Ind. Med.*, **22** (6), 825–834.

90 Bosetti, C., Boffetta, P., and La Vecchia, C. (2007) Occupational exposures to polycyclic aromatic hydrocarbons, and respiratory and urinary tract cancers: a quantitative review to 2005. *Ann. Oncol.*, **18** (3), 431–446.

91 Rybicki, B.A. *et al.* (2006) Prostate cancer risk from occupational exposure to polycyclic aromatic hydrocarbons interacting with the GSTP1 Ile105Val polymorphism. *Cancer Detect. Prev.*, **30** (5), 412–422.

92 Nebert, D.W. *et al.* (2004) Role of aryl hydrocarbon receptor-mediated induction of the CYP1 enzymes in environmental toxicity and cancer. *J. Biol. Chem.*, **279** (23), 23847–23850.

93 Guengerich, F.P. (1997) Comparisons of catalytic selectivity of cytochrome P450 subfamily enzymes from different species. *Chem. Biol. Interact.*, **106** (3), 161–182.

94 Weyand, E.H. and Wu, Y. (1995) Covalent binding of polycyclic aromatic hydrocarbon components of manufactured gas plant residue to mouse lung and forestomach DNA. *Chem. Res. Toxicol.*, **8** (7), 955–962.

95 Warshawsky, D., Barkley, W., and Bingham, E. (1993) Factors affecting carcinogenic potential of mixtures. *Fundam. Appl. Toxicol.*, **20** (3), 376–382.

96 Weyand, E.H. *et al.* (1995) Differences in the tumorigenic activity of a pure hydrocarbon and a complex mixture following ingestion: benzo[*a*]pyrene vs. manufactured gas plant residue. *Chem. Res. Toxicol.*, **8** (7), 949–954.

97 Jones, K. (1997) *Smith's Recognizable Patterns of Human Malformations*, 5th edn, WB Saunders, Philadelphia, PA.

98 Dodds, L. and Seviour, R. (2001) Congenital anomalies and other birth outcomes among infants born to women living near a hazardous waste site in Sydney, Nova Scotia. *Can. J. Public Health*, **92** (5), 331–334.

99 Dejmek, J. *et al.* (2000) The impact of polycyclic aromatic hydrocarbons and fine particles on pregnancy outcome. *Environ. Health Perspect.*, **108** (12), 1159–1164.

100 Perera, F. *et al.* (2002) *In utero* DNA damage from environmental pollution is associated with somatic gene mutation in newborns. *Cancer Epidemiol. Biomarkers Prev.*, **11** (10 Pt 1), 1134–1137.

101 Sram, R.J. *et al.* (1999) Adverse reproductive outcomes from exposure to environmental mutagens. *Mutat. Res.*, **428** (1–2), 203–215.

102 Preston, A.M. (1991) Cigarette smoking-nutritional implications. *Prog. Food Nutr. Sci.*, **15** (4), 183–217.

103 Perera, F.P. *et al.* (2005) Relationships among polycyclic aromatic hydrocarbon-DNA adducts, proximity to the World Trade Center, and effects on fetal growth. *Environ. Health Perspect.*, **113** (8), 1062–1067.

104 Sram, R.J. *et al.* (2005) Ambient air pollution and pregnancy outcomes: a review of the literature. *Environ. Health Perspect.*, **113** (4), 375–382.

105 Hansen, C., Asmussen, I., and Autrup, H. (1993) Detection of carcinogen-DNA adducts in human fetal tissues by the 32P-postlabeling procedure. *Environ. Health Perspect.*, **99**, 229–231.

106 Wang, D. and Yu, X. (2004) Morphological study on the neural tube defects caused by passive smoking. *Wei Sheng Yan Jiu*, **33** (2), 147–150.

107 Phillips, D.H. and Grover, P.L. (1994) Polycyclic hydrocarbon activation: bay regions and beyond. *Drug Metab. Rev.*, **26** (1–2), 443–467.

108 Alexandrov, K. *et al.* (2002) CYP1A1 and GSTM1 genotypes affect benzo[*a*]pyrene DNA adducts in smokers' lung: comparison with aromatic/hydrophobic adduct formation. *Carcinogenesis*, **23** (12), 1969–1977.

109 Gaspari, L. *et al.* (2003) Polycyclic aromatic hydrocarbon–DNA adducts in human sperm as a marker of DNA damage and infertility. *Mutat. Res.*, **535** (2), 155–160.

110 Hainaut, P. and Pfeifer, G.P. (2001) Patterns of p53 G $\rightarrow$ T transversions in lung cancers reflect the primary mutagenic signature of DNA-damage by tobacco smoke. *Carcinogenesis*, **22** (3), 367–374.

111 Uno, S. *et al.* (2004) Oral exposure to benzo[*a*]pyrene in the mouse: detoxication by inducible cytochrome P450 is more important than metabolic activation. *Mol. Pharmacol.*, **65** (5), 1225–1237.

112 Uno, S. *et al.* (2001) Benzo[*a*]pyrene-induced toxicity: paradoxical protection in Cyp1a1 (−/−) knockout mice having increased hepatic BaP-DNA adduct levels. *Biochem. Biophys. Res. Commun.*, **289** (5), 1049–1056.

113 Uno, S. *et al.* (2006) Oral benzo[*a*]pyrene in Cyp1 knockout mouse lines: CYP1A1 important in detoxication, CYP1B1 metabolism required for immune damage independent of total-body burden and clearance rate. *Mol. Pharmacol.*, **69** (4), 1103–1114.

114 Neal, J. and Rigdon, R.H. (1967) Gastric tumors in mice fed benzo(*a*)pyrene: a quantitative study. *Tex. Rep. Biol. Med.*, **25** (4), 553–557.

115 Brune, H. *et al.* (1981) Investigation of the tumorigenic response to benzo(*a*)pyrene in aqueous caffeine solution applied orally to Sprague–Dawley rats. *J. Cancer Res. Clin. Oncol.*, **102** (2), 153–157.

116 Hecht, S.S. *et al.* (2006) Comparison of polymorphisms in genes involved in polycyclic aromatic hydrocarbon metabolism with urinary phenanthrene metabolite ratios in smokers. *Cancer Epidemiol. Biomarkers Prev.*, **15** (10), 1805–1811.

117 Pavanello, S. *et al.* (2004) GSTM1 null genotype as a risk factor for anti-BPDE-DNA adduct formation in mononuclear white blood cells of coke-oven workers. *Mutat. Res.*, **558** (1–2), 53–62.

118 Iwanari, M. *et al.* (2002) Induction of CYP1A1, CYP1A2, and CYP1B1 mRNAs by nitropolycyclic aromatic hydrocarbons in various human tissue-derived cells: chemical-, cytochrome P450 isoform-, and cell-specific differences. *Arch. Toxicol.*, **76** (5), 287–298.

119 Pavanello, S. (2006) Biomarkers of toxicant susceptibility, in *Toxicologic Biomarkers* (ed. I.D. Anthony), Taylor & Francis.

120 Brockmoller, J. *et al.* (1996) Combined analysis of inherited polymorphisms in arylamine *N*-acetyltransferase 2, glutathione *S*-transferases M1 and T1, microsomal epoxide hydrolase, and cytochrome P450 enzymes as modulators of bladder cancer risk. *Cancer Res.*, **56** (17), 3915–3925.

121 Butkiewicz, D. *et al.* (2000) Polymorphisms of the GSTP1 and GSTM1 genes and PAH-DNA adducts in human mononuclear white blood cells. *Environ. Mol. Mutagen.*, **35** (2), 99–105.

122 McWilliams, J.E. *et al.* (1995) Glutathione *S*-transferase M1 (GSTM1) deficiency and lung cancer risk. *Cancer Epidemiol. Biomarkers Prev.*, **4** (6), 589–594.

123 Strange, R.C. *et al.* (1991) The human glutathione *S*-transferases: a case–control study of the incidence of the GST1 0 phenotype in patients with adenocarcinoma. *Carcinogenesis*, **12** (1), 25–28.

124 Vaury, C. *et al.* (1995) Human glutathione *S*-transferase M1 null genotype is associated with a high inducibility of cytochrome P450 1A1 gene transcription. *Cancer Res.*, **55** (23), 5520–5523.

125 Rojas, M. *et al.* (1998) High benzo[a] pyrene diol-epoxide DNA adduct levels in lung and blood cells from individuals with combined CYP1A1 MspI/Msp-GSTM1*0/*0 genotypes. *Pharmacogenetics*, **8** (2), 109–118.

126 Stucker, I. *et al.* (2002) Genetic polymorphisms of glutathione *S*-transferases as modulators of lung cancer susceptibility. *Carcinogenesis*, **23** (9), 1475–1481.

127 Matullo, G. *et al.* (2001) XRCC1, XRCC3, XPD gene polymorphisms, smoking and (32)P-DNA adducts in a sample of healthy subjects. *Carcinogenesis*, **22** (9), 1437–1445.

128 Hou, S.M. *et al.* (2002) The XPD variant alleles are associated with increased aromatic DNA adduct level and lung cancer risk. *Carcinogenesis*, **23** (4), 599–603.

129 Binkova, B. *et al.* (2007) PAH-DNA adducts in environmentally exposed

population in relation to metabolic and DNA repair gene polymorphisms. *Mutat. Res.*, **620** (1–2), 49–61.

130 Cebulska-Wasilewska, A. *et al.* (2007) Exposure to environmental polycyclic aromatic hydrocarbons: influences on cellular susceptibility to DNA damage (sampling Kosice and Sofia). *Mutat. Res.*, **620** (1–2), 145–154.

131 Skupinska, K., Misiewicz, I., and Kasprzycka-Guttman, T. (2004) Polycyclic aromatic hydrocarbons: physicochemical properties, environmental appearance and impact on living organisms. *Acta Pol. Pharm.*, **61** (3), 233–240.

132 Kleiner, H.E., Reed, M.J., and DiGiovanni, J. (2003) Naturally occurring coumarins inhibit human cytochromes P450 and block benzo[*a*]pyrene and 7,12-dimethylbenz[*a*]anthracene DNA adduct formation in MCF-7 cells. *Chem. Res. Toxicol.*, **16** (3), 415–422.

133 Niyogi, S. *et al.* (2001) Antioxidant enzymes in brackishwater oyster, *Saccostrea cucullata* as potential biomarkers of polyaromatic hydrocarbon pollution in Hooghly estuary (India): seasonality and its consequences. *Sci. Total Environ.*, **281** (1–3), 237–246.

134 Grimmer, G.D., Naujack, K.W., and Jacob, J. (1994) Relationship between inhaled PAH and urinary excretion of phenanthrene, pyrene and benzo[a] pyrene metabolites in coke plant workers. *Polycycl. Aromat. Comp.*, **5**, 269–277.

135 Jongeneelen, F.J. *et al.* (1988) 1-Hydroxypyrene in urine as a biological indicator of exposure to polycyclic aromatic hydrocarbons in several work environments. *Ann. Occup. Hyg.*, **32** (1), 35–43.

136 Keimig, S.D. *et al.* (1983) Identification of 1-hydroxypyrene as a major metabolite of pyrene in pig urine. *Xenobiotica*, **13** (7), 415–420.

137 Alexandrie, A.K. *et al.* (2000) CYP1A1 and GSTM1 polymorphisms affect urinary 1-hydroxypyrene levels after PAH exposure. *Carcinogenesis*, **21** (4), 669–676.

138 Pan, G. *et al.* (1998) A study of multiple biomarkers in coke oven workers: a cross-sectional study in China. *Carcinogenesis*, **19** (11), 1963–1968.

139 Siwinska, E., Mielzynska, D., and Kapka, L. (2004) Association between urinary 1-hydroxypyrene and genotoxic effects in coke oven workers. *Occup. Environ. Med.*, **61** (3), e10.

140 Sram, R.J. and Binkova, B. (2000) Molecular epidemiology studies on occupational and environmental exposure to mutagens and carcinogens, 1997–1999. *Environ. Health Perspect.*, **108** (Suppl. 1), 57–70.

141 Godschalk, R.W., Van Schooten, F.J., and Bartsch, H. (2003) A critical evaluation of DNA adducts as biological markers for human exposure to polycyclic aromatic compounds. *J. Biochem. Mol. Biol.*, **36** (1), 1–11.

142 Phillips, D.H. (2005) DNA adducts as markers of exposure and risk. *Mutat. Res.*, **577** (1–2), 284–292.

143 Angerer, J., Mannschreck, C., and Gundel, J. (1997) Biological monitoring and biochemical effect monitoring of exposure to polycyclic aromatic hydrocarbons. *Int. Arch. Occup. Environ. Health*, **70** (6), 365–377.

144 Shaham, J. and Ribak, J. (1996) The role of biomarkers in detecting early changes related to exposure to occupational carcinogens. *J. Occup. Health*, **38**, 170–178.

145 Boyd, J.A. and Barrett, J.C. (1990) Genetic and cellular basis of multistep carcinogenesis. *Pharmacol. Ther.*, **46** (3), 469–486.

146 Eriksson, H.L. *et al.* (2004) 32P-postlabeling of DNA adducts arising from complex mixtures: HPLC versus TLC separation applied to adducts from petroleum products. *Arch. Toxicol.*, **78** (3), 174–181.

147 Kondraganti, S.R. *et al.* (2003) Polycyclic aromatic hydrocarbon-inducible DNA adducts: evidence by 32P-postlabeling and use of knockout mice for Ah receptor-independent mechanisms of metabolic activation *in vivo*. *Int. J. Cancer*, **103** (1), 5–11.

148 Poirier, M.C. and Beland, F.A. (1994) DNA adduct measurements and tumor incidence during chronic carcinogen exposure in rodents. *Environ. Health Perspect.*, **102** (Suppl. 6), 161–165.

149 Tang, D. *et al.* (2001) Association between carcinogen–DNA adducts in white blood cells and lung cancer risk in the Physicians Health Study. *Cancer Res.*, **61** (18), 6708–6712.

150 Wilson, D.M., III and Thompson, L.H. (2007) Molecular mechanisms of sister-chromatid exchange. *Mutat. Res.*, **616** (1–2), 11–23.

151 Norppa, H. *et al.* (2006) Chromosomal aberrations and SCEs as biomarkers of cancer risk. *Mutat. Res.*, **600** (1–2), 37–45.

152 Nakatsuru, Y. *et al.* (2004) Dibenzo[*a,l*]pyrene-induced genotoxic and carcinogenic responses are dramatically suppressed in aryl hydrocarbon receptor-deficient mice. *Int. J. Cancer*, **112** (2), 179–183.

153 Wilding, C.S. *et al.* (2005) DNA repair gene polymorphisms in relation to chromosome aberration frequencies in retired radiation workers. *Mutat. Res.*, **570** (1), 137–145.

154 Lodovici, M. *et al.* (2004) Benzo(*a*)pyrene diolepoxide (BPDE)-DNA adduct levels in leukocytes of smokers in relation to polymorphism of CYP1A1, GSTM1, GSTP1, GSTT1, and mEH. *Cancer Epidemiol. Biomarkers Prev.*, **13** (8), 1342–1348.

155 Pavanello, S. and Clonfero, E. (2004) Individual susceptibility to occupational carcinogens: the evidence from biomonitoring and molecular epidemiology studies. *G. Ital. Med. Lav. Ergon.*, **26** (4), 311–321.

156 Salazar, A.M. *et al.* (2004) p53 expression in circulating lymphocytes of non-melanoma skin cancer patients from an arsenic contaminated region in Mexico. A pilot study. *Mol. Cell. Biochem.*, **255** (1–2), 25–31.

157 Whyatt, R.M. *et al.* (2000) Association between polycyclic aromatic hydrocarbon–DNA adduct levels in maternal and newborn white blood cells and glutathione *S*-transferase P1 and CYP1A1 polymorphisms. *Cancer Epidemiol. Biomarkers Prev.*, **9** (2), 207–212.

158 Perera, F.P. *et al.* (1992) Molecular and genetic damage in humans from environmental pollution in Poland. *Nature*, **360** (6401), 256–258.

159 Bonassi, S. *et al.* (2000) Chromosomal aberrations in lymphocytes predict human cancer independently of exposure to carcinogens. European Study Group on Cytogenetic Biomarkers and Health. *Cancer Res.*, **60** (6), 1619–1625.

160 Hagmar, L. *et al.* (1994) Cancer risk in humans predicted by increased levels of chromosomal aberrations in lymphocytes: Nordic Study Group on the Health Risk of Chromosome Damage. *Cancer Res.*, **54** (11), 2919–2922.

161 Kalina, I. *et al.* (1998) Cytogenetic monitoring in coke oven workers. *Mutat. Res.*, **417** (1), 9–17.

162 Burgaz, S. *et al.* (2002) Chromosomal damage in peripheral blood lymphocytes of traffic policemen and taxi drivers exposed to urban air pollution. *Chemosphere*, **47** (1), 57–64.

163 Park, S.Y. *et al.* (2006) Benzo[*a*]pyrene-induced DNA damage and p53 modulation in human hepatoma HepG2 cells for the identification of potential biomarkers for PAH monitoring and risk assessment. *Toxicol. Lett.*, **167** (1), 27–33.

164 Cariello, N.F. *et al.* (1994) Computer program for the analysis of mutational spectra: application to p53 mutations. *Carcinogenesis*, **15** (10), 2281–2285.

165 Binkova, B. *et al.* (2000) The effect of dibenzo[*a,l*]pyrene and benzo[*a*]pyrene on human diploid lung fibroblasts: the induction of DNA adducts, expression of p53 and p21 (WAF1) proteins and cell cycle distribution. *Mutat. Res.*, **471** (1–2), 57–70.

166 Mahadevan, B. *et al.* (2001) Effects of the (–)-anti-11*R*,12*S*-dihydrodiol-13*S*,14*R*-epoxide of dibenzo. *Carcinogenesis*, **22** (1), 161–169.

167 Brandt-Rauf, P.W. and Pincus, M.R. (1998) Molecular markers of carcinogenesis. *Pharmacol. Ther.*, **77** (2), 135–148.

168 Charuruks, N. *et al.* (2001) Clinical significance of p53 antigen and anti-p53 antibodies in the sera of hepatocellular carcinoma patients. *J. Gastroenterol.*, **36** (12), 830–836.

169 Krajewska, B., Lutz, W., and Pilacik, B. (1998) Determination of blood serum oncoprotein NEU and antioncoprotein p-53: molecular biomarkers in various types of occupational exposure. *Int. J. Occup. Med. Environ. Health*, **11** (4), 343–348.

170 Rossner, P., Jr., Binkova, B., and Sram, R.J. (2003) The influence of occupational exposure to PAHs on the blood plasma levels of p53 and p21WAF1 proteins. *Mutat. Res.*, **535** (1), 87–94.

171 Albinet, A. *et al.* (2007) Polycyclic aromatic hydrocarbons (PAHs), nitrated PAHs and oxygenated PAHs in ambient air of the Marseilles area (south of France): concentrations and sources. *Sci. Total Environ.*, **384** (1–3), 280–292.

172 Ciganek, M. *et al.* (2004) A combined chemical and bioassay analysis of traffic-emitted polycyclic aromatic hydrocarbons. *Sci. Total Environ.*, **334–335**, 141–148.

173 Hien, T.T. *et al.* (2007) Comparison of particle-phase polycyclic aromatic hydrocarbons and their variability causes in the ambient air in Ho Chi Minh City, Vietnam and in Osaka, Japan, during, 2005–2006. *Sci. Total Environ.*, **382** (1), 70–81.

174 Liu, L.B. *et al.* (2007) Determination of particle-associated polycyclic aromatic hydrocarbons in urban air of Beijing by GC/MS. *Anal. Sci.*, **23** (6), 667–671.

175 Sram, R.J. *et al.* (2007) Environmental exposure to carcinogenic polycyclic aromatic hydrocarbons: the interpretation of cytogenetic analysis by FISH. *Toxicol. Lett.*, **172** (1–2), 12–20.

176 Dubowsky, S.D., Wallace, L.A., and Buckley, T.J. (1999) The contribution of traffic to indoor concentrations of polycyclic aromatic hydrocarbons. *J. Exp. Anal. Environ. Epidemiol.*, **9** (4), 312–321.

177 Gevao, B. *et al.* (2007) Polycyclic aromatic hydrocarbons in indoor air and dust in Kuwait: implications for sources and nondietary human exposure. *Arch. Environ. Contam. Toxicol.*, **53** (4): 503–512.

178 Li, A. *et al.* (2005) Polycyclic aromatic hydrocarbons in residential air of ten Chicago area homes: concentrations and influencing factors. *Atmos. Environ.*, **39** (19), 3491–3501.

179 Li, C.-S. and Ro, Y.-S. (2000) Indoor characteristics of polycyclic aromatic hydrocarbons in the urban atmosphere of Taipei. *Atmos. Environ.*, **34** (4), 611–620.

180 Naumova, Y.Y. *et al.* (2002) Polycyclic aromatic hydrocarbons in the indoor and outdoor air of three cities in the U.S. *Environ. Sci. Technol.*, **36** (12), 2552–2559.

181 Ohura, T. *et al.* (2004) Polycyclic aromatic hydrocarbons in indoor and outdoor environments and factors affecting their concentrations. *Environ. Sci. Technol.*, **38** (1), 77–83.

182 Zhu, L. and Wang, J. (2003) Sources and patterns of polycyclic aromatic hydrocarbons pollution in kitchen air, China. *Chemosphere*, **50** (5), 611–618.

183 Black, W.V., Kosson, D.S., and Ahlert, R.C. (1989) Characterization and evaluation of environmental hazards in a large metropolitan landfill, in *Proceedings of the Industrial Waste Conference*, Lewis Publishers, Inc., Chelsea, MI.

184 Curtosi, A. *et al.* (2007) Polycyclic aromatic hydrocarbons in soil and surface marine sediment near Jubany Station (Antarctica). Role of permafrost as a low-permeability barrier. *Sci. Total Environ.*, **383** (1–3), 193–204.

185 Eom, I.C. *et al.* (2007) Ecotoxicity of a polycyclic aromatic hydrocarbon (PAH)-contaminated soil. *Ecotoxicol. Environ. Saf.*, **67** (2), 190–205.

186 Morillo, E. *et al.* (2007) Soil pollution by PAHs in urban soils: a comparison of three European cities. *J. Environ. Monit.*, **9** (9), 1001–1008.

187 Gu, S.H. *et al.* (2003) Source apportionment of PAHs in dated sediments from the Black River, Ohio. *Water Res.*, **37** (9), 2149–2161.

188 Kannan, K. *et al.* (2005) Spatial and temporal distribution of polycyclic aromatic hydrocarbons in sediments from Michigan inland lakes. *Environ. Sci. Technol.*, **39** (13), 4700–4706.

189 Neff, J.M., Stout, S.A., and Gunster, D.G. (2005) Ecological risk assessment of polycyclic aromatic hydrocarbons in sediments: identifying sources and

ecological hazard. *Integr. Environ. Assess. Manag.*, **1** (1), 22–33.

190 Su, M.C., Christensen, E.R., and Karls, J.F. (1998) Determination of PAH sources in dated sediments from Green Bay, Wisconsin, by a chemical mass balance model. *Environ. Pollut.*, **99** (3), 411–419.

191 Guo, W. *et al.* (2007) Distribution of polycyclic aromatic hydrocarbons in water, suspended particulate matter and sediment from Daliao River watershed, China. *Chemosphere*, **68** (1), 93–104.

192 Ke, R. *et al.* (2007) Comparison of the uptake of polycyclic aromatic hydrocarbons and organochlorine pesticides by semipermeable membrane devices and caged fish (*Carassius carassius*) in Taihu Lake, China. *Environ. Toxicol. Chem.*, **26** (6), 1258–1264.

193 Nagy, P., Fekete, J., and Sharma, V.K. (2007) Polycyclic aromatic hydrocarbons (PAHs) in surface waters of Rackevei-Soroksari Danube Branch, Hungary. *J. Environ. Sci. Health A*, **42** (3), 231–240.

194 Hartnik, T. *et al.* (2007) Bioassay-directed identification of toxic organic compounds in creosote-contaminated groundwater. *Chemosphere*, **66** (3), 435–443.

195 Rosenfeld, J. and Plumb, R. (1991) Ground water contamination at wood treatment facilities. *Ground Water Monitor. Rev.*, **11** (1), 133–140.

196 Kolarovic, L. and Traitler, H. (1982) Determination of polycyclic aromatic hydrocarbons in vegetable oils by caffeine complexation and glass capillary gas chromatography. *J. Chromatogr. A*, **237** (2), 263–272.

197 Moret, S. and Conte, L.S. (2000) Polycyclic aromatic hydrocarbons in edible fats and oils: occurrence and analytical methods. *J. Chromatogr. A*, **882** (1–2), 245–253.

198 Baumard, P. *et al.* (1998) Concentrations of PAHs (polycyclic aromatic hydrocarbons) in various marine organisms in relation to those in sediments and to trophic level. *Mar. Pollut. Bull.*, **36** (12), 951–960.

199 Mottier, P., Parisod, V., and Turesky, R.J. (2000) Quantitative determination of polycyclic aromatic hydrocarbons in barbecued meat sausages by gas chromatography coupled to mass spectrometry. *J. Agric. Food Chem.*, **48** (4), 1160–1166.

200 Krishnadasan, A. *et al.* (2007) Nested case–control study of occupational chemical exposures and prostate cancer in aerospace and radiation workers. *Am. J. Ind. Med.*, **50** (5), 383–390.

201 Zhao, Z.H., Quan, W.Y., and Tian, D.H. (1990) Urinary 1-hydroxypyrene as an indicator of human exposure to ambient polycyclic aromatic hydrocarbons in a coal-burning environment. *Sci. Total Environ.*, **92**, 145–154.

202 Granella, M. and Clonfero, E. (1993) Urinary excretion of 1-pyrenol in automotive repair workers. *Int. Arch. Occup. Environ. Health*, **65** (4), 241–245.

203 Lee, M.S. *et al.* (2007) 1-Hydroxypyrene as a biomarker of PAH exposure among subjects living in two separate regions from a steel mill. *Int. Arch. Occup. Environ. Health*, **80** (8), 671–678.

204 Chen, B. *et al.* (2007) Higher urinary 1-hydroxypyrene concentration is associated with cooking practice in a Chinese population. *Toxicol. Lett.*, **171** (3), 119–125.

205 Merlo, F. *et al.* (1998) Urinary excretion of 1-hydroxypyrene as a marker for exposure to urban air levels of polycyclic aromatic hydrocarbons. *Cancer Epidemiol. Biomarkers Prev.*, **7** (2), 147–155.

206 Bentsen-Farmen, R.K. *et al.* (1999) Detection of polycyclic aromatic hydrocarbon metabolites by high-pressure liquid chromatography after purification on immunoaffinity columns in urine from occupationally exposed workers. *Int. Arch. Occup. Environ. Health*, **72** (3), 161–168.

207 Jongeneelen, F.J. *et al.* (1986) Biological monitoring of polycyclic aromatic hydrocarbons. Metabolites in urine. *Scand. J. Work Environ. Health*, **12** (2), 137–143.

208 Chen, B. *et al.* (2007) Urinary 1-hydroxypyrene concentrations in Chinese coke oven workers relative to job category, respirator usage, and cigarette smoking. *Am. J. Ind. Med.*, **50** (9), 657–663.

21
Development of *In Vitro* Models to Assess Toxicity of Engineered Nanomaterials

Laura K. Braydich-Stolle, Richard C. Murdock, and Saber M. Hussain

21.1
Introduction

New chemical and pharmaceutical products are tested prior to human exposure in order to evaluate any potential hazard. *In vivo* animal models have been used to assess the toxicity of these new compounds, which has resulted in some criticism. To avoid testing in animals, systems that provide information on species-specific metabolism, pharmacokinetics, and toxicology are essential, and the development of *in vitro* models to assess toxicity may provide important tools for investigating specific mechanisms of toxicity.

One area that is receiving increased attention is the field of nanotechnology, and engineered nanomaterials are being incorporated into a wide range of products. Nanomaterials, by definition, have one dimension that falls into the 1–100 nm range, and are typically composed of metal or carbon [1]. Current medical applications include nanomaterial use in drug delivery, imaging, and bone composites [2–5]. Furthermore, in food and cosmetic industries, nanomaterials have been incorporated into products to enhance their quality [6]. Since there is a growing level of exposure to nanomaterials, it is important to answer questions regarding the toxicity of these materials. Currently, a wide range of cell lines are available which allow researchers to answer initial questions about the toxicological effects of these nanomaterials in a variety of target tissues and organs. However, proper characterization of test materials is important to ensure that results are reproducible and also to provide the basis for understanding the properties of nanoparticles that determine their biological effects [7].

Since nanomaterials can be inhaled, digested, or absorbed into the body and once inside the body they have been shown to colonize in a variety of organ systems, this chapter will focus on describing *in vitro* models to assess exposure in the lungs, skin, and liver. Perhaps even more interesting is that nanomaterials have also been shown to cross the blood–brain barrier and the blood–testis barrier, which most chemicals cannot. Therefore, the use of neuronal cell lines to study toxicity in the nerve cells, and the use of a novel spermatogonial stem cell line to study brain and testes toxicity, will

Principles and Practice of Mixtures Toxicology. Edited by Moiz Mumtaz

ISBN: 978-3-527-31992-3

also be discussed. While questions regarding mechanisms of nanomaterial toxicity need to be addressed, one must also realize that the unique properties of nanomaterials may be beneficial. Therefore, researchers also need to assess how non-toxic nanomaterials interact once inside a cell. An important avenue to evaluate is the interaction between nanomaterials and signaling proteins, and how this could be manipulated to label, tag, and turn off certain signaling proteins or genes. In addition, how nanomixtures can alter the dispersion, biocompatibility, and toxicity of certain nanomaterials will be addressed.

Furthermore, certain properties that nanoparticles possess differentiate them from their bulk counterparts, potentially leading to dissimilar biological effects. These characteristics which must be evaluated prior to nanoparticle studies include size, shape, dispersion, physical and chemical properties, surface area, and surface chemistry [8–10]. Many of these properties can be studied using dry, powdered nanomaterials; however, it is uncertain how these characteristics will change once placed into a physiologically compatible solution for *in vitro* dosing procedures. Therefore, one focus of this chapter will be on the characterization of these nanomaterials once they are dispersed in solution and prior to exposure. Overall, the applications of nanomaterials are very promising, and the use of *in vitro* systems will be useful in answering basic questions regarding characterization, toxicity, and interaction before moving to *in vivo* systems.

21.2 *In Vitro* Nanotoxicity Models

21.2.1 Lung Nanotoxicity

Since ultrafine particles are found in pollution which can be easily inhaled and absorbed systemically, the majority of nanotoxicity studies performed have focused on the lung exposure. Furthermore, studies have shown that nanoparticles are capable of penetrating the alveoli deep in the lung [9]. Therefore, the lung cell line types that researchers have focused on are the bronchial epithelial cells, alveolar epithelial cell lines, and alveolar macrophages, whose main function is to remove foreign material from the lungs.

In vitro studies, using human bronchial epithelial cells, have shown that SWNTs generate reactive oxygen species (ROS) indicating that oxidative stress is the predominant mechanism of acute toxicity [11, 12]. Similarly, TiO_2 particles induced oxidative damage in a human bronchial epithelial cell line in a size-dependent manner [13]. Additionally, the A549 cells, which are alveolar basal epithelial cells derived from a carcinoma have been used in a wide range of nanotoxicity studies. One study examining the size-dependent effects of crystalline silicon dioxide particles (15 and 46 nm) demonstrated that the toxicity was not size dependent but occurred in a dose-dependent manner along with increased levels of oxidative stress [14]. In addition, 20 nm cerium oxide also produces a dose-dependent cytotoxic response and increased levels of oxidative stress, which results in lipid peroxidation and cell membrane leakage [15]. Furthermore, 47 nm TiO_2 have been shown to be non-toxic, while silica nanospheres, 80 nm Al, and $<5\,\mu m$ quartz induce toxicity [16].

Similarly, aluminum nanoparticles have been shown to exhibit toxicity and significantly diminish the phagocytotic ability of macrophages in comparison to aluminum oxide nanoparticles which are not toxic [17]. Furthermore, silver nanoparticles have been shown to produce a size- and concentration-dependent toxicity effect [18]. In contrast, nanodiamonds have been shown to be biocompatible in rat alveolar macrophages [19], while carbon nanotubes induce the formation of ROS and decrease mitochondrial membrane potential in the rat macrophage cell line NR8383 and the A549 cells [20]. However, the toxicity of these nanotubes has been associated with metal traces from the synthesis stage [20].

21.2.2
Dermal Toxicity

Skin is the first line of defense for the human body against the external environment, and it provides a significant surface area for potential contamination. The skin is a complex organ that is comprised of the epidermis, dermis, and hypodermis. The epidermis is the most external layer, and does not contain any blood vessels. The epidermis contains several cell types which are keratinocytes, melanocytes, Langerhans cells, and Merkels cells. Between the epidermis and the dermis there is a very thin membrane, called the basement membrane, which attaches the epidermis firmly, though not rigidly, to the dermis. The dermis is the second layer of skin, which consists of connective tissue, and contains blood vessels, nerves, hair roots, and sweat glands. Below the dermis lies the hypodermis, which contains fibroblasts, adipose cells, and macrophages, as well as the subcutaneous fat, and larger blood vessels and nerves. The subcutaneous fat lies on the muscles and bones, to which the whole skin structure is attached by connective tissues.

The predominant *in vitro* model for assessing skin exposure has been keratinocyte cell lines since the outer layer of the skin is comprised primarily of keratinocytes. Using keratinocytes, studies have shown that different types of nanomaterials induce cytotoxicity and reactive oxygen species formation. For example, human epidermal keratinocytes treated with unmodified MWCNT demonstrate cytoxicity [21]. Furthermore, Nel *et al.* [22] reported that treating keratinocytes with high doses of SWCNTs results in ROS generation and oxidative stress. Furthermore, our laboratory has shown that mouse keratinocytes treated with MWCNT and SWCNT exhibit ROS generation, which is consistent with data demonstrating that MWCNTs and SWCNTs induce cytotoxicity in human epidermal keratinocytes [21, 23, 24]. In addition, our laboratory has also shown silicon dioxide and titanium dioxide nanoparticles can cause cytotoxicity and ROS production in mouse keratinocytes [25, 26].

In addition to transformed or immortalized cell lines, primary normal human keratinocytes (NHEK) are commercially available from companies such as PromoCell and Cell Applications, Inc. One study by Ryman-Rasmussen and colleagues has shown that in HEKs quantum dot uptake can occur regardless of surface coating;

however, the surface coating of the quantum dots will mediate the cytotoxicity [27]. Furthermore, fullerenes exposed to amino acid solutions at a concentration of 0.4–0.04 mg/ml showed decreases in cell viability and initiation of inflammatory responses in HEKs [28].

However, since these cell monolayers cannot fully depict what would happen in the entire tissue, companies have aimed to develop *in vitro* co-culture systems to help fully address dermal toxicity. Currently, the MatTek Corporation produces the EpiDerm™ Full Thickness (EpiDermFT™) model, which is a human dermal/epidermal skin equivalent with a fully developed basement membrane. This skin equivalent is prepared by culturing normal human epidermal keratinocytes with dermal fibroblasts to produce highly differentiated full-thickness skin tissue. As of now, there is no data evaluating nanotoxicity in this model.

21.2.3 Liver Nanotoxicity

In vivo studies have shown that the liver is a potential target for nanotoxicity [9, 29], which is to be expected since detoxicification is a major role of the liver. At present, a study using the rat BRL-3A liver cell line demonstrated that nanometals are differentially toxic. In the liver cells, Fe_3O_4 (30, 47 nm), Al (30, 103 nm), MnO_2 (1–2 μm), and W (27 μm) displayed little or no toxicity, while MoO_3 (30, 150 nm) was moderately toxic, and Ag (15, 100 nm) was highly toxic and oxidative stress was identified as the participating mechanism of Ag toxicity [30]. In addition to the BRL-3A rat liver cells, the HEP2G (human hepatoma cell line) liver cells provide a useful model to evaluate nanotoxicity.

So far, the literature is not as expansive on liver nanotoxicity, but since the liver plays such a crucial role in metabolism and detoxification, this system should not be ignored and future *in vitro* studies addressing liver nanotoxicity in our laboratory are underway.

21.2.4 Neuronal Neurotoxicity

Several studies have shown that intravenous and/or intra-abdominal administration of nanoparticles to mice results in their accumulation in the brain suggesting that they can penetrate the blood–brain barrier [31, 32], which is highly unexpected considering the fact that most of the pharmaceutical substances require careful modification to target the brain actually fail to cross the blood–brain barrier [33, 34]. The precise implications of this phenomenon remain unclear and warrant careful evaluation especially since researchers are using nanotechnology to deliver drugs and other agents to the brain [35].

The PC-12 cell line was derived from rat adrenal phenochomocytoma and can be induced to a neuronal phenotype using nerve growth factor [36]. One study using the PC-12 cells showed that Mn 40 nm particles induced dose-dependent dopamine depletion as well as depletion of dopamine metabolites [37], and further studies need to examine if dopamine depletion occurs *in vivo*. Moreover, PC-12 cells exposed to iron oxide nanoparticles have been shown to have decreased viability and inhibition of neurite extension in the presence of nerve growth factor in a dose-dependent manner [38]. Furthermore, the neuroblastoma cell line Neuro2A or N2A [39] has also been used to evaluate neurotoxicity. One study evaluating the effect of nanodiamonds on N2A cells has shown that these nanomaterials are non-toxic to the cells [19]. In addition to neurons, microglia, which are the immune cells in the brain, are also potential targets for neurotoxicity. These cells typically remain in a resting state until some neurodegenerative event occurs to trigger them into the active state where they will engulf the apoptotic neuron. Recently, it has been shown that TiO_2 produces reactive oxygen species formation in an immortalized brain microglia (BV2) [40].

Currently, these initial studies *in vitro* and *in vivo* provide evidence that nanoparticles are a potential neurological threat since they can cross the blood–brain barrier and interfere with physiological processes in cellular models of the brain.

21.2.5 Reproductive Nanotoxicity

Since the germline is responsible for passing genetic information on to future offspring, it is of paramount importance that we know how to protect the germline from environmental toxicants. There is still limited data available from toxicity studies of nanoparticles, in particular in adult stem cells. Recently, Kim and colleagues have shown that once 50 nm rhodamine-labeled magnetic nanoparticles cross the blood–testis barrier, they will colonize in the seminiferous epithelium where spermatogenesis takes places [41]. Furthermore, once inside the seminiferous epithelium, these nanoparticles incorporate into the developing germ cells and eventually leave the epithelium inside the mature sperm [41]. Therefore, certain nanoparticles could alter spermatogenesis and impact fertility. In particular, the cytotoxic effect of nanoparticles on the male germline needs to be addressed since this has the potential to impact future generations.

Current research has identified the C18-4 spermatogonial stem cell line [42] as a novel model for evaluating the reproductive toxicity of nanomaterials [43]. Furthermore, this study demonstrated that the silver-15 nm, aluminum-30 nm, and molybdenum oxide-30 nm nanoparticles were able to enter the C18-4 spermatogonial stem cells, disrupt proliferation, and damage the cell membranes. Furthermore, the different types of nanoparticles were toxic to the C18-4 cells in a dose-dependent manner. However, the direct mechanism by which these nanoparticles are toxic to the cells is still unknown. Since cell proliferation was disrupted and these cells are known to proliferate using a GDNF (glial cell line-derived neurotrophic factor) pathway [44], the impact of silver nanoparticles on this signaling pathway has been investigated. Specifically, once GDNF binds to its receptors GFRα-1 and Ret, this will activate the Ret receptor and allow it to phosphorylate the signaling protein Fyn (a member of the Src family kinases). Once Fyn is activated, this will allow the activation of the PI3/Akt pathway and the end result is that genes involved in cell cycle regulation are differentially expressed to increase the production of *N*-myc and promote proliferation (Figure 21.1a). Initial results have shown that treatment with silver nanoparticles at a non-lethal concentration will disrupt the activity of the Fyn kinase (Figure 21.1b). Further research will attempt to answer at which level this inhibition occurs. For example, the reduction in kinase activity could be occurring through several possible mechanisms: (1) GDNF being unable to bind its receptor, (2) the receptor being unable to be activated once GDNF is bound, or (3) the Fyn kinase being unable to be activated by the receptor (Figure 21.1c).

21.3 Toxicology of Nanomixtures

Nanoparticle combinations, or nanomixtures, are becoming more prevalent in industrial and commercial applications due to their unique individual and combined properties. TiO_2 and Ag nanoparticles have been combined for their increased antibacterial capabilities on everyday items, such as keyboards, and in medical applications, such as surgical masks [45]. In other cases, nanomixtures can create gels, liquids, or crystalline structures with unique properties different from the individual particles alone [46]. Studies have also been conducted to find out dispersion methods that yield homogeneous mixtures of nanomaterials for use in design of nanocomposites [47]. Also, nanomaterial mixtures, such as a combination of TiO_2 and Ag nanoparticles, can be used as an aid for photocatalytic reactions [48]. Understanding the possible changes in toxicity from individual particles to these nanoparticle mixtures is an important step for toxicology studies, especially to keep pace with commercial technology and to keep consumers safe. Therefore, the individual and combined toxicological effect of these nanomaterials needs to be addressed.

Studies have shown that copper (Cu) nanoparticles are highly toxic to PC-12 cells and alveolar macrophages [17]. In comparison, aluminum oxide (Al_2O_3) and silicon dioxide (SiO_2) nanoparticles are not toxic at low concentrations [17]. Therefore, the toxicity of Cu, Al_2O_3, and SiO_2 and mixtures of Cu, Al_2O_3, and SiO_2 were evaluated in rat alveolar macrophages to determine the individual and combined toxicity using cell proliferation and morphological changes as markers for toxicity. As expected, the cells treated with aluminum oxide and silicon dioxide alone showed no disruption of cell proliferation (Figure 21.2) and retained relatively normal

Figure 21.1 Model of GDNF signaling experiments in C18-4 cells. (a) GDNF signaling pathway in C18-4 cells. (b) Fyn kinase activity in C18-4 cells treated with varying sizes of Ag nanoparticles. Nanoparticles >25 nm completely abolish Fyn activity, whereas Ag 25 nm nanoparticles decrease Fyn activity, but still allow GDNF to stimulate an increase in the activity. (c) Hypothesis for GDNF disrupted signaling when C18-4 cells are treated with silver nanoparticles. Since disruption of kinase activity is seen, there are three possible mechanisms: (1) GDNF being unable to bind its receptor, (2) the receptor being unable to be activated once GDNF is bound, or (3) the Fyn kinase being unable to be activated by the receptor.

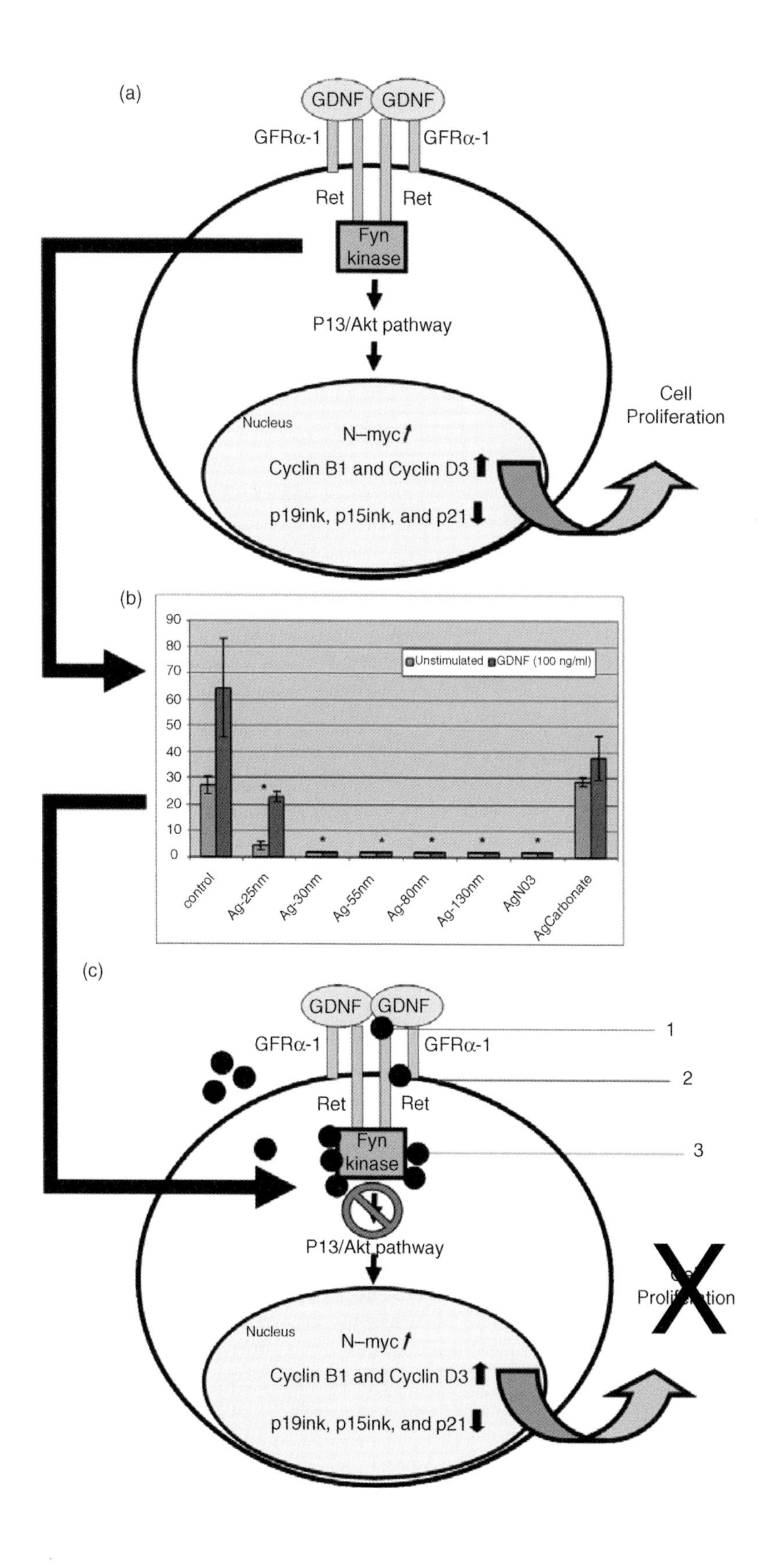

(a)
GDNF
GDNF
GFRα-1
GFRα-1
Ret
Ret
Fyn kinase
P13/Akt pathway
Nucleus
N–myc
Cyclin B1 and Cyclin D3
p19ink, p15ink, and p21
Cell Proliferation
(b)
Unstimulated
GDNF (100 ng/ml)
90
80
70
60
50
40
30
20
10
0
control
Ag-25nm
Ag-30nm
Ag-55nm
Ag-80nm
Ag-130nm
AgN03
AgCarbonate
(c)
GDNF
GDNF
GFRα-1
GFRα-1
1
2
Ret
Ret
Fyn kinase
3
P13/Akt pathway
Nucleus
N–myc
Cyclin B1 and Cyclin D3
p19ink, p15ink, and p21

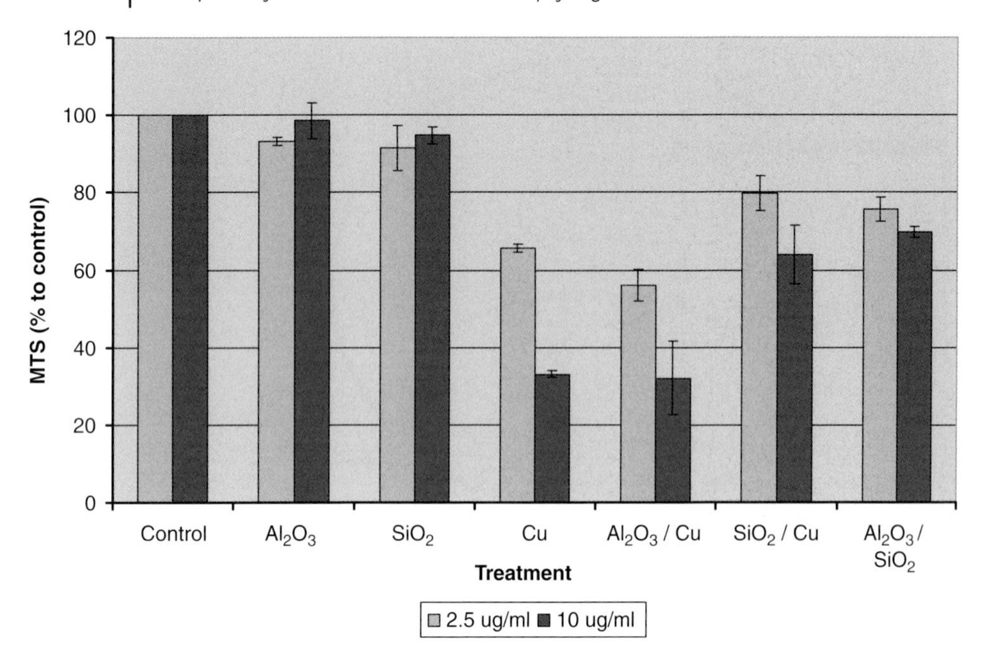

Figure 21.2 Cell viability assay for alveolar macrophages dosed with nanoparticles and nanomixtures. Cells exposed after 24 h growth and MTS performed 24 h postexposure time point. Individual nanoparticles dosed at 2.5 and 10 μg/ml. Mixtures contain 10 μg/ml of each nanomaterial listed.

morphology (Figure 21.3a–c). In contrast, the copper nanoparticles disrupted cell proliferation (Figure 21.2) and altered the morphology of the macrophages (Figure 21.3d). Interestingly, when Al_2O_3 and Cu were mixed together there was no change in the level of toxicity when compared to Cu alone; however, the SiO_2/Cu mixture showed less toxicity when compared to Cu alone, indicating that the SiO_2 may buffer some of the Cu toxicity. Surprisingly, Al_2O_3 and SiO_2 do not induce cytotoxicity, yet when the macrophages were treated with a combination of the Al_2O_3 and SiO_2, there was a mild disruption in cell proliferation (Figure 21.2). These data indicate that researchers should not limit their focus to just the individual effects of nanoparticles, but they should also be addressing the combined toxic effect since it is more likely that human exposure will occur at a mixtures level than at a pure particle level.

In addition, one challenge presented with *in vitro* studies of nanoparticles, is that these particles do not disperse homogeneously in solution and have a tendency to form aggregates. Portions of our research have been devoted to optimizing *in vitro* dosing procedures by evaluating different solvents for dispersion of nanoparticles. In the case of titanium dioxide particles (5, 50, and 223 nm), dispersion in water leads to the formation of aggregates that precipitate out of solution. To avoid these problems, we tried dispersing TiO_2 in polyethylene glycol (PEG) since it is known to have high biocompatibility, is soluble in water, and prevents the formation of

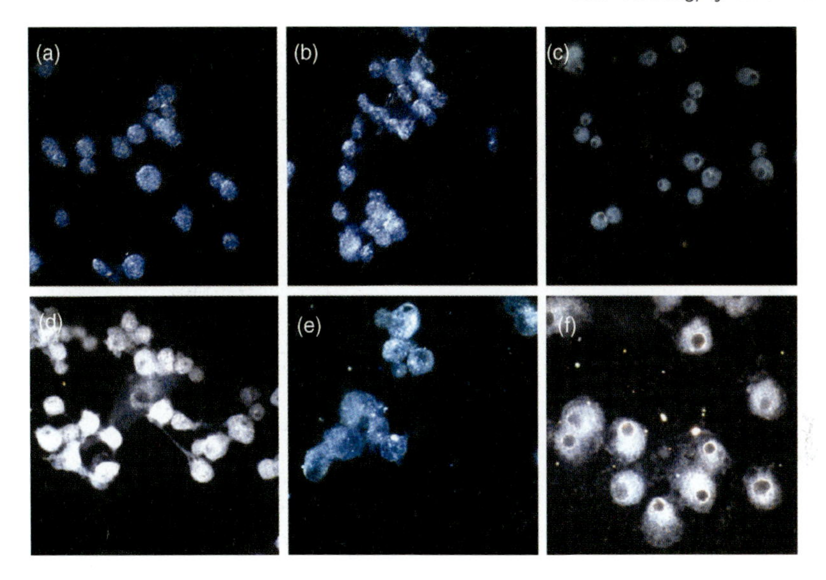

Figure 21.3 Alveolar macrophage morphology studied under ultraresolution imaging system. Images taken using CytoViva URI system at 96× magnification. (a) Control. (b) Macrophages dosed with Al_2O_3 30 nm. (c) Macrophages dosed with SiO_2 31 nm FITC. (d) Macrophages dosed with Cu 40 nm. (e) Macrophages dosed with Al_2O_3 30 nm and Cu 40 nm. (f) Macrophages dosed with SiO_2 31 nm FITC and Cu 40 nm.

aggregates [49–51]. Interestingly, when the solvent was varied the toxicity of the nanoparticles changed. When the nanoparticles were dispersed in sterile water, there was not a dose or size-dependent effect on cell proliferation after 24 h in mouse keratinocytes treated with TiO_2 nanoparticles. However, when the nanoparticles were dispersed in PEG, there was a dose-dependent decrease in cell proliferation (Figure 21.4), which could be for a variety of reasons. The particles could be acting as delivery vehicles for the PEG and the PEG could be disrupting proliferation; however, in cells that are treated with just PEG there is no significant decrease in cell proliferation. Another reason could be that because the PEG coats the particles and makes them hydrophilic, they are more homogeneously dispersed and therefore able to enter the cells more easily; therefore, more particles are entering the cells and this is why proliferation is disrupted. In contrast, keratinocytes treated with the nanoparticles dispersed in water, membrane leakage was two to three times higher compared to that of control keratinocytes. In comparison, cells treated with nanoparticles dispersed in PEG showed only a moderate increase in membrane leakage. This effect might be due to the fact that the particles did not agglomerate as much and therefore smaller agglomerates or single particles were being taken up by the cell and therefore less damage was done to the membrane. Furthermore, the PEG coating makes the particles more hydrophilic and this may have enabled them to use water channels or membrane channels to enter the keratinocytes and in turn cause less membrane disruption.

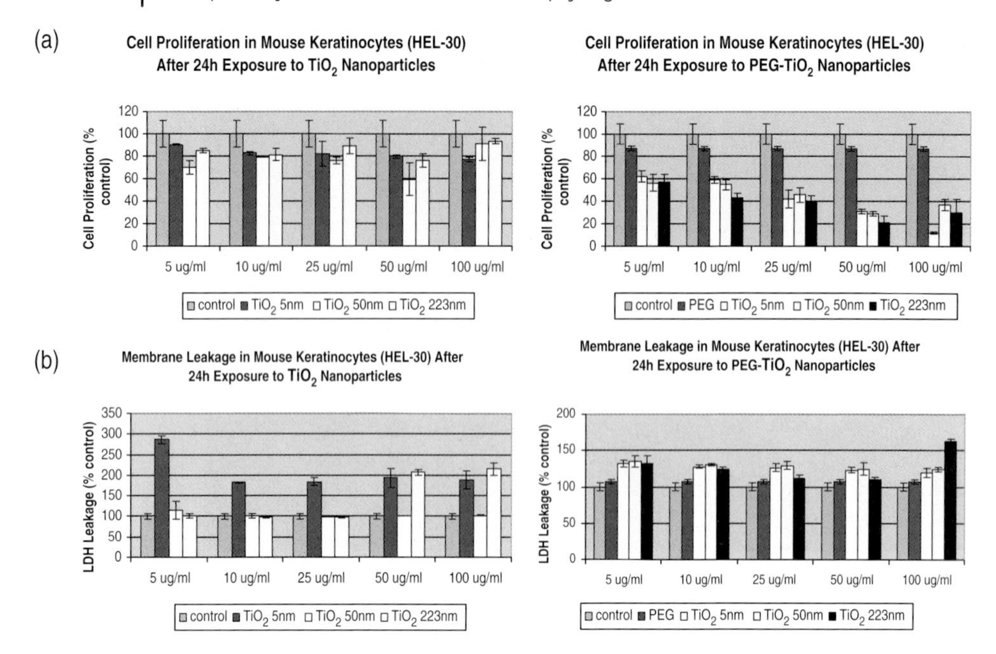

Figure 21.4 Changes in mouse keratinocyte toxicity after treatment with TiO_2 nanoparticles dispersed in water or PEG.

21.4 The *In Vitro* Debate

Since nanotoxicity is a relatively new field, a standard practice for testing these materials has yet to be created. *In vitro* models have received criticism for their lack of reproducibility, and a recent study by Sayes and colleagues has shown that in the case of lung exposure, *in vitro* models did not correlate well with *in vivo* models [52]. In order for an *in vitro* model to be used as a predictor of toxicity, the cells need to be validated as an accurate model for representing that tissue. For example, the C18-4 cell line has been extensively studied to illustrate the expression of the same key proteins expressed by spermatogonial stem cells [43, 53] and similar responses to key regulators of spermatogenesis [44, 45]. Important growth factor signaling studies have been performed in the C18-4 cells in parallel to the freshly isolated spermatogonial stem cells and organ cultures [45] to ensure the accuracy of using these cells as a model to understand the impact of nanoparticles on cell signaling pathways. Using uni-cellular *in vitro* models to represent complex tissues such as the lungs, skin, testes, and so on will not provide an exact depiction of how a multicellular tissue will respond; however, they provide a preliminary foundation for studies to assess dosing ranges, probable mechanisms of toxicity, and allow the refinement of techniques before progressing to costly *in vivo* studies.

Furthermore, *in vivo* studies have also yielded conflicting toxicity data. For example, *in vivo* SWNT studies by Lam and colleagues demonstrated that the SWNTs were more toxic than quartz in a dose-dependent manner in a mouse model [54], while Warheit and colleagues demonstrated dose-dependent responses with quartz and non-dose-dependent responses with the SWNTs in a rat model [55]. In addition, studies using rats that have examined the toxicity of nanosized TiO_2 and yielded conflicting results. One study concluded that ultrafine TiO_2 particles (29 nm) increased inflammation and altered macrophage chemotactic responses in rat lungs, when compared to TiO_2 particles that were 250 nm [56]. However, another study showed that in rats, exposure to nanoscale TiO_2 rods/dots produced inflammatory responses that were not different from pulmonary effects of larger TiO_2 particles [57], but rather toxicity was due to surface properties [58, 59]. These studies used the same delivery method, but these varied drastically in the nanoparticle source, amount of nanoparticles exposed, and the time of exposure that could account for differences. However, none of these studies attempted to characterize the nanomaterials once they were in aqueous solution and given that the properties of nanomaterials can change when they go from a dry environment to an aqueous environment, researchers need to evaluate these nanomaterials in solution prior to using them in toxicity studies. Based on this, questionable study designs and lack of characterization are the most likely source of the non-reproducible and potentially inaccurate results. Therefore, for future studies, there needs to be an optimization and validation of *in vitro* testing protocols and a strong focus needs to be placed on the characterization of these materials prior to, during, and after exposure to cells. Thus, the remainder of this chapter will focus on the characterization of nanomaterials.

21.5 Characterization of Nanomaterials

Researchers have shown that it is necessary to systematically and accurately define particle characteristics in correlation with potential toxicity of nanoparticles in a biological system [1] to ensure that results are reproducible and to determine which properties mediate the biological effects of nanoparticles [7]. In addition, previous research from our labs has shown that many factors can influence nanoparticle behavior, stressing the need to characterize these materials at each stage in the experimental process [60]. Furthermore, factors that have been shown to influence toxicity in both bacteria and mammalian cells include size/dimension, shape, crystal structure, porosity, and surface properties (e.g., area, charge, chemistry), which are all related to the concentration and duration of nanoparticle exposure [10, 61–66]. Therefore, before toxicity studies can be performed, it is imperative to perform a thorough characterization on the nanoparticles to evaluate attributes that contribute to the toxic response or otherwise effectively prevent cell injury. An example of a characterization plan for nanomaterials before and during toxicity testing is shown in Figure 21.5.

Once the nanoparticles have been received from the manufacturer, the first step is to confirm the average primary particle size and identify the size distribution using high-resolution transmission electron microscopy (HR-TEM) or scanning electron microscopy (SEM). The EM images will also provide information to address the morphology (shape, size, surface features, etc.) of the nanoparticles. Furthermore,

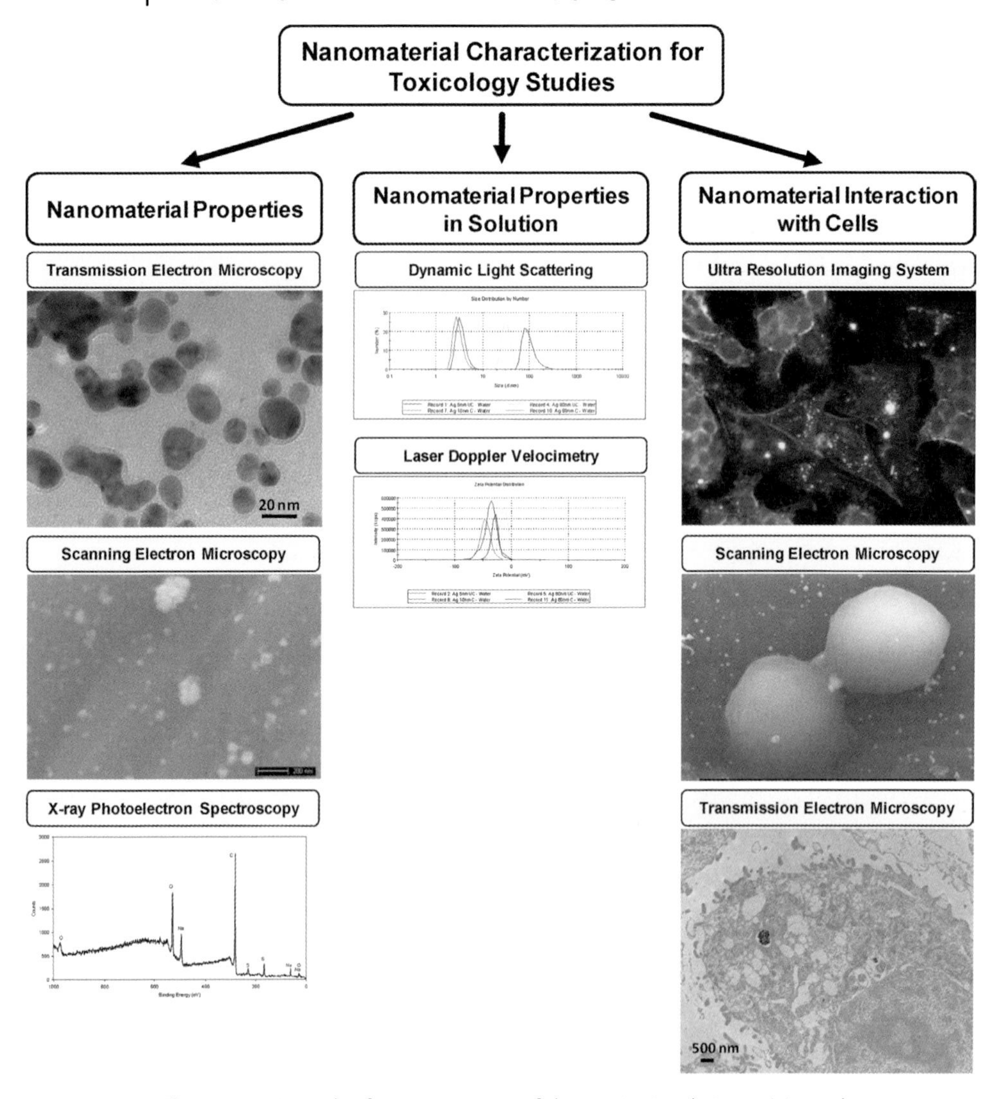

Figure 21.5 Example of necessary stages of characterization during toxicity studies.

examination of the surface coating and particle composition can be evaluated with SEM and HR-TEM in conjunction with the associated hardware that will allow collection of selected area diffraction patterns (SADPs) and energy dispersive X-ray (EDX) spectra. TEM and SEM can be performed to confirm uptake and localization of the nanoparticles in the cell. A precise size measurement and surface features of the nanoparticles can be evaluated using atomic force microscopy (AFM). The surface area of the particles can be determined by N_2 gas adsorption using the BET (Brunauer, Emmett, and Teller) method. In addition, inductively coupled plasma mass spectroscopy (ICP-MS) is used to evaluate the metal element concentrations of stock solutions, in order to determine if there are any impurities present in the samples, prior to exposure studies. Further chemical analysis can also be performed to elaborate on and confirm the elemental compositions, impurities, and surface chemistries using Raman and X-ray photoelectron spectroscopy (XPS).

After the nanoparticles have been dispersed in solution, their properties can change, and it has been shown that they will agglomerate in solution and that the degree of agglomeration varies depending on the solvent, or variation of contents in the solvent, used [60]. Therefore, once the nanoparticles have been dispersed in the different cell culture media for exposure, the size and the degree of agglomeration can be evaluated using dynamic light scattering (DLS). Characterization techniques to evaluate properties once the nanomaterial is in solution will be described in the next section.

21.5.1 Dynamic Light Scattering

As mentioned previously, there are numerous techniques to characterize dry nanomaterials: SEM, TEM, BET, AFM, and so on. However, no solid techniques for characterization of nanomaterials in solution have been established. One method, dynamic light scattering (DLS), also known as photon correlation spectroscopy (PCS), allows particles in a sample to be measured for their hydrodynamic size, with hydrodynamic being the key word. This allows real-time measurement of the particles in solution giving insight into the exact conditions under which they would be interacting with cells, tissues, or organs if placed in a biological fluid. DLS is not a new technique; in fact, it has been used in studies as early as 1975 as a simple method for determining suspension stability and measurement of particle hydrodynamic radius/diameter in solution [67–70]. DLS analyzes the velocity distribution of particle movement by measuring dynamic fluctuations of light scattering intensity, using Brownian motion. Brownian motion is the random movement of particles due to collisions at the molecular level and usually occurs only with particles smaller than 3 μm. This technique utilizes the Stokes–Einstein equation that determines the diffusion constant for a particle based on temperature of the solution, viscosity of the solvent, and radius of the particles.

There are a few drawbacks to the DLS and LDV methods. Usually, a 1 or 1.5 ml sample is needed for measurement, meaning a large amount of nanomaterial must be sacrificed. However, there are special adaptors available to lower this amount to near 300 μl. Also, it is not possible to obtain a visual image of the agglomerated structure while in solution, though DLS may be beneficial to know if the particles agglomerate in tight, compact structures or form loose chains of nanoparticles. A serious drawback to LDV measurements is the inability to measure samples in physiological solutions due to corrosion of the electrodes and varying protein charges. A similar solution in water can be used to mimic the higher salt concentrations and pH, but this still ignores the presence of proteins in the cell culture media.

21.5.2 Inductively Coupled Plasma Mass Spectrometry

To aid in quantifying cell uptake of nanomaterials, ICP-MS can be used to give an average uptake concentration. ICP-MS has been used since the early 1980s to determine metal composition, from samples in solution, in the parts per billion (ppb) or parts per trillion (ppt) ranges [71–73]. The ICP-MS uses an argon plasma torch to ionize liquid samples as they are atomized into it at normal atmospheric pressure. The ions then travel through a small hole into a vacuum chamber maintained by a pump system. An ion beam is formed and focused into a mass analyzer, typically a quadrupole analyzer, which can filter out specific ions depending on the mass/charge ratio of that ion and the voltage applied to the quadrupole rods. The selected ions travel through the mass analyzer and impact a detector, usually a cone- or horn-shaped tube or photomultiplier tube, where a signal of measurable amplitude is produced [71]. Ag, Au, and Cd can be detected down to 10–50 ppt while Mg, W, Co, and Mo are in the 50–100 ppt range. Cu and Mn and Zn and Ti are higher with 100–200 ppt and 400–500 ppt detection, respectively [74, 75].

As previously stated, ICP-MS can be particularly useful when trying to obtain an estimate of concentration of metal nanoparticle uptake by cells. This can be done by culturing the cells, exposing the cells with a known concentration of nanoparticles, incubating cells for a period of time, rinsing excess nanoparticles from cells, dissolving cells in an organic solvent, pelleting the nanomaterials via centrifugation, and resuspending nanoparticles in de-ionized water for analysis in ICP-MS. The concentration determined will yield an estimate of the nanoparticles significantly adhering to the cell membrane and/or of the nanoparticles taken in by the cell. Used in conjunction with other techniques, such as TEM sectioning of cells and SEM analysis of the surface of the cells, might help establish where, and in what quantity, the nanoparticles are interacting with the cells. The only foreseeable drawbacks are that this technique cannot be used with carbon nanomaterials, consistent sample preparation is difficult, and that it is an average of all the cells in a population. To expand on the last point, it is not correct to assume that all cells will uptake the same amount of nanomaterials and there is no way to distinguish between the amount interacting with the surface and the amount taken into the cells.

To summarize, a proper and complete characterization of nanomaterials used for toxicity studies, *in vitro* and *in vivo*, is an absolute necessity when determining what properties are responsible for toxic responses. It may not be possible for the laboratories conducting the toxicity studies to make these observations for each material; however, then it is necessary to have the manufacturer or supplier perform these measurements or contract an outside company to perform the measurements. Any one of the techniques listed helps add additional information to better understand the mechanisms for toxicity.

21.6 Conclusions

Nanotechnology is emerging as one of the most innovative technologies of the twenty-first century and promises to yield personal, social, and commercial benefits by being incorporated into medical, energy, materials, and microelectronics applications. However, the potential risks of nanomaterials need to be addressed before widespread incorporation of these materials occurs. Current preliminary studies have yielded challenges that must be addressed in the field of nanotoxicology, and *in vitro* cell models will be a valuable tool in addressing these challenges before progressing into much more complex *in vivo* models.

References

1 The Royal Society (2004) Nanoscience and Nanotechnologies: Opportunities and Uncertainties. RS Policy document 19/04.

2 Kim, D., Zhang, Y., Voit, W., Rao, K., Kehr, J., Bjelke, B., and Muhammed, M. (2001) Superparamagnetic iron oxide nanoparticles for bio-medical applications. *Scripta Mater.*, **44**, 1713–1717.

3 Murugan, R. and Ramakrishna, S. (2005) Development of nanocomposites for bone grafting. *Compos. Sci. Technol.*, **65**, 2385–2406.

4 Sinha, V. and Trehan, A. (2003) Biodegradable microspheres for protein delivery. *J. Controll. Release*, **90**, 261–280.

5 Wu, Y., Yang, W., Wang, C., Hu, J., and Fu, S. (2005) Chitosan nanoparticles as a novel delivery system for ammonium glycyrrhizinate. *Int. J. Pharm.*, **295**, 235–245.

6 Kimbrell, G. (2006) Nanomaterials in Personal Care Products and FDA Regulation. The International Center for Technology Assessment. www.icta.org, 13 September 2006.

7 Powers, K., Brown, S., Krishna, V., Wasdo, S., Moudgil, B., and Roberts, S. (2005) Research strategies for safety evaluation of nanomaterials. Part VI. Characterization of nanoscale particles for toxicological evaluation. *Toxicol. Sci.*, **90**, 296–303.

8 Bucher, J., Masten, S., Moudgil, B., Powers, K., Roberts, S., and Walker, N. (2004) Developing experimental approaches for the evaluation of toxicological interactions of nanoscale materials. Final Workshop Report 3-4 November 2004, 1–37; University of Florida, Gainesville, FL.

9 Oberdorster, G., Oberdorster, E., and Oberdorster, J. (2005) Nanotoxicology: an emerging discipline evolving from studies of ultrafine particles. *Environ. Health Perspect.*, **113**, 823–839.

10 Oberdorster, G., Maynard, A., Donaldson, K., Castranova, V., Fitzpatrick, J., Ausman, K., Carter, J., Karn, B., Kreyling, W., Lai, D. *et al.* (2005) Principles for characterizing the potential human health effects from exposure to nanomaterials: Elements of a screening strategy. *Particle Fibre Toxicol.*, **2**, 8.

11 Shvedova, A.A., Huang, H., Carlson, C., Schlager, J.J., Osawa, E., Hussian, S.M., and Dai, L. (2004) Cytotoxic and genotoxic effects of single wall carbon nanotube exposure on human keratinocytes and bronchial epithelial cells. 227th American Chemical Society National Meeting, 27 March–1 April 2004, American Chemical Society, IEC 20, Anaheim, CA. Washington, DC.

12 Shvedova, A.A., Kisin, E., Keshava, N., Murray, A.R., Gorelik, O., Arepalli, S. *et al.* (2004) Exposure of human bronchial cells to carbon nanotubes caused oxidative stress and cytotoxicity, in *Proceedings of the Meeting of the SFRR Europe 2004*, Taylor & Francis Group, Ioannina, Greece, Philadelphia, pp. 91–103.

13 Gurr, J.R., Wang, A.S., Chen, C.H., and Jan, K.Y. (2005) Ultrafine titanium dioxide particles in the absence of photoactivation can induce oxidative damage to human bronchial epithelial cells. *Toxicology*, **213** (1–2), 66–73.

14 Lin, W., Huang, Y.W., Zhou, X.D., and Ma, Y. (2006) *In vitro* toxicity of silica nanoparticles in human lung cancer cells. *Toxicol. Appl. Pharmacol.*, **217**, 252–259.

15 Weisheng, L., Yue-wern, H., Xiao-Dong, Z., and Yinfa, M. (2006) Toxicity of cerium oxide nanoparticles in human lung cancer cells. *Int. J. Toxicol.*, **25** (6), 451–457.

16 Erdos, G.W., Morgan, D., Palazuelos, M., and Powers, K. (2005) Cellular response to

nanoparticle exposure. NSTI-Nanotech, vol. 6, pp. 78–80.

17 Wagner, A.J., Bleckman, C.A., Murdock, R.C., Schrand, A.M., Schlager, J.J., and Hussain, S.M. (2007) Cellular interaction of different forms of aluminum nanoparticles in rat alveolar macrophages. *J. Phys. Chem. B*, **111** (25), 7353–7359.

18 Carlson, C., Hussain, S.M., Schrand, A., Braydich-Stolle, L.K., Hess, K., Jones, R., and Schlager, J.J. (2008) The cellular interaction of different sizes of silver nanoparticles in rat alveolar macrophages induced reactive oxygen species. *J. Phys. Chem. B.*, **112** (43), 13608–13619.

19 Schrand, A.M., Huang, H., Carlson, C., Schlager, J.J., Omacr Sawa, E., Hussain, S.M., and Dai, L. (2007) Are diamond nanoparticles cytotoxic? *J. Phys. Chem. B*, **111** (1), 2–7.

20 Pulskamp, K., Diabate, S., and Krug, H.F. (2007) Carbon nanotubes show no sign of acute toxicity but induce intracellular reactive oxygen species in dependence on contaminants. *Toxicol. Lett.*, **168** (1), 58–74.

21 Monteiro-Riviere, N.A., Nemanich, R.J., Inman, A.O., Wang, Y.Y., and Riviere, J.E. (2005) Multi-walled carbon nanotube interactions with human epidermal keratinocytes. *Toxicol. Lett.*, **155**, 377–384.

22 Nel, A., Xia, T., Madler, L., and Li, N. (2006) Toxic potential of materials at the nanolevel. *Science*, **311**, 622–627.

23 Allen, D.G., Riviere, J.E., and Monteiro-Riviere, N.A. (2001) Cytokine induction as a measure of cutaneous toxicity in primary and immortalized porcine keratinocytes exposed to jet fuels, and their relationship to normal human epidermal keratinocytes. *Toxicol. Lett.*, **119**, 209–217.

24 Shvedova, A.A., Castranova, V., Kisin, E.R., Schwegler-Berry, D., Murray, A.R., Gandelsman, V.Z. *et al.* (2003) Exposure to carbon nanotube material: assessment of nanotube cytotoxicity using human keratinocyte cells. *J. Toxicol. Environ. Health*, **A66**, 1909–1926.

25 Yu, K., Grabinski, C., Carlson, C., Murdock, R., Boyer, J., Wang, W., Gu, B., Schlager, J., and Hussain, S.M., Size and dose dependent toxicity of SiO_2 nanoparticles in keratinocytes. 47th Society of Toxicology National Meeting March 2007.

26 Braydich-Stolle, L.K., Hussain, S.M., Schlager, J.J., Jiang, J., and Biswas, P. (2007) The effect of titanium dioxide nanoparticles in mouse keratinocytes (HEL-30 cells). 47th Society of Toxicology National Meeting March 2007.

27 Ryman-Rasmussen, J.P., Riviere, J.E., and Monteiro-Riviere, N.A. (2007) Variables influencing interactions of untargeted quantum dot nanoparticles with skin cells and identification of biochemical modulators. *Nano. Lett.*, **7** (5), 1344–1348.

28 Rouse, J.G., Yang, J., Barron, A.R., and Monteiro-Riviere, N.A. (2006) Fullerene-based amino acid nanoparticle interactions with human epidermal keratinocytes. *Toxicol. In Vitro*, **20** (8), 1313–1320.

29 Chen, Z., Meng, H., Xing, G., Chen, C., Zhao, Y., Jia, G., Wang, T., Yuan, H., Ye, C., Zhao, F., Chai, Z., Zhu, C., Fang, X., Ma, B., and Wan, L. (2006) Acute toxicological effects of copper nanoparticles *in vivo*. *Toxicol. Lett.*, **115** (3), 259–265.

30 Hussain, S., Hess, K., Gearhart, J., Geiss, K., and Schlager, J. (2005) *In vitro* toxicity of nanoparticles in BRL3A rat liver cells. *Toxicol. In vitro*, **19** (7), 975–983.

31 Chen, Y., Xue, Z., Zheng, D., Xia, K., Zhao, Y., Liu, T., Long, Z., and Xia, J. (2003) Sodium chloride modified silica nanoparticles as a non-viral vector with a high efficiency of DNA transfer into cells. *Curr. Gene. Ther.*, **3**, 273–279.

32 Borm, P.J. and Kreyling, W. (2004) Toxicological hazards of inhaled nanoparticles – potential implications for drug delivery. *J. Nanosci. Nanotechnol.*, **4**, 521–531.

33 Costantino, L., Gandolfi, F., Tosi, G., Rivasi, F., Vandelli, M.A., and Forni, F. (2005) Peptide-derivatized biodegradable nanoparticles able to cross the blood–brain barrier. *J. Control Release*, **108** (1), 84–96.

34 Lockman, P.R., Mumper, R.J., Khan, M.A., and Allen, D.D. (2002) Nanoparticle technology for drug delivery across the blood–brain barrier. *Drug Dev. Ind. Pharm.*, **28** (1), 1–13.

35 Silva, G.A. (2007) Nanotechnology approaches for drug and small molecule delivery across the blood brain barrier. *Surg. Neurol.*, **67** (2), 113–116.

36 Goodman, R. and Herschman, H.R. (1978) Nerve growth factor-mediated induction of tyrosine hydroxylase in a clonal pheochromocytoma cell line. *Proc Natl. Acad. Sci. USA*, **75** (9), 4587–4590.

37 Hussain, S.M., Javorina, A.K., Schrand, A.M., Duhart, H.M., Ali, S.F., and Schlager, J.J. (2006) The interaction of manganese nanoparticles with PC-12 cells induces dopamine depletion. *Toxicol. Sci.*, **92** (2), 456–463.

38 Pisanic, T.R., 2nd, Blackwell, J.D., Shubayev, V.I., Finones, R.R., and Jin, S. (2007) Nanotoxicity of iron oxide nanoparticle internalization in growing neurons. *Biomaterials*, **28** (16), 2572–2581.

39 Klebe, R.J. and Ruddle, F.H. (1969) Neuroblastoma: cell culture analysis of a differentiating stem cell system. *J. Cell Biol.*, **43**, 69.

40 Long, T.C., Saleh, N., Tilton, R.D., Lowry, G.V., and Veronesi, B. (2006) Titanium dioxide (P25) produces reactive oxygen species in immortalized brain microglia (BV2): implications for nanoparticle neurotoxicity. *Environ. Sci. Technol.*, **40** (14), 4346–4352.

41 Kim, J.S., Yoon, T.J., Yu, K.N., Kim, B.G., Park, S.J., Kim, H.W., Lee, K.H., Park, S.B., Lee, J.K., and Cho, M.H. (2006) Toxicity and tissue distribution of magnetic nanoparticles in mice. *Toxicol. Sci.*, **89**, 338–347.

42 Hofmann, M.C., Braydich-Stolle, L., Dettin, L., Johnson, E., and Dym, M. (2005) Immortalization of mouse germ line stem cells. *Stem Cells*, **23**, 200–210.

43 Braydich-Stolle, L., Hussain, S., Schlager, J.J., and Hofmann, M.C. (2005) *In vitro* cytotoxicity of nanoparticles in mammalian germline stem cells. *Toxicol. Sci.*, **88**, 412–419.

44 Braydich-Stolle, L., Kostereva, N., Dym, M., and Hofmann, M.C. (2007) Role of Src family kinases and N-Myc in spermatogonial stem cell proliferation. *Dev. Biol.*, **304** (1), 34–45.

45 Li, Y., Leung, P., Yao, L., Song, Q., and Newton, E. (2006) Antimicrobial effect of surgical masks coated with nanoparticles. *J. Hosp. Infect.*, **62**, 58–63.

46 Martinez, C.J., Jiwen, L., Rhodes, S.K., Luijten, E., Eric, R., Weeks, E.R., and Lewis, J.A. (2005) Interparticle interactions and direct imaging of colloidal phases assembled from microsphere-nanoparticle mixtures. *Langmuir*, **21**, 9978–9989.

47 Wei, D., Dave, R., and Pfeffer, R. (2002) Mixing and characterization of nanosized powders: an assessment of different techniques. *J. Nanoparticle Res.*, **4**, 21–41.

48 Seery, M.K., George, R., Floris, P., and Pillai, S.C. (2007) Silver doped titanium dioxide nanomaterials for enhanced visible light photocatalysis. *J. Photochem. Photobiol. A*, **189**, 258–263.

49 Gerasimov, O.V. *et al.* Cytostatic Drug Delivery using pH and light sensitive liposomes (1999) *Adv. Drug Deliv. Rev.*, **38**, 317–338.

50 Cullis, P.R. *et al.* Recent advances in liposome technologies and their applications for systemic gene delivery. (1998) *Adv. Drug Deliv. Rev.*, **30**, 73–83.

51 Allen, C. *et al.* Controlling the physical behavior and biological performance of liposome formulations through use of surface grafted poly(ethylene glycol). (2002) *Biosci. Rep.*, **22**, 225–250.

52 Sayes, C.M., Reed, K.L., and Warheit, D.B. (2007) Assessing toxicity of fine and nanoparticles: comparing *in vitro* measurements to *in vivo* pulmonary toxicity profiles. *Toxicol. Sci.*, **7** (8), 2399–2406.

53 Hofmann, M.C., Braydich-Stolle, L., and Dym, M. (2005) Isolation of mouse germ line stem cells; influence of GDNF. *Dev. Biol.*, **279**, 114–124.

54 Lam, C.W., James, J.T., McCluskey, R., and Hunter, R.L. (2004) Pulmonary toxicity of single-wall carbon nanotubes in mice 7 and 90 days after intratracheal instillation. *Toxicol. Sci.*, **77** (1), 126–134.

55 Warheit, D.B., Laurence, B.R., Reed, K.L., Roach, D.H., Reynolds, G.A., and Webb, T.R. (2004) Comparative pulmonary toxicity assessment of single-wall carbon nanotubes in rats. *Toxicol. Sci.*, **77** (1), 117–125.

56 Renwick, L.C., Brown, D., Clouter, A., and Donaldson, K. (2004) Increased

inflammation and altered macrophage chemotactic responses caused by two ultrafine particle types. *Occup. Environ. Med.*, **61** (5), 442–447.

57 Warheit, D.B., Webb, T.R., Sayes, C.M., Colvin, V.L., and Reed, K.L. (2006) Pulmonary instillation studies with nanoscale TiO_2 rods and dots in rats: toxicity is not dependent upon particle size and surface area. *Toxicol. Sci.*, **91** (1), 227–236.

58 Warheit, D.B., Webb, T.R., Reed, K.L., Frerichs, S., and Sayes, C.M. (2007) Pulmonary toxicity study in rats with three forms of ultrafine-TiO_2 particles: differential responses related to surface properties. *Toxicology*, **230** (1), 90–104.

59 Warheit, D.B., Webb, T.R., Colvin, V.L., Reed, K.L., and Sayes, C.M. (2007) Pulmonary bioassay studies with nanoscale and fine-quartz particles in rats: toxicity is not dependent upon particle size but on surface characteristics. *Toxicol. Sci.*, **95** (1), 270–280.

60 Murdock, R.C., Braydich-Stolle, L., Schrand, A.M., Schlager, J.J., and Hussain, S.M. (2007) Characterization of nanomaterial dispersions in solution prior to *in vitro* exposure using dynamic light scattering technique. *Toxicol. Sci.*, **101** (2), 239–253.

61 Soto, K., Carrasco, A., Powell, T., Garza, K., and Murr, L. (2005) Comparative *in vitro* cytotoxicity assessment of some manufactured nanoparticulate materials characterized by transmission electron microscopy. *Nanoparticle Res.*, **7**, 145–169.

62 Pal, S., Tak, Y.K., and Song, J.M. (2007) Does antibacterial activity of silver nanoparticle depend on shape of nanoparticle? A study on Gram-negative *E. coli*. *Appl. Environ. Microbiol.*, **73**, 1712–1720.

63 Chithrani, B.D., Chan, W.C.W. (2007) Elucidating the mechanism of cellular uptake and removal of protein-coated gold nanoparticles of different sizes and shapes. *Nano Lett.*, **7** (6), 1542–1550.

64 Hoshino, A., Kujioka, F., Oku, T., Suga, M., Sasaki, Y.F., Ohta, T., Yasuhara, M., Suzuki, K., and Yamamoto, K. (2004) Physicochemical properties and cellular toxicity of nanocrystal quantum dots depend on their surface modification. *Nano Lett.*, **4** (11), 2163–2169.

65 Hohr, D., Steinfartz, Y., Schins, R.P.F., Knaapen, A.M., Martra, G., Fubini, B. *et al.* (2002) The surface area rather than the surface coating determines the acute inflammatory response after instillation of fine and ultrafine TiO_2 in the rat. *Int. J. Hyg. Environ. Health*, **205**, 239–244.

66 Rehn, B., Seiler, F., Rehn, S., Bruch, J., and Maier, M. (2003) Investigations on the inflammatory and genotoxic lung effects of two types of titanium dioxide: untreated and surface treated. *Toxicol. Appl. Pharmacol.*, **189**, 84–95;Berne, B.J., and Pecora, R. (1975) *Dynamic Light Scattering*, John Wiley & Sons, Inc., New York.

67 Berne, B.J. and Pecora, R. (1975). Dynamic Light Scattering. John Wiley, New York.

68 Simakov, S. and Tsur, Y. (2007) Surface stabilization of nano-sized titanium dioxide: improving the colloidal stability and the sintering morphology. *J. Nanoparticle Res.*, **9**, 403–417.

69 Williams, D., Ehrman, S., and Holoman, T. (2006) Evaluation of the microbial growth response to inorganic nanoparticles. *J. Nanobiotechnol.*, **4** (3), 1–8.

70 Wu, C. (1993) Simultaneous calibration of size exclusion chromatography and dynamic light scattering for the characterization of gelatin. *Macromolecules*, **26**, 5423–5426.

71 Jarvis, K.E., Gray, A.L., and Houk, R.S. (1992) *Handbook of Inductively Coupled Plasma Mass Spectrometry*, Chapman and Hall, New York.

72 Olesik, J.W. (1996) Fundamental research in ICP-OES and ICPMS. *Anal. Chem.*, **68** (1), 469A–474.

73 Worthy, W. (1988) Scope of ICP/MS expands to many fields. *Chem. Eng. News*, **66** (27), 33–34.

74 Newman, A. (1996) Elements of ICPMS. *Anal. Chem.*, **68** (1), 46A–51A.

75 B'Hymer, C., Brisbin, J.A., Sutton, K.L., and Caruso, J.A. (2000) New approaches for elemental speciation using plasma mass spectrometry. *Am. Lab.*, **32** (3), 17–32.

22
The Application of Physiologically Based Pharmacokinetics, Bayesian Population PBPK Modeling, and Biochemical Reaction Network Modeling to Chemical Mixture Toxicology

Raymond S.H. Yang, Arthur N. Mayeno, Michael A. Lyons, and Brad Reisfeld

This chapter relates the importance and significance of the application of biologically based computer modeling to chemical mixture toxicology. Here, the phrase, biologically based computer modeling includes physiologically based pharmacokinetic (PBPK) modeling, Bayesian population PBPK modeling, and biochemical reaction network (BRN) modeling – the three different types of technologies used by our Quantitative and Computational Toxicology Group at Colorado State University for dealing with the issues related to chemical mixture toxicology. These issues are discussed below in the form of "dialogue," where certain important questions are asked and discussed.

22.1
Why is Computer Simulation Not Only Important But Also Necessary for Chemical Mixture Toxicology?

In our modern society, computer usage is increasingly an integral and necessary part our daily lives. Much of this trend was the consequence of the extensive development and application of computer technologies in the physical sciences and engineering. Computer simulations are in every aspect of our lives, but most of us are not even aware of them. One of the most interesting examples to illustrate the maturity and utility of computer simulation is the modern jetliner. The Boeing 777 is the first jetliner to be 100% digitally designed using three-dimensional CAD/CAM (computer-aided design/computer-aided manufacturing) technology (http://www.boeing.com/commercial/777family/background.html). Throughout the design process, the airplane was "preassembled" on the computer, eliminating the need for a costly, full-scale mock-up. In contrast to this high level of computer applications in the physical sciences and engineering, the utilization of computational technology in the biomedical sciences has lagged far behind. It is our belief that coupling such advanced computer simulation technologies with sound biological and biochemical principles offer a new pathway to gaining insight into the complex field of chemical mixture toxicology.

Principles and Practice of Mixtures Toxicology. Edited by Moiz Mumtaz

ISBN: 978-3-527-31992-3

Of particular relevance to toxicology, computer simulation of biological processes is a realistic and workable method to replace, or at least reduce, animal usage and suffering in experimentation. Ethical reasons aside, the reality is that experimental toxicology cannot keep pace with the number of chemicals in commerce (about 70 000 to 80 000) plus the new synthetic organic chemicals coming on stream (about 2000/year) [1–3]. For instance, one of the largest toxicology programs in the world, the US National Toxicology Program and its predecessor, the National Cancer Institute Cancer Bioassay Program, has conducted chronic toxicity/carcinogenicity studies on only about 600 chemicals in their combined effort over the past 40 years. Considering further the potential presence of numerous chemical mixtures, it is clear that the present mode of dependence on experimental toxicology is inadequate at best and that the only reasonable alternative to meet these needs resides in efficient and thoughtful utilization of biologically based computer simulations [1].

To provide an illustration of the importance of predicting chemical mixture toxicity, we refer to two interesting studies by Lazarou *et al.* [4] and Jevtovic-Todorovic *et al.* [5]. Both of these studies are related to multiple drug interactions, an area on which there is inadequate information and understanding. Lazarou *et al.* [4] reported that in 1994 more than 2.2 million cases of serious adverse drug reactions (ADRs) occurred in hospital patients in the United States, and among these cases 106 000 were fatal. During their hospital stay, the patients were given an average of eight drugs, and alarmingly, ADRs became the fourth to sixth leading cause of death for that year in the United States. As a more concrete example, Jevtovic-Todorovic *et al.* [5] administered to 7-day old infant rats a combination of drugs commonly used in pediatric anesthesia (midazolam, nitrous oxide, and isoflurane) in doses sufficient to maintain a surgical plane of anesthesia for 6 h. These investigators observed that such a treatment caused widespread apoptotic neurodegeneration in the developing brain, deficits in hippocampal synaptic function, and persistent memory/learning impairments. These studies reflect the inadequate state of knowledge regarding drug–drug interactions. However, an even more urgent need is technologies and tools that can reliably predict the toxicity(ies) of mixtures of chemicals or drugs.

During the past 17 years or so, our research group has been working on advancing *in silico* approaches, coupled with targeted experimentation, to the complex field of chemical mixture toxicology. We believe that a "systems biology" approach, based on sound biological, biochemical, and mathematical principles, offers a promising path to gaining insight into the toxicology of chemical mixtures. Specifically, our aim is to integrate PBPK, Bayesian population PBPK modeling utilizing Markov Chain Monte Carlo (MCMC) simulation, biochemical reaction network modeling, other types of computer simulation, and the results of targeted experimentation into a unified, multiscale framework. The overall conceptual basis of our approach is illustrated in Figure 22.1.

Figure 22.1 represents the continuum from the time the toxicants (or chemical mixtures) enter the body or a population to the kinetic distribution into the organ/tissues, cells, receptors, to the molecular changes that manifest themselves in the form of organ/tissue toxicities, and further into organism/population toxicities. Embedded in this continuum is the multilevel biologically based computer modeling

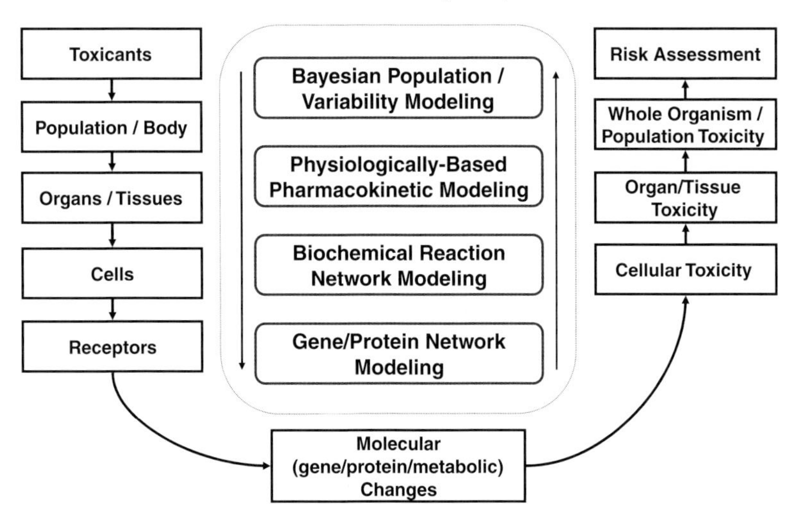

Figure 22.1 A multiscale computer simulation approach toward chemical mixture toxicology and risk assessment.

that, when developed fully, may not only be a predictive tool for chemical mixture toxicology but also be useful in the risk assessment process.

22.2 What Do We Mean by "Computer Simulation?" What Does it Entail?

If we consider the human body to be a chemical plant, we have pumps (heart, lung), pipes (blood vessels), reactors (liver, kidney, etc.), filtration apparatus (kidneys), and solid and liquid disposals. These individual "units" are operated under an extremely efficient and elegant physiological/biochemical system evolved through millions of years. In the past 200 years, we have accumulated a tremendous amount of knowledge on our body and these related "unit operations." Just as in the case of a chemical plant in which computer modeling of the plant enables operators to control the various chemical processes, computer simulations of the physiological/ biochemical processes in our body given a specific condition (e.g., the exposure of a chemical mixture) are feasible. Here, the words "computer simulation" denote the translation of the physiological/biochemical processes in our bodies to a simplified conceptual model that can be translated into mathematical notation and solved on a computer to yield properties of interest. In essence, a "virtual human" is created in the computer to simulate, for a given chemical or chemical mixture, the pharmacokinetic (what does the body do to the chemical) and pharmacodynamic (what the chemical does to the body) phenomena. Thus, the detailed time course behaviors and ADME (absorption, distribution, metabolism, and excretion) of the xenobiotics of interest are predicted. One of the most important prerequisites for this type of computer simulation is that all the computer modeling must be consistent with sound biological principles. Any given process must be biologically plausible.

22.3
What is Physiologically-Based Pharmacokinetic Modeling? How Does it work?

PBPK is a special type of pharmacokinetics in which the physiology and anatomy of the animal or human body, and the biochemistry of the chemical or chemicals of interest, are incorporated into a conceptual model for computer simulation. Unlike classical pharmacokinetics, PBPK modeling is a powerful tool useful for many types of extrapolation: interspecies, interroutes, interdoses, interlife stages, and so on.

The concept of PBPK had its embryonic development in the 1920s. PBPK modeling blossomed and flourished in the late 1960s and early 1970s in the chemotherapeutic area mainly due to the efforts of investigators with expertise in chemical engineering. In the mid-1980s, work on PBPK modeling of volatile solvents started yet another "revolution" in the toxicology and risk assessment arena. Today, there are more than 1000 publications directly related to PBPK modeling of industrial chemicals, drugs, environmental pollutants, and simple and complex chemical mixtures. A book on PBPK modeling was published in June 2005 from our laboratory in collaboration with others [6]; in this book, a specific chapter has been devoted to the PBPK modeling of chemical mixtures [7].

22.3.1
PBPK Model Structure and Parameters

A PBPK model, graphically illustrated in Figure 22.2, reflects the incorporation of basic physiology and anatomy. The compartments correspond to anatomic entities, such as the liver and lungs, while the blood circulation conforms to basic mammalian physiology. In the specific model in Figure 22.2, the exposure route of interest is inhalation and is represented by the lung and gas exchange compartments, with intake (CI) and exhalation (CX) vapor concentrations indicated. Depending on the need, other routes of exposures can be easily added. Some tissues are "lumped" together (e.g., richly (rapidly) or poorly (slowly) perfused tissues in Figure 22.2) when they are "kinetically" similar, for the specific chemical(s).

Three sets of parameters are needed for PBPK model building: physiological parameters (e.g., ventilation rates, cardiac output, organs as % body weight), thermodynamic parameters (e.g., tissue partition coefficients, protein binding), and biochemical parameters (e.g., K_m and V_{max}). Most, if not all, of the parameters for laboratory animals are available in the literature [9]. When information gaps exist, needed data can be obtained via experimentation or through allometric extrapolation, usually based on a power function of the body weight [10].

22.3.2
How Does PBPK Modeling Work?

The fundamentals of PBPK modeling are to identify the principal organs or tissues involved in the disposition of the chemical of interest and to correlate the chemical absorption, distribution, metabolism, and excretion within and among these organs

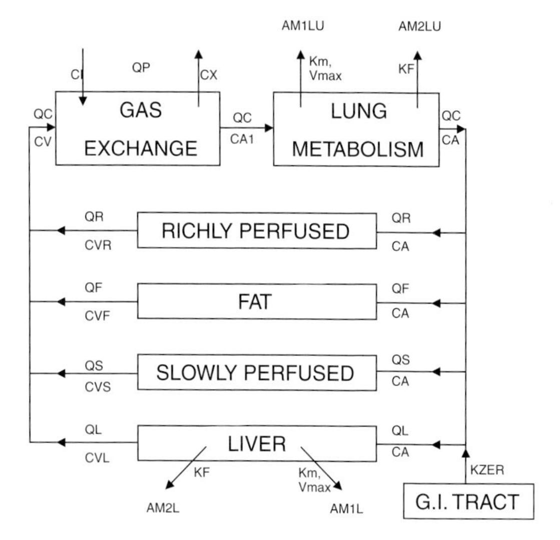

Figure 22.2 A graphical representation of a PBPK model for methylene chloride. Based on Andersen *et al.* [8], *Q* represents blood flow; thus, QC, QP, QR, QF, QS, and QL are cardiac output, pulmonary blood flow, and blood flows for richly perfused, fat, slowly perfused, and liver compartments, respectively. *C* represents concentration of the chemical we are studying; thus, CI, CX, CV, CA, CVR, CVF, CVS, and CVL are concentrations in inhaled breath, exhaled breath, venous blood, arterial blood, and venous blood for richly perfused, fat, slowly perfused, and liver compartments, respectively. K_m, V_{max}, and K_F are metabolic parameters representing affinity constant, maximum rate constant for Michaelis–Menten enzyme kinetics (saturation or nonlinear kinetics), and first-order rate constant (linear kinetics), respectively. AM1LU, AM2LU, AM1L, and AM2L are, respectively, amount of metabolite 1 or 2 for lung (LU) and liver (L).

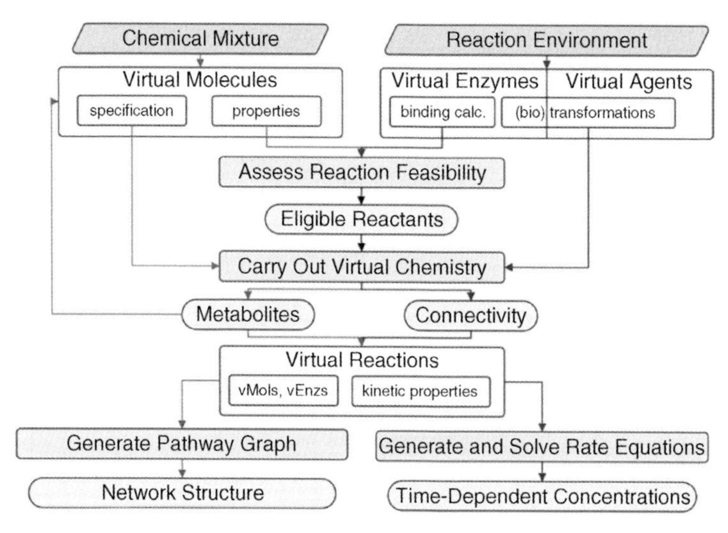

Figure 22.3 Information flow through the BioTRaNS.

and tissues in an integrated and biologically plausible manner. After a conceptual model is developed (e.g., Figure 22.2), time-dependent mass balance equations are written for each chemical in each compartment. These mass balances are essentially molecular accounting statements that include the rates at which molecules enter and leave the compartment, as well as the rates of reactions that produce or consume the chemical. For instance, a general equation, for chemical j in any tissue or organ i, is

$$V_i \frac{dC_{ij}}{dt} = Q_i(\mathrm{CA}_j - \mathrm{CV}_{ij}) - \mathrm{Metab}_{ij} - \mathrm{Elim}_{ij} + \mathrm{Absorp}_{ij} - \mathrm{Pr\ Binding}_{ij},$$

where V_i represents the volume of tissue group i, Q_i is the blood flow rate to tissue group i, CA_j is the concentration of chemical j in arterial blood, and C_{ij} and CV_{ij} are the concentrations of chemical j in tissue group i and in the effluent venous blood from tissue i, respectively. Metab_{ij} is the rate of metabolism for chemical j in tissue group i; liver, being the principal organ for metabolism, would have significant metabolic rates, while, with some exceptions, Metab_{ij} is usually equal to zero in other tissue groups. Elim_{ij} represents the rate of elimination from tissue group i (e.g., biliary excretion from the liver), Absorp_{ij} denotes uptake of the chemical from dosing (e.g., oral dosing), and $\mathrm{PrBinding}_{ij}$ represents protein binding of the chemical in the tissue. These terms are zero unless there is definitive knowledge that the particular organ/tissue of interest has such processes.

A set of such mass balance differential equations representing all of the interlinked compartments are formulated to express a mathematical representation, or model, of the biological system. This model can then be used for computer simulation to predict the time course behavior of any given chemical included in the model.

For the most well-studied chemicals or drugs, it is likely that the biochemical constants, such as K_m and V_{max}, are available in the literature. However, it must be noted that the K_m, V_{max}, and K_F (first-order rate constant) in a PBPK model (known as the *in vivo* K_m, V_{max}, K_F for a given chemical) are hybrid constants of all the saturable (K_m, V_{max}) or linear (K_F) metabolic pathways, for the chemical of interest in the body. They are different from the *in vitro* K_m, V_{max}, or K_F of a given pure enzyme. While they are not directly interchangeable, the *in vitro* constants in the literature may be used to estimate *in vivo* constants for modeling purposes [11, 12]. Also, for most well-known chemicals, it is likely that enough is known about the mechanism of toxicity to be incorporated into the model for computer simulation. Computer simulations may be made for any number of desired time course end points such as the blood levels of the parent compound, liver level of a reactive metabolite, and similar information on different species, at lower or higher dose levels, and/or via a different route of exposure. The experimental pharmacokinetic data may then be compared with a PBPK model simulation. If the model simulation does not agree with the measurements, the model might be deficient because critical scientific information might be missing or certain assumptions may be incorrect. The investigator, with knowledge of the chemical and a general understanding of the physiology and biochemistry of the animal species, can design and conduct critical experiments for refining the model to reach consistency with experimentation. This refinement process may be repeated again and again when necessary; such an iterative process is critically important for the development of a PBPK model. In that sense, PBPK modeling is a very good hypothesis-testing tool in toxicology, and it may be utilized to conduct different kinds of experiments on the computer (i.e., *in silico* toxicology). It should be noted that there is always the possibility that a good model may not be obtained at the time because of the limitation of our knowledge on the chemical. Validation of the PBPK model with data sets other than the working set (or training set) to develop the model is necessary. One should remember that a model is usually an oversimplification of reality; thus, "All models are wrong, some are useful," as George Box stated. The more the data sets against which a model is validated, the more robust is that model in its predictive capability. Once validated, the PBPK model is ready for extrapolation to other animal species, including humans.

22.3.3
PBPK Modeling of Chemical Mixtures

Since humans are rarely, if ever, exposed to a single chemical, a key feature of PBPK modeling is that it can be used to integrate information on toxicological interactions. The most ideal and scientifically defensible data requirement for establishing an interactive PBPK model is that an established, validated PBPK model is available for each component chemical in the mixture (it is rarely the case in reality). Furthermore, there need to be available many pharmacokinetic data sets both in laboratory animals and in humans for each of these component chemicals. We use the term "interactive PBPK model" to mean a PBPK model with the capability to simulate interactions between and among chemicals in a mixture. The interactive PBPK model is then built on the basis of known pharmacokinetic interactions. For instance, one chemical may inhibit the biotransformation of other mixture components. The individual PBPK models may then be linked together at the liver compartment by introducing competitive (or other) inhibition terms in the mass balance differential equations. In our opinion, the application of PBPK modeling to toxicological interactions of chemical mixtures is necessary in cumulative risk assessment. However, this area is very complex and it is still an emerging field. For a more thorough discussion, the readers may refer to a chapter on PBPK modeling of chemical mixtures [7], as well as a chapter devoted to the application of PBPK modeling in risk assessment [13]. It should be emphasized here that PBPK modeling will only handle part of the chemical mixture issue in cumulative risk assessment, namely, the pharmacokinetic interactions at the whole body level. It is necessary to integrate PBPK modeling with biochemical reaction network (BRN) modeling to reflect molecular interactions and, even further, to the linkage with toxic end points to fully address the chemical mixture issue in cumulative risk assessment.

22.4
What is Bayesian Inference and Population PBPK Modeling? What is Markov Chain Monte Carlo Simulation? Why Do We Need These Technologies?

A fundamental aspect of biological systems is the intrinsic heterogeneity (variability) of any population considered. PBPK models can be used to account for population variability by treating the model parameters (e.g., body weight, tissue volumes, and flow rates) as random variables with probability distributions descriptive of measured variability. Monte Carlo simulation [14, 15] can be used to calculate a distribution of target-tissue chemical concentration in the population by repeatedly evaluating the model with random draws from the assigned parameter distributions for a given exposure or dose.

The accuracy of Monte Carlo simulation depends, in part, on the proper assignment of probability distributions to the relevant parameters. Some model parameters, such as those related to metabolism, may be only poorly known; however, as such parameters have a definite biological interpretation, knowledge regarding possible ranges, central values, and measures of dispersion, as well as specific data from previous studies, can be used to define prior probability distributions for these parameters. Using additional observed concentration–time data, Bayesian inference [16] determines an updated, or posterior, probability distribution as the product of the prior distribution and the likelihood of the data for a given model (Bayes' theorem). Bayes' theorem can be applied to a hierarchical population model (which defines the statistical relationship between individual parameters and measurements, and population level parameters), in order to provide population estimates of the parameter distributions. The use of the posterior distributions for the PBPK model parameters results in a calibrated model that can then be used with Monte Carlo simulation to make population-based predictions of tissue concentrations for additional exposure/dose regimens.

For most cases of practical interest, the evaluation of Bayes' theorem is performed using numerical simulation; Markov Chain Monte Carlo (MCMC) methods being the current standard method [17]. MCMC simulation refers to a class of iterative simulations in which the random variables of interest are drawn from a sequence, or chain, of distributions that eventually converge to a stable posterior distribution. These chains can be determined by rejection sampling algorithms where a random draw is accepted or rejected on the basis of simple probabilistic rules. A major impediment faced in the use of MCMC methods in PBPK modeling has been the effort required to evaluate the often complicated posterior distributions. The relatively recent development of numerical algorithms and the availability of computational resources have led to a more widespread adoption of Bayesian methods for PBPK model calibration; in particular, MCSim [18] is a freely available, open-source program, developed to perform the necessary MCMC and MC simulations for hierarchical PBPK population modeling.

Bayesian population methods offer a number of advantages over more traditional likelihood-based methods, including [19] (1) the ability to incorporate and to update prior knowledge from previous experimental and modeling efforts for model calibration; (2) the ability to simultaneously account for multiple individual data sets, and to update multiple parameter values simultaneously using a hierarchical model; (3) the ability to separately consider parameter variability and uncertainty; and (4) the capacity to account for covariance of PBPK model parameters without the risk of incorrectly assuming all variables are independent.

High-level introductions to Bayesian PBPK population methods can be found in Jonsson and Johanson [20], and Hack [21]; more detailed accounts can be found in

Gelman *et al.* [22], Bernillon and Bois [23], and Covington *et al.* [24], with recent applications to risk assessment in Marino *et al.* [19] and David *et al.* [25].

22.5 What is Biochemical Reaction Network Modeling? Where Did It Come From? How Does It Work? Why Do We Need It for Chemical Mixture Toxicology?

22.5.1 Biochemical Reaction Network Modeling

Biochemical reaction network modeling has its origin in chemical and petroleum engineering. It was successfully employed in computer modeling and simulation of the complex processes in oil refineries. A reaction network model is a tool that is used to predict the amounts of reactants, intermediates, and products as a function of time for a series of coupled chemical reactions (potentially numbering in the tens of thousands of reactions). The simulation software for BRN modeling can be used to not only solve the kinetic equations of interest but also potentially generate the reaction mechanisms, rate constants, and reaction equations themselves. Reaction network modeling has been used very successfully in the fields of chemical and petroleum engineering over the past 25 years, utilizing development in the areas of group contribution methods [26], graph theory [27, 28], Monte Carlo techniques [29–31], and quantum chemistry [32]. To the best of our knowledge, we are the first to use reaction network modeling to study xenobiotic biotransformations [33–37].

In an attempt to describe hydrocarbon mixture properties, early models relied on the technique of lumping [38], where a relatively small number of lumps were used to describe the mixture. In these coarsely lumped kinetic models, the thousands of individual constituents in a complex feedstock were grouped into broad but measurable categories of compound classes or boiling range, with simplified reaction networks between the lumps. More recently, Quann and Jaffe [39] developed a more fine-grained lumping approach that they called "structure-oriented lumping" (SOL). Structure-oriented lumping was developed in response to the need for incorporating molecular detail in petroleum chemistry in order to predict product composition and properties. It is a group contribution method describing the structure of molecules, facilitating both molecular property estimation and description of process chemistry. It is also designed to be consistent with the limitations of analytical capability at the time to determine molecular detail. The concept central to the SOL approach is that any molecule can be described and represented by a set of certain structural features or groups. The SOL method organizes this set as a vector, with the elements of the vector representing the number of specific structural features sufficient to construct any molecule. Different molecules with the same set of structural groups, that is, certain isomers, are lumped and represented by the same vector. The structure vector provides a framework to enable rule-based generation of reaction networks and rate equations involving thousands of components and many thousands of reactions.

An even finer grain methodology was developed that used concepts of graph theory to represent species connectivity [32, 40, 41]. They also made use of computational quantum chemistry (CQC) and linear free energy relationships (LFER) [42] to automate the process of determining reaction rate constants. Computational quantum chemistry was applied to determine the optimal conformation and molecular properties, such as the electron affinity, electron density, bond order, and heat of formation, associated with each reactant and product structures. Linear free energy relationships [42] were then used to give a correlation of the rate or equilibrium constants with a property of a molecule or intermediate for a family of reactions. Thus, the computational quantum chemistry calculations ultimately provided an estimate of rate or equilibrium coefficients for reactions that had not been studied experimentally.

This general framework, which Klein [43] refers to as the Kinetic Modeler's Toolbox (KMT), allows the convenient construction and solution of even highly complex chemical reaction networks. Using this approach, various investigators have been able to encode complex hydrocarbon mixtures and create rule sets for a wide variety of reactions within the mixture. Results from model simulations [44] have shown good agreement with experimental observations in tracking the evolution of thousands of molecular components, and then predicting mixture properties such as normal boiling points, specific gravity, and narrow boiling cut yields. Joshi and coworkers [45] use the KMT for analyzing gas oil catalytic cracking and were able to derive optimized parameter values (activation energies and frequency factors) and lumped fractions that were in good agreement with experimental results reported in the literature. Along similar lines, Mizan and Klein [46] found good agreement in terms of product yields and yield profiles between simulation results and experimental data in the reaction network modeling of *n*-hexadecane hydroisomerization.

The idea of using reaction network modeling for biomedical applications was put forth by a joint effort of our laboratory and Rutgers University [34–36, 47]. The approach of biochemical reaction network modeling has great potential and represents groundbreaking work in predicting biotransformation of xenobiotics individually or as mixtures, especially in that metabolites will not only be predicted quantitatively but also be predicted in a time-dependent manner, and metabolic pathways will be interconnected via metabolites in common. This methodology should complement other systems biology approaches derived from advances in functional genomics, proteomics, and metabolomics.

22.5.2
BioTRaNS: A New Framework

Over the last few years, we have developed, *de novo*, a software package, BioTRaNS (*Bio*chemical *T*ool for *R*eaction *N*etwork *S*imulation) to address the shortcomings of SOL/KMT (which focuses on process chemical and petrochemical applications) and to satisfy the needs of biological systems analysis. BioTRaNS integrates modules of our own creation, as well as existing software and database tools, such as CORINA (molecule structure prediction), MOPAC7 (quantum chemical calculations), GraphViz (mathematical graph visualization), and Daylight and OpenEye toolkits (symbolic molecule manipulation and chemical/biochemical reaction transformations and prediction). As part of the BioTRaNS effort, we have also been developing methods (quantitative structure–activity relationships, decision trees) for the prediction of the probabilities of cytochrome P450 binding of chemicals.

A simplified description and information flow within BioTRaNS is illustrated in Figure 22.3 (see Mayeno *et al.* [37] for a more detailed description).

A chemical mixture (concentration of a single or multiple chemicals) and reaction environment (types and amounts of enzymes and other "agents") are input by the user. The chemical mixture is converted to a set of "virtual molecules" (vMols), each with its own specifications (i.e., a list of computed geometric, topological, energetic, and physicochemical properties). In general, properties are retrieved from associated databases when available or are computed as needed.

The reaction environment is translated into a set of "virtual enzymes" (vEnzs), representing the actual enzymes and "virtual agents" (vAgnts), representing non-enzymatic reactions. Each vEnz is endowed with (i) an appropriate binding calculator and (ii) "transforms" governing the biotransformations that the vEnz can mediate. The binding calculator computes the feasibility of a particular vMol binding to a vEnz, on the basis of quantitative structure–activity relationships (or decision tree) derived from properties of known substrates for each enzyme. Transformations are stored as a list of SMIRKS-based representations [48] (see Table 22.1 for examples). SMIRKS (*Smi*les *R*eac[*k*]tion *S*pecification) is a superset of the SMILES language that describes chemical transformations.

For any transformation, more than one SMIRKS can be written, depending on the specificity desired. A single SMIRKS (or transform) can perform a general

Table 22.1 Examples of transforms.

Chemical representation	**Name of transformation in the SMIRKS language**
 ═ ⟶ O	Epoxidation (alkene) [C:1]=[C:2]≫[C:1]1[C:2]O1
R-CHO → R-CH$_2$OH	Reduction (aldehyde to primary alcohol) [*:1][C:2](=[O:3])[H:99]≫[*:1][C:2]([O:3][H])([H:99])[H]
R-CH2OH → R-CHO	Oxidation (primary alcohol to aldehyde) [C:1][C:2]([O:3][H])([H:99])[H]≫[C:1][C:2](=[O:3])[H:99]
R-CHO → R-CH(OH)$_2$	Hydration (aldehyde) [*:1][C:2](=[O:3])[H:99]≫[*:1][C:2]([O:3][H])([H:99])O
R-CH(OH)$_2$ → R-CHO	Dehydration (gem-diol to aldehyde) [*:1][C:2]([O:3][H])([H:99])[OH]≫[*:1][C:2](=[O:3])[H:99]
R-COCl → R-COOH	Hydrolysis (acid chloride) [C:1](=O)Cl≫[C:1](=O)O
R-COCl → R-COSG	Acylation of glutathione (by acid chlorides) [C:1](=O)Cl≫OC(=O)[C@@H](N)CCC(=O) N[C@@H](CS[C:1](=O))C(=O)NCC(=O)O

reaction, such as converting any primary alcohol to an aldehyde. Moreover, our methodology is unique in implementing transforms that describe *steps of a reaction mechanism*, including those for enzyme-mediated reactions [49]. The use of mechanistic steps is key to automatically and accurately predicting metabolites (see below for an example). After the vMol is created, the binding calculator for each vEnz assesses binding feasibility; if feasible, the vMol becomes an eligible reactant. All eligible reactants then undergo appropriate virtual biotransformations, creating specific chemical reactions and converting the vMol into one or more metabolites, which also are checked for reaction feasibility. The information on the metabolites and the associated transform/agent interconnecting each substrate–product pair is converted to a "virtual reaction" (vRxn). Thus, each vRxn comprises the vMols and vEnzs from which it was derived, as well as appropriate kinetic properties.

The ability to delineate the metabolites and visualize where pathways intersect is key to gaining new insight into interrelationships among chemicals and their metabolic pathways. With this concept in mind, the aggregate information in the vRxns is used to construct a reaction network that captures these interrelationships. This network is generated in the form of a mathematical graph [50], wherein the metabolites constitute the vertices or nodes, and the reactions form the edges. Well described in discrete mathematics, graphs can be efficiently analyzed using algorithms of computer science [51]; graph theory allows efficient determination of characteristics such as common reactants between various pathways, the shortest path to a given metabolite, and common (or all) routes to given metabolites. These network graphs can be depicted at different levels of detail, from highly detailed, showing steps of a reaction mechanism, to overview, showing only key metabolites.

Simultaneous with the graph creation, kinetic properties in each vRxn are used to create the appropriate reaction rate equations (ordinary differential equations, ODE). These properties include (i) rate constants (e.g., Michaelis constant, K_m, and maximum velocity, V_{max}, for enzyme-catalyzed reactions, and k for nonenzymatic reactions); (ii) inhibitor constants, K_i; and (iii) modes of inhibition or "allosterism" [52–54]. The total set of rate equations and specified initial conditions forms an initial value problem that is solved by a stiff ODE equation solver for the concentrations of all species as a function of time. The essential output of BioTRaNS is a list of metabolites, their interconnections (through biotransformations), and their time course concentrations. This output can be expressed to the user in many forms, including graphs of the metabolic pathways, enumerations of all (or certain specified types of) metabolites and/or reactions, and plots of concentration versus time for species of interest.

22.5.3
BRN Modeling of Chemical Mixtures

As an example of the use of BRN modeling to examine interactions in multiple chemicals, we studied a mixture of four volatile organic chemicals (VOCs):

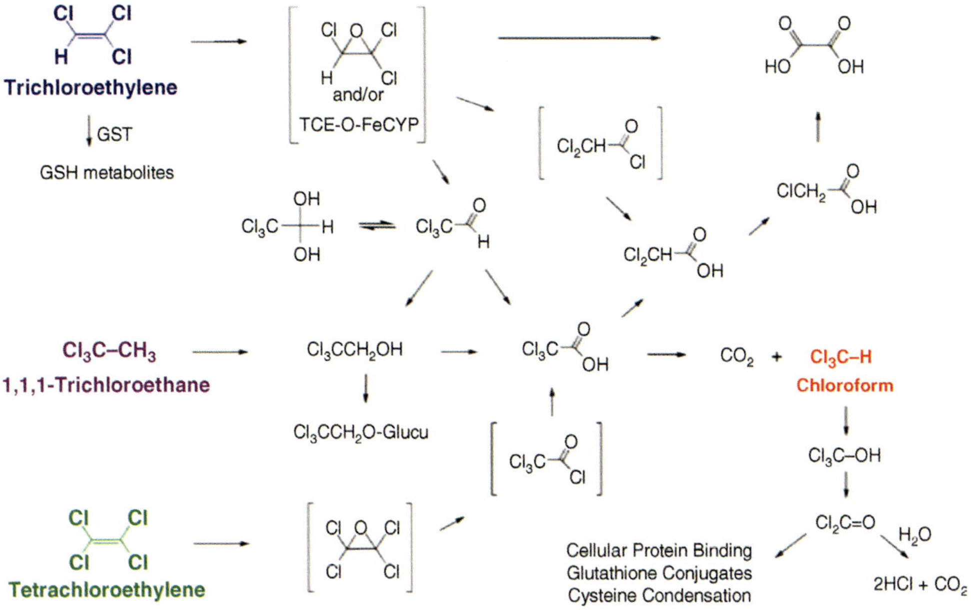

Figure 22.4 Combined metabolic pathways of trichloroethylene, chloroform, tetrachloroethylene, and 1,1,1-trichloroethane (only major pathways shown). Reproduced from Mayeno *et al.* [37] with permission.

trichloroethylene (TCE), chloroform (CF), tetrachloroethylene (Perc; perchlorethylene), and 1,1,1-trichloroethane (MC; methyl chloroform). All four are prevalent drinking water or groundwater contaminants and they are likely to be present in such media together. This study was reported in an earlier publication [37].

The interconnected biotransformation pathways of all four VOCs (Figure 22.4) illustrate the close relationship among the metabolic pathways of these chemicals, their shared enzyme systems, the potential of generating the same reactive species from different parent compounds, and the dynamic interactions among the linked pathways influencing the possible outcome of toxicities.

Our laboratory has completed the qualitative aspects of the biochemical reaction network modeling of the four volatile organic chemicals based on biochemical reaction mechanisms of the relevant CYP and related enzymes. Moreover, we have incorporated enzyme-reaction mechanisms to help predict metabolite formation. For instance, the mechanism-based biotransformation of TCE, as generated by BioTRaNS, is shown in Figure 22.5.

The first step of CYP2E1-catalyzed oxidation involves the formation of an intermediate between the high-valent iron-oxo complex, $(FeO)^{3+}$ of the CYP heme, and the alkene, as postulated by Miller and Guengerich [55]. This intermediate has been linked mechanistically to the 1,2-shifts of Cl (or H) leading to the formation of an aldehyde (or acid chloride), as shown in Figure 22.5. Details of these (bio) chemical processes and subsequent stepwise reaction mechanisms, with their

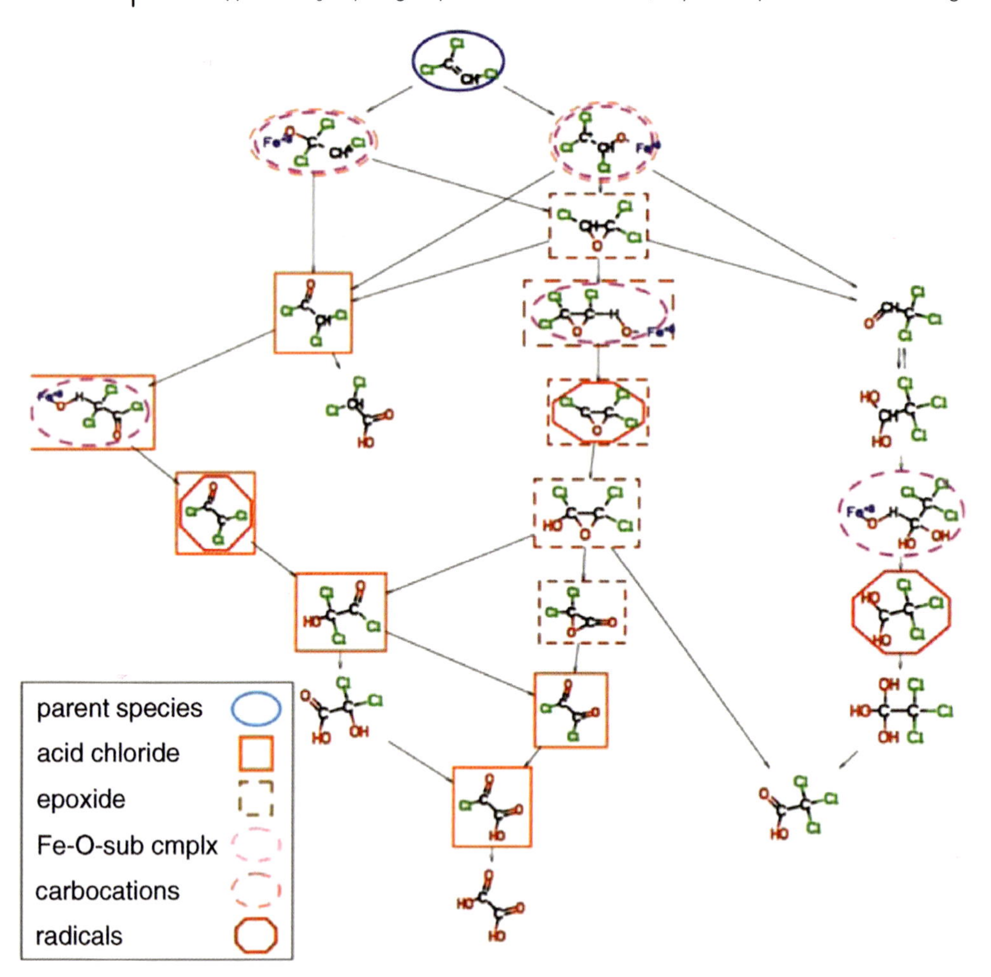

Figure 22.5 A portion of the BioTRaNS-generated biotransformation pathways using postulated *mechanisms* for CYP-mediated TCE oxidation [55]. Reactive metabolites are highlighted as follows: epoxides (brown, box, dashed); acid chlorides (orange, box, solid); and starting chemicals (blue, ellipse, solid). Reproduced from Mayeno *et al.* [37] with permission.

respective originally published sources, have been described in our recent paper [37]. Our computer-generated interconnected metabolic pathways for this set of four VOCs (Figure 22.6) demonstrate the metabolite inventory and interconnections.

We designate such a predicted biochemical reaction network on the basis of known enzymatic and chemical mechanisms "predictive xenobiotic metabolomics." The predicted pathways (Figure 22.6) match the known metabolic pathways (Figure 22.4) very well.

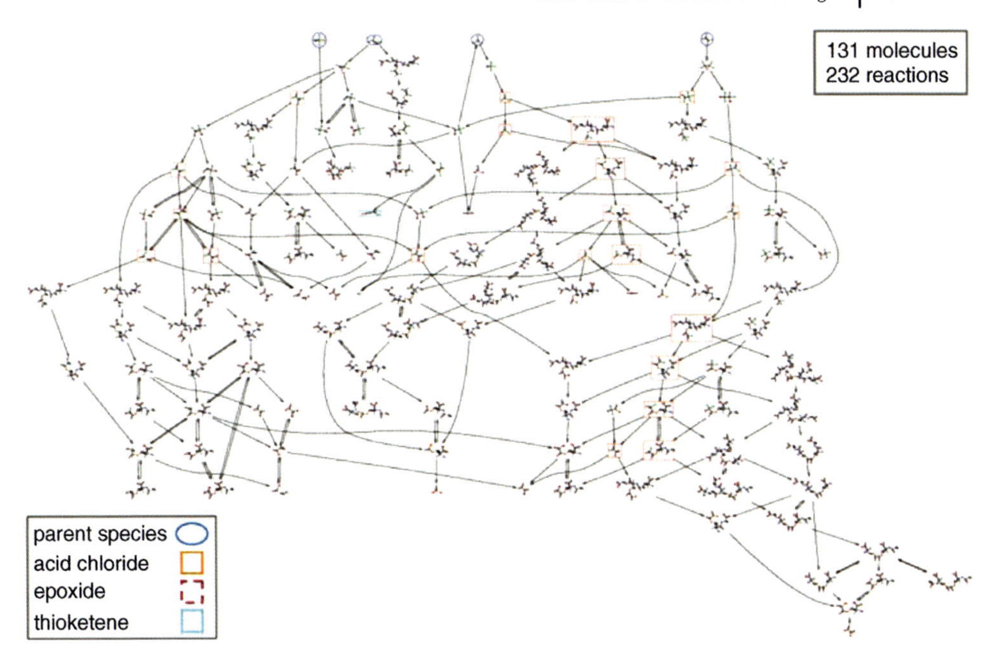

Figure 22.6 Our computer-generated biotransformation reaction network for four volatile organic chemicals: trichloroethylene, tetrachloroethylene, methyl chloroform, and chloroform. The computer automatically generated this figure based on reaction rules and interconnected the pathways via metabolites in common. Reactive species were automatically highlighted in red boxes, after substructures (SMARTS) of these species were input by the user.

22.6 What is "Multiscale Modeling?" How Do PBPK, Bayesian Population PBPK Modeling, and BRN modeling Fit Into "Multiscale Modeling?" Any Possible Inclusion of Other Types of Computer Modeling?

The biotransformation of a single xenobiotic within an organism may occur in several tissues and involve numerous enzymes [56] and yield a series of metabolites that can interact with cellular components near the site of generation or in distal tissues [57]. Within the cell, the enzymes can be localized in different subcellular compartments [58]. As such, a biological system is more than the sum of the parts; and to truly comprehend biological systems, an understanding of these dynamic interactions is required. "Systems biology" integrates computational and experimental sciences in an effort to describe and understand entire biological systems [59]. Figures 22.7 and Figure 22.1 are systems biology representations of the integration of biologically based computer modeling across a number of levels of biological organization.

In Figure 22.7, starting from the right, the ADME model refers to whole-body pharmacokinetics, as exemplified by PBPK modeling. In the middle, the metabolomic

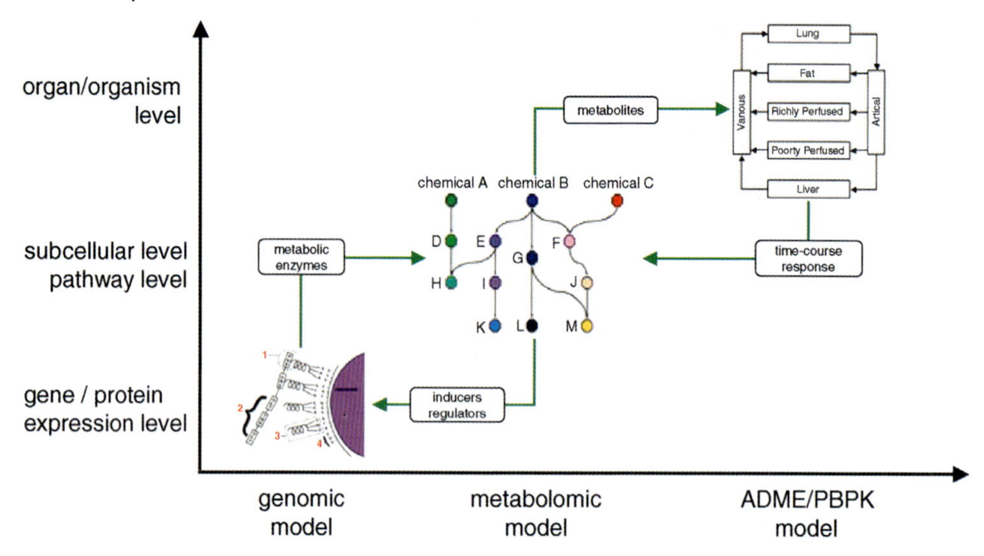

Figure 22.7 Multiscale modeling and systems biology approach toward global toxicological effects of a chemical or chemical mixtures.

(or metabonomic) model is exemplified by BRN modeling. On the left, the genomic model is a general representation of the gene and/or protein expressions related to the toxicological processes that we are studying. It is our belief that through appropriate models and linkages it will eventually be possible to capture the multiscale nature of toxicological interactions and effects. At present, it is possible to explore interactions at the ADME and metabolomic level through integration of PBPK and BRN models. However, the current state-of-the-science in genomics and proteomics modeling of chemical mixture toxicology is not yet at a stage where integration into the unified framework is possible.

22.7
Can We Predict Chemical Mixture Toxicities? What is the Potential Real-World Application of Such a "Multiscale Computer Simulation" Approach?

The ability to predict chemical mixture toxicities with a minimum of experimental data would represent an unprecedented breakthrough in the toxicology of chemical mixtures and cumulative risk assessment. Developing this ability has been a focus of our research within the Quantitative and Computational Toxicology Group for the past 17 years or so. With the recent development of biochemical reaction network modeling and PBPK modeling, we see a ray of hope in this area. Our approach can be explained in a stepwise manner as follows:

Step 1: Consider a given class of chemicals (e.g., volatile organic chemicals, polychlorinated biphenyls (PCBs), etc.,). Here, we designate such a class as "class A." As

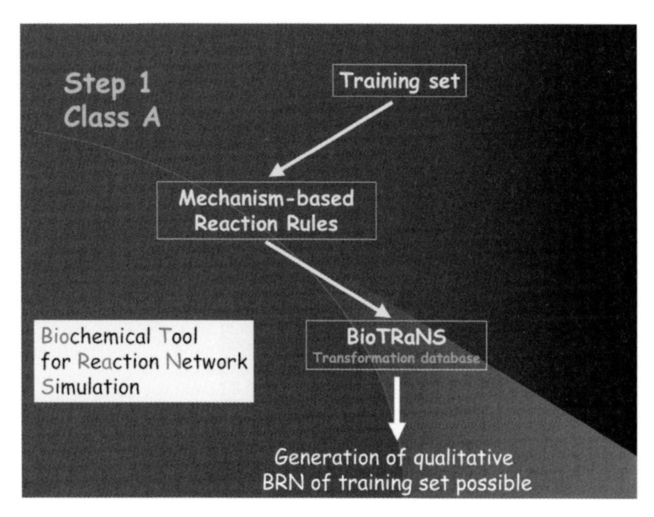

Figure 22.8 Biochemical reaction network modeling of a training set for a hypothetical class of chemicals: generation of qualitative biochemical reaction network.

shown in Figure 22.8, a biochemical reaction network model can be established for a training set (10–20 members) of class A chemicals in much the same way as described earlier for VOCs. When this is done, a *qualitative* biochemical reaction network for this training set will have been established. A *qualitative* BRN contains the predicted metabolic pathways for each member of class A chemicals, interconnections between these pathways, and metabolites and subpathways in common.

Step 2: Next, enzyme kinetic studies are conducted using commercially available recombinant human metabolic enzymes known to be involved in the metabolism of the chemicals in the training set of class A chemicals (Figure 22.9). The purpose of such studies is to generate reaction rate constants to be incorporated into BRN modeling for generation of the *quantitative* reaction network. A *quantitative* BRN contains predictions for the time rates of change of the concentrations of all chemicals comprising the network.

Step 3: Using quantitative structure–activity relationship (QSAR) modeling and other computational techniques (e.g., molecular modeling and computational quantum chemistry), the reaction rate constants of chemicals other than the training set in class A are calculated (Figure 22.10). At this stage, the generation of qualitative and quantitative biochemical reaction network for class A chemicals is possible.

Step 4: By integrating a generic PBPK model and BRN model for class A chemicals, pharmacokinetic information for toxicologically relevant species produced from the chemicals in class A can be predicted. This modeling effort is best carried out by an interdisciplinary team of scientists, including toxicologists, biological modelers, and chemists. In turn, such a team of scientists is in a position to be able to predict the possible outcome of toxicities for the mixture of class A chemicals, given that the

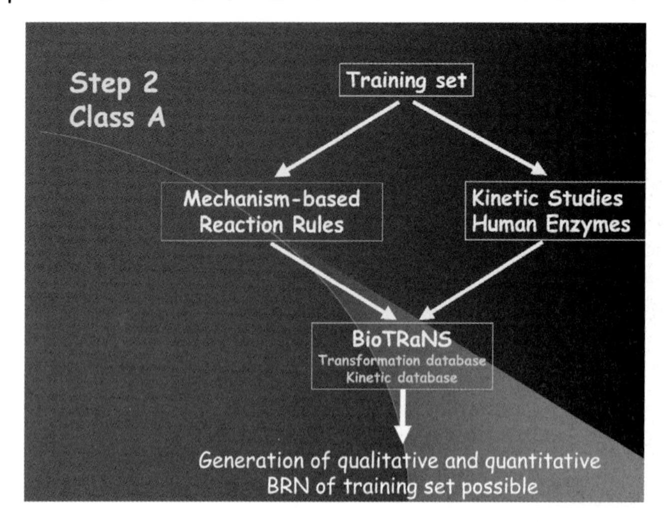

Figure 22.9 Biochemical reaction network modeling of a training set for a hypothetical class of chemicals: generation of qualitative and quantitative biochemical reaction networks.

mode(s) of action have been established for class A chemicals and their metabolites (Figure 22.11).

Step 5: Once predictions for class A chemicals are substantiated and the methodology validated, similar studies for other classes of chemicals should be possible (Figure 22.12), thus paving the way for a better understanding of the toxicities of a wide variety of chemical mixtures.

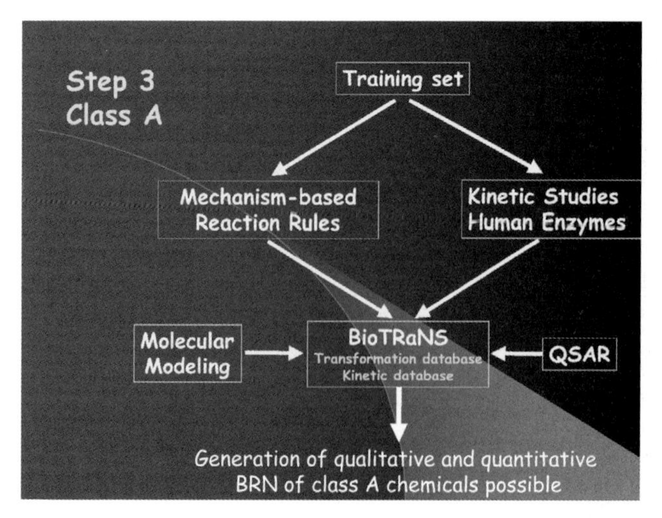

Figure 22.10 Biochemical reaction network modeling of a hypothetical class of chemicals: generation of qualitative and quantitative biochemical reaction networks for class A chemicals.

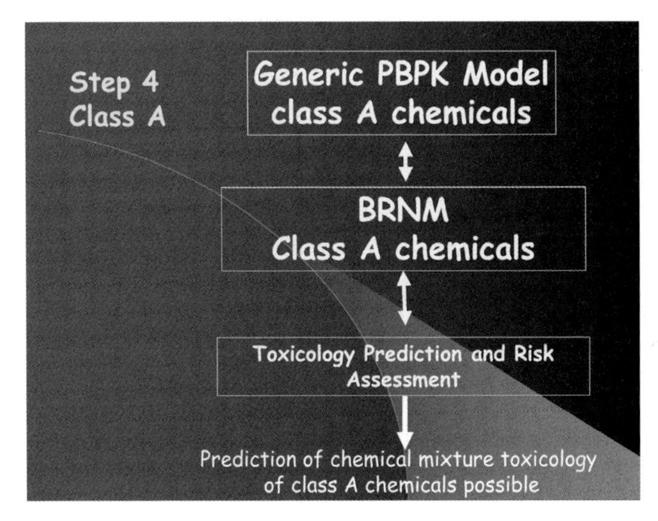

Figure 22.11 Integration of PBPK and biochemical reaction network modeling of class A and expert scientific assessment: prediction of chemical mixture toxicology of class A chemicals.

22.8
Concluding Remarks

In this chapter, we have provided a brief glimpse of some of the potential models and tools available to better understand important phenomena in chemical mixture toxicology. Each of these modeling areas represents the culmination of many years of devoted work by numerous investigators. Here, we have emphasized the concepts

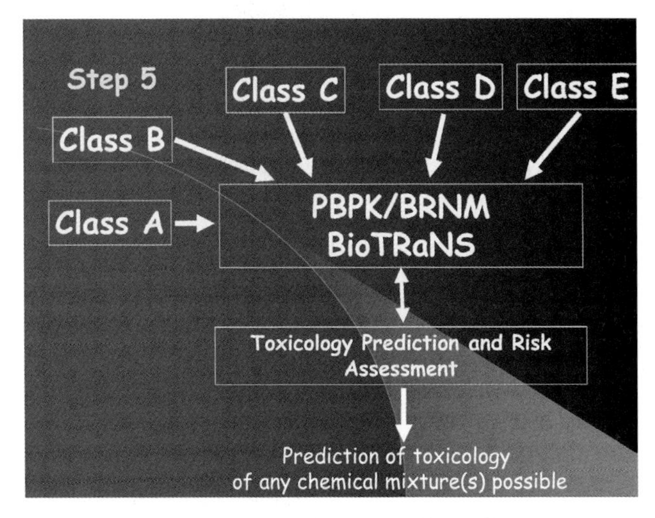

Figure 22.12 Prediction of toxicology for any chemical mixture(s).

and approaches, rather than focusing on the details of the modeling technologies. Therefore, we encourage interested readers to further study these and other modeling areas for advanced and detailed knowledge.

As indicated earlier, we see a ray of hope in the prediction of chemical mixture toxicities. This field of science is still in its infancy and there will be ample opportunities for significant contributions by those who are willing to undertake research in this important and complex area.

References

1 Yang, R.S.H., Thomas, R.S., Gustafson, D.L., Campain, J.A., Benjamin, S.A., Verhaar, H.J.M., and Mumtaz, M.M. (1998) Approaches to developing alternative and predictive toxicology based on PBPK/PD and QSAR modeling. *Environ. Health Perspect.*, **106** (Suppl. 6), 1385–1393.

2 Zeiger, E. and Margolin, B.H. (2000) The proportions of mutagens among chemicals in commerce. *Regul. Toxicol. Pharmacol.*, **32**, 219–225.

3 OTA (1995) Screening and Testing Chemicals in Commerce. Report No. OTA-BP-ENV-166. Office of Technology Assessment, Congress of the United States.

4 Lazarou, J., Pomeranz, B.H., and Corey, P.N. (1998) Incidence of adverse drug reactions in hospitalized patients: a meta-analysis of prospective studies. *JAMA*, **279** (15), 1200–1205.

5 Jevtovic-Todorovic, V., Hartman, R.E., Izumi, Y., Benshoff, N.D., Dikranian, K., Zorumski, C.F., Olney, J.W., and Wozniak, D.F. (2003) Early exposure to common anesthetic agents causes widespread neurodegeneration in the developing rat brain and persistent learning deficits. *J. Neurosci.*, **23** (3), 876–882.

6 Reddy, M.B., Yang, R.S.H., Clewell, H.J., 3rd, and Andersen, M.E. (2005) *Physiologically Based Pharmacokinetics: Science and Applications*, John Wiley & Sons, Inc., Hoboken, New Jersey.

7 Yang, R.S.H. and Andersen, M.E. (2005) Physiologically based pharmacokinetic modeling of chemical mixtures, in *Physiologically Based Pharmacokinetics: Science and Applications* (eds M.B. Reddy, R.S.H. Yang, H.J. Clewell, 3rd., and M.E. Andersen), John Wiley & Sons, Inc., New York, pp. 349–373.

8 Andersen, M.E., Clewell, H.J., 3rd, Gargas, M.L., Smith, F.A., and Reitz, R.H. (1987) Physiologically based pharmacokinetics and the risk assessment process for methylene chloride. *Toxicol. Appl. Pharmacol.*, **87** (2), 185–205.

9 Brown, R.P., Delp, M.D., Lindstedt, S.L., Rhomberg, L.R., and Beliles, R.P. (1997) Physiological parameter values for physiologically based pharmacokinetic models. *Toxicol. Ind. Health*, **13** (4), 407–484.

10 Lindstedt, S.L. (1987) Allometry: body size constraints in animal design, in *Pharmacokinetics in Risk Assessment, Drinking Water and Health*, vol. 8, National Academy Press, Washington, DC, pp. 65–79.

11 Lipscomb, J.C., Fisher, J.W., Confer, P.D., and Byczkowski, J.Z. (1998) *In vitro* to *in vivo* extrapolation for trichloroethylene metabolism in humans. *Toxicol. Appl. Pharmacol.*, **152** (2), 376–387.

12 Kedderis, G.L. and Lipscomb, J.C. (2001) Application of *in vitro* biotransformation data and pharmacokinetic modeling to risk assessment. *Toxicol. Ind. Health*, **17** (5–10), 315–321.

13 Yang, R.S.H. and Lu, Y. (2007) The application of physiologically based pharmacokinetic (PBPK) modeling to risk assessment, in *Risk Assessment for Environmental Health* (eds M. Robson and W. Toscano), John Wiley & Sons, Hoboken, NJ, pp. 85–120.

14 Portier, C.J. and Kaplan, N.L. (1989) Variability of safe dose estimates when using complicated models of the carcinogenic process. A case study: methylene chloride. *Fundam. Appl. Toxicol.*, **13** (3), 533–544.

15 Thomas, R.S., Lytle, W.E., Keefe, T.J., Constan, A.A., and Yang, R.S. (1996) Incorporating Monte Carlo simulation into physiologically based pharmacokinetic models using advanced continuous simulation language (ACSL): a computational method. *Fundam. Appl. Toxicol.*, **31** (1), 19–28.

16 Gelman, A., Carlin, J.B., Stern, H.S., and Rubin, D.B. (2004) *Bayesian Data Analysis*, Chapman & Hall/CRC.

17 Gilks, W.R., Richardson, S., and Spiegelhalter, D.J. (1996) *Markov Chain Monte Carlo in Practice*, Chapman & Hall/CRC.

18 Bois, F.Y. and Maszle, D.R. (1997) MCSim: a Monte Carlo simulation program. *J Stat Softw*, **2** (9), 1–60.

19 Marino, D.J., Clewell, H.J., Gentry, P.R., Covington, T.R., Hack, C.E., David, R.M., and Morgott, D.A. (2006) Revised assessment of cancer risk to dichloromethane. Part I. Bayesian PBPK and dose-response modeling in mice. *Regul. Toxicol. Pharmacol.*, **45** (1), 44–54.

20 Jonsson, F. and Johanson, G. (2003) The Bayesian population approach to physiological toxicokinetic–toxicodynamic models: an example using the MCSim software. *Toxicol. Lett.*, **138** (1–2), 143—150

21 Hack, C.E. (2006) Bayesian analysis of physiologically based toxicokinetic and toxicodynamic models. *Toxicology*, **221** (2–3), 241–248.

22 Gelman, A., Bois, F.Y., and Jiang, J. (1996) Physiological pharmacokinetic analysis using population modeling and informative prior distributions. *J. Am. Stat. Assoc.*, **91**, 1400–1412.

23 Bernillon, P. and Bois, F.Y. (2000) Statistical issues in toxicokinetic modeling: a Bayesian perspective. *Environ. Health Perspect.*, **108** (Suppl. 5), 883–893.

24 Covington, T.R., Gentry, P.R., Van Landingham, C.B., Andersen, M.E., Kester, J.E., and Clewell, H.J. (2007) The use of Markov chain Monte Carlo uncertainty analysis to support a public health goal for perchloroethylene. *Regul. Toxicol. Pharmacol.*, **47** (1), 1–18.

25 David, R.M., Clewell, H.J., Gentry, P.R., Covington, T.R., Morgott, D.A., and Marino, D.J. (2006) Revised assessment of cancer risk to dichloromethane II. Application of probabilistic methods to cancer risk determinations. *Regul. Toxicol. Pharmacol.*, **45** (1), 55–65

26 Lowry, T.H. and Richardson, K.S. (1987) *Mechanism and Theory in Organic Chemistry*, HarperCollins, New York.

27 Bonchev, D. and Rouvray, D.H. (1991) *Chemical Graph Theory: Introduction and Fundamentals*, Abacus Press, New York.

28 Biggs, N., Lloyd, E.K., and Wilson, R.J. (1998) *Graph Theory*, Oxford University Press, New York.

29 Campbell, D.M. and Klein, M.T. (1997) Construction of a molecular representation of a complex feedstock by Monte Carlo and quadrature methods. *Appl. Catal. A*, **160**, 41–54.

30 McDermott, J.B., Libanati, C., LaMarca, C., and Klein, M.T. (1990) Quantitative use of a model of compound information: Monte Carlo simulations of the reactions of complex macromolecules. *Ind. Eng. Chem. Res.*, **29**, 22–29.

31 Trauth, D.M., Stark, S.M., Petti, T.F., Neurock, M., and Klein, M.T. (1994) Representation of the molecular structure of petroleum resid through characterization and Monte Carlo modeling. *Energ. Fuel*, **8**, 576–580.

32 Broadbelt, L.J., Stark, S.M., and Klein, M.T. (1994) Computer generated reaction networks: on-the-fly calculation of species properties using computational quantum chemistry. *Chem. Eng. Sci.*, **49**, 4991–5010.

33 Klein, M.T., Hou, G., Quann, R., Wei, W., Liao, K.H., Yang, R.S.H., Campain, J.A., Mazurek, M., and Broadbelt, L.J. (2002) BioMOL: a computer-assisted biological modeling tool for complex chemical mixtures and biological processes at the molecular level. *Environ. Health Perspect.*, **110** (Suppl. 6), 1025–1029.

34 Liao, K.H. (2004) Development and validation of a hybrid reaction network/ physiologically based pharmacokinetic

model of benzo[*a*]pyrene and its metabolites. Ph.D. Dissertation, Colorado State University, p. 214.

35 Liao, K.H., Dobrev, I., Dennison, J.E., Andersen, M.E., Reisfeld, B., Reardon, K.F., Campain, J.A., Wei, W., Klein, M.T., Quann, R.J., and Yang, R.S.H. (2002) Application of biologically based computer modeling to simple or complex mixtures. *Environ. Health Perspect.*, **110** (Suppl. 6), 957–963.

36 Reisfeld, B. and Yang, R.S.H. (2004) A reaction network model for CYP2E1-mediated metabolism of toxicant mixtures. *Environ. Toxicol. Pharmacol.*, **18**, 173–179.

37 Mayeno, A.N., Yang, R.S.H., and Reisfeld, B. (2005) Biochemical reaction network modeling: a new tool for predicting metabolism of chemical mixtures. *Environ. Sci. Tech.*, **39**, 5363–5371.

38 Jacob, S.M., Gross, B., Voltz, S.E., and Weekman, V.W. (1976) A lumping and reaction scheme for catalytic cracking. *AIChE J.*, **22**, 701–713.

39 Quann, R.J. and Jaffe, S.B. (1992) Structure-oriented lumping: describing the chemistry of complex hydrocarbon mixtures. *Ind. Eng. Chem. Res.*, **31**, 2483–2497.

40 Broadbelt, L.J., Stark, S.M., and Klein, M.T. (1994) Computer generated pyrolysis modeling: on-the-fly generation of species, reactions, and rates. *Ind. Eng. Chem. Res.*, **33**, 790–799.

41 Broadbelt, L.J., Stark, S.M., and Klein, M.T. (1996) Computer generated reaction modeling: decomposition and encoding algorithms for determining species uniqueness. *Comput. Chem. Eng.*, **20**, 113–129.

42 Neurock, M. and Klein, M.T. (1993) When you can't measure – model. *CHEMTECH*, **23**, 26–32.

43 Klein, M.T. (2000) *Faculty Activities Publications*, Rutgers University, Piscataway, NJ.

44 Quann, R.J. and Jaffe, S.B. (1996) Building useful models of complex reaction systems in petroleum refining. *Chem. Eng. Sci.*, **51**, 1615–1635.

45 Joshi, P.V., Sadasivan, D.I., and Klein, M.T. (1998) Automatic mechanistic kinetic modeling of gas oil catalytic cracking. *Rev. Proc. Chem. Eng.*, **1**, 111–140.

46 Mizan, T.I. and Klein, M.T. (1999) Computer-assisted mechanistic modeling of *n*-hexadecane hydroisomerization over various bifunctional catalysts. *Catal. Today*, **50**, 159–170.

47 Klein, M.T., Hou, G., Quann, R.J., Wei, W., Liao, K.H., Yang, R.S., Campain, J.A., Mazurek, M.A., and Broadbelt, L.J. (2002) BioMOL: a computer-assisted biological modeling tool for complex chemical mixtures and biological processes at the molecular level. *Environ. Health Perspect.*, **110** (Suppl. 6), 1025–1029.

48 Daylight Chemical Information Systems, Inc. *Daylight Toolkits*. vol. 2004. Daylight Chemical Information Systems, Inc., Mission Viejo, CA.

49 Silverman, R.B. (2002) *Organic Chemistry of Enzyme-Catalyzed Reactions*, Academic Press, London.

50 Balaban, A.T. (1976) *Chemical Applications of Graph Theory*, Academic Press, New York.

51 Gross, J.L. and Yellen, J. (1999) *Graph Theory and Its Applications*, CRC Press, Boca Raton, FL.

52 Hlavica, P. and Lewis, D.F.V. (2001) Allosteric phenomena in cytochrome P450-catalyzed monooxygenations. *Eur. J. Biochem.*, **268** (18), 4817–4832.

53 Atkins, W.M., Wang, R.W., and Lu, A.Y.H. (2001) Allosteric behavior in cytochrome P450-dependent *in vitro* drug–drug interactions: a prospective based on conformational dynamics. *Chem. Res. Toxicol.*, **14** (4), 338–347.

54 Yoon, M.Y., Campbell, A.P., and Atkins, W.M. (2004) "Allosterism" in the elementary steps of the cytochrome P450 reaction cycle. *Drug Metab. Rev.*, **36** (2), 219–230.

55 Miller, R.E. and Guengerich, F.P. (1982) Oxidation of trichloroethylene by liver microsomal cytochrome P-450: evidence for chlorine migration in a transition state not involving trichloroethylene oxide. *Biochemistry*, **21**, 1090–1097.

56 Parkinson, A. (2001) Biotransformation of xenobiotics, in *Casarett and Doull's Toxicology: The Basic Science of Poisons*

(ed. C.D. Klaassen), McGraw-Hill, New York, pp. 133–224.

57 Elfarra, A.A., Krause, R.J., Last, A.R., Lash, L.H., and Parker, J.C. (1998) Species- and sex-related differences in metabolism of trichloroethylene to yield chloral and trichloroethanol in mouse, rat, and human liver microsomes. *Drug Metab. Dispos.*, **26**, 779–785.

58 Fiehn, O. and Weckwerth, W. (2003) Deciphering metabolic networks. *Eur. J. Biochem.*, **270**, 579–588.

59 Kitano, H. (2002) Computational systems biology. *Nature*, **420** (6912), 206–210.

23
Food Ingredients are Sometimes Mixtures

Mary E. LaVecchia, Paulette M. Gaynor, Negash Belay, Rebecca P. Danam, and Antonia Mattia

23.1
Introduction

The Center for Food Safety and Applied Nutrition (CFSAN) under the United States Food and Drug Administration (FDA) is responsible for promoting and protecting public health by making sure that the food supply is safe and wholesome. Its food safety mission is broad in scope and encompasses a variety of areas, including regulatory and research programs, to address health risks associated with foodborne chemical and biological contamination, proper labeling of foods, health claims, dietary supplements, food industry compliance, and international harmonization efforts.

Food primarily comes from plant or animal sources and contains carbohydrates, proteins, lipids, vitamins, minerals, and other nutrients. As such, food is a complex mixture of hundreds or thousands of chemicals. However, under the Federal Food, Drug, and Cosmetic Act (FD&C Act) whole foods are presumed to be safe. This follows from their history of common use. This presumption is not extended to ingredients added to food. Such ingredients must undergo a safety assessment and meet the safety standard of "reasonable certainty of no harm," as stated in the US Code of Federal Regulations (21 CFR 170.3(i)).

The term "food ingredients" includes food additives, color additives, and substances that are considered generally recognized as safe (GRAS) under specified conditions of use. These types of ingredients are intentionally added to food and include substances such as preservatives, sweeteners, colorants, and flavors. Materials used to package food are called food contact substances. Technically, food contact substances are food additives in spite of the fact that they are not intentionally added to food. CFSAN's Office of Food Additive Safety (OFAS) is responsible for assessing the safety of food ingredients and food contact substances.

In this chapter, FDA's framework for assessing the safety of ingredients added to conventional food will be explained. Distinctions will be made between food additives, color additives, GRAS substances, and food contact formulations with

Principles and Practice of Mixtures Toxicology. Edited by Moiz Mumtaz

ISBN: 978-3-527-31992-3

a discussion of unique examples that fall within each regulatory category. As described below, FDA ordinarily regulates ingredients individually added to food. However, in practical use, many food ingredients are mixtures rather than individual chemically defined substances. Some ingredients are racemic mixtures. In a limited number of cases, a condition of safe use for an ingredient may require that the ingredient be used in combination with another substance. FDA also evaluates food ingredients that are derived from plants that, depending on the part of the plant and other conditions, exhibit variable compositional complexity. For food packaging, multiple materials are formulated to produce polymeric substances, which may contain small amounts of residual contaminants from the manufacturing processes employed.

The diversity of substances that are the subject of submissions to FDA for consideration as food ingredients, and the distinction between food itself and what is added to it, presents a platform to describe the agency's regulatory framework and aspects that are related to mixtures. For the purpose of this chapter, mixtures will be considered as a combination or blend of two or more substances distinguishable by chemical analyses (not chemically bound to each other) that have the potential to influence the risk of chemical toxicity [1].

23.2
Safety Evaluation

Toxicological studies and related information submitted to FDA in food additive or color additive petitions, and GRAS or food contact substance notices are generally conducted and reported according to general requirements in FDA regulations (i.e., 21 CFR) and other guidance documents issued by the agency. Other approaches, if scientifically defensible, are acceptable too.

The FDA Redbook, or *Toxicological Principles for the Safety Assessment of Food Ingredients, Redbook 2000,* is a guidance document that is intended to assist petitioners and notifiers in developing and reporting information used in the safety assessment of food ingredients (see Redbook 2000 at http://www.fda.gov/~redbook/FoodRedbook). The Redbook includes guidelines for conducting various types of studies, including short- and long-term toxicity studies, metabolism and pharmacokinetic studies, reproductive and developmental toxicity studies, and immuno-toxicity, carcinogenicity, and neurotoxicity studies. The Redbook also describes clinical and epidemiological studies. The 1982 version of the Redbook [2] recommended minimum sets of toxicity tests by assigning ingredients to "concern levels" based on the chemical structure and dietary intake of the ingredient of interest. Typically, dietary intake has more weight than structure alert information. Testing recommendations based on concern levels are available at http://www.fda.gov/Food Toxicological Testing Summary.

In conducting safety tests, the substance tested should be the article of commerce. If the article of commerce is not the test substance, its relationship with the test substance must be established. In all cases, the test substance needs to be thoroughly characterized with respect to chemical identity, purity, stability, and other considerations such as solubility and formulation aides. In the case of mixtures, it is the individual substances that are combined in the blend that need to be properly identified. As part of the chemistry evaluation of a substance, the daily intake is estimated by considering the amount of a substance added to various foods and the amount of such foods generally consumed by the population at large over a lifetime. For complex mixtures, cumulative exposure estimates are complicated; such estimates are more easily derived when the components of the mixture are chemically similar and they have the same toxicological properties.

The article of commerce may be a mixture of a key ingredient and formulation aides such as diluents, antioxidants, and/or stabilizers (e.g., enzyme preparations). Often, formulation aides for which safety data and information are available are selected so that new safety data may be needed only for the key ingredient. If interactions occur between the components in a mixture, test results obtained using the mixture may not be applicable to the key ingredient.

Some ingredients, such as flavors and botanical extracts, are complex mixtures of numerous substances from various chemical classes. For these types of ingredients, safety tests may need to be conducted on a single component of the mixture if toxicity is either known or suspected. In cases of complex mixtures, compositional differences or contaminants may cause inconsistencies in the data reported rendering some test results unusable.

A food additive or color additive known to cause cancer in man or animals may not be intentionally added to food. On the other hand, exposure to an unintentional but potentially carcinogenic impurity in a food additive or color additive would be assessed via quantitative risk assessment to determine whether there is a safety concern or not.

FDA has provided guidance documents to assist individuals who wish to submit data for the safety assessment of a food ingredient. For detailed information on guidance documents used in the safety evaluation of food ingredients consult http://www.fda.gov/FoodRedbook.

23.3
Description of the Priority-Based Assessment of Food Additives

FDA identified a need for monitoring the post market safety of ingredients added to food when new information about an ingredient became available. The priority-based assessment of food additives (PAFA) is a system to rank food ingredients for reassessment based on concerns related to dietary exposure, chemical structure, results of toxicological tests conducted, and/or data gaps [3]. In the PAFA database, FDA created toxicological profiles, including Chemical Abstract Services (CAS) registry numbers and chemical structure categories, for over 2200 (including 1700 profiles for flavors) of approximately 3500 ingredients intentionally added to food in the United States [4].

The PAFA database is FDA's institutional memory for the toxicological effects of food ingredients. FDA has used the database to perform cyclic reviews of previously regulated food ingredients, but the database is also useful for other types of assessments. For example, the PAFA database is searchable for information on effects observed in a specific target organ. It can be searched for end points to design studies of structurally related compounds and it can be used to rank the relative toxicities of ingredients with similar structures or technical effects.

Since the PAFA database captures chemical structures and toxicity data, it has been useful in providing input into predictive toxicology software programs. In recent years, the FDA has been developing and evaluating such programs to support regulatory decision making. One such effort has focused on the use of quantitative structure–activity relationship modeling to predict the rodent carcinogenic potential of naturally occurring chemicals in the human diet [5]. Although computational safety analyses are limited to individual chemical substances, they can be applied to the components of new mixtures.

23.4 Food Additives

In 1958, Congress enacted the Food Additives Amendment to the FD&C Act. The amendment and/or supporting legislative documents defined the term "food additive" and required premarket approval for new uses of food additives. It also established the standard of review ("fair evaluation of the data . . ."), the standard of safety, and formal rule-making procedures for food additives. The term "safe," as it refers to food additives and ingredients (including food contact substances), is defined in 21 CFR 170.3(i) as a "reasonable certainty in the minds of competent scientists that a substance is not harmful under the intended conditions of use." The concept of safety involves the question whether a substance is hazardous to the health of man or animal and considers the reality that it is impossible to establish with complete certainty the absolute harmlessness of the use of any substance. Congress broadly defined "food additive" to include any substance the intended use of which results or may reasonably be expected to result, directly or indirectly, in its becoming a component or otherwise affecting the characteristics of food. Olestra, DL-alanine, DL-methionine, and high-intensity sweeteners are examples of food additives; these are discussed below in the context of mixtures.

On January 30, 1996, FDA approved olestra as a food additive intended to substitute for fats and oils in savory snacks, Federal Register (61 FR 3118). Olestra, also known as sucrose polyester, is a mixture of sucrose esters formed from the addition of six to eight fatty acids to the eight free hydroxyl moieties present in sucrose. Saturated fatty acids of medium and long chain lengths (C12 to C20 and higher) are used to manufacture olestra. In evaluating the safety of a mixture such as olestra, the range of substances produced has to be determined using appropriate analytical techniques and the manufacturing process has to be considered in detail.

DL-Alanine is a racemic mixture of D- and L-alanine isomers that may be safely used as a flavor enhancer for sweeteners in pickling mixtures at a level not to exceed 1% of the pickling spice that is added to the brine (21 CFR 172.540). DL-Methionine is another example of a racemic mixture that can be safely added to food as a nutrient in either the free, hydrated, or anhydrous form or as the hydrochloride, sodium, or potassium salt (21 CFR 172.320). Depending on the stability of the isomers, the composition of these additives could vary without raising a safety concern. This is in contrast to the use of DL-tartaric acid. L-Tartaric acid is a natural byproduct of wine with GRAS status in wine production. However, DL-tartaric acid is not interchangeable with L-tartaric acid for this use. DL-Tartaric acid is of potential toxicological concern because it is less soluble than L-tartaric acid.

High-intensity sweeteners are low or no-calorie food additives that can be used individually or blended with each other. For example, aspartame is blended with acesulfame-potassium to produce a more natural sugar taste and mask aftertaste.

Aspartame is composed of aspartic acid and phenylalanine, two amino acids commonly present in proteins and flavorings. It was first approved by the FDA in 1981 as a tabletop sweetener, and for use in gum, breakfast cereal, and other dry products. In subsequent years, aspartame was approved for use in soda and as a general-purpose sweetener in all foods and drinks [6]. Acesulfame-potassium (acesulfame-K) is the potassium salt of 6-methyl-1,2,3-oxathiazine-4(3*H*)-one-2,2-dioxide. The FDA approved acesulfame-K for specific uses, including use as a tabletop sweetener in 1988; 10 years later, the FDA approved acesulfame-K for use in beverages. It was approved for general use in food (e.g., use in baked goods, candies, cough drops, etc.) in 2003 (excluding use in meat or poultry). FDA has approved the use of aspartame and acesulfame-K individually; however, the agency has not approved specific combinations of these or other high-intensity sweeteners (e.g., sucralose). Such an approval is not necessary so long as the conditions of use for each ingredient conform to the conditions specified in the applicable regulation.

In 2000, the Food and Agriculture Organization (FAO)/World Health Organization (WHO) Joint Expert Committee on Food Additive (JECFA) evaluated an aspartame-acesulfame salt for use as a sweetener and flavor enhancer. The JECFA determined that the aspartame and acesulfame moieties are covered by acceptable daily intakes (ADIs) previously established by the committee for aspartame (0–40 mg/kg bw) and acesulfame potassium (0–15 mg/kg bw) [7]. Because JECFA had data to establish an ADI for each sweetener individually, it did not require new data in order to assess the safety of the aspartame-acesulfame salt as a replacement for aspartame and acesulfame-K in products that allow the use of both substances.

23.5 Color Additives

A couple of years after the passage of the Food Additives Amendment, Congress enacted the Color Additive Amendments to the FD&C Act. The Color Additive Amendments of 1960 defined "color additive" and required that only color additives (except coal-tar hair dyes) listed as "suitable and safe" for a given use could be used in foods, drugs, cosmetics, and medical devices. The 1960 Amendments prescribed both the factors that the FDA must consider in determining whether a proposed use of a color additive is safe and the specific conditions for safe use that must be included in the listing regulation. There is no GRAS exemption (see Section 23.6) to the definition of a color additive. Thus, a substance that imparts color is a color additive and is subject to premarket approval requirements unless the substance is used solely for a purpose other than coloring.

A color additive, as defined by regulation, is any dye, pigment, or other substance that can impart color to a food, drug, or cosmetic or to the human body (21 CFR 70.3 (f)). Color additives are classified as straight colors, lakes, and mixtures. Straight colors are color additives that have not been mixed or chemically reacted with any other substance (e.g., FD&C Green No. 3 or Green 3). Lakes are formed by chemically reacting straight colors with precipitants and substrata (e.g., Green 3 Lake). Lakes for food use must be made from certified batches of straight colors. (One exception is carmine, which is a lake made from cochineal extract.) Lakes for food use are made with aluminum cation as the precipitant and aluminum hydroxide as the substratum. Mixtures are a class of color additives that are formed by mixing one color additive with one or more other color additives or noncolored diluents, without a chemical reaction. As noted above, each of the "food ingredients" in the color additive mixture must conform to the conditions of use in the applicable regulation or be GRAS for the intended use in order for the color additive mixture to be lawful. Examples of color additive mixtures include food inks used to mark confectionery, and inks for marking fruits and vegetables and for coloring shell eggs.

In 1998, FDA published a notice announcing that a petition had been filed to amend the color additive regulations to allow the safe use of synthetic iron oxide and mica and titanium dioxide to color food at levels higher than the current limits. At the time of filing the notice for the color additive petition (CAP 8C0262), FDA considered the pigments to be color additive mixtures of synthetic iron oxide, mica, and titanium. During its review of the petition, FDA determined that the petitioned pigments are composite pigments, not color additive mixtures. Composite pigments are manufactured, for example, by coating the substrate mica with titanium salts, which precipitate on the surface as titanium hydroxide in basic solution. Upon heating, titanium hydroxide is converted to titanium dioxide resulting in titanium dioxide coated mica-based pearlescent pigments. Subsequently, FDA amended the filing notice for the petition to clarify that the petition proposed to amend the color additive regulations to allow the use of composite pigments prepared from synthetic iron oxide, mica, and titanium dioxide to color foods. In a partial response to the petition (71 FR 31927), only the composite pigments prepared from mica and titanium dioxide were approved to color food. FDA has listed mica-based pearlescent pigments for safe use in foods in 21 CFR 73.350. Mica-based pearlescent pigments are also listed for use in contact lenses (21 CFR 73.3128) and ingested drugs (21 CFR 73.1128). As color additives, the use of composite pigments is subject to premarket review and approval.

23.6
GRAS Substances

In defining a food additive, the FD&C Act states that "substances that are generally recognized, among experts qualified by scientific training and experience to evaluate their safety as having been adequately shown ... to be safe under the conditions of their intended use" are excluded from the definition. Put simply, substances that are GRAS under conditions of their intended use are not food additives and do not require premarket approval by FDA. Irrespective of whether a substance is deemed to be GRAS or if its safety is established through a premarket approval process, the safety determination is always limited to the substance's intended conditions of use.

The difference between a GRAS determination and a premarket approval relates to who has access to the scientific data and information and who has reviewed the scientific data and information. For a substance to be GRAS, the scientific data and information about the use of a substance must be widely known and there must be a consensus among qualified experts that those data and information establish that the substance is safe under the conditions of its intended use. GRAS determinations made in this manner are said to have been arrived at through scientific procedures, and any person can make a GRAS determination. For a food additive, proprietary data and information about the use of a substance are sent to FDA for safety evaluation under the conditions of its intended use (21 CFR 171.1). Thus, for a food additive, FDA determines the safety of the ingredient, whereas qualified nongovernment experts can determine that an ingredient is GRAS.

There is, however, an additional way that a GRAS determination can be made. For a substance used in food before 1958, a GRAS determination can be made through experience based on common use in food that requires a substantial history of consumption by a significant number of consumers (21 CFR 170.30(c) and 170.3(f)). This basis for GRAS determination is seldom relied on today.

The use of two specific long chain polyunsaturated fatty acid (LCPUFA) families in infant formula serves as an example of a "balanced" mixture. FDA has responded to several GRAS notices for a "balanced" use of an ingredient that is a source of arachidonic acid (ARA, a ω-6 family fatty acid) with an ingredient that is a source of docosahexaenoic acid (DHA, a ω-3 family fatty acid) in infant formula.

The source of the ARA and that of the DHA provides a ratio of the ARA to the DHA in the infant formula, thus balancing the ω-6 fatty acid with the ω-3 fatty acid. In the agency's evaluation of the notices for a certain source of ARA in combination with a certain source of DHA, FDA evaluated the data and information for the ARA source separately from that for the DHA source. Importantly, the GRAS determination was for the combination of these two ingredients. In the case of the GRAS Notices (GRNs) concerning the use of a certain source of ARA in combination with a certain source of DHA, the FDA responded by stating that the agency does not question the basis for the GRAS determinations. For details see response letters to GRNs 41 and 94 listed on the Internet at http://www.fda.gov/FoodGRASListing.

Certain other ingredients comprised of LCPUFAs (e.g., menhaden oil and substitutes for menhaden oil) are not intended for use in infant formula. These ingredients generally contain both DHA and eicosapentaenoic acid (EPA); DHA and EPA are both ω-3 fatty acids. Thus, the combined use of a balance of ARA with DHA (i.e., a specific ω-6 fatty acid with a specific ω-3 fatty acid) is a special consideration related to their intended use in infant formula that does not apply to other uses of these fatty acids as ingredients in food.

23.7 Flavorings

Individual flavoring substances, including many that are identical to chemicals found in nature as integral components of food, comprise a large class of food ingredients. They impart desirable organoleptic qualities and distinctive flavor characteristics in certain foods when present naturally or added intentionally. Most are added to food at low levels and are often combined for a technical effect in food. Most flavors do not have structural alerts for toxicity.

There are more than 2000 single-chemical entities commonly used as flavors in the United States and more worldwide to accommodate regional tastes. Nearly all of these flavors have been evaluated individually; however, in practical use, flavors are mixtures. Flavors are compounded to contain several individual flavoring substances. Specific formulations are generally well-guarded secrets.

FDA has published regulations for flavors in the US Code of Federal Regulations. Some flavors are listed in the food additive regulations (21 CFR 172.510 and 21 CFR 172.515) and some are listed in the GRAS regulations (21 CFR Part 182 and Part 184). In 1958, based on common use in food, the FDA published a GRAS list of 188 traditional herbs and spices followed by a second GRAS list of 27 synthetic, single chemical entities in 1960, based on their long history of "common use in food."

In the same year, the Flavor and Extracts Manufacturers Association (FEMA) of the United States established a panel of experts, qualified by scientific training and experience, for establishing the criteria and procedures to evaluate the safety of flavoring substances under the conditions of their intended use. Since 1965, FEMA has been publishing its GRAS determinations. At present, the most recent FEMA GRAS list (GRAS 24) was published in 2009 [8].

JECFA is an international expert scientific committee administered jointly by FAO and WHO charged with providing safety evaluations for food additives and contaminants. In 1995, at its 44th meeting, JECFA began an extensive evaluation of chemically related groups of flavorings using a decision-tree approach that is consistent with the FEMA approach and based in part on the scientific principles laid down and elaborated by Munro *et al.* [9, 15]. For each flavor evaluated, the JECFA procedure integrates information on human intake, chemical structure and structure–activity relationships, metabolism, and toxicity [10]. To date, JECFA has evaluated over 1800 flavors. The procedure has provided a practical and sound approach to evaluate the large number of flavoring substances in current use. In these evaluations, potential synergistic effects of various substances combined in flavor mixtures in practical use are not evaluated; however, JECFA does consider potential concern due to the combined intake of related substances.

23.8 Natural Flavor Complexes

FEMA has also developed guidelines for the safety evaluation of natural flavor complexes (NFCs) [11]. NFCs are mixtures of low molecular weight substances obtained from plants by physical processes such as distillation or extraction with water or organic solvents. The NFCs are similar to compounded flavors in that they contain several to many individual flavoring substances such as essential oils, extracts, and oleoresins. These materials may or may not be chemically well defined and characterized. In an evaluation, the components of an NFC are organized into congeneric groups that are expected to exhibit similar metabolic and toxicologic properties [12]. The approach builds upon the evaluations that have been completed by JECFA, in that the groupings are the same as the chemically related groups of flavors that JECFA has evaluated over the past 10 years. JECFA has not yet taken on the evaluation of NFCs to any great extent.

Citrus flavors are among the most popular fruit flavors for beverages. Some citrus varieties such as lemon and grapefruit contain one or two major flavor-impact compounds, whereas others, such as orange and mandarin, contain a mixture of several components responsible for their basic flavor.

Lemon peel oil is a good example of a natural flavor complex widely used in a variety of foods to provide the basic flavor and aroma of lemon. The quality of lemon oil has been based on the total aldehyde content, a measure of its main constituent, citral. Chemical analysis of lemon peel oil reveals that citral is a mixture of two isomers, neral and geranial, which together constitute the basic flavor-impact compound in lemon. However, other components such as β-pinene, γ-terpinene, bergamotene, and caryophyllene may modify the basic citral flavor [13].

On the other hand, orange flavor is the result of a combination of several classes of volatile components such as terpenes, aldehydes, esters, and alcohols that contribute to the distinct flavor of orange. Analysis shows that some of the flavor components are limonene, myrcene, α-pinene, valencene, citral, neral, geranial, α-sinensal, β-sinensal, and ethyl-3-hydroxyhexanoate. Flavorists try to mimic these natural flavors and use the components in different combinations to create new and better tastes. The best mixtures for providing orange flavor contain flavor-impact chemicals such as D-limonene, α-pinene, and citral and contributory chemicals such as ethyl butanoate, acetaldehyde, nonanal, and octanal. Varying the proportion of these components in a mixture determines the precise orange flavor.

23.9
Botanical Ingredients

Botanicals/herbs are another example of mixture-type substances used as ingredients in conventional food. Botanicals as food ingredients may be in various forms including whole plant, plant parts, or preparations from these materials such as extracts. These substances present a number of unique problems with respect to safety assessment, a key problem being their compositional complexity. There are a large number of broad categories and subcategories of plant constituents, some potentially toxic, produced as secondary metabolites. Examples include pyrrolizidine alkaloids and other alkaloids, phenolics, terpenes, steroids, carotenoids, flavonoids, and cyanogenic glycosides, and other glycosides. Information on these various chemical groups as constituents can be partly used to make a judgment about safety of a given "whole" botanical material intended for use as a food ingredient. Another challenge is the consideration of various additional issues related to composition and identity of a given botanical material. Examples of such issues include taxonomic uncertainty, seasonal variation of constituent levels, stability, effect of processing and manufacturing methods, nature and amount of excipients, and specificity of a chemical constituent to a certain plant part. Given the various complicating factors that arise in the safety assessment of these substances, a recent guidance document published by an expert group of the Natural Toxin Task Force of the European Branch of the International Life Sciences Institute proposed the use of a decision tree as a guidance tool in determining the information that needs to be considered [14]. This guidance is a useful tool for dealing with the challenges involved in safety assessments for botanicals.

The bulk of botanical substances that FDA has formally issued a ruling on with respect to use in conventional food are listed in flavor and spice regulations (21 CFR 172.510, 182.10, and 182.20). FDA's ruling on the safety of these substances is based primarily on the history of safe use in food and on the use being at low levels; essentially self-limiting quantities are involved when these substances are used in food consistent with good manufacturing practices. In the case of those botanicals that are known to produce a specific constituent with known toxicity, the regulations specify limitations on use conditions to address the hazard potential. Examples include, a thujone-free limitation for the finished food product in the case of use of Artemisia (wormwood), cedar (white), oak moss, tancy, and yarrow; a safrole-free limitation in the case of camphor tree and Sassafras leaves; and a Hypericin-free limitation in case of St. John's wort. These limitations were imposed because thujone is a neurotoxic substance, safrole is a suspected carcinogen, and hypericin is a photosensitizing agent. Plant part limitations and food category limitations also apply to some of the botanicals listed in the regulations.

At present, the most widespread use of botanicals/herbs is in dietary supplement products. However, it should be noted that the standard of safety and the burden of proof to demonstrate safety is different for ingredients used in dietary supplements and conventional foods. As stated above, the laws and regulations on food additives provide for a showing of safety before a substance can be intentionally added to a conventional food, unless the substance is GRAS. In the case of dietary supplements, the Dietary Supplement Health and Education Act (DSHEA) of 1994 excluded dietary ingredients in dietary supplements from the definition of a food additive and, hence, from any requirement for premarket approval by the FDA. Under DSHEA, the burden is placed on the FDA to prove that a dietary supplement presents a significant or unreasonable risk of illness or injury or, in the case of a new dietary ingredient (a dietary ingredient that was not marketed in the United States before October 15, 1994), that there is inadequate information to provide reasonable assurance that the ingredient does not present a significant or unreasonable risk of illness or injury.

Botanicals/herbs may also be used in the so-called "functional food" products. At present, FDA has neither a definition nor a specific regulatory rubric for foods being marketed as "functional foods." FDA regulates conventional foods being marketed as "functional foods" under the same regulatory framework as other conventional foods. The agency is confident that the existing provisions of the FD&C Act are adequate to ensure that conventional foods being marketed as "functional foods" are safe and lawful. Thus, any ingredient in a "functional food" needs to be safe and lawful. In conducting a safety assessment for a food ingredient, the purported benefits of a "functional" ingredient are not considered although health claims may be considered by FDA in a separate petition process.

23.10 Food Contact Substances/Formulations

In 1997, the Food and Drug Administration Modernization Act (FDAMA) amended the FD&C Act to streamline the way in which the FDA conducted business. One of the new procedures established to accomplish this goal was a notification process for food contact substances. The amended FD&C Act (Section 409(h)(6)) defined a food contact substance as "any substance intended for use as a component of materials used in manufacturing, packing, packaging, transporting, or holding food if such use is not intended to have any technical effect in such food." Examples of food contact substances include polymers (plastic packaging materials), pigments, and antioxidants used in polymers, can coatings, adhesives, materials used during the manufacture of paper and paperboard, slimicides and biocides (antimicrobial agents), and sealants for lids and caps.

A notification for a food contact substance formulation is only meant to verify that components of a food contact material may legally be used; they are not meant to authorize the use of a new food contact substance. The basis for determining compliance for each component in the formulation may be an effective food contact notification or an existing regulation. In either case, the component must meet the identity criteria, specifications, and limitations on the conditions of use already established.

Aqueous solutions of sodium chlorite provide an example of a food contact formulation used as an antimicrobial processing agent consisting of a mixture of individual chemicals that are applied to red meat (Food Contact Notification 450) and poultry (Food Contact Notification 739). The aqueous solution can be mixed with any GRAS acid to achieve a pH of 2.2–3.0, then the solution can be further diluted with a pH-elevating agent. When used according to the specifications and limitations of each notification, FDA allows its use.

23.11 Conclusions

Whole foods are presumed to be safe under the FD&C Act. FDA regulates ingredients that are added to conventional food individually. For a product that is composed of a mixture of several ingredients,FDA would consider the regulatory status of each ingredient. For an ingredient that is a food additive or color additive, the conditions of use must conform to the conditions for safe use set forth in the applicable regulation. If a GRAS determination has been made by either FDA or an independent party (and the intended use of an ingredient is truly GRAS), the ingredient can be lawfully added to conventional food. If each ingredient in a mixture is used within the limits of a food additive or color additive regulation or the intended use is GRAS, a firm that is marketing the mixture does not need to notify FDA. The FD&C Act prohibits marketing of conventional foods that contain ingredients that are not GRAS, approved food additives, or approved color additives. Such foods are subject to the adulteration and misbranding provisions of the FD&C Act.

Individuals who wish to determine if FDA has a regulation for a specific food additive can view a list of food additives and their corresponding regulations at www.fda.gov/FoodEAFus or www.fda.gov/~dms/FoodIndirectAdditive. The individual regulations for food additives and GRAS substances listed for use in the United States are located in Title 21 CFR Parts 172–186 and can be accessed through CFSAN's web site. The color additive regulations, listings, and specifications can be found in Title 21 CFR Parts 70–82. However, not all substances that are GRAS under specific conditions of use are listed in FDA's regulations. FDA has a program whereby companies may inform FDA of their own GRAS determination (GRAS Notification Program). A list of these GRAS notices, with FDA's response letter to the notifier, can be found at www.fda.gov/~rdb/FoodGRASListing.

The approach to assess the safety of food ingredients cannot be a stagnant process. For all ingredients added to food, safety must be reconsidered when significant advances in scientific data and testing methodology occur. For complex mixtures, future challenges extend to analytical analyses, safety testing strategies, and estimation of dietary intake. There is a need for more progress in the ability to assess the safety of mixtures, including simple or complex mixtures of ingredients that are intended for use in food.

Acknowledgments

The authors would like to thank Dr. Dennis Keefe of the Food and Drug Administration's Center for Food Safety and Applied Nutrition, for this cooperation by reviewing and commenting on this chapter. A special thanks to Dr. Antonia Mattia for focusing the content of this chapter in terms of the regulatory process for food ingredients and the challenges that mixtures present in that regulatory climate.

References

1 Jacobson-Kram, D. and Keller, K.A. (eds) (2006) *Toxicological Testing Handbook Principles, Applications, and Data Interpretation*, Informa Healthcare, New York.

2 US Food and Drug Administration (1982) *Toxicological Principles for the Safety Assessment of Direct Food Additives and Color Additives Used in*

Food (the Redbook), Food and Drug Administration, Washington, DC.

3 Rulis, A.M., Hattan, D.G., and Morgenroth, V.H., III (1984) FDA's priority-based assessment of food additives I. Preliminary results. *Regul. Toxicol. Pharmacol.*, **4**, 37–56.

4 Benz, R.D. and Irausquin, H. (1991) Priority-based assessment of food additives database of the U.S. Food and Drug Administration Center for Food Safety and Applied Nutrition. *Environ. Health Perspect.*, **96**, 85–89.

5 Valerio, L.G., Arvidson, K.B., Chanderbhan, R.F., and Contrera, J.F. (2007) Prediction of rodent carcinogenic potential of naturally occurring chemicals in the human diet using high-throughput QSAR predictive modeling. *Toxicol. Appl. Pharmacol.*, **222**, 1–16.

6 Magnuson, B.A., Burdock, G.A., Doull, J., Kroes, R.M. *et al.* (2007) Aspartame: a safety evaluation based on current use levels, regulations, and toxicological and epidemiological studies. *Crit. Rev. Toxicol.*, **37** (8), 629–727.

7 JECFA (2001) Evaluation of Certain Food Additives and Contaminants: Forty-Forth Report of the Joint FAO/WHO Expert Committee on Food Additives. WHO Technical Series No. 901, pp. 2–13.

8 Smith, R.L., Waddell, W.J., Cohen, S.M., Feron, V.J., Marnett, L.J., Portoghese, P.S., Rietiens, I., Adams, T.B., Lucas Gavin, C., McGowen, M.M., Taylor, S.V., and Williams, M.C. (2009) GRAS flavoring substances 24. *Food Technol.*, **63** (6): 46–81.

9 Munro, I.C., Kennepohl, E., and Kroes, R. (1999) A procedure for the safety evaluation of flavouring substances. *Food Chem. Toxicol.*, **37**, 207–232.

10 JECFA (1995) Evaluation of Certain Food Additives and Contaminants: Forty-Fourth Report of the Joint FAO/WHO Expert Committee on Food Additives, WHO Technical Report Series No. 859, pp. 1–53.

11 Smith, R.L., Cohen, S.M., Doull, J., Feron, V.J., Goodman, J.I., Marnett, L.J., Portoghese, P.S., Waddell, W.J., Wagner, B.M., Hall, R.L., Higley, N.A., Lucas-Gavin, C., and Adams, T.B. (2005) A procedure for the safety evaluation of natural flavor complexes used as ingredients in food: essential oils. *Food Chem. Toxicol.*, **43**, 345–363.

12 JECFA (2004) Evaluation of Certain Food Additives and Contaminants: Sixty-First Report of the Joint FAO/WHO Expert Committee on Food Additives, WHO Technical Report Series No. 922, pp. 1–164.

13 Shaw, P.E. (1991) Fruits, in *Volatile Compounds in Foods and Beverages* (ed. H. Maarse), Marcel Dekker, Inc., New York.

14 Schilter, B., Andersson, C., Anton, R., Constable, A., Kleiner, J., O'Brien, J., Renwick, A.G., Korver, O., Smit, F., and Walker, R. (2003) Guidance for the safety assessment of botanicals and botanical preparations for use in food and food supplements. *Food Chem. Toxicol.*, **4**, 1625–1649.

15 Munroe, I., Ford, R., Kennepohl, E., and Sprenger, J. (1996) Correlation of Structural Class with No-Observed-Effect Levels: A Proposal for Establishing a Threshold of Concern. *Food Chem. Toxicol.*, **34**, 829–867.

24
Biomonitoring[1)]

Richard Y. Wang, P. Barry Ryan, and Moiz Mumtaz

24.1
Introduction

Environmental contamination, whether through polluted air, contaminated food or water, or contact with contaminated soil, represents a significant source of exposure to individuals. Everyday activities result in exposure to a vast array of different chemical compounds (Table 24.1). Air may be contaminated with gaseous and particulate-borne pollutants. Water may be contaminated with industrial or agricultural runoff. Food may be contaminated with agricultural chemicals including pesticides and fertilizers. Soil to which we come in contact may be contaminated with various organic and metallic compounds. As outlined in Table 24.1, these contaminants are known to produce adverse health outcomes when examined individually. As mixtures, the effects may be purely additive or they may act synergistically or antagonistically to produce further health concerns. In order to assess the extent to which humans are affected by the exposure to these chemicals, it is necessary to quantify the amount of exposure. The assessment of human exposure to environmental chemicals may be accomplished by various mechanisms, including environmental monitoring, questionnaires, personal monitoring, and biomonitoring.

Biomonitoring is the assessment of human exposure to chemicals by measuring the internal dose of the chemical in the body [1]. The internal dose is the amount of the chemical that is absorbed by the body. It is commonly determined as a measurement of the chemical or its metabolite in blood or urine, although other biological matrices and analytes, such as biomolecular adducts, may be used. These analytes are known as biological markers (biomarkers), which can be broadly defined as almost any measurement reflecting an interaction between a biological system and an environmental agent, which may be chemical, physical, or biological. The focus of this chapter is on chemical agents, although the methods developed

1) The findings and conclusions in this chapter are those of the authors and do not necessarily represent the official position of Centers for Disease Control and Prevention/Agency for Toxic Substances and Disease Registry.

Principles and Practice of Mixtures Toxicology. Edited by Moiz Mumtaz

ISBN: 978-3-527-31992-3

Table 24.1 Selected examples of chemical mixtures in the environment from a common pathway.

Chemical mixture	Medium	Measurement system	Biological specimens or markers	Health outcomes or effects
Air pollution: benzene, toluene, ethyl benzene, and xylenes (BTEX)	Air	Passive systems, active measurements [41]	Breath, blood, and urine (parent chemical or metabolites) [42]	Carcinogenic effects [10, 11]
Air pollution: National Ambient Air Quality Standards (NAAQS) contaminants	Air	Numerous methods [43]	Generally, not available for gases and particulate matter. Blood for lead [43]	Local tissue irritation, respiratory effects, and cardiovascular effects [43]
Non-BTEX hazardous air pollutants (HAPs)	Air	Whole air sampling[b)]	Specific for the chemical[b)]	Various. Many have carcinogenic effects[a)]
Particulate matter	Air	Various. Usually pump and filter, followed by specific chemical or chemical class analysis [43, 44]	At present not well established. Tissue pathology shows uptake of particles [43]	Respiratory effects and cardiovascular effects [43]
Environmental tobacco smoke	Air	Various. Depends on the contaminant in ETS, such as particles and semivolatile and volatile chemicals. Nicotine is used as a marker of exposure [45]	Specific for the chemical. Carbon monoxide in exhaled breath is an indicator as is nicotine (cotinine) in blood, urine, and saliva. Particulate matter includes metals such as cadmium, which can be measured in blood and urine [46]	Various. Respiratory disorders include cancer, infections, and exacerbation of asthma. Cardiovascular effects include coronary heart disease [47]

Dietary intake: agricultural products	Food	Various. Depends on the contaminant, such as pesticides, polycyclic aromatic hydrocarbons, and phthalates from packing materials [48]	Generally, serum or urine (for metabolites) [36]	Neurological effects from organophosphorous and carbamate insecticides. Some chemicals have endocrine or hormonal activity [49, 50].
Dietary intake: methylmercury and persistent organic pollutants	Food: fish and marine mammals	Various. Depends on the contaminant, such as methylmercury, polychlorinated biphenyls, and related chemicals [51]	Specific for the chemical. Blood for mercury, and serum, human milk, or adipose tissue for PCBs and related chemicals [52]	Neurodevelopmental effects from methylmercury and PCBs. Dioxins, dioxin-like compounds, and PCBs have carcinogenic effects [10, 11, 53]
Dietary intake: persistent organic pollutants	Food: meats, dairy products, and human milk (infants)	Various. Depends on the contaminant, such as organochlorine insecticides, PCBs, polybrominated biphenyl ethers, and related compounds in human milk [54]	Serum and adipose tissue for PCBs and related chemicals. Direct sampling for human milk [55]	Dioxins, dioxin-like compounds, and PCBs have carcinogenic effects [10, 11]
Dietary intake: disinfectant byproducts	Drinking water	Disinfection byproducts including several trihalomethanes and other organic chemicals [56]	Exhaled breath and blood [57]	Some have carcinogenic effects [10, 11, 57]

The table includes identification of the mixture, its primary medium, potential measurement systems, potential biological markers or specimens, and potential health outcomes or effects associated with the exposure to the mixture.

a) There are 188 HAPs and they have specific characteristics. The reader may refer to the Agency for Toxic Substances and Disease Registry's Toxicological Profiles for specific measurements, biological markers, and health outcomes or effects associated with these chemicals [58].

here are applicable in other areas as well. Three classes of biomarkers are normally considered [2, 3].

- **Biomarkers of exposure** are exogenous substances, their metabolites, or the product of an interaction between a xenobiotic agent and some target molecule or cell that is usually measured in body fluid (blood, urine, and saliva) or, occasionally in less easily available tissues including adipose tissue and placenta.
- **Biomarkers of effect** are measurable biochemical, physiological, behavioral, or other alterations within an organism that, depending upon the magnitude, are recognized as being associated with an established or possible health impairment or disease (e.g., acetylcholinesterase inhibition, cognitive deficit, or neurological effect).
- **Biomarkers of susceptibility** are indicators of an inherent or acquired ability of an organism to respond to the challenge of exposure to a specific xenobiotic substance. Typically, these markers are identified through gene polymorphisms.

The first two definitions link exposures to health outcomes and can provide the basis for establishing a link between biological effect and exposure to contamination in the environment. The third refers to a modifier that influences the magnitude of the effect given a fixed magnitude of the exposure. All three types of biomarkers may be used to identify vulnerable individuals or populations.

A significant advantage of the internal dose over other measures of human exposure to environmental chemicals, such as an applied dose, is that it decreases the level of uncertainty in the assessment of the biologically effective dose and it can assess for multiple chemical exposures when the pathways are not well defined (Figure 24.1). For example, additional considerations (e.g., exposure route, bioavailability) needed to estimate the biologically effective dose from the applied dose will contribute to the variability in the final estimate compared to the estimate derived from the internal dose. Because the internal dose is a measurement of exposure from all routes and pathways, it is considered an integrated form of exposure assessment.

The various applications of biomonitoring include the characterization of the population's level of exposure to the chemical, the type of chemical exposure, and the characterization of risk associated with the exposure to chemicals with known health risks. Such information can be used to assess the efficacy of a change in the use of a chemical by monitoring the population's exposure to the chemical over time, such as lead [4, 5] and perfluoroalkyl chemicals [6], or to identify subgroups in general

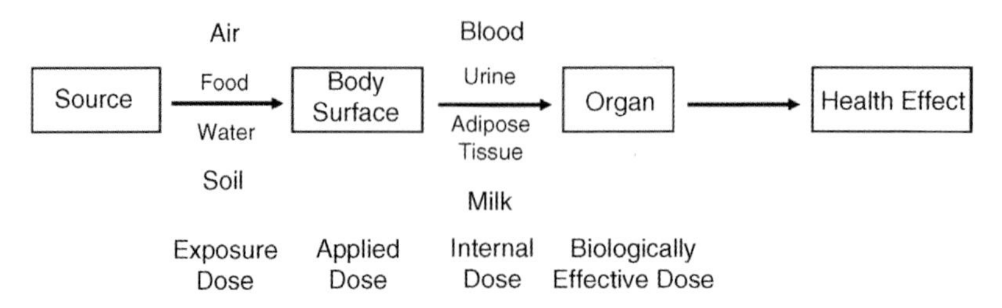

Figure 24.1 Schematic of the environmental exposure and health effect pathway.

population that have higher levels of exposure to a chemical than other subgroups. For example, Mexican Americans were found to have higher serum 1,1′-(2,2-dichloroethenylidene)-bis[4-chlorobenzene] (DDE) concentrations than non-Hispanic blacks and non-Hispanic whites in the population sampled for National Health and Nutrition Examination Survey (NHANES) in 2001–2002 [7]. NHANES is based on a complicated design that is representative for age, gender, and race/ethnicity (Mexican Americans, non-Hispanic blacks, and non-Hispanic whites) in general US population. In addition, biomonitoring can estimate the percentage of the population at risk from the exposure to a defined health threat, such as lead and mercury. For example, 2.2% of children aged 1–5 years had an elevated blood lead concentration (i.e., greater than 10 mcg/dl) [8], which represents a decrease from 4.4% in the early 1990s. In the general US population sampled from 1999 to 2002, 5.7% of women of reproductive age had a total blood mercury concentration within 10-fold of 5.8 mcg/dL, which is the concentration associated with neurological effects in the newborn from fetal exposure [9].

Because biomonitoring is an integrated form of exposure assessment, it is generally unable to discern exposure pathways, routes, or timing of exposure without additional information. The pathway and timing of exposure to a chemical are important for risk assessment because toxicity can differ by the route of exposure (i.e., bypass first-pass metabolism), and a measured exposure that is found to be “low” does not provide a complete picture as the value may be low after a large exposure source has ceased operations or low from the start. The temporal relation between the exposure and dose differs substantially in these instances. Thus, it is important to characterize the exposure profile for a chemical mixture along the primary exposure pathway when assessing the risk of exposure to multiple chemicals from biomonitoring data because this approach defines the relation between timing and dose based on the primary pathway, such as food, water, or air. Otherwise, the risk assessment of chemical mixtures with multiple routes of exposure and variable times of exposure can be very difficult. This is more of a challenge for chemicals with short biological half-life than those with a long half-life because of the possibility for frequent oscillations in the chemical concentration in the former setting. In situations when there is a constant level of background exposure to nonpersistent chemicals, the biological concentration may remain relatively steady. When the primary exposure pathway is uncertain for several chemicals, biomonitoring can be the most reliable approach to assess human exposure to chemical mixtures because it cannot be predicted.

The significance of environmental and personal monitoring in the assessment of human exposure to chemical mixtures is to characterize the movement of the chemicals from the source to the individual along the primary pathway. The magnitude, frequency, and duration of exposure are needed to assess the exposure profile for the chemical. These data will provide information on pathway and timing of exposure, which are not available from biomonitoring data. When dealing with mixtures of contaminants, the concept of implementation of exposure assessment strategies becomes more complex than for a single contaminant. Instead of an individual compound, one may consider a class of compounds with similar use, for example, organophosphate insecticides, or all contaminants coming from a particular source, say, motor vehicle exhaust. In examining the exposure profile for a mixture, one must be cognizant of the possibility that the composition of the mixture may change with time. Some examples of chemical mixtures that follow a common pathway existing in the environment that are of interest to exposure assessors are demonstrated in Table 24.1. They are discussed in detail from the perspective of an exposure assessor trying to understand the exposure pathway and the health impact of these mixtures in the following section.

24.1.1 Air Contaminants

Criteria air pollutants in the ambient atmosphere, such as ozone and nitrogen dioxide, display diurnal variability with ozone peaks delayed with respect to nitrogen dioxide peaks as the former is a secondary contaminant created by the interaction of nitrogen dioxide, sunlight, and other pollutants in the atmosphere. Furthermore, there may be seasonal variation. In the case just cited, nitrogen dioxide concentrations may be relatively constant over an annual cycle while ozone concentrations peak in the summer months. This is due to the changing intensity of sunlight, which is a necessary component for ozone formation. Motor vehicle exhaust is another example of a group of chemical mixtures from petroleum products. Combustion byproducts from gasoline and diesel engines are primarily air contaminants that are dispersed by advection from the tailpipe source of motor vehicles eventually reaching the breathing zone of an individual where they are subsequently inhaled and reach the lung epithelium. During their travel along this pathway, these chemicals encounter other materials and may react with them to be degraded, or due to their own physical properties may behave differentially, suggesting that just monitoring the tailpipe may be insufficient in deducing the composition of the "aged" mixture that reaches the lung epithelium. This change in the chemical composition in the atmosphere is critical in understanding the likely health impact of this mixture because, for example, benzene is a known human carcinogen, while the other components are not [10, 11]. Measurement is relatively simple for these chemicals as they are readily absorbed on charcoal and can be extracted easily for analysis. This suggests measurement at the individual level (i.e., personal monitor) using a simple sampler is a good strategy for assessing the exposure profile for these chemicals.

A mixture with great importance in indoor environments is environmental tobacco smoke (ETS). ETS is the combination of side-stream and exhaled mainstream tobacco smoke that contains literally hundreds of chemicals ranging from volatile gases through semivolatile compounds to particulate matter. These chemicals include carcinogens, such as benzene and polycyclic aromatic hydrocarbons (PAHs) (e.g., benzo[*a*]pyrene), metals (e.g., cadmium and polonium-210), and toxic gases (e.g., carbon monoxide and hydrogen cyanide). ETS is considered a known human carcinogen [10, 11]. In some sense, this treats the ETS mixture as a single pollutant, which it clearly is not, but considers only one potential health outcome, cancer. The noncancer diseases and disorders related to ETS include coronary heart disease, sudden infant death, respiratory infections, and asthma exacerbations in children [12], which may display other options for the risk assessor interested in determining the overall health risk to this complex mixture. Physical properties common to fresh ETS change with time, a process known as "aging." Residual ETS particulate matter deposited on surfaces is characterized by a yellowish sticky residue consistent with fairly high concentrations of nicotine and related compounds, and volatile gases remain in the air as do small particles with semivolatiles deposited on the outside. The risk characteristics associated with the aged mixture are different than those of fresh ETS. Thus, the exposure assessor interested in establishing the exposure to ETS must determine which "type" of ETS to monitor, fresh or aged, as monitoring methods differ for these components. Similarly, the risk assessor must evaluate which of these components is considered most relevant to the health assessment.

24.1.2
Food Contaminants

Food is another environmental medium that is in the exposure pathway for several chemicals, including pesticides, agricultural fertilizers, industrial chemicals, or materials contaminating the food due to packaging. The type of chemicals found in food depends on the type and origin of the food. For example, nonorganically produced agricultural products in the Unites States often contain residues from commonly used pesticides such as organophosphates and pyrethroids. Permethrin is a commonly used pyrethoid on agricultural crops in the United States and it is produced as a mixture of different isomers (e.g., *cis*- and *trans*-permethrin). The *cis* isomer of permethrin has more insecticidal activity than the *trans* isomer. While the concentrations of such residues in food are usually small, more than 60% of the NHANES participants sampled in 2001–2002 had measurable concentrations of dialkyl phosphates and 3-phenoxybenzoic acid in their urine [7]. Pyrethroids are also used to control pests in the residential setting and by public health programs, which may contribute to human exposure to these chemicals.

Chemical mixtures, such as PAHs and heterocyclic amines (e.g., 2-amino-3-methylimidazo[4,5-*f*]quinoline, 2-amino-3,4-dimethylimidazo[4,5-*f*]quinoline, 2-amino-3,8-dimethylimazo[4,5-*f*]quinoxaline, and 2-amino-1-methyl-6-phenylimidazo[4,5-*b*]pyridine), may also be introduced to foods during cooking and processing. Packaging materials, most notably plastic containers and plastic wrapping materials, may leach phthalates, which are plasticizers that make these products more flexible. Phthalates, such as di-2-ethylhexyl phthalate (DEHP), have hormonal activity that may be additive [13] and of concern to young children and women who are pregnant and breastfeeding [14]. The amount of these chemicals found in food depends on numerous factors, including source of the foods, shipping processes, packaging, and the nature of the food itself; high-fat foods tend to absorb more organic contaminants than do foods with high water content. In addition to lipid solubility, chemicals resistant to biological metabolism and environmental degradation tend to bioaccumulate in the food chain and are more likely to enter humans by this pathway. For example, up to 90% of human exposure to dioxin-like polychlorinated biphenyls (PCBs), polychlorinated-*p*-dibenzodioxins (PCDDs), and polychlorinated dibenzofurans (PCDFs) is from fatty foods, including meats, poultry, dairy products, and certain types of fish [15].

Fish has received a good deal of attention in the popular press and in the scientific community. Women of child-bearing age and, particularly, pregnant women are cautioned to limit their intake of large fishes (e.g., shark, swordfish, king mackerel, tilefish) due to the likely presence of mercury, which can lead to neurodevelopmental effects in newborns [16]. However, the risk assessment is more complicated because many of these same fish are a primary source of omega-3 oils (e.g., docosahexaenoic acid), which have been shown to benefit neural and retinal development in the fetus [17]. Marine mammals as a food source deserve the same consideration because they contain high concentrations of persistent organic pollutants (e.g., PCBs, PCDDs/PCDFs, and selected organochlorine insecticides) and are an essential source of high energy for certain communities.

Fish consumption may also lead to exposure to other chemicals with noncarcinogenic effects as they are in intimate contact with water and sediments, and they tend to bioaccumulate organic materials found therein. Contamination of the Hudson River in New York, the Housatonic River in Massachusetts, and the US Great Lakes with PCBs has resulted in increased concentration of these chemicals in fish. PCBs have been shown to modulate the neurodevelopment of pups by lowering serum thyroid hormones through *in utero* administration [18]. These findings may result from the interruption of thyroxine transport by the preferential binding of hydroxylated PCB metabolites to transthyretin. Because of these health concerns to the developing fetus, numerous locations in the United States have issued fish advisories on eating fish caught in contaminated waters.

Based on the above information, the exposure or risk assessor who wishes to understand the likely impact of dietary exposure to chemicals producing adverse health outcomes must appreciate the complexity of the mixtures found in food. Furthermore, they must be cognizant of which foods are likely to contain which chemicals and the appropriate chemical form or species. For example, inorganic arsenic may be prevalent in agricultural crops and species that come in contact with groundwater containing this naturally occurring mineral. Although inorganic arsenic is associated with peripheral vascular disease (blackfoot disease) and a known human carcinogen [10, 11], organic arsenic is relatively nontoxic and is commonly found in shellfish and certain other fish species. Thus, an exposure or risk assessor confusing these two species would substantially misclassify the risk associated with exposure to these chemicals.

24.1.3
Water Contaminants

Water is another environmental medium that contains mixtures of chemicals. Agricultural runoff, mainly pesticides and fertilizers, can contaminate surface water as can industrial runoff or runoff from city streets, which may contain industrial chemicals, metals, oils, and gasoline. The chemicals found in municipal water supplies may be from surficial contamination as mentioned above or through deepwater contamination, such as inorganic arsenic found most notably in Bangladesh and Taiwan, as well as in the western United States. Other solubilized metals, such as tungsten, may be evident depending on the location [19]. Perchlorate and nitrate are water contaminants that competitively block iodide uptake at the thyroidal sodium iodide symporter, which may reduce thyroid hormone production if there is sufficient inhibition of iodide uptake. In most municipal water supplies, microbial purification is effected through chlorination. However, residual free chlorine can interact with organic matter either in the raw water from the source or along the delivery route to produce disinfection by-products, including trihalomethanes, haloactic acids, haloacetonitriles, haloketones, and chlorophenols. The trihalomethanes, considered as reasonably anticipated human carcinogens, include bromodichloromethane and trichloromethane [10].

In each situation, the composition of the chemical mixture is determined by the "history" of the water, such as well water, surficial contamination, industrial contamination, or chlorine disinfection. For example, the factors determining the types of disinfectant byproducts and their concentrations measured in water include the chlorination process, the amount of disinfectant used, the amount of organic material present, the pH and the temperature of the environment where the disinfectant was used, the source of the water, and how the water was handled after it was dispensed. Each of these elements affects the complexity of the mixture in the water.

As can be seen from this discussion, the environmental pathway is an important component in understanding risk. While typical environmental contamination levels are low, the great variety of potential pathways increases the complexity of the assessment process. In addition, multiple chemicals with widely varying toxicological effects may be involved. The exposure assessor wishing to understand the impact of environmental contamination, or the risk assessor who wants to quantify the risk, must be aware of the complexities introduced by environmental contamination. The assessment of human exposure to chemical mixtures along a common primary pathway may be approached by characterizing the exposure profile by the pathway and integrating this information with biomonitoring data to assist in defining the temporal relation between the internal and the applied doses, which can be used in risk assessment. In situations when the exposure pathway is unclear, biomonitoring can serve as a useful means to assess the exposure to environmental chemicals, especially mixtures. The following section will discuss the considerations when conducting biomonitoring, such as the physical properties of the chemical, sampling, biological specimens, and data analysis.

24.2 Considerations for Biomonitoring

24.2.1 Environmental Chemicals

24.2.1.1 Persistent Organic Chemicals

These persistent organic chemicals refer to those with a long biological half-life, months to years, and they include chemicals such as the PCBs, PCDDs, PCDFs, organochlorine insecticides (e.g., dichlorodiphenyltrichloroethane, DDT), brominated flame retardants (e.g., polybrominated diphenyl ethers, PBDEs; polybrominated biphenyls), polychlorinated naphthalenes, and perfluoroalkyl chemicals (e.g., perfluorooctanesulfonic acid, PFOS; perfluorooctanoic acid, PFOA). These chemicals are readily measured in blood because of their ability to partition into subcompartments in this matrix. For example, the persistent organochlorine and organobrominated chemicals are found in blood lipids because of their lipid solubility, and the perfluorinated chemicals are found bound to blood proteins. The organochlorines (e.g., PCDDs) are more soluble in triglycerides than phospholipids because of the difference in polarity of these lipids, and the affinity to serum proteins appears to be inversely related to the length of the carbon chain of the perfluoroalkyl chemical [20]. Because neither proteins nor lipids are found in a substantial amount in urine, these chemicals are not routinely measured in this matrix. Lipid soluble chemicals can be detected in other tissues with high lipid content, such as milk and adipose tissue. These chemicals are more easily detected in tissues with high lipid content than in tissues with lower lipid content because of the higher chemical concentration in the former. For example, cord blood is of interest because it is indicative of fetal exposure, but a more sensitive analytical method is needed to measure these chemicals in this specimen compared to maternal blood because of the two- to threefold lower lipid concentration in cord blood. The route of elimination for the persistent organic chemicals from the body is biliary and includes lactation for those with increased lipid solubility.

Several of these chemicals exist as many possible congeners (e.g., 209 for PCBs, 75 for PCDDs, 135 for PCDFs, and 209 for PBDEs) and are present in the environment as mixtures because of production (e.g., PCBs, PBDEs) or they are by-products of combustion and synthetic reactions (e.g., brominated and chlorinated dioxins). In the United States, PCBs were produced as Aroclor mixtures, which varied in PCB content by order of the chlorinated homologue. For example, Aroclor 1260 contained approximately 80% of hexa- and hepta-chlorinated homologues and only 12% of penta-chlorinated homologues. The indicator PCBs (138 and158, 153, 180) are consistently detected in human populations and they have been estimated to contribute 40–60% of the total reported PCBs [21]. PCB153 has been used as a general indicator for the exposure to nondioxin-like PCBs and for total PCBs with a reported correlation of 0.90 [22]; however, in certain subgroups in the population, the observed general relation between noncoplanar and coplanar PCBs may not be as anticipated [23]. Environmental degradation and biological metabolism of higher to lower chlorinated homologues can account for differences in PCB concentrations among populations.

PBDEs are produced as mixtures, such as penta-, octa-, and deca-BDEs, which contain varying amounts of PBDE congeners by order of the brominated homologue. For example, the pentaBDE product contains mostly tetra- (i.e., PBDE 47) and penta- (i.e., PBDE 99, 100) brominated congeners, the octaBDE contains mostly hepta- and octa-brominated congeners, and the decaBDE primarily contains the deca-brominated congener, PBDE 209. The relative concentration for PBDE 99 to PBDE 47 is higher in the commercial mixture than that measured in the body, and these concentrations are generally one to two orders of magnitude higher in general US population than in European population. This difference in the PBDE concentration between these populations is attributed to the greater use of pentaBDE in the United States than in the European countries. Although the PBDEs are measured in the lipid compartment in the body, the primary pathway of human exposure to these chemicals is not attributed to food, which is contrary to the persistent organochlorine chemicals [24]. Dioxins and furans are contaminants in PCB and PBDE mixtures.

The toxicological effects of these chemicals can vary by congener, even though they belong to the same class of chemicals. For example, coplanar PCBs (e.g., 126, 169) and mono-*ortho*-substituted PCBs (e.g., 118, 156) have dioxin-like effects as defined by activity at the aryl hydrocarbon receptor (AhR), but the di- and tri-ortho-substituted PCBs do not [25]. The PCBs have additional activity through their hydroxylated metabolites, which compete with thyroxine for binding to transthyretin [26]. The dioxin-like activity for a chemical is compared with that of 2,3,7,8-tetrachlorinated dibenzo-*p*-dioxin (TCDD) by using the toxic equivalency factor (TEF), which allows the risk assessment of several chemicals with dioxin-like activity. Among the PCDD congeners, the TEF varies by several orders of magnitude. Thus, the measurement of specific congeners is more informative to the risk assessor than an aggregate value, such as an integrated or total PCB concentration.

The perfluoroalkyl chemicals are used in the synthesis of perfluorinated chemicals, such as fluoroelastomers and fluoropolymers (e.g., polytetrafluoroethylene), which are used as surfactants and surface protectants. There are several perfluoroalkyl chemicals and those that were detected in the serum of nearly all (>98%) of the NHANES participants (aged 12 years and older) sampled in 2003–2004 included PFOS, PFOA, perfluorohexane sulfonic acid, and perfluorononanoic acid [6]. Concentrations of these chemicals were significantly intercorrelated, suggesting a common exposure pathway. Environmental sources of exposure to perfluoroalkyl chemicals include production releases and environmental degradation. Developmental toxicity and various organ tumors have been reported in animal studies with PFOA and PFOS, but a common mechanism of action remains to be determined for this class of chemicals.

24.2.1.2 **Nonpersistent Organic Chemicals**

Nonpersistent chemicals have a short biological half-life, typically on the order of hours, and include contemporary pesticides (e.g., organophosphorous insecticides, carbamates, pyrethroids), plasticizers (e.g., phthalates, bisphenol A), polycyclic aromatic hydrocarbons, and volatile organic compounds (VOCs) (e.g., aromatic hydrocarbons, halogenated solvents), and environmental tobacco smoke (i.e., indexed by nicotine). Unlike the persistent organic chemicals, these chemicals may enter the body by inhalation and ingestion, and solvents may be absorbed by the body through the skin. These chemicals are typically measured in urine as metabolites because they are quickly metabolized and eliminated by the kidneys [7]. Thus, very little of the parent or original chemical can be detected in urine. More sensitive analytical methods are needed to measure these chemicals and their metabolites in blood than in urine because of their lower concentrations in the former matrix. VOCs can be measured in blood as the parent chemical, which is advantageous because some of the urinary metabolites are common to several VOCs. For example, hippurate is common to toluene and styrene, and *tert*-butyl alcohol is common to the gasoline oxygenators – methyl *tert*-butyl ether, ethyl *tert*-butyl ether, and *tert*-amyl methyl ether. Also, the source of the measured metabolite in a biological matrix needs to be distinguished from the environment as a degradate or the body as a metabolic by-product. Some of these chemicals, such as cotinine (metabolite of nicotine), have been detected at lower concentrations in saliva than in blood due to the binding to serum proteins. Phthalates are typically found in adhesives and cosmetics, and they are measured as monoester phthalates and as oxidized metabolites of the long chained dialkyl phthalates (i.e., di-2-ethylhexyl phthalate, DEHP) because of the concern for contamination during specimen collection or preparation. Organophosphorous insecticides can be measured as dialkyl phosphates and specific metabolites for certain insecticides, such as 3,5,6-trichloro-2-pyridinol for chlorpyrifos and chlorpyrifos-methyl.

The PAHs are byproducts from the combustion of fossil fuel and more than 150 of these chemicals have been found in tobacco smoke. These chemicals are converted by cytochrome P450 enzymes to several metabolites, including monohydroxylated derivatives, dihydrodiols, and reactive diolepoxides that form adducts with biomolecules, such as DNA and hemoglobin. The hydroxylated metabolites are measured in urine, such as phenanthrols and 1-hydroxypyrene, and the adducts (e.g., benzo[*a*]pyrene) are typically measured in blood components. PAHs with an increasing number of benzene rings are more likely to be a potential human carcinogen [10, 11]. Although pyrene is commonly found in PAH mixtures, the extent of its contribution to the mixture can vary by time and setting. Thus, the measurement of 1-hydroxypyrene alone may not accurately reflect the composition of the other PAHs in the mixture, such as those with carcinogenic effects. Apart from the parent chemical or metabolite, the measurement of biomolecular adducts (e.g., DNA, hemoglobin, albumin) for some of these chemicals (e.g., benzo[*a*]pyrene, benzene) can extend the sampling window to assess for exposure.

24.2.1.3 **Bioaccumulative Metals**

Lead, cadmium, and mercury are metals that are found at increasing concentrations with increased age in the population owing to their slow elimination from the body. These chemicals are typically measured in red blood cells because of their preferential binding to sulfhydryl groups. For example, more than 90% of lead in blood is found in erythrocytes. Bone is an alternative matrix to measure lead because the majority of the body's lead is stored in this compartment. The distribution of cadmium is similar to lead, and low iron stores in the body increases cadmium absorption from the gut. The measurement of mercury is typically speciated as inorganic and organic mercury because of their difference in toxicity. Inorganic mercury causes renal toxicity and organic or methyl mercury causes neurotoxicity. The common type of organic mercury in the general population is methyl mercury and its primary exposure pathway is through fish. The distribution of mercury in the blood depends on the species. Inorganic mercury distributes mostly to the kidneys and organic mercury distributes mostly to the brain. In the blood, methyl mercury distributes to a greater extent to red blood cells than inorganic mercury; thus, the majority (with approximation, 90%) of the total mercury concentration measured in the red blood cell is organic or methyl mercury. Mercury may be measured in hair specimens and this level represents the exposure to organic mercury. The kidneys are a route of elimination for cadmium, lead, and inorganic mercury, so these metals may be measured in urine.

24.2.2 Biological Sampling

For nonpersistent chemicals, the sampling period for the biological specimen can be important when assessing the exposure to these chemicals because of their relatively short residence time in the body. Thus, coordinating the biological sampling and exposure periods is preferred to improve the accuracy of representing the exposure to these chemicals. The sampling window for nonpersistent chemicals can be extended by measuring biomolecular adducts (e.g., PAH adducts), if these are formed for the chemical of interest. Independent of the environment, the sampling period of the biological specimen can be important because physiological variations can affect the concentrations of certain chemicals. For example, the specific gravity of the urine can vary throughout the day due to variable water excretion rates at the time of urine specimen collection, which can affect the concentration of the chemical. Depending on the exposure profile of chemicals with a short biological half-life, the urinary concentration can vary throughout the day, primarily because of episodic exposures to these chemicals throughout the day [27, 28]. Although 24-h urinary collections may be used to avoid this concern, this is not an easy or practical procedure when a large population is involved. The serum concentrations (whole weight) for persistent lipid soluble chemicals (e.g., PCBs, PCDDs/PCDFs) can be affected when the blood specimen is drawn soon after a meal. The increased serum triglyceride concentrations from a meal will cause a transiently elevated serum concentration (whole weight) for these chemicals because of the redistribution of these chemicals from adipose tissue. This variation can be corrected for by using lipid-adjusted units for these chemicals [29]. The sampling of human milk also needs to be considered because the lipid concentration varies by the maturity of the milk. Early postpartum milk contains a higher protein and lower lipid concentration than mature milk at 2–6 weeks postpartum.

It is advantageous to obtain biomonitoring data that represent a temporal exposure profile that is longer than the time it took to collect the specimen because this type of exposure pattern is more relevant to environmental exposures. For chemicals with long half-lives, biomolecular adducts, and certain biological matrices (i.e., meconium, hair), this may be possible but validation is needed. Although the long half-lives of persistent chemicals suggest only a single collection is necessary, at least two samplings at different periods are recommended to confirm the trend of the exposure level in the environment, especially if the individual has changed locations within 5 years of the assessment or sampling period. It is also important to define the temporal exposure profile for the chemical and biomonitoring profile for the biomarker in the specific environment of interest because this will establish the uniformity of the exposure level in the environment. If such information is to be correlated with a biological response or health effect, then that information also needs to be collected so the relation between these variables can be established.

24.2.3 Biological Specimens

The selection of the biological specimen depends on several factors, including the type of chemical, the purpose for the measurement, the host, and the nature of the exposure. Physical properties of chemicals, such as degree of ionization, lipid solubility, and molecular mass, can influence the disposition of chemicals and their metabolites in the body. Host factors to consider in the specimen selection include those affecting the ease of collection (e.g., age) and the disposition of chemicals in the body, including liver and renal function, serum lipid and protein concentrations, and body mass. For example, when evaluating pregnant women, it is technically easier to measure persistent lipophilic chemicals with low detection in serum during the latter half of gestation because of the physiological increase in serum lipid concentration. When exposures are acute or shortly sustained, blood or urine may be the better biological matrix to sample, depending on the purpose of the measurement and the time of sampling. For chemicals that have a short biological half-life and pharmacokinetic parameters are to be determined, then blood is the preferred specimen to use because it best approximates the steady-state concentration of the chemical in the body. However, if the collection of the specimen cannot be conducted immediately, then urine is the preferred specimen because it contains the metabolites of the chemical, which are continuously formed after the peak chemical concentration in the blood. For chemicals with a long biological half-life, such as persistent organic chemicals, blood is the preferred specimen to reconstruct the applied dose. In addition to serum, lipophilic chemicals can be detected in milk and adipose tissue because of their high lipid content, but these latter specimens may be more difficult to obtain than blood.

Blood is a complex matrix, containing an aqueous and lipid phase, and several components, such as proteins, lipids, and cellular components. Chemicals that are quickly metabolized to polar metabolites in the body are best detected in serum. Chemicals that are highly protein bound may be detected in the cellular fraction of the blood, such as mercury and lead. Lipid soluble chemicals are likely to partition or distribute to the lipid compartment and the likelihood for this to occur can be estimated by the octanol–water partition coefficient (K_{ow}) of the chemical. The lipid concentration in serum is about 0.6% and it consists of cholesterol, triglycerides, and phospholipids. The concentrations for these lipids can vary depending on the fasting state, physiological conditions, such as pregnancy, and pathologic states, such as familial hyperlipidemia. Besides the nonspecific binding of chemicals to proteins in the blood, reactive chemical metabolites may form covalent adducts with serum albumin, red blood cells, and DNA. Adducts are significant because they serve not only as an integrated marker of internal exposure but also as a biomolecular evidence of the chemical's clastogenic effect on the body. Because the duration of adducts in the blood is typically longer than that for the parent chemical or metabolite, it can be advantageous to use adducts to assess for exposure when the sampling period and the time of exposure cannot be synchronized. The approximate duration of adducts in the blood for albumin and erythrocytes are 21 days and 3 months, respectively. The duration of the DNA adduct appears to vary by the chemical and source of the DNA in the body [30]. The limitation of blood is the need to perform a venipuncture to obtain a specimen, which is an invasive procedure compared to the method of collection for other specimens, such as urine and saliva.

Urine offers several advantages compared to blood, including ease of collection and higher concentrations of nonpersistent chemicals than in blood. Chemicals metabolized to polar products are eliminated into the urine, typically by passive filtration; however, active tubular secretion may contribute to the amount of the chemical in the matrix. Unlike blood, chemicals in the urine may not be at steady state because chemicals are only eliminated into this matrix and the water content can vary throughout the day. Chemicals with a high lipid solubility or a high capacity for protein binding do not distribute well to urine from blood; thus, these chemicals are not detected at significant concentrations and urine would be a poor matrix. When the amount of biological specimen may be limited, urine may be preferred to blood because the former is more available.

Alternative biological matrices that are available for the detection of chemicals include human milk, adipose tissue, hair, meconium, and nail. Human milk and adipose tissue are used for measuring highly lipid soluble chemicals, such as PCBs, PCDDs/PCDFs, and PBDEs. When using lipid-adjusted units, the concentrations for these chemicals in these matrices are comparable to blood. The advantage of these matrices over blood is that a smaller quantity of the former is needed to measure the chemical because the lipid content is higher in these specimens (with approximation, milk 4% and adipose tissue 60–95%). Unique to human milk is that it can be used to assess an infant's exposure to these chemicals from breastfeeding. The extent to which the milk chemical content approximates the infant's exposure depends on the amount of breastfeeding as some mothers exclusively breastfeed and others use formula as well. Adipose tissue can be obtained only by surgical excision, which makes it less practical than blood as a biological specimen unless the procedure is already intended. Hair is a protein matrix and it can be used to assess the exposure to metals; however, the extent of its use to assess for internal dose is limited by environmental contamination. Methyl mercury is an example of a chemical that can be assessed using hair specimens. Meconium and cord blood are unique biological specimens because they assess for fetal exposure. Although both are collected at parturition, meconium can assess for fetal exposure since the second trimester [31].

24.2.4 Data Analysis and Evaluation

One of the several challenges when working with environmental chemicals is their low concentrations in the environment and even lower concentrations in the body.

Ultratrace analytical methods, including sophisticated instruments and specialized methodologies, are required to measure them. Despite these assets, the concentrations for certain chemicals can be below the limit of detection (LOD) of the method because of an inadequate amount of specimen submitted for measurement, the concentration of the chemical is lower than the detection limit of the method, or the chemical is not present in the specimen. Also, analytical methods that measure multiple chemicals in a single run tend to have a decreased sensitivity that can predispose to concentrations below the LOD. When a data set contains a significant proportion of values that are below the limit of detection, then a certain amount of uncertainty enters into estimates derived from the completed data set using estimation procedures. Left censored data can be filled with a fixed (e.g., zero, LOD, LOD/2, LOD/sqrt[2]) or variable value by using maximum likelihood estimation procedures that may include multiple imputation [32–34].

The other item to note regarding data is units. When using urine as a specimen, there are two commonly used units – the volumetric (chemical weight/urine volume) and the corrected for urinary creatinine (chemical weight/urinary creatinine weight). The latter is used to consider diurnal variations in urine specific gravity, which can contribute to variations in the concentration of the chemical. Although this approach may adjust for some of the variability in urine measurements, it may introduce a bias when comparing concentrations of chemicals between subgroups in the population [35]. The lower urinary creatinine concentration in children compared to adults is attributed to their lower skeletal muscle mass [36].

24.3 Interpretation

The interpretation of biomonitoring data is accomplished by considering the purpose of obtaining the measurement. For situations when concentrations are meant for comparisons, such as between subgroups in a population, has a population's level of exposure changed over time, or whether a worker has exceeded exposure levels at the workplace (e.g., biological exposure index; biological tolerance value, BAT), then descriptive analysis may suffice for these purposes. When using biomonitoring data to assess health risk, this is best accomplished for chemicals that have concentrations with defined health risks, such as lead and mercury. Otherwise, the traditional approach to assess risk (based on exposure or applied dose; minimal risk level, MRL; maximum contaminant level, MCL; and reference dose, RfD) may be used. If the desire is to use biomonitoring data to determine the applied dose, then the appropriate pharmacokinetic modeling may be considered for this purpose with human or animal biomonitoring data. The latter situation involving pharmacokinetic modeling will have added considerations for chemical mixtures compared to single compounds because of interactions (see below).

A reference range for a chemical that allows comparison of concentrations can be established from either a representative sample of a population (e.g., NHANES) or a comparable population to the individuals from whom the biomonitoring data were obtained. The Center for Disease Control and Prevention's National Report on Human Exposure to Environmental Chemicals presents reference ranges for several chemicals that are representative of general US population [7]. The sampling of the participants for the survey is based on a complex, stratified, multistage, probability cluster design that is representative for age, gender, and race/ethnicity (Mexican Americans, non-Hispanic blacks, and non-Hispanic whites). The concentrations are presented as percentiles and this survey data do not imply that these chemicals cause diseases. For many situations, the amount of effort needed to establish a representative reference range for a population is more than what may be available so a convenience sample may be adequate. In the latter situation, it is important to consider demographic variables that predict the concentrations of the chemical of interest. For example, increased age is associated with higher concentrations of PCBs/PCDDs/PCDFs. Similarly, smokers have higher concentrations of certain chemicals, such as cadmium, pyrene, and aromatic hydrocarbons (benzene, toluene, and xylenes). If these variables are not considered in the sampling design or data analysis, then unexpected biases may occur in the results. For chemicals with a long biological half-life, it is important to consider the length of stay at the current location if a regional comparison is important because the exposure to the chemical may have occurred at a former location. Nonpersistent chemicals or those with a short biological half-life and with episodic exposure patterns will require additional consideration when selecting a reference population. For example, the recruitment of participants during the winter months may yield a negative bias for pesticide exposure because these chemicals are typically used during the warmer seasons [37]. Thus, biomonitoring data for persistent chemicals are indicative of past exposures and those for nonpersistent chemicals are indicative of current exposures. Finally, occupational exposure to chemicals should be considered when establishing a comparable population for nonworkers.

Risk assessment based on biomonitoring data has been established for only a couple of chemicals, including lead and mercury. The basis for these chemicals included robust epidemiologic studies, sensitive and specific biomarkers of exposure, chemicals with long biological half-lives that are important because this provides a stable measure of exposure over time, knowledge of the exposure pathways, and a defined dose-response relation for a specific and adverse health end point. These criteria allowed assessment of risk by determining the applied dose using exposure assessment and biomonitoring data, which can be compared for consistency. The biomonitoring level of concern for lead is a blood concentration of 10 mcg/dL or greater in children. For mercury, the US EPA established an RfD based on a cord blood total mercury concentration of 5.8 mcg/dL.

When epidemiologic and exposure assessment data are lacking, pharmacokinetic modeling may be utilized to estimate the applied dose from biomonitoring data. For humans, blood and certain urine concentrations may be used to reconstruct the dose that can be compared with values determined by exposure assessment. When biomonitoring data are available from animal trials, then the applied dose may be derived or a dose-response relation may be constructed based on the internal dose. The latter situation would preclude the need to use pharmacokinetic modeling to determine the applied dose because the biological response is directly related to the internal dose. For human biomonitoring data, the estimation of the applied dose is based on the total body content, the total clearance, and the bioavailability, and the presumption of a steady-state condition and unchanging kinetics. The total body content for persistent organic chemicals, such as PCBs/PCDDs/PCDFs, is estimated from the body's lipid content because these chemicals reside mostly in the adipose tissue. For chemicals with a low lipid solubility, the total body content may be estimated from the volume of distribution. When metabolites are used as the biomarker of exposure, then the percentage of conversion from parent chemical to metabolite needs to be known, and it is preferred to use a major metabolite from the parent chemical [38]. Also, if the metabolite is present in the environment as a degradate, then that needs to be considered as well because that may contribute to elevated concentrations in the body. A limitation of using human biomonitoring data for dose reconstruction is that there is often only a single concentration from a participant, especially when studies with a large number of participants are being conducted. In most instances, a single measurement is indicative of current and not past exposure. Thus, it is advantageous to use biomonitoring data that is more representative of the exposure event.

If data from *in vivo* animal dosing or dose-response toxicity studies are available, then biomonitoring data may be derived from these models by using blood or urine chemical concentrations measured in these trials or estimating these concentrations with pharmacokinetic modeling, which should include appropriate selection of the biological matrix and time points for sampling of the biological matrix, type of analyte (parent chemical or metabolite), relevance to human biomonitoring data and exposure situation, and interactions from chemical mixtures when relevant. The significance of the latter is that these interactions can affect the concentration of the chemical in the body by altering its bioavailability, distribution, metabolism, or elimination [39]; thus, the applied dose at a given concentration for a chemical would be different between the exposure to the chemical as a single agent and as a mixture. Similarly, interactions among chemicals can lead to altered biological responses that are different from those from the exposure to a single chemical. These events are best discussed further in the following section by using classic examples with pharmaceutical agents [40] and a minor reference to environmental and occupational chemicals because the latter are discussed elsewhere in this text (refer to Chapter 12).

The interactions based on alterations in absorption involve vehicle effects, the chemical formation of poorly absorbed conjugates or complexes, the gastrointestinal (GI) flora, catabolic enzymes located in the GI epithelium, and decreases in GI motility. Examples of such effects have been noted for oral and dermal exposures. For example, the dermal absorption and consequent toxicity of 2,4,7,8-tetrachlorinated dibenzo-*p*-dioxin adsorbed on charcoal is considerably less than that of TCDD solubilized in a lipophilic medium. This is due to the reduced availability of the charcoal adsorbed TCDD for absorption by the gut. Conversely, dimethyl sulfoxide, a commonly used vehicle in dermal toxicity studies, is known to facilitate the absorption of many organic chemicals across the skin, thus causing apparent potentiation of the chemical's effect when compared to less lipophilic vehicles. A similar mechanism appears to be involved in the enhanced neurotoxicity of *O*-ethyl-*O*-4-nitrophenyl phenylphosphonothioate by various aliphatic hydrocarbons when dermally applied to hens. Similarly, the acute oral toxicity of many chemicals is substantially affected by the vehicle used, and these effects are probably due to differences in rate of absorption. For example, the gut absorption of 1,1-dichloroethylene is altered by the vehicle in which it is administered. As another example, the oral administration of clioquinol complexes with many metals and facilitates their absorption, and it has been implicated in an outbreak of heavy metal-induced subacute myelo-optic neuropathy in Japan. In contrast, there are examples of chemicals that form poorly absorbed complexes after oral administration, such as tetracycline and calcium carbonate. The amount of the absorbed chemical from the gut can be increased by altering intestinal bacteria (e.g., macrolide antibiotic and digoxin) or inhibiting catabolic enzymes located in the gut epithelium (e.g., monoamine oxidase inhibitors and tyramine). Chemicals with opioid or anticholinergic properties, such as codeine, morphine, and atropine, decrease the GI transit rate, which can increase or decrease the rate of absorption of orally administered chemicals. For the most part, such interactions usually lead to decreases in effects because of the slower rate of absorption, rather than increases in effects because of more complete absorption.

Interactions based on distribution can play a significant role if a more active agent is displaced from an inactive site to a primary receptor site by a less active or inactive agent. One of the best documented examples of this kind of activity is the displacement of anticoagulants from plasma proteins by compounds such as barbiturates, analgesics, antibiotics, or diuretics. Similarly, tri-*O*-tolyl phosphate decreases the binding of paraoxon to nonvital tissue in rat liver and plasma, consequently increasing the toxicity of paraoxon in rats. Because body fat represents a major nonvital storage site for many lipophilic xenobiotics, it may be anticipated that compounds that cause fat mobilization could result in similar potentiating effects. It has been shown that dithiocarbamates and tetramethylthiuram disulfide complex with lead and selectively increase the accumulation of lead in the brain that likely potentiates lead-induced central nervous system effects.

Altered patterns of chemical metabolism have been shown to be the basis of many toxicant interactions. A major metabolic enzyme system involved in such interactions is the liver microsomal mixed function oxidases (MFOs) that exist in multiple forms. The MFOs convert the parent chemical into biologically active or inactive metabolites. The types and levels of these enzymatic activities can be induced by agents, such as phenobarbital and mainstream tobacco smoke, and can be inhibited by agents, such as piperonyl butoxide. Thus, the chemical's concentration in the body and the overall joint toxicity outcome depend on the nature of the interaction at the MFO and the biological activity of the metabolites. An example of the former situation is the need to use a higher applied dose of theophylline to achieve the same serum concentration in patients who smoke compared to nonsmokers. Toxicant interactions involving the MFO can be complex and depend on both dose and duration of exposure, with some compounds causing an initial inhibition of enzymatic activity followed by a marked induction of activity. Although liver microsomal MFOs are the most commonly studied enzymes involved in toxicant interactions, they are also present in extrahepatic tissues and these sites may contribute to local toxicity by the production of biologically active metabolites. Also, there are several other enzyme systems, such as alcohol and aldehyde dehydrogenases, monamine and diamine oxidases, dehydrochlorinases, azo and nitro reductases, and hydrolases that play an important role in chemical metabolism. For instance, ethanol serves as an antagonist of the toxic effects of methanol by acting as a competitive inhibitor of alcohol dehydrogenase, thus prolonging the biological half-life of methanol and suppressing the formation of the toxic metabolites, formaldehyde and formic acid.

Interactions based on excretion involve chemicals and metabolites that are eliminated from the body organs, such as kidneys. For instance, probenecid or carinamide both competitively inhibit the elimination of penicillin, thus prolonging or potentiating its desirable therapeutic effect. Similarly, phenylbutazone inhibits the renal excretion of hydroxyhexamide, which can cause undesirably prolonged hypoglycemia. If a renally eliminated toxicant is a weak acid (e.g., phenobarbital), the administration of sodium bicarbonate to alkalinize the urine can enhance its urinary elimination by a mechanism known as "ion trapping." A less direct effect on renal elimination has been suggested by indomethacin's effect on the substantial potentiation of lithium toxicity. These investigators suggest that the potentiation is due to the inhibition of prostaglandin synthesis by indomethacin, which causes renal vasoconstriction and a decrease in the renal excretion of lithium. Several drugs and other chemicals are also able to compete for biliary excretion. Quinidine has a marked inhibitory effect on the presystemic elimination of ajmaline by the liver when both compounds are administered concurrently to rats; similar observations have been noted in humans.

The pharmacodynamic interactions occur at cellular receptor sites or among receptor sites. Examples of such interactions include the antagonistic effects of oxygen on carbon monoxide, atropine on cholinesterase inhibitors, and naloxone on morphine. Benzodiazepine on ethanol is an example of an agonistic effect. Pharmacodynamic interactions could also result as a consequence of two or more compounds acting on different receptor sites and causing opposite effects referred to as "functional antagonism." Interactions such as that of histidine and norepinephrine on vasodilation and blood pressure and the anticonvulsive effects of barbiturates on many compounds that cause convulsions are such examples. The sites of action for two compounds having the same type of activity may be different. This is the case when the effect can be caused either by a direct stimulation or by the annihilation of an inhibition. In both cases, the combination of two compounds, linked in parallel or in series, as it were, may well result in a synergistic effect. When the components of a combination possess different sites of action and different types of activity, no plausible prediction about the possibility of synergism can be made, unless their modes of action are well known. While examples of such interactions have not been well characterized in the literature, the potentiation of carbon tetrachloride by chlordecone may be at least partially mediated by an inhibition of hepatocellular repair. The increased risk for hepatocellular carcinoma from hepatitis B viral infection during the exposure to aflatoxin is another example.

24.4 Summary

Biomonitoring assesses human exposure to environmental chemicals by measuring the concentration of chemicals in the body. This information can assist in determining the extent of exposure in a population, a change in a population's level of exposure, subgroups in the population that are vulnerable to exposure, and the risk from the exposure to chemicals with known health risks. Biomonitoring reduces the level of uncertainty in these estimates compared to those derived by other means. Because there are very few chemicals with known health risks based on biomonitoring data, additional epidemiologic and toxicologic studies incorporating this approach in their designs are needed to better characterize the risk for other chemicals. Biomonitoring data can be utilized to assess risk by using pharmacokinetic modeling to estimate the applied dose and comparing it to an established dose for health safety or constructing a dose-response curve from animal trials or epidemiological studies. A similar approach may be used for chemical mixtures, depending on the level of complexity. When the primary exposure pathway is known for a chemical mixture, then the environmental medium can be monitored to characterize the relation between the exposure time and dose to assist in the interpretation of the biomonitoring data for the mixture. For more complicated scenarios, biomonitoring remains the best approach to assess for human exposure to chemical mixtures when the exposure pathway is unknown.

References

1 NRC (National Research Council), Committee on Human Biomonitoring for Environmental Toxicants, Board on Environmental Studies and Toxicology, National Academies of Sciences (2006) *Human Biomonitoring for Environmental Chemicals*, National Academy Press, Washington, DC.

2 NRC (National Research Council) (1987) Biological markers in environmental health research. *Environ. Health Perspect.*, **74**, 3–9.

3 WHO (World Health Organization) (1993) *Biomarkers and Risk Assessment: Concepts and Principles*, World Health Organization, Geneva.

4 Annest, J.L., Pirkle, J.L., Makuc, D., Neese, J.W., Bayse, D.D., and Kovar, M.G. (1983) Chronological trend in blood lead levels between 1976 and 1980. *N. Engl. J. Med.*, **308** (23), 1373–1377.

5 Pirkle, J.L., Kaufmann, R.B., Brody, D.J., Hickman, T., Gunter, E.W., and Paschal, D.C. (1998) Exposure of the U.S. population to lead, 1991–1994. *Environ. Health Perspect.*, **106** (11), 745–750.

6 Calafat, A.M., Wong, L.Y., Kuklenyik, Z., Reidy, J.A., and Needham, L.L. (2007) Polyfluoroalkyl chemicals in the U.S. population: data from the National Health and Nutrition Examination Survey (NHANES) 2003–2004 and comparisons with NHANES 1999–2000. *Environ. Health Perspect.*, **115** (11), 1596–1602.

7 CDC (Centers for Disease Control and Prevention) (2005) Third National Report on Human Exposure to

Environmental Chemicals. CDC, Atlanta, GA. http://www.cdc.gov/exposurereport/. Last accessed March 4, 2008.

8 Meyer, P.A., Pivetz, T., Dignam, T.A., Homa, D.M., Schoonover, J., and Brody, D. (2003) Centers for Disease Control and Prevention. Surveillance for elevated blood lead levels among children – United States, 1997–2001. *MMWR Surveill. Summ.*, **52** (10), 1–21.

9 CDC (Centers for Disease Control and Prevention) (2004) Blood mercury levels in young children and childbearing-aged women – United States, 1999–2002. *Morb. Mortal. Wkly. Rep.*, **53** (43), 1018–1020.

10 NTP (National Toxicology Program) (2005) Report on Carcinogens, Eleventh Edition. US Department of Health and Human Services, Public Health Service, National Toxicology Program. http://ntp.niehs.nih.gov/index.cfm?objectid=32BA9724-F1F6-975E-7FCE50709CB4C932. Accessed March 29, 2008.

11 IARC (International Agency for Research on Cancer) (2008) IARC Monographs. http://www.inchem.org/pages/iarc.html. Accessed March 29.

12 U.S. DHHS (U.S. Department of Health and Human Services) (2006) The Health Consequences of Involuntary Exposure to Tobacco Smoke: A Report of the Surgeon General. U.S. Department of Health and Human Services, Centers for Disease Control and Prevention, Coordinating Center for Health Promotion, National Center for Chronic Disease Prevention and Health Promotion, Office on Smoking and Health, Atlanta, GA. http://www.surgeongeneral.gov/library/secondhandsmoke/report/. Accessed March 29, 2008.

13 Howdeshell, K.L., Furr, J., Lambright, C.R., Rider, C.V., Wilson, V.S., and Gray, L.E. Jr. (2007) Cumulative effects of dibutyl phthalate and diethylhexyl phthalate on male rat reproductive tract development: altered fetal steroid hormones and genes. *Toxicol. Sci.*, **99** (1), 190–202.

14 NTP (National Toxicology Program) (2006) Brief on the potential human reproductive and developmental effects of di(2-ethylhexyl) phthalate. U.S. Department of Health and Human Services, Public Health Service, National Toxicology Program. Draft. http://cerhr.niehs.nih.gov/chemicals/dehp/dehp-eval.html. Accessed March 29, 2008.

15 NRC (National Research Council) (2003) *Committee on the Implications of Dioxin in the Food Supply*, National Academy Press, Washington, DC, p. 30.

16 FDA (Food and Drug Administration)/EPA (Environmental Protection Agency) (2006) FDA/EPA Advisory on Seafood Consumption. http://www.fda.gov/bbs/topics/NEWS/2006/NEW01382.html. Accessed on July 22, 2008.

17 Jacobson, J.L., Jacobson, S.W., Muckle, G., Kaplan-Estrin, M., Ayotte, P., and Dewailly, E. (2008) Beneficial effects of a polyunsaturated fatty acid on infant development: evidence from the inuit of arctic Quebec. *J. Pediatr.*, **152** (3), 356–364.

18 Goldey, E.S. and Crofton, K.M. (1998) Thyroxine replacement attenuates hypothyroxinemia, hearing loss, and motor deficits following developmental exposure to Aroclor 1254 in rats. *Toxicol. Sci.*, **45** (1), 94–105.

19 Rubin, C.S., Holmes, A.K., Belson, M.G., Jones, R.L., Flanders, W.D., Kieszak, S.M., Osterloh, J., Luber, G.E., Blount, B.C., Barr, D.B., Steinberg, K.K., Satten, G.A., McGeehin, M.A., and Todd, R.L. (2007) Investigating childhood leukemia in Churchill County, Nevada. *Environ. Health. Perspect.*, **115** (1), 151–157.

20 Jones, P.D., Hu, W., De Coen, W., Newsted, J.L., and Giesy, J.P. (2003) Binding of perfluorinated fatty acids to serum proteins. *Environ. Toxicol. Chem.*, **22** (11), 2639–2649.

21 Hansen, L.G. (1998) Stepping backward to improve assessment of PCB

congener toxicities. *Environ. Health Perspect.*, **106** (Suppl. 1), 171–189.

22 Grimvall, E., Rylander, L., Nilsson-Ehle, P., Nilsson, U., Stromberg, U., Hagmar, L., and Ostman, C. (1997) Monitoring of polychlorinated biphenyls in human blood plasma: methodological developments and influence of age, lactation, and fish consumption. *Arch. Environ. Contam. Toxicol.*, **32** (3), 329–336.

23 Dewailly, E., Ryan, J.J., Laliberte, C., Bruneau, S., Weber, J.P., Gingras, S., and Carrier, G. (1994) Exposure of remote maritime populations to coplanar PCBs. *Environ. Health Perspect.*, **102** (Suppl. 1), 205–209.

24 Lorber, M. (2008) Exposure of Americans to polybrominated diphenyl ethers. *J. Expos. Sci. Environ. Epidemiol.*, **18** (1), 2–19.

25 Van den Berg, M., Birnbaum, L.S., Denison, M., DeVito, M., Farland, W., Feeley, M., Fiedler, H., Hakansson, H. *et al.* (2006) The 2005 World Health Organization reevaluation of human and mammalian toxic equivalency factors for dioxins and dioxin-like compounds. *Toxicol. Sci.*, **93** (2), 223–241.

26 Purkey, H.E., Palaninathan, S.K., Kent, K.C., Smith, C., Safe, S.H., Sacchettini, J.C., and Kelly, J.W. (2004) Hydroxylated polychlorinated biphenyls selectively bind transthyretin in blood and inhibit amyloidogenesis: rationalizing rodent PCB toxicity. *Chem. Biol.*, **11** (12), 1719–1728.

27 Kissel, J.C., Curl, C.L., Kedan, G., Lu, C., Griffith, W., Barr, D.B., Needham, L.L., and Fenske, R.A. (2005) Comparison of organophosphorus pesticide metabolite levels in single and multiple daily urine samples collected from preschool children in Washington State. *J. Expos. Anal. Environ. Epidemiol.*, **15** (2), 164–171.

28 Calafat, A.M., Ye, X., Wong, L.Y., Reidy, J.A., and Needham, L.L. (2008) Exposure of the U.S. population to bisphenol A and 4-tertiary-octylphenol: 2003–2004. *Environ. Health Perspect.*, **116** (1), 39–44.

29 Phillips, D.L., Pirkle, J.L., Burse, V.W., Bernert, J.T., Jr., Henderson, L.O., and Needham, L.L. (1989) Chlorinated hydrocarbon levels in human serum: effects of fasting and feeding. *Arch. Environ. Contam. Toxicol.*, **18**, 495–500.

30 Nesnow, S., Ross, J., Nelson, G., Holden, K., Erexson, G., Kligerman, A., and Gupta, R.C. (1993) Quantitative and temporal relationships between DNA adduct formation in target and surrogate tissues: implications for biomonitoring. *Environ. Health Perspect.*, **101** (Suppl. 3), 37–42.

31 Bearer, C.F. (2003) Meconium as a biological marker of prenatal exposure. *Ambul. Pediatr.*, **3** (1), 40–43.

32 Lubin, J.H., Colt, J.S., Camann, D., Davis, S., Cerhan, J.R., Severson, R.K., Bernstein, L., and Hartge, P. (2004) Epidemiologic evaluation of measurement data in the presence of detection limits. *Environ. Health Perspect.*, **112** (17), 1691–1696.

33 Caudill, S.P., Wong, L.Y., Turner, W.E., Lee, R., Henderson, A., and Patterson, D.G., Jr. (2007) Percentile estimation using variable censored data. *Chemosphere*, **68** (1), 169–180.

34 Jain, R.B., Caudill, S.P., Wang, R.Y., and Monsell, E. (2008) Evaluation of maximum likelihood procedures to estimate left censored observations. *Anal. Chem.*, **80** (4), 1124–1132.

35 Tsuda, M., Hasunuma, R., Kawanishi, Y., and Okazaki, I. (1995) Urinary concentrations of heavy metals in healthy Japanese under 20 years of age: a comparison between concentrations expressed in terms of creatinine and of selenium. *Tokai J. Exp. Clin. Med.*, **20** (1), 53–64.

36 Barr, D.B., Bravo, R., Weerasekera, G., Caltabiano, L.M., Whitehead, R.D., Olsson, A.O., Caudill, S.A., Schober, S.E. *et al.* (2004) Concentrations of dialkyl phosphate metabolites of organophosphorus pesticides in the US population. *Environ. Health Perspect.*, **112** (2), 186–200.

37 MacIntosh, D.L., Kabiru, C., Echols, S.L., and Ryan, P.B. (2001) Dietary exposure to chlorpyrifos and levels of 3,5,6-trichloro-2-pyridinol in urine. *J. Expos. Anal. Environ. Epidemiol.*, **11**, 279–285.

38 Barr, D.B., Panuwet, P., Nguyen, J.V., Udunka, S., and Needham, L.L. (2007) Assessing exposure to atrazine and its metabolites using biomonitoring. *Environ. Health Perspect.*, **115** (10), 1474–1478.

39 Viau, C. (2002) Biological monitoring of exposure to mixtures. *Toxicol. Lett.*, **134** (1–3), 9–16.

40 Mumtaz, M.M. and Hertzberg, R.C. (1993) The status of interactions data in risk assessment of chemical mixtures, in *Hazard Assessment of Chemicals*, vol. 8 (ed. J. Saxena), Hemisphere Publishing Corporation, Washington, D.C., pp. 47–79.

41 Pratt, G., Bock, D., Stock, T., Morandi, M., Adgate, J., Ramachandran, G., Mongin, S., and Sexton, K. (2005) A field comparison of volatile organic compounds using passive organic vapor monitors and stainless steel canisters. *Environ. Sci. Technol.*, **39** (9), 3261–3268.

42 Paoletti, P. (1995) Application of biomarkers in population studies for respiratory nonmalignant diseases. *Toxicology*, **101** (1–2), 99–105.

43 U.S. EPA (2008) Criteria Documents and Integrated Science Assessments for Criteria Pollutants.

44 Moolgavkar, S.H. (2005) A review and critique of the EPA's rationale for a fine particle standard. *Reg. Toxicol. Pharm.*, **42** (1), 123–144.

45 Pershagen, G. (1986) Review of epidemiology in relation to passive smoking. *Arch. Toxicol. Suppl.*, **9**, 63–73.

46 Wallace, L.A. and O'Neill, I.K. (1987) Personal, biological and air monitoring for exposure to environmental tobacco smoke. *IARC Sci. Pub.*, **81**, 87–103.

47 Reardon, J.Z. (2007) Environmental tobacco smoke: respiratory and other health effects. *Clin. Chest Med.*, **28** (3), 559–573.

48 Pennington, J.A.T., Capar, S.G., Parfitt, C.H., and Edwards, C.W. (1996) History of the food and drug administration's total diet study. 2. 1987–1993. *J. AOAC Int.*, **79** (1), 163–170.

49 Eskenazi, B., Bradman, A., and Castorina, R. (1999) Exposures of children to organophosphate pesticides and their potential adverse health effects. *Environ. Health Perspect.*, **107**, 409–419.

50 Wolff, M.S. (2006) Endocrine disruptors: challenges for environmental research in the 21st century. *Ann. N.Y. Acad. Sci.*, **1076**, 228–238.

51 Hinck, J.E., Blazer, V.S., Denslow, N.D., Echols, K.R., Gross, T.S., May, T.W., Anderson, P.J., Coyle, J.J. *et al.* (2007) Chemical contaminants, health indicators, and reproductive biomarker responses in fish from the Colorado River and its tributaries. *Sci. Total Environ.*, **378** (3), 376–402.

52 Hinck, J.E., Blazer, V.S., Denslow, N.D., Echols, K.R., Gale, R.W., Wieser, C., Maya, T.W., Ellersiecke, M. *et al.* (2008) Chemical contaminants, health indicators, and reproductive biomarker responses in fish from rivers in the Southeastern United States. *Sci. Total Environ.*, **390** (2–3), 538–557.

53 Maycock, B.J. and Benford, D.J. (2007) Risk assessment of dietary exposure to methylmercury in fish in the UK. *Hum. Exp. Toxicol.*, **26** (3), 185–190.

54 Needham, L.L., Ryan, J.J., and Fürst, P. (2002) Technical Workshop on Human Milk Surveillance and Research on Environmental Chemicals in the United States. Guidelines for analysis of human milk for environmental chemicals. *J. Toxicol. Environ. Health A*, **65** (22), 1893–1908.

55 Wang, R.Y. and Needham, L.L. (2007) Environmental chemicals: from the environment to food, to breast milk, to the infant. *J. Toxicol. Environ. Health B*, **10** (8), 597–609.

56 Weisel, C.P., Kim, H., Haltmeier, P., and Klotz, J.B. (1999) Exposure

estimates to disinfection by-products of chlorinated drinking water. *Environ. Health Perspect.*, **107** (2), 103–110.

57 Arbuckle, T.E., Hrudey, S.E., Krasner, S.W., Nuckols, J.R., Richardson, S.D., Singer, P., Mendola, P., Dodds, L. *et al.* (2002) Assessing exposure in epidemiologic studies to disinfection by-products in drinking water: Report from an international workshop. *Environ. Health Perspect.*, **110**, 53–60.

58 ATSDR (Agency for Toxic Substances and Disease Registry) Toxicological Profiles. http://www.atsdr.cdc.gov/toxpro2.html. Accessed April 9, 2008.

25
Adverse Drug Reactions and Interactions[1)]

Patricia Ruiz, Moiz Mumtaz, and Chander Mehta

25.1
Introduction

A plethora of information appearing in various forms of media outlets and on the Internet has exposed patients to a whole spectrum of drug-related information and misinformation. The information includes the mechanism of the working of a drug, how long it takes to have its effects, or how similar it is to other available drugs. When the patient gets a prescription or refills, cautions and warnings to alert the patient about potential dangerous reactions are also provided, as are warnings about possible side effects, to better explain the drug's actions. The drug literature, given with the prescription, also describes a drug's interactions, so that the patient is informed if other drugs can or cannot be taken in combination, simultaneously, or sequentially. Because some interactions could be potentially dangerous, special information is also included – symptoms to watch for and advice to call a doctor if the patient is unsure about potential side effects. The literature accompanying the drug includes statements such as "Do not use over-the-counter cough, cold, or allergy medications"; "These products contain stimulant ingredients that can increase blood pressure"; and "This drug should be used with caution if you have kidney or liver disease, diabetes, hypothyroidism, severe heart disease, or a history of porphyria." Such information about possible side effects and the cautions and warnings can serve to aggravate a patient's existing anxieties.

Thus, considerable anxiety can be present when therapeutic drugs are ingested by a patient. For example, it is not uncommon to see a patient take multiple drugs for various chronic ailments and to experience some anxiety about interactions. Second, there exists a possibility of adverse drug reaction (ADR) due to a variety of patient, lifestyle, or environmental factors. Finally, a patient may experience a fear of harm simply by watching a television program about food–drug toxicity; for example, the airwaves have been filled recently with warnings about ingesting grapefruit juice

1) The findings and conclusions in this report are those of the author(s) and do not necessarily represent the official position of the Centers for Disease Control and Prevention/the Agency for Toxic Substances and Disease Registry.

Principles and Practice of Mixtures Toxicology. Edited by Moiz Mumtaz

ISBN: 978-3-527-31992-3

with an antihistamine such as terfenatine (Seldane) and the very serious toxic reactions, including ventricular arrhythmia resulting in death due to cardiac arrest, which can result.

ADRs have been considered a major public health problem; they might have played a role in the hospitalization of about three quarters of a million people each year, and they pose the possibility of permanent disability or even death [1–10]. The term *ADR* generally refers to unintended adverse responses to a therapeutic drug at normal therapeutic dosages in humans. ADRs can occur in connection with drug intake for prophylactic treatment to prevent a condition, for treatment of an illness, or for therapy. Since all drugs have the potential for causing adverse reactions, pragmatic physicians consider the benefits and risks of a drug before prescribing it.

In this chapter, an ADR is defined as a "response to a drug which is noxious and unintended and which occurs at doses normally used in man for prophylaxis, diagnosis, or therapy of disease or for the modification of physiologic function" [11]. In other words, an ADR is a harmful response caused by a therapeutic agent at a therapeutic dose-level during normal usage.

Pharmacists, in particular, have further defined ADR. The American Society of Health-System Pharmacists [12] defines an ADR as a response to a drug that requires changing medication, discontinuing the drug, altering the dose, increasing a patient's stay in a hospital, or resulting in temporary or permanent harm, disabling injury, or even death.

The incidence of ADR in hospitalized patient has been reported to be 6.7% [9], but it could be much higher if all patients in every clinical setting are considered.

Most of the clinical new drug trial studies fail to identify ADR as a potential problem with a drug because of the following reasons:

- The duration of the study of a drug is short. Therefore, an ADR that develops with chronic drug use or that has a long latency period is impossible to detect.
- Problems exist with a drug trial's sample population. Population samples in such trials do not include special groups (e.g., pediatric and geriatric patients) in a majority of the studies. In addition, the narrow population selected for such trials is not always representative of the population that may be exposed to the drug after approval.
- The small sample size of about 3000–4000 subjects so commonly used in clinical trials cannot usually identify a rare ADR that can occur in 1 in 100 000 cases.
- Drug trials generally focus on a narrow set of therapeutic indications for which the clinical efficacy is being investigated. Other future uses of a drug are not considered in a trial.

ADRs are not always attributable to active drug moieties. Various preservatives and additives such as coloring agent, vehicles, impurities, and metabolites are also capable of causing ADR.

There is no acceptable classification of ADRs, and hence different classification schemes are used. Many use an organ-based classification system for ADRs, such as hepatic, renal, hematological, cardiac, and the CNS ADRs. Another system used is based on the severity of ADRs; it classifies the responses as mild, moderate, severe, and lethal. A third method is to classify ADRs as involving either immune mechanisms or nonimmune mechanisms. Rawlins and Thompson's classification system, using such categories as predictable, unpredictable, long-term drug use-induced, and carcinogenic/teratogenic, is also frequently used [13].

Drugs used for therapeutic interventions either alone or in combination may sometimes cause unexpected toxicity to various critical organs of the body, including the liver, kidney, blood, and immune system, with the result of varying degrees of symptomatology, from mild discomfort and inconvenience to permanent damage and disability. The following section provides representative examples of drug toxicity effects for some important organs of the body.

25.2 Drug Toxicity in Major Body Organs

25.2.1 Hepatic Toxicity

Hepatic toxicity is the leading cause of adverse drug manifestations, and hepatotoxicity is a major cause of drug recall. A caregiver must be increasingly vigilant about detecting drug-induced hepatotoxicity since early detection and discontinuation results in less severe damage. Acute hepatotoxicity of a drug can be caused by direct damage to hepatocytes (cytotoxic) or by cholestasis (i.e., reduced bile flow resulting in icterus and pruritis). Since the occurrence of drug-induced hepatic injury ranges from 1 in 10 000 to 1 in 100 000, most of the clinical trials do not detect a hepatotoxic event due to its rarity [14]. More than 900 therapeutic agents, herbs, and toxins cause liver injury. Some injury can even result in fulminating hepatic toxicity. ADRs involving hepatotoxicity can be predictable or unpredictable. For example, chemicals such as chloroform, acetaminophen, carbon tetrachloride, and phosphorus cause predictable hepatotoxicity. On the other hand, unpredictable hepatotoxic response is neither dose-related nor does it have immediate timely manifestations; it can have an unpredictable latency period ranging from a few days to few months. Thus, hepatotoxicity can be classified as intrinsic or idiopathic. It is classified as intrinsic if a drug possesses a structural profile that causes potential liver damage. It is classified as idiopathic if the drug produces unpredictable hepatotoxicity in patients due to either host pharmacogenomic susceptibility or host hypersensitivity.

a. *Intrinsic hepatotoxicity* is direct when a chemical directly damages the normal hepatic cells. For example, direct or predictable hepatotoxicity is produced by acetaminophen if it is ingested in excessive amounts or at frequent intervals. It is also caused by environmental chemicals such as carbon tetrachloride and chloroform. Indirect hepatotoxicity occurs when a drug alters the normal or vital physiological functions such as secretion and metabolism of hepatocytes and causes damage. For example, anabolic steroids and oral contraceptives affect the metabolic functions, and tetracyclines, valproic acid, and isoniazid can affect the secretory function of the hepatocytes. Any disruption in the transport proteins involving the normal bile flow can result in loss of villous processes or interruption of transport pumps, producing cholestasis.

b. *Idiopathic hepatotoxicity* is indirect when it produces a hepatotoxic response that is not dose-related and is of variable latency that ranges from a few days to months. Drug-induced hepatotoxic response may involve several mechanisms, such as destruction of the hepatocytes and the release of various cell proteins that are covalently bound by the drug; such a process can result in the formation of altered antigenic moiety that can trigger hypersensitivity reaction. Methyldopa, phenytoin, sulfa drugs, nitrofurantoin, or halothane can cause hypersensitivity syndrome [15]. The inhibition of hepatic enzyme activity by a drug can also lead to hepatotoxicity, as observed in the antihistaminic drug terfenadine [16, 17].

25.2.2 Nephrotoxicity

Since the kidney handles 25% of cardiac output, it is susceptible to renal toxicity [18]. The acidic pH of the urine can cause precipitation of certain drugs whose solubility is altered by the acidic pH – for example, sulfa drugs that can cause crystalluria [19]. In addition, a drug or its metabolite that is protein bound may uncouple in the renal tubule so that the freed drug is more toxic than the protein-bound drug. A large concentration of sulfhydryl enzyme can precipitate heavy metal compounds, and some outdated tetracyclines are capable of producing Fanconi syndrome, resulting in tubular damage. Renal damage can involve structures outside the glomerulus; this is generally referred to as tubulointerstitial lesions. In these diseases, the glomeruli are not damaged; rather, the damage occurs to the interstitium of the kidney. Renal tubulointerstitial diseases, either acute or chronic, involve many etiologies and altered physiological processes. The most common form of tubulointerstitial inflammation is immunologic.

a. **Direct Toxicity to Kidney Structures:** Many drugs can cause direct damage to the glomerular apparatus or renal tubules – for example, aminoglycosides, cisplatin, some cephalosporins, and polymyxins [20].
b. **Acute and Chronic Tubulointerstitial Nephritis:** Acute tubulointerstitial nephritis is caused by hypersensitivity reactions induced by nonsteroidal anti-inflammatory agents (NSAID), sulfa drugs, and many antibiotics, such as nafcillin, methicillin, ampicillin, and rifampin. This type of damage is due to type III hypersensitivity reaction. Outdated tetracyclines can cause Fanconi syndrome, which involves tubular dysfunction. Chronic interstitial nephritis is caused by long-term use of many NSAIDs, lithium, tacrolimus, and cyclosporine.
c. **Acute and Chronic Interstitial Nephritis:** Therapeutic drugs associated with acute interstitial nephritis include phenytoin and many antibiotics, such as nafcillin, methicillin, ampicillin, and rifampin. This type of damage is due to type III hypersensitivity reaction. Chronic interstitial nephritis is caused by long-term use of many nonsteroidal anti-inflammatory drugs.

d. **Miscellaneous Obstructive Conditions:** Sulfa drugs with limited solubility in acidic medium can precipitate and cause crystalluria [21]. In addition, the uricosuric drug sulfinpyrazone, probenecid, methotrexate, and allopurinol can increase uric acid excretion in the kidney and cause crystalluria and subsequent renal damage.

25.2.3 Hematologic Toxicity

The hematopoiesis process, involving the maturation of blood cells from hematopoietic stem cells from the bone marrow (Figure 25.1), can be affected by some drugs, resulting in abnormal blood cells or alteration in their normal number. Drugs that affect bone marrow cells cause disruption in the formation of all blood cells. Other drugs target certain blood cells, damaging specific cell types [22].

a. **Leukopenia:** Therapeutic agents that cause agranulocytosis are phenylbutazone, zidovudine, and many anticancer agents [23]. Bone marrow suppression of myeloblasts causes agranulocytosis, a significant decrease in neutrophils, basophils, and eosinophils. Neutropenia is a condition in which neutrophils are

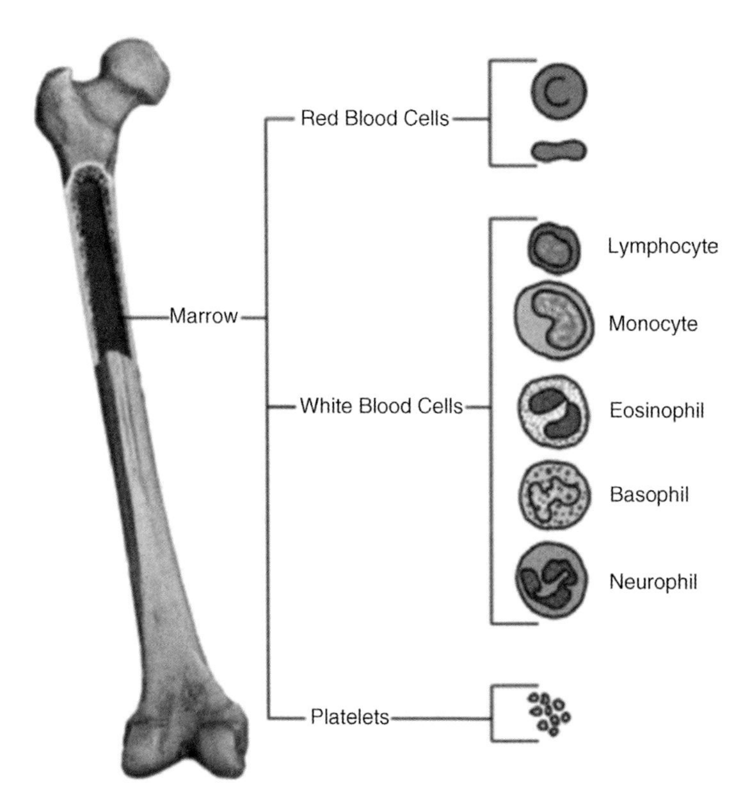

Figure 25.1 Maturation of blood cells.

significantly reduced. A patient can be advised that if he or she has sore throat (due to a decreased first line of defense involving low levels of neutrophils), fever, chills, and increased susceptibility to infections, then granulocytopenia may be suspected.

b. Anemia is caused by a significant decrease in red blood cell (RBC) counts. In a normal healthy person, RBCs are replaced every 120 days by new cells. In hemolytic anemia, this period is considerably shortened. The RBC death is hastened by oxidative stress or by hemolysis caused by immune reaction. Oxidative stress induced by certain oxidant drugs such as sulfonamides, antimalarial drugs, and nitrofurantoin can damage RBCs in patients with glucose-6-phosphate dehydrogenase (G6PD) deficiency. RBCs in many sickle-cell anemia patients are especially susceptible to oxidative drugs.

Immune hemolytic anemia involving antigen–antibody reaction is observed when certain drugs such as penicillin react with the RBC cell protein. The resultant antigen reacts with plasma protein molecules such as IgG antibodies, causing hemolysis.

Aplastic anemia is developed when a drug either suppresses or destroys the stem cells of the bone marrow, resulting in a deficiency of all cells of the blood. Chloramphenicol-induced aplastic anemia significantly reduces the usage of the antimicrobial agent [24].

Megaloblastic anemia is developed when the DNA synthesis of the RBC is altered, causing a slow down of the cell division and generation of large and immature RBCs whose normal functions are compromised.

Thrombocytopenia is caused by drugs that produce an abnormal reduction in platelet cell counts, either by destruction of platelet cells or by inhibition of the platelet cell synthesis [25]. Therapeutic agents that cause a decrease in the platelet function include ibuprofen, naproxen, indomethacin, tolmetin, imipramine, amitriptyline, nortriptyline, chlorpromazine, and nitrofurantoin.

25.2.4
Immunologic Toxicity

Many patients are wrongly classified as allergic or hypersensitive to certain medications [26], and in reality, these patients are simply not tolerating a medication, with the result that they exhibit nausea or vomiting.

The drug allergy symptoms involve a reproducible immunologically mediated series of reactions as described below:

a. Type I hypersensitivity reactions are of the immediate type, involving anaphylaxis, or IgE-mediated hypersensitivity responses, as in reactions to penicillin and cephalosporins [27] (especially those with, in addition to the beta-lactam structure, a similar side chain). Symptoms such as acute urticaria, bronchial spasm, or rhinitis may be observed, and cardiovascular collapse occurs in severe conditions that might need immediate medical intervention.
b. In Type II hypersensitivity reactions, the patient exhibits cytotoxic hypersensitivity involving IgG and IgM-antibodies. In addition, cytotoxicity involving

complement-dependent mechanisms can be observed. Patients taking penicillin or quinine can show this type of allergic reaction, manifested as hemolytic anemia.

c. Type III hypersensitivity reactions are caused by the deposition of the circulating immune complex of IgG and IgM antibodies with antigen complex in vessels or tissues, resulting in serum sickness symptoms. Isoniazid, phenytoin, penicillin, or cephalosporins may be involved [28].
d. Type IV hypersensitivity reactions are delayed, cell-mediated, or tuberculin types of hypersensitivity responses. Such hypersensitivity reactions are seen with topical, rather than systemic, therapy. A clinical rash is manifested as antigen reacts with sensitized T-cells; such a rash becomes increasingly worse with subsequent administrations of the causative agent.

Obviously, physicians need to consider a number of important potential drug-induced toxic effects and share information with patients. It is equally important for a physician to understand the pharmacodynamics and pharmacokinetics of drug–drug/drug–food/drug–herbal, and drug and environmental contaminant interactions, their potential impact on patient care, and the management strategies.

25.3 Drug Interactions

Most drug interactions involve either alterations in parameters of pharmacokinetics (changes in absorption, distribution, metabolism, and/or excretion processes of drugs) or pharmacodynamics (pharmacological responses involving the receptor mechanisms). Drug interaction can result in unexpected adverse effect(s), in increase in the activity of a drug, or in blocking of a drug response. In general, drug interaction is a reaction between two therapeutic agents – that is, a drug–drug interaction. However, drug interaction can be manifested in many other ways. For example, interactions between drug–food [29], drug–nutrient [30], drug–herbal [31], or drug–recreational drugs (alcohol or tobacco) can also be observed [32]. In addition, a pre-existing disease condition and drug interaction can be a significant clinical problem. For example, a patient suffering from chronic obstructive pulmonary disease (COPD) will be ill-advised to take a beta-blocking agent such as propranolol. Propranolol blocks both beta-1 receptors in the heart and the beta-2 receptors in the lungs. Blocking beta-2 receptors will adversely affect normal respiration by blocking bronchial smooth muscle cells.

Negative or adverse drug interactions usually are the focus of attention; however, positive or beneficial drug responses are also known to occur. For example, an epinephrine and lidocaine combination improves the bioavailability of the local anesthetic. In addition, an antagonistic action of naloxone is useful in treating opioid toxicity or the inhibition of such transport proteins as P-glycoproteins (Pgp), which belong to an ABC transporter superfamily and are capable of enhancing anticancer therapy responses. Still, negative or adverse drug interactions are the norm. For example, a hypertensive patient taking a nasal decongestant can suffer from a severe hypertensive episode. Because vasoconstriction action of decongestant in general is local, but at higher than therapeutic doses can cause systemic vasoconstriction thereby potentially aggravating hypertensive condition.

The following is a review of the major types of drug/chemical interactions and their effects.

25.3.1 Drug–Drug Interactions

Pharmacodynamic alterations during drug–drug interactions result in a change in the pharmacological activity of a drug (see also Chapter 13).

Additive drug interactions represent the combined effect of two concurrently administered agonist drugs on the same receptor – that is, of the $2 + 2 = 4$ value.
Potentiation drug interactions occur when a second drug binds to a site that is different from the active receptor site, resulting in a potentiation effect – that is, $2 + 0 = 4$. An example of a potentiation drug interaction is occupational or environmental exposure to isopropanol and carbon tetrachloride in combination [33]. Isopropanol *per se* is not hepatotoxic, but when it is combined with carbon tetrachloride, it causes increased hepatotoxicity of carbon tetrachloride. On the other hand, in colon cancer patients, 5-flurouracil was found to be very effective when used with leucovorine, and the combined therapeutic effect was greater than when either drug was used alone.
Synergistic drug interactions result in a final combined response that is larger than the sum of responses of each drug agonist administered alone – that is, $2 + 2 = 6$.
Antagonistic drug interactions produce diminished response when two drugs compete for the same receptors. For example, competitive antagonism is observed when morphine as an agonist is blocked at μ-receptors by naloxone [34], which *per se* has no activity except blocking opioid receptors. The antagonism effect of dimercaprol, a chelating agent that sequesters a number of metal ions, is useful in heavy metal poisoning. The binding of dimercaprol with metal renders the metal–dimercaprol complex inactive.

Pharmacokinetic drug–drug interactions include any antagonist-induced alteration in the absorption, distribution, metabolism, or excretion of an agonist drug. Some antagonists bind to drug molecules and inhibit GI absorption – for example, antacids can complicate and diminish GI absorption of tetracyclines [35]; similarly, cholestyramine resins can bind to cardiac glycosides, thiazides, and warfarin and decrease each drug's bioavailability.

Metabolism of a drug results in the decreased half-life of its parent compound and its metabolites. Enzymes responsible for drug metabolism are either induced or inhibited by drugs, causing either a significant enhancement or a diminution in the regulation of pharmacotherapeutic or adverse drug effects. The cytochrome P-450 (CYP) superfamily of enzymes exists in many isoforms to regulate the diverse structures of drug molecules. For example, grapefruit juice is a powerful inhibitor of CYP3A4 in the stomach. CYP3A4 is involved in first-pass metabolism of many therapeutic agents. Astemizole plasma levels are increased to dangerous levels when a drug is taken with grapefruit juice, and prolongation of the QT-interval can be fatal in some patients [36].

Another kind of drug–drug interaction is clearance alteration interaction. Clearance alteration interaction is best illustrated by the interaction of penicillin and probenecid. Early in the development of penicillin, a small quantity of the newly synthesized antibiotic had an extremely short half-life due to high renal excretion. Probenecid, as an inhibitor of urinary penicillin excretion, proved to be an important antagonistic agent that extended the half-life of the newly synthesized antibiotic [37].

Another example of drug–drug interactions is in HIV-infected patients using antiretroviral agents and other drugs. A 46-year-old African American who was doing well on antiretroviral therapy with tenofovir, emtricitabine, and ritonavir-boosted (/r) fosamprenavir, presented 1 day after falling off a motorcycle and receiving sutures for lacerations and analgesic treatment for arm and leg injuries at the emergency department. The man stated that he was unable to sleep or to concentrate at work because of pain, which was not responsive to ibuprofen. His physical examination showed multiple abrasions and contusions. He was prescribed acetaminophen with codeine since he showed no adverse effects to previous ingestion of codeine. Two days after his discharge, the patient called and reported vomiting and "feeling sick all day." The patient did not exhibit any response to the analgesic treatment. The most likely explanation for the patient's symptoms seemed to involve codeine, which is a prodrug and substrate for CYP2D6; codeine requires *in vivo* conversion to its active metabolite (morphine). If morphine is not formed, there is no pain relief. In an individual who has inhibited catabolism, codeine levels will accumulate, causing emesis without pain relief. The patient had no history of adverse events due to codeine on previous occasions. Most likely explanation of this ADR is that the patient taking many HIV medications was switched at some mid-point in the HIV therapy to ritonavir, a CYP2D6 inhibitor. The CYP2D6 inhibition reduces codeine metabolism, thereby raising codeine levels and producing codeine toxicity with no pain relief [38].

25.3.2
Drug–Food Interactions

The interaction between any medical agent and food can affect both the pharmacodynamic and the pharmacokinetic responses.

An example of a pharmacodynamic drug–food interaction is the interference of bananas, a potassium-rich fruit, with the effects of potassium-sparing diuretics. A second example is the reduction of the anticoagulant activity of warfarin when a patient eats vitamin K-rich food products such as broccoli, spinach, or Brussels sprouts [39, 40].

Pharmacokinetic drug–food interactions are more common than the pharmacodynamic type. Pharmacokinetic drug interactions with food are important because food affects pH, GI motility, alterations in the GI drug metabolism, and alterations in the drug transport across the GI mucosa, as well as chelation and adsorption to drug molecules. Some examples of drug–food interactions are as follows.

The oral absorption of tetracycline and quinolone antibiotics can be diminished by chelation with calcium-rich foods such as dairy products and calcium-containing drugs. The result is poor plasma blood levels and lower target tissue levels of the antimicrobials [41].

Grapefruit juice inhibits CYP3A4 enzymes in the GI tract, thereby improving the bioavailability of drugs that are normally destroyed by the enzyme – drugs such as astemizole, haloperidol, quinidine, simvastatin, vinvristine, and ritonavir, to name a few. In such instances, there is also a possibility of decreased plasma levels due to organic anilon-transporting polypeptide (OATP) inhibition [42–47].

A case of generalized dermatitis and itch induced by a possible drug–food interaction was reported in a young woman who was taking clomipramine for obsessive compulsive disorder (OCD). The 33-year-old woman, affected by anxiety symptoms, was undergoing pharmacological treatment with clomipramine (75–100 mg/day) plus alprazolam (0.5 mg/day). The patient developed a severe generalized urticaria with intense itch after 1 month of the treatment. A new anamnesis revealed that on the day before the development of the skin rash, no other drug was consumed, but the patient had eaten codfish. The treatment with clomipramine was gradually discontinued and changed to paroxetine (30 mg/day). The patient did not show any OCD-related symptom, and no adverse event associated with the paroxetine treatment was recorded. A possible interaction between clomipramine and codfish ingestion was postulated. The allergic potential of clomipramine was investigated. While clomipramine rechallenge induced a decrease in the skin rash, the drug rechallenge performed 1 month later did not induce any adverse event. In contrast, when the combined rechallenge of codfish and clomipramine was performed, urticaria was newly observed. The Naranjo probability scale score suggested a probable causal relationship between drug–food interaction and the skin rash [48].

25.3.3 Drug–Natural Products Interactions

Natural products such as herbs, nutraceuticals, plant alkaloids such as phytochemicals, and natural vitamin-containing products are increasingly ingested by patients [49–52]. These products also referred to as "alternative medicines" have been reported to be interacting with prescription and OTC medicinal agents. Since the ingredients of the alternative medicines are highly variable, both quantitatively and qualitatively, their interactions with medicinal agents can be very unpredictable. For example, St. John's wort (SJW), *Hypericum perforatum*, is generally used for the treatment of depression. The ingredients in SJW are capable of inducing CYP3A4, CYP1A2, the intestinal P-glycoprotein, and glutathione-*S*-transferase. Given these multiple actions of SJW, there has been increased interest in studying potential drug interactions. Failure of oral contraceptive activity in patients using SJW has been reported. In addition, the US Food and Drug Administration (FDA) has issued a public health advisory warning against the use of SJW for depression in HIV patients taking nonnucleoside reverse transcriptase inhibitors [53–55].

25.3.4
Drug–Environmental Contaminant Interactions

The harmful effects of cigarette smoking are well described in the literature [56–60]. The mechanisms of drug interactions with smoking can affect a drug therapy via both pharmacokinetic and pharmacodynamic mechanisms. For example, polycyclic aromatic hydrocarbons (PAHs) are some of the major lung carcinogens found in tobacco smoke. PAHs are potent inducers of the hepatic cytochrome P-450 isoenzymes 1A1, 1A2, and possibly 2E1. Drug therapy can be pharmacokinetically affected by polyaromatic hydrocarbons and pharmacodynamically by nicotine.

Cigarette smoking can reduce the efficacy of certain drugs or make drug therapy more unpredictable. Pharmacokinetic interactions may cause smokers to require a larger dosage of certain drugs through an increase in plasma clearance, a decrease in absorption, an induction of cytochrome P450 enzymes, or a combination of these factors. Pharmacokinetic drug interactions with cigarette smoking have been described for theophylline, tacrine, insulin, flecainide, propoxyphene, propranolol, and chlordiazepoxide. Pharmacodynamic interactions may increase the risk of adverse events in smokers with cardiovascular or peptic ulcer disease and in women who smoke and use oral contraceptives. Pharmacodynamic interactions have also been described for antihypertensive and antianginal agents, antilipidemics, and histamine2-receptor antagonists.

Other drugs showing interactions with smoking include pentazocine and phenylbutazone, both of which increase metabolism in smokers and may require administration in higher dosages. In addition, it is known that smokers experience less pain relief with propoxyphene than do nonsmokers. Codeine, meperidine, and morphine are pharmacokinetically unaffected by smoking, and the effect of smoking on acetaminophen is probably not clinically important. Carbamazepine, phenytoin, and phenobarbital are not influenced by smoking. Heparin metabolism, however, is elevated in smokers; this effect may necessitate modest increases in dosage. Warfarin clearance is increased in smokers, but the increase is not associated with a change in prothrombin time. The metabolism of estrogens to less active metabolites is increased in smokers. In addition, smoking is associated with decreased subcutaneous absorption of insulin and increased dosage requirements. Healing of GI ulcers with antiulcer medication may be compromised in smokers.

A 22-year-old woman developed severe hypotension and bradycardia after receiving a small dose of paracervical vasopressin. She was using a transdermal nicotine patch at the time of her surgery. The physicians suspected that her cardiac problems, and many recently reported cardiac events in other women receiving small doses of paracervical vasopressin, could have been caused by a synergism of the vasoconstrictive properties of nicotine and vasopressin. Therefore, the authors urge caution when vasopressin is to be administered to patients who smoke or use nicotine transdermal patches. In addition, healthcare professionals should always consider the negative impact of cigarette smoking in planning and assessing responses to drug therapy. Finally, drug interactions with cigarette smoking should be specifically studied in clinical trials of new drugs [61].

25.4 Conclusions

The body processes all chemicals and drugs – namely, biomolecules, natural chemicals, pharamaceuticals, and toxics – similarly through limited pathways inherently available to it. Thus, absorption, distribution, metabolism, and excretion (ADME) processes play a critical role in the disposition of the chemicals and therapeutic agents with which the body comes in contact. Essentially, these processes are a complex network of biochemical pathways and a gamut of enzymes that have taken years to coevolve as a function of encountered exposures. Resistance to drug action is a result of the adaptive capacity of the biological systems of microbes, insects, and humans to develop alternative mechanisms. There is a continuous evolution of and competition between the development of new efficacious drugs to achieve a desired effect and the body and biological systems to recognize an external factor and swiftly dispose it of.

In today's world, the demand of patients for quick recovery and for immediate satisfaction/gratification could subject patients to potentially more harmful drug-induced effects. The existing paucity of ADR or drug–drug interaction data is mainly due to lack of a data collection infrastructure; physicians or other healthcare professionals have not had a proper protocol of data reporting after they have noted any abnormal patient response to drugs. The data seem prolific now, mainly because of improved harvesting of the data from the field and the introduction of a number of novel drugs with complicated structures manufactured by newer biotechnology laboratories. What is lacking is a collection mechanism.

The existing problems of drug therapy in hospitalized patients, either receiving a single drug or a polypharmacy, in the United States are comparable to those occurring in other countries [62–67]. Although ADRs have been studied and reported for the past few years, lessons and experiences from these studies have not exactly been translated into effective problem management in any country. Further investigations are required to unearth the underlying deficiencies that are the cause of this failure in the current healthcare operating system. It is important to note, however, that age and gender are as important as the number of drugs prescribed as risk factors for an ADR among patients taking multiple drugs. This fact suggests the potential for improvement in the field of patient monitoring (i.e., laboratory testing) and prescription monitoring, especially if the monitoring is targeted at children, the elderly population, and pregnant/child-bearing-age women [63, 66, 67].

Acknowledgments

The authors acknowledge the editorial assistance rendered by Dr. Ernie Martin, and also acknowledge the assistance of Dr. Paula Burgess who provided a technical review to the chapter.

References

1 Onder, G., Pedone, C., Landi, F., Cesari, M., Della Vedova, C., Bernabei, R., and Gambassi, G. (2002) Adverse drug reactions as cause of hospital admissions: results from the Italian Group of Pharmaco-epidemiology in the Elderly (GIFA). *J. Am. Geriatr. Soc.*, **50**, 1962–1968.

2 Pirmohamed, M., James, S., Meakin, S., Green, C., Scott, A.K., Walley, T.J., Farrar, K., Park, B.K., and Breckenridge, A.M. (2004) Adverse drug reactions as cause of admission to hospital: prospective analysis of 18,820 patients. *BMJ*, **329**, 15–19.

3 Grymonpre, R.E., Mitenko, P.A., and Sitar, D.S. (1988) Drug associated hospital admissions in older medical patients. *J. Am. Geriatr. Soc.*, **36**, 1092–1098.

4 Laws, M.B. (2004) Adverse drug reactions as cause of admission to hospital: definition of adverse drug reactions needs to include overdose. *BMJ*, **329**, 459–460.

5 Brvar, M., Fokter, N., Bunc, M., and Mozina, M. (2009) The frequency of adverse drug reaction related admissions according to method of detection, admission urgency and medical department specialty. *BMC Clin. Pharmacol.*, **4**, 9–18.

6 van der Hooft, C.S., Sturkenboom, M.C., van Grootheest, K., Kingma, H.J., and Sticker, B.H. (2006) Adverse drug reaction-related hospitalizations: a nationwide study in the Netherlands. *Drug Saf.*, **29**, 161–168.

7 Koh, Y., Fatimah, M.K., and Li, S.C. (2003) Therapy related hospital admission in patients on polypharmacy in Singapore: a pilot study. *Pharm. World Sci.*, **25**, 135–137.

8 Lagnaoui, R., Moore, N., and Fach, J. (2000) Adverse drug reactions in a department of systemic diseases-oriented internal medicine: prevalence, incidence, direct costs and avoidability. *Eur. J. Clin. Pharmacol.*, **6**, 181–186.

9 Lazarou, J., Pomeranz, B.H., and Corey, P.N. (1998) Incidence of adverse drug reactions in hospitalized patients: a meta-analysis of prospective studies. *JAMA*, **279**, 1200–1205.

10 van der Hooft, C.S., Dieleman, J.P., Siemes, C., Aarnoudse, A.J., Verhamme, K.M., Stricker, B.H., and Sturkenboom, M.C. (2008) Adverse drug reaction-related hospitalizations: a population-based cohort study. *Pharmacoepidemiol. Drug Saf.*, **17**, 365–371.

11 World Health Organization (WHO) (1975) Requirements for adverse reaction reporting, Geneva, Switzerland.

12 American Society of Health-System Pharmacists (1995) ASHP guidelines on adverse drug reaction monitoring and reporting. *Am. J. Health-Syst. Pharm.*, **52**, 417–419.

13 Rawlins, M. and Thompson, W. (1991) Mechanisms of adverse drug reactions, in *Textbook of Adverse Drug Reactions* (ed. D. Davies), Oxford University Press, New York, pp. 18–45.

14 Navarro, V.J. and Senior, J.R. (2006) Drug-related hepatotoxicity. *NEJM*, **354**, 731–739.

15 Knowles, S.R., Uetrecht, J., and Shear, N.H. (2000) Idiosyncratic drug reactions: the reactive metabolite syndromes. *Lancet*, **356**, 1587–1591.

16 Honig, P.K., Woosley, R.I., Zamani, K., Conner, D.P., and Cantilena, L.R., Jr. (1992) Changes in the pharmacokinetics and electrocardiographic pharmacodynamics of terfenadine with concomitant administration of

erythromycin. *Clin. Pharmacol. Ther.*, **52**, 231–238.

17 Yun, C.H., Okerholm, R.A., and Guengerich, F.P. (1993) Oxidation of the antihistaminic drug terfenadine in human liver microsomes: role of cytochrome P-450 3A4 in *N*-dealkylation and *C*-hydroxylation. *Drug Metab. Disp.*, **21**, 403–409.

18 Servais, H., Ortiz, A., Devuyst, O., Denamur, S., Tulkens, P.M., and Mingeot-Leclercq, M.P. (2008) Renal cell apoptosis induced by nephrotoxic drugs: cellular and molecular mechanisms and potential approaches to modulation. *Apoptosis*, **13**, 11–32.

19 de la Prada Alvarez, F.J., Prados Gallardo, A.M., Tugores Vázquez, A., Uriol Rivera, M., and Morey Molina, A. (2007) Acute renal failure due to sulfadiazine crystalluria. *Ann. Med. Intern.*, **24**, 235–238.

20 Lau, A.H. (1999) Apoptosis induced by cisplatin nephrotoxic injury. *Kidney Int.*, **56**, 1295–1298.

21 Kuhlmann, M.K., Burkhardt, G., and Köhler, H. (1997) Insights into potential cellular mechanisms of cisplatin nephrotoxicity and their clinical application. *Nephrol. Dial. Transplant.*, **12**, 2478–2480.

22 Acharya, S. and Bussel, J.B. (2000) Hematologic toxicity of sodium valproate. *J. Pediatr. Hematol. Oncol.*, **22**, 62–65.

23 Antoniou, T., Gough, K., Yoong, D., and Arbess, G. (2004) Severe anemia secondary to a probable drug interaction between zidovudine and valproic acid. *Clin. Infect. Dis.*, **38** (5), e38–e40.

24 Yuan, Z.R. and Shi, Y. (2008) Chloramphenicol induces abnormal differentiation and inhibits apoptosis in activated T cells. *Cancer Res.*, **68**, 4875–4881.

25 Kimland, B., Höjeberg, B., and von Euler, M. (2004) Levetiracetam-induced thrombocytopenia. *Epilepsia*, **45**, 877–878.

26 Mendelson, L. (1998) Adverse reactions to β-lactam antibiotics. *Immunol. Allergy Clin. North Am.*, **18**, 745–756.

27 Annè, S. and Reisman, R.E. (1995) Risk of administering cephalosporin antibiotics to patients with histories of penicillin allergy. *Ann. Allergy Asthma Immunol.*, **74**, 167–170.

28 Antunez, C., Blanca-Lopez, N., and Torres, M.J. (2006) Immediate allergic reactions to cephalosporins: evaluation of cross-reactivity with a panel of penicillins and cephalosporins. *J. Allergy Clin. Immunol.*, **117**, 404–410.

29 Schmidt, L.E. and Dalhoff, K. (2002) Food drug interactions. *Drugs*, **62**, 1481–1502.

30 Lourenco, R. (2001) Enteral feeding: drug/nutrient interaction. *Clin. Nutr.*, **20**, 187–193.

31 Scott, G.N. and Elmer, G.W. (2002) Update on natural product–drug interactions. *Am. J. Health Syst. Pharm.*, **59**, 339–347.

32 Shoaf, S.E. and Linnolia, S. (1991) Interaction of ethanol and smoking on the pharmacokinetics and pharmacodynamics of psychotropic medications. *Psychopharmacol. Bull.*, **27**, 577–594.

33 Ueng, T.H., Moore, L., Elves, R.G., and Alvares, A.P. (1983) Isopropanol enhancement of cytochrome P-450-dependent monooxygenase activities and its effects on carbon tetrachloride intoxication. *Toxicol. Appl. Pharmacol.*, **71**, 204–214.

34 Schaefer, G.J. and Michael, R.P. (1990) Interactions of naloxone with morphine, amphetamine and phencyclidine on fixed interval responding for intracranial self-stimulation in rats. *Psychopharmacology (Berl.)*, **102**, 263–268.

35 Xu, Z., Fan, J., Zheng, S., Ma, F., and Yin, D. (2009) On the adsorption of tetracycline by calcined magnesium-aluminum hydrotalcites. *J. Environ. Qual.*, **38**, 1302–1310.

36 Lilja, J.J., Kivisto, K.T., and Neuvonen, P.J. (2000) Duration of effect of grapefruit juice on the pharmacokinetics of the CYP3A4 substrate simvastatin. *Clin. Pharmacol. Ther.*, **68**, 384–390.

37 Catlin, B.W. (1984) Probenecid: antibacterial action against *Neisseria gonorrhoeae* and interaction with benzylpenicillin. *Antimicrob. Agents Chemother.*, **25**, 676–682.

38 Aberg, J.A. (2008) Drug–drug interactions with newer antiretroviral agents. *Top. HIV Med.*, **16**, 146–150.

39 Hanslik, T. and Prinseau, J. (2004) The use of vitamin K in patients on anticoagulant therapy: a practical guide. *Am. J. Cardiovasc. Drugs*, **4**, 43–55.

40 Cambria-Kiely, J.A. (2002) Effect of soy milk on warfarin efficacy. *Ann. Pharmacother.*, **36**, 1893–1896.

41 Jung, H., Peregrina, A.A., Rodriguez, J.M., and Moreno-Esparza, R. (1997) The influence of coffee with milk and tea with milk on the bioavailability of tetracycline. *Biopharm. Drug Dispos.*, **18**, 459–463.

42 Fuhr, U., Maier-Bruggemann, A., Blume, H., Muck, W., Unger, S., and Kuhlmann, J. (1998) Grapefruit juice increases oral nimodipinebioavailability. *Int. J. Clin. Pharmacol. Ther.*, **36**, 126–132.

43 Lilja, J.J., Kivisto, K.T., Backman, J.T., and Neuvonen, P.J. (2000) Effect of grapefruit juice dose on grapefruit juice–triazolam interaction: repeated consumption prolongs triazolam half-life. *Eur. J. Clin. Pharmacol.*, **56**, 411–415.

44 Kupferschmidt, H.H., Fattinger, K.E., Ha, H.R., Follath, F., and Krahenbuhl, S. (1998) Grapefruit juice enhances the bioavailability of the HIV protease inhibitor saquinavir in man. *Br. J. Clin. Pharmacol.*, **45**, 355–359.

45 Lundahl, J., Regardh, C.G., Edgar, B., and Johnsson, G. (1997) Effects of grapefruit juice ingestion: pharmacokinetics and haemodynamics of intravenously and orally administered felodipine in healthy men. *Eur. J. Clin. Pharmacol.*, **52**, 139–145.

46 Clifford, C.P., Adams, D.A., Murray, S., Taylor, G.W., Wilkins, M.R., and Boobis, A.R. (1997) The cardiac effects of terfenadine after inhibition of its metabolism by grapefruit juice. *Eur. J. Clin. Pharmacol.*, **52**, 311–315.

47 Kupferschmidt, H.H., Ha, H.R., Ziegler, W.H., Meier, P.J., and Krahenbuhl, S. (1995) Interaction between grapefruit juice and midazolam in humans. *Clin. Pharmacol. Ther.*, **58**, 20–28.

48 Gallelli, L., De Fazio, S., Corace, E., De Sarro, G., Garcia, C.S., and De Fazio, P. (2006) Generalized urticaria in a young woman treated with clomipramine and after ingestion of codfish. *Pharmacopsychiatry*, **39**, 154–156.

49 Williamson, E.M. (2003) Drug interactions between herbal and prescription medicines. *Drug Saf.*, **26**, 1075–1092.

50 Ernst, E. (2002) The risk–benefit profile of commonly used herbal therapies: ginkgo, St. John's wort, ginseng, echinacea, saw palmetto, and kava. *Ann. Intern. Med.*, **136**, 42–53.

51 Budzinski, J.W., Foster, B.C., Vandenhoek, S., and Arnason, J.T. (2000) An *in vitro* evaluation of human cytochrome P450 3A4 inhibition by selected commercial herbal extracts and tinctures. *Phytomedicine*, **7**, 273–282.

52 Gorski, J.C., Huang, S.M., and Zaheer, N. (2003) The effect of echinacea on CYP3A activity *in vivo*. *Clin. Pharmacol. Ther.*, 73–78.

53 Izzo, A.A. (2004) Drug interactions with St. John's wort (*Hypericum perforatum*): a review of the clinical evidence. *Int. J. Clin. Pharmacol. Ther.*, **42**, 139–148.

54 Wang, Z., Gorski, J.C., Hamman, M.A., Huang, S.M., Lesko, L.J., and Hall, S.D. (2001) The effects of St. John's wort (*Hypericum perforatum*) on humancytochrome P450 activity. *Clin. Pharmacol. Ther.*, **70**, 317–326.

55 Wang, Z., Hamman, M.A., Huang, S.M., Lesko, L.J., and Hall, S.D. (2002) Effect of St John's wort on the pharmacokinetics of fexofenadine. *Clin. Pharmacol. Ther.*, **71**, 414–420.

56 Schein, J.R. (1995) Cigarette smoking and clinically significant drug interactions. *Ann. Pharmacother.*, **29**, 1139–1148.

57 Desai, H.D., Seabolt, J., and Jann, M.W. (2001) Smoking in patients receiving psychotropic medications: a pharmacokinetic perspective. *CNS Drugs*, **15**, 469–494.

58 Lee, B.L., Benowitz, N.L., and Jacob, P., III (1987) Influence of tobacco abstinence on the disposition kinetics and effects of nicotine. *Clin. Pharmacol. Ther.*, **41**, 474–479.

59 Benowitz, N.L. and Jacob, P., III (1993) Nicotine and cotinine elimination pharmacokinetics in smokers and nonsmokers. *Clin. Pharmacol. Ther.*, **53**, 316–323.

60 Zevin, S. and Benowitz, N.L. (1999) Drug interactions with tobacco smoking. An update. *Clin. Pharmacokinet.*, **36**, 425–438.

61 Groudine, S.B. and Morley, J.N. (1996) Recent problems with paracervical vasopressin: a possible synergistic reaction with nicotine. *Med. Hypotheses*, **47**, 19–21.

62 Kishore, P.V., Palaian, S., Ojha, P., and Shankar, P.R. (2008) Pattern of adverse drug reactions experienced by tuberculosis patients in a tertiary care teaching hospital in Western Nepal. *Pak. J. Pharm. Sci.*, **21**, 51–56.

63 Loya, A.M., González-Stuart, A., and Rivera, J.O. (2009) Prevalence of polypharmacy, polyherbacy, nutritional supplement use and potential product interactions among older adults living on the United States–Mexico border: a descriptive, questionnaire-based study. *Drugs Aging*, **26**, 423–436.

64 Divac, N., Jasović-Gasić, M., Samardzić, R., Lacković, M., and Prostran, M. (2007) Antipsychotic polypharmacy at the University Psychiatric Hospital in Serbia. *Pharmacoepidemiol. Drug Saf.*, **16**, 1250–1251.

65 Wunsch, M.J., Nakamoto, K., Behonick, G., and Massello, W. (2009) Opioid deaths in rural Virginia: a description of the high prevalence of accidental fatalities involving prescribed medications. *Am. J. Addict.*, **18**, 5–14.

66 Ay, P., Akici, A., and Harmanc, H. (2005) Drug utilization and potentially inappropriate drug use in elderly residents of a community in Istanbul, Turkey. *Int. J. Clin. Pharmacol. Ther.*, **43**, 195–202.

67 Flaherty, J.H., Liu, M.L., Ding, L., Dong, B., Ding, Q., Li, X., and Xiao, S. (2007) China: the aging giant. *J. Am. Geriatr. Soc.*, **55**, 1295–1300.

Index

Principles and Practice of Mixtures Toxicology. Edited by Moiz Mumtaz

ISBN: 978-3-527-31992-3

e

g

h

i